SYSTEMS ENGINEERING COMPETENCY ASSESSMENT GUIDE

SYSTEMS ENGINEERING COMPETENCY ASSESSMENT GUIDE

A combined INCOSE Systems Engineering Competency Framework (SECF) and associated Systems Engineering Competency Assessment Guide (SECAG) document

---

INCOSE

Compiled and Edited by:

Ian Presland CEng, FIET, ESEP (Primary Editor)
Clifford Whitcomb, Ph.D., INCOSE Fellow
Lori Zipes, ESEP

This edition first published 2023

Registered Office
John Wiley & Sons, Inc., 111 River Street, Hoboken, NJ 07030, USA

For details of our global editorial offices, customer services, and more information about Wiley products visit us at www.wiley.com.

Wiley also publishes its books in a variety of electronic formats and by print-on-demand. Some content that appears in standard print versions of this book may not be available in other formats.

Library of Congress Cataloging-in-Publication Data Applied for:
ISBN: 9781119862550 (hardback)

Cover Design: Wiley
Cover Image: © metamorworks/Shutterstock

Set in 10/12pt and TimesLTStd by Straive, Pondicherry, India

DISCLAIMER

Reasonable endeavors have been used throughout its preparation to ensure that the Systems Engineering Competency Framework is complete and correct as [illegible]. INCOSE, its officers, and members shall not be held liable for the consequences of decisions based on any information contained in or excluded from this framework. Inclusion or exclusion of references to any organization, individual, or service shall not be construed as endorsement or the converse. Where value judgements are expressed, these are the consensus view of expert and experienced members of INCOSE.

CONTENTS

LIST OF SECF TABLES

LIST OF SECF FIGURES

LIST OF SECAG TABLES

LIST OF SECAG FIGURES

INCOSE NOTICES

This product was prepared by the International Competency Working Group of the International Council on Systems Engineering (INCOSE). It is approved by the INCOSE Technical Operations for release as an INCOSE Technical Product.

ACKNOWLEDGEMENTS

We would like to acknowledge the contribution of INCOSE UK Advisory Board members who participated in the UK Working group that created the INCOSE UK Systems Engineering Competencies Framework (Issue 3, 2010) which was used as the original basis for the first edition of the INCOSE Systems Engineering Competency Framework. A full list of contributing UK organizations and individuals is included in the main body of the INCOSE Systems Engineering Competency Framework (First Edition 2018).

We would like to thank all other past and current members of the INCOSE Competency Working Group for their support, ideas, and comments.

INTRODUCTION

The goal of the INCOSE International Competency Working Group is to define a global standard for those competencies regarded as central to the practice and profession of Systems Engineering, together with a set of indicators which can be used to verify attainment of those competencies.

This document is an output from that working group.

This book comprises two self-contained elements: (1) A definition of the framework itself (referred to as the "Systems Engineering Competency Framework"), and (2) an accompanying assessment guide (referred to as the "Systems Engineering Competency Assessment Guide") which provides guidance in assessing each of the competencies defined within that framework.

PURPOSE

The purpose of this document is to provide a set of competencies for Systems Engineering within a framework that provides guidance for both beneficiaries and practitioners to identify knowledge, skills, abilities and behaviors important to Systems Engineering effectiveness in the application domain (e.g. space, transportation, medical) for which the competency model is applied.

SCOPE

This document consists of two parts.

Part I is the Systems Engineering Competency Framework (SECF), a generic systems engineering framework. This framework can be applied in the context of any application, project, organization or enterprise for both individual and/or organizational assessment and/or development. The framework is expected to be tailored to suit the application domain in which it is applied, combining competencies identified herein with others taken from complimentary frameworks (e.g. Program Management, Human Resources, Aerospace, Medical), or generated organizationally, to define the required knowledge, skills and behaviors appropriate to an area or role.

Part II is the Systems Engineering Competency Assessment Guide (SECAG) designed to guide those assessing systems engineering competencies characterized within the SECF. The SECAG is a general document which is expected to be tailored to reflect organizational SECF tailoring, and further tailored to reflect local or organizational specifics as defined herein.

PART I: SYSTEMS ENGINEERING COMPETENCY FRAMEWORK

The INCOSE Systems Engineering Competency Framework (Second Edition) is a collaborative product generated from a series of meetings of the INCOSE International Competency Working Group between 2018 and 2021. While it builds upon experience gained in applying the First Edition (originally published by INCOSE in July 2018), the primary changes made in this edition are due to the extensive work done by the group in developing the accompanying Systems Engineering Competency Assessment Guide (SECAG). The Competency Working Group is Chaired by Cliff Whitcomb and Co-Chaired by Lori Zipes. The SECAG guide is published separately through an agreement between INCOSE and John Wiley and Sons, Inc.

Compiled and Edited by:

Ian Presland, CEng, FIET, ESEP

Primary Contributing Authors (alphabetical order):

Juan P. Amenabar, ESEP
Jonas Hallqvist, ESEP
Ian Presland, CEng, FIET, ESEP
Clifford Whitcomb, Ph.D., INCOSE Fellow
Lori Zipes, ESEP

Reviewers and additional contributors (alphabetical order):

We would like to acknowledge the contributions of the following people in the development and review of this update to the framework:

Richard Beasley, ESEP
Ray Dellefave
Suja Joseph-Malherbe, CSEP

Systems Engineering Competency Assessment Guide: A combined INCOSE Systems Engineering Competency Framework (SECF) and associated Systems Engineering Competency Assessment Guide (SECAG) document, First Edition. INCOSE.

Ruediger Kaffenberger, CSEP
Rabia Khan, ASEP
Kirk Michealson
Susan Plano-Faber, ASEP
Philip Quan, CSEP
Sven-Olaf Schulze, CSEP
Corina White, CSEP

SECF ACKNOWLEDGEMENTS

This INCOSE Systems Engineering Competency Framework is a new collaborative product based on several sources.

We would also like to acknowledge the INCOSE UK Advisory Board who participated in the UK Working group and created the INCOSE UK Systems Engineering Competency Framework Issue 3, used to create the INCOSE Framework first edition in 2018 (INCOSE 2018b).

We would like to thank all other members of the INCOSE International Competency Working Group for their contribution and support to the ideas and concepts within this framework.

DISCLAIMER

Reasonable endeavors have been used throughout its preparation to ensure that the Systems Engineering Competency Framework is as complete and correct as is practical. INCOSE, its officers, and members shall not be held liable for the consequences of decisions based on any information contained in or excluded from this framework. Inclusion or exclusion of references to any organization, individual, or service shall not be construed as endorsement or the converse. Where value judgements are expressed, these are the consensus view of expert and experienced members of INCOSE.

COPYRIGHT

SECF INTRODUCTION

SECF SCOPE

This document is the second edition of the INCOSE Systems Engineering Competency Framework (SECF). The SECF is designed to be a source of competencies for systems engineering. The SECF is a general document which is expected to be tailored to reflect organizational SECF tailoring, and further tailored to reflect local or organizational specifics as defined herein.

This second edition of the INCOSE Systems Engineering Competency Framework (SECF) captures updates to the first edition which have resulted from three activities:

- Work performed on the Systems Engineering Competency Assessment Guide (SECAG).
- Deployment of the SECF by various organizations.
- Reorganization of the original SECF to publish the SECAG as a companion document to the SECF.

The first two of these activities have resulted in generally minor technical changes. The third resulted in several document structural changes including updates to text and references throughout the document.

A summary and more detailed rationale of the changes made in this edition can be found in the SECF Annex A.

SECF PURPOSE

The purpose of this document is to provide a resource on how to establish systems engineering competencies defined in the framework and how to differentiate between proficiency at each of the five levels defined within the document.

SECF CONTEXT

The context of this document is well represented in the definition of Systems Engineering as published by the International Council of Systems Engineering (INCOSE) and reflected in standards used across industry today. This Competency Framework specifically aligns with these standards in the areas of terminology and concepts to ensure

Systems Engineering Competency Assessment Guide: A combined INCOSE Systems Engineering Competency Framework (SECF) and associated Systems Engineering Competency Assessment Guide (SECAG) document, First Edition. INCOSE.

using organizations have the ability to use these as complementary resources and to ensure the framework is consistent with industry standards.

In developing the Systems Engineering competencies in this document, the working group considered the following sources:

- Atlas 1.1: An Update to the Theory of Effective Systems Engineers (Hutchison et al. 2018).
- ISO/IEC/IEEE 15288:2015 Systems and Software engineering – System life cycle processes (ISO/IEC JTC 1/SC 7 Software and Systems Engineering Technical Committee 2015).
- Capability Maturity Model Integration V1.3 (CMMI® Institute 2010).
- EIA731 (Electronic Industries Alliance 2002).
- INCOSE Systems Engineering Body of Knowledge (INCOSE 2017).
- *INCOSE Systems Engineering Handbook*, Fourth Edition (INCOSE 2015).
- *NASA Systems Engineering Handbook*, Rev 2 (National Aeronautics and Space Administration (NASA) 2016).
- EE/BCS Safety Competency Guidelines (Institution of Engineering and Technology 2013).
- US DoD's Better Buying Power 3.0 Implementation Plan (Kendall 2014).
- Defense Acquisition University Competency Model (Defense Acquisition University 2013).
- US Navy's Systems Engineering Competency Career Model (SECCM) (Whitcomb et al. 2014).
- INCOSE Systems Engineering Professional Certification Program (INCOSE 2018a).

SECF OBJECTIVE

The objective of the SECF is to leverage existing competency frameworks and competency models in order to:

- Capitalize on feedback received on existing frameworks and models from a decade of practical use globally.
- Improve alignment with other INCOSE initiatives.
- Address content and language to widen its appeal, recognizing the growth of a systems approach within several new domains.
- Reflect the latest collective intelligence of industry as reflected in the data, descriptions, and standards available as learning benchmarks globally.

SECF DOCUMENT OVERVIEW

The SECF Introduction summarizes the context and objective and provides a competency overview, explains the framework structure, describes the format of the competence proficiency levels recognized within the framework, explains the framework language standardization, and describes the competency area table format.

The Using the Competency Framework section provides introductory guidance as to how to use the framework, describes tailoring of assessments, and explains the relationship between roles, job descriptions, and competencies.

The SECF Acronyms and Abbreviations section provides a table for acronyms and abbreviations.

The SECF Glossary section provides definitions of the terminology used in the framework.

SECF Annex A contains a summary of the key changes made between the last formal release of this framework (July 2018) and this release.

SECF Annex B contains background information as to how the framework aligns with INCOSE and other initiatives.

SECF Annex C contains information on defining roles and how to describe the typical structure of a generic role statement within an organization.

SECF Annex D forms the main technical body of this document and contains the formal Competency Framework definition.

Guidance on how to perform competency assessment using this framework is addressed in the INCOSE Systems Engineering Competency Assessment Guide (SECAG) Part II of this book.

The authors welcome feedback on this document. SECF Annex E is a comment form provided for this purpose.

INCOSE SE COMPETENCY FRAMEWORK DEFINITION

COMPETENCY OVERVIEW

The terms "competency" and "competencies" focus on the personal attributes or inputs of an individual. They can be defined as the technical attributes and behaviors that individuals must have, or must acquire, to perform effectively at work.

"Competence" and "competences" are broader concepts that encompass demonstrable performance outputs as well as behavior inputs and may relate to a system or set of minimum standards required for effective performance at work (Chartered Institute of Personnel Development (CIPD) 2021b).

Competency is "a measure of an individual's ability in terms of their knowledge, skills, and behavior to perform a given role (p. xvi)." (Holt and Perry 2011).

Competency is distinct from competence, defined by Merriam-Webster as the ability to do something well (Merriam-Webster 2018). Competence, then, reflects the total characteristics of the individual, while "competency" reflects a single area; the sum of an individual's competencies makes up their competence (Holt and Perry 2011).

Although these terms are now regularly used interchangeably, in this document, the term "competency" is used to define the need ("requirement") and "competence" is used to characterize the "outcome" ("validation").

A "Competency Framework" is a structure that sets out and defines each individual competency required by individuals working in an organization or part of an organization (Chartered Institute of Personnel Development (CIPD) 2021b). This document defines a framework of competencies for the Systems Engineering discipline.

Note that "competence" differs from "capability." Capability is an organizational or organizational team attribute that refers to the ability to execute the organization's mission (i.e. deliver a product or service). Engineering capabilities are defined in terms of people (competencies), processes, facilities, and equipment, the integration of which leads to the ability to produce an engineered product or service.

Indeed, to be effective in any Systems Engineering task, an individual will normally require in addition:

- Supporting Skills and Techniques.
- Domain Knowledge.

Systems Engineering Competency Assessment Guide: A combined INCOSE Systems Engineering Competency Framework (SECF) and associated Systems Engineering Competency Assessment Guide (SECAG) document, First Edition. INCOSE.

Supporting skills and techniques help an individual perform a task effectively within a context. Skills and techniques may be organization, project, or role specific. For instance, the ability to use a company-standard "Requirements Management" tool may be central to performing a given role effectively within one organization, but this tool-specific skill will not help in another organization if it uses a different tool. However, possessing the knowledge, understanding, and experience (i.e. possessing competence) to perform Requirements Management effectively as a task goes beyond the specifics of using any one tool or technique. Indeed, competence may be gained through using a variety of tools. However, an individual organization is likely to require additional "tool-specific skills" for any practitioner to operate effectively within it. The definition of such required skills will form part of the organization-specific tailoring (i.e. the contextualization of the framework).

In a similar way, "domain knowledge" exists within an application or industrial context (e.g. automotive, healthcare, and space). This requires specialization appropriate to the domain addressing areas such as the commercial or organizational environment, the supply chain, and domain-specific technical standards/protocols.

Competence supports the ability to carry out a process, but it is not just about executing a defined process. Competence includes capitalizing on a wider understanding of the relationship between processes, understanding an individual's specific role in supporting their execution within the wider organizational structure, and the behavioral skills required to ensure that process activities are executed effectively. This document includes a mapping of competencies to *INCOSE Systems Engineering Handbook* processes (see SECF Figure 3) which clearly demonstrates the relationship is not one-to-one.

This framework excludes competencies associated with domain knowledge. Users of this framework can use the ideas defined herein, together with resources available within their domain to extend the scope of this framework by adding domain-specific competencies applicable to their project, business, organization, or domain area.

FRAMEWORK STRUCTURE

Competencies predominately associated with Systems Engineering have been identified and grouped into five themes which are summarized in SECF Figure 1. All competence areas are fully defined in a series of tables forming Annex D of this document.

Note that Annex D only defines the "requirements" which express competence at a particular proficiency level. Validation that requirements have been met (i.e. confirming competence at levels characterized in Annex D) needs to be performed by formally assessing an individual against these indicators.

Guidelines for competency assessment are defined in the INCOSE Systems Engineering Competency Assessment Guide (SECAG). Thus, the SECAG Guidelines can be interpreted as guidelines for "verification and validation" evidence required for SECF Annex D.

Although the specific nature of evidence provided will differ from organization and individual, each defined indicator has at least one possible element of potential evidence associated with it.

COMPETENCE PROFICIENCY LEVELS

Five "levels" of increasing competence have been defined in terms of levels of knowledge and experience for each competency area:

- **Awareness**

  The person displays knowledge of key ideas associated with the competency area and understands key issues and their implications. They ask relevant and constructive questions on the subject. This level characterizes engineers new to the competency area. It could also characterize an individual outside Systems Engineering who requires an understanding of the competency area to perform their role.
- **Supervised Practitioner**

  The person displays an understanding of the competency area and have either limited or historical experience. They require regular guidance and supervision.

Core competencies		Professional competencies		Management competencies		Technical competencies	
Core competencies underpin engineering as well as systems engineering.		Behavioral competencies well established within the Human Resources (HR) domain. To facilitate alignment with existing HR frameworks, where practicable, competency definitions have been taken from well-established, internationally recognized definitions rather than partial or complete reinvention by INCOSE.		The ability to perform tasks associated with controlling and managing Systems Engineering activities. This includes tasks associated with the Management Processes identified in the INCOSE SE Handbook.		The ability to perform tasks associated primarily with the suite of Technical Processes identified in the *INCOSE SE Handbook*.	
Systems Thinking (ST)	The application of the fundamental concepts of systems thinking to systems engineering.	Communications (CC)	The dynamic process of transmitting or exchanging information.	Planning (PL)	Producing, coordinating, and maintaining effective and workable plans across multiple disciplines.	Requirements Definition (RD)	To analyze the stakeholder needs and expectations to establish the requirements for a system.
Life Cycles (LC)	Selection of the appropriate life cycles in the realization of a system.	Ethics and Professionalism (EP)	The personal, organizational, and corporate standards of behavior expected of systems engineers.	Monitoring and Control (MC)	Assessment of an ongoing project to see if the current plans are aligned and feasible.	System Architecting (SA)	The definition of the system structure, interfaces, and associated derived requirements to produce a solution that can be implemented.
Capability Engineering (CP)	An appreciation of the role the system of interest plays in the system of which it is a part.	Technical Leadership (TL)	The application of technical knowledge and experience in systems engineering together with appropriate professional competencies.	Decision Management (DM)	The structured, analytical framework for objectively identifying, characterizing, and evaluating a set of alternatives.	Design for... (DF)	Ensuring that the requirements of all life cycle stages are addressed at the correct point in the system design.
General Engineering (GE)	Foundational concepts in mathematics, science, and engineering and their application.	Negotiation (NE)	Dialogue between two or more parties intended to reach a beneficial outcome where difference exist between them.	Concurrent Engineering (CE)	A work methodology based on the parallelization of tasks.	Integration (IN)	The logical process for assembling a set of system elements and aggregates into the realized system, product, or service.
Critical Thinking (CT)	The objective analysis and evaluation of a topic in order to form a judgement.	Team Dynamics (TD)	The unconscious, psychological forces that influence the direction of a team's behavior and performance.	Business and Enterprise Integration (BE)	The consideration of needs and requirements of other internal stakeholders as part of the system development.	Interfaces (IF)	The identification, definition, and control of interactions across system or system element boundaries.
Systems Modeling and Analysis (SM)	Provision of rigorous data and information including the use of modeling to support technical understanding and decision-making.	Facilitation (FA)	The act of helping others to deal with a process, solve a problem, or reach a goal without getting directly getting involved.	Acquisition and Supply (AS)	Obtaining or providing a product or service in accordance with requirements.	Verification (VE)	A formal process of obtaining objective evidence that a system fulfills its specified requirements and characteristics.
		Emotional Intelligence (EI)	The ability to monitor one's own and others' feelings and use this information to guide thinking and action.	Information Management (IM)	Addresses activities associated with all aspects of information, to provide designated stakeholders with appropriate levels of timeliness, accuracy, and security.	Validation (VA)	A formal process of obtaining objective evidence that the system achieves its intended use in its intended operational environment.
		Coaching and Mentoring (ME)	Development approaches based on the use of one-to-one conversations to enhance an individual's skills, knowledge, or work performance.	Configuration Management (CM)	Ensuring the overall coherence of system functional, performance, and physical characteristics throughout its life cycle.	Transition (TR)	Integration of a verified system into its operational environment including the wider system of which it forms a part.
				Risk and Opportunity Management (RO)	The identification and reduction in the probability of uncertain events, or maximizing the potential of opportunities provided by them.	Utilization and Support (US)	When the system is used to deliver its capabilities and is sustained over its lifetime.
						Retirement (RE)	The final stage of a system life cycle, where the existence of a system is ended for a specific use, through controlled activities.
Integrating competencies	This competency group recognizes Systems Engineering as an integrating discipline, joining activities and thinking from specialists in other disciplines to create a coherent whole.	Project Management (PM)	Identification, planning, and coordinating activities to deliver a satisfactory system, product, service of appropriate quality.	Logistics (LO)	The support and sustainment of a product once it is transitioned to the end user.		
		Finance (FI)	Estimating and tracking costs associated with the project.	Quality (QU)	Achieving customer satisfaction through the control of key product characteristics.		

SECF FIGURE 1 Complete listing of competencies in the Systems Engineering Competency.

This level addresses two categories of individuals:

- Individuals new to the competency who are "in-training" or inexperienced in the competency area.
- Individuals who gained experience as a Practitioner (or higher) level in the competency area in the past, but who have not been operating as a day-to-day Practitioner for an extended period. These individuals are currently not able to operate without some degree of regular supervision. Clearly, over time, their learning may progress at a faster pace than those in-training for the first time as skills, tools, and techniques are relearned.
- **Practitioner**

  The person displays both knowledge and practical experience of the competency area and can function without supervision on a day-to-day basis. They are also capable of providing guidance and advice to less-experienced practitioners.
- **Lead Practitioner**

  The person displays extensive and substantial practical knowledge and experience of the competency area and provides guidance to others including practitioners encountering unusual situations. Typically, this level is associated with an individual who is the "go-to" person for advice and to determine best practice within the competency area within an organization or business unit.
- **Expert**

  In addition to extensive and substantial practical experience and applied knowledge of the competency area, this individual contributes to and is recognized beyond the organizational or business boundary. Typically, this level is associated with an individual contributing to and defining regional or international best practices within the competency area.

  During a formal assessment of competence, in addition, there may be a need to record a level of
- "**Unaware**" meaning that the criteria for "awareness" have not been reached;
- "**Not Applicable**" which would be appropriate if the competence area is tailored out and not applicable to the business; and
- "**Not Assessed**" meaning that for whatever reason, the competency area was not covered as part of the assessment activity.

The assessment of competence is covered in the INCOSE SECAG.

LANGUAGE STANDARDIZATION WITHIN THE FRAMEWORK

In order to allow a common interpretation of indicators across the SECF, a common vocabulary has been defined and used in defining competency indicators and assessment evidence.

Where practicable, the terms use standard dictionary definitions although in some cases, these have needed to be adapted to reflect the activity within the context of the framework.

A list of key vocabulary used to define framework competency indicators is contained in SECF Table 1.

SECF TABLE 1 Key vocabulary used to define SECF and SECAG indicators

Verb	Definition
Acts	Behaves in the way specified.
Adapts	Make (something) suitable for a new use, or new conditions.
Advises	Offers suggestions about the best course of action.
Analyzes	Examines (something) methodically and in detail, typically in order to explain and interpret it.

SECF TABLE 1 (Continued)

Verb	Definition
Arbitrates	Reaches an authoritative judgement or settlement in a dispute or decision. Note: This term is only used in the SECAG Assessment Guide and is included here only for completeness.
Assesses	Evaluates the nature, ability, or quality of something (or someone).
Authorizes	Gives official (e.g. project or organizational) permission for or approval to. Note: This term is only used in the SECAG Assessment Guide and is included here only for completeness.
Challenges	Dispute the truth or validity of.
Champions	Vigorously supports or defends the cause of (a cause, idea, technique, tool, method, etc.).
Coaches	Trains or instructs (a team member or individual) using a set of defined professional skills in this area.
Collaborates	Works jointly on an activity or project. Note: This term is only used in the SECAG Assessment Guide and is included here only for completeness.
Collates	Collects and combines (texts, information, or data).
Communicates	Shares or exchanges information, news, or ideas.
Compares	Notes or measures the similarity or dissimilarity between two or more items.
Complies	Acts in accordance with certain specified standard or guidance.
Conducts	Organizes, and carries out.
Coordinates	Regulates, adjusts, or combines the actions of others to attain coherence or harmony.
Creates	Brings something into existence, causes something to happen as a result of one's actions.
Defines	Characterizes the essential qualities or meaning of an item.
Describes	Gives an account in words, including all the relevant characteristics, qualities, or events.
Determines	Ascertains or establishes specific values of an item by research or through calculation.
Develops	Grows or causes something to grow and become more mature, advanced, or elaborate.
Elicits	Evokes or draws out (a need or requirement, reaction, answer, or fact) from someone.
Ensures	Makes certain that (something) will occur or be the case.
Evaluates	Makes a judgement as a result of informed analysis. Note: This term is only used in the SECAG Assessment Guide and is included here only for completeness.
Explains	Makes an idea, situation, or problem clear to someone by describing it in more detail or revealing relevant facts.
Follows	Acts or adheres to instructions as ordered or required.
Fosters	Encourages the development of (something, especially something desirable).
Guides	Directs or influences the behavior or development of someone.
Identifies	Establishes the identity or complete scope of an item.
Influences	Has an effect on someone or something.
Judges	Forms an opinion or conclusion about.
Liaises	Acts as a link to assist communication between (people or groups). Note: This term is only used in the SECAG Assessment Guide and is included here only for completeness.

(Continued)

SECF TABLE 1 (Continued)

Verb	Definition
Lists	Names a set of items that form a meaningful grouping.
Maintains	Keeps in an existing state.
Monitors	Observes and checks the progress or quality of (something) over a period of time; keeps under systematic review.
Negotiates	Confers or discusses with others with a view to reaching agreement.
Performs	Carries out, accomplishes, or fulfills (an action, task, or function).
Persuades	Causes someone to do something through reasoning or argument.
Prepares	Makes (something) ready for use or consideration.
Produces	Causes something to come into existence, by intellectual effort.
Promotes	Supports or actively encourages further the progress of (a cause, idea, technique, tool, method, etc.).
Proposes	Puts forward (an idea or plan) for consideration or discussion by others.
Provides	Makes something available for use; supplies. Note: This term is only used in the SECAG Assessment Guide and is included here only for completeness.
Reacts	Acts in response to something; responds in a particular way.
Recognizes	Acknowledges or takes notice of in some definite way.
Records	Registers or sets down in writing.
Reviews	Formally assesses a topic or issue, proposing any necessary or desirable changes.
Selects	Carefully chooses as being the best or most suitable.
Trains	Teaches a particular skill or type of behavior through sustained practice or instruction. Note: This term is only used in the SECAG Assessment Guide and is included here only for completeness.
Uses	Employs something as a means of accomplishing a purpose or achieving a result.

COMPETENCY AREA TABLE FORMAT

SECF Annex D contains the formal definition of all competencies within the framework.

Each competency has associated with it a single competency table, which provides:

- **Description** explains what the competency is and provides meaning behind the title. Each title can mean different things to different individuals and enterprises;
- **Why it matters** indicates the importance of the competency and the problems that may be encountered in the absence of that competency; and
- **Effective indicators of knowledge and experience** table contains a list of evidence-based indicators for five defined levels of competence, which are themselves defined below. Indicators provide a definition of what is expected to be demonstrated in order that competence can be assumed at the level indicated. Note that the indicators defined are "entry level" requirements for each level. The timing of when experience was gained should be taken into consideration when assessing current competence: individuals may have had competence at a defined level historically.

The text for each competency indicator is followed by a unique indicator identifier contained within two square brackets thus [. . .], and constructed as follows:

[Group][Competency Area][Level][nn], where

- **[Group]** is a single alpha character referring to one of the competency groups identified in Figure 1, as follows:
  - "C" = "Core" competency group
  - "P" = "Professional" competency group
  - "T" = "Technical" competency group
  - "M" = "Management" competency group
  - "I" = "Integrating" competency group.
- **[Competency Area]** is a two alpha character referencing one of the 37 defined competency areas identified in Figure 2.
- **[Level]** is a single alpha character identifying one of the proficiency levels. Proficiency levels are discussed in the following section.
  - "A" = Awareness proficiency level
  - "S" = Supervised Practitioner proficiency level

Competency area name	Reference code	Competency area name	Reference code
Acquisition and Supply	AS	Logistics	LO
Business and Enterprise Integration	BE	Monitoring and Control	MC
Capability Engineering	CP	Negotiation	NE
Coaching and Mentoring	ME	Planning	PL
Communications	CC	Project Management	PM
Concurrent Engineering	CE	Quality	QU
Configuration Management	CM	Requirements Definition	RD
Critical Thinking	CT	Retirement	RE
Decision Management	DM	Risk and Opportunity Management	RO
Design for…	DF	System Architecting	SA
Emotional Intelligence	EI	Systems Modeling and Analysis	SM
Ethics and Professionalism	EP	Systems Thinking	ST
Facilitation	FA	Team Dynamics	TD
Finance	FI	Technical Leadership	TL
General Engineering	GE	Transition	TR
Information Management	IM	Utilization and Support	US
Integration	IN	Validation	VA
Interfaces	IF	Verification	VE
Life Cycles	LC		

SECF FIGURE 2 Codes used in competency indicator index creation.

  - "P" = Practitioner proficiency level
  - "L" = Lead Practitioner proficiency level
  - "E" = Expert proficiency level.
- **[nn]** is a two-digit numeric which uniquely references a competency identifier within a competency area proficiency level.

For example, the reference [CSTP05] refers to the "Core" group of competencies, the "Systems Thinking" area within that group, the "Practitioner" level of proficiency within that competency area, and the fifth indicator i.e. the text "Explains why the boundary of a system needs to be managed."

USING THE COMPETENCY FRAMEWORK

TYPICAL USAGE SCENARIOS

Organizations and individuals have numerous ways in which they can use the INCOSE Systems Engineering Competency Framework to their advantage.

Organizations use Competency Frameworks for human resource management, as described in the *INCOSE Systems Engineering Handbook* Fourth Edition (INCOSE 2015). This may include using competence assessment in recruiting and selecting candidates for employment; for appraisals, promotions, and compensation decisions; for aligning organizational structures to maximize organizational capability; and to identify workforce training requirements that can be communicated to internal or external training providers who can develop and tailor content that will deliver the required competencies (Holt and Perry 2011; SFIA Foundation 2021).

Individuals may self-assess their competence levels for career planning and identifying needs for personal and professional development; comparing self-assessed competence levels against competency-based vacancy announcements also helps individuals to identify opportunities which match their skills and experience (Holt and Perry 2011; SFIA Foundation 2021). SFIA notes that use of Competency Frameworks in job postings reduces risks both to the individual and the organization, reducing churn induced when individuals feel "the job is not what they thought it would be" and minimizing situations in which the organization discovers they do not have the right set of skills for effective mission execution.

Educational institutions and training providers use industry- and discipline-specific Competency Frameworks to align their offerings to provide graduates the knowledge they need to develop their skills at the right level (SFIA Foundation 2021).

Several "standard" usage scenarios were developed in the Universal Competency Assessment Model (UCAM) (Holt and Perry 2011). These scenarios are not described in detail here. A detailed analysis can be found in the paper "Use Cases for the INCOSE Competency Frameworks" (Hahn and Whitcomb 2017).

However, in summary, the framework can be used to support any of the usage scenarios defined in SECF Table 2.

This document only defines a framework of systems engineering competencies. It does not define how competence can be assessed. The INCOSE Systems Engineering Competency Assessment Guide (SECAG) provides extensive information and guidelines for the assessment of the competencies defined in this framework.

Systems Engineering Competency Assessment Guide: A combined INCOSE Systems Engineering Competency Framework (SECF) and associated Systems Engineering Competency Assessment Guide (SECAG) document, First Edition. INCOSE.

SECF TABLE 2 Typical usage scenarios for the INCOSE SE Competency Framework

Usage	Users	Description
To define required competence outcomes from educational courses	Educators, Employers	A company recruiter or capability manager interacts with a representative of an educational institution to define the competencies expected from those leaving the educational institution. This helps align program content to better prepare graduates for company employment.
To assure employers that students completing a course will have acquired specific knowledge and skills	Educators, Employers	A company recruiter or capability manager interacts with a representative of an educational institution to assess and recruit prequalified students against a set of competency needs for a company pipeline programs.
To align course curricula for external accreditation purposes	Educators, Accreditors	An educational curriculum provider interacts with curriculum sponsors and/or accreditation agencies to assess the effectiveness of an educational course/module in delivering stated outcomes against predefined accreditation objectives. This might be through assessment of learning objectives against competency needs, and outcomes against competence acquired or those attending the course.
To create (or maintain) role descriptions	Employers	An employer defines the needs for an organizational role in terms of competencies and their minimum required levels. This use case is elaborated further in the section on role definitions elsewhere in this document. A competency-driven job definition can also help ensure that the requirements for a role are based upon ability rather than age and thus aligns with age-discrimination legislation in areas such as the European Union (GOV.UK 2017).
To create job vacancies	Employers	An employer publishes the requirements for an organizational role in terms of competencies and their minimum required levels – as defined above. Candidates and recruiting agencies can compare this against their own (or their candidates') competences to determine their suitability for the position. It also supports candidate preparation as it provides an insight into the evidence they may be asked to provide during their application and/or interview.
To support candidate recruitment	Employers	Having defined the requirements for a role in terms of competencies, an employer can assess candidate competence against the required competencies using the Competency Framework Assessment Guide. This helps to provide an objective (and repeatable) assessment of candidates at interview.
To support employee performance assessments and rating	Employers, Employees	An employer sets targets for individual competence attainment in one or more competencies, and provides opportunities for competency development to occur. The competence assessment activity can be used to formally gauge competence level attainment against the targets set, as an input to their overall performance rating.
To define career path models	Employers, Employees	An employer can link career paths within the organization to differing expected combinations of competencies and associated minimum competence levels. This can be used to provide insight to employees as to the competence needs for differing career development paths. This indicates the competencies and levels necessary to progress a selected career path – informing employee career development choices along the way.

SECF TABLE 2 (Continued)

Usage	Users	Description
To self-assess supporting personal career development planning	Employees	An employee can "self-assess" their skills against the Competency Framework, using the assessment guidance provided. This helps inform their career development choices – whether as part of a job application or more generally as part of a personal career path development.
To perform workforce risk analyses or mission/business case analysis	Organizations, Acquirers	An organization can use information gathered through individual employee competency assessment against the framework to analyze organizational capability within a specific application domain of Systems Engineering, or more generally. This could be driven by current or future business aspirations. Acquirers (i.e. organizations placing contracts) could mandate minimum organizational capability requirements for those supporting a contract/task as a risk reduction strategy – requesting capability data based upon competency assessments using the framework rather than traditional more generalized experiential statements from a business.
To target training investment	Organizations	An organization gathers enterprise-wide data through individual employee competence assessment against the framework and uses this to assess organizational-level strengths and weaknesses. This enables training investment to be focused on areas deemed organizationally (and individually) in areas where it is needed most.

TAILORING THE FRAMEWORK

The INCOSE Competency Framework should be tailored as part of its deployment.

The framework is structured so that organizations can tailor it to develop competency models ideally suited to their organizational needs and workforce. The framework contains the fundamental Systems Engineering competencies that can support almost any Systems Engineering role. Using organizations can tailor this Competency Framework to derive a bespoke competency model by:

- Adding or deleting competencies as needed.
- Revising or only using a subset of the competencies.
- Adding, deleting, or revising the proficiency level indicators for any of the five levels for any of the competencies.
- Developing a bespoke set of Systems Engineering roles associated with the necessary supporting competencies.
- Developing their own unique set of use cases for the competency models they derive from the Competency Framework (Gelosh et al. 2017).

Systems Engineering is a broad discipline that interacts with all other engineering disciplines and as such can be deployed in a variety of ways. To support this, the INCOSE Competency Framework can be tailored to make it relevant and appropriate to a specific use. The terminology used in this document for different levels of competence may be relabeled as needed (e.g. to remove any reference to specific roles). The range of competencies encompassed by the Systems Engineering framework is very large and it is not expected that an individual will be operating at the "Expert" or even "Lead Practitioner" level in more than a few of these competencies.

It is important therefore that this framework be used as the common starting point or baseline for tailoring the description of Systems Engineering relevant to an organization and individual. It is expected that an organization will have a set of roles, each with a profile against these competencies (or a tailored subset), with different levels of competence needed. These roles may well include requirements for expertise in other engineering skills and application domain-specific knowledge/experience. An important check for the enterprise will be to ensure that the roles are balanced (expertise not diluted and all key competencies covered) and the means of communication and integration of the roles understood – so that the "team" is appropriately competent in Systems Engineering (Beasley 2013).

Individual Professional Development

Individuals may decide to tailor the framework based on the systems engineering requirements for specific or proposed future roles, and their current level of competence. This allows them to identify career progression exploiting their identified strengths and identification of personal development plans.

Enterprise Capability Development

To use the Competency Framework, an enterprise will need to review their requirements for the different competencies and competence levels and generate a scope for the skills required across the enterprise, in generic roles, within teams and at an individual level. These role specifications can then be mapped to existing and potential employees. These competencies provide a framework for career development and recruitment processes by describing the Systems Engineering skill requirements for a role.

This may require some specific tailoring of competencies to suit the needs of the enterprise. This tailoring can include:

- Combining competencies into definitions relevant/appropriate to the enterprise.
- Utilizing a subset of competencies depending on the specific activities of the organization.

An organization may wish to trace the tailoring back to this original framework, to enable benchmarking against other organizations, and to update in-line with changes to this source framework.

This framework can be adapted and integrated with other frameworks to describe specific roles in the organization. It is important that roles are profiled to define requirements, and then individuals assessed against/matched to the role, rather than starting with the individual.

A general test for completeness of role definitions is to check whether the full scope of Systems Engineering competencies defined within this framework is covered somewhere within the set of enterprise role definitions.

Other Tailoring Approaches

The *INCOSE Systems Engineering Handbook* Fourth Edition (INCOSE 2015), Chapter 8, "Tailoring Process and Application of Systems Engineering," describes several methods whereby organizations can tailor SE processes.

These methods and approaches can also be used to help analyze and tailor this Competency Framework.

The technical report entitled *Atlas 1.1: An Update to the Theory of Effective Systems Engineers* (Hutchison et al. 2018), Section 5.2 Tailoring the Proficiency Framework, provides a general description and two examples of how a using organization can tailor the Atlas Proficiency Framework. These same tailoring approaches could be applied to this Competency Framework to derive a unique competency model to satisfy user needs.

Within INCOSE there have been several published papers and presentations relating to the tailoring and assessment of the framework for systems engineers which can be used to provide additional support and guidance. INCOSE members are able to download these directly from the INCOSE website.

Several organizations other than INCOSE have produced additional generic information which can support the tailoring and general application of frameworks within organizations. One specific example is the Skills Framework for the Information Age Foundation (SFIA) which has been defining information technology skills requirements for over 20 years. In 2021, they published the latest version (Version 8) of their information age skills framework (SFIA Foundation 2021). This provides some good general advice not only in the tailoring of frameworks, but also in their organizational application.

THE RELATIONSHIP BETWEEN ROLES, JOB DESCRIPTIONS, AND COMPETENCIES

A competency framework can be used to define roles and job descriptions. This section summarizes how this can be achieved.

Roles and job descriptions are not the same thing. Merriam-Webster online (Merriam Webster 2018) defines each as follows:

- "Job"
  - A regular remunerative position.
  - A specific duty.
- "Role"
  - A function or part performed, especially in a particular operation or process.

One way to understand this distinction is through analogy. Imagine a hot dog vendor at a ballgame. Their "job" is to sell hot dogs. To do this, they are required to fulfill several distinct roles: driver, cook, server, cashier, salesperson (as well as accountant, business developer, marketing, legal representative, procurement, etc.). To be effective in their job, individuals need to be competent in each of the many roles which make up their overall job scope.

It is often the case that some specific competencies required to discharge one role may be similar to those required by another role. Equally, some roles may demand a higher level of competence than others. For instance, the "cashier" and "salesperson" roles in the example above are likely to both require good "communications" skills, while the "accountant" and cashier roles both need some form of "math" competence, although the level expected from the accountant role is likely to be considerably higher than the cashier role.

Job descriptions also reflect the nature and scope of the enterprise and the way that it is set up and operated. In a small enterprise, such as that of the hot dog vendor, many roles will be performed by a single individual. In large enterprises, roles may be performed by different people within the organization. The situation is the same when defining job descriptions for Systems Engineers in an enterprise, project or team.

A competency-based role statement defines a role not only in terms of what the person performing the role does (e.g. the processes followed) but the competencies required to execute those tasks. Looking at the level of involvement in each of the processes allows definition of the level of each competence needed.

Considerations When Defining Competency-Based Role Statements

Every organization is different and will have different – and possibly incomplete – definitions of processes and competency as well as different organizational structures in which these are deployed. In addition, organizations evolve over time, requiring competencies and processes to evolve to match. The specifics of an individual role statement and which roles are appropriate (or not) within an organization is context dependent.

The SECF Annex C discusses this topic in more detail. It lists several important considerations when defining role statements. In summary, these are:

- Take account of the way the organization is structured.
- Consider the visibility of Systems Engineering within an organization and its relationship with other disciplines.
- Add any application domain-specific competencies to ensure requirements reflect the domain.
- Adjust language and terminology to reflect organizational norms.
- Take account of the purpose of the role statement (e.g. role definition and career development).

SECF ACRONYMS AND ABBREVIATIONS

SECF Table 3 summarizes the acronyms and abbreviations used in the SECF.

SECF TABLE 3 INCOSE SE Competency Framework acronyms and abbreviations

ACRONYM	MEANING
ARCIFE	Accountable, Responsible, Consulted, Informed, Facilitator, Expert
BCS	British Computer Society
BKCASE	Body of Knowledge and Curriculum to Advance Systems Engineering
CIO	Chief Information Officer
CIPD	Chartered Institute of Personnel Development (United Kingdom)
CMMI®	Capability Maturity Model® Integration (CMMI Institute)
CPD	Continued Professional Development
CSEP	Certified Systems Engineering Professional
CSEP-Acq	Certified Systems Engineering Professional (with Acquisition extension)
DAU	Defense Acquisition University (United States)
DoD	Department of Defense (United States)
DoDAF	Department of Defense Architecture Framework
EIA	Electronic Industries Alliance
ESEP	Expert Systems Engineering Professional
IEC	International Electrotechnical Commission

(Continued)

Systems Engineering Competency Assessment Guide: A combined INCOSE Systems Engineering Competency Framework (SECF) and associated Systems Engineering Competency Assessment Guide (SECAG) document, First Edition. INCOSE.

SECF TABLE 3 (Continued)

ACRONYM	MEANING
IEE	Institution of Electronics Engineering (now renamed Institution of Engineering and Technology [IET])
IEEE	Institute of Electrical and Electronics Engineers
INCOSE	International Council on Systems Engineering
INCOSE UK Ltd.	UK Chapter of the International Council on Systems Engineering
ISO	International Organization for Standardization
JTC	Joint Technical Committee
KSA	Knowledge, Skills, and Abilities
MBSE	Model-Based Systems Engineering
MOD	Ministry of Defence (United Kingdom)
NASA	National Aeronautics and Space Administration (United States)
NPS	Naval Postgraduate School
OPM	Office of Personnel Management (United States)
RACI	Responsible, Accountable, Consulted, and Informed
SE	Systems Engineering
SEBoK	Systems Engineering Body of Knowledge
SECAG	Systems Engineering Competency Assessment Guide
SECCM	Systems Engineering Competency Career Model (US Navy)
SECF	Systems Engineering Competency Framework
SEP	(INCOSE) Systems Engineering Professional
SFIA	Skills Framework for the Information Age
UCAM	Universal Competency Assessment Model
UK	United Kingdom
US	United States

SECF GLOSSARY

The glossary in SECF Table 4 defines words/phrases in the context of use within the Systems Engineering Competency Framework. Several sources have been provided in some cases to aid explanation.

Where multiple potential definitions exist, the glossary definition is based upon definitions from the following sources, in the priority listed below:

1. *INCOSE Systems Engineering Handbook* (Fourth Edition) (INCOSE 2015).
2. INCOSE Systems Engineering Body of Knowledge (SEBoK) (BKCASE Project 2017).
3. INCOSE UK Ltd. Systems Engineering Competencies Framework (Issue 3) (INCOSE UK 2010).
4. Other well-established internationally recognized sources.
5. INCOSE Competency Working Group.

SECF TABLE 4 INCOSE SECF glossary

Glossary term	Definition
Ability	A term used in human resource management denoting an acquired or natural capacity or talent that enables an individual to perform a particular task successfully (BusinessDictionary.com 2018).
Architecture	(System) fundamental concepts or properties of a system in its environment embodied in its elements, relationships, and in the principles of its design and evolution (INCOSE 2015).
Architecting	Process of conceiving, defining, expressing, documenting, communicating, certifying proper implementation of, maintaining and improving an architecture throughout a system's life cycle (ISO/IEC/IEEE 2011). The architecting process sometime involves the use of heuristics to establish the form of architectural options before quantitative analyses can be applied. Heuristics are design principles learned from experience (Rechtin 1990).
Authored	Wrote the document or work product (i.e. did not just sign the front page) (INCOSE UK 2010).

(Continued)

Systems Engineering Competency Assessment Guide: A combined INCOSE Systems Engineering Competency Framework (SECF) and associated Systems Engineering Competency Assessment Guide (SECAG) document, First Edition. INCOSE.

SECF TABLE 4 (Continued)

Glossary term	Definition
Behavior	The way in which one acts or conducts oneself, especially towards others (Oxford Dictionaries 2018).
Best practice	A procedure that has been shown by research and experience to produce optimal results and that is established or proposed as a standard suitable for widespread adoption (Merriam-Webster 2018).
Capability	An expression of a system, product, function, or process ability to achieve a specific objective under stated conditions (INCOSE 2015). The ability to achieve a desired effect under specified (performance) standards and conditions through combinations of ways and means (activities and resources) to perform a set of activities (DoD 2009).
Coaching	Helping, supporting, advising, explaining, demonstrating, instructing, and directing others resulting in transfer of knowledge and skills (INCOSE 2015).
Competence	The measure of specified ability (INCOSE 2015).
Competence-based assessment	An evidence-based activity, where an individual is independently assessed (or self-assesses) in one or more defined areas, to determine whether they are able to demonstrate that their knowledge, skills, and experience meet a defined or required level of proficiency ("competence level").
Competency	An observable, measurable set of skills, knowledge, abilities, behaviors, and other characteristics an individual needs to successfully perform work roles or occupational functions. Competencies are typically required at different levels of proficiency depending on the specific work role or occupational function. Competencies can help ensure individual and team performance aligns with the organization's mission and strategic direction (U.S. Office of Personnel Management (OPM) 2015). A measure of an individual's ability in terms of their knowledge, skills, and behavior to perform a given role (Holt and Perry 2011).
Competent	Having a specified level of competence (INCOSE UK 2010).
Complexity (of a system)	A measure of how difficult it is to understand how a system will behave or to predict the consequences of changing it (Sheard and Mostashari 2009). The degree to which a system's design or code is difficult to understand because of numerous components or relationships among components (ISO/IEC 2009). In a complex system, the exact relationship between elements is either unknown and possibly unknowable.
Complicated system	A system comprising many interacting elements where the overall behavior of the system is both knowable and deterministic (c.f. complex).
Configuration Management	Configuration Management is the discipline of identifying and formalizing the functional and physical characteristics of a configuration item at discrete points in the product evolution for the purpose of maintaining the integrity of the product system and controlling changes to the baseline (INCOSE 2017).
Discipline	Area of expertise, e.g. systems, software, hardware, program management, quality assurance, etc. (INCOSE UK 2010).
Domain	A "problem space" (IEEE 2010).
Element	The level below the system of interest (INCOSE 2015). A system that is part of a larger system (Merriam-Webster 2018).

SECF TABLE 4 (Continued)

Glossary term	Definition
Enterprise	One or more organizations sharing a definite mission, goals, and objectives to offer an output such as a product or service (ISO 2000). An organization (or cross-organizational entity) supporting a defined business scope and mission that includes interdependent resources (people, organizations, and technologies) that must coordinate their functions and share information in support of a common mission (or set of related missions) (CIO Council 1999). The term enterprise can be defined in one of two ways. The first is when the entity being considered is tightly bounded and directed by a single executive function. The second is when organizational boundaries are less well defined and where there may be multiple owners in terms of direction of the resources being employed. The common factor is that both entities exist to achieve specified outcomes (MOD 2004). A complex, (adaptive) socio-technical system that comprises interdependent resources of people, processes, information, and technology that must interact with each other and their environment in support of a common mission (Giachetti 2010).
Enterprise Asset	(As applied to a person) known by reputation to be a leader in the field, highly valued, highly regarded. Recognized by the community outside employer organization (e.g. asked to be on conference panel, government advisory board, etc.) (INCOSE UK 2010).
Framework	A basic conceptional structure (as of ideas) (Merriam-Webster 2018). Broad overview, outline, or skeleton of interlinked items which supports a particular approach to a specific objective and serves as a guide that can be modified as required by adding or deleting items (BusinessDictionary.com 2018).
Information	Information is an item of data concerning something or someone. Information includes technical, project, organizational, agreement, and user information.
Interface	A point where two or more entities interact. Interactions may involve systems, system elements including their environment, organizations, disciplines, humans (users, operators, maintainers, developers, etc.), or some combination thereof.
Job	A job is a recognized organizational position, usually performed in exchange for payment. Historically the terms "job" and "role" have been used interchangeably, more recently a distinction has appeared. A job comprises all, or parts, of one or more defined "roles" which govern organizational processes and activities. An individual may remain in the same "job" for a long period, but during this time will usually perform multiple roles. This topic is discussed in further detail in Annex C of this document.
Knowledge	(In the context of KSA) A body of information applied directly to the performance of a function. US Office of Personnel Management (Wikipedia 2017).
Mentoring	Mentoring is a relationship where a more experienced colleague shares their greater knowledge to support development of a less experienced member of staff. It uses many of the techniques associated with coaching. One key distinction is that mentoring relationships tend to be longer term than coaching arrangements.
Organization	A group of people and facilities with an arrangement of responsibilities, authorities, and relationships (INCOSE 2015).
Professional Development	A structured approach to learning to help ensure competence to practice, taking in knowledge, skills, and practical experience. Continued Professional Development (CPD) can involve any relevant learning activity, whether formal and structured or informal and self-directed (INCOSE UK Chapter 2017).

(Continued)

SECF TABLE 4 (Continued)

Glossary term	Definition
Program	A group of related projects managed in a coordinated way to obtain benefits and control not available from managing them individually. Programs may include elements of related work outside of the scope of the discrete projects in the program (PMI 2008).
Project	An endeavor with defined start and finish criteria undertaken to create a product or service in accordance with specified resources and requirements (INCOSE 2015).
Recent	(In the context of competency assessment) Within the last five years (INCOSE UK 2010).
Role	A role is a recognized organizational position, usually performed in exchange for payment. Historically the terms "job" and "role" have been used interchangeably, more recently a distinction has appeared. A job comprises all, or parts, of one or more defined "roles" which govern organizational processes and activities. An individual may remain in the same "job" for a long period, but during this time will usually perform multiple roles. This topic is discussed in further detail in Annex C of this document.
Skills	(In the context of KSA) An observable competence to perform a learned psychomotor act. US Office of Personnel Management (Wikipedia 2017).
Specialty Engineering	(Verb) Analysis of specific features of a system that requires special skills to identify requirements and assess their impact on the system life cycle (INCOSE 2015). (Noun) Specialty engineering is the collection of those narrow disciplines that are needed to engineer a complete system (Elowitz 2006).
Stakeholder	A party having a right, share, claim, or vested interest in a system or in its possession of characteristics that meet that party's needs and expectations (INCOSE 2015).
Stage (Life Cycle Stage)	A period within the life cycle of an entity that relates to the state of its description or realization (ISO/IEC/IEEE 2015).
Sub System	See Element.
Super system	The level above the system of interest (INCOSE 2015). A system that is made up of systems (Merriam-Webster 2018).
System	A system is an arrangement of parts or elements that together exhibit behavior or meaning that the individual constituents do not (INCOSE 2019).
Systems Engineering	Systems Engineering is a transdisciplinary and integrative approach to enable the successful realization, use, and retirement of engineered systems, using systems principles and concepts, and scientific, technological, and management methods (INCOSE 2019). *INCOSE uses the terms* "engineering" *and* "engineered" *in their widest sense:* "the action of working artfully to bring something about." "Engineered systems" *may be composed of any or all of people, products, services, information, processes, and natural elements.*
System Element	A member of a set of elements that constitutes a system. A system element is a discrete part of a system that can be implemented to fulfill specified requirements. A system element can be hardware, software, data, humans, processes (e.g. processes for providing service to users), procedures (e.g. operator instructions), facilities, materials, and naturally occurring entities (e.g. water, organisms, minerals), or any combination (ISO/IEC/IEEE 2015). A system element can be a subsystem.
System of Interest	The system whose life cycle is under consideration (INCOSE 2015).

SECF TABLE 4 (Continued)

Glossary term	Definition
System of Systems	A System of Interest whose system elements are themselves systems; typically, these entail large-scale interdisciplinary problems with multiple, heterogeneous, distributed systems (INCOSE 2015).
Tailoring	Tailoring a process adapts the process description for a particular end. For example, a project creates its defined process by tailoring the organization's set of standard processes to meet the objectives, constraints, and environment of the project (Adapted from notes/discussion of "tailoring guide") (ISO/IEC/IEEE 2009).
Team	A group of individuals who work together to achieve a common goal.
Verification	Provision of objective evidence that the system or system element fulfills its specified requirements and characteristics (INCOSE 2015).
Validation	Provision of objective evidence that the system, when in use, fulfills its business or mission objectives and stakeholder requirements, achieving its intended use in its intended operational environment (INCOSE 2015).

SECF BIBLIOGRAPHY

Beasley, R. (2013). The need to tailor competency models – with a use case from Rolls-Royce. *INCOSE International Symposium IS2013,* Philadelphia, PA (24–27 June 2013).

BKCASE Project (2017). SEBoK v1.9 section 5: enabling systems engineering. http://sebokwiki.org/wiki/Enabling_Systems_Engineering (accessed 30 November 2018).

Brackett, M.A., Rivers, S.E., and Salovey, P. (2011). Emotional intelligence: implications for personal, social, academic, and workplace success. *Social and Personality Psychology Compass* 5: 88–103. https://doi.org/10.1111/j.1751-9004.2010.00334.

BusinessDictionary.com (2018). Framework definition. http://www.businessdictionary.com/definition/framework.html (accessed April 2018).

Chartered Institute of Personnel Development (CIPD) (2021a). What are coaching and mentoring? Coaching and Mentoring Factsheet. https://www.cipd.co.uk/knowledge/fundamentals/people/development/coaching-mentoring-factsheet (accessed 13 July 2022).

Chartered Institute of Personnel Development (CIPD) (2021b). Competence or competency? https://www.cipd.co.uk/knowledge/fundamentals/people/performance/competency-factsheet (accessed 13 July 2022).

CIO Council (1999). *Federal Enterprise Architecture Framework (FEAF).* Washington, DC, USA: Chief Information Officer (CIO) Council.

CMMI® Institute (2010). *Capability Maturity Model Integration (CMMI®) V1.3.* CMMI® Institute.

Defense Acquisition University (2013). *SPRDE Competency Model.* Defense Acquisition University (DAU)/U.S. Department of Defense June 12.

DoD (2009). *DoD Architecture Framework (DoDAF), version 2.0.* Washington, DC, USA: U.S. Department of Defense (DoD).

Electronic Industries Alliance (2002). *Systems Engineering Capability Model EIA731_1:2002.* EIA.

Elowitz, M. (2006). Specialty engineering as an element of systems engineering. Albuquerque, NM, USA: Presentation to Enchantment Chapter, INCOSE. Albuquerque, NM, USA, (9 August).

Gelosh, D., Heisey, M., Snoderly, J., and Nidiffer, K.. (2017). Version 0.75 of the proposed INCOSE systems engineering competency framework. *Proceedings of the International Symposium 2017 of the International Council on Systems Engineering (INCOSE),* Adelaide, Australia (15–20 July 2017).

Giachetti, R.E. (2010). *Design of Enterprise Systems: Theory, Architecture, and Methods.* Boca Raton, FL, USA: CRC Press, Taylor and Francis Group.

Systems Engineering Competency Assessment Guide: A combined INCOSE Systems Engineering Competency Framework (SECF) and associated Systems Engineering Competency Assessment Guide (SECAG) document, First Edition. INCOSE.

GOV.UK (2017). Equality act 2010: guidance. https://www.gov.uk/guidance/equality-act-2010-guidance#age-discrimination (accessed 31 December 2017).

Hahn, H.A. and Whitcomb, C.A.. (2017). Use Cases for the INCOSE Competencies Framework. Unpublished.

Holt, J. and Perry, S. (2011). *A Pragmatic Guide to Competency Tools, Frameworks and Assessment*. BCS.

Hutchison, N., Verma, D., Burke, P. et al. (2018). Atlas 1.1: an update to the theory of effective systems engineers. *Technical Report SERC-2018-TR-101A*. Report, Hoboken, NJ: Systems Engineering Research Center, Stevens Institute of Technology. https://sercuarc.org/publication/?id=186&pub-type=Technical-Report&publication=SERC-2018-TR-101-A-Atlas+1.1%3A+An+Update+to+the+Theory+of+Effective+Systems+Engineers (accessed 1 July 2018).

IEEE (2010). *IEEE 1570-2010 Standards for Information Technology Standard*. New York, NY, USA: Institute of Electrical and Electronics Engineers.

INCOSE (2021). *Engineering Solutions for a Better World Vision 2035*. San Diego, CA: INCOSE.

INCOSE (2018a). Certification. https://www.incose.org/certification (accessed 2018).

INCOSE (2017). SEBoK v1.9 systems engineering body of knowledge. http://www.sebokwiki.org/wiki/Guide_to_the_Systems_Engineering_Body_of_Knowledge_(SEBoK) (accessed 30 November 2017).

INCOSE (2015). *Systems Engineering Handbook – A Guide for System Life Cycle Processes and Activities*, 4e. San Diego, CA: Wiley.

INCOSE UK Chapter (2017). U3: what is continuing professional development (CPD). https://incoseonline.org.uk/Documents/uGuides/dot107_U-guide_U3_WEB.pdf (accessed February 2018).

INCOSE UK (2010). *INCOSE UK SE Competency Framework/Guide*. INCOSE UK.

INCOSE (2004). What is systems engineering? http://www.incose.org/practice/whatissystemseng.aspx (accessed April 2018).

INCOSE (2018b). INCOSE Systems Engineering Competency Framework. *INCOSE INCOSE-TP-2018-002-01.0*.

INCOSE (2019). INCOSE website SE and SE definitions. https://www.incose.org/about-systems-engineering/system-and-se-definition (accessed November 2021).

INCOSE (2022). Systems Engineering Competency Assessment Guide (SECAG). *INCOSE-TPP-2020-38*.

INCOSE, OMG (2018). MBSE wiki. http://www.omgwiki.org/MBSE/doku.php (accessed April 2018).

Institution of Engineering and Technology (2013). Competency framework for independent safety assessors (ISAs), November 17. https://www.theiet.org/factfiles/isa/index.cfm (accessed 19 February 2018).

ISO. 2000. ISO 15704. 2000 (2000). *Industrial Automation Systems – Requirements for Enterprise-Reference Architectures and Methodologies Standard*. Geneva, Switzerland: International Organization for Standardization (ISO).

ISO/IEC (2009). *ISO/IEC 24765. Systems and Software Engineering Vocabulary (SEVocab) Standard*. Geneva, Switzerland: International Organization for Standardization (ISO)/International Electrotechnical Commission (IEC) [database online].

ISO/IEC JTC 1/SC 7 Software and Systems Engineering Technical Committee (2015). *ISO/IEC/IEEE 15288:2015 Systems and software engineering – System life cycle processes*. ISO/IEC.

ISO/IEC/IEEE (2009). *ISO/IEC/IEEE 24765:2009 Systems and Software Engineering – System and Software Engineering Vocabulary*. Geneva: International Organization for Standardization (ISO)/ International Electrotechnical Commission (IEC)/ Institute of Electrical and Electronics Engineers (IEEE).

ISO/IEC/IEEE (2011). *ISO/IEC/IEEE 42010 Systems and software engineering – Architecture Description*. Geneva, Switzerland: International Organization for Standardization (ISO)/International Electrotechnical Commission (IEC)/Institute of Electrical and Electronics Engineers (IEEE).

ISO/IEC/IEEE (2015). *ISO15288: Systems and Software Engineering – System Life Cycle Processes*. Geneva, Switzerland: International Organization for Standardization/International Electrotechnical Commissions / Institute of Electrical and Electronics Engineers. ISO/IEC/IEEE.

Kendall, F. (2014). US DoD's better buying power 3.0 implementation plan. White Paper, Office of the Under Secretary of Defense Acquisition, Technology and Logistics.

Mayer, J.D. and Salovey, P. (1997). What is emotional intelligence? In: *Emotional Development and Emotional Intelligence: Educational implications* (ed. P. Salovey and D.J. Sluyter), 3–31. New York: Harper Collins.

Merriam Webster (2018). Dictionary by Merriam Webster. https://www.merriam-webster.com (accessed 2018).

MOD (2004). *Ministry of Defence Architecture Framework (MODAF), Version 2*. London, UK: U.K. Ministry of Defence.

National Aeronautics and Space Administration (NASA) (2016). *NASA Systems Engineering Handbook (SP-2016-6105, Rev 2)*. NASA.

Oxford Dictionaries (2018). Definition: behaviour (US: behavior). https://en.oxforddictionaries.com/definition/behaviour (accessed April 2018).

PMI (2008). *A Guide to the Project Management Body of Knowledge (PMBOK® Guide)*, 4e. Newtown Square, PA, USA: Project Management Institute (PMI).

Prati, M., Douglas, F., and Ammeter, B. (2003). Emotional intelligence, leadership Effectiveness, and team outcomes. *The International Journal of Organizational Analysis* 11 (9): 21–40.

Presland, I. (2017). Comparison of SEP Technical Areas with Version 4 Framework Competencies performed for Certification Advisory Group. Unpublished – Internal INCOSE Document, June.

Rechtin, E. (1990). *Systems Architecting: Creating & Building Complex Systems*. Upper Saddle River, NJ, USA: Prentice-Hall.

Rumsfeld, D. (2012). There are known knowns. Briefing at United States Defence Department. Wikipedia (accessed 2018).

Salovey, P. and Mayer, J.D. (1990). Emotional intelligence. *Imagination, Cognition and Personality* 9 (3): 185–211.

SFIA Foundation (2021). Skills framework for the information age (V8). https://sfia-online.org/en/sfia-8 (accessed 31 December 2021).

Sheard, S. and Mostashari, A. (2009). Principles of complex systems for systems engineering. *Systems Engineering* 12 (4): 295–311.

Sheard, S. (1996). 12 Systems engineering roles. *INCOSE International Symposium*. Boston MA, USA: INCOSE.

U.S. Office of Personnel Management (OPM) (2015). Human capital assessment and accountability framework (HCAAF) resource center, "glossary". http://www.opm.gov/hcaaf_resource_center/glossary.asp (accessed June 2015).

Whitcomb, C.A., Khan, R.H., and White, C.L. (2014). Development of a System Engineering Competency Career Model: An Analytical Approach Using Bloom's Taxonomy. Technical Report, US Naval Postgraduate School.

Wikipedia (2017). Knowledge, skills and abilities. https://en.m.wikipedia.org/wiki/Knowledge,_Skills,_and_Abilities (accessed 31 December 2017).

Williams, C.R. and Reyes, A. (2012). Developing Systems Engineers at NASA. *Global Journal of Flexible Systems Management* 13 (3): 159–164.

SECF ANNEX A: SUMMARY AND RATIONALE FOR CHANGES IN THE SECF SECOND EDITION

This Annex presents a summary of the changes made from the 2018 edition of the SECF.

SECF STRUCTURAL CHANGES

The most significant document structure change is the removal of matter which solely relates to the assessment of competency, which is published in the Systems Engineering Competency Assessment Guide. The motivation for this is to limit the scope of this document to the framework and its definition, rather than the application of the framework. As a result of this change, text for many items has been reworked and associated sections, tables, and figures updated and/or renumbered.

In addition, a list of verbs used in the definition of competency indicators has been added.

PRIMARY TECHNICAL CHANGES

The most significant technical change is the creation of a NEW (37th) competency through the split of the "Operation and Support" technical competency into two distinct technical competencies: "Utilization and Support" and "Retirement." This split recognizes the distinct set of competencies required in order to perform system retirement and better reflects the underlying ISO 15288 standard in this area.

Although changes have been made to many of the indicators within competency areas, the majority of these changes are minor in nature and have been made in order try to improve consistency across the competency suite. They include:

- Standardization of terminology usage throughout (e.g. trying to ensure the same term is used throughout for a similar type of activity).
- Related to the above, there has been additional work to ensure all competencies express similar ideas in a similar manner. We believe this helps standardize the assessment approach and methods.

Systems Engineering Competency Assessment Guide: A combined INCOSE Systems Engineering Competency Framework (SECF) and associated Systems Engineering Competency Assessment Guide (SECAG) document, First Edition. INCOSE.

- Inclusion of a requirement across all competency areas to perform continuous professional development at all levels, from Awareness to Expert.
- A greater focus on diversity and inclusion in the Ethics and Professionalism competency area.
- A greater focus on sustainability and the environment, primarily in the Ethics and Professionalism competency area, but also more generally.
- Clarification of the relationship between Systems Engineering and its "project management" activities such as planning, monitoring, and control and the "Project Management" integrating competency.
- Clarification of other integrating competencies to align with this approach.

SECF ANNEX B: ALIGNMENT WITH INCOSE AND OTHER INITIATIVES

INCOSE SYSTEMS ENGINEERING HANDBOOK FOURTH EDITION

The *INCOSE Systems Engineering Handbook*, Fourth Edition (INCOSE 2015) describes the discipline, the practice, and the processes that need to be carried out when doing Systems Engineering. Competencies are the specific abilities needed by individuals to perform the processes and conduct "good" Systems Engineering. Competency is the specific ability needed by an individual to support/contribute to the execution of Systems Engineering processes.

Competency provides a different perspective to Systems Engineering to the process view. Combining process and competency gives a more complete view of what is needed by both organizations and individuals to achieve the benefits of Systems Engineering.

There are many Systems Engineering processes, and they are performed by teams of individuals. Understanding the relationships among the processes and competencies helps to assign process activities and the aligned supporting competencies to the individuals in those teams.

The general relationships between competencies defined in this framework and processes defined in the INCOSE Systems Engineering Handbook, Fourth Edition (2015) are shown in SECF Figure 3. This figure is primarily intended for those seeking guidance in this area as the mappings are somewhat subjective: many of the relationships are not one-to-one; competencies may underpin more than one process and their linkage is complex.

In SECF Figure 3, the following key applies:

- ✓✓✓ The framework competency definition has a significant relationship with the reference handbook process.
- ✓ The framework competency definition has a relationship to the reference handbook process, but this is somewhat limited.

Note that SECF Figure 3 does not attempt to map professional competencies. These are addressed in Chapter 2 of the INCOSE Systems Engineering Handbook, Fourth Edition (2015).

Systems Engineering Competency Assessment Guide: A combined INCOSE Systems Engineering Competency Framework (SECF) and associated Systems Engineering Competency Assessment Guide (SECAG) document, First Edition. INCOSE.

Columns 1–14: Technical processes

SECF competency areas		Business or mission analysis	Stakeholder needs and requirements definition	System requirements definition	Architecture definition	Design definition	System analysis	Implementation	Integration	Verification	Transition	Validation	Operation	Maintenance	Disposal
Core SE principles	Systems thinking	✓✓✓	✓✓✓	✓✓✓	✓✓✓	✓	✓	✓✓✓	✓✓✓	✓	✓✓✓	✓✓✓	✓✓✓	✓✓✓	✓✓✓
	Life cycles	✓	✓	✓	✓		✓	✓✓✓	✓	✓✓✓	✓	✓	✓	✓	✓
	Capability engineering	✓	✓✓✓	✓	✓	✓	✓		✓✓✓	✓	✓	✓✓✓	✓	✓	✓
	General engineering	✓	✓✓✓	✓	✓✓✓	✓✓✓	✓✓✓	✓	✓	✓	✓	✓	✓	✓	✓
	Critical thinking	✓	✓✓✓	✓	✓✓✓	✓✓✓	✓	✓	✓	✓	✓	✓	✓	✓	✓
	Systems modeling and analysis	✓✓✓	✓✓✓	✓✓✓	✓✓✓	✓✓✓	✓✓✓	✓	✓	✓	✓	✓	✓	✓	✓
Professional competencies	ALL	Not directly addressed in *INCOSE SE Handbook*													
Technical competencies	Requirements definition	✓✓✓	✓✓✓	✓✓✓	✓	✓				✓		✓	✓	✓	✓
	System architecting		✓		✓✓✓	✓✓✓									
	Design for...					✓✓✓	✓✓✓								
	Integration				✓	✓		✓✓✓	✓✓✓						
	Interfaces		✓	✓	✓	✓			✓						
	Verification			✓✓✓						✓✓✓					
	Validation	✓	✓✓✓									✓✓✓			
	Transition										✓✓✓				
	Utilization and support		✓✓✓	✓✓✓	✓✓✓	✓✓✓							✓✓✓	✓✓✓	✓✓✓
	Retirement		✓✓✓	✓✓✓	✓	✓								✓	✓
SE management competencies	Planning	✓	✓	✓		✓	✓	✓	✓✓✓	✓	✓✓✓	✓✓✓			
	Monitoring and control							✓	✓	✓	✓	✓		✓	
	Decision management				✓✓✓	✓✓✓		✓							
	Concurrent engineering			✓	✓	✓	✓	✓	✓	✓					
	Business and enterprise integration	✓✓✓	✓					✓			✓		✓	✓	✓
	Acquisition and supply			✓✓✓				✓✓✓	✓						
	Information management	✓	✓	✓	✓	✓	✓	✓	✓	✓✓✓		✓	✓	✓	✓
	Configuration management				✓	✓	✓	✓✓✓	✓	✓✓✓	✓	✓	✓	✓	
	Risk and opportunity management	✓	✓		✓	✓				✓		✓			
Integrating competencies	Project management		✓												
	Finance	✓													
	Logistics												✓	✓	
	Quality	✓	✓	✓											

Columns 15–30: Technical management processes (15–22), Agreement processes (23–24), Organizational project-enabling processes (25–30)

SECF competency areas		Project planning	Project assessment and control	Decision management	Risk management	Configuration management	Information management	Measurement	Quality assurance	Acquisition	Supply	Life cycle model management	Infrastructure management	Portfolio management	Human resource management	Quality management	Knowledge management
Core SE principles	Systems thinking	✓✓✓	✓	✓	✓		✓	✓	✓			✓✓✓	✓	✓	✓	✓	✓
	Life cycles	✓✓✓	✓	✓	✓	✓	✓			✓	✓	✓✓✓		✓			
	Capability engineering	✓	✓	✓	✓	✓	✓		✓			✓	✓	✓✓✓	✓		✓
	General engineering	✓		✓	✓✓✓		✓										✓
	Critical thinking	✓	✓✓✓	✓✓✓	✓✓✓		✓	✓✓✓	✓✓✓							✓✓✓	✓
	Systems modeling and analysis			✓	✓✓✓							✓	✓				✓
Professional competencies	ALL	Not directly addressed in *INCOSE SE Handbook*															
Technical competencies	Requirements definition				✓		✓		✓✓✓	✓	✓					✓✓✓	
	System architecting					✓✓✓											
	Design for...								✓			✓				✓✓✓	
	Integration																
	Interfaces																
	Verification								✓✓✓								
	Validation								✓✓✓								
	Transition									✓	✓	✓✓✓					
	Utilization and support			✓									✓				
	Retirement			✓									✓				
SE management competencies	Planning	✓✓✓			✓							✓		✓			
	Monitoring and control		✓✓✓	✓	✓			✓✓✓	✓								
	Decision management		✓	✓✓✓			✓	✓	✓								
	Concurrent engineering	✓	✓				✓✓✓										
	Business and enterprise integration						✓						✓✓✓	✓	✓✓✓	✓✓✓	✓✓✓
	Acquisition and supply			✓	✓	✓				✓✓✓	✓✓✓		✓				
	Information management	✓		✓		✓	✓✓✓										✓
	Configuration management	✓		✓		✓✓✓	✓			✓✓✓	✓✓✓						
	Risk and opportunity management	✓✓✓	✓✓✓	✓	✓✓✓	✓	✓		✓					✓✓✓			
Integrating competencies	Project management	✓✓✓	✓✓✓							✓	✓	✓	✓	✓✓✓	✓		
	Finance									✓	✓		✓✓✓	✓✓✓	✓✓✓		
	Logistics					✓	✓			✓	✓✓✓		✓				
	Quality								✓✓✓	✓✓✓	✓✓✓					✓✓✓	

SECF FIGURE 3 Mapping of SE Handbook processes to framework competencies.

INCOSE SYSTEMS ENGINEERING PROFESSIONAL (SEP) CERTIFICATION PROGRAM

The INCOSE Systems Engineering Professional (SEP) Certification program (INCOSE 2018a) is another important initiative for alignment.

At "Practitioner" level status, to acquire Certified Systems Engineering Professional or CSEP designation, this program requires applicants to demonstrate knowledge and experience in 14 Systems Engineering "Technical Areas." This is supplemented by a "knowledge examination" based upon the *INCOSE Systems Engineering Handbook* (INCOSE 2015).

To acquire Expert Systems Engineering Professional (ESEP) designation, the requirements for demonstrable experience are made significantly broader and are supplemented with an additional requirement to demonstrate technical leadership within the Systems Engineering profession. (NOTE: There is currently no "knowledge" examination at this level.)

Clearly, these concepts of demonstrating "knowledge and experience" bear a resemblance to demonstrating competence, although the SEP program does not currently employ the assessment-based approach to validating competence identified in this framework.

As part of a Certification Advisory Group (CAG) initiative to improve alignment of the SEP program to other INCOSE products, in June 2017, the SEP Technical Areas were compared against the 36 competencies contained within the 2018 framework (Presland 2017). This initial comparison has now been updated to reflect the revised set of 37 competencies and further review by CWG participants. The results of this updated comparison are shown in Figure 4.

In Figure 4, the following key applies:

- ✓✓✓ The SEP technical area has an extensive or significant overlap with the framework competency area.
- ✓ The SEP technical area has a more limited, or implicit relationship to the framework competency area or is encompassed by multiple areas.

Grayed-out cells indicate professional competencies covered formally only as part of Expert-level INCOSE designation. These competencies are not expected to be demonstrated at the "Practitioner" (i.e. CSEP) level.

SECF Figure 4 is a primarily intended subjective mapping provided for those seeking guidance in this area. However, it suggests all SEP Technical Areas are covered by at least one of the competencies within the Competency Framework. Further work on mapping competencies, especially the Professional competencies is still required.

NOTE: The SEP Technical Area "Other" is designed primarily to capture experience not classifiable by an applicant into one of the other 13 SEP Technical Areas. Such experience (if deemed to be "Systems Engineering") will normally be reclassified by application reviewers into one or more of the other Technical Areas, meaning that coverage of "other" as a technical area is not required.

INCOSE VISION 2035 ROLES AND COMPETENCIES

The INCOSE publication Engineering Solutions for a Better World Systems Engineering Vision 2035 (INCOSE 2021) contains a section devoted to "Roles and Competencies". This Vision 2035 section highlights that "systems engineers must be holistic thinkers, strong communicators, and maintain a broad view of systems and how they are being used". The section goes on to say that systems engineers "must develop a breadth of knowledge, and a balance of skills". The section includes the full list of SECF systems engineering competency areas as the source that covers much of the systems engineering discipline. The section goes on to state that no one person may need expertise in all of the competencies, that the composite expertise of systems engineering teams should provide the requisite skills. The Vision 2035 also includes a Top-Level Roadmap that includes competencies in a path to "Realizing the Vision 2035". The identified 2035 goal is that "systems engineering should be embedded at all educational levels, and across disciplines, supported by innovative education and training approaches". This INCOSE Systems Engineering Competency Framework update aligns well with the intent of the Vision 2035 document.

SEP program technical areas \ SECF competency areas	Core SE principles						Professional competencies							
	Systems thinking	Life cycles	Capability engineering	General engineering	Critical thinking	Systems modeling and analysis	Communications	Ethics and professionalism	Technical leadership	Negotiation	Team dynamics	Facilitation	Emotional intelligence	Coaching and mentoring
Requirements engineering	✓✓✓	✓	✓✓✓	✓	✓	✓✓✓	✓	✓	✓✓✓	✓✓✓	✓	✓✓✓	✓	✓
Systems and decision analysis	✓✓✓		✓	✓	✓	✓✓✓	✓	✓	✓✓✓		✓	✓✓✓	✓	✓
Architecture/design development	✓✓✓		✓✓✓	✓	✓	✓✓✓	✓	✓	✓✓✓	✓	✓	✓✓✓	✓	✓
Systems integration	✓		✓✓✓	✓	✓	✓✓✓	✓	✓	✓✓✓			✓		✓
Verification and validation	✓		✓	✓	✓	✓✓✓	✓	✓	✓✓✓	✓✓✓		✓		✓
System operation and maintenance	✓	✓	✓	✓	✓	✓	✓	✓	✓✓✓			✓		✓
Technical planning	✓✓✓		✓	✓	✓		✓	✓	✓✓✓	✓	✓✓✓	✓✓✓	✓	✓✓✓
Technical monitoring and control	✓✓✓		✓	✓	✓	✓	✓	✓	✓✓✓	✓	✓✓✓	✓✓✓	✓✓✓	✓✓✓
Acquisition and supply	✓	✓	✓✓✓	✓	✓		✓	✓	✓✓✓	✓✓✓	✓	✓✓✓	✓✓✓	✓
Information and configuration management	✓	✓	✓	✓	✓		✓	✓	✓✓✓			✓		✓
Risk and opportunity management	✓✓✓		✓✓✓	✓	✓	✓✓✓	✓	✓	✓✓✓	✓	✓	✓	✓	✓
Life cycle process definition and management	✓	✓✓✓	✓	✓	✓		✓	✓	✓✓✓			✓		✓
Specialty engineering	✓✓✓		✓	✓	✓	✓✓✓	✓	✓	✓✓✓	✓	✓	✓	✓	✓
Organizational project enabling activities	✓✓✓		✓	✓	✓		✓	✓	✓✓✓	✓	✓	✓	✓	✓✓✓
Other														

SEP program technical areas \ SECF competency areas	Technical competencies									
	Requirements definition	System architecting	Design for...	Integration	Interfaces	Verification	Validation	Transition	Utilization and support	Retirement
Requirements engineering	✓✓✓	✓✓✓	✓✓✓	✓✓✓	✓✓✓	✓✓✓	✓✓✓	✓	✓✓✓	✓✓✓
Systems and decision analysis	✓✓✓	✓	✓						✓	✓
Architecture/design development	✓	✓✓✓	✓✓✓	✓✓✓	✓✓✓	✓	✓		✓	✓
Systems integration	✓	✓✓✓	✓	✓✓✓	✓✓✓	✓✓✓	✓✓✓	✓		✓
Verification and validation	✓✓✓	✓✓✓	✓✓✓	✓✓✓	✓✓✓	✓✓✓	✓✓✓	✓✓✓	✓✓✓	✓✓✓
System operation and maintenance			✓		✓			✓✓✓	✓✓✓	✓✓✓
Technical planning	✓	✓		✓✓✓	✓	✓	✓	✓	✓✓✓	✓✓✓
Technical monitoring and control			✓						✓✓✓	✓✓✓
Acquisition and supply	✓✓✓	✓✓✓	✓	✓✓✓	✓✓✓	✓✓✓	✓✓✓		✓	✓
Information and configuration management	✓	✓		✓	✓✓✓	✓✓✓	✓✓✓	✓✓✓	✓✓✓	✓✓✓
Risk and opportunity management	✓✓✓	✓✓✓	✓✓✓	✓✓✓	✓✓✓	✓✓✓	✓✓✓	✓✓✓	✓✓✓	✓✓✓
Life cycle process definition and management			✓✓✓	✓		✓	✓	✓	✓	✓
Specialty engineering	✓	✓	✓✓✓	✓	✓	✓✓✓	✓✓✓	✓	✓	✓✓✓
Organizational project enabling activities	✓			✓	✓				✓✓✓	✓✓✓
Other										

SEP program technical areas \ SECF competency areas	SE management competencies									Integrating competencies			
	Planning	Monitoring and control	Decision management	Concurrent engineering	Business and enterprise integration	Acquisition and supply	Information management	Configuration management	Risk and opportunity management	Project management	Finance	Logistics	Quality
Requirements engineering	✓✓✓	✓	✓		✓	✓✓✓	✓	✓	✓✓✓	✓		✓	✓
Systems and decision analysis	✓	✓	✓✓✓						✓	✓	✓	✓	✓
Architecture/design development			✓	✓	✓	✓✓✓	✓	✓	✓✓✓				
Systems integration			✓	✓	✓	✓✓✓	✓	✓	✓✓✓				✓
Verification and validation	✓	✓	✓	✓	✓	✓✓✓	✓	✓	✓✓✓	✓		✓✓✓	✓✓✓
System operation and maintenance	✓✓✓	✓✓✓	✓	✓✓✓	✓	✓	✓✓✓	✓✓✓	✓	✓		✓✓✓	✓✓✓
Technical planning	✓✓✓	✓	✓	✓✓✓	✓	✓✓✓	✓	✓	✓✓✓	✓✓✓	✓✓✓	✓	✓
Technical monitoring and control	✓✓✓	✓✓✓	✓	✓	✓	✓			✓	✓✓✓	✓✓✓	✓	✓
Acquisition and supply	✓✓✓	✓✓✓	✓	✓✓✓		✓✓✓	✓✓✓	✓✓✓	✓✓✓	✓	✓	✓	✓✓✓
Information and configuration management		✓✓✓	✓		✓	✓✓✓	✓✓✓	✓✓✓	✓	✓✓✓		✓	✓
Risk and opportunity management	✓✓✓	✓✓✓	✓✓✓	✓✓✓	✓	✓✓✓			✓✓✓	✓✓✓	✓✓✓	✓	✓
Life cycle process definition and management	✓✓✓	✓	✓	✓✓✓	✓	✓	✓	✓	✓	✓✓✓		✓	✓
Specialty engineering	✓	✓	✓	✓	✓	✓✓✓	✓	✓	✓			✓✓✓	✓✓✓
Organizational project enabling activities	✓✓✓	✓✓✓	✓	✓	✓✓✓		✓	✓	✓	✓✓✓	✓✓✓	✓✓✓	✓✓✓
Other													

SECF FIGURE 4 Comparison of SEP Technical Areas to SECF framework competencies.

INCOSE MODEL-BASED SYSTEMS ENGINEERING (MBSE) INITIATIVE

The INCOSE Model-Based Systems Engineering (MBSE) initiative is an important direction for Systems Engineering and is a key plank of the INCOSE strategic direction. An overview of this initiative can be found at INCOSE, OMG (2018). This Competency Framework fully supports this initiative in two key ways:

First, Systems Modeling and Analysis has been added as a core competency in the framework, recognizing its importance to both current and future direction of Systems Engineering. Within the competence definition, it is recognized that modeling is more than just the skills to run modeling software; it includes the "modeling mindset" and the ability not only to produce appropriate models, but also to understand their purpose and to use their output effectively.

Second, INCOSE recognizes that models of many types are used increasingly to support all aspects of Systems Engineering: concept generation, requirements elicitation, use cases, architecture, detailed analysis, integration, verification, validation, and in support of operation and maintenance. Sometimes these models are distinct and used in isolation. Others are increasingly linked together to form an integrated approach.

We expect Systems Engineers to be required to apply aspects of MBSE within their organization on a day-to-day basis, although the specific tools, processes, and scope of any MBSE implementation will vary greatly from organization to organization.

However, irrespective of the scope of MBSE within an organization, the expectation will always be that an individual understands and can apply the systems engineering principles implemented rather than the tools in isolation. As such, the SECF focuses on Systems Engineering technical principles which should have been captured in any MBSE approach. The scope and specifics of tools will then form part of organizational tailoring as described elsewhere in this document.

Furthermore, Systems Engineering Management competencies will be required to control the scope and production of models through planned activities, and Professional Competencies will assist in the creation of effective models and communication of results to all parties.

ATLAS PROFICIENCY MODEL

The technical report entitled Atlas 1.1: An Update to the Theory of Effective Systems Engineers (Hutchison et al. 2018) describes the Atlas Proficiency Model developed by the Systems Engineering Research Center. Section 5.4 of the technical report compares the Atlas Proficiency Model with the draft INCOSE Competency Framework Version 0.75. The set of competencies in this final Competency Framework are the same as in Version 0.75 so the comparison remains valid. Section 5.4 also states that: "Overall, the Helix team found that though the different approaches taken led to different grouping of knowledge, skills, and abilities, the INCOSE 0.75 Competency Model and the Atlas proficiency model aligned well."

SECF ANNEX C: DEFINING ROLES USING THE FRAMEWORK

DEFINING ROLES - INTRODUCTION

The purpose of this Annex is to describe the typical structure of a generic role statement within an organization.

It uses a competency-based approach, utilizing the Systems Engineering competencies defined in this framework, and their relationship to the Systems Engineering processes defined in the *INCOSE Systems Engineering Handbook* (INCOSE 2015).

The resulting role statements can be used for recruitment, promotion, planned, and/or individual development as described earlier. The Annex is generic to help deployment within a wide range of organizations.

A competency-based role statement can be considered as a set of "requirements" for individuals (components) who make up the enterprise (the system). It defines the role not only in terms of what the person in the role does (e.g. the processes followed) but also the competencies required to follow those processes. Looking at the level of involvement in the processes allows definition of the level of each competence needed. Further information on the topic is available from Section 5 of BKCASE "Enabling Systems Engineering" (BKCASE Project 2017).

The structure of the rest of this section is as follows:

- A list of key considerations. Major points, issues, and assumptions to address when defining competency-based role statements.
- The recommended structure for a role statement. This is built from the Competency Framework and defined Systems Engineering processes.
- Guidance on defining competence levels within a role statement. Competence levels depend upon the level of engagement with the generic Systems Engineering processes needed for the role.
- Guidance on the tailoring of generic role statements to real organizations. This extends the generic approach to real-world organizations, where competencies and processes may not be those included in the INCOSE Competency Framework and the *INCOSE Systems Engineering Handbook*.

Systems Engineering Competency Assessment Guide: A combined INCOSE Systems Engineering Competency Framework (SECF) and associated Systems Engineering Competency Assessment Guide (SECAG) document, First Edition. INCOSE.

CONSIDERATIONS WHEN DEFINING ROLE STATEMENTS

Every organization is different and will have different – and possibly incomplete – definitions of processes and competencies as well as different organizational structures in which these are deployed. In addition, organizations evolve over time, requiring competencies and processes to evolve to match. Development of generic role descriptions should bear these points in mind. We are after all dealing with the most variable element in an organization – its people.

Role statements can be used for a variety of purposes. There is no right or wrong purpose and no right or wrong roles within an organization. Both are context dependent.

Whatever the purpose, the following should be noted:

A job description is different from a role statement. A role statement is part of an organization design, whereas a job description may comprise a combination of roles adapted to the specific needs of the organization and the strengths of any individual. An individual's job may require them to perform several different roles.

- Ideally, roles defined within an organization need to form an integrated whole. However, while organizational structures vary, some organizations will have a Systems Engineering department; others make Systems Engineering competency core to all roles. Organizational structure will affect the way roles interact with each other.
- As Systems Engineering is an "integrating" discipline, any role statement covering Systems Engineering is likely to contain domain-specific competencies.
- Tailoring of the INCOSE Competency Framework and of the terminology within it needs to reflect the needs, language, approaches, and priorities appropriate to an organization is to be expected. A good example of such tailoring is to be found in the paper "The need to tailor competency models – with a use case from Rolls-Royce" (Beasley 2013) which was produced for the INCOSE Competency Working Group.
- Role statements can be used to support many different purposes; the purpose of creating a role statement must be clear when it is created. In some organizations, role statements are "entry level" (i.e. competencies are required to be, mostly, "met" to get the job), or they can define "stretch targets" challenging an individual to perform the role beyond expectations (i.e. they are used as guidance supporting ongoing professional development).
- The relationship between role, reward, and position within the organization needs to be carefully considered. Linking competence directly to reward can be dangerous (make honest self-assessment hard). Different roles may be undertaken during the development of a career (e.g. some may provide essential breadth of experience to enable an individual to take on a more senior role). Some valuable additional information on this topic can be obtained from the Helix (Hutchison, et al. 2018) research program.

In summary, when defining role statement, we need to ensure

- their purpose is understood;
- the way the organization integrates Systems Engineering with other disciplines (e.g. Program Management, Sales, etc.) is understood; and
- the role definition aligns with organizational HR policies and terminology.

ROLE STATEMENT STRUCTURE

It is recommended that a typical generic role statement should include the following elements:

- **Role name** – The title of the role, which should describe the role succinctly.
- **Role purpose** – This should provide the primary aim of the role, ideally in just one or two sentences.

- **Activities performed** – This should be an accountability statement for the role.
  - List the process activities that the role does. Any role may do a range of activities, or focus (specialize) on just one. Other, non-Systems Engineering processes used in the organization should be included as needed.
  - The level of accountability should be defined by applying (or consulting) a responsibility and accountability matrix (e.g. a RACI or similar).
  - Generic activities that can be defined external to the process or specific applications or instances of the process can be included.
- **Competencies required (highest level)**
  - These should be divided into the groups defined in this framework.
  - The overall level should be built by assessing the competencies needed for each activity, taking the highest level from all activities within each competence area.
- **Other constraints or preferences for role**
  - For example, licenses, qualifications, specific experiences, or domain knowledge needed.
  - The contents of this section will be specific to the organization.

ASSIGNING COMPETENCIES TO A ROLE STATEMENT

This section provides generic guidance in the definition of competencies, and associated competence levels for a role statement. Any real-world implementation will also have to address organizational concerns. This is addressed later.

SECF Figure 3 shows relationships between INCOSE SE Competencies and life cycle processes. Any role involving systems engineering will require its holder to be competent in several of these processes. The competencies associated with each process can be derived from this figure. The level of competence expected within a role will be determined by considering the nature of the involvement in the process using SECF Figure 3 in combination with SECF Figure 5, the ARCIFE model.

	Aware	Supervised practitioner	Practitioner	Lead practitioner	Expert
Accountable (A)		See note *			
Responsible (R)					
Consulted (C)					
Informed (I)					
Facilitator (F)					
Expert (E)					

** Note:*
Those accountable for a process may not necessarily have the highest competency level. While they may, in the past, have operated at a higher level of competency – and will draw on this experience - as a leader, their role has moved beyond "doing" to "leadership" such as setting the culture and environment for the activity to be done correctly. Of course, if a role involves both accountability and responsibility, a higher level of competency will normally be required.

SECF FIGURE 5 ARCIFE levels mapped to competency levels.

ARCIFE (Accountable, Responsible, Consulted, Informed, Facilitator, Expert) analysis is a refinement on the standard RACI (Responsible, Accountable, Consulted, Informed) analysis for engagement in an activity. This considers some of the specifics of the nature of aspects of Systems Engineering (notably the integration of specialist glue role referred to by Sheard in 12 Systems Engineering Roles; Sheard 1996). It is not uncommon for some Systems Engineering roles to be heavily involved with helping other Engineers do Systems Engineering activities.

The ARCIFE terms are defined below:

- **A – Accountable** Leadership, making sure the activity is done, and done right. Often the accountable person delegates the actual doing. There should only be one accountable person for a specific activity/issue.
- **R – Responsible** People (maybe different roles) that do the activity (so if the output of the activity is a report, these are the authors). Responsibility can be shared in a team.
- **C – Consulted** Engaged in the work; may provide input, or (more likely) either apply specific technical/domain knowledge to assist with the activity, or they use/act on the outcome of the process and influence it.
- **I – Informed** Needs to know either that the work product is produced, or the outcome (or part of outcome and decision from process activity.

The next two terms are in a different dimension to the level of descending importance above. To embed Systems Engineering, individuals may have to help the team apply specific SE techniques or define and develop best practice.

- **F – Facilitator/Coach** – Lead workshops or discussions applying Systems Approach (with people from other skills) and build consolidated and agreed models. This aspect includes sufficient expertise and knowledge in Systems Approach (process and techniques) to select the most appropriate for the situation, considering both the nature of the system of interest, and for the "systems" competence and experience of the team they are working with.
- **E – Expert** – Develops/explains or teaches methods and process in this area and advances the state of the art. Considered a specialist in the competency.

  [Note the term "Expert" is capturing organizational role expectations and should not be taken as an assumption that the individual is operating at an "Expert" level of proficiency.]

Once the level of ARCIFE involvement is determined, then the competency level can be determined using Figure 5 as a starting point.

SECF Figure 5 gives a range for the required competence level. Judgement based on the level of difficulty and criticality of the activities the role undertakes is needed to finalize the level required.

Since several processes may call for the same competencies, a role statement can be completed by taking the highest required competency level defined when all the processes the role is involved with have been analyzed.

ROLE TAILORING AND ORGANIZATION

This INCOSE Competency Framework is generic, and care needs to be taken during its application to ensure the specific purpose and scope of a role definition, and the level to which Systems Engineering is integrated into the wider organization. Systems Engineering is both implemented and described differently in every organization (e.g. some competencies or processes may be grouped together, or specific terminology applied). It is common for each organization to have specific additional processes and competencies that will need to be included in role descriptions. This subsection expands the simple generic mapping described earlier to allow this tailoring and adjustment.

SECF Figure 6 shows nine steps that are needed to integrate organization roles, any existing competency processes and competency definition, the processes defined in the *INCOSE Systems Engineering Handbook* Fourth Edition (INCOSE 2015), and the contents of this framework. The individual steps are described and explained below.

The overall intent of the steps in SECF Figure 6 is to map organization roles to the processes they support, to map organization definitions and terminology (for process and competency) to INCOSE definitions, to allow use of the INCOSE definition of how competencies support process, and to finalize with organization-specific role definitions. The outputs produced in Steps 2 and 4 should be retained and used to help map INCOSE terminology for process and competency to any organization-specific language. Each step is explained in more detail below.

- **STEP 1** Perform ARCIFE analysis for the complete matrix for existing organization processes and roles. This allows determination (in Step 9) of elements of the process to include in the "activities performed" part of the role statement (Note: This should be prioritized to keep role statement brief). Depending on the scope of the roles being defined, this activity may allow some processes to be ignored hereafter.
- **STEP 2** Map organization processes onto *INCOSE Systems Engineering Handbook* processes. If there are additional processes that cannot be mapped, these should be handled separately by extending SECF Figure 5 (see Step 5). One benefit of this activity is that it may highlight overall "gaps" in local process definitions, in terms of INCOSE-recommended processes.
- **STEP 3** Combine the outputs of Steps 1 and 2 to produce a mapping of organization roles to *INCOSE Systems Engineering Handbook* processes. This step should retain any additional organization processes defined. Note this mapping is still in terms of the ARCIFE level of engagement by roles to support the process.
- **STEP 4** Map INCOSE competency area definitions to organization competency definitions. This may include tailoring of INCOSE competencies, and inclusion of additional competencies required by the organization for its roles.
- **STEP 5** Expand the mapping between *INCOSE Systems Engineering Handbook* processes and INCOSE competency area definitions in SECF Figure 4 to include any organization processes identified in Step 2 and organization competencies identified in Step 4.
- **STEP 6** Use the mapping between processes and competencies produced in Step 5 and the mapping of organization roles to *INCOSE Systems Engineering Handbook* processes (output of Step 3) to determine the INCOSE-defined competencies relevant to each organization role. Judgement will be needed to map the impact of competency on process and the ARCIFE analysis (i.e. engagement of role in the performance of role).
- **STEP 7** Use the mapping of INCOSE organization to organization competencies (produced in Step 4) to translate each organization role to the equivalent INCOSE Competency output (produced in Step 6). This produces a set of roles mapped to competency in organization terms.
- **STEP 8** Translate the definition of competencies needed by each role (in terms of ARCIFE and competency impact on process to the level of competency needed using SECF Figure 6 as a guide. During this analysis, it will be recognized that some competencies are needed to support many processes at different levels. These should be merged together taking the highest level of competency needed from the range defined for the role as the output.
- **STEP 9** Complete the generic role profile for each organization role, taking the output from Step 1 as "activities performed" and the output from Step 8 as required competency.

The above analysis activity is not straightforward and the individual or team performing the mapping will need to use their judgement, an extensive knowledge of applicable INCOSE products, and knowledge of their own organization's definitions and terminologies to complete the analysis successfully.

ACTIVITY PRIORITIZATION AND ROLE TAILORING

In the above analysis, a relatively simple relationship exists between competencies and roles: If a competency is required to complete any activity or process, then it is required for the role. If the same competency is required to complete several activities or processes, the highest proficiency level required for any activity or process is the required

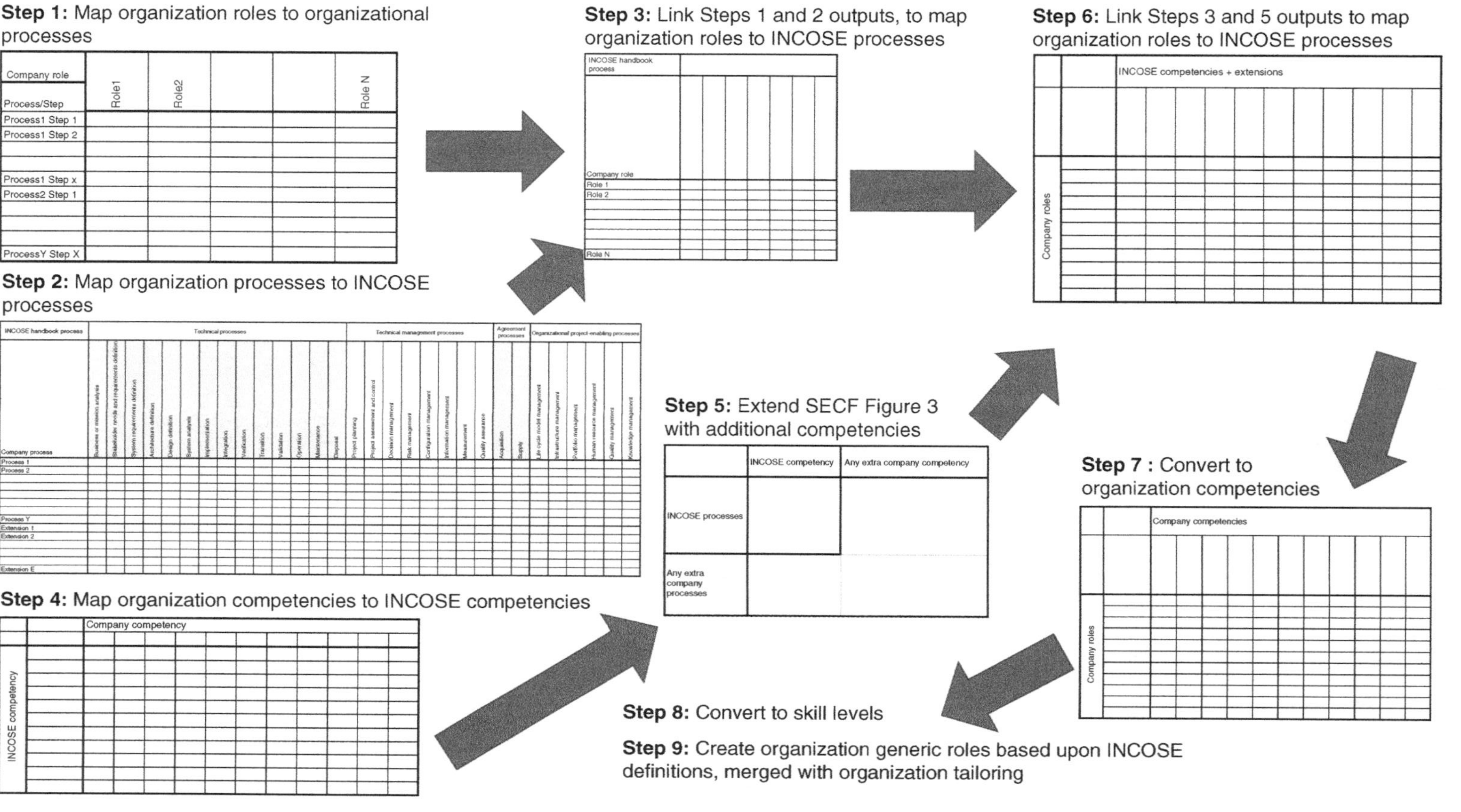

SECF FIGURE 6 Steps required to create organization generic role profile using INCOSE Competency Framework.

proficiency level for the role. A "perfect" role holder will hold all the required competencies at the required proficiency levels.

But in reality, individuals – particularly those new to a role – may not (yet) have acquired all the experience necessary to match every single competency at the required level (or above). The areas where they lack would normally be the subject of a personal development plan or similar.

However, in reality, a role holder may not be required to perform every activity or process defined for their role on a day-to-day basis. Some activities or processes will be common, while others will be very rare. Furthermore, it may be critical that specific activities or processes are performed perfectly, whereas the impact of effectiveness on others may have less impact (e.g. a safety process would normally be critical; a weekly reporting process may be important but less impactful if performed poorly).

In other words, we can optimize a role description by highlighting certain key or critical activities or processes and their associated competencies and proficiency levels, while still recognizing additional competencies may be required to effectively perform ALL allocated activities or processes in the role (or existing competencies may require higher proficiency levels to perform all activities or processes in a role). On a day-to-day basis, the competency expectations of a role holder may differ from theoretical requirements and an individual could successfully discharge the role on a day-to-day basis without ever having to display the full range of competencies/competency proficiency levels.

NOTE: This relationship is likely to vary from role to role and from organization to organization. Factors to consider might be the frequency with which the activity or process is performed in the role, the support available in performing the activity or process, or the criticality of successful execution of the activity to the organization.

With this approach, by prioritizing activities or processes expected to be required regularly (or which are deemed critical) for a defined role and using this prioritized list to identify "key" activities and processes and their associated competencies and levels, the required proficiency levels can be adjusted from that defined for the role in its complete form (i.e. as defined in the ARCIFE or RACI), to reflect key tasks an individual needs to perform on a day-to-day basis. Those with a higher proficiency level will be able to evidence more of the prioritized activities or processes and indeed some of the lower-priority activities or processes as well.

In the analysis below, we have used Pareto-like assumptions:

- If an individual can demonstrate the competencies (and levels) necessary to complete the top 20% of key or critical activities or processes associated with a role, while they will require day-to-day guidance, they will have a degree of effectiveness in their role.
- If an individual can demonstrate competencies (and levels) necessary to complete the top 80% of key or critical activities or processes associated with a role, they will be for most purposes be operating effectively on a day-to-day basis, requiring guidance only for unusual situations (i.e. the remaining 20%).
- This approach can be applied at competency levels above Practitioner, since at these levels, individuals tend to demonstrate their higher levels of competency by correctly applying their accrued experience and knowledge to unusual or new situations which even they may not have experienced themselves.

While this approach can be used to improve granularity when defining role requirements (or measuring role performance), it requires a tailored definition of proficiency levels to help in its application.

The following guidance demonstrates one way in which proficiency levels for competencies associated with a role might be assessed using this approach.

- As a minimum, individuals operating at "Awareness" level in a competency area should be able to demonstrate through evidence that they have the knowledge and understand the principles behind all knowledge indicators defined for the competency area.

Awareness-level practitioners in a competency area may or may not have started to perform day-to-day activities in this area.

While organizations may choose to tailor the knowledge to meet their domain needs, basic knowledge is always a central aspect of competency.

- Individuals operating at the Supervised Practitioner level in a competency area level are expected to be able to demonstrate through evidence provided that they are currently applying Awareness-level knowledge without supervision on a day-to-day basis, to perform the top 20% of key activities or processes identified for their role.

  Supervised Practitioners in a competency area may be able to perform other common activities or processes associated with this competency area (with some additional guidance from those with a higher proficiency level) but this is not an expectation.

- Individuals operating at the Practitioner level in a competency area are expected to be able to demonstrate through evidence provided that they are currently applying their knowledge and accrued experience to operate without supervision on a day-to-day basis to perform the top 80% of key activities or processes identified for their role.

  Practitioners in a competency area may be able to perform other less common competency-related activities or processes (with some additional guidance from those with a higher proficiency level) but this is not an expectation.

- Individuals operating at the Lead Practitioner level in a competency area are expected to be able to demonstrate through evidence provided, that they are currently applying their knowledge and accrued experience to operate without supervision on a day-to-day basis to perform all activities or processes identified for their role.

  In addition to the above, Lead Practitioners in a competency area are also expected to be able to demonstrate through evidence provided also that they are currently operating as an organizational point of reference for the competency area, applying their knowledge and experience to provide guidance or solutions to problems involving the competency area, or to improve or standardize organizational best practices within the competency area.

  In practice, an individual may not be able to provide evidence of having experienced every complex situation expected for their role but should still be able to successfully demonstrate to assessors through evidence provided that they have nonetheless acquired sufficient knowledge and experience to enable them to produce effective outcomes for any of the 20% of activities not explicitly covered within the Practitioner-level definition.

- Expert-level Practitioners in a competency area are expected to be able to demonstrate through evidence provided that they are currently applying their knowledge and accrued experience to operate without supervision on a day-to-day basis in competency-related activities reaching beyond the organizational boundary, and working with other organizations to research, improve, or standardize global best practices for the competency.

  In addition, Expert-level practitioners in a competency area are expected to be able to demonstrate through evidence provided that they have acquired sufficient knowledge and experience to enable them to produce effective outcomes for all activities associated with the Lead Practitioner level in the competency area, as described above.

  In practice, an individual may not be able to provide evidence of having experienced every complex situation expected for their role but should still be able to successfully demonstrate to assessors through evidence provided that they have nonetheless acquired sufficient knowledge and experience to enable them to produce effective outcomes for all activities identified.

  As with the Lead Practitioner, while the assessment of an individual against an Expert-level proficiency should include confirmation of experience in a role as a Practitioner or Lead Practitioner and having proven this to the assessor's satisfaction, the bulk of the assessment is expected to focus on confirming evidence of activities beyond the organizational boundary.

Please note:

- The percentage figures used above are provided as guidance and may vary between organizations. Organizational tailoring should define the precise requirements for assessment of indicators.

- The above analysis is to support role tailoring only. The role requirements for competencies and levels resulting from applying the above approach will produce a set of indicators which will form the basis of the assessment.
- %-ages do not relate to how many competency indicators in SECF tables are to be met (e.g. "20% of SP indicators. . .", etc.). The precise relationship is more complex as explained above.
- The "Supervised Practitioner" level spans a continuum, connecting Awareness and Practitioner levels and there is no "time" element. An individual assessed as operating at the Supervised Practitioner level may have to bridge a small or a large "gap" in order to demonstrate the higher level of proficiency but depending on the organizational opportunities available, this may take a short or long time to achieve.
- There is no expectation that an individual will automatically progress from Practitioner to Lead Practitioner or indeed Expert level over time. Organizational need and the opportunities to acquire and display competence that arise in an organization will determine if higher levels are (ever) achieved.

Notwithstanding the above, instructions for assessing competency at any level within an organization should be documented as part of assessment tailoring guidance for the organization.

SECF ANNEX D: INCOSE SYSTEMS ENGINEERING COMPETENCY FRAMEWORK

This annex defines each of the competencies identified as forming INCOSE SECF.

It should be read in conjunction with the SECAG, Part II of this book, which provides an assessment guide for each competency area.

In summary, the SECF represents the "requirements" for proficiency in 37 competencies at five levels of competence, while the SECAG defines the evidence expected in order to verify that proficiency at a particular level has been achieved.

Systems Engineering Competency Assessment Guide: A combined INCOSE Systems Engineering Competency Framework (SECF) and associated Systems Engineering Competency Assessment Guide (SECAG) document, First Edition. INCOSE.

Competency area – Core: Systems Thinking

Description
The application of the fundamental concepts of systems thinking to Systems Engineering. These concepts include understanding what a system is, its context within its environment, its boundaries and interfaces, and that it has a life cycle. Systems thinking applies to the definition, development, and production of systems within an enterprise and technological environment and is a framework for curiosity about any system of interest.

Why it matters
Systems thinking is a way of dealing with increasing complexity. The fundamental concepts of systems thinking involve understanding how actions and decisions in one area affect another, and that the optimization of a system within its environment does not necessarily come from optimizing the individual system components. Systems thinking is conducted within an enterprise and technological context. These contexts impact the life cycle of the system and place requirements and constraints on the systems thinking being conducted. Failing to meet such constraints can have a serious effect on the enterprise and the value of the system.

Effective indicators of knowledge and experience

Awareness	Supervised Practitioner	Practitioner	Lead Practitioner	Expert
Explains what "systems thinking" is and explains why it is important. [CSTA01]	Defines the properties of a system. [CSTS01]	Identifies and manages complexity using appropriate techniques. [CSTP01]	Creates enterprise-level policies, procedures, guidance, and best practice for systems thinking, including associated tools. [CSTL01]	Communicates own knowledge and experience in systems thinking in order to improve best practice beyond the enterprise boundary. [CSTE01]
Explains what "emergence" is, why it is important, and how it can be "positive" or "negative" in its effect upon the system as a whole. [CSTA02]	Explains how system behavior produces emergent properties. [CSTS02]	Uses analysis of a system functions and parts to predict resultant system behavior. [CSTP02]	Judges the suitability of project-level systems thinking on behalf of the enterprise, to ensure its validity. [CSTL02]	Influences individuals and activities beyond the enterprise boundary to support the systems thinking approach of their own enterprise. [CSTE02]
Explains what a "system hierarchy" is and why it is important. [CSTA03]	Uses the principles of system partitioning within system hierarchy on a project. [CSTS03]	Identifies the context of a system from a range of viewpoints including system boundaries and external interfaces. [CSTP03]	Persuades key enterprise-level stakeholders across the enterprise to support and maintain the technical capability and strategy of the enterprise. [CSTL03]	Advises organizations beyond the enterprise boundary on the suitability of their approach to systems thinking. [CSTE03]
Explains what "system context" is for a given system of interest and describes why it is important. [CSTA04]	Defines system characteristics in order to improve understanding of need. [CSTS04]	Identifies the interaction between humans and systems, and systems and systems. [CSTP04]	Adapts existing systems thinking practices on behalf of the enterprise to accommodate novel, complex, or difficult system situations or problems. [CSTL04]	Advises organizations beyond the enterprise boundary on complex or sensitive systems thinking issues. [CSTE04]
Explains why it is important to be able to identify and understand what interfaces are. [CSTA05]	Explains why the boundary of a system needs to be managed. [CSTS05]	Identifies enterprise and technology issues affecting the design of a system and addresses them using a systems thinking approach. [CSTP05]	Persuades project stakeholders across the enterprise to improve the suitability of project technical strategies in order to maintain their validity. [CSTL05]	Champions the introduction of novel techniques and ideas in systems thinking beyond the enterprise boundary, in order to develop the wider Systems Engineering community in this competency. [CSTE05]

Explains why it is important to recognize interactions among systems and their elements. [CSTA06]	Explains how humans and systems interact and how humans can be elements of systems. [CSTS06]	Uses appropriate systems thinking approaches to a range of situations, integrating the outcomes to get a full understanding of the whole. [CSTP06]	Persuades key stakeholders to address enterprise-level issues identified through systems thinking. [CSTL06]	Coaches individuals beyond the enterprise boundary in systems thinking techniques, in order to further develop their knowledge, abilities, skills, or associated behaviors. [CSTE06]
Explains why it is important to understand purpose and functionality of a system of interest. [CSTA07]	Identifies the influence of wider enterprise on a project. [CSTS07]	Identifies potential enterprise improvements to enable system development. [CSTP07]	Coaches or mentors practitioners across the enterprise in systems thinking in order to develop their knowledge, abilities, skills, or associated behaviors. [CSTL07]	Maintains expertise in this competency area through specialist Continual Professional Development (CPD) activities. [CSTE07]
Explains how business, enterprise, and technology can each influence the definition and development of the system and vice versa. [CSTA08]	Uses systems thinking to contribute to enterprise technology development activities. [CSTS08]	Guides team systems thinking activities in order to ensure current activities align to purpose. [CSTP08]	Promotes the introduction and use of novel techniques and ideas in systems thinking across the enterprise, to improve enterprise competence in this area. [CSTL08]	
Explains why it may be necessary to approach systems thinking in different ways, depending on the situation, and provides examples. [CSTA09]	Develops their own systems thinking insights to share thinking across the wider project (e.g. working groups and other teams). [CSTS09]	Develops existing case studies and examples of systems thinking to apply in new situations. [CSTP09]	Develops expertise in this competency area through specialist Continual Professional Development (CPD) activities. [CSTL09]	
	Develops own understanding of this competency area through Continual Professional Development (CPD). [CSTS10]	Guides new or supervised practitioners in systems thinking techniques in order to develop their knowledge, abilities, skills, or associated behaviors. [CSTP10]		
		Maintains and enhances own competence in this area through Continual Professional Development (CPD) activities. [CSTP11]		

Competency area – Core: Life Cycles

Description

The selection of appropriate life cycles in the realization of a system. Systems and their constituent elements have individual life cycles, characterizing the nature of their evolution. Each life cycle is itself divided into a series of stages, marking key transition points during the evolution of that element. As different system elements may have different life cycles, the relationship between life cycle stages on differing elements is complex, varying depending on the scope of the project, characteristics of the wider system of which it forms a part, the stakeholder requirements, and perceived risk.

Why it matters

Life cycles form the basis for project planning and estimating. Selection of the appropriate life cycles and their alignment has a large impact on and may be crucial to project success. Ensuring coordination between related life cycles at all levels is critical to the realization of a successful system.

Effective indicators of knowledge and experience

Awareness	Supervised Practitioner	Practitioner	Lead Practitioner	Expert
Identifies different life cycle types and summarizes the key characteristics of each. [CLCA01]	Describes Systems Engineering life cycle processes. [CLCS01]	Explains the advantages and disadvantages of different types of systems life cycle and where each might be used advantageously. [CLCP01]	Creates enterprise-level policies, procedures, guidance, and best practice for life cycle definition and management, including associated tools. [CLCL01]	Communicates own knowledge and experience in life cycle definition and management in order to improve best practice beyond the enterprise boundary. [CLCE01]
Explains why selection of life cycle is important when developing a system solution. [CLCA02]	Identifies the impact of failing to consider future life cycle stages in the current stage. [CLCS02]	Creates a governing project life cycle, using enterprise-level policies, procedures, guidance, and best practice. [CLCP02]	Judges life cycle selections across the enterprise, to ensure they meet the needs of the project. [CLCL02]	Advises organizations beyond the enterprise boundary on the suitability of life cycle tailoring or life cycle definitions. [CLCE02]
Explains why it is necessary to define an appropriate life cycle process model and the key steps involved. [CLCA03]	Prepares inputs to life cycle definition activities at system or system element level. [CLCS03]	Identifies dependencies aligning life cycles and life cycle stages of different system elements accordingly. [CLCP03]	Adapts standard life cycle models on behalf of the enterprise, to address complex or difficult situations or to resolve conflicts between life cycles where required. [CLCL03]	Advises organizations beyond the enterprise boundary on complex, concurrent, or sensitive projects. [CLCE03]
Explains why differing engineering approaches are required in different life cycle phases and provides examples. [CLCA04]	Complies with a governing project system life cycle, using appropriate processes and tools to plan and control their own activities. [CLCS04]	Acts to influence the life cycle of system elements beyond boundary of the system of interest, to improve the development strategy. [CLCP04]	Identifies work or issues relevant to the current life cycle phase by applying knowledge of life cycles to projects across the enterprise. [CLCL04]	Champions the introduction of novel techniques and ideas in life cycle management, beyond the enterprise boundary, in order to develop the wider Systems Engineering community in this competency. [CLCE04]
Explains how different life cycle characteristics relate to the system life cycle. [CLCA05]	Describes the system life cycle in which they are working on their project. [CLCS05]	Prepares plans addressing future life cycle phases to take into consideration their impact on the current phase, improving current activities accordingly. [CLCP05]	Persuades key stakeholders across the enterprise to support activities required now in order to address future life cycle stages. [CLCL05]	Coaches individuals beyond the enterprise boundary in life cycle management techniques, in order to further develop their knowledge, abilities, skills, or associated behaviors. [CLCE05]

	Develops own understanding of this competency area through Continual Professional Development (CPD). [CLCS06]	Prepares plans governing transitions between life cycle stages to reduce project impact at those transitions. [CLCP06]	Coaches or mentors practitioners across the enterprise in life cycle definition and management in order to develop their knowledge, abilities, skills, or associated behaviors. [CLCL06]	Maintains expertise in this competency area through specialist Continual Professional Development (CPD) activities. [CLCE06]
		Guides new or supervised practitioners in Systems Engineering life cycles in order to develop their knowledge, abilities, skills, or associated behaviors. [CLCP07]	Promotes the introduction and use of novel techniques and ideas in life cycle definition and management across the enterprise, to improve enterprise competence in this area. [CLCL07]	
		Maintains and enhances own competence in this area through Continual Professional Development (CPD) activities. [CLCP08]	Develops expertise in this competency area through specialist Continual Professional Development (CPD) activities. [CLCL08]	

Competency area – Core: Capability Engineering

Description

A system or enterprise "capability" relates to the delivery of a desired effect (outcome) rather than delivery of a desired performance level (output). Capability is achieved by combining differing systems, products and services, using a combination of people, processes, information, as well as equipment. Capability-based systems are enduring systems, adapting to changing situations rather than remaining static against a fixed performance baseline.

Why it matters

Understanding the difference between capability and product is important in order to be able to separate the desired effect from any preconceived expectations as to its delivery mechanism. Even if the system of interest only delivers part of a wider capability, obtaining a good understanding of the wider intent helps keep options option, facilitating innovation and creativity in solutions. Failure to do this will generally result in suboptimization and lost opportunities.

Effective indicators of knowledge and experience

Awareness	Supervised Practitioner	Practitioner	Lead Practitioner	Expert
Explains the concept of capability and how its use can prove beneficial. [CCPA01]	Explains how project capability and environment are linked. [CCPS01]	Identifies capability issues of the wider (super) system which will affect the design of own system and translates these into system requirements. [CCPP01]	Creates enterprise-level policies, procedures, guidance, and best practice for capability engineering, including associated tools. [CCPL01]	Communicates own knowledge and experience in capability engineering in order to improve best practice beyond the enterprise boundary. [CCPE01]
Explains how capability requirements can be satisfied by integrating several systems. [CCPA02]	Identifies capability issues from the wider system, which will affect the design of a system of interest. [CCPS02]	Reviews proposed system solutions to ensure their ability to deliver the capability required by a wider system, making changes as necessary. [CCPP02]	Judges the suitability of capability solutions and the planned approach on projects across the enterprise. [CCPL02]	Advises organizations beyond the enterprise boundary on the suitability of their approach to capability engineering. [CCPE02]
Explains how super system capability needs impact on the development of each system that contributes to the capability. [CCPA03]	Prepares inputs to technology planning activities required in order to provide capability. [CCPS03]	Prepares technology plan that includes technology innovation, risk, maturity, readiness levels, and insertion points into existing capability. [CCPP03]	Identifies impact and changes needed in super system environment as a result of the capability development on behalf of the enterprise. [CCPL03]	Advises organizations beyond the enterprise boundary on their handling of complex or sensitive capability engineering strategy issues. [CCPE03]
Describes the difficulties of translating capability needs of the wider system into system requirements. [CCPA04]	Prepares information that supports the embedding or utilization of capability. [CCPS04]	Creates an operational concept for a capability (what it does, why, how, where, when, and who). [CCPP04]	Identifies improvements required to enterprise capabilities on behalf of the enterprise. [CCPL04]	Advises organizations beyond the enterprise boundary on differences and relationships between capability and product-based systems. [CCPE04]
	Identifies different elements that make up capability. [CCPS05]	Reviews existing capability to identify gaps relative to desired capability, documenting approaches which reduce or eliminate this deficit. [CCPP05]	Coaches or mentors practitioners across the enterprise in capability engineering in order to develop their knowledge, abilities, skills, or associated behaviors. [CCPL05]	Assesses capability engineering in multiple domains beyond the enterprise boundary in order to develop or improve capability solutions within own enterprise. [CCPE05]

	Prepares multiple views that focus on value, purpose, and solution for capability. [CCPS06]	Prepares information that supports improvements to enterprise capabilities. [CCPP06]	Promotes the introduction and use of novel techniques and ideas in Capability Engineering across the enterprise, to improve enterprise competence in the area. [CCPL06]	Champions the introduction of novel techniques and ideas in capability engineering, beyond the enterprise boundary, in order to develop the wider Systems Engineering community in this competency. [CCPE06]
	Develops own understanding of this competency area through Continual Professional Development (CPD). [CCPS07]	Identifies key "pinch points" in the development and implementation of specific capability. [CCPP07]	Develops expertise in this competency area through specialist Continual Professional Development (CPD) activities. [CCPL07]	Coaches individuals beyond the enterprise boundary in capability engineering, in order to further develop their knowledge, abilities, skills, or associated behaviors. [CCPE07]
		Uses multiple views to analyze alignment, balance, and trade-offs in and between the different elements (in a level) ensuring that capability performance is not traded out. [CCPP08]		Maintains expertise in this competency area through specialist Continual Professional Development (CPD) activities. [CCPE08]
		Guides new or supervised practitioners in Capability engineering in order to develop their knowledge, abilities, skills, or associated behaviors. [CCPP09]		
		Maintains and enhances own competence in this area through Continual Professional Development (CPD) activities. [CCPP10]		

Competency area – Core: General Engineering

Description
Foundational concepts in mathematics, science, and engineering and their application.

Why it matters
Systems Engineering is performed in a technical scientific environment and as a result, a good understanding of mathematics, science coupled with a sound appreciation of core engineering principles is a critical foundation for effective Systems Engineering. Without this, systems engineers cannot communicate effectively and efficiently with engineers from other domains.

Effective indicators of knowledge and experience

Awareness	Supervised Practitioner	Practitioner	Lead Practitioner	Expert
Explains core principles of science and mathematics applicable to engineering. [CGEA01]	Uses scientific and mathematical knowledge when performing engineering tasks. [CGES01]	Selects software-based tools together with the products they create to facilitate and progress engineering tasks. [CGEP01]	Creates enterprise-level policies, procedures, guidance, and best practice for general engineering, including associated tools. [CGEL01]	Communicates own knowledge and experience in general engineering in order to improve best practice beyond the enterprise boundary. [CGEE01]
Explains fundamentals of engineering as a discipline. [CGEA02]	Uses appropriate engineering approaches when performing engineering tasks. [CGES02]	Uses selected software-based tools together with the products they create to facilitate and progress engineering tasks. [CGEP02]	Adapts mathematics and engineering principles so that they can be applied to specific engineering situations on behalf of the enterprise. [CGEL02]	Advises organizations beyond the enterprise boundary on the suitability of their approach to general engineering activities. [CGEE02]
Explains why probability and statistics are both relevant to engineering. [CGEA03]	Explains the concept of "variation" and its effect in engineering tasks. [CGES03]	Selects scientific and mathematical methods to be used in support of tasks and justifies their selection. [CGEP03]	Communicates the difference between scientific and engineering approaches in order to engage with pure scientific advances on behalf of the enterprise. [CGEL03]	Develops new applications of mathematical methods to engineering practices applicable beyond the enterprise boundary. [CGEE03]
Explains why analytical methods and sound judgment are central to engineering decisions. [CGEA04]	Uses proven analytical methods when performing engineering tasks, while appreciating the limitations of their applicability. [CGES04]	Selects relevant engineering approaches and methods to be used in support of engineering tasks and justifies their selection. [CGEP04]	Assesses items across the enterprise for the appropriate level of understanding of impact of variation and uncertainty on engineering outcomes. [CGEL04]	Advises organizations beyond the enterprise boundary on their handling of complex general engineering challenges. [CGEE04]
Explains the characteristics of an engineered system. [CGEA05]	Uses software-based tools, together with the products they create, to facilitate and progress engineering tasks. [CGES05]	Uses probability and statistics in engineering tasks recognizing benefits and limitations on results obtained. [CGEP05]	Advises stakeholders across the enterprise on issues requiring the application of engineering judgment. [CGEL05]	Maintains own awareness of developments in sciences, technologies, and related engineering disciplines beyond the enterprise boundary, recognizing areas where new developments might be applicable within their own discipline or enterprise. [CGEE05]

Describes engineered systems that are physical, software, and socio-technical systems or combinations thereof. [CGEA06]	Explains why the value of a mathematical approach can be limited in a human-centric or human-originated system, with examples. [CGES06]	Determines the level of variation or probability appropriate to the current task and justifies this decision. [CGEP06]	Fosters creative or innovative approaches to performing general engineering activities across the enterprise. [CGEL06]	Fosters creative or innovative approaches to performing general engineering activities beyond the enterprise boundary. [CGEE06]
Explains how different sciences impact the technology domain and the Systems Engineering discipline. [CGEA07]	Acts creatively or innovatively when performing own activities. [CGES07]	Uses practical and proven engineering principles to structure engineering tasks. [CGEP07]	Advises stakeholders across the enterprise on issues affecting the "broader" engineering approach to engineering activities. [CGEL07]	Champions the introduction of novel techniques and ideas in general engineering, beyond the enterprise boundary, in order to develop the wider Systems Engineering community in this competency. [CGEE07]
Explains why uncertainty is an important factor in engineering and explains how it might arise from many sources. [CGEA08]	Develops own understanding of this competency area through Continual Professional Development (CPD). [CGES08]	Reviews situations using well-established engineering principles and uses these assessments to make sound engineering judgements. [CGEP08]	Judges the quality of engineering judgments made by others across the enterprise. [CGEL08]	Coaches individuals beyond the enterprise boundary in general engineering techniques in order to further develop their knowledge, abilities, skills, or associated behaviors. [CGEE08]
		Adapts engineering approaches to take account of human-centric aspects of systems development. [CGEP09]	Coaches or mentors practitioners across the enterprise in general engineering techniques in order to develop their knowledge, abilities, skills, or associated behaviors. [CGEL09]	Maintains expertise in this competency area through specialist Continual Professional Development (CPD) activities. [CGEE09]
		Uses creative or innovative approaches when performing project activities. [CGEP10]	Promotes the introduction and use of novel techniques and ideas in general engineering across the enterprise, to improve enterprise competence in this area. [CGEL10]	
		Guides new or supervised practitioners in Core Engineering principles in order to develop their knowledge, abilities, skills, or associated behaviors. [CGEP11]	Develops expertise in this competency area through specialist Continual Professional Development (CPD) activities. [CGEL11]	
		Maintains and enhances own competence in this area through Continual Professional Development (CPD) activities. [CGEP12]		

Competency area – Core: Critical Thinking

Description
The objective analysis and evaluation of a topic in order to form a judgement.

Why it matters
Artifacts produced during the conduct of Systems Engineering need to be defendable. Critical thinking helps improve the quality of input information, assumptions, and decisions. A failure to apply critical thinking may lead to invalid outputs including decisions and the solutions from which they are derived.

Effective indicators of knowledge and experience

Awareness	Supervised Practitioner	Practitioner	Lead Practitioner	Expert
Explains why conclusions and arguments made by others may be based upon incomplete, potentially erroneous, or inadequate information, with examples. [CCTA01]	Collates evidence constructing arguments needed in order to make informed decisions. [CCTS01]	Reviews work performed for the quality of critical thinking applied in deriving outcomes. [CCTP01]	Creates enterprise-level policies, procedures, guidance, and best practice for critical thinking, including associated tools. [CCTL01]	Communicates own knowledge and experience in critical thinking in order to improve best practice beyond the enterprise boundary. [CCTE01]
Identifies logical steps in an argument or proposition and the information needed to justify each. [CCTA02]	Uses logical relationships and dependencies between propositions to develop an argument. [CCTS02]	Reviews the impact of assumptions or weak logic in order to locate substantive arguments. [CCTP02]	Uses a range of techniques and viewpoints to critically evaluate assumptions, approaches, arguments, conclusions, and decisions made across the enterprise. [CCTL02]	Advises organizations beyond the enterprise boundary on the suitability of their approach to critical thinking activities. [CCTE02]
Explains why assumptions are important and why there is a need to ensure that they are based upon sound information. [CCTA03]	Uses critical thinking techniques to review own work to test logic, assumptions, arguments, approach, and conclusions. [CCTS03]	Develops robust arguments when responding to critical thinking analyses. [CCTP03]	Identifies alternative approaches to an existing approach to problem solving, to address flawed thinking and its results. [CCTL03]	Advises organizations beyond the enterprise boundary on complex or sensitive assumptions, approaches, arguments, conclusions, and decisions. [CCTE03]
Explains the relationship between assumptions and risk and why assumptions need to be validated. [CCTA04]	Prepares robust arguments in order to respond to critical thinking within a collaborate environment. [CCTS04]	Uses a range of different critical thinking approaches to challenge conclusions of others. [CCTP04]	Judges impact of weak, incomplete or flawed arguments, conclusions, and decisions made across the enterprise. [CCTL04]	Advises organizations beyond the enterprise boundary on the resolution of weak, incomplete, or flawed approaches impacting arguments, conclusions, and decisions made. [CCTE04]
Explains why ideas, arguments, and solutions need to be critically evaluated. [CCTA05]	Reviews ideas from others in order to improve the quality of their own approach, decisions, or conclusions. [CCTS05]	Guides new or supervised practitioners in systems thinking in order to develop their knowledge, abilities, skills, or associated behaviors. [CCTP05]	Produces logical and clear explanations in support of the resolution of intricate or difficult situations across the enterprise. [CCTL05]	Develops own critical thinking expertise through regular review and analysis of critical thinking successes and failures documented beyond the enterprise boundary. [CCTE05]

Lists common techniques and approaches used to propose or define arguments. [CCTA06]	Identifies weaknesses and assumptions in their own arguments. [CCTS06]	Maintains and enhances own competence in this area through Continual Professional Development (CPD) activities. [CCTP06]	Assesses uncertainty in situational assessment made across the enterprise, recommending approaches to address the impact of this. [CCTL06]	Champions the introduction of novel techniques and ideas in critical thinking, beyond the enterprise boundary, in order to develop the wider Systems Engineering community in this competency. [CCTE06]
Explains how own perception of arguments from others may be biased and how this can be recognized. [CCTA07]	Uses common techniques to propose defining or challenging arguments, and conclusions, with guidance. [CCTS07]		Uses own experiences to inform the critical examination of novel scenarios or domains across the enterprise. [CCTL07]	Coaches individuals beyond the enterprise boundary in critical thinking in order to further develop their knowledge, abilities, skills, or associated behaviors. [CCTE07]
Explains how different stakeholders' experiences may cause arguments to be presented in an incomplete or biased manner and how this can be overcome. [CCTA08]	Identifies potential limitations in others' which may impact arguments made regarding proposals or ideas. [CCTS08]		Judges aspects of decision-making which require deeper critical review on behalf of the enterprise. [CCTL08]	Maintains expertise in this competency area through specialist Continual Professional Development (CPD) activities. [CCTE08]
	Identifies own perspective for potential cognitive bias for or against arguments made by others and modifies approach accordingly. [CCTS09]		Develops own critical thinking approaches through a regular analysis of both personal experiences and the experiences of others across the enterprise. [CCTL09]	
	Develops own understanding of this competency area through Continual Professional Development (CPD). [CCTS10]		Coaches or mentors practitioners across the enterprise in critical thinking in order to develop their knowledge, abilities, skills, or associated behaviors. [CCTL10]	
			Promotes the introduction and use of novel techniques and ideas in critical thinking across the enterprise, to improve enterprise competence in this area. [CCTL11]	
			Develops expertise in this competency area through specialist Continual Professional Development (CPD) activities. [CCTL12]	

Competency area – Core: Systems Modeling and Analysis

Description
Modeling is a physical, mathematical, or logical representation of a system entity, phenomenon, or process. System analysis provides a rigorous set of data and information to aid technical understanding and decision-making across the life cycle. A key part of systems analysis is modeling.

Why it matters
Modeling, analysis, and simulation can provide early, cost-effective, indications of function and performance, thereby driving the solution design, enabling risk mitigation and supporting the verification and validation of a solution. Modeling and simulation also allow the exploration of scenarios outside the normal operating parameters of the system.

Effective indicators of knowledge and experience

Awareness	Supervised Practitioner	Practitioner	Lead Practitioner	Expert
Explains why system representations are required and the benefits they can bring to developments. [CSMA01]	Uses modeling and simulation tools and techniques to represent a system or system element. [CSMS01]	Identifies project-specific modeling or analysis needs that need to be addressed when performing modeling on a project. [CSMP01]	Creates enterprise-level policies, procedures, guidance, and best practice for systems modeling and analysis definition and management, including associated tools. [CSML01]	Communicates own knowledge and experience in Systems Modeling and Analysis in order to improve best practice beyond the enterprise boundary. [CSME01]
Explains the scope and limitations of models and simulations, including definition, implementation, and analysis. [CSMA02]	Analyzes outcomes of modeling and analysis and uses this to improve understanding of a system. [CSMS02]	Creates a governing process, plan, and associated tools for systems modeling and analysis in order to monitor and control systems modeling and analysis activities on a system or system element. [CSMP02]	Judges the correctness of tailoring of enterprise-level modeling and analysis processes to meet the needs of a project, on behalf of the enterprise. [CSML02]	Advises organizations beyond the enterprise on the appropriateness of their selected approaches in any given level of complexity and novelty. [CSME02]
Explains different types of modeling and simulation approaches. [CSMA03]	Analyzes risks or limits of a model or simulation. [CSMS03]	Determines key parameters or constraints, which scope or limit the modeling and analysis activities. [CSMP03]	Advises stakeholders across the enterprise, on systems modeling and analysis. [CSML03]	Advises organizations beyond the enterprise boundary on the modeling and analysis of complex or novel systems, or system elements. [CSME03]
Explains how the purpose of modeling and simulation affect the approach taken. [CSMA04]	Uses systems modeling and analysis tools and techniques to verify a model or simulation. [CSMS04]	Uses a governing process and appropriate tools to manage and control their own system modeling and analysis activities. [CSMP04]	Coordinates modeling or analysis activities across the enterprise in order to determine appropriate representations or analysis of complex system or system elements. [CSML04]	Advises organizations beyond the enterprise boundary on the model or analysis validation issues and risks. [CSME04]
Explains why functional analysis and modeling is important in Systems Engineering. [CSMA05]	Prepares inputs used in support of model development activities. [CSMS05]	Analyzes a system, determining the representation of the system or system element, collaborating with model stakeholders as required. [CSMP05]	Adapts approaches used to accommodate complex or challenging aspects of a system of interest being modeled or analyze on projects across the enterprise. [CSML05]	Advises organizations beyond the enterprise boundary on the suitability of their approach to systems modeling and analysis. [CSME05]

Explains the relevance of outputs from systems modeling and analysis, and how these relate to overall system development. [CSMA06]	Uses different types of models for different reasons. [CSMS06]	Selects appropriate tools and techniques for system modeling and analysis. [CSMP06]	Assesses the outputs of systems modeling and analysis across the enterprise to ensure that the results can be used for the intended purpose. [CSML06]	Advises organizations beyond the enterprise boundary on complex or sensitive systems modeling and analysis issues. [CSME06]
Explains the difference between modeling and simulation. [CSMA07]	Uses system analysis techniques to derive information about the real system. [CSMS07]	Defines appropriate representations of a system or system element. [CSMP07]	Advises stakeholders across the enterprise on selection of appropriate modeling or analysis approach across the enterprise. [CSML07]	Champions the introduction of novel techniques and ideas in systems modeling and analysis, beyond the enterprise boundary, in order to develop the wider Systems Engineering community in this competency. [CSME07]
Describes a variety of system analysis techniques that can be used to derive information about a system. [CSMA08]	Develops own understanding of this competency area through Continual Professional Development (CPD). [CSMS08]	Uses appropriate representations and analysis techniques to derive information about a real system. [CSMP08]	Coordinates the integration and combination of different models and analyses for a system or system element across the enterprise. [CSML08]	Coaches individuals beyond the enterprise boundary in systems modeling and analysis, in order to further develop their knowledge, abilities, skills, or associated behaviors. [CSME08]
Explains why the benefits of modeling can only be realized if choices made in defining the model are correct. [CSMA09]		Ensures the content of models that are produced within a project are controlled and coordinated. [CSMP09]	Coaches or mentors practitioners across the enterprise in systems modeling and analysis in order to develop their knowledge, abilities, skills, or associated behaviors. [CSML09]	Maintains expertise in this competency area through specialist Continual Professional Development (CPD) activities. [CSME09]
Explains why models and simulations have a limit of valid use, and the risks of using models and simulations outside those limits. [CSMA10]		Uses systems modeling and analysis tools and techniques to validate a model or simulation. [CSMP10]	Promotes the introduction and use of novel techniques and ideas in Systems Modeling and Analysis across the enterprise, to improve enterprise competence in this area. [CSML10]	
		Guides new or supervised practitioners in modeling and systems analysis to operation in order to develop their knowledge, abilities, skills, or associated behaviors. [CSMP11]	Develops expertise in this competency area through specialist Continual Professional Development (CPD) activities. [CSML11]	
		Maintains and enhances own competence in this area through Continual Professional Development (CPD) activities. [CSMP12]		

Competency area – Professional: Communications				
Description The dynamic process of transmitting or exchanging information using various principles such as verbal, speech, body-language, signals, behavior, writing, audio, video, graphics, language, etc. Communication includes all interactions between individuals, individuals and groups, or between different groups.				
Why it matters Communication plays a fundamental role in all facets of business within an organization, in order to: transfer information between individuals and groups to develop a common understanding and build and maintain relationships and other intangible benefits. Ineffective communication has been identified as the root cause of problems on projects.				
Effective indicators of knowledge and experience				
Awareness	Supervised Practitioner	Practitioner	Lead Practitioner	Expert
Explains communications in terms of the sender, the receiver, and the message and why these three parameters are central to the success of any team communication. [PCCA01]	Follows guidance received (e.g. from mentors) when using communications skills to plan and control their own communications activities. [PCCS01]	Uses a governing communications plan and appropriate tools to control communications. [PCCP01]	Creates enterprise-level policies, procedures, guidance, and best practice for systems engineering communications, including associated tools. [PCCL01]	Communicates own knowledge and experience in Communications Techniques in order to improve best practice beyond the enterprise boundary. [PCCE01]
Explains why there is a need for clear and concise communications. [PCCA02]	Uses appropriate communications techniques to ensure a shared understanding of information with peers. [PCCS02]	Uses appropriate communications techniques to ensure a shared understanding of information with all project stakeholders. [PCCP02]	Uses best practice communications techniques to improve the effectiveness of Systems Engineering activities across the enterprise. [PCCL02]	Advises organizations beyond the enterprise boundary on the suitability of their approach to communications. [PCCE02]
Describes the role communications has in developing positive relationships. [PCCA03]	Fosters positive relationships through effective communications. [PCCS03]	Uses appropriate communications techniques to ensure positive relationships are maintained. [PCCP03]	Maintains positive relationships across the enterprise through effective communications in challenging situations, adapting as necessary to achieve communications clarity or to improve the relationship. [PCCL03]	Fosters a collaborative learning, listening atmosphere among key stakeholders beyond the enterprise boundary. [PCCE03]
Explains why employing the appropriate means for communications is essential. [PCCA04]	Uses appropriate communications techniques to interact with others, depending on the nature of the relationship. [PCCS04]	Uses appropriate communications techniques to express alternate points of view in a diplomatic manner using the appropriate means of communication. [PCCP04]	Uses effective communications techniques to convince stakeholders across the enterprise to reach consensus in challenging situations. [PCCL04]	Advises organizations beyond the enterprise boundary on complex or sensitive communications-related matters affecting Systems Engineering. [PCCE04]
Explains why openness and transparency in communications matters. [PCCA05]	Fosters trust through openness and transparency in communication. [PCCS05]	Fosters a communicating culture by finding appropriate language and communication styles, augmenting where necessary to avoid misunderstanding. [PCCP05]	Uses a proactive style, building consensus among stakeholders across the enterprise using techniques supporting the verbal messages (e.g. nonverbal communication). [PCCL05]	Champions the introduction of novel techniques and ideas in "communications," beyond the enterprise boundary, in order to develop the wider Systems Engineering community in this competency. [PCCE05]

Explains why systems engineers need to listen to stakeholders' point of view. [PCCA06]	Uses active listening techniques to clarify understanding of information or views. [PCCS06]	Uses appropriate communications techniques to express own thoughts effectively and convincingly in order to reinforce the content of the message. [PCCP06]	Adapts communications techniques or expresses ideas differently to improve effectiveness of communications to stakeholders across the enterprise, by changing language, content, or style. [PCCL06]	Coaches individuals beyond the enterprise boundary in Communications techniques, in order to further develop their knowledge, abilities, skills, or associated behaviors. [PCCE06]
	Develops own understanding of this competency area through Continual Professional Development (CPD). [PCCS07]	Uses full range of active listening techniques to clarify information or views. [PCCP07]	Reviews ongoing communications across the enterprise, anticipating and mitigating potential problems. [PCCL07]	Maintains expertise in this competency area through specialist Continual Professional Development (CPD) activities. [PCCE07]
		Uses appropriate feedback techniques to verify success of communications. [PCCP08]	Fosters the wider enterprise vision, communicating it successfully across the enterprise. [PCCL08]	
		Guides new or supervised Systems Engineering practitioners in Communications techniques in order to develop their knowledge, abilities, skills, or associated behaviors. [PCCP09]	Coaches or mentors practitioners across the enterprise, or those new to this competency are in order to develop their knowledge, abilities, skills, or associated behaviors. [PCCL09]	
		Maintains and enhances own competence in this area through Continual Professional Development (CPD) activities. [PCCP10]	Develops expertise in this competency area through specialist Continual Professional Development (CPD) activities. [PCCL10]	

Competency area – Professional: Ethics and Professionalism

Description

Professional ethics encompass the personal, organizational, and corporate standards of behavior expected of systems engineers. Professional ethics also encompasses the use of specialist knowledge and skills by systems engineers when providing a service to the public. Overall, competence in ethics and professionalism can be summarized by a personal commitment to professional standards, recognizing obligations to society, the profession, and the environment.

Why it matters

Systems engineers are routinely trusted to apply their skills, make judgements, and to reach unbiased, informed, and potentially significant decisions because of their specialized knowledge and skills. It is important that the professional systems engineer always acts ethically, in order to maintain trust, ensure professional standards are upheld, and that their wider obligations to society and the environment are met.

Effective indicators of knowledge and experience

Awareness	Supervised Practitioner	Practitioner	Lead Practitioner	Expert
Explains why Systems Engineering has a social significance. [PEPA01]	Complies with applicable codes of professional conduct within the enterprise. [PEPS01]	Follows governing ethics and professionalism guidance, adapting as required to address new situations if required. [PEPP01]	Promotes best practice ethics and professionalism across the enterprise. [PEPL01]	Communicates own knowledge and experience in ethics and professionalism in order to improve best practice beyond the enterprise boundary. [PEPE01]
Describes applicable codes of conduct for professional systems engineers including institutional or company codes of conduct. [PEPA02]	Follows safe systems principles at work, by interpreting relevant health, safety, and welfare processes, legislation, and standards seeking guidance if required. [PEPS02]	Acts to ensure safe systems are used at work, by interpreting relevant health, safety, and welfare legislation and standards. [PEPP02]	Judges compliance with relevant workplace social and employment legislation and regulatory framework on behalf of the enterprise. [PEPL02]	Persuades legislative and regulatory framework stakeholders beyond the enterprise to follow a particular path for in support of improving professionalism and ethics within Systems Engineering. [PEPE02]
Lists typical safety standards and requirements. [PEPA03]	Follows systems security principles at work, by interpreting relevant security processes, legislation, and standards seeking guidance if required. [PEPS03]	Acts to promote consideration and elimination of security issues or threats across project activities. [PEPP03]	[PEPL03] - ITEM DELETED	Persuades stakeholders beyond the enterprise boundary to improve health, safety, and welfare issues, systems, or safety culture in their activities. [PEPE03]
Explains why security has become increasingly important general requirement in the development of systems and provides examples. [PEPA04]	Acts to ensure their own activities are performed in a way that contributes to sustainable development. [PEPS04]	Ensures compliance with relevant workplace social and employment legislation and regulatory framework across the project. [PEPP04]	Judges the security of systems across the organization, including compliance with requirements, security risk management, and security awareness culture, on behalf of the enterprise. [PEPL04]	Persuades stakeholders beyond the enterprise boundary to address security issues, systems, or security culture in their activities. [PEPE04]
Explains why there is a need to undertake engineering activities in a way that contributes to sustainable, environmentally sound development and the relationship these have with the economic sustainability of a system. [PEPA05]	Acts to ensure their own activities are conducted in a way that reduces their environmental impact. [PEPS05]	Fosters a sustainable development, taking personal responsibility to promote this area in project activities. [PEPP05]	Promotes the goal of performing engineering activities in a sustainable manner across the enterprise. [PEPL05]	Persuades stakeholders beyond the enterprise boundary to address relevant employment and social regulatory compliance issues within their activities. [PEPE05]

Explains why there is a need to undertake engineering activities in a way that considers diversity, equality, and inclusivity, and provides examples. [PEPA06]	Acts to take on personal responsibility for ensuring their own activities consider diversity, equality, and inclusivity. [PEPS06]	Fosters an environmentally sound approach to project activities, taking personal responsibility to promote environmental and community considerations in project activities. [PEPP06]	Promotes the goal of performing engineering activities in an environmentally sound manner across the enterprise. [PEPL06]	Champions the development of a sustainable and environmentally sound approach to systems engineering beyond the enterprise boundary. [PEPE06]
Explains why it is necessary to develop, plan, carry out, and record Continued Professional Development (CPD) in order to maintain and enhance competence in own area of practice. [PEPA07]	Proposes changes to the project or organization which maintain and enhance the quality of the environment and community and meet financial objectives. [PEPS07]	Acts to address own professional development needs in order to maintain and enhance professional competence in own area of practice, evaluating outcomes against any plans made. [PEPP07]	Judges continual professional development planning activities at enterprise level to ensure they maintain and enhance organizational and individual competencies. [PEPL07]	Advises organizations beyond the enterprise boundary on the suitability of their approach to ethics and professionalism. [PEPE07]
Explains why Systems Engineering has a relationship to ethics and professionalism. [PEPA08]	Maintains personal continual development records and plans. [PEPS08]	Acts to ensure all members of the project/team operate with integrity and in an ethical manner. [PEPP08]	Coaches or mentors practitioners across the enterprise in matters relating to ethics and professionalism, including career development planning, in order to develop their knowledge, abilities, skills, or associated behaviors. [PEPL08]	Champions an ethical and professional culture beyond the enterprise boundary. [PEPE08]
	Acts with integrity when fulfilling own responsibilities. [PEPS09]	Acts in an ethical manner when fulfilling their own responsibilities, without support of guidance. [PEPP09]	Promotes the introduction and use of novel techniques and ideas in ethics and professionalism across the enterprise, to improve enterprise competence in this area. [PEPL09]	Champions the introduction of novel techniques and ideas in Systems Engineering ethics and professionalism, beyond the enterprise boundary, in order to develop the wider Systems Engineering community in these competencies. [PEPE09]
	Acts ethically when fulfilling own responsibilities. [PEPS10]	Guides new or supervised practitioners in matters relating to ethics and professionalism, including career development planning, in order to develop their knowledge, abilities, skills, or associated behaviors. [PEPP10]	Develops expertise in this competency area through specialist Continual Professional Development (CPD) activities. [PEPL10]	Coaches individuals beyond the enterprise boundary in ethics and professionalism, including career development planning in order to further develop their knowledge, abilities, skills, or associated behaviors. [PEPE10]
	Develops own understanding of this competency area through Continual Professional Development (CPD). [PEPS11]	Maintains and enhances own competence in this area through Continual Professional Development (CPD) activities. [PEPP11]		Maintains expertise in this competency area through specialist Continual Professional Development (CPD) activities. [PEPE11]

Competency area – Professional: Technical Leadership

Description

Systems Engineering technical leadership is the combination of the application of technical knowledge and experience in Systems Engineering with appropriate professional competencies. This encompasses an understanding of customer need, problem solving, creativity and innovation skills, communications, team building, relationship management, operational oversight, and accountability skills coupled with core Systems Engineering competency and engineering instinct.

Why it matters

The complexity of modern system designs, the severity of their constraints and the need to succeed in a high tempo, high-stakes environment where competitive advantage matters, demands the highest levels of technical excellence and integrity throughout the life cycle. Systems Engineering technical leadership helps teams meet these challenges.

Effective indicators of knowledge and experience

Awareness	Supervised Practitioner	Practitioner	Lead Practitioner	Expert
Explains the role of technical leadership within Systems Engineering. [PTLA01]	Follows guidance received (e.g. from mentors), to plan and control their own technical leadership activities or approaches. [PTLS01]	Follows guidance received to develop their own technical leadership skills, using leadership techniques and tools as instructed. [PTLP01]	Uses best practice technical leadership techniques to guide, influence, and gain trust from systems engineering stakeholders across the enterprise. [PTLL01]	Communicates own knowledge and experience in technical leadership in order to improve best practice beyond the enterprise boundary. [PTLE01]
Defines the terms "vision," "strategy," and "goal" terms explaining why each is important in leadership. [PTLA02]	Acts to gain trust in their Systems Engineering leadership activities. [PTLS02]	Acts with integrity in their leadership activities, being trusted by their team. [PTLP02]	Reacts professionally and positively to constructive criticism received from others across the enterprise. [PTLL02]	Advises organizations beyond the enterprise boundary on the suitability of their approach to technical leadership issues. [PTLE02]
Explains why understanding the strategy is central to Systems Engineering leadership. [PTLA03]	Complies with a project, or wider, vision in performing Systems Engineering leadership activities. [PTLS03]	Guides and actively coordinates Systems Engineering activities across a team, combining appropriate professional and technical competencies, with demonstrable success. [PTLP03]	Uses appropriate communications techniques to offer constructive criticism to others across the enterprise. [PTLL03]	Guides and actively coordinates the progress of Systems Engineering activities beyond the enterprise boundary, combining appropriate professional competencies with technical knowledge and experience. [PTLE03]
Explains why fostering collaboration is central to Systems Engineering. [PTLA04]	Uses team and project to guide direction, thinking strategically, holistically, and systemically when performing own Systems Engineering leadership activities. [PTLS04]	Develops technical vision for a project team, influencing and integrating the viewpoints of others in order to gain acceptance. [PTLP04]	Fosters stakeholder collaboration across the enterprise, sharing ideas and knowledge, and establishing mutual trust. [PTLL04]	Guides and actively coordinates the progress of collaborative activities beyond the enterprise boundary, establishing mutual trust. [PTLE04]
Explains why the art of communications is central to Systems Engineering including the impact of poor communications. [PTLA05]	Recognizes constructive criticism from others following guidance to improve their SE leadership. [PTLS05]	Identifies a leadership strategy to support of project goals, changing as necessary, to ensure success. [PTLP05]	Fosters the empowerment of individuals across the enterprise, by supporting, facilitating, promoting, giving ownership, and supporting them in their endeavors. [PTLL05]	Fosters empowerment of others beyond the enterprise boundary. [PTLE05]
Explains how technical analysis, problem-solving techniques, and established best practices can be used to improve the excellence of Systems Engineering solutions. [PTLA06]	Uses appropriate mechanisms to offer constructive criticism to others on the team. [PTLS06]	Recognizes constructive criticism from others within the enterprise following guidance to improve their SE leadership. [PTLP06]	Acts with creativity and innovation, applying problem-solving techniques to develop strategies or resolve complex project or enterprise technical leadership issues. [PTLL06]	Advises organizations beyond the enterprise boundary on complex or sensitive team leadership problems or issues, applying creativity and innovation to ensure successful delivery. [PTLE06]

Explains how creativity, ingenuity, experimentation, and accidents or errors, often lead to technological and engineering successes and advances and provides examples. [PTLA07]	Elicits viewpoints from others when developing solutions as part of their Systems Engineering leadership role. [PTLS07]	Uses appropriate communications techniques to offer constructive criticism to others on the team. [PTLP07]	Coaches or mentors practitioners across the enterprise in technical and leadership issues in order to develop their knowledge, abilities, skills, or associated behaviors. [PTLL07]	Uses their extended network and influencing skills to gain collaborative agreement with key stakeholders beyond the enterprise boundary in order to progress project or their own enterprise needs. [PTLE07]
Explains how different sciences impact the technology domain and the engineering discipline. [PTLA08]	Uses appropriate communications mechanisms to reinforce their Systems Engineering leadership activities. [PTLS08]	Fosters a collaborative approach in their Systems Engineering leadership activities. [PTLP08]	Promotes the introduction and use of novel techniques and ideas in SE technical leadership across the enterprise, to improve enterprise competence in this area. [PTLL08]	Champions the introduction of novel techniques and ideas in Systems Engineering technical leadership, beyond the enterprise boundary, in order to develop the wider Systems Engineering community in this competency. [PTLE08]
Explains how complexity impacts the role of the engineering leader. [PTLA09]	Acts creatively and innovatively in their SE leadership activities. [PTLS09]	Fosters the empowerment of team members, by supporting, facilitating, promoting, giving ownership, and supporting them in their endeavors. [PTLP09]	Develops expertise in this competency area through specialist Continual Professional Development (CPD) activities. [PTLL09]	Coaches individuals beyond the enterprise boundary, in technical leadership techniques in order to further develop their knowledge, abilities, skills, or associated behaviors. [PTLE09]
	Identifies concepts and ideas in sciences, technologies, or engineering disciplines beyond their own discipline, applying them to benefit their own Systems Engineering leadership activities on a project. [PTLS10]	Uses best practice communications techniques in their leadership activities, in order to express their ideas clearly and effectively. [PTLP10]		Maintains expertise in this competency area through specialist Continual Professional Development (CPD) activities. [PTLE10]
	Develops own understanding of this competency area through Continual Professional Development (CPD). [PTLS11]	Develops strategies for leadership activities or the resolution of team issues, using creativity and innovation. [PTLP11]		
		Guides new or supervised practitioners in matters relating to technical leadership in Systems Engineering, in order to develop their knowledge, abilities, skills, or associated behaviors. [PTLP12]		
		Maintains and enhances own competence in this area through Continual Professional Development (CPD) activities. [PTLP13]		

Competency area – Professional: Negotiation

Description
Negotiation is a dialogue between two or more parties intended to reach a beneficial outcome over one or more issues where differences exist with respect to at least one of these issues. This beneficial outcome can be for all parties involved, or just for one or some of them. Negotiation aims to resolve points of difference, to gain advantage for an individual or collective, or to craft outcomes to satisfy various interests. It is often conducted by putting forward a position and making small concessions to achieve an agreement.

Why it matters
Systems Engineers are the "glue" that hold elements of a complex system development together. To achieve success, they need to involve themselves in many aspects of a project, interacting with different types of stakeholders and organizations. This necessitates resolution of many different types of issue in order to gain agreement between differing groups of stakeholders. Good negotiation skills are central to this activity.

Effective indicators of knowledge and experience

Awareness	Supervised Practitioner	Practitioner	Lead Practitioner	Expert
Explains key terminology associated with negotiation. [PNEA01]	Develops good working level relationships with counterparts by negotiating to resolve routine issues. [PNES01]	Follows established best practice strategies for negotiation in terms of preparation, approach, strategy, tactics, and style. [PNEP01]	Promotes best practice negotiation techniques across the enterprise to improve the effectiveness of systems engineering negotiations. [PNEL01]	Communicates own knowledge and experience in negotiation skills in order to improve best practice beyond the enterprise boundary. [PNEE01]
Describes situations where it may be necessary to negotiate and why. [PNEA02]	Collates data from a range of sources through research and analysis to provide useful input to a negotiation team. [PNES02]	Negotiates successfully with internal and external project stakeholders. [PNEP02]	Judges the suitability of the planned approach or strategy for negotiations affecting Systems Engineering across the enterprise. [PNEL02]	Influences stakeholders beyond the enterprise boundary in support of negotiations activities affecting Systems Engineering. [PNEE02]
Explains how different stakeholders hold different positions and bargaining power. [PNEA03]	Identifies stakeholders with different bargaining power on a project. [PNES03]	Acts to ensure buy-in and gain trust with internal stakeholders prior to and during negotiations. [PNEP03]	Guides and actively coordinates the direction of negotiation teams across the enterprise, accepting accountability for final negotiation outcomes whether successful or not. [PNEL03]	Guides and actively coordinates the direction of negotiations beyond the enterprise boundary, on complex or strategic decisions. [PNEE03]
Identifies situations which do or do not require negotiation, to support negotiating strategies. [PNEA04]	Describes key stakeholders' negotiation positions of these stakeholders. [PNES04]	Communicates negotiation developments to internal stakeholders in order to manage expectations while keeping all parties informed. [PNEP04]	Adapts personal positions and style quickly if circumstances change favorably and unfavorably. [PNEL04]	Advises organizations beyond the enterprise boundary on the suitability of their negotiating strategies. [PNEE04]
	Prepares inputs to the review of a negotiation, covering the broad implications and unintended consequences of a negotiation decision. [PNES05]	Analyzes data from a range of sources to make robust fact-based statements during negotiations, to make available choices clear and simple to stakeholders. [PNEP05]	Acts on behalf of the wider enterprise during tough, challenging negotiating situations with both external and internal stakeholders. [PNEL05]	Champions the introduction of novel techniques and ideas in negotiation techniques, beyond the enterprise boundary, in order to develop the wider Systems Engineering community in this competency. [PNEE05]

	Maintains own confidence in the face of objections during negotiations. [PNES06]	Reacts positively when handling objections or points of view expressed by others challenging these views without damaging stakeholder relationship. [PNEP06]	Acts on behalf of the wider enterprise to gain credibility and gains trust and respect of all parties during difficult negotiations. [PNEL06]	Coaches individuals beyond the enterprise boundary, in negotiation techniques in order to further develop their knowledge, abilities, skills, or associated behaviors. [PNEE06]
	Develops own understanding of this competency area through Continual Professional Development (CPD). [PNES07]	Reviews the immediate results, broad implications, and unintended consequences of a negotiation decision to ensure decision is sound. [PNEP07]	Acts positively when handling objections or points of view expressed by senior enterprise stakeholders challenging views without damaging stakeholder relationship and persuading them to change their mind. [PNEL07]	Maintains expertise in this competency area through specialist Continual Professional Development (CPD) activities. [PNEE07]
		Acts with political awareness when negotiating with key decision-makers. [PNEP08]	Persuades third-party decision-makers to move toward wider enterprise goals, using good political awareness. [PNEL08]	
		Acts to gain credibility and gains trust and respect of all parties to negotiations. [PNEP09]	Acts to accept accountability for final negotiation outcomes on behalf of the enterprise, whether successful or not. [PNEL09]	
		Guides new or supervised practitioners in negotiation techniques, in order to develop their knowledge, abilities, skills, or associated behaviors. [PNEP10]	Coaches or mentors practitioners across the enterprise in negotiation techniques in order to develop their knowledge, abilities, skills, or associated behaviors. [PNEL10]	
		Maintains and enhances own competence in this area through Continual Professional Development (CPD) activities. [PNEP11]	Promotes the introduction and use of novel techniques and ideas in negotiation across the enterprise, to improve enterprise competence in this area. [PNEL11]	
			Develops expertise in this competency area through specialist Continual Professional Development (CPD) activities. [PNEL12]	

Competency area – Professional: Team Dynamics				
Description Team dynamics are the unconscious, psychological forces that influence the direction of a team's behavior and performance. Team dynamics are created by the nature of the team's work, the personalities within the team, their working relationships with other people, and the environment in which the team works.				
Why it matters Team dynamics can be good – for example, when they improve overall team performance and/or get the best out of individual team members. They can also be bad – for example, when they cause unproductive conflict, demotivation, and prevent the team from achieving its goals.				
Effective indicators of knowledge and experience				
Awareness	Supervised Practitioner	Practitioner	Lead Practitioner	Expert
Lists different types of team and the role of each team within the project or organization. [PTDA01]	Identifies when and when not to identify own positions, roles, and responsibilities within different teams within the project, or organization. [PTDS01]	Acts collaboratively with other teams to accomplish interdependent project or organizational goals. [PTDP01]	Uses best practice team dynamics techniques to improve the effectiveness of Systems Engineering activities across the enterprise. [PTDL01]	Communicates own knowledge and experience in negotiation skills in order to improve best practice beyond the enterprise boundary. [PTDE01]
Explains the different stages of team development and how they affect team dynamics and performance. [PTDA02]	Uses team dynamics to improve their effectiveness in performing team goals. [PTDS02]	Recognizes the dynamic of their team and applies best practice to improve this as necessary. [PTDP02]	Judges the dynamic of teams across the enterprise, advising where improvement is necessary. [PTDL02]	Advises organizations beyond the enterprise boundary on the suitability of their approach to team dynamics. [PTDE02]
Explains the positive and negative features of cooperation and competition within teams. [PTDA03]	Identifies the stage (e.g. forming, Storming, and Norming) at which each of the teams within which they participate is operating and provides rationale. [PTDS03]	Fosters a common understanding of an assignment in line with organizational intent within their team. [PTDP03]	Advises stakeholders across the enterprise, on the selection of measurable group goals, communication, or interpersonal actions designed to improve team performance. [PTDL03]	Advises organizations beyond the enterprise boundary on the selection and interpretation of goals used to challenge, measure, and assess team performance. [PTDE03]
Explains how the effectiveness of communications affects team dynamics. [PTDA04]	Explains the building blocks of successful team performance and why they affect performance. [PTDS04]	Fosters cooperation and pride within the team through strategies focused on group goals, communication, and interpersonal actions. [PTDP04]	Challenges negative behaviors of key enterprise stakeholders, with measurable success. [PTDL04]	Advises organizations beyond the enterprise boundary on how team members can be rewarded to act cooperatively. [PTDE04]
Explains the differing nature of disagreement, conflict, and criticism in teams, and core strategies for resolving conflict. [PTDA05]	Explains how team goals, communication, and interpersonal actions are affected by competitive behaviors. [PTDS05]	Identifies negative behaviors within the team, challenging these to create positive outcomes. [PTDP05]	Advises stakeholders across the enterprise on different best practice team dynamics techniques across the enterprise depending on the situation and decision required. [PTDL05]	Challenges negative behaviors of beyond the enterprise boundary, with measurable success. [PTDE05]
Explains why team building can help form effective teams, what it involves, and its key challenges. [PTDA06]	Identifies competitive behaviors within a team and their potential cause (e.g. cultural, personal, and organizational reasons). [PTDS06]	Uses communications skills to offer constructive feedback to improve team performance, managing emotions as an important aspect of team's communications. [PTDP06]	Fosters communication across the wider enterprise building trust through the application of team dynamics techniques. [PTDL06]	Influences key stakeholders beyond the enterprise boundary to follow a revised path to improve team dynamics across or beyond the enterprise. [PTDE06]

Identifies different types of team-building activities, their aims, and provides examples. [PTDA07]	Describes different potential types of team conflict and the differing techniques available to resolve them. [PTDS07]	Recognizes conflict in a team in order to resolve it. [PTDP07]	Influences key stakeholders across the enterprise to follow a revised path to improve a project or enterprise team dynamics. [PTDL07]	Champions the introduction of novel techniques and ideas in team dynamics, beyond the enterprise boundary, in order to develop the wider Systems Engineering community in this competency. [PTDE07]
	Explains how team dynamics affect decision-making. [PTDS08]	Fosters an open team dynamic within the team so that all team members can express their opinions and feelings. [PTDP08]	Uses different types of team-building activities depending on the team context, to improve team dynamics across the enterprise. [PTDL08]	Coaches individuals beyond the enterprise boundary, in team dynamics in order to further develop their knowledge, abilities, skills, or associated behaviors. [PTDE08]
	Develops own understanding of this competency area through Continual Professional Development (CPD). [PTDS09]	Uses best practice team dynamics techniques to obtain team consensus when making decisions. [PTDP09]	Coaches or mentors practitioners across the enterprise in team dynamics techniques in order to develop their knowledge, abilities, skills, or associated behaviors. [PTDL09]	Maintains expertise in this competency area through specialist Continual Professional Development (CPD) activities. [PTDE09]
		Uses team-building activities to improve team dynamics. [PTDP10]	Promotes the introduction and use of novel techniques and ideas in team dynamics across the enterprise, to improve enterprise competence in this area. [PTDL10]	
		Guides new or supervised practitioners in negotiation techniques, in order to develop their knowledge, abilities, skills, or associated behaviors. [PTDP11]	Develops expertise in this competency area through specialist Continual Professional Development (CPD) activities. [PTDL11]	
		Maintains and enhances own competence in this area through Continual Professional Development (CPD) activities. [PTDP12]		

Competency area – Professional: Facilitation				
Description The act of helping others to deal with a process, solve a problem, or reach a goal without getting directly involved. The goal is set by the individuals or groups, not by the facilitator.				
Why it matters Modern systems engineers must perform successfully in environments where accountability expectations are increasing, but where the use of direct authority may not achieve the desired results. Numerous sources indicate that an alternative form of leadership can address these seemingly contradictory conditions. This form of leadership has been named "facilitative leadership," and is the ability to lead without controlling, while making it easier for everyone in the organization to achieve agreed-upon goals.				
Effective indicators of knowledge and experience				
Awareness	Supervised Practitioner	Practitioner	Lead Practitioner	Expert
Explains the concept of "facilitation," summarizing its key characteristics and techniques. [PFAA01]	Acts as a neutral servant of a group performing a facilitated task. [PFAS01]	Identifies rules of conduct between individuals within a facilitated group. [PFAP01]	Uses best practice facilitation techniques to improve the effectiveness of Systems Engineering activities across the enterprise. [PFAL01]	Communicates own knowledge and experience in facilitation skills in order to improve best practice beyond the enterprise boundary. [PFAE01]
Explains how facilitation can help individuals and groups to achieve their goals. [PFAA02]	Identifies members of a group in order to perform a facilitated task. [PFAS02]	Guides a facilitated group problem-solving session. [PFAP02]	Creates a plan for a facilitated enterprise-level activity, defining methods to be used, coordinating the logistics of the meeting arrangements. [PFAL02]	Advises organizations beyond the enterprise boundary on the suitability of their approach to facilitation. [PFAE02]
Describes why the effectiveness of facilitation can differ during different stages of group formation. [PFAA03]	Identifies rules of conduct between individuals within a facilitated group, with guidance. [PFAS03]	Acts to ensure own views and feelings remain hidden, when facilitating group activities on a project. [PFAP03]	Selects the most appropriate style of facilitation based upon enterprise-level facilitated group maturity. [PFAL03]	Persuades key stakeholders beyond the enterprise boundary to support facilitated group activities. [PFAE03]
Describes how different facilitation skills can help resolve different forms of conflict and dissent in a group to mitigate their impact. [PFAA04]	Acts as impartial observer focused on facilitated group activities, with guidance. [PFAS04]	Acts to facilitate self-improvement of the performance of a group. [PFAP04]	Uses facilitation skills to ensure that an enterprise-level facilitated group clarifies its goals. [PFAL04]	Reviews the suitability of facilitation programs affecting Systems Engineering beyond the enterprise boundary. [PFAE04]
Describes how facilitation skills supplement different approaches to problem solving and the patterns of thinking associated with each. [PFAA05]	Acts in support of facilitation of a group problem-solving session. [PFAS05]	Acts to engage individuals to improve performance of a group. [PFAP05]	Uses facilitation skills to facilitate an enterprise-level facilitated group toward achieving its objectives. [PFAL05]	Identifies alternative ways of working to reinforce collaboration within the context of a facilitated group with membership beyond the enterprise boundary. [PFAE05]
	Conducts a small, facilitated group problem-solving session on their team. [PFAS06]	Acts to protect individuals and their ideas from attack within a facilitated group. [PFAP06]	Acts as a referee in times of conflict, disagreement, or tension within an enterprise-level facilitated group. [PFAL06]	Advises organizations beyond the enterprise boundary on complex or sensitive matters, conflict disagreement, or tension affecting facilitated group. [PFAE06]

	Develops own understanding of this competency area through Continual Professional Development (CPD). [PFAS07]	Adapts strategy if a facilitated group requires a change of direction. [PFAP07]	Fosters systematic patterns of thinking during the facilitated enterprise-level group problem-solving process. [PFAL07]	Acts to anticipate and mitigate potential problems in facilitation of a group extending beyond the enterprise boundary. [PFAE07]
		Guides and actively coordinates a facilitated group problem-solving session on the project. [PFAP08]	Coaches or mentors practitioners across the enterprise in facilitation techniques in order to develop their knowledge, abilities, skills, or associated behaviors. [PFAL08]	Fosters open communications which surface prevailing mental models and challenge a facilitation group extending beyond the enterprise boundary to build a shared vision. [PFAE08]
		Guides new or supervised practitioners in facilitation techniques, in order to develop their knowledge, abilities, skills, or associated behaviors. [PFAP09]	Promotes the introduction and use of novel techniques and ideas in facilitation across the enterprise, to improve enterprise competence in this area. [PFAL09]	Champions the introduction of novel techniques and ideas in facilitation, beyond the enterprise boundary, in order to develop the wider Systems Engineering community in this competency. [PFAE09]
		Maintains and enhances own competence in this area through Continual Professional Development (CPD) activities. [PFAP10]	Develops expertise in this competency area through specialist Continual Professional Development (CPD) activities. [PFAL10]	Coaches individuals beyond the enterprise boundary, in facilitation techniques in order to further develop their knowledge, abilities, skills, or associated behaviors. [PFAE10]
				Maintains expertise in this competency area through specialist Continual Professional Development (CPD) activities. [PFAE11]

Competency area – Professional: Emotional Intelligence

Description

Emotional intelligence is the ability to monitor one's own and others' feelings, to discriminate among them, and to use this information to guide thinking and action. This is usually broken down into four distinct but related proposed abilities: perceiving, using, understanding, and managing emotions.

Why it matters

Emotional intelligence is regularly cited as a critical competency for effective leadership and team performance in organizations. It influences the success with which individuals in organizations interact with colleagues, the approaches they use to manage conflict and stress, and their overall job performance. As Systems Engineering involves interacting with many diverse stakeholders, emotional intelligence is critical to its success.

Effective indicators of knowledge and experience

Awareness	Supervised Practitioner	Practitioner	Lead Practitioner	Expert
Explains why the perception of emotion is important including differentiating one's own emotions from those of others. [PEIA01]	Identifies emotions in one's physical states, feelings, and thoughts. [PEIS01]	Uses Emotional Intelligence techniques to interpret meanings and origins of emotions and acts accordingly. [PEIP01]	Uses best practice emotional intelligence techniques to improve the effectiveness of Systems Engineering activities across the enterprise. [PEIL01]	Communicates own knowledge and experience in emotional intelligence in order to improve best practice beyond the enterprise boundary. [PEIE01]
Explains how emotions can be used to facilitate thinking such as reasoning, problem solving, and interpersonal communication and explains why this is important. [PEIA02]	Uses emotional intelligence techniques to identify the emotions of others via verbal and nonverbal cues. [PEIS02]	Uses Emotional Intelligence techniques to identify needs related to emotional feelings. [PEIP02]	Guides others across the enterprise in controlling their own emotional responses. [PEIL02]	Uses emotional intelligence to influence beyond the enterprise boundary. [PEIE02]
Explains why it is important to be able to understand and analyze emotions. [PEIA03]	Explains the language used to label emotions. [PEIS03]	Uses emotional intelligence techniques to monitor their own emotions in relation to others. [PEIP03]	Uses emotional intelligence techniques in tough, challenging situations with both external and internal stakeholders, with demonstrable results. [PEIL03]	Advises organizations beyond the enterprise boundary on the suitability of their approach to emotional intelligence awareness and its utilization. [PEIE03]
Explains why managing and regulating emotions in both oneself and in others is important. [PEIA04]	Develops own understanding of this competency area through Continual Professional Development (CPD). [PEIS04]	Acts to capitalize fully upon changing moods in order to best fit the task at hand. [PEIP04]	Uses emotional intelligence to influence key stakeholders within the enterprise. [PEIL04]	Advises beyond the enterprise boundary on complex or sensitive emotionally charged issues. [PEIE04]
		Acts to remain open to feelings, both those that are pleasant and those that are unpleasant. [PEIP05]	Coaches or mentors practitioners across the enterprise in emotional intelligence techniques in order to develop their knowledge, abilities, skills, or associated behaviors. [PEIL05]	Champions the introduction of novel techniques and ideas in the application of emotional intelligence, beyond the enterprise boundary, in order to develop the wider Systems Engineering community in this competency. [PEIE05]

		Acts to control own emotion by preventing, reducing, enhancing, or modifying an emotional response. [PEIP06]	Promotes the introduction and use of novel techniques and ideas in Emotional Intelligence techniques across the enterprise, to improve enterprise competence in this area. [PEIL06]	Coaches individuals beyond the enterprise boundary, in emotional intelligence techniques in order to further develop their knowledge, abilities, skills, or associated behaviors. [PEIE06]
		Guides new or supervised practitioners in emotional intelligence techniques, in order to develop their knowledge, abilities, skills, or associated behaviors. [PEIP07]	Develops expertise in this competency area through specialist Continual Professional Development (CPD) activities. [PEIL07]	Maintains expertise in this competency area through specialist Continual Professional Development (CPD) activities. [PEIE07]
		Maintains and enhances own competence in this area through Continual Professional Development (CPD) activities. [PEIP08]		

Competency area – Professional: Coaching and Mentoring

Description

Coaching and mentoring are development approaches based on the use of one-to-one conversations to enhance an individual's skills, knowledge, or work performance. Coaching is a nondirective form of development aiming to produce optimal performance and improvement at work. It focuses on specific skills and goals, although may impact an individual's personal attributes. The process typically lasts for a defined period. Mentoring is a relationship where a more experienced colleague shares their greater knowledge to support development of a less experienced member of staff. It uses many of the techniques associated with coaching. One key distinction is that mentoring relationships tend to be longer term than coaching arrangements.

Why it matters

Coaching and mentoring play an important role in the development of Systems Engineering professionals, providing targeted development and guidance, organizational and cultural insights. They represent learning opportunities for both parties, encouraging sharing and learning across generations and/or between roles. In addition, an organization may benefit through greater retention of staff, improved skills and productivity, improved communication, etc.

Effective indicators of knowledge and experience

Awareness	Supervised Practitioner	Practitioner	Lead Practitioner	Expert
Describes key characteristics and personal attributes of coach and mentor roles, and how both approaches help to develop individual potential. [PMEA01]	Identifies areas of own skills, knowledge, or experience which could be improved. [PMES01]	Coaches (or mentors) others on the project as part of an enterprise coaching and mentoring program. [PMEP01]	Promotes the use of best practice coaching and mentoring techniques to improve the effectiveness of Systems Engineering activities across the enterprise. [PMEL01]	Communicates own knowledge and experience in coaching and mentoring skills in order to improve best practice beyond the enterprise boundary. [PMEE01]
Explains how those undergoing coaching and mentoring need to act in order to benefit from the activity. [PMEA02]	Identifies personal challenges through various perspectives. [PMES02]	Creates career development goals and objectives with individuals. [PMEP02]	Judges the suitability of planned coaching and mentoring programs affecting Systems Engineering within the enterprise. [PMEL02]	Persuades key stakeholders beyond the enterprise boundary to follow a particular path for coaching and mentoring activities affecting Systems Engineering. [PMEE02]
Explains why listening to an individual's goals and objectives is important. [PMEA03]	Prepares information supporting the development of others within the team. [PMES03]	Develops individual career development paths based on development goals and objectives. [PMEP03]	Defines the direction of enterprise coaching and mentoring program development. [PMEL03]	Advises organizations beyond the enterprise boundary on the suitability of their approach to coaching and mentoring. [PMEE03]
Lists enterprise goals and describes the influence mentoring may have on meeting those goals. [PMEA04]	Develops own understanding of this competency area through Continual Professional Development (CPD). [PMES04]	Uses available coaching and mentoring opportunities to develop individuals within the enterprise. [PMEP04]	Guides and actively coordinates the implementation of an enterprise-level coaching and mentoring program. [PMEL04]	Advises organizations beyond the enterprise boundary on the development of coaching and mentoring programs. [PMEE04]
Explains why taking a comprehensive approach to assess an individual's challenge is important. [PMEA05]		Develops individuals within their team by supporting them in solving their individual challenges. [PMEP05]	Assesses career development path activities for individuals across the enterprise, providing regular feedback. [PMEL05]	Assesses the effectiveness of a mentoring program for an organization beyond the enterprise boundary, providing regular feedback. [PMEE05]

Describes the design and operation of the enterprise's coaching and mentoring program. [PMEA06]		Guides new or supervised practitioners in coaching and mentoring techniques, in order to develop their knowledge, abilities, skills, or associated behaviors. [PMEP06]	Advises stakeholders across the enterprise on individual coaching and mentoring issues with demonstrable success. [PMEL06]	Advises organizations beyond the enterprise boundary on complex or challenging coaching and mentoring issues. [PMEE06]
		Maintains and enhances own competence in this area through Continual Professional Development (CPD) activities. [PMEP07]	Coaches or mentors practitioners across the enterprise in coaching and mentoring techniques in order to develop their knowledge, abilities, skills, or associated behaviors. [PMEL07]	Champions the introduction of novel techniques and ideas in coaching and mentoring, beyond the enterprise boundary, in order to develop the wider Systems Engineering community in this competency. [PMEE07]
			Promotes the introduction and use of novel techniques and ideas in Coaching and Mentoring across the enterprise, to improve enterprise competence in this area. [PMEL08]	Coaches individuals beyond the enterprise boundary, in coaching and mentoring techniques in order to further develop their knowledge, abilities, skills, or associated behaviors. [PMEE08]
			Develops expertise in this competency area through specialist Continual Professional Development (CPD) activities. [PMEL09]	Maintains expertise in this competency area through specialist Continual Professional Development (CPD) activities. [PMEE09]

Competency area – Technical: Requirements Definition

Description
To analyze the stakeholder needs and expectations to establish the requirements for a system.

Why it matters
The requirements of a system describe the problem to be solved (its purpose, how it performs, how it is to be used, maintained and disposed of, and what the expectations of the stakeholders are).

Effective indicators of knowledge and experience

Awareness	Supervised Practitioner	Practitioner	Lead Practitioner	Expert
Describes what a requirement is, the purpose of requirements, and why requirements are important. [TRDA01]	Uses a governing process using appropriate tools to manage and control their own requirements definition activities. [TRDS01]	Creates a strategy for requirements definition on a project to support SE project and wider enterprise needs. [TRDP01]	Creates enterprise-level policies, procedures, guidance, and best practice for requirements elicitation and management, including associated tools. [TRDL01]	Communicates own knowledge and experience in Systems Engineering requirements definition in order to promote best practice beyond the enterprise boundary. [TRDE01]
Describes different types of requirements and constraints that may be placed on a system. [TRDA02]	Identifies examples of internal and external project stakeholders highlighting their sphere of influence. [TRDS02]	Creates a governing process, plan, and associated tools for Requirements Definition, which reflect project and business strategy. [TRDP02]	Judges the tailoring of enterprise-level requirements elicitation and management processes to meet the needs of a project. [TRDL02]	Persuades key stakeholders beyond the enterprise boundary to address identified requirements definition issues to reduce project risk. [TRDE02]
Explains why there is a need for good quality requirements. [TRDA03]	Elicits requirements from stakeholders under guidance, in order to understand their need and ensuring requirement validity. [TRDS03]	Uses plans and processes for requirements definition, interpreting, evolving, or seeking guidance where appropriate. [TRDP03]	Advises on complex or challenging requirements from across the enterprise to ensure completeness and suitability. [TRDL03]	Advises organizations beyond the enterprise boundary on the suitability of their approach to requirements definition. [TRDE03]
Identifies major stakeholders and their needs. [TRDA04]	Describes the characteristics of good quality requirements and provides examples. [TRDS04]	Elicits requirements from stakeholders ensuring their validity, to understand their need. [TRDP04]	Defines strategies for requirements resolution in situations across the enterprise where stakeholders (or their requirements) demand unusual or sensitive treatment. [TRDL04]	Advises organizations beyond the enterprise boundary on the handling of complex or sensitive Systems Engineering requirements definition issues. [TRDE04]
Explains why managing requirements throughout the life cycle is important. [TRDA05]	Describes different mechanisms used to gather requirements. [TRDS05]	Develops good quality, consistent requirements. [TRDP05]	Persuades key stakeholders across the enterprise to address identified enterprise-level requirements elicitation and management issues to reduce enterprise-level risk. [TRDL05]	Champions the introduction of novel techniques and ideas in the requirements definition, beyond the enterprise boundary, in order to develop the wider Systems Engineering community in this competency. [TRDE05]

Describes the relationship between requirements, testing, and acceptance. [TRDA06]	Defines acceptance criteria for requirements, under guidance. [TRDS06]	Determines derived requirements. [TRDP06]	Coaches or mentors practitioners across the enterprise in Systems Engineering Requirements Definition in order to develop their knowledge, abilities, skills, or associated behaviors. [TRDL06]	Coaches individuals beyond the enterprise boundary, in requirements definition in order to further develop their knowledge, abilities, skills, or associated behaviors. [TRDE06]
	Explains why there may be potential requirement conflicts within a requirement set. [TRDS07]	Creates a system to support requirements management and traceability. [TRDP07]	Promotes the introduction and use of novel techniques and ideas in Requirements Definition across the enterprise, to improve enterprise competence in this area. [TRDL07]	Maintains expertise in this competency area through specialist Continual Professional Development (CPD) activities. [TRDE07]
	Explains how requirements affect design and vice versa. [TRDS08]	Determines acceptance criteria for requirements. [TRDP08]	Develops expertise in this competency area through specialist Continual Professional Development (CPD) activities. [TRDL08]	
	Defines (or maintains) requirements traceability information. [TRDS09]	Negotiates agreement in requirement conflicts within a requirement set. [TRDP09]		
	Reviews developed requirements. [TRDS10]	Analyzes the impact of changes to requirements on the solution and program. [TRDP10]		
	Develops own understanding of this competency area through Continual Professional Development (CPD). [TRDS11]	Maintains requirements traceability information to ensure source(s) and test records are correctly linked over the life cycle. [TRDP11]		
		Guides new or supervised practitioners in Systems Engineering Requirements Definition to develop their knowledge, abilities, skills, or associated behaviors. [TRDP12]		
		Maintains and enhances own competence in this area through Continual Professional Development (CPD) activities. [TRDP13]		

Competency area – Technical: System Architecting				
Description The definition of the system structure, interfaces, and associated derived requirements to produce a solution that can be implemented to enable a balanced and optimum result that considers all stakeholder requirements (business, technical, . . .). This includes the early generation of potential system concepts that meet a set of needs and demonstration that one or more credible, feasible options exist.				
Why it matters Effective architectural design enables systems to be partitioned into realizable system elements which can be brought together to meet the requirements. Failure to explore alternative conceptual options as part of architectural analysis may result in a nonoptimal system. There may be no viable option (e.g. technology not available).				
Effective indicators of knowledge and experience				
Awareness	Supervised Practitioner	Practitioner	Lead Practitioner	Expert
Describes the principles of architectural design and its role within the life cycle. [TSAA01]	Uses a governing process using appropriate tools to manage and control their own system architectural design activities. [TSAS01]	Creates a strategy for system architecting on a project to support SE project and wider enterprise needs. [TSAP01]	Creates enterprise-level policies, procedures, guidance, and best practice for system architectural design including associated tools. [TSAL01]	Communicates own knowledge and experience in Systems Architecting in order to promote best practice beyond the enterprise boundary. [TSAE01]
Describes different types of architecture and provides examples. [TSAA02]	Uses analysis techniques or principles used to support an architectural design process. [TSAS02]	Creates a governing process, plan, and associated tools for systems architecting, which reflect project and business strategy. [TSAP02]	Assesses the tailoring of enterprise-level system architectural design processes to meet the needs of a project. [TSAL02]	Persuades key stakeholders beyond the enterprise boundary in order to facilitate the system architectural design. [TSAE02]
Explains why architectural decisions can constrain and limit future use and evolution and provides examples. [TSAA03]	Develops multiple different architectural solutions (or parts thereof) meeting the same set of requirements to highlight different options available. [TSAS03]	Uses plans and processes for system architecting, interpreting, evolving, or seeking guidance where appropriate. [TSAP03]	Advises stakeholders across the enterprise on selection of architectural design and functional analysis techniques to ensure effectiveness and efficiency of approach. [TSAL03]	Advises organizations beyond the enterprise boundary on the suitability of their approach to system architectural design. [TSAE03]
Explains why there is a need to explore alternative and innovative ways of satisfying the requirements. [TSAA04]	Produces traceability information linking differing architectural design solutions to requirements. [TSAS04]	Creates alternative architectural designs traceable to the requirements to demonstrate different approaches to the solution. [TSAP04]	Judges the suitability of architectural solutions across the enterprise in areas of complex or challenging technical requirements or needs. [TSAL04]	Advises organizations beyond the enterprise boundary on improving their handling of complex or sensitive Systems Architecting issues. [TSAE04]
Explains why alternative discipline technologies can be used to satisfy the same requirement and provides examples. [TSAA05]	Uses different techniques to develop architectural solutions. [TSAS05]	Analyzes options and concepts in order to demonstrate that credible, feasible options exist. [TSAP05]	Assesses system architectures across the enterprise, to determine whether they meet the overall needs of individual projects. [TSAL05]	Advises organizations beyond the enterprise boundary on improving their concept generation activities. [TSAE05]

Describes the process and key artifacts of functional analysis. [TSAA06]	Compares the characteristics of different concepts to determine their strengths and weaknesses. [TSAS06]	Uses appropriate analysis techniques to ensure different viewpoints are considered. [TSAP06]	Persuades key stakeholders across the enterprise to address identified enterprise-level Systems Engineering architectural design issues to reduce project cost, schedule, or technical risk. [TSAL06]	Champions the introduction of novel techniques and ideas in systems architecting, beyond the enterprise boundary, in order to develop the wider Systems Engineering community in this competency. [TSAE06]
Explains why there is a need for functional models of the system. [TSAA07]	Prepares a functional analysis using appropriate tools and techniques to characterize a system. [TSAS07]	Elicits derived discipline specific architectural constraints from specialists to support partitioning and decomposition. [TSAP07]	Coaches or mentors practitioners across the enterprise in Systems Architecting in order to develop their knowledge, abilities, skills, or associated behaviors. [TSAL07]	Coaches individuals beyond the enterprise boundary in Systems Architecting, in order to further develop their knowledge, abilities, skills, or associated behaviors. [TSAE07]
Explains how outputs from functional analysis relate to the overall system design and provides examples. [TSAA08]	Prepares architectural design work products (or parts thereof) traceable to the requirements. [TSAS08]	Uses the results of system analysis activities to inform system architectural design. [TSAP08]	Promotes the introduction and use of novel techniques and ideas in Systems Architecting across the enterprise, to improve enterprise competence in this area. [TSAL08]	Maintains expertise in this competency area through specialist Continual Professional Development (CPD) activities. [TSAE08]
	Develops own understanding of this competency area through Continual Professional Development (CPD). [TSAS09]	Identifies the strengths and weaknesses of relevant technologies in the context of the requirement and provides examples. [TSAP09]	Develops expertise in this competency area through specialist Continual Professional Development (CPD) activities. [TSAL09]	
		Monitors key aspects of the evolving design solution in order to adjust architecture, if appropriate. [TSAP10]		
		Guides new or supervised practitioners in Systems Architecting to develop their knowledge, abilities, skills, or associated behaviors. [TSAP11]		
		Maintains and enhances own competence in this area through Continual Professional Development (CPD) activities. [TSAP12]		

Competency area – Technical: Design for. . .

Description

Ensuring that the requirements of all life cycle stages are addressed at the correct point in the system design. During the design process, consideration should be given to the design attributes such as manufacturability, testability, reliability, maintainability, affordability, safety, security, human factors, environmental impacts, robustness and resilience, flexibility, interoperability, capability growth, disposal, cost, natural variations, etc. Includes the need to design for robustness. A robust system is tolerant of misuse, out of spec scenarios, component failure, environmental stress, and evolving needs.

Why it matters

Failure to design for these attributes at the correct point in the development life cycle may result in the attributes never being achieved or achieved at escalated cost. In particular, a robust system provides greater availability during operation.

Effective indicators of knowledge and experience

Awareness	Supervised Practitioner	Practitioner	Lead Practitioner	Expert
Explains why there is a need to accommodate the requirements of all life cycle stages when determining a solution. [TDFA01]	Uses a governing process using appropriate tools and techniques to manage and control their own specialty engineering activities, interpreting, evolving, or seeking guidance where appropriate. [TDFS01]	Creates a strategy for "Designing for. . ." specialties on a project to support SE project and wider enterprise needs. [TDFP01]	Creates enterprise-level policies, procedures, guidance, and best practice relating to specialty engineering including associated tools. [TDFL01]	Communicates own knowledge and experience in specialty engineering, in order to promote best practice beyond the enterprise boundary. [TDFE01]
Identifies design attributes and explains why attributes must be balanced using trade-off studies. [TDFA02]	Explains the concept of design attributes, explaining how they influence the design. [TDFS02]	Creates a governing process, plan, and associated tools for Specialty engineering which reflect project and business strategy. [TDFP02]	Judges the tailoring of enterprise-level system specialty engineering processes to meet the needs of a project. [TDFL02]	Persuades key stakeholders beyond the enterprise boundary to accept recommendations in support of specialty engineering activities. [TDFE02]
Identifies different design specialties and describes their role and key activities. [TDFA03]	Selects design attributes in order to balance differing specialty engineering needs. [TDFS03]	Uses governing plans and processes to ensure Specialty engineering is accommodated in the evolving design, interpreting, evolving, or seeking guidance where appropriate. [TDFP03]	Judges the strategy to be adopted on projects across the enterprise to ensure required specialty engineering characteristics are met. [TDFL03]	Advises organizations beyond the enterprise boundary on the suitability of their approach to specialty engineering including its organization and integration across their enterprise. [TDFE03]
Explains why it is important to integrate design specialties into the solution and how this can be a potential source of conflict with requirements. [TDFA04]	Reviews design attributes with specialists to ensure they are addressed. [TDFS04]	Selects design attributes throughout the design process balancing these to support specialty engineering needs. [TDFP04]	Judges the adequacy of sensitivity analysis made for specialty engineering criteria across the enterprise. [TDFL04]	Advises organizations beyond the enterprise boundary on complex or sensitive specialty engineering planning issues. [TDFE04]
Explains how design specialties can affect the cost of ownership and provides examples. [TDFA05]	Identifies conflicting demands from differing design specialties and records these in trade studies in order to compare alternative solutions. [TDFS05]	Identifies the appropriate specialists to ensure it addresses design attributes effectively and at the correct time. [TDFP05]	Judges the suitability of plans across the enterprise, for the incorporation of all life cycle design attributes at the correct point within the design process. [TDFL05]	Coordinates activities beyond the enterprise boundary which cover multiple specialties. [TDFE05]

Explains how the design, throughout the life cycle, affects the robustness of the solution. [TDFA06]	Elicits operational environment characteristics from specialty engineers in support of specialty engineering activities. [TDFS06]	Analyzes demands from differing design specialties highlighted in trade studies, resolving identified conflicts as necessary. [TDFP06]	Judges selected solutions from across the enterprise, against key specialty engineering design parameters to support the decision-making process. [TDFL06]	Identifies conflicts involving specialty engineering issues that extend beyond the enterprise boundary in order to enable the project to progress. [TDFE06]
Describes the relationship between reliability, availability, maintainability, and safety. [TDFA07]	Compares specialty characteristics for proposed solutions and records these in trade studies in order to compare differing solutions. [TDFS07]	Produces a sensitivity analysis on specialty engineering trade-off criteria. [TDFP07]	Persuades key stakeholders to address identified enterprise-level specialty-related design issues to reduce project cost, schedule, or technical risk. [TDFL07]	Advises organizations beyond the enterprise boundary on how evolving needs impacts specialty engineering. [TDFE07]
	Describes how design integrity affects their project and provides examples. [TDFS08]	Uses appropriate techniques to characterize the operational environment in order to support specialty engineering activities. [TDFP08]	Coaches or mentors practitioners across the enterprise in "designing for. . ." specialties, in order to develop their knowledge, abilities, skills, or associated behaviors. [TDFL08]	Champions the introduction of novel techniques and ideas in the "Design for. . ." area, beyond the enterprise boundary, in order to develop the wider Systems Engineering community in this competency. [TDFE08]
	Identifies constraints placed on the system because of the needs of design specialties. [TDFS09]	Uses appropriate techniques and trade studies to determine and characterize specialty characteristics of proposed solutions. [TDFP09]	Promotes the introduction and use of novel techniques and ideas in "designing for. . ." specialties across the enterprise, to improve enterprise competence in this area. [TDFL09]	Coaches individuals beyond the enterprise boundary in designing for specialty engineering, in order to further develop their knowledge, abilities, skills, or associated behaviors. [TDFE09]
	Uses specialty engineering techniques and tools to ensure delivery of designs meeting specialty needs. [TDFS10]	Ensures specialty engineering experts and specialty engineering activities are fully integrated into Systems Engineering development activities. [TDFP10]	Develops expertise in this competency area through specialist Continual Professional Development (CPD) activities. [TDFL10]	Maintains expertise in this competency area through specialist Continual Professional Development (CPD) activities. [TDFE10]
	Develops own understanding of this competency area through Continual Professional Development (CPD). [TDFS11]	Identifies constraints on a system which reflect the needs of different design specialties. [TDFP11]		
		Guides new or supervised practitioners in the area of specialty engineering to develop their knowledge, abilities, skills, or associated behaviors. [TDFP12]		
		Maintains and enhances own competence in this area through Continual Professional Development (CPD) activities. [TDFP13]		

Competency area – Technical: Integration

Description

Systems Integration is the logical process for assembling a set of system elements and aggregates into the realized system, product, or service that satisfies system requirements, architecture, and design. Systems integration focuses on the testing of interfaces, data flows, and control mechanisms, checking that realized elements and aggregates perform as predicted by their design and architectural solution, since it may not always be practicable or cost-effective to confirm these lower-level aspects at higher levels of system integration.

Why it matters

Systems Integration should be planned so that system elements are brought together in a logical sequence to avoid wasted effort. Systematic and incremental integration makes it easier to find, isolate, diagnose, and correct problems. A system or system element that has not been integrated systematically cannot be relied on to meet its requirements.

Effective indicators of knowledge and experience

Awareness	Supervised Practitioner	Practitioner	Lead Practitioner	Expert
Explains why integration is important and how it confirms the system design, architecture, and interfaces. [TINA01]	Uses a governing process using appropriate tools to manage and control their own integration activities. [TINS01]	Creates a strategy for system integration on a project to support SE project and wider enterprise needs. [TINP01]	Creates enterprise-level policies, procedures, guidance, and best practice for integration, including associated tools for a project. [TINL01]	Communicates own knowledge and experience in Systems Integration in order to promote best practice beyond the enterprise boundary. [TINE01]
Explains why it is important to integrate the system in a logical sequence. [TINA02]	Prepares inputs to integration plans based upon governing standards and processes including identification of method and timing for each activity to meet project requirements. [TINS02]	Creates a governing process, plan, and associated tools for systems integration, which reflect project and business strategy. [TINP02]	Judges the tailoring of enterprise-level integration processes to meet the needs of a project. [TINL02]	Persuades key stakeholders beyond the enterprise boundary to accept recommendation associated with integration activities. [TINE02]
Explains why planning and management of systems integration is necessary. [TINA03]	Prepares plans which address integration for system elements (or noncomplex systems) in order to define or scope that activity. [TINS03]	Uses governing plans and processes to plan and execute system integration activities, interpreting, evolving, or seeking guidance where appropriate. [TINP03]	Judges the suitability of integration plans from projects across the enterprise, to ensure project success. [TINL03]	Advises organizations beyond the enterprise boundary on the suitability of their approach to integration to support enterprise needs. [TINE03]
Explains the relationship between integration and verification. [TINA04]	Records the causes of simple faults typically found during integration activities in order to communicate with stakeholders. [TINS04]	Performs rectification of faults found during integration activities. [TINP04]	Judges detailed integration procedures from projects across the enterprise, to ensure project success. [TINL04]	Advises organizations beyond the enterprise boundary on evidence generated during integration to support enterprise needs. [TINE04]
	Collates evidence during integration in support of downstream test and acceptance activities. [TINS05]	Prepares evidence obtained during integration in support of downstream test and acceptance activities. [TINP05]	Judges integration evidence generated by projects across the enterprise, to ensure adequacy of information. [TINL05]	Advises organizations beyond the enterprise boundary on complex or sensitive integration-related issues to support enterprise needs. [TINE05]

	Identifies an integration environment to facilitate system integration activities. [TINS06]	Guides and actively coordinates integration activities for a system. [TINP06]	Guides and actively coordinates integration activities on complex systems or across multiple projects from projects across the enterprise. [TINL06]	Champions the introduction of novel techniques and ideas in systems integration, beyond the enterprise boundary, in order to develop the wider Systems Engineering community in this competency. [TINE06]
	Develops own understanding of this competency area through Continual Professional Development (CPD). [TINS07]	Identifies a suitable integration environment. [TINP07]	Persuades key stakeholders to address identified enterprise-level system integration issues to reduce project cost, schedule, or technical risk. [TINL07]	Coaches individuals beyond the enterprise boundary in Systems Integration, in order to further develop their knowledge, abilities, skills, or associated behaviors. [TINE07]
		Creates detailed integration procedures. [TINP08]	Coaches or mentors practitioners across the enterprise in systems integration in order to develop their knowledge, abilities, skills, or associated behaviors. [TINL08]	Maintains expertise in this competency area through specialist Continual Professional Development (CPD) activities. [TINE08]
		Guides new or supervised practitioners in Systems integration to develop their knowledge, abilities, skills, or associated behaviors. [TINP09]	Promotes the introduction and use of novel techniques and ideas in systems integration across the enterprise, to improve enterprise competence in this area. [TINL09]	
		Maintains and enhances own competence in this area through Continual Professional Development (CPD) activities. [TINP10]	Develops expertise in this competency area through specialist Continual Professional Development (CPD) activities. [TINL10]	

Competency area – Technical: Interfaces

Description
Interfaces occur where system elements interact, for example human, mechanical, electrical, thermal, data, etc. Interface Management comprises the identification, definition, and control of interactions across system or system element boundaries.

Why it matters
Poor interface definition and management can result in incompatible system elements (either internal to the system or between the system and its environment) which may ultimately result in system failure or project overrun.

Effective indicators of knowledge and experience

Awareness	Supervised Practitioner	Practitioner	Lead Practitioner	Expert
Defines key concepts within interface definition and management. [TIFA01]	Uses a governing process to manage and control their own interface management activities. [TIFS01]	Creates a strategy for interface definition and management on a project to support SE project and wider enterprise needs. [TIFP01]	Creates enterprise- level policies, procedures, guidance, and best practice for interface definition and management, including associated tools. [TIFL01]	Communicates own knowledge and experience in Systems Engineering interface definition and management in order to improve best practice beyond the enterprise boundary. [TIFE01]
Explains how interface definition and management affects the integrity of the system solution. [TIFA02]	Identifies the properties of simple interfaces in order to define them. [TIFS02]	Creates a governing process, plan, and associated tools for interface definition and management, which reflect project and business strategy. [TIFP02]	Judges the tailoring of enterprise-level interface definition and management processes to meet the needs of a project. [TIFL02]	Influences key stakeholders beyond the enterprise boundary in interface definition and management to support enterprise needs. [TIFE02]
Identifies possible sources of complexity in interface definition and management. [TIFA03]	Explains the potential consequences of changes on system interfaces to coordinate and control ongoing development. [TIFS03]	Uses interface management techniques and governing processes, to manage and control their own interface management activities. [TIFP03]	Judges the suitability and completeness of interfaces and associated management practices used on projects across the enterprise. [TIFL03]	Advises organizations beyond the enterprise boundary on the suitability of their approach to Systems Engineering Interface Management and Control. [TIFE03]
Explains how different sources of complexity affect interface definition and management. [TIFA04]	Maintains technical parameters associated with an interface to ensure continued stability of definition. [TIFS04]	Maintains interfaces over time to ensure continued coherence and alignment with project need. [TIFP04]	Identifies conflicts in the definition or management of interfaces requiring resolution on projects across the enterprise. [TIFL04]	Advises organizations beyond the enterprise boundary on their handling of complex or sensitive Systems Engineering Interface management issues. [TIFE04]
	Develops own understanding of this competency area through Continual Professional Development (CPD). [TIFS05]	Explains the effect of complexity on interface definition and management. [TIFP05]	Acts to arbitrate when there are conflicts in the definition of interfaces or their management on projects across the enterprise. [TIFL05]	Champions the introduction of novel techniques and ideas in interface management, beyond the enterprise boundary, in order to develop the wider Systems Engineering community in this competency. [TIFE05]

		Negotiates interfaces between interface stakeholders to facilitate system development. [TIFP06]	Coaches or mentors practitioners across the enterprise in Systems Engineering interface management in order to develop their knowledge, abilities, skills, or associated behaviors. [TIFL06]	Coaches individuals beyond the enterprise boundary in Interface Management, in order to further develop their knowledge, abilities, skills, or associated behaviors. [TIFE06]
		Identifies impact on interface definitions as a result of wider changes. [TIFP07]	Promotes the introduction and use of novel techniques and ideas in interface management across the enterprise, to improve enterprise competence in this area. [TIFL07]	Maintains expertise in this competency area through specialist Continual Professional Development (CPD) activities. [TIFE07]
		Guides new or supervised practitioners in Systems Engineering interface management in order to develop their knowledge, abilities, skills, or associated behaviors. [TIFP08]	Develops expertise in this competency area through specialist Continual Professional Development (CPD) activities. [TIFL08]	
		Maintains and enhances own competence in this area through Continual Professional Development (CPD) activities. [TIFP09]		

Competency area – Technical: Verification

Description
Verification is the formal process of obtaining objective evidence that a system or system element, product, or service fulfills its specified requirements and characteristics. Verification includes formal testing of the system against the system requirements; including qualification against the super system environment (e.g. electromagnetic compatibility, thermal, vibration, humidity, fungus growth, etc.). Put simply, it answers the question "Did we build the system right?"

Why it matters
System verification should be planned so that system elements are tested in a logical sequence to avoid wasted effort. Systematic and incremental verification makes it easier to find, isolate, diagnose, and correct problems. A system or system element that has not been verified cannot be relied on to meet its requirements. Systems verification is an essential prerequisite to customer acceptance and certification.

Effective indicators of knowledge and experience

Awareness	Supervised Practitioner	Practitioner	Lead Practitioner	Expert
Explains what verification is, the purpose of verification, and why verification against the system requirements is important. [TVEA01]	Complies with a governing process and appropriate tools to plan and control their own verification activities. [TVES01]	Creates a strategy for system verification on a project to support SE project and wider enterprise needs. [TVEP01]	Creates enterprise-level policies, procedures, guidance, and best practice for verification, including associated tools. [TVEL01]	Communicates own knowledge and experience in Systems Engineering verification in order to improve best practice beyond the enterprise boundary. [TVEE01]
Explains why there is a need to verify the system in a logical sequence. [TVEA02]	Prepares inputs to verification plans. [TVES02]	Creates a governing process, plan, and associated tools for systems verification, which reflect project and business strategy. [TVEP02]	Judges the tailoring of enterprise-level verification processes to meet the needs of a project. [TVEL02]	Advises organizations beyond the enterprise boundary on the suitability of their approach to Systems Engineering verification. [TVEE02]
Explains why planning for system verification is necessary. [TVEA03]	Prepares verification plans for smaller projects. [TVES03]	Uses governing plans and processes for System verification, interpreting, evolving, or seeking guidance where appropriate. [TVEP03]	Judges the suitability of verification plans, from multiple projects, on behalf of the enterprise. [TVEL03]	Advises organizations beyond the enterprise boundary on their Systems Engineering Verification plans or practices on complex systems or projects. [TVEE03]
Explains how traceability can be used to establish whether a system meets requirements. [TVEA04]	Performs verification testing as part of system verification activities. [TVES04]	Prepares verification plans for systems or projects. [TVEP04]	Advises on verification approaches on complex or challenging systems or projects across the enterprise. [TVEL04]	Advises organizations beyond the enterprise boundary on complex or sensitive verification-related issues. [TVEE04]
Describes the relationship between verification, validation, qualification, certification, and acceptance. [TVEA05]	Identifies simple faults found during verification through diagnosis and consequential corrective actions. [TVES05]	Reviews project-level system verification plans. [TVEP05]	Judges detailed verification procedures from multiple projects, on behalf of the enterprise. [TVEL05]	Champions the introduction of novel techniques and ideas in systems verification, beyond the enterprise boundary, in order to develop the wider Systems Engineering community in this competency. [TVEE05]

	Collates evidence in support of verification, qualification, certification, and acceptance. [TVES06]	Reviews verification results, diagnosing complex faults found during verification activities. [TVEP06]	Judges verification evidence generated from multiple projects on behalf of the enterprise. [TVEL06]	Coaches individuals beyond the enterprise boundary in Systems Verification, in order to further develop their knowledge, abilities, skills, or associated behaviors. [TVEE06]
	Reviews verification evidence to establish whether a system meets requirements. [TVES07]	Prepares evidence obtained during verification testing to support system verification or downstream qualification, certification, and acceptance activities. [TVEP07]	Guides and actively coordinates verification activities for complex systems or projects across the enterprise. [TVEL07]	Maintains expertise in this competency area through specialist Continual Professional Development (CPD) activities. [TVEE07]
	Selects a verification environment to ensure requirements can be fully verified. [TVES08]	Monitors the traceability of verification requirements and tests to system requirements and vice versa. [TVEP08]	Coaches or mentors practitioners across the enterprise in systems verification in order to develop their knowledge, abilities, skills, or associated behaviors. [TVEL08]	
	Develops own understanding of this competency area through Continual Professional Development (CPD). [TVES09]	Identifies a suitable verification environment. [TVEP09]	Promotes the introduction and use of novel techniques and ideas in verification across the enterprise, to improve enterprise competence in this area. [TVEL09]	
		Creates detailed verification procedures. [TVEP10]	Develops expertise in this competency area through specialist Continual Professional Development (CPD) activities. [TVEL10]	
		Performs system verification activities. [TVEP11]		
		Prepares evidence obtained during verification testing to support downstream verification testing, integration, or validation activities. [TVEP12]		
		Guides new or supervised practitioners in Systems verification in order to develop their knowledge, abilities, skills, or associated behaviors. [TVEP13]		
		Maintains and enhances own competence in this area through Continual Professional Development (CPD) activities. [TVEP14]		

Competency area – Technical: Validation

Description

The purpose of validation is to provide objective evidence that the system, product, or service when in use, fulfills its business or mission objectives and stakeholder requirements, achieving its intended use in its intended operational environment. Put simply, validation checks that the needs of the customer/end user have been met and answers the question "Did we build the right system?"

Why it matters

Validation is used to check that the system meets the needs of the customer/end user. Failure to satisfy the customer will impact future business. Validation provides some important inputs to future system development.

Effective indicators of knowledge and experience

Awareness	Supervised Practitioner	Practitioner	Lead Practitioner	Expert
Explains what validation is, the purpose of validation, and why validation is important. [TVAA01]	Complies with a governing process and appropriate tools to plan and control their own validation activities. [TVAS01]	Creates a strategy for system validation on a project to support SE project and wider enterprise needs. [TVAP01]	Creates enterprise-level policies, procedures, guidance, and best practice for validation, including associated tools. [TVAL01]	Communicates own knowledge and experience in Systems Engineering validation in order to improve best practice beyond the enterprise boundary. [TVAE01]
Explains why there is a need for early planning for validation. [TVAA02]	Prepares inputs to validation plans. [TVAS02]	Creates a governing process, plan, and associated tools for system validation, which reflect project and business strategy. [TVAP02]	Judges the tailoring of enterprise-level validation processes to meet the needs of a project. [TVAL02]	Advises organizations beyond the enterprise boundary on the suitability of their approach to Systems Engineering validation. [TVAE02]
Describes the relationship between validation, verification, qualification, certification, and acceptance. [TVAA03]	Prepares validation plans for smaller projects. [TVAS03]	Uses governing plans and processes for System validation, interpreting, evolving, or seeking guidance where appropriate. [TVAP03]	Judges the suitability of validation plans from multiple projects, on behalf of the enterprise. [TVAL03]	Advises organizations beyond the enterprise boundary on their handling of complex or sensitive Systems Engineering validation issues. [TVAE03]
Describes the relationship between traceability and validation. [TVAA04]	Performs validation testing as part of system validation or system acceptance. [TVAS04]	Communicates using the terminology of the customer while focusing on customer need. [TVAP04]	Advises on validation approaches on complex or challenging systems or projects across the enterprise. [TVAL04]	Advises organizations beyond the enterprise boundary on complex or sensitive validation-related issues. [TVAE04]
	Identifies simple faults found during validation through diagnosis and consequential corrective actions. [TVAS05]	Prepares validation plans for systems or projects. [TVAP05]	Judges detailed validation procedures from multiple projects, on behalf of the enterprise. [TVAL05]	Champions the introduction of novel techniques and ideas in system validation, beyond the enterprise boundary, in order to develop the wider Systems Engineering community in this competency. [TVAE05]

	Collates evidence in support of validation, qualification, certification, and acceptance. [TVAS06]	Reviews project-level system validation plans. [TVAP06]	Judges validation evidence generated from multiple projects on behalf of the enterprise. [TVAL06]	Coaches individuals beyond the enterprise boundary in Systems validation, in order to further develop their knowledge, abilities, skills, or associated behaviors. [TVAE06]
	Reviews validation evidence to establish whether a system will meet the operational need. [TVAS07]	Reviews validation results, diagnosing complex faults found during validation activities. [TVAP07]	Guides and actively coordinates validation activities on complex systems or projects across the enterprise. [TVAL07]	Maintains expertise in this competency area through specialist Continual Professional Development (CPD) activities. [TVAE07]
	Selects a validation environment to ensure requirements can be fully validated. [TVAS08]	Identifies a suitable validation environment. [TVAP08]	Coaches or mentors practitioners across the enterprise in systems validation in order to develop their knowledge, abilities, skills, or associated behaviors. [TVAL08]	
	Develops own understanding of this competency area through Continual Professional Development (CPD). [TVAS09]	Creates detailed validation procedures. [TVAP09]	Promotes the introduction and use of novel techniques and ideas in validation across the enterprise, to improve enterprise competence in this area. [TVAL09]	
		Performs system validation activities. [TVAP10]	Develops expertise in this competency area through specialist Continual Professional Development (CPD) activities. [TVAL10]	
		Prepares evidence obtained during validation testing to support certification and acceptance activities. [TVAP11]		
		Monitors the traceability of validation requirements and tests to system requirements and vice versa. [TVAP12]		
		Guides new or supervised practitioners in System Validation in order to develop their knowledge, abilities, skills, or associated behaviors. [TVAP13]		
		Maintains and enhances own competence in this area through Continual Professional Development (CPD) activities. [TVAP14]		

Competency area – Technical: Transition

Description

Transition is the integration of a verified system, product, or service into its operational environment including the wider ("Super") system of which it forms a part.
Transition is performed in accordance with stakeholder agreements and includes support activities and provision of relevant enabling systems (e.g. production and volume manufacturing, site preparation, support and logistics systems, and operator training). Transition is used at each level in the system structure.

Why it matters

Incorrectly transitioning the system into operation can lead to misuse, failure to perform, and customer or end user dissatisfaction. Failure to plan for transition to operation may result in a system that is delayed into service or market with a consequential impact on the customer or business. Failure to satisfy the customer will impact future business.

Effective indicators of knowledge and experience

Awareness	Supervised Practitioner	Practitioner	Lead Practitioner	Expert
Explains why there is a need to carry out transition to operation. [TTRA01]	Uses a governing process and appropriate tools to plan and control their own transition activities. [TTRS01]	Creates a strategy for transitioning a system, which supports project and wider enterprise needs. [TTRP01]	Creates enterprise-level policies, procedures, guidance, and best practice for system transition, including associated tools. [TTRL01]	Communicates own knowledge and experience in Systems transition in order to improve best practice beyond the enterprise boundary. [TTRE01]
Explains key activities and work products required for transition to operation. [TTRA02]	Performs transitioning of a system into production or operation in order to meet the requirements of a plan. [TTRS02]	Creates a governing process, plan, and associated tools for systems transition activities that reflect project and business strategy. [TTRP02]	Judges the tailoring of enterprise-level transition processes to meet the needs of a project. [TTRL02]	Advises organizations beyond the enterprise boundary on the suitability of their approach to Systems Engineering Transition activities. [TTRE02]
	Describes the system's contribution to the wider system (super-system) of which it forms a part. [TTRS03]	Uses governing plans and processes to plan and execute system transition activities, interpreting, evolving, or seeking guidance where appropriate. [TTRP03]	Judges the adequacy of transition approaches and procedures on complex or challenging systems or projects across the enterprise. [TTRL03]	Advises organizations beyond the enterprise boundary on the handling of complex or sensitive Systems transition issues. [TTRE03]
	Develops own understanding of this competency area through Continual Professional Development (CPD). [TTRS04]	Performs a system transition to production and operation taking into consideration its contribution to the wider (super) system. [TTRP04]	Judges the suitability of transition plans to ensure wider enterprise needs are addressed. [TTRL04]	Champions the introduction of novel techniques and ideas in system transition, beyond the enterprise boundary, in order to develop the wider Systems Engineering community in this competency. [TTRE04]
		Communicates transition activities using user terminology to ensure clear communications. [TTRP05]	Persuades key stakeholders to address identified enterprise-level transition issues to reduce project and business risk. [TTRL05]	Coaches individuals beyond the enterprise boundary in Systems Transition, in order to further develop their knowledge, abilities, skills, or associated behaviors. [TTRE05]

		Acts to ensure system transition addresses export control and licensing obligations. [TTRP06]	Guides the transition of a complex system or projects across the enterprise, into service. [TTRL06]	Maintains expertise in this competency area through specialist Continual Professional Development (CPD) activities. [TTRE06]
		Acts to ensure system transition activities gain customer approval. [TTRP07]	Coaches or mentors practitioners across the enterprise in systems transition in order to develop their knowledge, abilities, skills, or associated behaviors. [TTRL07]	
		Guides new or supervised practitioners in System transition to operation in order to develop their knowledge, abilities, skills, or associated behaviors. [TTRP08]	Promotes the introduction and use of novel techniques and ideas in transition management across the enterprise, to improve enterprise competence in this area. [TTRL08]	
		Maintains and enhances own competence in this area through Continual Professional Development (CPD) activities. [TTRP09]	Develops expertise in this competency area through specialist Continual Professional Development (CPD) activities. [TTRL09]	

Competency area – Technical: Utilization and Support

Description
Utilization is the stage of development when a system, product, or service is used (operated) in its intended environment to deliver its intended capabilities. The support stage of a system life cycle encompasses the activities required to sustain operation of the system, product, or service over time, such as maintaining the system to continue or extend its operational life, address performance issues, evolving needs, obsolescence, and technology upgrades and changes. Support entails monitoring system performance, addressing system failures and performance issues, and updating the system to accommodate evolving needs and technology.

Why it matters
The Utilization and support stages of a system, product, or service typically account for the largest portion of the total life cycle cost. Proactive and systematic responses to operational issues contribute significantly to user satisfaction and operational cost management.

Effective indicators of knowledge and experience

Awareness	Supervised Practitioner	Practitioner	Lead Practitioner	Expert
Explains why a system needs to be supported during operation. [TUSA01]	Uses a governing process and appropriate tools to plan and control their own operations and support activities. [TUSS01]	Creates a strategy for system utilization and support, which reflects wider project and business strategies. [TUSP01]	Creates enterprise-level policies, procedures, guidance, and best practice for utilization and support, including associated tools. [TUSL01]	Communicates own knowledge and experience in systems utilization and support in order to improve best practice beyond the enterprise boundary. [TUSE01]
Describes the difference between preventive and corrective maintenance. [TUSA02]	Identifies operational data in order to assess system performance. [TUSS02]	Creates a governing process, plan, and associated tools for system utilization and support, which reflect wider project and business plans. [TUSP02]	Judges the tailoring of enterprise-level utilization and support processes to meet the needs of a project. [TUSL02]	Advises organizations beyond the enterprise boundary on the suitability of their approach to system utilization and support. [TUSE02]
Explains why it is necessary to address failures, parts obsolescence, and evolving user requirements during system operation. [TUSA03]	Reviews system failures or performance issues, proposing design changes to rectify such failures. [TUSS03]	Uses governing plans and processes for System Utilization and support, interpreting, evolving, or seeking guidance where appropriate. [TUSP03]	Advises across the enterprise on the application of advanced practices to improve the effectiveness of project-level system support or operations. [TUSL03]	Advises organizations beyond the enterprise boundary on the handling of complex or sensitive operations, maintenance, and support-related issues. [TUSE03]
Lists the different levels of repair capability and describes the characteristics of each. [TUSA04]	Performs rectification of system failures or performance issues. [TUSS04]	Guides and actively coordinates in-service support activities for a system. [TUSP04]	Advises across the enterprise on technology upgrade implementations in order to improve the cost–benefit ratio of an upgraded design solution. [TUSL04]	Champions the introduction of novel techniques and ideas in systems utilization and support, beyond the enterprise boundary, in order to develop the wider Systems Engineering community in this competency. [TUSE04]
Explains the impact of operations and support on specialty engineering areas. [TUSA05]	Reviews the feasibility and impact of evolving user need on operations, maintenance, and support. [TUSS05]	Identifies data to be collected in order to assess system operational performance. [TUSP05]	Persuades key stakeholders across the enterprise to address identified operation, maintenance, and support issues to reduce project or wider enterprise risk. [TUSL05]	Coaches individuals beyond the enterprise boundary in System Utilization and support, in order to further develop their knowledge, abilities, skills, or associated behaviors. [TUSE05]

	Prepares inputs to concept studies to document the impact or feasibility of new technologies or possible system updates. [TUSS06]	Reviews system failures or performance issues in order to initiate design change proposals rectifying these failures. [TUSP06]	Coaches or mentors practitioners across the enterprise in systems utilization and support in order to develop their knowledge, abilities, skills, or associated behaviors. [TUSL06]	Maintains expertise in this competency area through specialist Continual Professional Development (CPD) activities. [TUSE06]
	Prepares inputs to obsolescence studies to identify obsolescent components and suitable replacements. [TUSS07]	Identifies system elements approaching obsolescence and conducts studies to identify suitable replacements. [TUSP07]	Promotes the introduction and use of novel techniques and ideas in systems operations and support across the enterprise, to improve enterprise competence in this area. [TUSL07]	
	Prepares updates to technical data (e.g. procedures, guidelines, checklists, and training materials) to ensure operations and maintenance activities and data are current. [TUSS08]	Maintains system elements and associated documentation following their replacement due to obsolescence. [TUSP08]	Develops expertise in this competency area through specialist Continual Professional Development (CPD) activities. [TUSL08]	
	Identifies potential changes to system operational environment or external interfaces. [TUSS09]	Monitors the effectiveness of system support or operations. [TUSP09]		
	Develops own understanding of this competency area through Continual Professional Development (CPD). [TUSS10]	Reviews the timing of technology upgrade implementations in order to improve the cost–benefit ratio of an upgraded design solution. [TUSP10]		
		Reviews potential changes to the system operational environment or external interfaces. [TUSP11]		
		Reviews technical support data (e.g. procedures, guidelines, checklists, training, and maintenance materials) to ensure it is current. [TUSP12]		
		Guides new or supervised practitioners in System operation, support, and maintenance, in order to develop their knowledge, abilities, skills, or associated behaviors. [TUSP13]		
		Maintains and enhances own competence in this area through Continual Professional Development (CPD) activities. [TUSP14]		

Competency area – Technical: Retirement

Description
The retirement stage of a system is the final stage of a system life cycle, where the existence of a system or product is ended for a specific use, through appropriate and often controlled handling or recycling. System engineering activities in this stage primarily focus on ensuring that disposal requirements are addressed. They could also concern preparing for the next generation of a system.

Why it matters
Retirement is an essential part of a system overall concept as experience has shown that failure to plan for retirement (e.g. disposal, recycling, and reuse) early can be both costly and time-consuming. Indeed, many countries have changed their laws to insist that the developer is now responsible for ensuring the proper end-of-life disposal of all system components.

Effective indicators of knowledge and experience

Awareness	Supervised Practitioner	Practitioner	Lead Practitioner	Expert
Explains why the needs of system retirement need to be considered, even as part of the original system design concept. [TREA01]	Uses a governing process and appropriate tools to plan and control their retirement activities. [TRES01]	Creates a strategy for system retirement which reflects wider project and business strategies. [TREP01]	Creates enterprise-level policies, procedures, guidance, and best practice for retirement, including associated tools. [TREL01]	Communicates own knowledge and experience in System retirement, in order to improve best practice beyond the enterprise boundary. [TREE01]
Identifies areas requiring special consideration when determining retirement requirements across each of the life cycle stages. [TREA02]	Identifies required retirement requirements or design changes in order to address system retirement needs. [TRES02]	Uses governing plans and processes for system retirement, interpreting, evolving, or seeking guidance where appropriate. [TREP02]	Judges the tailoring of enterprise-level retirement processes to meet the needs of a project. [TREL02]	Advises organizations beyond the enterprise boundary on the suitability of their approach to system retirement. [TREE02]
Explains how evolving user requirements could affect retirement. [TREA03]	Reviews the feasibility and impact of evolving user need on system retirement. [TRES03]	Complies with governing plans and processes for system retirement, interpreting, evolving, or seeking guidance where appropriate. [TREP03]	Advises across the enterprise on the application of advanced practices to improve the effectiveness of project-level retirement activities. [TREL03]	Advises organizations beyond the enterprise boundary on their handling of complex or sensitive retirement-related issues. [TREE03]
	Prepares updates to technical data (e.g. procedures, guidelines, checklists, and training) to ensure retirement activities and data are current. [TRES04]	Guides and actively coordinates the retirement of a system at end-of-life. [TREP04]	Persuades key stakeholders across the enterprise to address identified retirement issues to reduce project and business risk. [TREL04]	Champions the introduction of novel techniques and ideas in system retirement, beyond the enterprise boundary, in order to develop the wider Systems Engineering community in this competency. [TREE04]
	Prepares inputs to obsolescence studies to identify impact on retirement of obsolescent components and their replacements. [TRES05]	Determines the data to be collected in order to assess system retirement performance. [TREP05]	Coaches or mentors practitioners across the enterprise in system retirement in order to develop their knowledge, abilities, skills, or associated behaviors. [TREL05]	Coaches individuals beyond the enterprise boundary, in System retirement in order to further develop their knowledge, abilities, skills, or associated behaviors. [TREE05]

	Identifies potential changes to system retirement environment or external interfaces as a result of system evolution or other technology change. [TRES06]	Monitors the implementation of changes to system retirement environment or external interfaces. [TREP06]	Promotes the introduction and use of novel techniques and ideas in system retirement across the enterprise, to improve enterprise competence in this area. [TREL06]	Maintains expertise in this competency area through specialist Continual Professional Development (CPD) activities. [TREE06]
	Identifies potential changes to system retirement process or interfaces as a result of changes to interfacing systems or usage changes. [TRES07]	Reviews retirement technical support data (e.g. procedures, guidelines, checklists, training, and materials) to ensure it is current. [TREP07]	Develops expertise in this competency area through specialist Continual Professional Development (CPD) activities. [TREL07]	
	Develops own understanding of this competency area through Continual Professional Development (CPD). [TRES08]	Guides new or supervised practitioners in system retirement in order to develop their knowledge, abilities, skills, or associated behaviors. [TREP08]		
		Maintains and enhances own competence in this area through Continual Professional Development (CPD) activities. [TREP09]		

Competency area – Systems Engineering Management: Planning

Description

The purpose of planning is to produce, coordinate, and maintain effective and workable plans across multiple disciplines. Systems Engineering planning includes planning the way the engineering of the system will be performed and managed, tailoring generic engineering processes to address specific project context, technical activities, and identified risks. This includes estimating the effort, resources, and timescales required to complete the project to the required quality level. Planning is performed in association with the Project Manager. Plans and estimates may need updating to reflect changes or to overcome unexpected issues encountered during the development process.

Why it matters

It is important to identify the full scope and timing of all Systems Engineering activities and their associated resource needs and to link this with task effort and cost estimation through controlled planning. Alignment between Systems Engineering planning and estimation is vital to ensure that assumptions made when developing a plan, such as ways of working and process tailoring are taken into consideration. Failure to plan correctly will mean inadequate visibility of progress and is likely to cause ongoing problems with time, budget, and quality.

Effective indicators of knowledge and experience

Awareness	Supervised Practitioner	Practitioner	Lead Practitioner	Expert
Identifies key planning and estimating terms and acronyms and the relationships between them. [MPLA01]	Follows a defined governing process, using appropriate tools, to guide their Systems Engineering planning activities. [MPLS01]	Creates a strategy for performing Systems Engineering life cycle activities considering the wider project plan to ensure integration and coherence across the development. [MPLP01]	Creates enterprise-level policies, procedures, guidance, and best practice for Systems Engineering planning, including associated tools to improve organizational effectiveness. [MPLL01]	Communicates own knowledge and experience in Systems Engineering planning, in order to promote best practice beyond the enterprise boundary. [MPLE01]
Explains why planning Systems Engineering activities is important and how planning interacts across disciplines and organizations. [MPLA02]	Explains the role of Systems Engineering planning and its relationship to wider project planning and management. [MPLS02]	Creates a governing Systems Engineering management plan, which reflect project and business strategy and all development constraints. [MPLP02]	Judges tailoring of enterprise-level Systems Engineering planning processes to balance the needs of the project and business. [MPLL02]	Advises organizations beyond the enterprise boundary on the suitability of their approach to Systems Engineering planning and estimation. [MPLE02]
Identifies key areas that need to be addressed in a project Systems Engineering plan. [MPLA03]	Prepares information required in order to create an SE plan, in order to control the management of a system development. [MPLS03]	Develops effort, resource, and schedule estimates to scope Systems Engineering life cycle activities. [MPLP03]	Creates Systems Engineering plans integrating multiple diverse projects or a complex system to ensure coherence across the development. [MPLL03]	Advises organizations beyond the enterprise boundary on the handling of complex or sensitive Systems Engineering planning issues. [MPLE03]
Explains the principles of Systems Engineering process tailoring including its benefits and potential issues. [MPLA04]	Prepares inputs to SE management plan. [MPLS04]	Selects key design parameters required to track critical aspects of the design during development. [MPLP04]	Judges Systems Engineering plans from across the enterprise for their suitability in meeting both the needs of the project and the wider enterprise. [MPLL04]	Champions the introduction of novel techniques and ideas in Systems Engineering planning, beyond the enterprise boundary, in order to develop the wider Systems Engineering community in this competency. [MPLE04]

Identifies key potential sources of change on a project and why the impact of such changes needs to be carefully assessed and planned. [MPLA05]	Explains how Systems Engineering estimates are compiled in order to scope the size of a development. [MPLS05]	Negotiates successfully with others to secure identified future Systems Engineering needs of a project. [MPLP05]	Judges Systems Engineering effort, resource, and schedule estimates from across the enterprise for their quality. [MPLL05]	Coaches individuals beyond the enterprise boundary in Systems Engineering planning, in order to further develop their knowledge, abilities, skills, or associated behaviors. [MPLE05]
Explains the relationship between life cycle reviews and planning. [MPLA06]	Prepares inputs to Systems Engineering work packages to support the scoping of Systems Engineering tasks. [MPLS06]	Negotiates successfully with project management to secure identified future Systems Engineering needs of a project. [MPLP06]	Reviews engineering changes from across the enterprise to establish the impact on both the project itself and the wider enterprise. [MPLL06]	Maintains expertise in this competency area through specialist Continual Professional Development (CPD) activities. [MPLE06]
	Prepares inputs to Systems Engineering replanning activities in order to implement engineering changes. [MPLS07]	Prepares updates required to Systems Engineering plan to address internal or external changes. [MPLP07]	Persuades key stakeholders to address identified enterprise-level Systems Engineering project planning issues to reduce project cost, schedule, or technical risk. [MPLL07]	
	Prepares updates to Systems Engineering plans to reflect authorized changes. [MPLS08]	Coordinates implementation of updates to Systems Engineering plan to address internal or external changes. [MPLP08]	Coaches or mentors practitioners across the enterprise in Systems Engineering planning in order to develop their knowledge, abilities, skills, or associated behaviors. [MPLL08]	
	Identifies development lessons learned performing SE planning to inform future projects. [MPLS09]	Guides new or supervised practitioners in Systems Engineering planning to develop their knowledge, abilities, skills, or associated behaviors. [MPLP09]	Promotes the introduction and use of novel techniques and ideas in Systems Engineering planning across the enterprise, to improve enterprise competence in this area. [MPLL09]	
	Develops own understanding of this competency area through Continual Professional Development (CPD). [MPLS10]	Maintains and enhances own competence in this area through Continual Professional Development (CPD) activities. [MPLP10]	Develops expertise in this competency area through specialist Continual Professional Development (CPD) activities. [MPLL10]	

Competency area – Systems Engineering Management: Monitoring and Control

Description

Monitoring and control assesses the project to see if the current plans are aligned and feasible; determines the status of a project, including its technical and process performance, and directs execution to ensure that performance is according to plans and schedule, within project budgets, and satisfies technical objectives.

Why it matters

Failure to adequately assess and monitor performance against the plan prevents visibility of progress and, in consequence, appropriate corrective actions may not be identified and/or taken when project performance deviates from that required.

Effective indicators of knowledge and experience

Awareness	Supervised Practitioner	Practitioner	Lead Practitioner	Expert
Explains why monitoring and controlling Systems Engineering activities is important. [MMCA01]	Follows a defined governing process, using appropriate tools, to guide their Systems Engineering monitoring and control activities. [MMCS01]	Creates a strategy for Monitoring and Control on a project to support project and wider enterprise needs. [MMCP01]	Creates enterprise-level policies, procedures, guidance, and best practice for Systems Engineering monitoring and control, including associated tools to improve organizational effectiveness. [MMCL01]	Communicates own knowledge and experience in Systems Engineering monitoring and control in order to improve best practice beyond the enterprise boundary. [MMCE01]
Explains how Systems Engineering monitoring and control fits within the wider execution and control of a project. [MMCA02]	Records technical data identified as requiring monitoring or control in plans to facilitate analysis. [MMCS02]	Creates a governing process, plan, and associated tools for systems decision management, which reflect project and business strategy. [MMCP02]	Assesses tailoring of enterprise-level Systems Engineering monitoring and control processes to balance the needs of the project and business. [MMCL02]	Advises organizations beyond the enterprise boundary on the handling of complex or sensitive Systems Engineering monitoring and control issues. [MMCE02]
Explains the purpose of reviews and decision gates and their relationship to the monitoring and control of Systems Engineering tasks. [MMCA03]	Monitors key data parameters against expectations to determine deviations. [MMCS03]	Complies with governing plans and processes for system monitoring and control, interpreting, evolving, or seeking guidance where appropriate. [MMCP03]	Assesses ongoing Systems Engineering projects at enterprise-level to ensure they are being monitored and controlled successfully. [MMCL03]	Advises organizations beyond the enterprise boundary on the suitability of their approach to Systems Engineering monitoring and control. [MMCE03]
Explains how Systems Engineering metrics and measures contribute to monitoring and controlling Systems Engineering on a project. [MMCA04]	Identifies potential corrective actions to control and correct deviations from expectations. [MMCS04]	Monitors Systems Engineering activities in order to determine and report progress against estimates and plans on a project. [MMCP04]	Judges the suitability of management and trading of design technical margins to satisfy the needs of the project, on ongoing projects across the enterprise. [MMCL04]	Champions the introduction of novel techniques and ideas in Systems Engineering monitoring and control, beyond the enterprise boundary, in order to develop the wider Systems Engineering community in this competency. [MMCE04]

Explains how communications support the successful monitoring and control of a Systems Engineering project. [MMCA05]	Monitors technical margins both horizontally and vertically through the project hierarchy and over time, to control and monitor design evolution. [MMCS05]	Monitors Systems Engineering activities by processing measurement data in order to determine deviations or trends against plans. [MMCP05]	Assesses proposals for preventative or remedial actions when assessment indicates a trend toward deviation on multiple distinct projects across the enterprise or a complex project. [MMCL05]	Coaches individuals beyond the enterprise boundary in Systems Engineering monitoring and control, in order to further develop their knowledge, abilities, skills, or associated behaviors. [MMCE05]
	Identifies development lessons learned performing monitoring and control to inform future projects. [MMCS06]	Analyzes measurement and assessment data to determine and implement necessary remedial corrective actions in order to control SE activities. [MMCP06]	Analyzes monitoring and control data from multiple diverse projects or a complex system to provide enterprise level coordination of SE. [MMCL06]	Maintains expertise in this competency area through specialist Continual Professional Development (CPD) activities. [MMCE06]
	Develops own understanding of this competency area through Continual Professional Development (CPD). [MMCS07]	Prepares recommendations for updates to existing monitoring and control plans to address internal or external changes. [MMCP07]	Persuades key stakeholders to address identified enterprise-level Systems Engineering monitoring and control issues to reduce project cost, schedule, or technical risk. [MMCL07]	
		Reviews technical margins both horizontally and vertically through the project hierarchy to maintain overall required margins. [MMCP08]	Coaches or mentors practitioners across the enterprise in Systems Engineering monitoring and control in order to develop their knowledge, abilities, skills, or associated behaviors. [MMCL08]	
		Guides new or supervised practitioners in Systems Engineering monitoring and control in order to develop their knowledge, abilities, skills, or associated behaviors. [MMCP09]	Promotes the introduction and use of novel techniques and ideas in monitoring and control of Systems Engineering, across the enterprise, to improve enterprise competence in this area. [MMCL09]	
		Maintains and enhances own competence in this area through Continual Professional Development (CPD) activities. [MMCP10]	Develops expertise in this competency area through specialist Continual Professional Development (CPD) activities. [MMCL10]	

Competency area – Systems Engineering Management: Risk and Opportunity Management

Description

Risk is an uncertain event or condition that, if it occurs, has a positive or negative effect on project or enterprise objectives. The purpose of risk and opportunity management is to reduce potential risks to an acceptable level before they occur, or maximize the potential of any opportunity, throughout the life of the project. Risk and opportunity management is a continuous, forward-looking process that is applied to anticipate and avert risks that may adversely impact the project and can be considered both a project management and a Systems Engineering activity.

Why it matters

Every new system (or existing system modification) has inherent risk but is also based upon the pursuit of an opportunity. Risk and opportunity are both present throughout the life cycle of systems and the primary objective of managing these areas as part of Systems Engineering activities is to balance the allocation of resources to achieve greatest risk mitigation (or opportunity benefits).

Effective indicators of knowledge and experience

Awareness	Supervised Practitioner	Practitioner	Lead Practitioner	Expert
Describes the distinction between risk, issue, and opportunity, and can provide examples of each. [MROA01]	Follows a governing process and appropriate tools to plan and control their own risk and opportunity management activities. [MROS01]	Creates a strategy for risk and opportunity management on a project to support SE project and wider enterprise needs. [MROP01]	Creates enterprise-level policies, procedures, guidance, and best practice for Systems Engineering risk and opportunity management, including associated tools. [MROL01]	Communicates own knowledge and experience in Systems Engineering risk and opportunity management, in order to best practice beyond the enterprise boundary. [MROE01]
Identifies key factors associated with good risk management and why these factors are important. [MROA02]	Identifies potential risks and opportunities on a project. [MROS02]	Creates a governing process, plan, and associated tools for risk and opportunity management, which reflect project and business strategy. [MROP02]	Judges the tailoring of enterprise-level risk and opportunity management processes and associated work products to meet the needs of a project. [MROL02]	Influences key stakeholders beyond the enterprise boundary in support of risk and opportunity management. [MROE02]
Identifies different classes of risk and can provide examples of each. [MROA03]	Identifies action plans to treat risks and opportunities on a project. [MROS03]	Develops a project risk and opportunity profile including context, probability, consequences, thresholds, priority, and risk action and status. [MROP03]	Guides and actively coordinates Systems Engineering risk and opportunity management across multiple diverse projects or across a complex system, with proven success. [MROL03]	Advises organizations beyond the enterprise boundary on the handling of complex or sensitive risk and opportunity issues. [MROE03]
Identifies different types of risk treatment available and can provide examples of each. [MROA04]	Develops own understanding of this competency area through Continual Professional Development (CPD). [MROS04]	Analyzes risks and opportunities for likelihood and consequence in order to determine magnitude and priority for treatment. [MROP04]	Produces an enterprise-level risk profile including context, probability, consequences, thresholds, priority, and risk action and status. [MROL04]	Advises organizations beyond the enterprise boundary on the suitability of their approach to risk and opportunity management. [MROE04]
Identifies different types of opportunity and can provide examples of each. [MROA05]		Monitors Systems Engineering risks and opportunities during project execution. [MROP05]	Judges on the treatment of risks and opportunities across multiple diverse projects or a complex project, with proven success. [MROL05]	Champions the introduction of novel techniques and ideas in Systems Engineering risk and opportunity management, beyond the enterprise boundary, in order to develop the wider Systems Engineering community in this competency. [MROE05]

Describes a typical high-level process for risk and opportunity management. [MROA06]		Analyzes risks and opportunities effectively, considering alternative treatments and generating a plan of action when thresholds exceed certain levels. [MROP06]	Persuades key enterprise stakeholders to address identified enterprise-level project risks and opportunities to reduce enterprise-level risks. [MROL06]	Coaches individuals beyond the enterprise boundary in Systems Engineering risk and opportunity management, in order to further develop their knowledge, abilities, skills, or associated behaviors. [MROE06]
Explains how risk is typically assessed and can provide examples. [MROA07]		Communicates risk and opportunity status to affected stakeholders. [MROP07]	Coaches or mentors practitioners across the enterprise in Systems Engineering risk and opportunity management in order to develop their knowledge, abilities, skills, or associated behaviors. [MROL07]	Maintains expertise in this competency area through specialist Continual Professional Development (CPD) activities. [MROE07]
		Guides new or supervised practitioners in risk and opportunity Management techniques in order to develop their knowledge, abilities, skills, or associated behaviors. [MROP08]	Promotes the introduction and use of novel techniques and ideas in Systems Engineering risk and opportunity management across the enterprise, to improve enterprise competence in this area. [MROL08]	
		Maintains and enhances own competence in this area through Continual Professional Development (CPD) activities. [MROP09]	Develops expertise in this competency area through specialist Continual Professional Development (CPD) activities. [MROL09]	

Competency area – Systems Engineering Management: Decision Management

Description

Decision management provides a structured, analytical framework for objectively identifying, characterizing, and evaluating a set of alternatives for a decision at any point in the life cycle in order to select the most beneficial course of action.

Why it matters

System development entails an array of interrelated decisions that require the holistic perspective of the Systems Engineering discipline. Decisions include selection of preferred solution at every level of the system, including technology option selection, architecture selection, make-or-buy decisions, strategy selection for maintenance, disposal. While some low-value decisions can be made and recorded simply, key decisions which might affect the long-term success and delivery of the project's desired value need to be controlled using a formalized decision management process.

Effective indicators of knowledge and experience

Awareness	Supervised Practitioner	Practitioner	Lead Practitioner	Expert
Identifies the Systems Engineering situations where a structured decision is, and is not, appropriate. [MDMA01]	Follows a governing process and appropriate tools to plan and control their own decision management activities. [MDMS01]	Creates a strategy for decision management on a project to support SE project and wider enterprise needs. [MDMP01]	Creates enterprise-level policies, procedures, guidance, and best practice for decision management and communication, including associated tools. [MDML01]	Communicates own knowledge and experience in Systems Engineering decision management, in order to improve best practice beyond the enterprise boundary. [MDME01]
Explains why there is a need to select a preferred solution. [MDMA02]	Identifies potential decision criteria and performance parameters for consideration. [MDMS02]	Creates a governing process, plan, and associated tools for systems decision management, which reflect project and business strategy. [MDMP02]	Judges the tailoring of enterprise-level decision management processes and associated work products to meet the needs of a project. [MDML02]	Influences key decision stakeholders beyond the enterprise boundary. [MDME02]
Describes the relevance of comparative techniques (e.g. trade studies, and make/buy) to assist decision processes. [MDMA03]	Identifies tools and techniques for the decision process. [MDMS03]	Complies with governing plans and processes for system decision management, interpreting, evolving, or seeking guidance where appropriate. [MDMP03]	Coordinates decision management and trade analysis using different techniques, across multiple diverse projects or across a complex system, with proven success. [MDML03]	Advises organizations beyond the enterprise boundary on complex or sensitive decision management or trade-off issues. [MDME03]
Explains how to frame, tailor, and structure a decision including its objectives and measures and outlines the key characteristics of a structured decision-making approach. [MDMA04]	Prepares information in support of decision trade studies. [MDMS04]	Develops governing decision management plans, processes, and appropriate tools and uses these to control and monitor decision management activities. [MDMP04]	Persuades key stakeholders across the enterprise to address identified enterprise-level decision management issues. [MDML04]	Advises organizations beyond the enterprise boundary on the suitability of their approach to decision management. [MDME04]
Explains how uncertainty impacts on decision-making. [MDMA05]	Monitors the decision process to catalog actions taken and their supporting rationale. [MDMS05]	Guides and actively coordinates ongoing decision management activities to ensure successful outcomes with decision management stakeholders. [MDMP05]	Negotiates complex trades on behalf of the enterprise. [MDML05]	Identifies strategies for organizations beyond the enterprise boundary, in order to resolve their issues with complex system trade-offs. [MDME05]

Explains why there is a need for communication and accurate recording in all aspects of the decision-making process. [MDMA06]	Develops own understanding of this competency area through Continual Professional Development (CPD). [MDMS06]	Determines decision selection criteria, weightings of the criteria, and assess alternatives against selection criteria. [MDMP06]	Judges decisions affecting solutions and the criteria for making the solution across the enterprise. [MDML06]	Champions the introduction of novel techniques and ideas in Systems Engineering decision management, beyond the enterprise boundary, in order to develop the wider Systems Engineering community in this competency. [MDME06]
		Selects appropriate tools and techniques for making different types of decision. [MDMP07]	Coaches or mentors practitioners across the enterprise in Systems Engineering decision management in order to develop their knowledge, abilities, skills, or associated behaviors. [MDML07]	Coaches individuals beyond the enterprise boundary in Systems Engineering decision management, in order to further develop their knowledge, abilities, skills, or associated behaviors. [MDME07]
		Prepares trade-off analyses and justifies the selection in terms that can be quantified and qualified. [MDMP08]	Promotes the introduction and use of novel techniques and ideas in decision resolution and management across the enterprise, to improve enterprise competence in this area. [MDML08]	Maintains expertise in this competency area through specialist Continual Professional Development (CPD) activities. [MDME08]
		Assesses sensitivity of selection criteria through a sensitivity analysis, reporting as required. [MDMP09]	Develops expertise in this competency area through specialist Continual Professional Development (CPD) activities. [MDML09]	
		Guides new or supervised practitioners in Decision management techniques in order to develop their knowledge, abilities, skills, or associated behaviors. [MDMP10]		
		Maintains and enhances own competence in this area through Continual Professional Development (CPD) activities. [MDMP11]		

Competency area – Systems Engineering Management: Concurrent Engineering

Description
Concurrent engineering is a work methodology based on the parallelization of tasks (i.e. performing tasks concurrently). It refers to an approach used in Systems Engineering in which functions of design and development engineering, manufacturing engineering, and other enterprise functions are integrated to reduce the elapsed time required to bring a new system, product, or service to market.

Why it matters
Systems Engineering life cycles involve multiple, concurrent processes and activities which must be coordinated to mitigate risk and prevent unnecessary work, paralysis, and a lack of convergence to an effective solution. Concurrency may be the only approach capable of meeting the customer schedule or gaining a competitive advantage. Performance can be constrained unnecessarily by allowing individual system elements to progress too quickly.

Effective indicators of knowledge and experience

Awareness	Supervised Practitioner	Practitioner	Lead Practitioner	Expert
Explains how Systems Engineering life cycle processes and activities and the development of systems elements can be concurrent and provides examples. [MCEA01]	Describes Systems Engineering life cycle processes in place on their project and how concurrency issues may impact its successful execution. [MCES01]	Creates governing concurrency management strategies and uses these to perform concurrent engineering on a project. [MCEP01]	Creates enterprise-level policies, procedures, guidance, and best practice for concurrent engineering, including associated tools. [MCEL01]	Communicates own knowledge and experience in Systems Engineering concurrent engineering in order to improve best practice beyond the enterprise boundary. [MCEE01]
Describes the advantages and disadvantages of concurrent engineering. [MCEA02]	Coordinates concurrent engineering activities on a Systems Engineering project. [MCES02]	Identifies elements which can be developed concurrently on a Systems Engineering project. [MCEP02]	Judges the tailoring of enterprise-level concurrency processes and associated work products to meet the needs of a project. [MCEL02]	Influences key stakeholders beyond the enterprise boundary in order to resolve concurrent engineering issues. [MCEE02]
	Prepares inputs to concurrency-related inputs to management plans for a Systems Engineering project. [MCES03]	Identifies concurrent interactions within a Systems Engineering life cycle on a project. [MCEP03]	Coordinates concurrent activities and deals with emerging issues across multiple diverse projects, or across a complex system, with proven results. [MCEL03]	Advises organizations beyond the enterprise boundary on the suitability of their approach to concurrent engineering developments. [MCEE03]
	Develops own understanding of this competency area through Continual Professional Development (CPD). [MCES04]	Coordinates concurrent activities and deals with emerging issues on a Systems Engineering project. [MCEP04]	Guides and actively coordinates interactions within Systems Engineering concurrency issues across multiple diverse projects, or across a complex system. [MCEL04]	Advises organizations beyond the enterprise boundary on complex or sensitive concurrency issues. [MCEE04]
		Identifies concurrency-related aspects of appropriate management plans for a Systems Engineering project. [MCEP05]	Judges on concurrency issues and risks across multiple diverse projects, or on a complex system. [MCEL05]	Develops new strategies for concurrent engineering for use beyond the enterprise boundary. [MCEE05]

		Analyzes concurrency issues and risks on a Systems Engineering project. [MCEP06]	Judges the suitability of plans for concurrent system developments from across the enterprise. [MCEL06]	Advises organizations beyond the enterprise boundary on concurrency issues and risks. [MCEE06]
		Guides new or supervised practitioners in concurrent engineering principles in order to develop their knowledge, abilities, skills, or associated behaviors. [MCEP07]	Persuades key stakeholders to address identified enterprise-level concurrent engineering issues. [MCEL07]	Champions the introduction of novel techniques and ideas in concurrency management, beyond the enterprise boundary, in order to develop the wider Systems Engineering community in this competency. [MCEE07]
		Maintains and enhances own competence in this area through Continual Professional Development (CPD) activities. [MCEP08]	Coaches or mentors practitioners across the enterprise in Systems Engineering concurrent engineering in order to develop their knowledge, abilities, skills, or associated behaviors. [MCEL08]	Coaches individuals beyond the enterprise boundary, in the concurrent engineering of Systems Engineering projects in order to further develop their knowledge, abilities, skills, or associated behaviors. [MCEE08]
			Promotes the introduction and use of novel techniques and ideas in concurrent engineering management across the enterprise, to improve enterprise competence in this area. [MCEL09]	Maintains expertise in this competency area through specialist Continual Professional Development (CPD) activities. [MCEE09]
			Develops expertise in this competency area through specialist Continual Professional Development (CPD) activities. [MCEL10]	

Competency area – Systems Engineering Management: Business and Enterprise Integration

Description

Businesses and Enterprises are systems in their own right. Systems Engineering is just one of many activities that must occur in order to bring about a successful system development meeting the needs of all its stakeholders. Systems Engineering addresses the needs of all other internal business and enterprise stakeholders, covering areas such as infrastructure, portfolio management, human resources, knowledge management, quality, information technology, production, sales, marketing, commercial, legal, and finance, within and beyond the local enterprise.

Why it matters

As businesses and enterprises become larger, more complex, and the functions within the enterprise more insular, the interdependencies between individual enterprise functions should be engineered using a systems approach at an enterprise level in order to meet the demands of increased business effectiveness and efficiency.

Effective indicators of knowledge and experience

Awareness	Supervised Practitioner	Practitioner	Lead Practitioner	Expert
Explains why a business or enterprise is a system in its own right and describes the business or enterprise "system" using Systems Engineering ideas and terminology. [MBEA01]	Follows a governing process and appropriate tools to plan and control their own business and enterprise integration activities. [MBES01]	Creates a strategy for Business and Enterprise integration on a project to support SE project and wider enterprise needs. [MBEP01]	Creates enterprise-level policies, procedures, guidance, and best practice for business and enterprise integration, including associated tools. [MBEL01]	Communicates own knowledge and experience in business and enterprise integration, in order to best practice beyond the enterprise boundary. [MBEE01]
Lists other business or enterprise functions and provides examples. [MBEA02]	Describes Systems Engineering work products that support business and enterprise infrastructure and provides examples of why this is the case. [MBES02]	Identifies the needs of the wider business and enterprise in order to ensure integration of ongoing activities (e.g. portfolio sustainment, infrastructure, HR, and knowledge management). [MBEP02]	Judges the tailoring of enterprise-level business and enterprise processes and associated work products to meet the needs of a project. [MBEL02]	Influences stakeholders beyond the enterprise boundary on business or enterprise issues. [MBEE02]
Lists work products from business and enterprise functions to the Systems Engineering process and vice versa and can provide examples. [MBEA03]	Describes work products produced elsewhere in the enterprise, which impact Systems Engineering activities and provides examples of why this is the case. [MBES03]	Creates Systems Engineering work products needed by to manage infrastructure across business or enterprise objectives. [MBEP03]	Coordinates business and enterprise integration across multiple diverse projects or across a complex system, with proven success. [MBEL03]	Advises organizations beyond the enterprise boundary on the suitability of their approach to business or enterprise integration. [MBEE03]
	Prepares inputs to Systems Engineering work products required by other business and enterprise functions. [MBES04]	Creates Systems Engineering work products needed to initiate or sustain the needs of the wider business or enterprise project portfolio. [MBEP04]	Judges Systems Engineering work products created for use by other parts of the business and enterprise. [MBEL04]	Advises organizations beyond the enterprise boundary on the effectiveness of the business or enterprise as a system. [MBEE04]
	Analyzes the impact of work products produced by other business and enterprise functions for their impact on Systems Engineering activities, in order to improve integration across the project or enterprise. [MBES05]	Creates Systems Engineering work products needed by Human Resources in order to meet business or enterprise objectives. [MBEP05]	Advises stakeholders across the enterprise regarding activities and work products affecting Systems Engineering. [MBEL05]	Advises organizations beyond the enterprise boundary on developing a Systems Engineering capability within a business or enterprise context. [MBEE05]

	Develops own understanding of this competency area through Continual Professional Development (CPD). [MBES06]	Creates Systems Engineering work products needed by to manage knowledge across business or enterprise objectives in order to support enterprise knowledge management, reuse, or exploitation. [MBEP06]	Advises stakeholders across the enterprise regarding the use of Systems Engineering techniques to contribute to the definition of the business/enterprise. [MBEL06]	Advises organizations beyond the enterprise boundary on the impact of inputs from other business/ enterprise functions on the Systems Engineering process. [MBEE06]
		Complies with governing plans and processes for business and enterprise integration, interpreting, evolving, or seeking guidance where appropriate. [MBEP07]	Persuades key stakeholders to address identified enterprise-level business and enterprise integration issues and constraints to reduce project cost, schedule, or technical risk. [MBEL07]	Advises organizations beyond the enterprise boundary on the impact of inputs from other business/ enterprise functions on the Systems Engineering process. [MBEE07]
		Uses Systems Engineering techniques to contribute to the definition of the business or enterprise. [MBEP08]	Coaches or mentors practitioners across the enterprise in business and enterprise integration in order to develop their knowledge, abilities, skills, or associated behaviors. [MBEL08]	Champions the introduction of novel techniques and ideas in business and enterprise integration, beyond the enterprise boundary, in order to develop the wider Systems Engineering community in this competency. [MBEE08]
		Reviews work products produced by other enterprise functions for their impact on Systems Engineering activities. [MBEP09]	Promotes the introduction and use of novel techniques and ideas in business and enterprise integration across the enterprise, to improve the enterprise competence in this area. [MBEL09]	Coaches individuals beyond the enterprise boundary in business and enterprise integration techniques, in order to further develop their knowledge, abilities, skills, or associated behaviors. [MBEE09]
		Identifies constraints placed on the Systems Engineering process by the business or enterprise. [MBEP10]	Develops expertise in this competency area through specialist Continual Professional Development (CPD) activities. [MBEL10]	Maintains expertise in this competency area through specialist Continual Professional Development (CPD) activities. [MBEE10]
		Guides new or supervised practitioners in Business and Enterprise integration in order to develop their knowledge, abilities, skills, or associated behaviors. [MBEP11]		
		Maintains and enhances own competence in this area through Continual Professional Development (CPD) activities. [MBEP12]		

Competency area – Systems Engineering Management: Acquisition and Supply

Description

The purpose of Acquisition is to obtain a product or service in accordance with the Acquirer's requirements. The purpose of Supply is to provide an Acquirer with a product or service that meets agreed needs.

Why it matters

All system solutions require agreements between different organizations under which one party acquires or supplies products or services from the other. Systems Engineering helps facilitate the successful acquisition and supply of products or services, in order to ensure that the need is defined accurately, to evaluate the supplier against complex criteria, to monitor the ongoing agreement especially when technical circumstances change, and to support formal acceptance of the product or service.

Effective indicators of knowledge and experience

Awareness	Supervised Practitioner	Practitioner	Lead Practitioner	Expert
Describes the key stages in the acquisition of a system. [MASA01]	Follows a governing process and appropriate tools to plan and control their own acquisition and supply activities. [MASS01]	Creates a strategy for Acquisition or Supply on a project to support SE project and wider enterprise needs. [MASP01]	Creates enterprise-level policies, procedures, guidance, and for acquisition and supply, including associated tools. [MASL01]	Communicates own knowledge and experience in acquisition and supply in order to best practice beyond the enterprise boundary. [MASE01]
Describes the key stages in the supply of a system. [MASA02]	Prepares inputs to work products associated with acquisition of a system. [MASS02]	Creates governing plans, processes, and appropriate tools and uses these to control and monitor Acquisition or Supply on a project. [MASP02]	Judges the tailoring of enterprise-level acquisition and supply processes and associated work products to meet the needs of a project. [MASL02]	Influences key acquisition and supply stakeholders beyond the enterprise boundary. [MASE02]
Describes legal and ethical obligations associated with acquisition and supply, and provides examples. [MASA03]	Identifies potential acquirers of organization systems, products, and services on their program. [MASS03]	Develops a tender document requesting the supply of a system. [MASP03]	Coordinates acquisition and supply across multiple diverse projects or across a complex system, with proven success. [MASL03]	Advises organizations beyond the enterprise boundary on complex or sensitive acquisition and supply issues. [MASE03]
	Prepares inputs to work products associated with supply of a system. [MASS04]	Identifies potential suppliers using criteria to judge their suitability. [MASP04]	Identifies opportunities, arising from projects across the enterprise, to supply systems, products, or services in accordance with wider enterprise goals. [MASL04]	Advises organizations beyond the enterprise boundary on the suitability of their approach to acquisition and supply. [MASE04]
	Develops own understanding of this competency area through Continual Professional Development (CPD). [MASS05]	Reviews supplier responses to a tender document and makes formal recommendations. [MASP05]	Persuades key stakeholders to address identified enterprise-level acquisition and supply issues in order to reduce risk or eliminate issues. [MASL05]	Champions the introduction of novel techniques and ideas in acquisition and supply, beyond the enterprise boundary, in order to develop the wider Systems Engineering community in this competency. [MASE05]

		Reviews acquirer requests and works with key internal stakeholders to propose a solution that meets acquirer needs. [MASP06]	Coaches or mentors practitioners across the enterprise in acquisition and supply in order to develop their knowledge, abilities, skills, or associated behaviors. [MASL06]	Coaches individuals beyond the enterprise boundary in system or element acquisition and supply, in order to further develop their knowledge, abilities, skills, or associated behaviors. [MASE06]
		Negotiates an agreement with a supplier for a system including acceptance criteria. [MASP07]	Promotes the introduction and use of novel techniques and ideas in Acquisition and Supply, across the enterprise, to improve enterprise competence in this area. [MASL07]	Maintains expertise in this competency area through specialist Continual Professional Development (CPD) activities. [MASE07]
		Negotiates an agreement with an acquirer for a system, including acceptance criteria. [MASP08]	Develops expertise in this competency area through specialist Continual Professional Development (CPD) activities. [MASL08]	
		Monitors supplier adherence to terms of agreement to ensure compliance. [MASP09]		
		Maintains an agreement with a supplier to reflect changes on a project. [MASP10]		
		Maintains an agreement with an acquirer maintaining in accordance with agreement terms and conditions. [MASP11]		
		Guides new or supervised practitioners in Acquisition and Supply in order to develop their knowledge, abilities, skills, or associated behaviors. [MASP12]		
		Maintains and enhances own competence in this area through Continual Professional Development (CPD) activities. [MASP13]		

Competency area – Systems Engineering Management: Information Management

Description

Information Management addresses activities associated with the generation, obtaining, confirming, transforming, retaining, retrieval, dissemination, and disposal of information to designated stakeholders with appropriate levels of timeliness, accuracy, and security. Information Management plans, executes, and controls the provision of information to designated stakeholders that is unambiguous, complete, verifiable, consistent, modifiable, traceable, and presentable. Information includes technical, project, organizational, agreement, and user information.

Why it matters

System Engineering requires relevant, timely, and complete information during and after the system life cycle to support all aspects of the development; from the analysis of future concepts to the ultimate archiving and potential subsequent retrieval of project data. Information also supports decision-making across every aspect of the development including suppliers and agreements. Information security and assurance are crucial parts of Information Management: ensuring only designated individuals are able to access certain data; while protecting intellectual property and making sure information is available as required in line with the sender's intent.

Effective indicators of knowledge and experience

Awareness	Supervised Practitioner	Practitioner	Lead Practitioner	Expert
Describes various types of information required to be managed in support of Systems Engineering activities and provides examples. [MIMA01]	Follows a governing process and appropriate tools to plan and control information management activities. [MIMS01]	Creates a strategy for Information Management on a project to support SE project and wider enterprise needs. [MIMP01]	Creates enterprise-level policies, procedures, guidance, and best practice for information management, including associated tools. [MIML01]	Communicates own knowledge and experience in information management, in order to best practice beyond the enterprise boundary. [MIME01]
Describes various types of information assets that may need to be managed within a project or system. [MIMA02]	Prepares inputs to a data dictionary and technical data library. [MIMS02]	Creates governing plans, processes, and appropriate tools and uses these to control and monitor information management and associated communications activities. [MIMP02]	Judges the tailoring of enterprise-level information management processes and associated work products to meet the needs of a project. [MIML02]	Influences individuals beyond the enterprise boundary to adopt appropriate information management techniques or approaches. [MIME02]
Identifies different classes of risk to information integrity and can provide examples of each. [MIMA03]	Identifies valid sources of information and associated authorities on a project. [MIMS03]	Maintains a data dictionary, technical data library appropriate to the project. [MIMP03]	Coordinates information management across multiple diverse projects or across a complex system, with proven success. [MIML03]	Advises organizations beyond the enterprise boundary on complex or sensitive information management issues recommending appropriate solutions. [MIME03]
Describes the relationship between information management and configuration change management. [MIMA04]	Maintains information in accordance with integrity, security, privacy requirements, and data rights. [MIMS04]	Identifies valid sources of information and designated authorities and responsibilities for the information. [MIMP04]	Advises on appropriate information management solutions to be used on projects across the enterprise. [MIML04]	Advises organizations beyond the enterprise boundary on the suitability of their approach to information management. [MIME04]
Describes potential scenarios where information may require modification. [MIMA05]	Identifies information or approaches which requires replanning in order to implement engineering changes on a project. [MIMS05]	Maintains information artifacts in accordance with integrity, security, privacy requirements, and data rights. [MIMP05]	Influences key stakeholders to address identified enterprise-level information management issues. [MIML05]	Advises organizations beyond the enterprise boundary on security, data management, data rights, privacy standards, and regulations. [MIME05]

Explains how data rights may affect information management on a project. [MIMA06]	Identifies designated information requiring archiving in compliance with the requirements on a project. [MIMS06]	Determines formats and media for capture, retention, transmission, and retrieval of information, and data requirements for the sharing of information. [MIMP06]	Communicates Systems Engineering lessons learned gathered from projects across the enterprise. [MIML06]	Champions the introduction of novel techniques and ideas in information management, beyond the enterprise boundary, in order to develop the wider Systems Engineering community in this competency. [MIME06]
Describes the legal and ethical responsibilities associated with access to and sharing of enterprise and customer information and summarizes regulations regarding information sharing. [MIMA07]	Identifies information requiring disposal such as unwanted, invalid, or unverifiable information in accordance with requirements on a project. [MIMS07]	Selects information archival requirements reflecting legal, audit, knowledge retention, and project closure obligations. [MIMP07]	Coaches or mentors practitioners across the enterprise in information management in order to develop their knowledge, abilities, skills, or associated behaviors. [MIML07]	Coaches individuals beyond the enterprise boundary in information management, in order to further develop their knowledge, abilities, skills, or associated behaviors. [MIME07]
Describes what constitutes personal data and why its protection and management is important. [MIMA08]	Prepares information management data products to support management reporting at organizational level. [MIMS08]	Prepares managed information in support of organizational configuration management and knowledge management requirements (e.g. sharing lessons learned). [MIMP08]	Promotes the introduction and use of novel techniques and ideas in Information Management, across the enterprise, to improve enterprise competence in this area. [MIML08]	Maintains expertise in this competency area through specialist Continual Professional Development (CPD) activities. [MIME08]
	Prepares inputs to plans and work products addressing information management and its communication. [MIMS09]	Follows security, data management, privacy standards, and regulations applicable to the project. [MIMP09]	Develops expertise in this competency area through specialist Continual Professional Development (CPD) activities. [MIML09]	
	Records lessons learned and shares beyond the project boundary. [MIMS10]	Selects and implements information management solutions consistent with project security and privacy requirements, data rights, and information management standards. [MIMP10]		
	Develops own understanding of this competency area through Continual Professional Development (CPD). [MIMS11]	Guides new or supervised practitioners in Information Management to develop their knowledge, abilities, skills, or associated behaviors. [MIMP11]		
		Maintains and enhances own competence in this area through Continual Professional Development (CPD) activities. [MIMP12]		

Competency area – Systems Engineering Management: Configuration Management

Description

Configuration Management (CM) manages and controls system elements and configurations over the program life cycle, ensuring the overall coherence of the "evolving" design of a system is maintained in a verifiable manner, throughout the life cycle, and retains the original intent. The CM activity includes planning, identification, change management and control, reporting, and auditing.

Why it matters

CM ensures that the product functional, performance, and physical characteristics are properly identified, documented, validated, and verified to establish product integrity; that changes to these product characteristics are properly identified, reviewed, approved, documented, and implemented; and that the products produced against a given set of documentation are known. Without CM, loss of control over the evolving design, development, and operation of a product will occur.

Effective indicators of knowledge and experience

Awareness	Supervised Practitioner	Practitioner	Lead Practitioner	Expert
Explains why the integrity of the design needs to be maintained and how configuration management supports this. [MCMA01]	Follows a governing configuration and change management process and appropriate tools to plan and control their own activities relating to maintaining design integrity. [MCMS01]	Creates a strategy for Configuration Management on a project to support SE project and wider enterprise needs. [MCMP01]	Creates enterprise-level policies, procedures, guidance, and best practice for configuration management, including associated tools. [MCML01]	Communicates own knowledge and experience in configuration management in order to best practice beyond the enterprise boundary. [MCME01]
Describes the key characteristics of a configuration item (CI) including how configuration items are selected and controlled. [MCMA02]	Prepares information for configuration management work products. [MCMS02]	Creates governing configuration and change management plans, processes, and appropriate tools, and uses these to control and monitor design integrity during the full life cycle of a project or system. [MCMP02]	Judges the tailoring of enterprise-level configuration and change management processes and associated work products to meet the needs of a project. [MCML02]	Influences individuals beyond the enterprise boundary regarding configuration and change management issues. [MCME02]
Identifies key baselines and baseline reviews in a typical development life cycle. [MCMA03]	Describes the need to identify configuration items and why this is done. [MCMS03]	Identifies required remedial actions in the presence of baseline inconsistencies. [MCMP03]	Coordinates configuration management across multiple diverse projects or across a complex system, with proven success. [MCML03]	Advises organizations beyond the enterprise boundary on the suitability of their approach to configuration management. [MCME03]
Describes the process for changing baselined information and a typical life cycle for an engineering change. [MCMA04]	Prepares information in support of configuration change control activities. [MCMS04]	Coordinates changes to configuration items understanding the potential scope within the context of the project. [MCMP04]	Influences key stakeholders to address identified enterprise-level configuration management issues. [MCML04]	Advises organizations beyond the enterprise boundary on complex or sensitive configuration and change management issues. [MCME04]

Lists key activities performed as part of configuration management and can outline the key activities involved in each. [MCMA05]	Prepares material in support of change control decisions and associated review meetings. [MCMS05]	Identifies selection of configuration items and associated documentation by working with design teams justifying the decisions reached. [MCMP05]	Advises stakeholders across the enterprise on remedial actions to address baseline inconsistencies for projects of various size and complexity. [MCML05]	Champions the introduction of novel techniques and ideas in configuration management, beyond the enterprise boundary, in order to develop the wider Systems Engineering community in this competency. [MCME05]
Explains why change occurs and why changes need to be carefully managed. [MCMA06]	Produces management reports in support of configuration item status accounting and audits. [MCMS06]	Coordinates change control review activities in conjunction with customer representative and directs resolutions and action items. [MCMP06]	Advises stakeholders across the enterprise on major changes and influences them to reduce impact of such changes. [MCML06]	Coaches individuals beyond the enterprise boundary in Configuration Management, in order to further develop their knowledge, abilities, skills, or associated behaviors. [MCME06]
Describes the processes and work products used to assist in Change Management. [MCMA07]	Identifies applicable standards, regulations, and enterprise level processes on their project. [MCMS07]	Coordinates configuration status accounting reports and audits. [MCMP07]	Coaches or mentors practitioners across the enterprise in configuration management in order to develop their knowledge, abilities, skills, or associated behaviors. [MCML07]	Maintains expertise in this competency area through specialist Continual Professional Development (CPD) activities. [MCME07]
Describes the meaning of key terminology and acronyms used within Change Management and their relationships. [MCMA08]	Identifies and reports baseline inconsistencies. [MCMS08]	Guides new or supervised practitioners in configuration management to develop their knowledge, abilities, skills, or associated behaviors. [MCMP08]	Promotes the introduction and use of novel techniques and ideas in Configuration Management, across the enterprise, to improve enterprise competence in this area. [MCML08]	
	Develops own understanding of this competency area through Continual Professional Development (CPD). [MCMS09]	Maintains and enhances own competence in this area through Continual Professional Development (CPD) activities. [MCMP09]	Develops expertise in this competency area through specialist Continual Professional Development (CPD) activities. [MCML09]	

Competency area – Integrating Competencies: Project Management

Description

Project Management identifies, plans, and coordinates activities required in order to deliver a satisfactory system, product, service of appropriate quality, within the constraints of schedule, budget, resources, infrastructure, available staffing, and technology. Project Management includes development engineering but covers the complete project (i.e. beyond the engineering boundary), encompassing disciplines such as sales, business development, finance, commercial, legal, human resources, production, procurement and supply chain management, and logistics.

Why it matters

Good project management reduces risk, maximizes opportunity, cut system, product, or service costs, and improves both the success rate and the return on investment of projects.

Effective indicators of knowledge and experience

Awareness	Supervised Practitioner	Practitioner	Lead Practitioner	Expert
Explains the role the project management function plays in developing a successful system product or service. [IPMA01]	Follows a governing process in order to interface successfully to project management activities. [IPMS01]	Follows governing project management plans and processes, and uses appropriate tools to control and monitor project management-related Systems Engineering tasks, interpreting as necessary. [IPMP01]	Creates enterprise-level policies, procedures, guidance, and best practice in order to ensure Systems Engineering project management activities integrate with enterprise-level Project Management goals. [IPML01]	Communicates own knowledge and experience in the integration of project management with Systems Engineering, in order to improve Systems Engineering best practice beyond the enterprise boundary. [IPME01]
Explains the meaning of commonly used project management terms and applicable standards. [IPMA02]	Prepares inputs to work products which interface to project management stakeholders to ensure Systems Engineering work aligns with wider project management activities. [IPMS02]	Identifies Systems Engineering tasks ensuring that these tasks integrate successfully with project management activities. [IPMP02]	Assesses enterprise-level project management processes and tailoring to ensure they integrate with Systems Engineering needs. [IPML02]	Advises organizations beyond the enterprise boundary on complex or sensitive project management-related issues affecting Systems Engineering. [IPME02]
Explains the relationship between cost, schedule, quality, and performance and why this matters. [IPMA03]	Identifies potential issues with interfacing work products received from project management Stakeholders or produced by Systems Engineering for project management stakeholders taking appropriate action. [IPMS03]	Identifies activities required to ensure integration of project management planning and estimating with Systems Engineering planning and estimating. [IPMP03]	Assesses project management information produced across the enterprise using appropriate techniques for its integration with Systems Engineering data. [IPML03]	Advises organizations beyond the enterprise boundary on the suitability of their approach to project management plans affecting Systems Engineering activities. [IPME03]
Describes the role and typical responsibilities of a project manager on a project team, within the wider project management function. [IPMA04]	Prepares Systems Engineering information for project management in support of wider project initiation activities. [IPMS04]	Develops inputs to a project management plan for a complete project beyond those required for Systems Engineering planning to support wider project or business project management. [IPMP04]	Judges appropriateness of enterprise-level project management decisions in a rational way to ensure alignment with Systems Engineering needs. [IPML04]	Champions the introduction of novel techniques and ideas to improve the integration of Systems Engineering and project management functions, beyond the enterprise boundary, in order to develop the wider Systems Engineering community in this competency. [IPME04]

Describes the differences between performing project management and Systems Engineering management on that project. [IPMA05]	Prepares Systems Engineering Work Breakdown Structure (WBS) information for project management in support of their creation of a wider project WBS. [IPMS05]	Develops Systems Engineering inputs for project management status reviews to enable informed decision-making. [IPMP05]	Judges conflicts between project management needs and Systems Engineering needs on behalf of the enterprise, arbitrating as required. [IPML05]	Coaches individuals beyond the enterprise boundary, in the relationship between Systems Engineering and project management, to further develop their knowledge, abilities, skills, or associated behaviors. [IPME05]
Describes the key interfaces between project management stakeholders within the enterprise and the project team. [IPMA06]	Prepares Systems Engineering Work Package definitions and estimating information for project management in support of their work creating project-level Work Packages and estimates. [IPMS06]	Develops project initiation information required to support Project Start-up by project management on a project. [IPMP06]	Guides and actively coordinates complex or challenging relationships with key stakeholders affecting Systems Engineering. [IPML06]	Maintains expertise in this competency area through specialist Continual Professional Development (CPD) activities. [IPME06]
Describes the wider program environment within which the system is being developed, and the influence each can have on this other. [IPMA07]	Follows a governing process in order to interface successfully to project management activities. [IPMS07]	Develops Systems Engineering information required to support termination of a project by senior management. [IPMP07]	Persuades key project management stakeholders to address identified enterprise-level project management issues affecting Systems Engineering. [IPML07]	
	Prepares information used in project management contract reviews for project management on a project. [IPMS08]	Creates working groups extending beyond Systems Engineering. [IPMP08]	Coaches or mentors practitioners across the enterprise in the integration of project management with Systems Engineering, in order to develop their knowledge, abilities, skills, or associated behaviors. [IPML08]	
	Prepares Systems Engineering information for project management in support of wider project termination activities. [IPMS09]	Guides new or supervised practitioners in finance and its relationship to Systems Engineering, to develop their knowledge, abilities, skills, or associated behaviors. [IPMP09]	Promotes the introduction and use of novel techniques and ideas across the enterprise, which improve the integration of Systems Engineering and project management functions. [IPML09]	
	Develops own understanding of this competency area through Continual Professional Development (CPD). [IPMS10]	Maintains and enhances own competence in this area through Continual Professional Development (CPD) activities. [IPMP10]	Develops expertise in this competency area through specialist Continual Professional Development (CPD) activities. [IPML10]	

Competency area – Integrating Competencies: Finance

Description

Finance is the area of estimating and tracking costs associated with the project. It also includes understanding of the financial environment in which the project is being executed.

Why it matters

Appropriate funding is the life blood of any system development project. It is important for systems engineers to recognize the importance of cost estimation, budgeting, and controlling project finances and to support the finance discipline in its activities.

Effective indicators of knowledge and experience

Awareness	Supervised Practitioner	Practitioner	Lead Practitioner	Expert
Explains the role the finance function plays in developing a successful system product or service. [IFIA01]	Follows a governing process in order to interface successfully to financial management activities. [IFIS01]	Follows governing finance plans, processes, and uses appropriate tools to control and monitor finance-related Systems Engineering tasks, interpreting as necessary. [IFIP01]	Creates enterprise-level policies, procedures, guidance, and best practice in order to ensure Systems Engineering finance-related activities integrate with enterprise financial goals, including associated tools. [IFIL01]	Communicates own knowledge and experience in the integration of finance needs with Systems Engineering, in order to improve Systems Engineering best practice beyond the enterprise boundary. [IFIE01]
Explains the meaning of commonly used financial terms and applicable standards. [IFIA02]	Prepares inputs to work products which interface to financial stakeholders to ensure Systems Engineering work aligns with wider financial management activities. [IFIS02]	Prepares work products required by financial stakeholders to ensure Systems Engineering work aligns with wider financial management activities. [IFIP02]	Assesses enterprise-level financial management materials to ensure they integrate with Systems Engineering needs. [IFIL02]	Advises organizations beyond the enterprise boundary on the suitability of financial management plans affecting Systems Engineering activities. [IFIE02]
Explains how business financial decisions may impact a product or service through its entire life cycle, and vice versa. [IFIA03]	Identifies potential issues with interfacing work products received from Financial Stakeholders or produced by Systems Engineering for financial stakeholders taking appropriate action. [IFIS03]	Creates detailed cost estimating work products required by financial stakeholders to scope the financial aspects of a project. [IFIP03]	Judges tailoring required for enterprise-level Systems Engineering processes in order to ensure that the needs of financial stakeholders are fully integrated. [IFIL03]	Advises organizations beyond the enterprise boundary on complex or sensitive Financial matters and their effect on Systems Engineering. [IFIE03]
Explains primary interfaces between the finance function and the Systems Engineering team. [IFIA04]	Prepares inputs to financial cost estimation work products for financial stakeholders ensuring Systems Engineering work aligns with wider financial management activities. [IFIS04]	Analyzes activity costs and scheduling as required by financial stakeholders in order to develop project funding requirements and a cost management plan. [IFIP04]	Judges appropriateness of enterprise-level financial decisions in a rational way to ensure alignment with Systems Engineering needs. [IFIL04]	Champions the introduction of novel techniques and ideas to improve the integration of Systems Engineering with the finance function, beyond the enterprise boundary, in order to develop the wider Systems Engineering community in this competency. [IFIE04]

Describes the key work products exchanged between finance stakeholders and the Systems Engineering team. [IFIA05]	Uses cost aggregation and analysis techniques to communicate funding information for financial stakeholders during creation or approval of funding requests. [IFIS05]	Analyzes system life cycle cost issues and decisions as required by financial stakeholders in order to make recommendations. [IFIP05]	Assesses financial information produced across the enterprise using appropriate techniques for its integration with Systems Engineering data. [IFIL05]	Coaches individuals beyond the enterprise boundary, in the relationship between Systems Engineering and finance, to further develop their knowledge, abilities, skills, or associated behaviors. [IFIE05]
Describes the difference between performing financial management on a project or wider enterprise and managing financial resources as part of Systems Engineering activities. [IFIA06]	Uses system life cycle cost analysis techniques to communicate cost information to financial stakeholders on a project. [IFIS06]	Analyzes project performance and expenditures as required by financial stakeholders in order to determine variance from plans. [IFIP06]	Persuades key financial stakeholders to address identified enterprise-level financial management issues affecting Systems Engineering. [IFIL06]	Maintains expertise in this competency area through specialist Continual Professional Development (CPD) activities. [IFIE06]
Explains how financial management concerns relate to Systems Engineering. [IFIA07]	Uses project performance and expenditure tracking techniques to communicate performance and expenditure tracking information to financial stakeholders on a project. [IFIS07]	Analyzes variances to budget tolerance as required by financial stakeholders in order to identify and implement corrective actions. [IFIP07]	Coaches or mentors practitioners across the enterprise in the integration of finance with Systems Engineering to develop their knowledge, abilities, skills, or associated behaviors. [IFIL07]	
	Uses financial variance and tolerance data to communicate budget or financial variances to financial stakeholders on a project. [IFIS08]	Guides new or supervised practitioners in finance and its relationship to Systems Engineering, to develop their knowledge, abilities, skills, or associated behaviors. [IFIP08]	Promotes the introduction and use of novel techniques and ideas across the enterprise, which improve the integration of Systems Engineering and finance functions. [IFIL08]	
	Develops own understanding of this competency area through Continual Professional Development (CPD). [IFIS09]	Maintains and enhances own competence in this area through Continual Professional Development (CPD) activities. [IFIP09]	Develops expertise in this competency area through specialist Continual Professional Development (CPD) activities. [IFIL09]	

Competency area – Integrating Competencies: Logistics

Description
Logistics focuses on the support and sustainment of the product once it is transitioned to the end user. It includes areas such as life cycle cost analysis, supportability analysis, sustainment engineering, maintenance planning and execution, training, spares and inventory control, associated facilities and infrastructure, packaging, handling and shipping, and support equipment for the system and its elements.

Why it matters
Factoring logistics considerations such as availability, storage and transport, and training needs early in the design effort can significantly reduce total life cycle cost for the system.

Effective indicators of knowledge and experience

Awareness	Supervised Practitioner	Practitioner	Lead Practitioner	Expert
Explains the role the logistics function plays in developing a successful system, product, or service. [ILOA01]	Follows a governing process in order to interface successfully to logistics management activities. [ILOS01]	Follows governing logistics plans, processes, and uses appropriate tools to control and monitor logistics-related Systems Engineering tasks, interpreting as necessary. [ILOP01]	Creates enterprise-level policies, procedures, guidance, and best practice in order to ensure Systems Engineering logistics-related activities integrate with enterprise logistics goals, including associated tools. [ILOL01]	Communicates own knowledge and experience in the integration of logistics needs with Systems Engineering, in order to improve Systems Engineering best practice beyond the enterprise boundary. [ILOE01]
Explains the meaning of commonly used logistics terms and applicable standards. [ILOA02]	Identifies potential issues with interfacing work products received from logistics Stakeholders or produced by Systems Engineering for logistics stakeholders taking appropriate action. [ILOS02]	Prepares work products required by logistics stakeholders to ensure Systems Engineering work aligns with wider logistics management activities. [ILOP02]	Assesses enterprise-level logistics management processes to ensure they integrate with Systems Engineering needs. [ILOL02]	Advises organizations beyond the enterprise boundary on the suitability of their approach to logistics management within Systems Engineering. [ILOE02]
Describes key logistics activities and why they are important to the success of a system. [ILOA03]	Prepares inputs to a supportability analysis on a project to assist logistics stakeholders. [ILOS03]	Prepares supportability analysis information required by logistics stakeholders to meet project and enterprise requirements. [ILOP03]	Judges the appropriateness of enterprise-level logistics decisions in a rational way to ensure alignment with Systems Engineering needs. [ILOL03]	Assesses the suitability of Logistics Management Plans affecting Systems Engineering activities. [ILOE03]
Explains primary interfaces between the logistics function and the Systems Engineering team. [ILOA04]	Explains how different concepts for maintenance may have different life cycle costs. [ILOS04]	Develops maintenance concepts required by logistics stakeholders to ensure alignment with system engineering activities. [ILOP04]	Judges the supportability strategies and supportability decisions across the enterprise to ensure they align with Systems Engineering performance, readiness, and life cycle cost needs. [ILOL04]	Advises organizations beyond the enterprise boundary on complex or sensitive logistics-related issues and its effect on Systems Engineering. [ILOE04]
Describes the key work products exchanged between logistics stakeholders and the Systems Engineering team. [ILOA05]	Uses recognized analysis techniques to calculate spares, repairs, or supply-related information for logistics stakeholders on a project. [ILOS05]	Develops spares and repair concepts required by logistics stakeholders to ensure alignment with system engineering activities. [ILOP05]	Judges logistics plans and decisions across the enterprise to ensure they align with Systems Engineering performance, readiness, and life cycle cost needs. [ILOL05]	Champions the introduction of novel techniques and ideas to improve the integration of Systems Engineering and logistics functions, beyond the enterprise boundary, in order to develop the wider Systems Engineering community in this competency. [ILOE05]

Explains the concept and value of life cycle cost and how this affects both the system solution and logistics. [ILOA06]	Uses recognized analysis techniques to produce facilities and infrastructure operation and maintenance information for logistics stakeholders on a project. [ILOS06]	Develops facilities infrastructure concepts required by logistics stakeholders to support operation and maintenance of a system across its life cycle. [ILOP06]	Assesses enterprise-level logistics work products for their alignment with Systems Engineering. [ILOL06]	Coaches individuals beyond the enterprise boundary, in the relationship between Systems Engineering and logistics, to further develop their knowledge, abilities, skills, or associated behaviors. [ILOE06]
Describes the wider logistics environment within which the system is being developed, and the influence each can have on this other. [ILOA07]	Uses recognized techniques to produce system engineering information in support of operator or personnel training or simulation activities for logistics stakeholders on a project. [ILOS07]	Develops logistics training products required by logistics stakeholders to maximize the effectiveness of operators and personnel sustaining the system at lowest life cycle cost. [ILOP07]	Persuades key logistics stakeholders to address identified enterprise-level logistics management issues affecting Systems Engineering. [ILOL07]	Maintains expertise in this competency area through specialist Continual Professional Development (CPD) activities. [ILOE07]
	Uses recognized techniques to produce system operation and maintenance information for logistics stakeholders on a project. [ILOS08]	Develops concepts for support equipment in collaboration with logistics stakeholders to sustain the operation and maintenance of a system across its life cycle. [ILOP08]	Coaches or mentors practitioners across the enterprise in the integration of logistics with Systems Engineering in order to develop their knowledge, abilities, skills, or associated behaviors. [ILOL08]	
	Uses recognized techniques to produce system installation, operation, maintenance, and sustainment information for logistics stakeholders on a project. [ILOS09]	Develops packaging, handling, storage, and transportation required by logistics stakeholders to ensure safe and secure transportation of a system. [ILOP09]	Promotes the introduction and use of novel techniques and ideas across the enterprise, which improve the integration of Systems Engineering and logistics functions. [ILOL09]	
	Uses recognized techniques to produce system packaging, handling, storage, and transportation information for logistics stakeholders on a project. [ILOS10]	Develops work products required by logistics stakeholders in order to support the installation, operation, maintenance, and sustainment of the system. [ILOP10]	Develops expertise in this competency area through specialist Continual Professional Development (CPD) activities. [ILOL10]	
	Develops own understanding of this competency area through Continual Professional Development (CPD). [ILOS11]	Guides new or supervised practitioners in logistics and its relationship to Systems Engineering, to develop their knowledge, abilities, skills, or associated behaviors. [ILOP11]		
		Maintains and enhances own competence in this area through Continual Professional Development (CPD) activities. [ILOP12]		

Competency area – Integrating Competencies: Quality

Description
Quality focuses on customer satisfaction via the control of key product characteristics and corresponding key manufacturing process characteristics.

Why it matters
Proactive quality management improves both the quality of the system, product, or service provided, as well as the quality of the project's management processes.

Effective indicators of knowledge and experience

Awareness	Supervised Practitioner	Practitioner	Lead Practitioner	Expert
Explains the role the quality function plays in developing a successful system product or service. [IQUA01]	Follows a governing process in order to interface successfully to quality management activities. [IQUS01]	Follows governing quality plans and processes, and uses appropriate tools to control and monitor quality-related Systems Engineering tasks, interpreting as necessary. [IQUP01]	Creates enterprise-level policies, procedures, guidance, and best practice in order to ensure Systems Engineering quality-related activities integrate with enterprise-level quality goals including associated tools. [IQUL01]	Communicates own knowledge and experience in the integration of quality function needs with Systems Engineering, in order to improve Systems Engineering best practice beyond the enterprise boundary. [IQUE01]
Explains the meaning of commonly used quality-related terms and applicable standards. [IQUA02]	Prepares inputs to work products which interface to quality stakeholders to ensure Systems Engineering work aligns with wider quality management activities. [IQUS02]	Prepares work products required by quality stakeholders to ensure Systems Engineering work aligns with wider quality management activities. [IQUP02]	Assesses enterprise-level quality management processes to ensure they integrate with Systems Engineering needs. [IQUL02]	Advises organizations beyond the enterprise boundary on the suitability of their approach to Quality Management and the effect of their plans on Systems Engineering activities. [IQUE02]
Explains primary interfaces between the quality management function and the Systems Engineering team. [IQUA03]	Identifies potential issues with interfacing work products received from quality Stakeholders or produced by Systems Engineering for quality taking appropriate action. [IQUS03]	Identifies alternative mechanisms for measuring quality to support the quality function in achieving the targeted standard of excellence on a project. [IQUP03]	Judges appropriateness of enterprise-level quality decisions in a rational way to ensure alignment with Systems Engineering needs. [IQUL03]	Fosters a culture of continuous quality improvement beyond the enterprise boundary. [IQUE03]
Describes the key work products exchanged between quality management stakeholders and the Systems Engineering team. [IQUA04]	Identifies measures of quality which ensure an appropriate standard of excellence is targeted on a project in support of quality function activities. [IQUS04]	Identifies mechanisms measuring process performance to support the quality function in achieving the targeted standard of excellence on a project. [IQUP04]	Persuades quality stakeholders to address identified enterprise-level quality management issues affecting Systems Engineering. [IQUL04]	Advises organizations beyond the enterprise boundary on complex or sensitive quality-related issues affecting Systems Engineering. [IQUE04]
Explains the difference between quality assurance and quality control. [IQUA05]	Identifies quality characteristics which ensure an appropriate standard of excellence is targeted on a project in support of quality function activities. [IQUS05]	Guides and actively coordinates Systems Engineering process improvement activities to enable the quality function to achieve its targeted standard of Systems Engineering excellence on a project. [IQUP05]	Assesses quality information produced across the enterprise using appropriate techniques for its integration with Systems Engineering data. [IQUL05]	Champions the introduction of novel techniques and ideas to improve the integration of Systems Engineering and quality functions, beyond the enterprise boundary, in order to develop the wider Systems Engineering community in this competency. [IQUE05]

Explains how project-level decisions can impact the quality of a system. [IQUA06]	Monitors process adherence on a project in support of quality function activities. [IQUS06]	Analyzes design information or test (e.g. verification) results for a product or project to confirm conformance to standards. [IQUP06]	Reviews quality audit outcomes at enterprise level to establish their impact on system engineering across the enterprise. [IQUL06]	Coaches individuals beyond the enterprise boundary, in the relationship between Systems Engineering and quality management, to further develop their knowledge, abilities, skills, or associated behaviors. [IQUE06]
Explains the difference between performing quality management on a project or wider enterprise and managing quality as part of Systems Engineering activities. [IQUA07]	Uses recognized techniques to support verification of product or system conformity for quality stakeholders on a project. [IQUS07]	Analyzes the root-cause analysis of failures, determining appropriate corrective actions in support of quality function needs. [IQUP07]	Promotes continuous improvement in Systems Engineering at the enterprise level to support quality management function initiatives. [IQUL07]	Maintains expertise in this competency area through specialist Continual Professional Development (CPD) activities. [IQUE07]
Describes the wider quality environment within which the system is being developed, and the influence each can have on this other. [IQUA08]	Uses recognized techniques to perform system root-cause analysis and failure elimination for quality stakeholders on a project. [IQUS08]	Conducts an audit of project practices against recognized quality or project standards to support quality function needs. [IQUP08]	Assesses quality management plans from projects across the enterprise for their impact on Systems Engineering activities. [IQUL08]	
	Identifies measures of quality which ensure an appropriate standard of excellence is targeted on a project in support of quality function activities. [IQUS09]	Reviews the results of Quality Management Plans affecting Systems Engineering activities. [IQUP09]	Fosters a culture of continuous quality improvement in projects across the enterprise. [IQUL09]	
	Complies with required quality standards to support the quality function in auditing ongoing projects. [IQUS10]	Guides new or supervised practitioners in quality and its relationship to Systems Engineering, to develop their knowledge, abilities, skills, or associated behaviors. [IQUP10]	Coaches or mentors practitioners across the enterprise in the integration of quality with Systems Engineering in order to develop their knowledge, abilities, skills, or associated behaviors. [IQUL10]	
	Develops own understanding of this competency area through Continual Professional Development (CPD). [IQUS11]	Maintains and enhances own competence in this area through Continual Professional Development (CPD) activities. [IQUP11]	Promotes the introduction and use of novel techniques and ideas across the enterprise, which improve the integration of Systems Engineering and quality management functions. [IQUL11]	
			Develops expertise in this competency area through specialist Continual Professional Development (CPD) activities. [IQUL12]	

SECF ANNEX E: SECF COMMENT FORM

Please submit feedback comment form information to SECFCompetencyWG@incose.net.

Reviewed document:	INCOSE Systems Engineering Competency Framework (SECF)
Name of submitter:	Given name, family name (e.g. Jo DOE)
Date of submission:	YYYY-MM-DD (e.g. 2018-04-09)
Contact Info:	Email address (e.g. jo.DOE@anywhere.com)
Type of submission:	Group, individual
Group name and number of contributors	Group name if applicable (e.g. INCOSE XYZ Working group)
Comments	Please provide comment details including precise reference to the document section, paragraph, or line item requiring change. Ideally, comments should be formatted as shown in the table below.

Comment ID	**Category**	**Section number**	**Specific reference**	**Issue, comment, and rationale**	**Proposed change or new text (mandatory)**
Unique Identifier	G, E, TH, TL As follows: • G = general • E = editorial • TH = technical comment, high priority • TL = technical comment, low priority	E.g. section n, table m	E.g. paragraph, line	Please provide rationale so that comment is clear and supportable.	Good quality new or revised text will increase odds of acceptance.

Systems Engineering Competency Assessment Guide: A combined INCOSE Systems Engineering Competency Framework (SECF) and associated Systems Engineering Competency Assessment Guide (SECAG) document, First Edition. INCOSE.

PART II: SECAG - SYSTEMS ENGINEERING COMPETENCY ASSESSMENT GUIDE

The INCOSE Systems Engineering Competency Assessment Guide (SECAG) is a collaborative product from a series of meetings of the INCOSE International Competency Working Group, held at INCOSE International Workshops and on conference calls between January 2019 and December 2021. The Competency Working Group is Chaired by Cliff Whitcomb and Co-Chaired by Lori Zipes.

Compiled and edited by:

Ian Presland CEng, FIET, ESEP (Primary Editor)
Clifford Whitcomb, Ph.D., INCOSE Fellow
Lori Zipes, ESEP

Primary contributing authors (alphabetical order):

Juan P. Amenabar, ESEP
Jonas Hallqvist, ESEP
Ian Presland, CEng, FIET, ESEP
Clifford Whitcomb, Ph.D., INCOSE Fellow
Lori Zipes, ESEP

Reviewers and additional contributors (alphabetical order):

Richard Beasley, ESEP
Ray Dellefave
Suja Joseph-Malherbe, CSEP
Ruediger Kaffenberger, CSEP
Rabia Khan, ASEP
Kirk Michealson

Systems Engineering Competency Assessment Guide: A combined INCOSE Systems Engineering Competency Framework (SECF) and associated Systems Engineering Competency Assessment Guide (SECAG) document, First Edition. INCOSE.

Susan Plano-Faber, ASEP
Philip Quan, CSEP
Sven-Olaf Schulze, CSEP
Corina White, CSEP

SECAG ACKNOWLEDGEMENTS

We would like to acknowledge the contribution of INCOSE UK Advisory Board members who participated in the UK Working group that created the INCOSE UK Systems Engineering Competencies Framework (Issue 3, 2010) which was used as the original basis for the first edition of the INCOSE Systems Engineering Competency Framework (INCOSE 2018b). A full list of contributing UK organizations and individuals is included in the main body of the INCOSE Systems Engineering Competency Framework (First Edition) (INCOSE 2018b).

We would like to thank all other past and current members of the INCOSE Competency Working Group for their support, ideas, and comments.

DISCLAIMER

COPYRIGHT

SECAG INTRODUCTION

SECAG SCOPE

This document is the *INCOSE Systems Engineering Competency Assessment Guide* (SECAG).

The SECAG is an Assessment Guide for users of the *INCOSE Systems Engineering Competency Framework* (SECF), Part I of this book, designed to guide those assessing systems engineering competencies characterized within that framework.

The SECAG is a general document which is expected to be tailored to reflect organizational SECF tailoring, and further tailored to reflect local or organizational specifics as defined herein.

SECAG PURPOSE

The purpose of this document is to provide guidance on how to evaluate individuals to establish their proficiency in the competencies defined in the framework and how to differentiate between proficiency at each of the five levels defined within that document.

SECAG CONTEXT

The content of this document aligns with INCOSE technical products and initiatives published by end 2021.

In particular, the document uses the *INCOSE Systems Engineering Handbook* (Fourth Edition) (INCOSE 2015) as its primary technical knowledge base as well as extracts taken from the INCOSE Systems Engineering Body of Knowledge (SEBoK) wiki (Ref: https://www.sebokwiki.org/wiki) during its production.

The INCOSE SECF was originally published in 2018 by INCOSE (2018a). The SECAG is aligned to an updated version of that document which we have included as Annex D in Part I of this book in order to ensure completeness.

Systems Engineering Competency Assessment Guide: A combined INCOSE Systems Engineering Competency Framework (SECF) and associated Systems Engineering Competency Assessment Guide (SECAG) document, First Edition. INCOSE.

SECAG OBJECTIVE

The objective of this assessment guide is to provide detailed guidance to users of the competency framework in order to:

- Create a benchmark standard for each level of proficiency within each competence area.
- Define a set of standardized terminology for competency indicators to promote like-for-like comparison.
- Provide typical non-domain-specific indicators of evidence which may be used to confirm experience in each competency area.

LINK TO COMPETENCY-BASED CERTIFICATION WITHIN THE INCOSE SEP PROGRAM

In 2020, INCOSE piloted a "Practitioner" level "competency-based" equivalent as part of their INCOSE Systems Engineering Professional (SEP) Certification program. For simplicity, the INCOSE pilot used an older version of the framework than that defined in this document.

However, the success of the pilot means that INCOSE is now considering adopting a version of this assessment guide for a future "competence-based" equivalent. Please contact INCOSE for more details.

SECAG DOCUMENT OVERVIEW

The SECAG Introduction summarizes the scope, purpose, context, objective, and describes the link to competency-based certification within the INCOSE SEP program.

The Tailoring the Assessment Guide section explains terminology tailoring and provides general guidance for tailoring assessment of proficiency levels.

The Applying the Assessment Guide section explains assessing the assessors.

The Explanation of Assessment Guide Tables section explains assessment guide language usage, describes sub-indicator classifications, and explains assessing evidence sub-indicator types.

The SECAG Bibliography section provides a list of the documents referenced.

The SECAG Acronyms and Abbreviations section provides a table for acronyms and abbreviations.

The SECAG Glossary section provides definitions of the terminology used in the assessment guide.

SECAG Annex A contains the competency framework assessment guidance.

SECAG Annex B provides a set of examples of usage of both the framework and assessment guide documents.

The authors welcome feedback on this document. SECAG Annex C is a comment form provided for this purpose.

TAILORING THE ASSESSMENT GUIDE

The INCOSE Systems Engineering Competency Assessment Guide will be applicable to a variety of organizations and institutions covering many different domains, in different countries worldwide. In addition, the framework can be applied in support of a number of goals from recruitment to individual and organizational capability development.

Every organization, institution, domain, and country operate in a unique *context* – having its own constraints, terminology, regulations, standards, tools, procedures, standard practices, etc.

Therefore, it is critical that this Assessment Guide is *tailored* before use to reflect the operational context.

Tailoring should align the terminology and the content of the guide to the specific needs of the organization or enterprise using the guide. Tailoring the guide will help maximize the value of the assessments performed. Tailoring will typically include:

- Removal of unnecessary or less important element(s)
- Modification of element(s) (e.g. to reflect terminology)
- Addition of new element(s)

An "element" could be a competence area, a competency indicator, or an evidence indicator.

Tailoring helps simplify deployment by increasing alignment with other organizational systems and with domain-specific language and concepts.

An approach to tailoring is defined in the core framework definition document. However, some additional points are described below.

TERMINOLOGY TAILORING

Terminology is often organization specific. We have endeavored to use the same terms consistently throughout. Organizations are free to modify the terminology to reflect their own organizational definitions.

For example, the term "system" has a standard INCOSE definition which is consistently used in this guide. However, some organizations applying the guide may prefer to adopt an alternative word (such as "product," "service," or even "project") depending on the nature of the organization.

Systems Engineering Competency Assessment Guide: A combined INCOSE Systems Engineering Competency Framework (SECF) and associated Systems Engineering Competency Assessment Guide (SECAG) document, First Edition. INCOSE.

"LEAD PRACTITIONER" VS "LEAD ROLE" ASSESSMENT

We have tried to avoid assuming any organizational hierarchy in the framework (e.g. a requirement to be an appointed "team lead" or "project lead") although clearly, a "Lead Practitioner" will generally be in an organizational role with greater seniority (e.g. responsibility, authority, and accountability) than a "Practitioner" due to a requirement for having accrued increased experience. (The same argument applies with other level comparisons.)

There is however no implication that an individual *is required* to be in a "team leader" role or "project leader" role when being assessed as a "Lead Practitioner." For instance, they may be operating in an organizational role which identifies them as part of a team led by another individual. Clearly, their additional experience will help the team, and the individual may be underutilized in their role, but the individual's underlying competence is not affected.

"EXPERT" PRACTITIONER VS "EXPERT ROLE"

Many organization's use the term "Expert" in defining roles. Some organizations even grant the title "Expert" to reflect that organizations valuation of particular experience or leadership contributions. Indeed, INCOSE uses the term "Expert" in its Systems Engineering Certification program for its "ESEP designation (ESEP = Expert Systems Engineering Professional).

However, within this framework, as with the "Lead Practitioner" level, there is no expectation that an individual is *required* to have an organizational (or INCOSE) designation of "expert" in order to be assessed as an Expert Practitioner in a competence area.

GENERAL GUIDANCE FOR TAILORING ASSESSMENT OF PROFICIENCY LEVELS – ACCUMULATED EVIDENCE

With five increasing levels of competency defined for each competency area, a question remains about assessment of indicators for higher levels of competency.

Some have argued that assessing an individual for competence at any proficiency level above Awareness should require explicit confirmation that all indicators at all lower proficiency levels should also be confirmed through evidence.

However, the INCOSE Working Group members have concluded that this is neither beneficial nor cost-effective. Furthermore, there are arguments that at higher proficiency levels, assessments should change to increase their focus on an individual's effectiveness in applying their accrued knowledge and experience to resolve increasingly complex challenges, rather than their accumulated experience alone.

With this in mind, the following provides guidance on how to select appropriate lower-level evidence when assessing an individual's proficiency in a competency above Awareness level.

- Assessment of proficiency at "Awareness" level should validate the knowledge indicators defined for the competency area.

  While organizations may choose to tailor the knowledge to meet their domain needs, basic knowledge is always a central aspect of competency.
- For Supervised Practitioner level, a reconfirmation of Awareness level (knowledge) indicators may still be reasonable. However, at this level, assessment should focus on new knowledge indicators defined for the level and on validating the experiential evidence showing Awareness knowledge applied.

Assessors may choose to include revalidation of some Awareness knowledge in areas where they have doubt as to the other evidence provided or for knowledge formally identified by the organization as key to success.

- When assessing a competency proficiency level at the Practitioner level (and above), adopting a "100% compliant" approach is unlikely to be beneficial or cost-effective. Furthermore, the nature of competency indicators is such that assessing an individual at this level will usually (directly or indirectly) confirm the experiential and knowledge expectations of both Supervised Practitioner and Awareness levels. As a result, explicit assessment of all indicators for Supervised Practitioner and Awareness levels during an assessment at the Practitioner level should be left to the discretion of the assessor (e.g. to confirm a specific item within evidence provided, or to cover an area wholly omitted through the evidence provided).

  Practitioner level indicators are generally an elaboration of those defined at Supervised Practitioner level and thus repeating assessment of Supervised Practitioner indicators will not be beneficial. However, there may be merit in explicitly confirming some selected knowledge indicators, especially those formally identified by the organization as key to success or where the assessors have doubts as to the depth of experience or knowledge provided.
- When assessing a competency proficiency level at the Lead Practitioner level, the degree to which lower-level competency indicators need to be explicitly confirmed during an assessment will be driven primarily by organizational needs. But in principle, assessors should focus on assessing the evidence for all Lead Practitioner proficiency level indicators since usually this will directly or indirectly highlight deficiencies of experience and core knowledge.

  There is merit in supplementing the above with explicit confirmation of some selected lower-level indicators, especially those formally identified by the organization as key to success or where the assessors have doubts as to the depth of experience or knowledge provided for those indicators. The degree to which this is done will be at the discretion of the assessors, based upon their belief as to whether or not the candidate has demonstrated the necessary experience and knowledge required.
- Assessing a competency proficiency level at the Expert level may (directly or indirectly) confirm some aspects of the expectations of the Lead Practitioner level (and by implication, all other lower levels). Assessment of proficiency at the Expert level will normally be focused on examining an individual's wider contributions to the competency area (e.g. beyond the organizational boundary).

  There is merit in supplementing the above with explicit confirmation of some selected lower-level indicators, especially those formally identified by the organization as key to success or where the assessors have doubts as to the depth of experience or knowledge evidence provided for those indicators. The degree to which this is done will be at the discretion of the assessors, based upon their belief as to whether or not the candidate has demonstrated the necessary experience and knowledge required.

Notwithstanding the above, instructions for assessing a competency at any level within an organization should be documented as part of assessment tailoring guidance for the organization.

EVIDENCE INDICATOR TAILORING

Work products (e.g. documents, models, and drawings) are often domain, organization, or project specific. They will also be unique to the processes applied to create them.

As a result, our stated "evidence indicators" are examples only. Organizations are free to modify the terminology in evidence indicators to reflect their own organizational definitions.

ASSESSMENT APPROACH TAILORING

Competence assessment can be implemented with any one of three approaches (or combinations of these). These are analyzed in SECAG Table 1. The pros and cons of each basic approach need to be considered when determining the deployment approach for an organization.

ATLAS 1.1 PROFICIENCY ASSESSMENT

The Atlas 1.1 technical report (Hutchison et al. 2018), Section 5.3 Proficiency Assessments, describes another method to assess proficiency. The Atlas model assesses proficiency across five levels: Fundamental Awareness, Novice, Intermediate, Advanced, and Expert. The Atlas model uses radar maps to illustrate the evolving proficiency level profiles of individuals against the six areas. Using organizations could also use radar maps to assess the proficiency levels of individuals against the competency areas identified in the ATLAS model, or against a subset of the 37 competencies described in the INCOSE SECF.

SECAG TABLE 1 Comparison between competency assessment régimes

Method	Description	Benefits	Disadvantages
Self-assessment	Individuals are provided with a formal description of each competence area and competence level indicator by the organization, and independently determine what they believe their competence levels to be.	• Reasonably cost-effective. • Effort of assessment is split across the organization. • Requires little support from the organization. • Individuals can include experience which may not be. Well-known within their current organization. • Can be used by individuals in organizations for career planning where Systems Engineering is not well established.	• Individuals may not understand the full scope of the competence, leading to potential overstating of their own competence. • Equally individuals may also understate their competence, due to lack of self-confidence in competency areas. • Organizational consistency is hard as individuals may self-assess to different standards. • Individuals reluctant to accept Systems Engineering may overscore themselves to demonstrate they do not need development in this area.

SECAG TABLE 1 (Continued)

Method	Description	Benefits	Disadvantages
Manager assessment	Managers are provided with a formal description of each competence area and competence level indicator, and independently determine the individual competence levels of staff members for the purpose of training or job assignments. This could be with or without interview of the individual concerned.	• Cost-effective. • If managers understand the competencies and know their staff well, assessment can be a quick process. • If managers use an interview-based technique, this can be very accurate. • The consistency of organizational assessment can be good, if managers are prepared well. • Can be aligned well to organizational strategies – implemented through managers.	• May be a burden to managers with large numbers of staff, or if the manager formally interviews individuals. • If managers do not understand the full scope of the competencies, errors in ratings can occur (e.g. if a manager is not an expert in Systems Engineering). • If managers do not know their staff well, assessment can be erroneous. • Individuals may feel uncomfortable admitting a lack of competence to their manager or may feel stressed at the idea of assessment, which may influence accuracy. • Managers may exhibit bias for/against an individual influencing outcome. Managers may not be aware of experience gained by individuals before they worked for the manager.
Independent assessment	Independent trained assessors (from inside or outside the organization) formally interview individual staff members to assess their competence. This is commonly deployed with two assessors to provide consistency and results analysis but can be achieved with just one or indeed three.	• The use of trained assessors ensures candidates are put at ease, helping to ensure complete and honest responses. • The use of trained assessors ensures an accurate reflection of the scope of competence areas and indicators in the framework. • There is unlikely to be any subjective bias from knowing the history or circumstances of a candidate. The assessment is fact-based. • Two or more assessors can further ensure consistency against the defined standards than one alone.	• Can be quite expensive, especially if assessors are formally trained internally as part of the initiative to ensure their full understanding of the framework. • Administration required to set up interviews can be time-consuming. • Individuals may feel uncomfortable admitting a lack of competence if they feel they are being "judged" or may feel stressed at the idea of independent assessment, both of which may influence accuracy.

USING THE ASSESSMENT GUIDE

While the INCOSE Systems Engineering Competency Framework defines an overarching framework set of "requirements" for 37 competencies at each of the five levels of competence, this assessment guide completes that definition by defining a method of "verification" of the competency indicators listed in the SECF and examples of documents and other items which could be produced as supporting evidence. These elements will be tailored as required within an organization and as described in the tailoring section of this document.

This section provides additional guidance supporting the deployment and application of the framework and assessment guide. This guidance is based upon experience from real applications in INCOSE organizations. These examples implement particular competency framework use cases identified in the framework.

ASSESSING THE ASSESSORS

Successful competence assessment relies upon those carrying out the assessment ("assessors") being competent themselves. How can this be achieved within an organization "new" to the discipline of Systems Engineering? Here are some general points to consider.

- Assessors should be trained in how to perform a competency-based interview. Competency-based assessment is not the same as merely asking questions about statements made in a résumé provided by the interviewee. Competency-based interviews search for evidence, not just for affirmations by the candidate. Put simply, one of the key skills of this form of assessment is to examine evidence of competency by asking "open" questions. (An open question is one which does not immediately elicit the response (" Yes" or "No"). Follow-up questions can probe further detail as necessary.
- Assessors should be selected for their ability to conduct the assessment in an open and constructive manner. This helps candidates to relax and should help ensure their responses are open and honest as well.
- Lack of a full understanding of the full scope and meaning of the competence and its associated competence indicators will undermine the assessment approach. Even a well-written competence framework can only help so far; it is necessary to understand, interpret, and contextualize the framework for an organizational need.
  - Assessors may be highly regarded experts within the organization's business domain, but this does not guarantee they are experts in Systems Engineering.

- There may be merit in seeking out training in the scope of the SECF from Systems Engineering experts, to ensure that each competence area is properly understood by both assessors.
- There may be merit in selecting assessors based upon formal Systems Engineering credentials. For example, INCOSE offers Certification at three levels. Ensuring that assessors have attained "Practitioner" or "Expert" level accreditation (i.e. "CSEP," "ESEP" levels) within the INCOSE Systems Engineering Professional (SEP) program (INCOSE 2018b) would help ensure their knowledge and understanding of the full scope of Systems Engineering as defined within the framework was adequate to interpret the framework correctly.

FRAMEWORK AND ASSESSMENT USE CASE EXAMPLES

A set of SECF and SECAG application examples are included in SECAG Annex B.

EXPLANATION OF ASSESSMENT GUIDE TABLES

Each of the 37 competencies identified in the INCOSE SECF has five tabular sections in the SECAG, one for each of the defined proficiency levels (i.e. Awareness, Supervised Practitioner, Practitioner, Lead Practitioner, and Expert).

The general layout and interpretation of each competency area is similar and is defined in SECAG Table 2.

SECAG TABLE 2 Assessment guide table structure

Assessment table item	Where found	Explanation
Competency Area	All competence area headers, all levels	The name of the competency area as taken from the INCOSE Systems Engineering Competency Framework main body.
Description	As above	Summary of the scope of the competency area, as taken from the INCOSE Systems Engineering Competency Framework main body.
Why it matters	As above	Rationale for defining the competency area, as taken from the INCOSE Systems Engineering Competency Framework main body.
Possible contributory types of evidence	As above	This introduces the nature of information that might be useful to evidence competence. Evidence examples are provided in detail within each section against the indicators listed. Evidence will be organization and domain driven and therefore needs tailoring. Examples provided in this framework are not intended to be definitive or exhaustive, merely illustrative. It is the responsibility of the organization applying the framework and/or the assessors performing an assessment to determine the validity of any evidence items provided in support of competence. Further detail is provided later in this table.

SECAG TABLE 2 (Continued)

Assessment table item	Where found	Explanation
Learning and Development	As above	This section is currently a placeholder to accommodate potential future expansion. A general reference is made to the "INCOSE Professional Development Portal (PDP)." The INCOSE PDP is a web-based facility currently under development within INCOSE. It is expected that in due course the PDP will provide a validated source of personal development guidance covering both how to gain an initial awareness of a competency area and options for developing competence thereafter.
ID	All competence area tables, all levels	A sequential identifier designed to aid in referencing indicators within a competence area.
Indicators of Competence	As above	This is the text of the competence indicators defined in the INCOSE Systems Engineering Competency Framework (SECF). Competency indicators are high-level statements each summarizing one particular aspect of a competency at a particular level of proficiency. *For information only, the structure of the identifier is as follows:* *[Group][Competency name][Proficiency]* *Where:* *Group = A Competency Group identifier (e.g. Core (C), Professional (P), Technical (T), Management (M) or Integrating (I))* *Competency name = A unique two alphanumeric competency area identifier (e.g. Systems Thinking = "ST")* *Proficiency = A level identifier (e.g. Awareness (A), Supervised Practitioner (S) etc.)* *Indicator = A two-digit identifier for a specific framework indicator in the SECF (e.g. 01, 02, 03. etc.)*
Sub ID	As above	A sequential identifier for each of the competency sub-identifiers (i.e. (K), (A), (P) sub-indicators) within a competency indicator. Sub-indicators (A), (K), and (P) are explained in detail elsewhere in this document.
Ref ID	As above	A reference used internally by INCOSE. However, this ID can also be used as a location reference when making comments on the document. *For information only, the structure of the identifier is as follows:* *[Group][Competency name][Proficiency]-[Indicator][Assessment table line item]* *Where:* *Group = A Competency Group identifier (e.g. Core (C), Professional (P), Technical (T), Management (M), or Integrating (I))* *Competency name = A unique two alphanumeric competency area identifier (e.g. Systems Thinking = "ST")* *Proficiency = A level identifier (e.g. Awareness (A), Supervised Practitioner (S), etc.)* *Indicator = A two-digit identifier for a specific framework indicator in the SECF (e.g. 01, 02, 03, etc.)* *Assessment table line item = A unique two-digit reference identifier for a line item within the current table (e.g. 10, 20, 30, etc.)*

(Continued)

SECAG TABLE 2 (Continued)

Assessment table item	Where found	Explanation
Relevant knowledge sub-indicators	All "Awareness Level" tables	This column addresses items which form the evidence of competence at this level. In the INCOSE SE Competency Framework at Awareness level, competence is demonstrated solely by "knowledge." Because the competency indicators identified in the INCOSE SE Competency Framework represent distinct combinations of knowledge, experience, and behavior, evidence of competence at all proficiency levels has been subdivided into a demonstration of knowledge (K), evidence of required activities performed (A), or required professional behaviors displayed (P). Since "Awareness" level is limited to knowledge, all indicators at this level are marked with "(K)" and represent items of knowledge that are expected to be known at this level. We have tried to make the definitions of knowledge agnostic to domains. However, in different organizations, aspects of knowledge will be different (e.g. naming conventions for documents, review names, or domain ideas) but the underlying knowledge scope will be similar.
Relevant knowledge, experience, and/or behaviors	All tables *excluding* "Awareness" tables, i.e. tables from "Supervised Practitioner" to "Expert" levels	This column addresses items which characterize competence at this level in detail. This column attempts to break down the higher-level "Indicators of Competence" into a set of constituent Sub-Indicators which are themselves somewhat more verifiable. (Note, some sub-indicators are identical to their indicators.) The competency indicators identified in the INCOSE SE Competency Framework represent combinations of knowledge, experience, and behavior, and so evidence of competence at each proficiency level is subdivided into more verifiable demonstrations of knowledge (K), activities performed (A), or professional behaviors displayed (P) through a set of sub-indicators. Sub-indicators (A), (K), and (P) are explained in detail elsewhere in this document. We have tried to make the definitions of all items agnostic to domains. However, in different organizations, knowledge, activities, and occasionally behaviors will be different, but we believe the underlying scope will be broadly similar.
Possible examples of objective evidence of personal contribution to activities performed, or professional behaviors applied	All tables excluding "Awareness" tables, i.e. from "Supervised Practitioner" to "Expert" levels	This column is designed to provide some representative examples of possible objective evidence that might be used to formally verify an individual has the required competency indicator. Other examples may well exist both generally and specifically within an organization. Evidence items may be physical (e.g. documents and drawings) but could equally be digital artifacts (e.g. contained within software applications and software models). Ideally, these will be accessible to those assessing competence, in order to confirm its validity and content. However, the extent to which any proof is required will be determined by the nature or purpose of the assessment. Determining appropriate evidence items and access requirements for assessment forms part of the tailoring activity at this level. *NOTE: This column represents example evidence at the INDICATOR level, not SUB-INDICATOR level. There is no implied correspondence between the knowledge, although clearly some example evidence items may relate better to one sub-indicator rather than another.*

ASSESSMENT GUIDE LANGUAGE USAGE

Verbs used in the SECAG to define indicators and sub-indicators (e.g. "*Describes* . . ." or "(A) *Identifies*. . ." at sublevel) have been standardized in order to assist in the understanding of the intent of each of the indicators and sub-indicators. The nature of evidence expected in order to demonstrate competence for an indicator will differ depending on whether the indicator relates to concerns knowledge, activities, professional behaviors, and the proficiency level at which the verb is being applied. SECAG Table 3 explains the meaning of the standardized set of verbs used.

SECAG TABLE 3 Indicator language definitions

Verb	Definition	Additional notes on interpretation
Acts	Behaves in the way specified.	When referenced as a Professional skill or behavior . . . Written or oral commentary should demonstrate the identified behavior, how this was performed, the challenges involved and the skills used, coupled with any supporting knowledge used, thinking applied or activities performed to achieve a successful outcome.
Adapts	Make (something) suitable for a new use, or new conditions.	When referenced as an activity . . . Written or oral commentary should identify the knowledge used, thinking applied or activities performed for the identified adaptation. When referenced as a Professional skill and behavior . . . Written or oral commentary should demonstrate what needed to be adapted, how this was performed, the challenges involved and the skills used, coupled with any supporting knowledge used, thinking applied or activities performed to achieve a successful outcome.
Advises	Offers suggestions about the best course of action.	When referenced as an activity . . . Written or oral commentary should identify the knowledge used, thinking applied or activities performed for the identified adaptation. When referenced as a Professional skill and behavior . . . Written or oral commentary should demonstrate the advice required, how this was performed, the challenges involved and the skills used, coupled with any supporting knowledge used, thinking applied or activities performed to achieve a successful outcome.
Analyzes	Examines (something) methodically and in detail, typically in order to explain and interpret it.	When referenced as an activity . . . Written or oral commentary should identify the knowledge used, thinking applied, or activities performed in order to complete the identified analysis.
Arbitrates	Reaches an authoritative judgement or settlement in a dispute or decision.	When referenced as a Professional skill and behavior . . . Written or oral commentary should demonstrate what needed to be arbitrated, how this was performed, the challenges involved and the skills used, coupled with any supporting knowledge used, thinking applied, or activities performed to achieve a successful outcome.
Assesses	Evaluates the nature, ability, or quality of something (or someone).	When referenced as an activity . . . Written or oral commentary should identify the knowledge used, thinking applied, or activities performed to complete the identified assessment.

(*Continued*)

SECAG TABLE 3 (Continued)

Verb	Definition	Additional notes on interpretation
Authorizes	Gives official (e.g. project or organizational) permission for or approval to.	When referenced as an activity . . . Written or oral commentary should identify the knowledge used, thinking applied, or activities performed to complete the identified authorization.
Challenges	Dispute the truth or validity of.	When referenced as a Professional skill and behavior . . . Written or oral commentary should demonstrate why the challenge was needed, how this was performed, the challenges involved and the skills used, coupled with any supporting knowledge used, thinking applied, or activities performed to achieve a successful outcome.
Champions	Vigorously supports or defends the cause of (a cause, idea, technique, tool, method, etc.).	When referenced as a Professional skill and behavior . . . Written or oral commentary should demonstrate why the identified support was needed, how this was performed, the challenges involved and the skills used, coupled with any supporting knowledge used, thinking applied, or activities performed to achieve a successful outcome.
Coaches	Trains or instructs (a team member or individual) using a set of defined professional skills in this area.	When referenced as a Professional skill and behavior . . . Written or oral commentary should demonstrate why the identified instruction was needed, how this was performed, the challenges involved and the professional skills used, coupled with any supporting knowledge used, thinking applied. or activities performed to achieve a successful outcome.
Collaborates	Works jointly on an activity or project.	When referenced as a Professional skill and behavior . . . Written or oral commentary should demonstrate why the identified collaboration activity was needed how this was performed, the challenges involved and the skills used, coupled with any supporting knowledge used, thinking applied. or activities performed to achieve a successful outcome.
Collates	Collects and combines (texts, information, or data).	When referenced as an activity . . . Written or oral commentary should identify the knowledge used, thinking applied, or activities performed to collect or combine the identified information.
Communicates	Shares or exchanges information, news, or ideas.	When referenced as an activity . . . Written or oral commentary should identify the knowledge used, thinking applied, or activities performed to achieve the identified communication result. When referenced as a Professional skill and behavior . . . Written or oral commentary should demonstrate what needed to be communicated, how this was performed, the challenges involved and the skills used, coupled with any supporting knowledge used, thinking applied or activities performed to achieve a successful outcome.

SECAG TABLE 3 (Continued)

Verb	Definition	Additional notes on interpretation
Compares	Notes or measures the similarity or dissimilarity between two or more items.	When referenced as an activity . . . Written or oral commentary should identify the knowledge used, thinking applied, or activities performed to compare the identified information.
Complies	Acts in accordance with certain specified standard or guidance.	When referenced as an activity . . . Written or oral commentary should identify the knowledge used, thinking applied, or activities performed to comply with the identified guidance or standard.
Conducts	Organizes, and carries out.	When referenced as an activity . . . Written or oral commentary should identify the knowledge used, thinking applied, or activities performed to conduct the identified activity.
Coordinates	Regulates, adjusts, or combines the actions of others to attain coherence or harmony.	When referenced as an activity . . . Written or oral commentary should identify the knowledge used, thinking applied, or activities performed to coordinate the identified activity.
Creates	Brings something into existence, causes something to happen as a result of one's actions.	When referenced as an activity . . . Written or oral commentary should identify the knowledge used, thinking applied, or activities performed to create the identified work product(s).
Defines	Characterizes the essential qualities or meaning of an item.	When referenced as a knowledge item . . . Written or oral commentary which demonstrate that the essential qualities or scope of the referenced item is known and comprehended.
Describes	Gives an account in words, including all the relevant characteristics, qualities, or events.	When referenced as a knowledge item . . . Written or oral commentary which demonstrate that the essential all the relevant characteristics, qualities, or events associated with the referenced item are known and comprehended.
Determines	Ascertains or establishes specific values of an item by research or through calculation.	When referenced as an activity . . . Written or oral commentary should identify the knowledge used, thinking applied, or activities performed to establish or ascertain the identified item.
Develops	Grows or causes something to grow and become more mature, advanced, or elaborate.	When referenced as an activity . . . Written or oral commentary should identify the knowledge used, thinking applied, or activities performed to advance the maturity of the identified item.
Elicits	Evokes or draws out (a need or requirement, reaction, answer, or fact) from someone.	When referenced as an activity . . . Written or oral commentary should identify the knowledge used, thinking applied, or activities performed to draw out the identified information (e.g. requirement or need).

(Continued)

SECAG TABLE 3 (Continued)

Verb	Definition	Additional notes on interpretation
Ensures	Makes certain that (something) will occur or be the case,	When referenced as an activity . . . Written or oral commentary should identify the knowledge used, thinking applied, or activities performed to make certain that the identified item will be correct.
Evaluates	Makes a judgement as a result of informed analysis.	When referenced as an activity . . . Written or oral commentary should identify the knowledge used, thinking applied, or activities performed to judge the identified item.
Explains	Makes an idea, situation, or problem clear to someone by describing it in more detail or revealing relevant facts.	When referenced as a knowledge item . . . Written or oral commentary which provide additional detail and facts regarding the identified item or situation and display an understanding and comprehension of its scope or complexity.
Follows	Acts or adheres to instructions as ordered or required,	When referenced as an activity . . . Written or oral commentary should identify the knowledge used, thinking applied, or activities performed to follow the identified instructions.
Fosters	encourages the development of (something, especially something desirable),	When referenced as a Professional skill and behavior . . . Written or oral commentary should identify the individual or item requiring development encouragement, how this was performed, the challenges involved and the skills used, coupled with any supporting knowledge used, thinking applied, or activities performed to achieve a successful outcome.
Guides	Directs or influences the behavior or development of someone,	When referenced as a Professional skill and behavior . . . Written or oral commentary should identify the nature of the individuals requiring guidance or development, how guidance or development was performed, the challenges involved and the skills used, coupled with any supporting knowledge used, thinking applied, or activities performed to achieve a successful outcome.
Identifies	Establishes the identity or complete scope of an item,	When referenced as a knowledge item . . . Written or oral commentary which provide additional detail and facts regarding the referenced item or situation and display an understanding and comprehension of its scope or complexity. When referenced as an activity . . . Written or oral commentary should identify the knowledge used, thinking applied, or activities performed to establish the identity and scope of the referenced item.
Influences	Has an effect on someone or something.	*NOTE: "Influences" is only used within the main SECF framework. In the SECAG, this term is decomposed into associated activities and behaviors.*
Judges	Forms an opinion or conclusion about.	When referenced as an activity . . . Written or oral commentary should identify the knowledge used, thinking applied, or activities performed to establish the identity and scope of the referenced item.

SECAG TABLE 3 (Continued)

Verb	Definition	Additional notes on interpretation
Liaises	Acts as a link to assist communication between (people or groups).	When referenced as a Professional skill and behavior . . . Written or oral commentary should identify the nature of the individuals requiring liaison for the identified activity, how liaison was performed, the challenges involved and the skills used, coupled with any supporting knowledge used, thinking applied, or activities performed to achieve a successful outcome.
Lists	Names a set of items that form a meaningful grouping.	When referenced as a knowledge item . . . Written or oral commentary which names the full group in the referenced item, or if the group is large, its most common or key constituent members.
Maintains	Keeps in an existing state.	When referenced as an activity . . . Written or oral commentary should identify the knowledge used, thinking applied, or activities performed to keep the referenced item in its required state.
Monitors	Observes and checks the progress or quality of (something) over a period of time; keeps under systematic review.	When referenced as an activity . . . Written or oral commentary should identify the knowledge used, thinking applied, or activities performed to observe and check the identified item(s) over time.
Negotiates	Confers or discusses with others with a view to reaching agreement.	When referenced as a Professional skill and behavior . . . Written or oral commentary should identify the nature of the agreement required, the nature of the individuals involved, how negotiation was performed, the challenges involved and the skills used, coupled with any supporting knowledge used, thinking applied, or activities performed to achieve a successful outcome.
Performs	Carries out, accomplishes, or fulfills (an action, task, or function).	When referenced as an activity . . . Written or oral commentary should identify the knowledge used, thinking applied or activities, tasks, or functions carried out.
Persuades	Causes someone to do something through reasoning or argument.	When referenced as a Professional skill and behavior . . . Written or oral commentary should identify the nature of the decision, or change in opinion required, the nature of the individuals concerned and arguments used, the challenges involved and the skills used, coupled with any supporting knowledge used, thinking applied, or activities performed to achieve a successful outcome.
Prepares	Makes (something) ready for use or consideration.	When referenced as an activity . . . Written or oral commentary should identify the knowledge used, thinking applied, or activities performed to make the referenced item ready for use or consideration.
Produces	Causes something to come into existence, by intellectual effort.	When referenced as an activity . . . Written or oral commentary should identify the knowledge used, thinking applied. or activities performed to cause the referenced item to come into existence.

(Continued)

SECAG TABLE 3 (Continued)

Verb	Definition	Additional notes on interpretation
Promotes	Supports or actively encourages further the progress of (a cause, idea, technique, tool, method, etc.).	*NOTE: "Promotes" is only used within the main SECF framework. In the SECAG, this term is decomposed into associated activities and behaviors.*
Proposes	Puts forward (an idea or plan) for consideration or discussion by others.	When referenced as an activity . . . Written or oral commentary should identify the knowledge used, thinking applied, or activities performed to put forward the referenced item(s).
Provides	Makes something available for use; supplies.	When referenced as an activity . . . Written or oral commentary should identify the knowledge used, thinking applied, or activities performed to make available the referenced item(s).
Reacts	Acts in response to something; responds in a particular way.	When referenced as a Professional skill and behavior . . . Written or oral commentary should identify the nature of the situation requiring the identified response, the nature of the individuals involved, the challenges and skills used, coupled with any supporting knowledge, and thinking applied or activities performed to achieve a successful outcome.
Recognizes	Acknowledges or takes notice of in some definite way.	When referenced as a Professional skill and behavior . . . Written or oral commentary should identify the references item requiring acknowledgement, the nature of individuals involved, the challenges and skills used, coupled with any supporting knowledge, and thinking applied or activities performed to achieve a successful outcome.
Records	Registers or sets down in writing.	When referenced as an activity . . . Written or oral commentary should identify the knowledge used, thinking applied. or activities performed to register or set the referenced item in writing.
Reviews	Formally assesses a topic or issue, proposing any necessary or desirable changes.	When referenced as an activity . . . Written or oral commentary should identify the knowledge used, thinking applied, or activities performed to examine the referenced item.
Selects	Carefully chooses as being the best or most suitable.	When referenced as an activity . . . Written or oral commentary should identify the knowledge used, thinking applied, or activities performed to choose the referenced item.
Trains	Teaches a particular skill or type of behavior through sustained practice or instruction.	When referenced as an activity . . . Written or oral commentary should identify the knowledge used, thinking applied, or activities performed to ensure that the referenced skill or behavior has been successfully passed on the recipients.

SECAG TABLE 3 (Continued)

Verb	Definition	Additional notes on interpretation
Uses	Employs something as a means of accomplishing a purpose or achieving a result.	When referenced as an activity . . . Written or oral commentary should identify the referenced knowledge used, thinking applied, or activities performed to accomplish the referenced result. When referenced as a Professional skill and behavior . . . Written or oral commentary should identify the referenced outcome to be accomplished, the nature of individuals involved, the challenges and skills used, coupled with any supporting knowledge, and thinking applied or activities performed to achieve a successful outcome.

SUB-INDICATOR CLASSIFICATIONS ("K", "A," AND "P")

Evidence indicators defined in the SECF can often be complex and thus hard to assess. In order to help assessors in their understanding this complexity, competency indicators defined in the framework have been broken down into combinations of one or more "sub-indicators" which will be classified as of types:

- "K" – indicating a competence sub-indicator defining an item of knowledge which is required in order to demonstrate the competency indicator.
- "A" – indicating a competence sub-indicator defining an activity to be performed (or "ability" in order to demonstrate the competency indicator.
- "P" – indicating a competence sub-indicator defining a professional attitude or behavior to be demonstrated in order to demonstrate the competency indicator.

Most competencies are a combination of several sub-indicators.

A review of the competency assessment tables in the guide will show that competence indicators at the "Awareness" proficiency level are always defined as knowledge only (type "K"), in order to reflect that fact that the "Awareness" proficiency level is aligned to the acquisition of knowledge but no experience.

Above "Awareness" proficiency level, additional knowledge *may* be required (type "K"); however, in general, higher-level proficiency levels focus mostly on demonstrable experience of performing activities often with demonstrated behaviors.

ASSESSING EVIDENCE SUB-INDICATOR TYPES ("K", "A," AND "P")

There are many ways in which the possession of an aspect of competence can be assessed. Examples of how each of the three types identified in this assessment guide might be assessed are listed in SECAG Table 4. Organizational tailoring guidance for assessors should define the precise methods used within any organization.

SECAG TABLE 4 Assessment mechanisms for the different indicator types

Indicator type	Possible assessment mechanisms
"Knowledge" (K)	• Responses to "open" questions ("Tell me about topic xxx . . .", "List items which . . .") containing specific words or phrases. • Examination (e.g. INCOSE Systems Engineering Professional [SEP] knowledge examination – although this currently does not address all areas in the Competency Framework). • Validation of certificates from activities assessing relevant knowledge (e.g. prior assessment(s), accredited or internally approved course completion certificates). • Evidence produced indirectly as part of a response to the assessment of another competence area.
"Activity" or "Ability" (A)	• Validation of physical evidence items (e.g. documents, drawings, reports, and minutes). • Validation of digital artifacts evidence items (e.g. data held within software applications such as tools, models. or repositories). • Attestations by approved or independent individuals regarding their contribution, signature pages, management records, etc. • Responses to "open" questions ("Can you give me an example of when you performed activity xxx . . .") and their own contribution and impact on the identified activities, challenges overcome, and learning gained. • Evidence produced indirectly as part of a response to the assessment of another competence area.
"Professional Behavior" (P)	• Validation of physical evidence items (e.g. documents, drawings, reports, and minutes). • Validation of digital artifacts evidence items (e.g. items held within software applications such as data repositories). • Attestations by approved or independent individuals regarding their behavior or responses to certain scenarios (e.g. human resources, customers, and suppliers). • Reponses to "open" questions ("Tell me about a situation where you had to act/respond to someone else's actions . . .") regarding demonstration of specific behaviors, behavioral responses, challenges, and learning gained. • Evidence produced indirectly as part of a response to the assessment of another competence area.

A key point about all evidence produced as part of an assessment is that it should reflect an individual's PERSONAL contribution to a system development rather than just being a document produced by others on a project.

Individuals who only originated a small part of the document or digital artifact, produced a complete document or digital artifact with little "help," reviewed a document or digital artifact originated wholly by others, or who have merely "overseen" the origination and review of a document or digital artifact without any direct technical contribution to its technical content should all highlight this aspect since it affects the experience they will have gained in that document's production.

SECAG ACRONYMS AND ABBREVIATIONS

SECAG Table 5 summarizes the acronyms and abbreviations used in the SECAG.

SECAG TABLE 5 Acronyms and abbreviations

ACRONYM	MEANING
8D	"8D" is a technique for problem diagnosis and resolution
A3	A term used for a particular system architecture representation
BKCASE	Body of Knowledge and Curriculum to Advance Systems Engineering
CCB	Change Control Board
CM	Configuration Management
CMM	Capability Maturity Model
COTS	Commercial Off the Shelf
CPD	Continued (Continuous) Professional Development
CPI	Cost Performance Index
DMSMS	Diminishing Manufacturing Sources and Material Shortages
DoD	Department of Defense
DoDAF	Department of Defense Architectural Framework
ECP	Engineering Change Proposal
ECR	Engineering Change Request
ESEP	Expert Systems Engineering Professional

(*Continued*)

Systems Engineering Competency Assessment Guide: A combined INCOSE Systems Engineering Competency Framework (SECF) and associated Systems Engineering Competency Assessment Guide (SECAG) document, First Edition. INCOSE.

SECAG TABLE 5 (Continued)

ACRONYM	MEANING
FMEA	Failure Mode and Effects Analysis
FMECA	Failure Modes, Effects, and Criticality Analysis
FTA	Fault Tree Analysis
GROW	Goal, Reality, Options, What-next
HCI	Human Computer Interface
HR	Human Resources
ICD	Interface Control Document
ID	Identifier
IDD	Interface Definition (or Description) Document
IEEE	Institute of Electrical and Electronics Engineers
IM	Information Management
INCOSE	International Council on Systems Engineering
ISO	International Organization for Standardization
ISO/IEC	International Organization for Standardization/International Electrotechnical Commission
ISR	In-Service Review
IT	Information Technology
LRU	Line Replaceable Unit
MBSE	Model-Based Systems Engineering
MIL-STD	Military Standard
MoDAF	Ministry of Defence Architectural Framework (UK)
OBS	Organizational Breakdown Structure
OEM	Original Equipment Manufacturer
OODA	A Quality management approach (Observe, Orient, Decide, and Act)
PDCA	A Quality Management approach (Plan, Do, Check, Act)
PDR	Preliminary Design Review
PHS&T	Packaging Handling Storage and Transportation
PMP	Project (or Program) Management Plan
QFD	Quality Function Deployment
QMP	Quality Management Plan
RMP	Risk Management Plan
RPR	Rapid Problem Resolution
SE	Systems Engineering
SEP	Systems Engineering Professional

SECAG TABLE 5 (Continued)

ACRONYM	MEANING
SEBOK	Systems Engineering Body of Knowledge
SECF	Systems Engineering Competency Framework
SECAG	Systems Engineering Competency Assessment Guide
SEIT	Systems Engineering Integration Team
SEMP	Systems Engineering Management Plan
SFIA	Skills Framework for the Information Age
SFR	System (or Solution) Functional Review
SME	Subject Matter Expert
SPI	Schedule Performance Index
STEM	Science, Technology, Engineering, and Mathematics
SysML	Systems Modeling Language
TCF	Tailored Competency Framework
TLCC	Total Life Cycle Cost
TOGAF	The Open Group Architectural Framework
TQM	Total Quality Management
TRIZ	A problem-solving, analysis and forecasting tool
TRR	Test Readiness Review
UK	United Kingdom
US	United States
WBS	Work Breakdown Structure

SECAG GLOSSARY

This glossary defines words/phrases in the context of use within the Systems Engineering Competency Assessment Guide (SECAG). Several sources have been provided in some cases to aid explanation.

Where multiple potential definitions exist, the glossary definition is based upon definitions from the following sources, in the priority listed below:

1. *INCOSE Systems Engineering Handbook* (Fourth Edition) (INCOSE 2015).
2. INCOSE Systems Engineering Body of Knowledge (SEBoK) (BKCASE Project 2017).
3. INCOSE UK Ltd. Systems Engineering Competencies Framework (Issue 3) (INCOSE UK 2010).
4. Other well-established internationally recognized sources.
5. INCOSE Competency Working Group.

Glossary Term	Definition
Ability	A term used in human resource management denoting an acquired or natural capacity or talent that enables an individual to perform a particular task successfully (BusinessDictionary.com 2018).
Architecture	(System) fundamental concepts or properties of a system in its environment embodied in its elements, relationships, and in the principles of its design and evolution (INCOSE 2015).
Architecting	Process of conceiving, defining, expressing, documenting, communicating, certifying proper implementation of, maintaining, and improving an architecture throughout a system's life cycle (ISO/IEC/IEEE 2011). The architecting process sometimes involves the use of heuristics to establish the form of architectural options before quantitative analyses can be applied. Heuristics are design principles learned from experience (Rechtin 1990).
Authored	Wrote the document or work product (i.e. did not just sign the front page) (INCOSE UK 2010).
Behavior	The way in which one acts or conducts oneself, especially towards others (Oxford Dictionaries 2018).

(*Continued*)

Systems Engineering Competency Assessment Guide: A combined INCOSE Systems Engineering Competency Framework (SECF) and associated Systems Engineering Competency Assessment Guide (SECAG) document, First Edition. INCOSE.

Glossary Term	Definition
Best practice	A procedure that has been shown by research and experience to produce optimal results and that is established or proposed as a standard suitable for widespread adoption (Merriam-Webster 2018).
Capability	An expression of a system, product, function, or process ability to achieve a specific objective under stated conditions (INCOSE 2015). The ability to achieve a desired effect under specified (performance) standards and conditions through combinations of ways and means (activities and resources) to perform a set of activities (DoD 2009).
Coaching	Helping, supporting, advising, explaining, demonstrating, instructing, and directing others resulting in transfer of knowledge and skills (INCOSE 2015).
Competence	The measure of specified ability (INCOSE 2015).
Competence-based assessment	An evidence-based activity, where an individual is independently assessed (or self-assesses) in one or more defined areas, to determine whether they are able to demonstrate that their knowledge, skills, and experience meet a defined or required level of proficiency ("competence level").
Competency	An observable, measurable set of skills, knowledge, abilities, behaviors, and other characteristics an individual needs to successfully perform work roles or occupational functions. Competencies are typically required at different levels of proficiency depending on the specific work role or occupational function. Competencies can help ensure individual and team performance aligns with the organization's mission and strategic direction (U.S. Office of Personnel Management (OPM) 2015). A measure of an individual's ability in terms of their knowledge, skills, and behavior to perform a given role (Holt and Perry 2011).
Competent	Having a specified level of competence (INCOSE UK 2010).
Complexity (of a system)	A measure of how difficult it is to understand how a system will behave or to predict the consequences of changing it (Sheard and Mostashari 2009). The degree to which a system's design or code is difficult to understand because of numerous components or relationships among components (ISO/IEC 2009). In a complex system, the exact relationship between elements is either unknown and possibly unknowable.
Complicated system	A system comprising many interacting elements where the overall behavior of the system is both knowable and deterministic (cf. complex).
Configuration Management	Configuration management is the discipline of identifying and formalizing the functional and physical characteristics of a configuration item at discrete points in the product evolution for the purpose of maintaining the integrity of the product system and controlling changes to the baseline (INCOSE 2017).
Discipline	Area of expertise, e.g. systems, software, hardware, program management, quality assurance, etc. (INCOSE UK 2010).
Domain	A "problem space" (IEEE 2010).
Element	The level below the system of interest (INCOSE 2015). A system that is part of a larger system (Merriam-Webster 2018).
Enterprise	One or more organizations sharing a definite mission, goals, and objectives to offer an output such as a product or service (ISO 2000). An organization (or cross-organizational entity) supporting a defined business scope and mission that includes interdependent resources (people, organizations, and technologies) that must coordinate their functions and share information in support of a common mission (or set of related missions) (CIO Council 1999). The term enterprise can be defined in one of two ways. The first is when the entity being considered is tightly bounded and directed by a single executive function. The second is when organizational boundaries are less well defined and where there may be multiple owners in terms of direction of the resources being employed. The common factor is that both entities exist to achieve specified outcomes (MOD 2004). A complex, (adaptive) socio-technical system that comprises interdependent resources of people, processes, information, and technology that must interact with each other and their environment in support of a common mission (Giachetti 2010).

Glossary Term	Definition
Enterprise Asset	(As applied to a person) known by reputation to be a leader in the field, highly valued, highly regarded. Recognized by the community outside employer organization (e.g. asked to be on conference panel, government advisory board, etc.) (INCOSE UK 2010).
Framework	A basic conceptional structure (as of ideas) (Merriam-Webster 2018). Broad overview, outline, or skeleton of interlinked items which supports a particular approach to a specific objective and serves as a guide that can be modified as required by adding or deleting items (BusinessDictionary.com 2018).
Information	Information is an item of data concerning something or someone. Information includes technical, project, organizational, agreement, and user information.
Interface	A point where two or more entities interact. Interactions may involve systems, system elements including their environment, organizations, disciplines, humans (users, operators, maintainers, developers, etc.), or some combination thereof.
Job	A job is a recognized organizational position, usually performed in exchange for payment. Historically, the terms "job" and "role" have been used interchangeably, more recently a distinction has appeared. A job comprises all, or parts, of one or more defined "roles" which govern organizational processes and activities. An individual may remain in the same "job" for a long period, but during this time will usually perform multiple roles. This topic is discussed in further detail in Annex C of this document.
Knowledge	(In the context of KSA) A body of information applied directly to the performance of a function. US Office of Personnel Management (Wikipedia 2017).
Mentoring	Mentoring is a relationship where a more experienced colleague shares their greater knowledge to support development of a less-experienced member of staff. It uses many of the techniques associated with coaching. One key distinction is that mentoring relationships tend to be longer term than coaching arrangements.
Organization	A group of people and facilities with an arrangement of responsibilities, authorities, and relationships (INCOSE 2015).
Professional Development	A structured approach to learning to help ensure competence to practice, taking in knowledge, skills, and practical experience. Continued Professional Development (CPD) can involve any relevant learning activity, whether formal and structured or informal and self-directed (INCOSE UK Chapter 2017).
Project	An endeavor with defined start and finish criteria undertaken to create a product or service in accordance with specified resources and requirements (INCOSE 2015).
Program	A group of related projects managed in a coordinated way to obtain benefits and control not available from managing them individually. Programs may include elements of related work outside of the scope of the discrete projects in the program (PMI 2008).
Recent	(In the context of competency assessment) Within the last five years (INCOSE UK 2010).
Role	A role is a recognized organizational position, usually performed in exchange for payment. Historically, the terms "job" and "role" have been used interchangeably, more recently a distinction has appeared. A job comprises all, or parts, of one or more defined "roles" which govern organizational processes and activities. An individual may remain in the same "job" for a long period, but during this time will usually perform multiple roles. This topic is discussed in further detail in Annex C of this document.
Skills	(In the context of KSA) An observable competence to perform a learned psychomotor act. US Office of Personnel Management (Wikipedia 2017).
Specialty Engineering	(Verb) Analysis of specific features of a system that requires special skills to identify requirements and assess their impact on the system life cycle (INCOSE 2015). (Noun) Specialty engineering is the collection of those narrow disciplines that are needed to engineer a complete system (Elowitz 2006).

(*Continued*)

Glossary Term	Definition
Stakeholder	A party having a right, share, claim, or vested interest in a system or in its possession of characteristics that meet that party's needs and expectations (INCOSE 2015).
Stage (life cycle Stage)	A period within the life cycle of an entity that relates to the state of its description or realization (ISO/IEC/IEEE 2015).
System	A system is an arrangement of parts or elements that together exhibit behavior or meaning that the individual constituents do not (INCOSE 2019).
Systems Engineering	Systems Engineering is a transdisciplinary and integrative approach to enable the successful realization, use, and retirement of engineered systems, using systems principles and concepts, and scientific, technological, and management methods (INCOSE 2019). *INCOSE uses the terms "engineering" and "engineered" in their widest sense: "the action of working artfully to bring something about." "Engineered systems" may be composed of any or all of people, products, services, information, processes, and natural elements.*
System Element	A member of a set of elements that constitutes a system. A system element is a discrete part of a system that can be implemented to fulfill specified requirements. A system element can be hardware, software, data, humans, processes (e.g. processes for providing service to users), procedures (e.g. operator instructions), facilities, materials, and naturally occurring entities (e.g. water, organisms, and minerals), or any combination (ISO/IEC/IEEE 2015). A system element can be a subsystem.
Sub System	See element
Super system	The level above the system of interest (INCOSE 2015). A system that is made up of systems (Merriam-Webster 2018).
System of Interest	The system whose life cycle is under consideration (INCOSE 2015).
System of Systems	A System of Interest whose system elements are themselves systems; typically, these entail large-scale interdisciplinary problems with multiple, heterogeneous, distributed systems (INCOSE 2015).
Tailoring	Tailoring a process adapts the process description for a particular end. For example, a project creates its defined process by tailoring the organization's set of standard processes to meet the objectives, constraints, and environment of the project. (Adapted from notes/discussion of "tailoring guide") (ISO/IEC/IEEE 2009).
Team	A group of individuals who work together to achieve a common goal.
Verification	Provision of objective evidence that the system or system element fulfills its specified requirements and characteristics (INCOSE 2015).
Validation	Provision of objective evidence that the system, when in use, fulfills its business or mission objectives and stakeholder requirements, achieving its intended use in its intended operational environment (INCOSE 2015).

SECAG BIBLIOGRAPHY

Beasley, R. (2013). The need to tailor competency models – with a use case from Rolls-Royce. *INCOSE International Symposium IS2013,* Philadelphia, PA (24–27 June 2013).

BKCASE Project (2017). SEBoK v1.9 section 5: enabling systems engineering. http://sebokwiki.org/wiki/Enabling_Systems_Engineering (accessed 30 November 2018).

Brackett, M.A., Rivers, S.E., and Salovey, P. (2011). Emotional intelligence: implications for personal, social, academic, and workplace success. *Social and Personality Psychology Compass* 5: 88–103. https://doi.org/10.1111/j.1751-9004.2010.00334.

BusinessDictionary.com (2018). Framework definition. http://www.businessdictionary.com/definition/framework.html (accessed April 2018).

Chartered Institute of Personnel Development (CIPD) (2021a). What are coaching and mentoring?. Coaching and Mentoring Factsheet. https://www.cipd.co.uk/knowledge/fundamentals/people/development/coaching-mentoring-factsheet (accessed 2018).

Chartered Institute of Personnel Development (CIPD) (2021b). Competence or competency? https://www.cipd.co.uk/knowledge/fundamentals/people/performance/competency-factsheet.

CIO Council (1999). *Federal Enterprise Architecture Framework (FEAF).* Washington, DC, USA: Chief Information Officer (CIO) Council.

CMMI® Institute (2010). *Capability Maturity Model Integration (CMMI®) V1.3.* CMMI® Institute.

Defense Acquisition University (2013). *SPRDE Competency Model.* Defense Acquisition University (DAU)/U.S. Department of Defense June 12.

DoD (2009). *DoD Architecture Framework (DoDAF), Version 2.0.* Washington, DC, USA: U.S. Department of Defense (DoD).

Electronic Industries Alliance (2002). *Systems Engineering Capability Model EIA731_1:2002.* EIA.

Elowitz, M. 2006. Specialty engineering as an element of systems engineering. Presentation to Enchantment Chapter, INCOSE, Albuquerque, NM, USA (9 August).

Gelosh, D., Heisey, M., Snoderly, J. and Nidiffer, K. (2017). Version 0.75 of the proposed INCOSE systems engineering competency framework. *Proceedings of the International Symposium 2017 of the International Council on Systems Engineering (INCOSE),* Adelaide, Australia (15–20 July 2017).

Giachetti, R.E. (2010). *Design of Enterprise Systems: Theory, Architecture, and Methods.* Boca Raton, FL, USA: CRC Press, Taylor and Francis Group.

Systems Engineering Competency Assessment Guide: A combined INCOSE Systems Engineering Competency Framework (SECF) and associated Systems Engineering Competency Assessment Guide (SECAG) document, First Edition. INCOSE.

GOV.UK (2017). Equality act 2010: guidance. https://www.gov.uk/guidance/equality-act-2010-guidance#age-discrimination (accessed 31 December 2017).

Hahn, H.A. and Whitcomb, C.A. (2017). Use cases for the INCOSE competencies framework. Unpublished.

Holt, J. and Perry, S. (2011). *A Pragmatic Guide to Competency Tools, Frameworks and Assessment*. BCS.

Hutchison, N., Verma, D., Burke, P. et al. (2018). Atlas 1.1: An Update to the Theory of Effective Systems Engineers. *Technical Report SERC-2018-TR-101A*. Report, Hoboken, NJ: Systems Engineering Research Center, Stevens Institute of Technology. http://www.dtic.mil/dtic/tr/fulltext/u2/1046509.pdf (accessed 2018).

IEEE (2010). *IEEE 1570-2010 Standards for Information Technology Standard*. New York, NY, USA: Institute of Electrical and Electronics Engineers.

INCOSE (2004). What is systems engineering? http://www.incose.org/practice/whatissystemseng.aspx (accessed April 2018).

INCOSE (2021). *Engineering Solutions for a Better World Vision 2035*. San Diego, CA: INCOSE.

INCOSE (2015). *Systems Engineering Handbook – A Guide for System Life Cycle Processes and Activities*, 4e. San Diego, CA: Wiley.

INCOSE (2017). SEBoK v1.9 systems engineering body of knowledge. http://www.sebokwiki.org/wiki/Guide_to_the_Systems_Engineering_Body_of_Knowledge_(SEBoK) (accessed 30 November).

INCOSE (2018a). INCOSE Systems Engineering Competency Framework. *INCOSE INCOSE-TP-2018-002-01.0*.

INCOSE (2018b). Certification. https://www.incose.org/certification (accessed 2018).

INCOSE (2019). INCOSE website SE and SE definitions. https://www.incose.org/about-systems-engineering/system-and-se-definition (accessed November 2021).

INCOSE UK Chapter (2017). U3: what is continuing professional development (CPD). https://incoseonline.org.uk/Documents/uGuides/dot107_U-guide_U3_WEB.pdf (accessed February 2018).

INCOSE UK (2010). *INCOSE UK SE Competency Framework/Guide*. INCOSE UK.

INCOSE, OMG (2018). MBSE wiki. http://www.omgwiki.org/MBSE/doku.php (accessed April 2018).

Institution of Engineering and Technology (2013). Competency framework for independent safety assessors (ISAs),. November 17. https://www.theiet.org/factfiles/isa/index.cfm (accessed 19 February 2018).

ISO 2000. ISO 15704. 2000 (2000). *Industrial Automation Systems – Requirements for Enterprise-Reference Architectures and Methodologies Standard*. Geneva, Switzerland: International Organization for Standardization (ISO).

ISO/IEC (2009). *ISO/IEC 24765. Systems and Software Engineering Vocabulary (SEVocab) Standard*. Geneva, Switzerland: International Organization for Standardization (ISO)/International Electrotechnical Commission (IEC) [database online].

ISO/IEC JTC 1/SC 7 Software and Systems Engineering Technical Committee (2015). *ISO/IEC/IEEE 15288:2015 Systems and Software Engineering – System Life Cycle Processes*. ISO/IEC.

ISO/IEC/IEEE (2009). *ISO/IEC/IEEE 24765:2009 Systems and Software Engineering – System and Software Engineering Vocabulary*. Geneva, Switzerland: International Organization for Standardization (ISO)/ International Electrotechnical Commission (IEC)/ Institute of Electrical and Electronics Engineers (IEEE).

ISO/IEC/IEEE (2011). *ISO/IEC/IEEE 42010 Systems and software engineering – Architecture Description*. Geneva: International Organization for Standardization (ISO)/International Electrotechnical Commission (IEC)/Institute of Electrical and Electronics Engineers (IEEE).

ISO/IEC/IEEE (2015). *ISO15288: Systems and Software Engineering -- System Life Cycle Processes*. Geneva, Switzerland: International Organization for Standardization/International Electrotechnical Commissions/Institute of Electrical and Electronics Engineers. ISO/IEC/IEEE.

Kendall, F. (2014). US DoD's better buying power 3.0 implementation plan. White Paper, Office of the Under Secretary of Defense Acquisition, Technology and Logistics.

Mayer, J.D. and Salovey, P. (1997). What is emotional intelligence? In: *Emotional Development and Emotional Intelligence: Educational implications* (ed. P. Salovey and D.J. Sluyter), 3–31. New York: Harper Collins.

Merriam Webster (2018). Dictionary by Merriam Webster. https://www.merriam-webster.com/ (accessed 2018).

MOD (2004). *Ministry of Defence Architecture Framework (MODAF), Version 2*. London, UK: U.K. Ministry of Defence.

National Aeronautics and Space Administration (NASA) (2016). *NASA Systems Engineering Handbook (SP-2016-6105, Rev 2)*. NASA.

Oxford Dictionaries (2018). Definition: behaviour (US: behavior). https://en.oxforddictionaries.com/definition/behaviour (accessed April 2018).

PMI (2008). *A Guide to the Project Management Body of Knowledge (PMBOK® Guide)*, 4e. Newtown Square, PA, USA: Project Management Institute (PMI).

Prati, L.M., Douglas, C., Ferris, G.R. et al. (2003). Emotional intelligence, leadership effectiveness, and team outcomes. *The International Journal of Organizational Analysis* 11 (9): 21–40.

Presland, I. 2017. Comparison of SEP technical areas with version 4 framework competencies performed for certification advisory group. Unpublished – Internal INCOSE Document, June.

Rechtin, E. (1990). *Systems Architecting: Creating & Building Complex Systems*. Upper Saddle River, NJ, USA: Prentice-Hall.

Rumsfeld, D. (2012). There are known knowns. Briefing at United States Defence Department, 12 February 2002, Wikipedia. (accessed 2018).

Salovey, P. and Mayer, J.D. (1990). Emotional intelligence. *Imagination, Cognition and Personality* 9 (3): 185–211.

SFIA Foundation (2021). Skills framework for the information age (V8). https://sfia-online.org/en/sfia-8 (accessed 31 December 2021).

Sheard, S.A. and Mostashari, A. (2009). Principles of complex systems for systems engineering. *Systems Engineering* 12 (4): 295–311.

Sheard, S. (1996). 12 systems engineering roles. *INCOSE International Symposium* (7–11 July 1996). Boston MA: INCOSE.

U.S. Office of Personnel Management (OPM) (2015). Human capital assessment and accountability framework (HCAAF) resource center, "glossary". http://www.opm.gov/hcaaf_resource_center/glossary.asp (accessed June 2015).

Whitcomb, C., Khan, R. and White, C. (2014). Development of a System Engineering Competency Career Model: An Analytical Approach Using Bloom's Taxonomy. Technical Report, US Naval Postgraduate School.

Wikipedia (2017). Knowledge, skills and abilities. https://en.m.wikipedia.org/wiki/Knowledge,_Skills,_and_Abilities (accessed 31 December 2017).

Williams, C.R. and Reyes, A. (2012). Developing Systems Engineers at NASA. *Global Journal of Flexible Systems Management* 13 (3): 159–164.

SECAG ANNEX A: SYSTEMS ENGINEERING COMPETENCY ASSESSMENT GUIDE TABLES

This annex provides a set of tables providing detailed assessment criteria associated with each of the competence areas identified in the Systems Engineering Competency Framework.

Systems Engineering Competency Assessment Guide: A combined INCOSE Systems Engineering Competency Framework (SECF) and associated Systems Engineering Competency Assessment Guide (SECAG) document, First Edition. INCOSE.

Competency area – Core: Systems Thinking
Description The application of the fundamental concepts of systems thinking to Systems Engineering. These concepts include understanding what a system is, its context within its environment, its boundaries and interfaces, and that it has a life cycle. System thinking applies to the definition, development, and production of systems within an enterprise and technological environment and is a framework for curiosity about any system of interest.
Why it matters Systems thinking is a way of dealing with increasing complexity. The fundamental concepts of systems thinking involve understanding how actions and decisions in one area affect another, and that the optimization of a system within its environment does not necessarily come from optimizing the individual system components. Systems Thinking is conducted within an enterprise and technological context. These contexts impact the life cycle of the system and place requirements and constraints on the systems thinking being conducted. Failing to meet such constraints can have a serious effect on the enterprise and the value of the system.
Possible contributory types of evidence Any combination of the types of evidence may be acceptable (depending on how the Framework is tailored and used). The evidence items identified at each level indicate example work products only. Contributions to work products will generally differ at each proficiency level.
Learning and development The INCOSE Professional Development Portal provides example guidance on how to gain an initial awareness of a competency area and options for developing further competence thereafter.

Awareness – Core: Systems Thinking		
ID	**Indicators of Competence**	**Relevant knowledge sub-indicators**
1	Explains what "systems thinking" is and explains why it is important. [CSTA01]	(K) Explains why systems are more than interfaced collections of parts. [CSTA01-10K] (K) Lists static and dynamic properties of systems. [CSTA01-20K] (K) Lists different perspectives on systems (viewpoints). [CSTA01-30K] (K) Defines the concepts of abstraction, interaction, and emergence. [CSTA01-40K]
2	Explains what "emergence" is, why it is important, and how it can be "positive" or "negative" in its effect upon the system as a whole. [CSTA02]	(K) Explains why the whole may be different (not just greater) than the sum of the parts. [CSTA02-10K] (K) Lists examples of attributes or performance characteristics that are only associated with the whole rather than the parts. [CSTA02-20K] (K) Explains how emergence may be unanticipated and only found late in system development and how pre-work and anticipation can help prevent this. [CSTA02-30K]
3	Explains what a "system hierarchy" is and why it is important. [CSTA03]	(K) Defines what a system hierarchy is. [CSTA03-10K] (K) Explains why a "system hierarchy" implies more than just a decomposition. [CSTA03-20K] (K) Describes the idea of levels of detail. [CSTA03-30K]

		(K) Explains the relationship between system hierarchy and context, super system, system of interest, system elements (subsystems), and lower-level elements. [CSTA03-40K]
4	Explains what "system context" is for a given system of interest and describes why it is important. [CSTA04]	(K) Defines context as part of a hierarchical view. [CSTA04-10K] (K) Explains why context is important when considering systems. [CSTA04-20K]
5	Explains why it is important to be able to identify and understand what interfaces are. [CSTA05]	(K) Defines the concept of a system boundary. [CSTA05-10K] (K) Explains how the system interacts across its boundary. [CSTA05-20K] (K) Explains how interfaces may be external or internal to a system. [CSTA05-30K]
6	Explains why it is important to recognize interactions among systems and their elements. [CSTA06]	(K) Explains how different parts of a system may affect each other just by being there. [CSTA06-10K] (K) Explains how a part of a system is affected as much by being within the system as the system is by the part itself. [CSTA06-20K]
7	Explains why it is important to understand purpose and functionality of a system of interest. [CSTA07]	(K) Explains how an abstract, solution-free description of the purpose (and functions) of a system can help ensure developers think more widely about potential solutions. [CSTA07-10K] (K) Explains how improving our understanding of a system's purpose helps improve our understanding of the context of the development and the underlying needs to be addressed during the development. [CSTA07-20K]
8	Explains how business, enterprise, and technology can each influence the definition and development of the system and vice versa. [CSTA08]	(K) Lists business, enterprise and technology influences that may affect the system development and how. [CSTA08-10K] (K) Explains why stakeholder needs may influence the agreement of stakeholders. [CSTA08-20K] (K) Describes potential risks of mandating a particular technology. [CSTA08-30K] (K) Explains the risk of relying on technology innovation to provide solutions. [CSTA08-40K] (K) Explains how technology availability and maturity affects system development. [CSTA08-50K] (K) Explains why the system may have an effect on the enterprise (e.g. facilities and number of staff). [CSTA08-60K] (K) Explains why business, enterprise, and technology effects may not be apparent in the early stages of system development. [CSTA08-70K] (K) Explains how a new system development may reinforce or broaden an enterprise's understanding of its technology base when internally sourced or may do the same for a supplier when outsourced. [CSTA08-80K] (K) Explains why determining the supplier of a technology (e.g. in house or external) is an enterprise-level strategic issue. [CSTA08-90K]
9	Explains why it may be necessary to approach systems thinking in different ways, depending on the situation, and provides examples. [CSTA09]	(K) Explains why systems thinking is not a rigid process but a way of thinking (or a thinking "toolkit"). [CSTA09-10K] (K) Explains how systems thinking can facilitate investigation into a system solution. [CSTA09-20K]

Supervised Practitioner – Core: Systems Thinking			
ID	**Indicators of competence *(in addition to those at awareness level)***	**Relevant knowledge, experience, and/or behaviors**	**Possible examples of objective evidence of personal contribution to activities performed, or professional behaviors applied**
1	Defines the properties of a system. [CSTS01]	(A) Defines the following system properties for a system they have worked on life cycle, context, hierarchy, sum of parts, purpose, boundary, interactions. [CSTS01-10A]	Concept map or other model of a system. [CSTS01-E10] System concept document defining system life cycle, context, hierarchy, sum of parts, purpose, boundary, and key interactions. [CSTS01-E20]
2	Explains how system behavior produces emergent properties. [CSTS02]	(A) Identifies interactions between system elements, which may cause emergent properties of a system to appear. [CSTS02-10A] (K) Explains how emergent properties may be desirable or undesirable, planned or unplanned. [CSTS02-20K]	Emergent properties of different types identified in analysis work. [CSTS02-E10]
3	Uses the principles of system partitioning within system hierarchy on a project. [CSTS03]	(K) Explains how system partitioning may be carried out through various techniques such as an analysis of scenarios, functional decomposition, physical decomposition, interface reduction, and heritage. [CSTS03-10K] (K) Explains how the process of system partitioning deals with complexity by breaking down the system into realizable system elements. [CSTS03-20K] (K) Explains how partitioning moves from an understanding of high-level purpose, through analysis, to an eventual allocation of identified functions to elements within the system. [CSTS03-30K] (K) Explains the challenges of system partitioning. [CSTS03-40K] (K) Explains the relative merits of different system partitioning approaches. [CSTS03-50K] (K) Explains why hierarchy and partitions are merely constructs but how they impact our solution. [CSTS03-60K] (A) Uses partitioning principles to support system decomposition. [CSTS03-70A]	Contributions to a functional analysis or other system decomposition model. [CSTS03-E10]
4	Defines system characteristics in order to improve understanding of need. [CSTS04]	(K) Defines what a function is using active verb–noun combinations. [CSTS04-10K] (A) Prepares information characterizing a system in order to improve understanding of need. [CSTS04-20A]	System function descriptions/definitions and characteristics created. [CSTS04-E10]

		(A) Uses functional analysis to create descriptions/definitions of system functions defining functional performance requirements, suggesting implementation methods for functions in system design. [CSTS04-30A]	
5	Explains why the boundary of a system needs to be managed. [CSTS05]	(K) Explains how system boundaries are identified. [CSTS05-10K] (K) Defines the concepts of physical and functional boundaries, illustrating why these may differ. [CSTS05-20K] (K) Describes the potential impact of failing to manage boundaries they identified on a project. [CSTS05-30K] (A) Identifies system external interfaces in order to bound a system, using appropriate methods or tools. [CSTS05-40A]	Functional partitioning or interface work contributions. [CSTS05-E10] Physical boundaries of a system of interest defined. [CSTS05-E20] Functional boundaries of a system of interest defined. [CSTS05-E30]
6	Explains how humans and systems interact and how humans can be elements of systems. [CSTS06]	(K) Explains the difference between humans in the loop and human activity systems. [CSTS06-10K] (K) Explains how a human can be regarded as a part of the system, or as another system interacting with the system of interest. [CSTS06-20K] (K) Explains the importance of human factors and their effect on the wider system. [CSTS06-30K] (A) Analyzes human factor considerations for elements of a system. [CSTS06-40A]	System human factors analysis contributions. [CSTS06-E10]
7	Identifies the influence of wider enterprise on a project. [CSTS07]	(K) Lists potential influences at enterprise level including how they affect behavior (part and future), e.g. markets, products, policies, and finance. [CSTS07-10K] (K) Explains how cultural barriers and norms can be important when dealing with soft systems. [CSTS07-20K] (A) Identifies the part played by the enterprise in a system design. [CSTS07-30A] (A) Uses one or more appropriate systems methods to support their enterprise-level analysis work. [CSTS07-40A]	Contribution to a project/program analysis highlighting enterprise issues. [CSTS07-E10]
8	Uses systems thinking to contribute to enterprise technology development activities. [CSTS08]	(K) Explains how technology plans cover technologies required and how they are obtained. [CSTS08-10K] (K) Identifies various technology factors that could influence a system. [CSTS08-20K]	Comments made on an enterprise technology plan. [CSTS08-E10] Contribution to an enterprise technology plan. [CSTS08-E20]

(Continued)

ID	Indicators of competence *(in addition to those at awareness level)*	Relevant knowledge, experience, and/or behaviors	Possible examples of objective evidence of personal contribution to activities performed, or professional behaviors applied
		(K) Explains how technology is sometimes part of another system and the effect of this. [CSTS08-30K] (A) Uses systems thinking in support of enterprise technology planning. [CSTS08-40A] (A) Identifies technology (or capability) required to enable a system to be realized. [CSTS08-50A] (A) Prepares work products documenting the influence of technology development on a project. [CSTS08-60A]	Technology (or capability) recommendations made as part of enterprise planning. [CSTS08-E30]
9	Develops their own systems thinking insights to share thinking across the wider project (e.g. working groups and other teams). [CSTS09]	(A) Records own insights gained from considering a situation from a systems perspective. [CSTS09-10A] (A) Prepares inputs to a Project Team or Working group, which document their thinking. [CSTS09-20A]	Contributions to a cross-discipline (or similar) team. [CSTS09-E10] Systems thinking reflections (mental, pictures, or electronic), modeling insights. [CSTS09-E20]
10	Develops own understanding of this competency area through Continual Professional Development (CPD). [CSTS10]	(A) Identifies potential gaps in own knowledge or development needs in this area, identifying opportunities to address these through continual professional development activities. [CSTS10-10A] (A) Performs continual professional development activities to improve their knowledge and understanding in this area. [CSTS10-20A] (A) Records continual professional development activities undertaken including learning or insights gained. [CSTS10-30A]	Records of Continual Professional Development (CPD) performed and learning outcomes. [CSTS10-E10]

Practitioner – Core: Systems Thinking			
ID	**Indicators of Competence *(in addition to those at Supervised Practitioner level)***	**Relevant knowledge, experience, and/or behaviors**	**Possible examples of objective evidence of personal contribution to activities performed, or professional behaviors applied**
1	Identifies and manages complexity using appropriate techniques. [CSTP01]	(A) Identifies sources of both complication and complexity on a project. [CSTP01-10A] (A) Performs system partitioning on projects/programs, explaining the choices made. [CSTP01-20A] (A) Uses different simplification techniques, discussing the relative merits of each. [CSTP01-30A] (A) Uses simplification techniques to appreciate there are complexity overheads to partitioning and other forms of "simplifying" complexity. [CSTP01-40A]	System studies tackling complication/complexity and recommending suitable approaches. [CSTP01-E10]
2	Uses analysis of a system functions and parts to predict resultant system behavior. [CSTP02]	(A) Analyzes a system functions and parts in a model (e.g. a systems model). [CSTP02-10A] (A) Uses model analysis to predict resultant system behavior. [CSTP02-20A]	System analysis used to predict behavior. [CSTP02-E10]
3	Identifies the context of a system from a range of viewpoints including system boundaries and external interfaces. [CSTP03]	(A) Creates system boundaries. [CSTP03-10A] (A) Creates external system interfaces. [CSTP03-20A]	System boundary analysis document. [CSTP03-E10] System interface definition document. [CSTP03-E20]
4	Identifies the interaction between humans and systems, and systems and systems. [CSTP04]	(A) Uses human factors modeling/task analysis, ergonomic models, or other modeling techniques to characterize human interactions. [CSTP04-10A] (A) Uses system analysis, simulation and modeling to determine and understand interactions between systems to characterize human interactions. [CSTP04-20A]	System interface control document (human factors aspects). [CSTP04-E10] Human factors analysis reports. [CSTP04-E20] Human Computer Interface (HCI) models. [CSTP04-E30] System analysis reports (human factors aspects). [CSTP04-E40]
5	Identifies enterprise and technology issues affecting the design of a system and addresses them using a systems thinking approach. [CSTP05]	(A) Elicits enterprise and technology issues on a project from stakeholder dialogue to capture potential needs. [CSTP05-10A] (A) Creates system requirements to capture enterprise technology requirements. [CSTP05-20A]	System requirements taking enterprise and technology considerations into account. [CSTP05-E10] Records from enterprise/technology projects recognizing the needs of the systems to be used to achieve the needs of the ultimate system. [CSTP05-E20]
6	Uses appropriate systems thinking approaches to a range of situations, integrating the outcomes to get a full understanding of the whole. [CSTP06]	(A) Selects an appropriate systems thinking methodology (from a broad portfolio) based upon understanding the most relevant to the situation. [CSTP06-10A]	Records of contribution to enquiry into a system situation addressing techniques and insights gained. [CSTP06-E10]

(*Continued*)

ID	Indicators of Competence *(in addition to those at Supervised Practitioner level)*	Relevant knowledge, experience, and/or behaviors	Possible examples of objective evidence of personal contribution to activities performed, or professional behaviors applied
		(A) Uses appropriate systems thinking methodologies to increase understanding of a system. [CSTP06-20A]	
7	Identifies potential enterprise improvements to enable system development. [CSTP07]	(A) Identifies and plans what is needed (in the organization) to enable the system development. [CSTP07-10A]	Records of contribution to activities improving "system capability" for realization of systems in an enterprise (i.e. not enterprise output "products" but the systems needed to create them). [CSTP07-E10]
8	Guides team systems thinking activities in order to ensure current activities align to purpose. [CSTP08]	(A) Defines appropriate approach and methods to use based upon the need. [CSTP08-10A] (P) Uses insights gained from team systems thinking activities to guide and actively coordinate actions performed. [CSTP08-20P]	Records of applying systems thinking activities at team level to guide other team actions or activities. [CSTP08-E10] Records of applying systems thinking activities at team level to guide own actions or activities. [CSTP08-E20]
9	Develops existing case studies and examples of systems thinking to apply in new situations. [CSTP09]	(A) Analyzes the approach used in the existing systems thinking, identifying the rationale for the approach taken. [CSTP09-10A] (A) Maintains systems thinking approach to reflect new circumstances. [CSTP09-20A] (A) Reviews approach over time, updating as required to overcome unforeseen challenges or issues (e.g. as understanding emerges). [CSTP09-30A]	Lessons learned from personal application of systems thinking and how experience on one system situation informed approach to another. [CSTP09-E10] Lessons learned from a given situation that made certain aspects more or less easy to do and relevant to the situation. [CSTP09-E20]
10	Guides new or supervised practitioners in system thinking techniques in order to develop their knowledge, abilities, skills, or associated behaviors. [CSTP10]	(P) Guides new or supervised practitioners in executing activities that form part of this competency. [CSTP10-10P] (A) Trains individuals to an "Awareness" level in this competency area. [CSTP10-20A]	Organizational Breakdown Structure showing responsibility for technical supervision in this area. [CSTP10-E10] Records of on-the-job training objectives/guidance etc. provided. [CSTP10-E20] Coaching or mentoring assignment records. [CSTP10-E30] Records highlighting their impact on another individual in terms of improvement or professional development in this competency. [CSTP10-E40]
11	Maintains and enhances own competence in this area through Continual Professional Development (CPD) activities. [CSTP11]	(A) Identifies potential development needs in this area, identifying opportunities to address these through continual professional development activities. [CSTP11-10A] (A) Performs continual professional development activities to maintain and enhance their competency in this area. [CSTP11-20A] (A) Records continual professional development activities undertaken, including learning or insights gained. [CSTP11-30A]	Records of Continual Professional Development (CPD) performed and learning outcomes. [CSTP11-E10]

Lead Practitioner – Core: Systems Thinking			
ID	**Indicators of competence** ***(in addition to those at practitioner level)***	**Relevant knowledge, experience, and/or behaviors**	**Possible examples of objective evidence of personal contribution to activities performed, or professional behaviors applied**
1	Creates enterprise-level policies, procedures, guidance, and best practice for systems thinking, including associated tools. [CSTL01]	(A) Analyzes enterprise need for systems thinking policies, processes, tools, or guidance. [CSTL01-10A]	Records showing their role in embedding systems thinking into enterprise policies (e.g. guidance introduced at enterprise level, enterprise-level review minutes). [CSTL01-E10]
		(A) Creates enterprise policies, procedures, or guidance for systems thinking activities. [CSTL01-20A]	Procedures they have written. [CSTL01-E20]
		(A) Selects and acquires appropriate tools supporting systems thinking activities. [CSTL01-30A]	Records of support for tool introduction. [CSTL01-E30]
2	Judges the suitability of project-level systems thinking on behalf of the enterprise to ensure its validity. [CSTL02]	(A) Reviews systems thinking approaches based on a deep understanding of systems thinking in assessing suitability of systems solutions. [CSTL02-10A]	Records of role as independent reviewer in the assessment of project- or enterprise-level systems thinking (e.g. ideas, proposals, decisions, and their underlying justification). [CSTL02-E10]
		(P) Communicates conclusions made based on a deep understanding of systems thinking in assessing suitability of systems solutions. [CSTL02-20P]	
3	Persuades key enterprise-level stakeholders across the enterprise to support and maintain the technical capability and strategy of the enterprise. [CSTL03]	(A) Uses systems thinking to analyze capability and strategy issues at enterprise level. [CSTL03-10A]	Dialog (e.g. meeting minutes, documents, or communications) advising with key enterprise stakeholders on technical capability or strategy at enterprise level. [CSTL03-E10]
		(A) Uses systems thinking to maintain the technical capability and strategy of an enterprise. [CSTL03-20A]	Records indicating changed stakeholder thinking from personal intervention. [CSTL03-E20]
		(P) Persuades key stakeholders to address identified enterprise-level systems thinking issues. [CSTL03-30P]	
4	Adapts existing systems thinking practices on behalf of the enterprise to accommodate novel, complex, or difficult system situations or problems. [CSTL04]	(A) Develops updates to systems thinking in complex or difficult system situations or where situations require a novel approach. [CSTL04-10A]	Records of changes made to stakeholder thinking, which identify the novelty/difficulty of situation, how the practice was applied, and how understanding was gained. [CSTL04-E10]
5	Persuades project stakeholders across the enterprise to improve the suitability of project technical strategies in order to maintain their validity. [CSTL05]	(A) Reviews local technical strategy from a systems thinking perspective. [CSTL05-10A]	Records of local guidance and improvement plans for the practice, which include reflections on application of systems thinking. [CSTL05-E10]

(Continued)

ID	Indicators of competence *(in addition to those at practitioner level)*	Relevant knowledge, experience, and/or behaviors	Possible examples of objective evidence of personal contribution to activities performed, or professional behaviors applied
		(P) Persuades key stakeholders to accept improvements to technical strategy in order to address identified systems thinking issues. [CSTL05-20P]	Records of contributions to system strategy planning. [CSTL05-E20]
6	Persuades key stakeholders to address enterprise-level issues identified through systems thinking. [CSTL06]	(A) Reviews systems thinking in organization, not just on specific system or product development. [CSTL06-10A]	Records of enterprise-wide role (e.g. minutes, reports, and communications) in resolving differing system thinking issues (i.e. not just project role). [CSTL06-E10]
		(P) Persuades key stakeholders to address identified enterprise-level systems thinking issues. [CSTL06-20P]	Records showing senior/organization/enterprise leadership engagement in systems thinking problems, not just project leaders. [CSTL06-E20]
7	Coaches or mentors practitioners across the enterprise in Systems thinking in order to develop their knowledge, abilities, skills, or associated behaviors. [CSTL07]	(P) Coaches or mentors practitioners across the enterprise in competency-related techniques, recommending development activities. [CSTL07-10P]	Coaching or mentoring assignment records. [CSTL07-E10]
		(A) Develops or authorizes enterprise training materials in this competency area. [CSTL07-20A]	Records of formal training courses, workshops, seminars, and authored training material supported by successful post-training evaluation data. [CSTL07-E20]
		(A) Provides enterprise workshops/seminars or training in this competency area. [CSTL07-30A]	Listing as an approved organizational trainer for this competency area. [CSTL07-E30]
8	Promotes the introduction and use of novel techniques and ideas in systems thinking across the enterprise, to improve enterprise competence in this area. [CSTL08]	(A) Analyzes different approaches across different domains through research. [CSTL08-10A]	Research records. [CSTL08-E10]
		(A) Defines novel approaches that could potentially improve the SE discipline within the enterprise. [CSTL08-20A]	Published papers in refereed journals/company literature. [CSTL08-E20]
		(P) Fosters awareness of these novel techniques within the enterprise. [CSTL08-30P]	Records showing introduction of enabling systems supporting the new techniques or ideas. [CSTL08-E30]
		(P) Collaborates with enterprise stakeholders to facilitate the introduction of techniques new to the enterprise. [CSTL08-40P]	Published papers (or similar) at enterprise level. [CSTL08-E40]
		(A) Monitors new techniques after their introduction to determine their effectiveness. [CSTL08-50A]	Records of improvements made against a recognized process improvement model. [CSTL08-E50]
		(A) Adapts approach to reflect actual enterprise performance improvements. [CSTL08-60A]	

9	Develops expertise in this competency area through specialist Continual Professional Development (CPD) activities. [CSTL09]	(A) Identifies own needs for further professional development in order to increase competence beyond practitioner level. [CSTL09-10A] (A) Performs professional development activities to move competence toward expert level. [CSTL09-20A] (A) Records continual professional development activities undertaken including learning or insights gained. [CSTL09-30A]	Records of Continual Professional Development (CPD) performed and learning outcomes. [CSTL09-E10]

NOTES	In addition to items above, enterprise-level or independent 3rd-Party generated evidence may be used to amplify other evidence presented and may include: a. Formally recognized by senior management in current organization as an expert in this competency area b. Evidence of role as Product/System Design Authority or Technical Authority on a complex project with responsibilities in this area or where skills within this competency area were used c. Recognized as an authorizing signatory on behalf of enterprise for formal documentation in this competency area (e.g. policies, processes, and deliverables) d. Formal commendation or award within own enterprise for contribution or item of work successfully performed, which required proficiency in this competency area e. Customer, Supplier, or other external project-specific key Stakeholder accolades for specific work performed in this competency area f. Independently assessed or accredited work in this competency area (e.g. for independent publication or use) g. Formal organizational HR records positively highlighting any specific professional competencies or behaviors identified (if applicable) plus any of the evidence indicators listed at Expert level below

Expert – Core: Systems Thinking			
ID	**Indicators of Competence *(in addition to those at Lead Practitioner level)***	**Relevant knowledge, experience, and/or behaviors**	**Possible examples of objective evidence of personal contribution to activities performed, or professional behaviors applied**
1	Communicates own knowledge and experience in systems thinking in order to improve best practice beyond the enterprise boundary. [CSTE01]	(A) Produces papers, seminars, or presentations outside own enterprise for publication in order to share own ideas and improve industry best practices in this competence area. [CSTE01-10A]	Published papers or books etc. on new technique in refereed journals/company literature. [CSTE01-E10]
		(P) Fosters incorporation of own ideas into industry best practices in this area. [CSTE01-20P]	Published papers in refereed journals or internal literature proposing new practices in this competence area (or presentations, tutorials, etc.). [CSTE01-E20]
		(P) Develops guidance materials identifying new (or updating existing) best practice in this competence area. [CSTE01-30P]	Own proposals adopted as industry best practices in this competence area. [CSTE01-E30]
2	Influences individuals and activities beyond the enterprise boundary to support the systems thinking approach of their own enterprise. [CSTE02]	(A) Reviews systems thinking approaches proposed by organizations beyond the enterprise boundary. [CSTE02-10A]	Records of review, including personal role played and review outcome, including recommendations and changes resulting. [CSTE02-E10]
		(A) Advises organizations on changes required to their systems thinking strategies. [CSTE02-20A]	Records of advice requested and provided on the suitability of systems solutions or approaches. [CSTE02-E20]
		(A) Produces guidance and recommendations of best practice for use by external organizations. [CSTE02-30A]	Records of own involvement in external company reviews and recommendations into System Thinking. [CSTE02-E30]
		(P) Persuades individuals beyond the enterprise boundary to adopt approaches, which help underpin the systems thinking of their own enterprise. [CSTE02-40P]	
3	Advises organizations beyond the enterprise boundary on the suitability of their approach to systems thinking. [CSTE03]	(A) Reviews systems thinking strategies. [CSTE03-10A]	Records of review, including personal role played, and review outcome, including recommendations and changes resulting. [CSTE03-E10]
		(A) Advises on systems thinking strategies of changes required. [CSTE03-20A]	Records of advice requested and provided on the suitability of systems solutions or approaches. [CSTE03-E20]
		(A) Produces guidance and recommendations or best practice for external stakeholders. [CSTE03-30A]	Records of own involvement in external company reviews and recommendations into a systems approach taken. [CSTE03-E30]

4	Advises organizations beyond the enterprise boundary on complex or sensitive systems thinking issues. [CSTE04]	(A) Advises stakeholders beyond the enterprise on their systems thinking handling of complex or sensitive issues. [CSTE04-10A]	Records of systems thinking advice provided together with evidence that the issues advised on were by their nature either complex or sensitive. [CSTE04-E10]
		(P) Conducts sensitive negotiations on a highly complex or sensitive system thinking issue, making limited use of specialized, technical terminology. [CSTE04-20P]	Records from negotiations demonstrating awareness of customer's background and knowledge. [CSTE04-E20]
		(A) Uses a systems thinking approach to complex issue resolution including balanced, rational arguments on way forward. [CSTE04-30A]	Records of role as meeting or activity arbiter determining the most appropriate approach to take. [CSTE04-E30]
5	Champions the introduction of novel techniques and ideas in systems thinking beyond the enterprise boundary, in order to develop the wider Systems Engineering community in this competency. [CSTE05]	(A) Analyzes different approaches across different domains through research. [CSTE05-10A]	Records of activities promoting research and need to adopt novel technique or ideas. [CSTE05-E10]
		(A) Produces reports for the wider SE community on the effectiveness of new techniques after their introduction. [CSTE05-20A]	Records of improvements made to process and appraisal against a recognized process improvement model. [CSTE05-E20]
		(P) Collaborates with those introducing novel techniques within the wider SE community. [CSTE05-30P]	Research records. [CSTE05-E30]
		(A) Defines novel approaches that could potentially improve the wider SE discipline. [CSTE05-40A]	Published papers in refereed journals/company literature. [CSTE05-E40]
		(P) Fosters awareness of these novel techniques within the wider SE community. [CSTE05-50P]	Records showing introduction of enabling systems supporting the new techniques or ideas. [CSTE05-E50]
6	Coaches individuals beyond the enterprise boundary in systems thinking techniques, in order to further develop their knowledge, abilities, skills, or associated behaviors. [CSTE06]	(P) Coaches or mentors individuals beyond the enterprise boundary, in competency-related techniques, recommending development activities. [CSTE06-10P]	Coaching or mentoring assignment records. [CSTE06-E10]
		(A) Develops or authorizes training materials in this competency area, which are subsequently successfully delivered beyond the enterprise boundary. [CSTE06-20A]	Records of formal training courses, workshops, seminars, and authored training material supported by successful post-training evaluation data. [CSTE06-E20]
		(A) Provides workshops/seminars or training in this competency area for practitioners or lead practitioners beyond the enterprise boundary (e.g. conferences and open training days). [CSTE06-30A]	Records of training/workshops/seminars created supported by successful post-training evaluation data. [CSTE06-E30]

(Continued)

ID	Indicators of Competence *(in addition to those at Lead Practitioner level)*	Relevant knowledge, experience, and/or behaviors	Possible examples of objective evidence of personal contribution to activities performed, or professional behaviors applied
7	Maintains expertise in this competency area through specialist Continual Professional Development (CPD) activities. [CSTE07]	(A) Reviews research, new ideas, and state of the art to identify relevant new areas requiring personal development in order to maintain expertise in this competency area. [CSTE07-10A] (A) Performs identified specialist professional development activities in order to maintain or further develop competence at expert level. [CSTE07-20A] (A) Records continual professional development activities undertaken including learning or insights gained. [CSTE07-30A]	Records of documents reviewed and insights gained as part of own research into this competency area. [CSTE07-E10] Records of Continual Professional Development (CPD) performed and learning outcomes. [CSTE07-E20]

NOTES	In addition to items above, enterprise-level or independent 3rd-Party-generated evidence may be used to amplify other evidence presented and may include: a. Formally recognized by a reputable external organization as an expert in this competency area b. Evidence of role as independent assessor or reviewer on project outside own organization where skills in this competency area were used c. Evidence of invitation(s) from wider community for contribution of systems engineering expertise in this area (e.g. industry conference panel, government advisory board etc. cross-industry working groups, partnerships, accredited advanced university courses or research, or as part of professional institute) d. Formal commendation beyond the enterprise (e.g. by INCOSE or other recognized authority) for work performed in this competency area e. Independently assessed or accredited work product in this competency area (e.g. for independent publication or use) f. Accolades of expertise in this area from recognized industry leaders

Competency area – Core: Life Cycles
Description The selection of appropriate life cycles in the realization of a system. Systems and their constituent elements have individual life cycles, characterizing the nature of their evolution. Each life cycle is itself divided into a series of stages, marking key transition points during the evolution of that element. As different system elements may have different life cycles, the relationship between life cycle stages on differing elements is complex, varying depending on the scope of the project, characteristics of the wider system of which it forms a part, the stakeholder requirements, and perceived risk.
Why it matters Life cycles form the basis for project planning and estimating. Selection of the appropriate life cycles and their alignment has a large impact on and may be crucial to project success. Ensuring coordination between related life cycles at all levels is critical to the realization of a successful system.
Possible contributory types of evidence Any combination of the types of evidence may be acceptable (depending on how the Framework is tailored and used). The evidence items identified at each level indicate example work products only. Contributions to work products will generally differ at each proficiency level.
Learning and development The INCOSE Professional Development Portal provides example guidance on how to gain an initial awareness of a competency area and options for developing further competence thereafter.

Awareness – Core: Life Cycles

ID	Indicators of competence	Relevant knowledge sub-indicators
1	Identifies different life cycle types and summarizes the key characteristics of each. [CLCA01]	(K) Defines key characteristics of the following life cycle types: Waterfall, Vee, Iterative, Incremental, Evolutionary/Spiral. [CLCA01-10K] (K) Lists the different life cycle stages from concept top retirement identified in ISO/IEC 15288. [CLCA01-20K] (K) Defines the difference between a project and a system life cycle (from Planning). [CLCA01-30K] (K) Defines the relationship between different project life cycle types (e.g. "bid," "development," "support") and different system life cycle stages or activities (from Planning). [CLCA01-40K] (K) Defines the stages involved and the similarities and differences between a system or product "Acquisition" by a customer or 3rd party and system or product development by a supplier. [CLCA01-50K]
2	Explains why selection of life cycle is important when developing a system solution. [CLCA02]	(K) Defines the relationship between different stages of the life cycle. [CLCA02-10K] (K) Explains why different life cycles may suit different types of development. [CLCA02-20K] (K) Explains why activities within different life cycle stages may interact, requiring trade-offs in solutions, such as development cost versus operating costs. [CLCA02-30K]

(Continued)

ID	Indicators of competence	Relevant knowledge sub-indicators
3	Explains why it is necessary to define an appropriate life cycle process model and the key steps involved. [CLCA03]	(K) Describes what a life cycle process model is and how it is used to manage and control a project development. [CLCA03-10K] (K) Describes different life cycle process models (e.g. the Project model vs. product model). [CLCA03-20K] (K) Explains how differing project or program characteristics (e.g. size of project/program, experience of staff, cycle time, acceptable defect levels) may affect life cycle models. [CLCA03-30K] (K) Describes the concept of "tailoring" and (K) Explains how this affects life cycle process definition. [CLCA03-40K]
4	Explains why differing engineering approaches are required in different life cycle phases and provides examples. [CLCA04]	(K) Explains why early stages may require more investigation or concept proving (iterative, spiral) or could be waterfall if requirements are few and well-defined. [CLCA04-10K] (K) Explains why later stages of program may become more predictable as design stabilizes, but how this may also be offset by the presence of "legacy systems" and how such factors may influence the selected approach. [CLCA04-20K] (K) Explains how while a software-intensive system may benefit from a software-optimized life cycle (e.g. agile, incremental) even this approach may be limited to underlying hardware development life cycle or constraints. [CLCA04-30K] (K) Explains why tailoring defines but also optimizes the SE process and how too little or too much may increase rather than decrease cost or risk. [CLCA04-40K]
5	Explains how different life cycle characteristics relate to the system life cycle. [CLCA05]	(K) Explains how different life cycle types (such as "Waterfall," "Vee," "Spiral," "Iterative," "Incremental," "Evolutionary") relate to the stages of a system development identified in the ISO15288 model. [CLCA05-10K]

Supervised Practitioner – Core: Life Cycles			
ID	**Indicators of competence *(in addition to those at awareness level)***	**Relevant knowledge, experience, and/or behaviors**	**Possible examples of objective evidence of personal contribution to activities performed, or professional behaviors applied**
1	Describes Systems Engineering life cycle processes. [CLCS01]	(K) Describes Systems Engineering processes applicable to their project (e.g. Requirements capture, requirements analysis, design, build, integration, verification, validation, operation, or disposal). [CLCS01-10K]	Systems Engineering processes followed on a Project or Program. [CLCS01-E10]
		(A) Uses system engineering life cycle processes within a project/program life cycle. [CLCS01-20A]	System engineering life cycle processes developed. [CLCS01-E20]
2	Identifies the impact of failing to consider future life cycle stages in the current stage. [CLCS02]	(K) Describes why the impact of future life cycle stages needs to be considered within the current life cycle stage. [CLCS02-10K]	Records of life cycle considerations taken into account in life cycle stages. [CLCS02-E10]
		(A) Identifies how future life cycle needs might be accommodated in activities they perform. [CLCS02-20A]	Records showing the impact of omission of future life cycle considerations on an outcome. [CLCS02-E20]
3	Prepares inputs to life cycle definition activities at system or system element level. [CLCS03]	(K) Explains how life cycle tailoring can address variables such as nature of the customer, cost, schedule, quality of trade-offs, technical difficulty, and experience of the people implementing the process. [CLCS03-10K]	Contributions to project process life cycle tailoring. [CLCS03-E10]
		(K) Explains why a fully integrated project/program team should be involved in tailoring the process. [CLCS03-20K]	Contributions to project/program definition or tailoring. [CLCS03-E20]
		(A) Identifies a life cycle within the enterprise including the rationale for life cycle decisions made. [CLCS03-30A]	
4	Complies with a governing project system life cycle, using appropriate processes and tools to plan and control their own activities. [CLCS04]	(K) Explains the purpose and key activities of each stage of the system life cycle in which they are working. [CLCS04-10K]	Contributions to project/program definition or tailoring. [CLCS04-E10]
		(K) Explains the model used within the business, which handles life cycle realization and maintenance processes. [CLCS04-20K]	
		(K) Explains actual or potential limitations of the life cycle and tailoring process they are using and how this might be addressed. [CLCS04-30K]	
		(A) Complies with the processes, tools, and reviews defined within a project life cycle. [CLCS04-40A]	

(Continued)

ID	Indicators of competence *(in addition to those at awareness level)*	Relevant knowledge, experience, and/or behaviors	Possible examples of objective evidence of personal contribution to activities performed, or professional behaviors applied
5	Describes the system life cycle in which they are working on their project. [CLCS05]	(K) Describes the different stages of the life cycle being used on their current project. [CLCS05-10K] (K) Identifies the life cycle stage in which they are working on their project (e.g. concept, development, and production) and the primary goals of the stage in which they are working. [CLCS05-20K] (A) Identifies the Systems Engineering life cycle processes they are following and where they are identified on their project (e.g. Systems Engineering Management Plan (SEMP)). [CLCS05-30A] (A) Identifies key life cycle review gates on their project. [CLCS05-40A]	Contributions to project/program definition or tailoring. [CLCS05-E10] Records of Systems Engineering life cycles used on projects/programs to which they have contributed. [CLCS05-E20]
6	Develops own understanding of this competency area through Continual Professional Development (CPD). [CLCS06]	(A) Identifies potential gaps in own knowledge or development needs in this area, identifying opportunities to address these through continual professional development activities. [CLCS06-10A] (A) Performs continual professional development activities to improve their knowledge and understanding in this area. [CLCS06-20A] (A) Records continual professional development activities undertaken including learning or insights gained. [CLCS06-30A]	Records of Continual Professional Development (CPD) performed and learning outcomes. [CLCS06-E10]

Practitioner – Core: Life Cycles			
ID	**Indicators of competence *(in addition to those at supervised practitioner level)***	**Relevant knowledge, experience, and/or behaviors**	**Possible examples of objective evidence of personal contribution to activities performed, or professional behaviors applied**
1	Explains the advantages and disadvantages of different types of systems life cycle and where each might be used advantageously. [CLCP01]	(K) Identifies different types of system life cycle (e.g. Waterfall, Spiral, Iterative, Incremental, and Evolutionary), their key characteristics, and the pros and cons of each type. [CLCP01-10K]	Records showing rationale for life cycle selection(s) made on a project. [CLCP01-E10]
		(K) Explains why characteristics of complex systems mean that different life cycle approaches (e.g. probe sense respond c.f. sense analyze respond) may be required to optimize development. [CLCP01-20K] (K) Identifies factors that affect selection and definition of life cycle such as customer program life cycle, complexity, requirements stability, milestones, delivery dates, technology insertion/readiness, standards, internal policy/process requirements, product life cycle, availability of tools. [CLCP01-30K] (A) Selects life cycles to address identified project-specific development factors. [CLCP01-40A]	Process tailoring performed on projects/programs (e.g. in a systems engineering management plan). [CLCP01-E20]
2	Creates a governing project life cycle, using enterprise-level policies, procedures, guidance, and best practice. [CLCP02]	(A) Uses enterprise policies and procedures when selecting life cycles for a project. [CLCP02-10A]	Gaining approval for a project life cycle, which selects appropriate enterprise policies and procedures (e.g. a systems engineering management plan). [CLCP02-E10]
		(A) Uses enterprise guidelines and best practices when selecting life cycles for a project. [CLCP02-20A] (A) Creates life cycle definition information for the project. [CLCP02-30A] (P) Communicates definition to affected stakeholders. [CLCP02-40P]	Gaining approval for a project life cycle, which interprets selection guidelines (e.g. a systems engineering management plan). [CLCP02-E20]
3	Identifies dependencies aligning life cycles and life cycle stages of different system elements accordingly. [CLCP03]	(A) Identifies the life cycle of different system elements in a system and potential dependencies caused. [CLCP03-10A]	Contributions made to documents identifying life cycle dependencies between system elements (e.g. a systems engineering management plan, or its review). [CLCP03-E10]
		(A) Identifies the constraints and risks on the system elements as a result of dependencies between system elements. [CLCP03-20A]	Contributions made to documents identifying development constraints and risks due to differing life cycles of system elements (e.g. a systems engineering management plan, or its review). [CLCP03-E20]
4	Acts to influence the life cycle of system elements beyond boundary of the system of interest, to improve the development strategy. [CLCP04]	(A) Identifies super system elements related to system of interest and their life cycles. [CLCP04-10A]	Minutes of meetings with the Customer / Acquirer discussing super system and system life cycle-related issues. [CLCP04-E10]
		(A) Identifies the potential dependencies, constraints, and risks on the target system from super-system elements leading to changes in life cycle. [CLCP04-20A]	Document they have contributed to highlighting dependencies, constraints, and risks of differing life cycles. [CLCP04-E20]

(Continued)

ID	Indicators of competence *(in addition to those at supervised practitioner level)*	Relevant knowledge, experience, and/or behaviors	Possible examples of objective evidence of personal contribution to activities performed, or professional behaviors applied
5	Prepares plans addressing future life cycle phases to take into consideration their impact on the current phase, improving current activities accordingly. [CLCP05]	(K) Explains how activities in future life cycle phases may affect activities in current phase. [CLCP05-10K] (A) Analyzes the needs of future life cycle phases when planning current life cycle phase. [CLCP05-20A]	Document they have contributed to identifying needs of future life cycle phases. [CLCP05-E10] Document they have contributed to demonstrating that needs of future life cycle activities have influenced or changed activities performed in current phase. [CLCP05-E20]
6	Prepares plans governing transitions between life cycle stages to reduce project impact at those transitions. [CLCP06]	(A) Identifies key life cycle stages and associated reviews on their project. [CLCP06-10A] (A) Identifies key elements addressed when transitioning between different life cycle stages. [CLCP06-20A]	Contribution to end of stage review, which identifies needs of next stage. [CLCP06-E10] Document they have contributed, which identified and prepared for changed needs in the next stage of development (e.g. systems engineering management plan or life cycle review minutes). [CLCP06-E20]
7	Guides new or supervised practitioners in Systems Engineering Life cycles in order to develop their knowledge, abilities, skills, or associated behaviors. [CLCP07]	(P) Guides new or supervised practitioners in executing activities that form part of this competency. [CLCP07-10P] (A) Trains individuals to an "Awareness" level in this competency area. [CLCP07-20A]	Organizational Breakdown Structure showing their responsibility for technical supervision in this area. [CLCP07-E10] On-the-job training records. [CLCP07-E20] Coaching or mentoring assignment records. [CLCP07-E30] Records highlighting their impact on another individual in terms of improvement or professional development in this competency. [CLCP07-E40]
8	Maintains and enhances own competence in this area through Continual Professional Development (CPD) activities. [CLCP08]	(A) Identifies potential development needs in this area, identifying opportunities to address these through continual professional development activities. [CLCP08-10A] (A) Performs continual professional development activities to maintain and enhance their competency in this area. [CLCP08-20A] (A) Records continual professional development activities undertaken including learning or insights gained. [CLCP08-30A]	Records of Continual Professional Development (CPD) performed and learning outcomes. [CLCP08-E10]

Lead Practitioner – Core: Life Cycles

ID	Indicators of competence *(in addition to those at practitioner level)*	Relevant knowledge, experience, and/or behaviors	Possible examples of objective evidence of personal contribution to activities performed, or professional behaviors applied
1	Creates enterprise-level policies, procedures, guidance, and best practice for life cycle definition and management, including associated tools. [CLCL01]	(A) Analyzes enterprise need for life cycle policies, processes, tools, or guidance. [CLCL01-10A]	Records showing their role in embedding Life cycle definition into enterprise policies (e.g. guidance introduced at enterprise level, enterprise -level review minutes). [CLCL01-E10]
		(A) Creates enterprise policies, procedures, or guidance for life cycle activities. [CLCL01-20A]	Procedures they have written. [CLCL01-E20]
		(A) Selects and acquires appropriate tools supporting life cycle activities. [CLCL01-30A]	Records of support for tool introduction. [CLCL01-E30]
2	Judges life cycle selections across the enterprise to ensure they meet the needs of the project. [CLCL02]	(A) Reviews program life cycles and process tailoring. [CLCL02-10A]	Review comments on selections made by others. [CLCL02-E10]
3	Adapts standard life cycle models on behalf of the enterprise to address complex or difficult situations or to resolve conflicts between life cycles where required. [CLCL03]	(K) Explains how the variability of life cycle stages of different elements of a system impacts on standard life cycle model selection. [CLCL03-10K]	Documentation of SE life cycles on many occasions. [CLCL03-E10]
		(K) Explains how the variability of life cycle of elements of a system at different levels impacts on standard life cycle model selection. [CLCL03-20K]	Records of life cycle adaptations made. [CLCL03-E20]
		(A) Analyzes standard life cycle models to ensure that they address project-specific context. [CLCL03-30A]	Life cycle definition documents they have written which addressed life cycle challenges between elements. [CLCL03-E30]
		(A) Adapts standard life cycle models to address project-specific life cycle constraints. [CLCL03-40A]	Life cycle plans for different products demonstrating different life cycle approaches. [CLCL03-E40]
		(A) Adapts life cycle stages of both the system-of-interest and its elements to define or refine life cycle-related plans. [CLCL03-50A]	Risks and mitigations they applied across the stages of the systems engineering "V" diagram. [CLCL03-E50]
4	Identifies work or issues relevant to the current life cycle phase by applying knowledge of life cycles to projects across the enterprise. [CLCL04]	(A) Uses knowledge of life cycles to identify activities required in current life cycle stage. [CLCL04-10A]	Life cycle definition documents, which identify life cycle activities required. [CLCL04-E10]
		(A) Uses knowledge of life cycles to identify challenges or potential issues in current life cycle stage. [CLCL04-20A]	Life cycle definition documents, which address identified life cycle challenges. [CLCL04-E20]

(Continued)

ID	Indicators of competence *(in addition to those at practitioner level)*	Relevant knowledge, experience, and/or behaviors	Possible examples of objective evidence of personal contribution to activities performed, or professional behaviors applied
5	Persuades key stakeholders across the enterprise to support activities required now in order to address future life cycle stages. [CLCL05]	(A) Identifies and engages with key stakeholders where life cycle challenges exist. [CLCL05-10A]	Records of advice provided, which resulted in a life cycle process change. [CLCL05-E10]
		(P) Fosters agreement between key stakeholders at enterprise level to resolve life cycle issues by promoting a holistic viewpoint. [CLCL05-20P]	Minutes of key stakeholder review meetings and actions taken. [CLCL05-E20]
		(P) Persuades key stakeholders to address identified enterprise-level transition issues. [CLCL05-30P]	Records indicating changed stakeholder thinking from personal intervention. [CLCL05-E30]
6	Coaches or mentors practitioners across the enterprise in life cycle definition and management in order to develop their knowledge, abilities, skills, or associated behaviors. [CLCL06]	(P) Coaches or mentors practitioners across the enterprise in competency-related techniques, recommending development activities. [CLCL06-10P]	Coaching or mentoring assignment records. [CLCL06-E10]
		(A) Develops or authorizes enterprise training materials in this competency area. [CLCL06-20A]	Records of formal training courses, workshops, seminars, and authored training material supported by successful post-training evaluation data. [CLCL06-E20]
		(A) Provides enterprise workshops/seminars or training in this competency area. [CLCL06-30A]	Listing as an approved organizational trainer for this competency area. [CLCL06-E30]
7	Promotes the introduction and use of novel techniques and ideas in life cycle definition and management across the enterprise to improve enterprise competence in this area. [CLCL07]	(A) Analyzes different approaches across different domains through research. [CLCL07-10A]	Research records. [CLCL07-E10]
		(A) Defines novel approaches that could potentially improve the SE discipline within the enterprise. [CLCL07-20A]	Published papers in refereed journals/company literature. [CLCL07-E20]
		(P) Fosters awareness of these novel techniques within the enterprise. [CLCL07-30P]	Records showing introduction of enabling systems supporting the new techniques or ideas. [CLCL07-E30]
		(P) Collaborates with enterprise stakeholders to facilitate the introduction of techniques new to the enterprise. [CLCL07-40P]	Published papers (or similar) at enterprise level. [CLCL07-E40]
		(A) Monitors new techniques after their introduction to determine their effectiveness. [CLCL07-50A]	Records of improvements made against a recognized process improvement model. [CLCL07-E50]
		(A) Adapts approach to reflect actual enterprise performance improvements. [CLCL07-60A]	Health metrics records tracking a return to program health. [CLCL07-E60]

8	Develops expertise in this competency area through specialist Continual Professional Development (CPD) activities. [CLCL08]	(A) Identifies own needs for further professional development in order to increase competence beyond practitioner level. [CLCL08-10A] (A) Performs professional development activities in order to move own competence toward expert level. [CLCL08-20A] (A) Records continual professional development activities undertaken including learning or insights gained. [CLCL08-30A]	Records of Continual Professional Development (CPD) performed and learning outcomes. [CLCL08-E10]

NOTES	In addition to items above, enterprise-level or independent 3rd-Party-generated evidence may be used to amplify other evidence presented and may include: a. Formally recognized by senior management in current organization as an expert in this competency area b. Evidence of role as Product/System Design Authority or Technical Authority on a complex project with responsibilities in this area or where skills within this competency area were used c. Recognized as an authorizing signatory on behalf of enterprise for formal documentation in this competency area (e.g. policies, processes, and deliverables) d. Formal commendation or award within own enterprise for contribution or item of work successfully performed, which required proficiency in this competency area e. Customer, Supplier, or other external project-specific key Stakeholder accolades for specific work performed in this competency area f. Independently assessed or accredited work in this competency area (e.g. for independent publication or use) g. Formal organizational HR records positively highlighting any specific professional competencies or behaviors identified (if applicable) plus any of the evidence indicators listed at Expert level below

Expert – Core: Life Cycles			
ID	**Indicators of competence** ***(in addition to those at Lead Practitioner level)***	**Relevant knowledge, experience, and/or behaviors**	**Possible examples of objective evidence of personal contribution to activities performed, or professional behaviors applied**
1	Communicates own knowledge and experience in life cycle definition and management in order to improve best practice beyond the enterprise boundary. [CLCE01]	(A) Produces papers, seminars, or presentations outside own enterprise for publication in order to share own ideas and improve industry best practices in this competence area. [CLCE01-10A]	Published papers or books etc. on new technique in refereed journals/company literature. [CLCE01-E10]
		(P) Fosters incorporation of own ideas into industry best practices in this area. [CLCE01-20P]	Published papers in refereed journals or internal literature proposing new practices in this competence area (or presentations, tutorials, etc.). [CLCE01-E20]
		(P) Develops guidance materials identifying new (or updating existing) best practice in this competence area. [CLCE01-30P]	Proposals adopted as industry best practices. [CLCE01-E30]
2	Advises organizations beyond the enterprise boundary on the suitability of life cycle tailoring or life cycle definitions. [CLCE02]	(A) Advises stakeholders beyond the enterprise boundary on program life cycles and process tailoring. [CLCE02-10A]	Records of advice provided. [CLCE02-E10]
		(A) Advises stakeholders beyond the enterprise boundary on issues or implications of life cycle definitions. [CLCE02-20A]	Peer-reviewed paper on life cycle usage or adaptation on complex or concurrent projects. [CLCE02-E20]
3	Advises organizations beyond the enterprise boundary on complex, concurrent, or sensitive projects. [CLCE03]	(A) Advises stakeholders beyond enterprise boundary on how their separate life cycles inter-relate and best practices to ensure a balanced solution. [CLCE03-10A]	Records of life cycle advice provided together with evidence that the issues advised on were by their nature either complex or sensitive. [CLCE03-E10]
		(A) Advises on complex or sensitive life cycle issues relating to assumptions, approaches, arguments, conclusions, and decisions with successful outcome. [CLCE03-20A]	Review comments. [CLCE03-E20]
		(A) Advises external organizations on complex or sensitive situation demanding the application of their life cycle knowledge. [CLCE03-30A]	
4	Champions the introduction of novel techniques and ideas in life cycle management, beyond the enterprise boundary, in order to develop the wider Systems Engineering community in this competency. [CLCE04]	(A) Analyzes different approaches across different domains through research. [CLCE04-10A]	Records of activities promoting research and need to adopt novel technique or ideas. [CLCE04-E10]
		(A) Produces reports for the wider SE community on the effectiveness of new techniques after their introduction. [CLCE04-20A]	Records of improvements made to process and appraisal against a recognized process improvement model. [CLCE04-E20]

		(P) Collaborates with those introducing novel techniques within the wider SE community. [CLCE04-30P]	Research records. [CLCE04-E30]
		(A) Defines novel approaches that could potentially improve the wider SE discipline. [CLCE04-40A]	Published papers in refereed journals/company literature. [CLCE04-E40]
		(P) Fosters awareness of these novel techniques within the wider SE community. [CLCE04-50P]	Records showing introduction of enabling systems supporting the new techniques or ideas. [CLCE04-E50]
5	Coaches individuals beyond the enterprise boundary in life cycle management techniques, in order to further develop their knowledge, abilities, skills, or associated behaviors. [CLCE05]	(P) Coaches or mentors individuals beyond the enterprise boundary, in competency-related techniques, recommending development activities. [CLCE05-10P]	Coaching or mentoring assignment records. [CLCE05-E10]
		(A) Develops or authorizes training materials in this competency area, which are subsequently successfully delivered beyond the enterprise boundary. [CLCE05-20A]	Records of formal training courses, workshops, seminars, and authored training material supported by successful post-training evaluation data. [CLCE05-E20]
		(A) Provides workshops/seminars or training in this competency area for practitioners or lead practitioners beyond the enterprise boundary (e.g. conferences and open training days). [CLCE05-30A]	Records of Training/ workshops/ seminars created supported by successful post-training evaluation data. [CLCE05-E30]
6	Maintains expertise in this competency area through specialist Continual Professional Development (CPD) activities. [CLCE06]	(A) Reviews research, new ideas, and state of the art to identify relevant new areas requiring personal development in order to maintain expertise in this competency area. [CLCE06-10A]	Records of documents reviewed and insights gained as part of own research into this competency area. [CLCE06-E10]
		(A) Performs identified specialist professional development activities in order to maintain or further develop competence at expert level. [CLCE06-20A]	Records of Continual Professional Development (CPD) performed and learning outcomes. [CLCE06-E20]
		(A) Records continual professional development activities undertaken including learning or insights gained. [CLCE06-30A]	

NOTES	In addition to items above, enterprise-level or independent 3rd-Party-generated evidence may be used to amplify other evidence presented and may include: a. Formally recognized by a reputable external organization as an expert in this competency area b. Evidence of role as independent assessor or reviewer on project outside own organization where skills in this competency area were used c. Evidence of invitation(s) from wider community for contribution of systems engineering expertise in this area (e.g. industry conference panel, government advisory board etc. cross-industry working groups, partnerships, accredited advanced university courses or research, or as part of professional institute) d. Formal commendation beyond the enterprise (e.g. by INCOSE or other recognized authority) for work performed in this competency area e. Independently assessed or accredited work product in this competency area (e.g. for independent publication or use) f. Accolades of expertise in this area from recognized industry leaders

Competency area – Core: Capability Engineering
Description A system or enterprise "capability" relates to the delivery of a desired effect (outcome) rather than delivery of a desired performance level (output). Capability is achieved by combining differing systems, products, and services, using a combination of people, processes, information, as well as equipment. Capability-based systems are enduring systems, adapting to changing situations rather than remaining static against a fixed performance baseline.
Why it matters Understanding the difference between capability and product is important in order to be able to separate the desired effect from any preconceived expectations as to its delivery mechanism. Even if the system of interest only delivers part of a wider capability, obtaining a good understanding of the wider intent helps keep options open, facilitating innovation and creativity in solutions. Failure to do this will generally result in sub-optimization and lost opportunities.
Possible contributory types of evidence Any combination of the types of evidence may be acceptable (depending on how the Framework is tailored and used). The evidence items identified at each level indicate example work products only. Contributions to work products will generally differ at each proficiency level.
Learning and development The INCOSE Professional Development Portal provides example guidance on how to gain an initial awareness of a competency area and options for developing further competence thereafter.

Awareness – Core: Capability Engineering		
ID	**Indicators of competence**	**Relevant knowledge sub-indicators**
1	Explains the concept of capability and how its use can prove beneficial. [CCPA01]	(K) Explains why capability includes people, information, organization, strategic goals, and the technical systems needed to achieve the aims of the super system owner. [CCPA01-10K] (K) Describes the concept of capability and its relationship to system requirements. [CCPA01-20K] (K) Explains how the use of capabilities to characterize systems can prove beneficial. [CCPA01-30K] (K) Describes the concept of "levels" within a system and a system hierarchy. [CCPA01-40K]
2	Explains how capability requirements can be satisfied by integrating several systems. [CCPA02]	(K) Describes the term systems of systems. [CCPA02-10K] (K) Explains how different organizations/teams may develop the individual systems. [CCPA02-20K]
3	Explains how super system capability needs impact on the development of each system that contributes to the capability. [CCPA03]	(K) Explains how different levels of a system interact and why this interaction may not be fully captured in an interface alone, i.e. the system affects the super system and vice versa. [CCPA03-10K] (K) Explains how the super system constraints/impacts the system of interest. [CCPA03-20K]
4	Describes the difficulties of translating capability needs of the wider system into system requirements. [CCPA04]	(K) Describes the conceptual mapping between capability and lower-level requirements. [CCPA04-10K] (K) Explains why there may be difficulties which might arise when trying to translate capability needs of the wider system into system requirements. [CCPA04-20K] (K) Explains why there may be a need for modeling or simulation to assist the mapping of super system capabilities to system requirements. [CCPA04-30K]

Supervised Practitioner – Core: Capability Engineering			
ID	**Indicators of competence** ***(in addition to those at awareness level)***	**Relevant knowledge, experience, and/or behaviors**	**Possible examples of objective evidence of personal contribution to activities performed, or professional behaviors applied**
1	Explains how project capability and environment are linked. [CCPS01]	(K) Explains what "emergence" is and how project environment may cause "emergence" to arise downstream on a project. [CCPS01-10K] (A) Identifies the context of a system of interest and its relationship to a super system. [CCPS01-20A] (A) Defines the nature of an "environment" of a project. [CCPS01-30A] (A) Identifies project capabilities and their associated requirements. [CCPS01-40A]	Project context diagram identifying super system and environmental elements. [CCPS01-E10] Project capabilities identifies and their associated requirements. [CCPS01-E20]
2	Identifies capability issues from the wider system, which will affect the design of a system of interest. [CCPS02]	(K) Explains how elements of the super system may be at different stages of the life cycle. [CCPS02-10K] (A) Identifies capability interfaces and interactions with the super system. [CCPS02-20A] (A) Identifies the mapping between a system of interest and the super system and vice versa. [CCPS02-30A]	System context definition document. [CCPS02-E10]
3	Prepares inputs to technology planning activities required in order to provide capability. [CCPS03]	(K) Explains how technology obsolescence requires planning and how it impacts requirements, design, and testing. [CCPS03-10K] (K) Explains why technology refresh may require planning and how it impacts requirements, design, and testing. [CCPS03-20K] (A) Identifies project technology readiness levels and how emerging technology impacts requirements, design, and testing. [CCPS03-30A] (A) Prepares project technology refresh or obsolescence for use in capability planning. [CCPS03-40A]	Records demonstrating Technology Plan contributions. [CCPS03-E10]
4	Prepares information that supports the embedding or utilization of capability. [CCPS04]	(K) Explains how capability utilization can affect the wider enterprise. [CCPS04-10K] (K) Explains why the concept of capability needs to be embedded at enterprise level to ensure utilization. [CCPS04-20K]	Organizational or Enterprise Capability Utilization plans. [CCPS04-E10] Organizational or Enterprise Capability Development Plans. [CCPS04-E20]

(Continued)

ID	Indicators of competence *(in addition to those at awareness level)*	Relevant knowledge, experience, and/or behaviors	Possible examples of objective evidence of personal contribution to activities performed, or professional behaviors applied
		(K) Explains how capability utilization can be improved through changes at enterprise level. [CCPS04-30K] (A) Prepares information that embeds or improves capability utilization. [CCPS04-40A]	
5	Identifies different elements that make up capability. [CCPS05]	(K) Explains how distinct parts or aspects of a project may combine to deliver own project or own enterprise capability. [CCPS05-10K] (A) Identifies own Organizational or Enterprise Capability Breakdown role and how responsibilities can contribute to a capability. [CCPS05-20A]	Organizational or Enterprise Capability Breakdown document. [CCPS05-E10] Records showing overlap between Capability definition and their own Role Definition. [CCPS05-E20]
6	Prepares multiple views that focus on value, purpose, and solution for capability. [CCPS06]	(A) Uses multiple views/focus on value, purpose, and solution for capability (rather than any one of these). [CCPS06-10A] (A) Prepares information documenting multiple views for a capability. [CCPS06-20A]	Records demonstrating contribution to the different viewpoints in capability. [CCPS06-E10] Trade-off studies or report contributions. [CCPS06-E20]
7	Develops own understanding of this competency area through Continual Professional Development (CPD). [CCPS07]	(A) Identifies potential gaps in own knowledge or development needs in this area, identifying opportunities to address these through continual professional development activities. [CCPS07-10A] (A) Performs continual professional development activities to improve their knowledge and understanding in this area. [CCPS07-20A] (A) Records continual professional development activities undertaken including learning or insights gained. [CCPS07-30A]	Records of Continual Professional Development (CPD) performed and learning outcomes. [CCPS07-E10]

Practitioner – Core: Capability Engineering

ID	Indicators of competence *(in addition to those at supervised practitioner level)*	Relevant knowledge, experience, and/or behaviors	Possible examples of objective evidence of personal contribution to activities performed, or professional behaviors applied
1	Identifies capability issues of the wider (super) system, which will affect the design of own system and translates these into system requirements. [CCPP01]	(A) Identifies the context in which a system operated in order to deliver a specific super system capability. [CCPP01-10A]	System Requirements Documentation showing capabilities identified. [CCPP01-E10]
		(A) Identifies the effect upon the design of the system changed super systems contexts. [CCPP01-20A]	User/system Capability Review contributions. [CCPP01-E20]
		(A) Develops a capability change successfully, dealing with implementation issues raised. [CCPP01-30A]	Capability Technical Reports or Review contributions. [CCPP01-E30]
		(A) Analyzes potential conflicts between capabilities at different levels and resolves issues found. [CCPP01-40A]	Capability Development or Enhancement Plan contributions. [CCPP01-E40]
		(A) Identifies the translation of requirements set against clear statements of capability. [CCPP01-50A]	
2	Reviews proposed system solutions to ensure their ability to deliver the capability required by a wider system, making changes as necessary. [CCPP02]	(A) Analyzes the extent to which the proposed system solution meets the super system capability. [CCPP02-10A]	Trade studies generated. [CCPP02-E10]
		(A) Identifies trade-offs to improve capability delivery if required. [CCPP02-20A]	Review contributions made. [CCPP02-E20]
		(A) Reviews and comments on capability trade-offs. [CCPP02-30A]	Technical reports generated. [CCPP02-E30]
3	Prepares technology plan that includes technology innovation, risk, maturity, readiness levels, and insertion points into existing capability. [CCPP03]	(A) Ensures that innovative technology (or technology innovation) is incorporated into existing capability. [CCPP03-10A]	Technology Plans, Readiness Plans, Obsolescence Plans. [CCPP03-E10]
		(K) Explains how factors such as risk, technology maturity, and readiness levels affected the implementation. [CCPP03-20K]	Technology risk mitigation items and actions completed from Risk Register. [CCPP03-E20]
		(A) Ensures that technology refresh requirements are incorporated into existing capability. [CCPP03-30A]	Capability Enhancement Plan contributions. [CCPP03-E30]
		(A) Uses existing plans to address technology obsolescence. [CCPP03-40A]	Technology planning Review records. [CCPP03-E40]

(Continued)

ID	Indicators of competence *(in addition to those at supervised practitioner level)*	Relevant knowledge, experience, and/or behaviors	Possible examples of objective evidence of personal contribution to activities performed, or professional behaviors applied
4	Creates an operational concept for a capability (what it does, why, how, where, when, and who). [CCPP04]	(A) Analyzes an operational concept addressing the "what," "why," "where," "when," "who," and "how" of the concept. [CCPP04-10A]	Technical Reports generated. [CCPP04-E10]
		(A) Prepares work products documenting a capability. [CCPP04-20A]	Operational Concept or Concept of Operations documentation. [CCPP04-E20]
5	Reviews existing capability to identify gaps relative to desired capability, documenting approaches which reduce or eliminate this deficit. [CCPP05]	(A) Identifies and addresses current or future gaps in technology capability to reduce the deficit. [CCPP05-10A]	Technology or Capability Enhancement Plans implemented. [CCPP05-E10]
		(A) Determines solutions that address gaps in technology capability. [CCPP05-20A]	Review contributions. [CCPP05-E20]
			Technical reports generated. [CCPP05-E30]
6	Prepares information that supports improvements to enterprise capabilities. [CCPP06]	(A) Identifies potential improvements to an enterprise capability. [CCPP06-10A]	Technical Reports generated. [CCPP06-E10]
		(A) Prepares work products in order to maintain an existing enterprise capability. [CCPP06-20A]	Review contributions. [CCPP06-E20]
			Technology Plans, Readiness Plans, Obsolescence Plans implemented. [CCPP06-E30]
7	Identifies key "pinch points" in the development and implementation of specific capability. [CCPP07]	(A) Identifies key "pinch points" in the development and implementation of specific capability. [CCPP07-10A]	Records demonstrating contribution to capability engineering improvements in this area. [CCPP07-E10]
8	Uses multiple views to analyze alignment, balance, and trade-offs in and between the different elements (in a level) ensuring that capability performance is not traded out. [CCPP08]	(A) Analyzes trade-offs between the different elements (in a level) to ensure alignment and balance. [CCPP08-10A]	Records demonstrating contribution to capability engineering improvements in this area. [CCPP08-E10]
		(A) Analyzes trade-offs between the different elements (in a level) to ensure that capability performance is not traded out. [CCPP08-20A]	Trade-off studies or report contributions. [CCPP08-E20]
		(A) Analyzes multiple views/focus on value, purpose, and solution for capability (rather than any one of these). [CCPP08-30A]	

9	Guides new or supervised practitioners in Capability engineering in order to develop their knowledge, abilities, skills, or associated behaviors. [CCPP09]	(P) Guides new or supervised practitioners in executing activities that form part of this competency. [CCPP09-10P] (A) Trains individuals to an "Awareness" level in this competency area. [CCPP09-20A]	Organizational Breakdown Structure showing their responsibility for technical supervision in this area. [CCPP09-E10] On-the-job training objectives/guidance. [CCPP09-E20] Coaching or mentoring assignment records. [CCPP09-E30] Records highlighting their impact on another individual in terms of improvement or professional development in this competency. [CCPP09-E40]
10	Maintains and enhances own competence in this area through Continual Professional Development (CPD) activities. [CCPP10]	(A) Identifies potential development needs in this area, identifying opportunities to address these through continual professional development activities. [CCPP10-10A] (A) Performs continual professional development activities to maintain and enhance their competency in this area. [CCPP10-20A] (A) Records continual professional development activities undertaken including learning or insights gained. [CCPP10-30A]	Records of Continual Professional Development (CPD) performed and learning outcomes. [CCPP10-E10]

Lead Practitioner – Core: Capability Engineering			
ID	**Indicators of competence *(in addition to those at practitioner level)***	**Relevant knowledge, experience, and/or behaviors**	**Possible examples of objective evidence of personal contribution to activities performed, or professional behaviors applied**
1	Creates enterprise-level policies, procedures, guidance, and best practice for capability engineering, including associated tools. [CCPL01]	(A) Analyzes enterprise need for capability policies, processes, tools, or guidance. [CCPL01-10A]	Records showing their role in embedding capability engineering into enterprise policies (e.g. guidance introduced at enterprise level, enterprise -level review minutes). [CCPL01-E10]
		(A) Creates enterprise policies, procedures, or guidance for capability activities. [CCPL01-20A]	Procedures they have written. [CCPL01-E20]
		(A) Selects and acquires appropriate tools supporting capability activities. [CCPL01-30A]	Records of support for tool introduction. [CCPL01-E30]
2	Judges the suitability of capability solutions and the planned approach on projects across the enterprise. [CCPL02]	(A) Evaluates the attributes of a capability strategy in the context of the project/program/domain/business to ensure a successful outcome. [CCPL02-10A]	Records demonstrating role as internal or external reviewer or consultant in the relevant areas. [CCPL02-E10]
		(A) Evaluates capability risks and proposes mitigation strategies to reduce project or enterprise risk. [CCPL02-20A]	
3	Identifies impact and changes needed in super system environment as a result of the capability development on behalf of the enterprise. [CCPL03]	(A) Identifies impact and changes needed in super system environment as a result of the capability development. [CCPL03-10A]	Records demonstrating contribution to capability engineering improvements in this area. [CCPL03-E10]
4	Identifies improvements required to enterprise capabilities on behalf of the enterprise. [CCPL04]	(A) Analyzes enterprise capability against current and future needs. [CCPL04-10A]	Technical Reports generated. [CCPL04-E10]
		(A) Evaluates changes required in order to maintain or develop an existing enterprise capability. [CCPL04-20A]	Review contributions. [CCPL04-E20]
		(A) Proposes changes to an existing enterprise capability to align with enterprise needs. [CCPL04-30A]	Technology Plans, Readiness Plans, Obsolescence Plans implemented. [CCPL04-E30]
5	Coaches or mentors practitioners across the enterprise in capability engineering in order to develop their knowledge, abilities, skills, or associated behaviors. [CCPL05]	(P) Coaches or mentors practitioners across the enterprise in competency-related techniques, recommending development activities. [CCPL05-10P]	Coaching or mentoring assignment records. [CCPL05-E10]
		(A) Develops or authorizes enterprise training materials in this competency area. [CCPL05-20A]	Records of formal training courses, workshops, seminars, and authored training material supported by successful post-training evaluation data. [CCPL05-E20]
		(A) Provides enterprise workshops/seminars or training in this competency area. [CCPL05-30A]	Listing as an approved organizational trainer for this competency area. [CCPL05-E30]

6	Promotes the introduction and use of novel techniques and ideas in Capability Engineering across the enterprise, to improve enterprise competence in the area. [CCPL06]	(A) Analyzes different approaches across different domains through research. [CCPL06-10A]	Research records. [CCPL06-E10]
		(A) Defines novel approaches that could potentially improve the SE discipline within the enterprise. [CCPL06-20A]	Published papers in refereed journals/company literature. [CCPL06-E20]
		(P) Fosters awareness of these novel techniques within the enterprise. [CCPL06-30P]	Records showing introduction of enabling systems supporting the new techniques or ideas. [CCPL06-E30]
		(P) Collaborates with enterprise stakeholders to facilitate the introduction of techniques new to the enterprise. [CCPL06-40P]	Published papers (or similar) at enterprise level. [CCPL06-E40]
		(A) Monitors new techniques after their introduction to determine their effectiveness. [CCPL06-50A]	Records of improvements made against a recognized process improvement model in this area. [CCPL06-E50]
		(A) Adapts approach to reflect actual enterprise performance improvements. [CCPL06-60A]	
7	Develops expertise in this competency area through specialist Continual Professional Development (CPD) activities. [CCPL07]	(A) Identifies own needs for further professional development in order to increase competence beyond practitioner level. [CCPL07-10A]	Records of Continual Professional Development (CPD) performed and learning outcomes. [CCPL07-E10]
		(A) Performs professional development activities in order to move own competence toward expert level. [CCPL07-20A]	
		(A) Records continual professional development activities undertaken including learning or insights gained. [CCPL07-30A]	

NOTES	In addition to items above, enterprise-level or independent 3rd-Party-generated evidence may be used to amplify other evidence presented and may include: a. Formally recognized by senior management in current organization as an expert in this competency area b. Evidence of role as Product/System Design Authority or Technical Authority on a complex project with responsibilities in this area or where skills within this competency area were used c. Recognized as an authorizing signatory on behalf of enterprise for formal documentation in this competency area (e.g. policies, processes, and deliverables) d. Formal commendation or award within own enterprise for contribution or item of work successfully performed, which required proficiency in this competency area e. Customer, Supplier, or other external project-specific key Stakeholder accolades for specific work performed in this competency area f. Independently assessed or accredited work in this competency area (e.g. for independent publication or use) g. Formal organizational HR records positively highlighting any specific professional competencies or behaviors identified (if applicable) plus any of the evidence indicators listed at Expert level below

Expert – Core: Capability Engineering			
ID	**Indicators of competence** ***(in addition to those at Lead Practitioner level)***	**Relevant knowledge, experience, and/or behaviors**	**Possible examples of objective evidence of personal contribution to activities performed, or professional behaviors applied**
1	Communicates own knowledge and experience in capability engineering in order to improve best practice beyond the enterprise boundary. [CCPE01]	(A) Produces papers, seminars, or presentations outside own enterprise for publication in order to share own ideas and improve industry best practices in this competence area. [CCPE01-10A]	Published papers or books etc. on new technique in refereed journals/company literature. [CCPE01-E10]
		(P) Fosters incorporation of own ideas into industry best practices in this area. [CCPE01-20P]	Published papers in refereed journals or internal literature proposing new practices in this competence area (or presentations, tutorials, etc.). [CCPE01-E20]
		(P) Develops guidance materials identifying new (or updating existing) best practice in this competence area. [CCPE01-30P]	Own proposals adopted as industry best practices in this competence area. [CCPE01-E30]
2	Advises organizations beyond the enterprise boundary on the suitability of their approach to capability engineering. [CCPE02]	(A) Advises on the suitability of capability solutions or derived needs for the systems that make up the capability. [CCPE02-10A]	Records demonstrating internal or external advisory or consultative role in capability engineering and management. [CCPE02-E10]
		(A) Advises on capability strategies. [CCPE02-20A]	
3	Advises organizations beyond the enterprise boundary on their handling of complex or sensitive capability engineering strategy issues. [CCPE03]	(A) Advises stakeholders beyond enterprise boundary to improve their handling of complex or sensitive capability strategy issues, arbitrating as necessary. [CCPE03-10A]	Records demonstrating internal or external advisory or consultative role in capability engineering and management. [CCPE03-E10]
			Records of advice provided together with evidence that the issues advised on were by their nature either complex or sensitive. [CCPE03-E20]
4	Advises organizations beyond the enterprise boundary on differences and relationships between capability and product-based systems. [CCPE04]	(A) Advises on differences and relationships between capability and product-based systems. [CCPE04-10A]	Records demonstrating internal or external advisory or consultative role in capability engineering and management. [CCPE04-E10]
5	Assesses capability engineering in multiple domains beyond the enterprise boundary in order to develop or improve capability solutions within own enterprise. [CCPE05]	(A) Analyzes different approaches across different domains through research. [CCPE05-10A]	Research records. [CCPE05-E10]
		(A) Develops improvements in capabilities within the enterprise to incorporate knowledge from other systems or domains. [CCPE05-20A]	Published papers in refereed journals on this topic. [CCPE05-E20]
			Published articles or books etc. [CCPE05-E30]
			Records demonstrating the successful application of capability engineering research in other domains to own domain. [CCPE05-E40]

6	Champions the introduction of novel techniques and ideas in capability engineering, beyond the enterprise boundary, in order to develop the wider Systems Engineering community in this competency. [CCPE06]	(A) Analyzes different approaches across different domains through research. [CCPE06-10A]	Records of activities promoting research and need to adopt novel technique or ideas. [CCPE06-E10]
		(A) Produces reports for the wider SE community on the effectiveness of new techniques after their introduction. [CCPE06-20A]	Records of improvements made to process and appraisal against a recognized process improvement model. [CCPE06-E20]
		(P) Collaborates with those introducing novel techniques within the wider SE community. [CCPE06-30P]	Research records. [CCPE06-E30]
		(A) Defines novel approaches that could potentially improve the wider SE discipline. [CCPE06-40A]	Published papers in refereed journals/company literature. [CCPE06-E40]
		(P) Fosters awareness of these novel techniques within the wider SE community. [CCPE06-50P]	Records showing introduction of enabling systems supporting the new techniques or ideas. [CCPE06-E50]
7	Coaches individuals beyond the enterprise boundary in capability engineering, in order to further develop their knowledge, abilities, skills, or associated behaviors. [CCPE07]	(P) Coaches or mentors individuals beyond the enterprise boundary, in competency-related techniques, recommending development activities. [CCPE07-10P]	Coaching or mentoring assignment records. [CCPE07-E10]
		(A) Develops or authorizes training materials in this competency area, which are subsequently successfully delivered beyond the enterprise boundary. [CCPE07-20A]	Records of formal training courses, workshops, seminars, and authored training material supported by successful post-training evaluation data. [CCPE07-E20]
		(A) Provides workshops/seminars or training in this competency area for practitioners or lead practitioners beyond the enterprise boundary (e.g. conferences and open training days). [CCPE07-30A]	Records of training/workshops/seminars created supported by successful post-training evaluation data. [CCPE07-E30]
8	Maintains expertise in this competency area through specialist Continual Professional Development (CPD) activities. [CCPE08]	(A) Reviews research, new ideas, and state of the art to identify relevant new areas requiring personal development in order to maintain expertise in this competency area. [CCPE08-10A]	Records of documents reviewed and insights gained as part of own research into this competency area. [CCPE08-E10]
		(A) Performs identified specialist professional development activities in order to maintain or further develop competence at expert level. [CCPE08-20A]	Records of Continual Professional Development (CPD) performed and learning outcomes. [CCPE08-E20]
		(A) Records continual professional development activities undertaken including learning or insights gained. [CCPE08-30A]	

NOTES	In addition to items above, enterprise-level or independent 3rd-Party-generated evidence may be used to amplify other evidence presented and may include: a. Formally recognized by a reputable external organization as an expert in this competency area b. Evidence of role as independent assessor or reviewer on project outside own organization where skills in this competency area were used c. Evidence of invitation(s) from wider community for contribution of systems engineering expertise in this area (e.g. industry conference panel and government advisory board cross-industry working groups, partnerships, accredited advanced university courses or research, or as part of professional institute) d. Formal commendation beyond the enterprise (e.g. by INCOSE or other recognized authority) for work performed in this competency area e. Independently assessed or accredited work product in this competency area (e.g. for independent publication or use) f. Accolades of expertise in this area from recognized industry leaders

Competency area – Core: General Engineering
Description Foundational concepts in mathematics, science, and engineering and their application. Provides depth and breadth subject understanding beyond typical specialty depth.
Why it matters Systems Engineering is performed in a technical scientific environment and as a result, a good understanding of mathematics, science coupled with a sound appreciation of core engineering principles is a critical foundation for effective Systems Engineering. Without this, systems engineers cannot communicate effectively and efficiently with engineers from other domains. Typical systems engineering profile is of someone with depth in a particular area and general top-level awareness across other areas. General engineering is someone with depth across several areas providing greater knowledge breath.
Possible contributory types of evidence Any combination of the types of evidence may be acceptable (depending on how the Framework is tailored and used). The evidence items identified at each level indicate example work products only. Contributions to work products will generally differ at each proficiency level.
Learning and development The INCOSE Professional Development Portal provides example guidance on how to gain an initial awareness of a competency area and options for developing further competence thereafter.

Awareness – Core: General Engineering		
ID	**Indicators of competence**	**Relevant knowledge sub-indicators**
1	Explains core principles of science and mathematics applicable to engineering. [CGEA01]	(K) Describes basic mathematical theories apply to physics and engineering. [CGEA01-10K] (K) Describes shortcomings of theoretical foundations and need for experimental, stochastic, numerical solutions. [CGEA01-20K] (K) Describes a central mathematical theory and its possible application to engineering principles. [CGEA01-30K]
2	Explains fundamentals of engineering as a discipline. [CGEA02]	(K) Lists key engineering fundamentals of a selected engineering area. [CGEA02-10K] (K) Describes the context of a selected engineering application and theories applicable to this application. [CGEA02-20K] (K) Describes key interfaces and links between engineering functions and stakeholders in the wider enterprise outside engineering (e.g. infrastructure management, human resource management, quality management, knowledge management, portfolio management, and life cycle model management). [CGEA02-30K]
3	Explains why probability and statistics are both relevant to engineering. [CGEA03]	(K) Describes aspects of probability and statistics applicable to core engineering activities. [CGEA03-10K] (K) Describes some shortcomings of theoretical foundations and the consequential need for experimental, stochastic, numerical solutions. [CGEA03-20K]

4	Explains why analytical methods and sound judgment are central to engineering decisions. [CGEA04]	(K) Describes the principles of an analytical method applicable to a selected engineering discipline. [CGEA04-10K] (K) Describes the principal foundations of a mathematical or scientific theory and summarizes the understanding it provides for problem solving and reasoning. [CGEA04-20K] (K) Explains how sound judgement would normally be required in addition to mathematical results within the engineering discipline. [CGEA04-30K]
5	Explains the characteristics of an engineered system. [CGEA05]	(K) Defines key concepts and underlying differences between "engineered" and "natural" systems. [CGEA05-10K] (K) Explains how different engineering disciplines relate to differing types of systems and how they interact with each other. [CGEA05-20K]
6	Describes engineered systems that are physical, software, and socio-technical systems or combinations thereof. [CGEA06]	(K) Defines key concepts and underlying differences between "physical," "software," and "socio-technical" systems. [CGEA06-10K] (K) Lists examples of physical, software, and socio-technical engineered systems and combinations of these types. [CGEA06-20K]
7	Explains how different sciences impact the technology domain and the Systems Engineering discipline. [CGEA07]	(K) Explains how different sciences impact the technology domain, such as cybersecurity or artificial intelligence. [CGEA07-10K]
8	Explains why uncertainty is an important factor in engineering and explains how it might arise from many sources. [CGEA08]	(K) Describes potential sources of uncertainty in differing types of system. [CGEA08-10K] (K) Describes potential sources of uncertainty in differing life cycle types (e.g. spiral, waterfall, incremental, iterative, agile). [CGEA08-20K] (K) Explains how uncertainty might be reduced or eliminated (e.g. through additional analysis, mockups, simulation, expert opinion, and other techniques). [CGEA08-30K]

Supervised Practitioner – Core: General Engineering			
ID	**Indicators of competence** ***(in addition to those at awareness level)***	**Relevant knowledge, experience, and/or behaviors**	**Possible examples of objective evidence of personal contribution to activities performed, or professional behaviors applied**
1	Uses scientific and mathematical knowledge when performing engineering tasks. [CGES01]	(A) Uses underlying engineering principles to achieve desired outcomes. [CGES01-10A]	Contributions to work products containing scientific or mathematical study and analysis. [CGES01-E10]
		(A) Uses basic analytical methods and stochastic data analysis methods and techniques. [CGES01-20A]	Contributions to scientific or mathematical plans and reports. [CGES01-E20]
2	Uses appropriate engineering approaches when performing engineering tasks. [CGES02]	(K) Explains how shortcomings of basic theoretical foundations may impact the activities they perform and the steps taken to address this. [CGES02-10K]	Contributions to and analysis work products which used engineering approaches to support a task. [CGES02-E10]
		(K) Describes phenomenological aspects and extension to theory applicable to engineering activities they have performed. [CGES02-20K]	Contributions to a study plan or report where they have used engineering approaches to support a task. [CGES02-E20]
		(A) Selects engineering approaches appropriate to different types of engineering problem. [CGES02-30A]	
		(A) Uses selected engineering approach to address different types of engineering problem. [CGES02-40A]	
3	Explains the concept of "variation" and its effect in engineering tasks. [CGES03]	(K) Explains how probability and random process theory may affect the work they are performing, and the steps taken to address this. [CGES03-10K]	Contributions to a stochastic data analysis. [CGES03-E10]
		(A) Identifies variability between theory and practice on a project. [CGES03-20A]	Contributions to study plans and reports dealing with variability. [CGES03-E20]
			Modeling and simulation performed where variability was a factor. [CGES03-E30]
4	Uses proven analytical methods when performing engineering tasks, while appreciating the limitations of their applicability. [CGES04]	(A) Uses basic analytical methods to support their work. [CGES04-10A]	Contributions to stochastic data analysis. [CGES04-E10]
		(A) Identifies variations from theory in real-world applications. [CGES04-20A]	Contributions to study plans and reports dealing with variability. [CGES04-E20]
		(A) Uses phenomenological principles to study events. [CGES04-30A]	Reports highlighting limitation of Modeling and simulation analysis. [CGES04-E30]

5	Uses software-based tools, together with the products they create, to facilitate and progress engineering tasks. [CGES05]	(K) Explains how software-based tools and products can be used in support of the resolution of a variety of different engineering problems. [CGES05-10K] (A) Uses software-based tools or products to address specific engineering problems. [CGES05-20A] (A) Uses software-based tools and products in support of engineering problems and their potential or actual limitations. [CGES05-30A]	Software-based modeling and simulation outputs. [CGES05-E10]
6	Explains why the value of a mathematical approach can be limited in a human-centric or human-originated system, with examples. [CGES06]	(K) Explains how human-originated systems may have limitations, which produce variation from mathematical theory to real-world practice. [CGES06-10K] (A) Identifies variation from mathematical theory to real-world practice due to the effect of humans within the system. [CGES06-20A]	Contributions to mathematical modeling or simulation where theory and reality differed. [CGES06-E10]
7	Acts creatively or innovatively when performing own activities. [CGES07]	(A) Uses creative or innovative approaches when performing own activities. [CGES07-10A]	Project records, agendas, organization charts, minutes, or reports demonstrating leading projects with creativity or innovation. [CGES07-E10]
8	Develops own understanding of this competency area through Continual Professional Development (CPD). [CGES08]	(A) Identifies potential gaps in own knowledge or development needs in this area, identifying opportunities to address these through continual professional development activities. [CGES08-10A] (A) Performs continual professional development activities to improve their knowledge and understanding in this area. [CGES08-20A] (A) Records continual professional development activities undertaken including learning or insights gained. [CGES08-30A]	Records of Continual Professional Development (CPD) performed and learning outcomes. [CGES08-E10]

Practitioner – Core: General Engineering			
ID	**Indicators of competence** ***(in addition to those at supervised practitioner level)***	**Relevant knowledge, experience, and/or behaviors**	**Possible examples of objective evidence of personal contribution to activities performed, or professional behaviors applied**
1	Selects software-based tools together with the products they create to facilitate and progress engineering tasks. [CGEP01]	(A) Identifies software-based tools available for selection on the project. [CGEP01-10A] (A) Explains how specific software-based tools were selected as appropriate including the criteria used. [CGEP01-20A]	Trade study analysis reports. [CGEP01-E10] Reports commissioned on tools or techniques. [CGEP01-E20] Tool output work products. [CGEP01-E30]
2	Uses selected software-based tools together with the products they create to facilitate and progress engineering tasks. [CGEP02]	(K) Explains how software tools used in support of engineering can facilitate progress. [CGEP02-10K] (A) Uses a variety of software tools in support of engineering, summarizing their differing characteristics and the progress they have facilitated. [CGEP02-20A]	Trade study analysis reports. [CGEP02-E10] Reports commissioned on tools or techniques. [CGEP02-E20] Tool output work products. [CGEP02-E30]
3	Selects scientific and mathematical methods to be used in support of tasks and justifies their selection. [CGEP03]	(A) Identifies scientific and mathematical models available to support specific engineering tasks. [CGEP03-10A] (A) Reviews the theoretical strengths and weaknesses of available scientific and mathematical models in the context of the task for which they were being considered. [CGEP03-20A] (A) Reviews the practical strengths and weaknesses of available scientific and mathematical models in the context of the task for which they were being considered. [CGEP03-30A] (A) Selects appropriate specific scientific and mathematical models from those available to support specific engineering tasks, justifying the selection. [CGEP03-40A]	Trade study analysis reports. [CGEP03-E10] Reports commissioned on tools or techniques. [CGEP03-E20]
4	Selects relevant engineering approaches and methods to be used in support of engineering tasks and justifies their selection. [CGEP04]	(A) Identifies engineering approaches and methods available to support specific engineering tasks. [CGEP04-10A] (A) Reviews the strengths and weaknesses of available engineering approaches and methods in the context of the task for which they were being considered. [CGEP04-20A] (A) Reviews the practical strengths and weaknesses of available engineering approaches and methods in the context of the task for which they were being considered. [CGEP04-30A]	Systems Engineering plan. [CGEP04-E10] Trade studies for different approaches. [CGEP04-E20] Test results, analysis reports, corrective action findings. [CGEP04-E30]

		(A) Adapts specific engineering approaches and methods to address project need. [CGEP04-40A]	
		(A) Selects appropriate engineering approaches and methods from those available to support specific engineering tasks, justifying their selection. [CGEP04-50A]	
5	Uses probability and statistics in engineering tasks recognizing benefits and limitations on results obtained. [CGEP05]	(A) Reviews the strengths and weaknesses of specific probability and statistical methods in the context of the task for which they were being considered. [CGEP05-10A]	Trade study reports. [CGEP05-E10]
		(A) Selects appropriate probabilistic and statistical from those available to support specific engineering tasks, justifying the selection. [CGEP05-20A]	Engineering reports or work products. [CGEP05-E20]
		(A) Uses selected probability and statistics in support of specific engineering tasks. [CGEP05-30A]	Data analysis reviews. [CGEP05-E30]
		(A) Reviews data analysis results for accuracy. [CGEP05-40A]	Test results, analysis reports, corrective action findings. [CGEP05-E40]
6	Determines the level of variation or probability appropriate to the current task and justifies this decision. [CGEP06]	(A) Identifies how probabilistic or statistical variability exists within the context of a task they have performed. [CGEP06-10A]	Engineering reports or work products. [CGEP06-E10]
		(A) Creates probabilistic and statistical variability limits for specific engineering tasks, justifying the limits defined. [CGEP06-20A]	Data analysis reviews. [CGEP06-E20]
			Test results, analysis reports, corrective action findings. [CGEP06-E30]
7	Uses practical and proven engineering principles to structure engineering tasks. [CGEP07]	(A) Identifies goals and constraints from their project and how they may have impacted selection of engineering approaches and methods. [CGEP07-10A]	Contributions to Systems Engineering Management Plan (SEMP) or similar document. [CGEP07-E10]
		(K) Explains how engineering tasks were selected, planned, and executed in a particular way in order to address the specific needs, constraints, or challenges of the project. [CGEP07-20K]	Trade studies for different approaches. [CGEP07-E20]
			Test strategies or test Plans. [CGEP07-E30]
8	Reviews situations using well-established engineering principles and uses these assessments to make sound engineering judgments. [CGEP08]	(K) Identifies day-to-day project situations that could require an engineering technical review. [CGEP08-10K]	Study reports, whitepapers. [CGEP08-E10]
		(K) Identifies engineering technical review principles. [CGEP08-20K]	Trade studies for different approaches. [CGEP08-E20]
		(A) Reviews engineering situations using well-established engineering review principles to achieve a successful outcome. [CGEP08-30A]	Review minutes. [CGEP08-E30]

(*Continued*)

ID	Indicators of competence *(in addition to those at supervised practitioner level)*	Relevant knowledge, experience, and/or behaviors	Possible examples of objective evidence of personal contribution to activities performed, or professional behaviors applied
		(A) Records engineering review activities and judgments made. [CGEP08-40A] (A) Records judgments made. [CGEP08-50A]	Team briefings on key decisions made. [CGEP08-E40]
9	Adapts engineering approaches to take account of human-centric aspects of systems development. [CGEP09]	(A) Identifies the impact of human factors on an engineering task. [CGEP09-10A] (A) Uses human-factors standards or guidelines to adapt their own development activities. [CGEP09-20A]	Human factors requirements, analysis, or design adapted. [CGEP09-E10] Trade studies for different approaches. [CGEP09-E20] Review minutes. [CGEP09-E30]
10	Uses creative or innovative approaches when performing project activities. [CGEP10]	(P) Acts to support creative or innovative approaches in others when performing project activities. [CGEP10-10P] (P) Acts creatively or innovatively to overcome wider project-level challenges. [CGEP10-20P]	Project records, agendas, organization charts, minutes, or reports demonstrating leading projects with creativity or innovation. [CGEP10-E10] Records of actions implemented creatively or innovatively to overcome wider project-level challenges. [CGEP10-E20]
11	Guides new or supervised practitioners in Core Engineering principles in order to develop their knowledge, abilities, skills, or associated behaviors. [CGEP11]	(P) Guides new or supervised practitioners in executing activities that form part of this competency. [CGEP11-10P] (A) Trains individuals to an "Awareness" level in this competency area. [CGEP11-20A]	Organizational Breakdown Structure showing their responsibility for technical supervision in this area. [CGEP11-E10] On-the-job training records. [CGEP11-E20] Coaching or mentoring assignment records. [CGEP11-E30] Records highlighting their impact on another individual in terms of improvement or professional development in this competency. [CGEP11-E40]
12	Maintains and enhances own competence in this area through Continual Professional Development (CPD) activities. [CGEP12]	(A) Identifies potential development needs in this area, identifying opportunities to address these through continual professional development activities. [CGEP12-10A] (A) Performs continual professional development activities to maintain and enhance their competency in this area. [CGEP12-20A] (A) Records continual professional development activities undertaken including learning or insights gained. [CGEP12-30A]	Records of Continual Professional Development (CPD) performed and learning outcomes. [CGEP12-E10]

Lead Practitioner – Core: General Engineering

ID	Indicators of Competence *(in addition to those at Practitioner level)*	Relevant knowledge, experience, and/or behaviors	Possible examples of objective evidence of personal contribution to activities performed, or professional behaviors applied
1	Creates enterprise-level policies, procedures, guidance, and best practice for general engineering, including associated tools. [CGEL01]	(A) Analyzes enterprise need for general engineering policies, processes, tools, or guidance. [CGEL01-10A]	Records showing their role in embedding general engineering approaches into enterprise policies (e.g. guidance introduced at enterprise level, enterprise-level review minutes). [CGEL01-E10]
		(A) Creates enterprise policies, procedures, or guidance for general engineering activities. [CGEL01-20A]	Procedures they have written. [CGEL01-E20]
		(A) Selects and acquires appropriate tools supporting general engineering activities. [CGEL01-30A]	Records of support for tool introduction. [CGEL01-E30]
2	Adapts mathematics and engineering principles so that they can be applied to specific engineering situations on behalf of the enterprise. [CGEL02]	(A) Reviews standard mathematical and engineering principles and modifies for their applicability within the project-specific context. [CGEL02-10A]	Documentation demonstrating mathematics and engineering principles adapted to address project need. [CGEL02-E10]
		(A) Adapts standard mathematical and engineering principles to address project-specific conflicts where appropriate. [CGEL02-20A]	
3	Communicates the difference between scientific and engineering approaches in order to engage with pure scientific advances on behalf of the enterprise. [CGEL03]	(P) Communicates the differences between scientific and engineering approaches to stakeholders. [CGEL03-10P]	Studies and analysis performed. [CGEL03-E10]
		(A) Proposes the investigation of new scientific advance(s). [CGEL03-20A]	Records of recommendations made to perform scientific investigations. [CGEL03-E20]
4	Assesses items across the enterprise for the appropriate level of understanding of impact of variation and uncertainty on engineering outcomes. [CGEL04]	(A) Determines an appropriate level of variability or uncertainty in assessing potential system performance and engineering outcomes. [CGEL04-10A]	Records showing contribution to projects in the area of understanding variation and uncertainty. [CGEL04-E10]
		(A) Uses an appropriate level of stochastic analysis to improve understanding of system performance and engineering outcomes. [CGEL04-20A]	
5	Advises stakeholders across the enterprise on issues requiring the application of engineering judgment. [CGEL05]	(A) Uses engineering knowledge in order to advise the wider organization on issues requiring the application of engineering judgment where this advice has been formally acknowledged. [CGEL05-10A]	Records of advice provided on engineering judgments made at project and enterprise levels. [CGEL05-E10]

(Continued)

ID	Indicators of Competence *(in addition to those at Practitioner level)*	Relevant knowledge, experience, and/or behaviors	Possible examples of objective evidence of personal contribution to activities performed, or professional behaviors applied
6	Fosters creative or innovative approaches to performing general engineering activities across the enterprise. [CGEL06]	(P) Fosters creatively or innovatively in general engineering activities across the enterprise. [CGEL06-10P]	Enterprise-level records, agendas, organization charts, minutes, or reports demonstrating leading projects with creativity or innovation. [CGEL06-E10]
		(P) Acts creatively or innovatively to overcome wider enterprise-level challenges. [CGEL06-20P]	Records of actions implemented creatively or innovatively to overcome wider enterprise-level challenges. [CGEL06-E20]
7	Advises stakeholders across the enterprise on issues affecting the "broader" engineering approach to engineering activities. [CGEL07]	(A) Uses engineering knowledge in order to advise the wider organization on broader engineering issues and approaches made. [CGEL07-10A]	Records documenting advice provided on broader engineering approach to be used at project and enterprise levels. [CGEL07-E10]
8	Judges the quality of engineering judgments made by others across the enterprise. [CGEL08]	(A) Reviews the quality of general engineering judgments made across the enterprise. [CGEL08-10A]	Review comments made on engineering judgments made by others. [CGEL08-E10]
		(P) Challenges the quality of general engineering judgments made across the enterprise, resulting in improvement. [CGEL08-20P]	Changes made as a result of assessments made. [CGEL08-E20]
9	Coaches or mentors practitioners across the enterprise in general engineering techniques in order to develop their knowledge, abilities, skills, or associated behaviors. [CGEL09]	(P) Coaches or mentors practitioners across the enterprise in competency-related techniques, recommending development activities. [CGEL09-10P]	Coaching or mentoring assignment records. [CGEL09-E10]
		(A) Develops or authorizes enterprise training materials in this competency area. [CGEL09-20A]	Records of formal training courses, workshops, seminars, and authored training material supported by successful post-training evaluation data. [CGEL09-E20]
		(A) Provides enterprise workshops/seminars or training in this competency area. [CGEL09-30A]	Listing as an approved organizational trainer for this competency area. [CGEL09-E30]
10	Promotes the introduction and use of novel techniques and ideas in general engineering across the enterprise to improve enterprise competence in this area. [CGEL10]	(A) Analyzes different approaches across different domains through research. [CGEL10-10A]	Research records. [CGEL10-E10]
		(A) Defines novel approaches that could potentially improve the SE discipline within the enterprise. [CGEL10-20A]	Published papers in refereed journals/company literature. [CGEL10-E20]
		(P) Fosters awareness of these novel techniques within the enterprise. [CGEL10-30P]	Records showing introduction of enabling systems supporting the new techniques or ideas. [CGEL10-E30]

		(P) Collaborates with enterprise stakeholders to facilitate the introduction of techniques new to the enterprise. [CGEL10-40P]	Published papers (or similar) at enterprise level. [CGEL10-E40]
		(A) Monitors new techniques after their introduction to determine their effectiveness. [CGEL10-50A]	Records of improvements made against a recognized process improvement model. [CGEL10-E50]
		(A) Adapts approach to reflect actual enterprise performance improvements. [CGEL10-60A]	
11	Develops expertise in this competency area through specialist Continual Professional Development (CPD) activities. [CGEL11]	(A) Identifies own needs for further professional development in order to increase competence beyond practitioner level. [CGEL11-10A] (A) Performs professional development activities in order to move own competence toward expert level. [CGEL11-20A] (A) Records continual professional development activities undertaken including learning or insights gained. [CGEL11-30A]	Records of Continual Professional Development (CPD) performed and learning outcomes. [CGEL11-E10]

NOTES

In addition to items above, enterprise-level or independent 3rd-Party-generated evidence may be used to amplify other evidence presented and may include:

a. Formally recognized by senior management in current organization as an expert in this competency area
b. Evidence of role as Product/System Design Authority or Technical Authority on a complex project with responsibilities in this area or where skills within this competency area were used
c. Recognized as an authorizing signatory on behalf of enterprise for formal documentation in this competency area (e.g. policies, processes, and deliverables)
d. Formal commendation or award within own enterprise for contribution or item of work successfully performed, which required proficiency in this competency area
e. Customer, Supplier, or other external project-specific key Stakeholder accolades for specific work performed in this competency area
f. Independently assessed or accredited work in this competency area (e.g. for independent publication or use)
g. Formal organizational HR records positively highlighting any specific professional competencies or behaviors identified (if applicable) plus any of the evidence indicators listed at Expert level below

Expert – Core: General Engineering			
ID	**Indicators of Competence** ***(in addition to those at Lead Practitioner level)***	**Relevant knowledge, experience, and/or behaviors**	**Possible examples of objective evidence of personal contribution to activities performed, or professional behaviors applied**
1	Communicates own knowledge and experience in general engineering in order to improve best practice beyond the enterprise boundary. [CGEE01]	(A) Produces papers, seminars, or presentations outside own enterprise for publication in order to share own ideas and improve industry best practices in this competence area. [CGEE01-10A] (P) Fosters incorporation of own ideas into industry best practices in this area. [CGEE01-20P] (P) Develops guidance materials identifying new (or updating existing) best practice in this competence area. [CGEE01-30P]	Published papers or books etc. on new technique in refereed journals/company literature. [CGEE01-E10] Published papers in refereed journals or internal literature proposing new practices in this competence area (or presentations, tutorials, etc.). [CGEE01-E20] Own proposals adopted as industry best practices in this competence area. [CGEE01-E30]
2	Advises organizations beyond the enterprise boundary on the suitability of their approach to general engineering activities. [CGEE02]	(A) Advises on the suitability of general engineering solutions. [CGEE02-10A] (A) Advises on general engineering strategies. [CGEE02-20A]	Records demonstrating internal or external advisory or consultative role in general engineering and management. [CGEE02-E10]
3	Develops new applications of mathematical methods to engineering practices applicable beyond the enterprise boundary. [CGEE03]	(A) Produces research to understand the application of mathematical methods to general engineering practice in different domains. [CGEE03-10A]	Published papers in refereed journals on this topic. [CGEE03-E10] Published articles or books etc. [CGEE03-E20]
4	Advises organizations beyond the enterprise boundary on their handling of complex general engineering challenges. [CGEE04]	(A) Advises stakeholders on their general engineering requirements and issues. [CGEE04-10A] (P) Communicates effectively on highly complex system making limited use of specialized, technical terminology. [CGEE04-20P] (A) Uses a holistic approach to complex issue resolution including balanced, rational arguments on way forward. [CGEE04-30A]	Records showing internal or external advisory or consultative role in general engineering and management. [CGEE04-E10]
5	Maintains own awareness of developments in sciences, technologies, and related engineering disciplines beyond the enterprise boundary, recognizing areas where new developments might be applicable within their own discipline or enterprise. [CGEE05]	(A) Analyzes applicable research developments in sciences, technologies, and related engineering to evaluate their potential applicability to general engineering practice in different domains. [CGEE05-10A]	Papers, publications, books. [CGEE05-E10] Records showing Society participation. [CGEE05-E20]
6	Fosters creative or innovative approaches to performing general engineering activities beyond the enterprise boundary. [CGEE06]	(P) Fosters creatively or innovatively in general engineering activities beyond the enterprise boundary. [CGEE06-10P] (P) Acts creatively or innovatively to overcome wider general engineering challenges beyond the enterprise boundary. [CGEE06-20P]	Records showing creativity or innovation support beyond enterprise boundary. [CGEE06-E10] Records showing successful creative or innovative ventures beyond enterprise boundary. [CGEE06-E20]

7	Champions the introduction of novel techniques and ideas in general engineering, beyond the enterprise boundary, in order to develop the wider Systems Engineering community in this competency. [CGEE07]	(A) Analyzes different approaches across different domains through research. [CGEE07-10A]	Records of activities promoting research and need to adopt novel technique or ideas. [CGEE07-E10]
		(A) Produces reports for the wider SE community on the effectiveness of new techniques after their introduction. [CGEE07-20A]	Records of improvements made to process and appraisal against a recognized process improvement model. [CGEE07-E20]
		(P) Collaborates with those introducing novel techniques within the wider SE community. [CGEE07-30P]	Research records. [CGEE07-E30]
		(A) Defines novel approaches that could potentially improve the wider SE discipline. [CGEE07-40A]	Published papers in refereed journals/company literature. [CGEE07-E40]
		(P) Fosters awareness of these novel techniques within the wider SE community. [CGEE07-50P]	Records showing introduction of enabling systems supporting the new techniques or ideas. [CGEE07-E50]
8	Coaches individuals beyond the enterprise boundary in general engineering techniques in order to further develop their knowledge, abilities, skills, or associated behaviors. [CGEE08]	(P) Coaches or mentors individuals beyond the enterprise boundary, in competency-related techniques, recommending development activities. [CGEE08-10P]	Coaching or mentoring assignment records. [CGEE08-E10]
		(A) Develops or authorizes training materials in this competency area, which are subsequently successfully delivered beyond the enterprise boundary. [CGEE08-20A]	Records of formal training courses, workshops, seminars, and authored training material supported by successful post-training evaluation data. [CGEE08-E20]
		(A) Provides workshops/seminars or training in this competency area for practitioners or lead practitioners beyond the enterprise boundary (e.g. conferences and open training days). [CGEE08-30A]	Records of training/ workshops/ seminars created supported by successful post-training evaluation data. [CGEE08-E30]
9	Maintains expertise in this competency area through specialist Continual Professional Development (CPD) activities. [CGEE09]	(A) Reviews research, new ideas, and state of the art to identify relevant new areas requiring personal development in order to maintain expertise in this competency area. [CGEE09-10A]	Records of documents reviewed and insights gained as part of own research into this competency area. [CGEE09-E10]
		(A) Performs identified specialist professional development activities in order to maintain or further develop competence at expert level. [CGEE09-20A]	Records of Continual Professional Development (CPD) performed and learning outcomes. [CGEE09-E20]
		(A) Records continual professional development activities undertaken including learning or insights gained. [CGEE09-30A]	

NOTES	In addition to items above, enterprise-level or independent 3rd-Party-generated evidence may be used to amplify other evidence presented and may include: a. Formally recognized by a reputable external organization as an expert in this competency area b. Evidence of role as independent assessor or reviewer on project outside own organization where skills in this competency area were used c. Evidence of invitation(s) from wider community for contribution of systems engineering expertise in this area (e.g. industry conference panel, government advisory board etc. cross-industry working groups, partnerships, accredited advanced university courses or research, or as part of professional institute) d. Formal commendation beyond the enterprise (e.g. by INCOSE or other recognized authority) for work performed in this competency area e. Independently assessed or accredited work product in this competency area (e.g. for independent publication or use) f. Accolades of expertise in this area from recognized industry leaders

Competency area – Core: Critical Thinking
Description The objective analysis and evaluation of a topic in order to form a judgement.
Why it matters Artifacts produced during the conduct of Systems Engineering need to be defendable. Critical thinking helps improve the quality of input information, assumptions, and decisions. A failure to apply critical thinking may lead to invalid outputs including decisions and the solutions from which they are derived.
Possible contributory types of evidence Any combination of the types of evidence may be acceptable (depending on how the Framework is tailored and used). The evidence items identified at each level indicate example work products only. Contributions to work products will generally differ at each proficiency level.
Learning and development The INCOSE Professional Development Portal provides example guidance on how to gain an initial awareness of a competency area and options for developing further competence thereafter.

Awareness – Core: Critical Thinking		
ID	**Indicators of Competence**	**Relevant knowledge sub-indicators**
1	Explains why conclusions and arguments made by others may be based upon incomplete, potentially erroneous, or inadequate information, with examples. [CCTA01]	(K) Explains how incomplete, inadequate or erroneous information can cause false conclusions or poor outcomes to be drawn. [CCTA01-10K]
2	Identifies logical steps in an argument or proposition and the information needed to justify each. [CCTA02]	(K) Describes each of the steps to a logical argument (e.g. organize information, structure reasoning, consider all evidence, identify assumptions, evaluate arguments, communicate outcome). [CCTA02-10K]
3	Explains why assumptions are important and why there is a need to ensure that they are based upon sound information. [CCTA03]	(K) Explains how assumptions used to underpin the validity of any argument can affect an outcome. [CCTA03-10K] (K) Explains why "context" is important in assessing the validity of an argument. [CCTA03-20K] (K) Explains why the test of a solid argument is how good the evidence to underpin the claims made. [CCTA03-30K]
4	Explains the relationship between assumptions and risk and why assumptions need to be validated. [CCTA04]	(K) Explains why risk management is based upon an expectation that the risk being assessed is fully understood. [CCTA04-10K] (K) Explains how a failure to understand the underlying problem causing a risk may lead to incorrect assumptions about risk probabilities and possible outcomes, and therefor incorrect risk assessment. [CCTA04-20K] (K) Explains why additional analysis helps ensure assumptions underpinning a risk are valid. [CCTA04-30K]
5	Explains why ideas, arguments, and solutions need to be critically evaluated. [CCTA05]	(K) Explains why failure to critically evaluate an argument, proposal, or solution (e.g. leaping to first answer found) may result in an incomplete or inadequate result or fail to deliver maximum value to stakeholders. [CCTA05-10K] (K) Explains why it is easy to gather information, but the key is to present it effectively for consumption of others. [CCTA05-20K]

		(K) Describes common fallacies in logical arguments (e.g. ad hominem, strawman, bandwagon, false premise, appeal to authority, false dilemma, correlation/causation, anecdotal evidence, to quoque, appeal to emotion, black and white/dichotomy, burden of proof…). [CCTA05-30K] (K) Explains why arguments may be presented in a way which supports particular stakeholder perspectives and that ideally arguments need to be reasoned and unbiased or at least recognize the perspective they represent. [CCTA05-40K]
6	Lists common techniques and approaches used to propose or define arguments. [CCTA06]	(K) Explains why presenting results with clarity is essential to their acceptance when analyzing complex situations. [CCTA06-10K] (K) Explains the four common ways of presenting an argument: Deductive; Inductive; Abductive; Analogy. [CCTA06-20K] (K) Explains the terms: proposition, premise, inference, conclusion. [CCTA06-30K]
7	Explains how own perception of arguments from others may be biased and how this can be recognized. [CCTA07]	(K) Explains why while an opinion is reasonable, it is not necessarily substantiated by adequate supporting rationale and thus needs to be tested. [CCTA07-10K] (K) Explains why good critical thinkers need to be open-minded. [CCTA07-20K] (K) Explains how personal experience scan affect the interpretation of arguments from others and how this bias can be overcome. [CCTA07-30K]
8	Explains how different stakeholders, experiences may cause arguments to be presented in an incomplete or biased manner and how this can be overcome. [CCTA08]	(K) Explains how different stakeholders experiences may cause arguments to be presented in an incomplete or biased manner and how this can be overcome. [CCTA08-10K]

Supervised Practitioner – Core: Critical Thinking			
ID	**Indicators of Competence** ***(in addition to those at Awareness level)***	**Relevant knowledge, experience, and/or behaviors**	**Possible examples of objective evidence of personal contribution to activities performed, or professional behaviors applied**
1	Collates evidence constructing arguments needed in order to make informed decisions. [CCTS01]	(A) Collates supporting analysis for project decisions, such as the nature of the analysis required, the information, the arguments presented. [CCTS01-10A] (A) Records arguments in support of project decisions. [CCTS01-20A] (A) Identifies improvements they would make next time they presented similar arguments. [CCTS01-30A]	Inputs created to facilitate project decisions. [CCTS01-E10] Project decision data. [CCTS01-E20]
2	Uses logical relationships and dependencies between propositions to develop an argument. [CCTS02]	(K) Describes the relationship between differed parts on an argument. [CCTS02-10K] (K) Identifies the links between propositions made within an argument. [CCTS02-20K] (A) Uses relationships between parts of an argument and links between propositions within an argument to develop an argument. [CCTS02-30A]	An argument they have written having multiple parts and several related propositions. [CCTS02-E10]
3	Uses critical thinking techniques to review own work to test logic, assumptions, arguments, approach, and conclusions. [CCTS03]	(A) Uses critical thinking techniques to review own work, testing their own logic, assumptions, arguments, approach, or conclusions. [CCTS03-10A] (A) Uses critical thinking techniques to adjust own thinking as a result of testing their own logic, assumptions, arguments, approach, or conclusions. [CCTS03-20A]	Authored notes/reports testing validity of own decisions or results. [CCTS03-E10] Authored notes/reports adjusting own decisions or results through critical thinking. [CCTS03-E20]
4	Prepares robust arguments in order to respond to critical thinking within a collaborate environment. [CCTS04]	(A) Uses critical thinking techniques to support the construction of arguments from other team members. [CCTS04-10A] (A) Uses critical thinking techniques to challenge aspects of arguments made by others in order to improve the overall outcome. [CCTS04-20A]	Complex arguments supporting key project decisions. [CCTS04-E10] Reports challenging arguments decisions or propositions on a project. [CCTS04-E20] Records of positive outcomes to complex scenarios resulting from critical thinking applied. [CCTS04-E30]

5	Reviews ideas from others in order to improve the quality of their own approach, decisions, or conclusions. [CCTS05]	(A) Reviews their chosen analysis approach, decisions, or conclusions as a result of the input from others. [CCTS05-10A] (A) Produces updates to their analysis, decisions, or conclusions, improving the outcome. [CCTS05-20A]	Comments made on reviews from others on one's own work or thinking. [CCTS05-E10] Unilateral updates on arguments or decisions as a result of comments from others. [CCTS05-E20]
6	Identifies weaknesses and assumptions in their own arguments. [CCTS06]	(A) Identifies weaknesses in an existing piece of own analysis or arguments. [CCTS06-10A] (A) Maintains an existing piece of own analysis as a result of identifying weaknesses. [CCTS06-20A]	Unilateral updates of own arguments or decisions to address weaknesses detected by self. [CCTS06-E10] Records of updates of own arguments or decisions. [CCTS06-E20]
7	Uses common techniques to propose defining or challenging arguments, and conclusions, with guidance. [CCTS07]	(A) Uses appropriate techniques to challenge arguments and conclusions made by others, with guidance. [CCTS07-10A]	Work products created to challenge arguments or conclusions of others. [CCTS07-E10]
8	Identifies potential limitations in others' which may impact arguments made regarding proposals or ideas. [CCTS08]	(A) Identifies potential limitations in arguments or ideas made by others. [CCTS08-10A] (A) Develops arguments and ideas using critical thinking. [CCTS08-20A] (A) Prepares documentation for review by others. [CCTS08-30A]	Comments made on limitations detected in others' reports/documentation. [CCTS08-E10] Review comments critiquing analysis performed by others. [CCTS08-E20]
9	Identifies own perspective for potential cognitive bias for or against arguments made by others and modifies approach accordingly. [CCTS09]	(A) Identifies potential cognitive bias in their own arguments or ideas. [CCTS09-10A] (A) Uses awareness gained to modify own arguments or ideas positively. [CCTS09-20A]	Comments highlighting potential cognitive bias in others' work. [CCTS09-E10] Updates made to own work or thinking following detection of potential cognitive bias. [CCTS09-E20] Report or analysis contributions addressing potential cognitive bias. [CCTS09-E30]
10	Develops own understanding of this competency area through Continual Professional Development (CPD). [CCTS10]	(A) Identifies potential gaps in own knowledge or development needs in this area, identifying opportunities to address these through continual professional development activities. [CCTS10-10A] (A) Performs continual professional development activities to improve their knowledge and understanding in this area. [CCTS10-20A] (A) Records continual professional development activities undertaken including learning or insights gained. [CCTS10-30A]	Records of Continual Professional Development (CPD) performed and learning outcomes. [CCTS10-E10]

Practitioner – Core: Critical Thinking			
ID	**Indicators of Competence** ***(in addition to those at Supervised Practitioner level)***	**Relevant knowledge, experience, and/or behaviors**	**Possible examples of objective evidence of personal contribution to activities performed, or professional behaviors applied**
1	Reviews work performed for the quality of critical thinking applied in deriving outcomes. [CCTP01]	(A) Reviews assumptions, approaches, arguments, conclusions, and decisions made by colleagues using critical thinking. [CCTP01-10A] (A) Prepares balanced critique of decisions made. [CCTP01-20A]	Reports or technical papers to which they have contributed. [CCTP01-E10]
2	Reviews the impact of assumptions or weak logic in order to locate substantive arguments. [CCTP02]	(A) Reviews assumptions or logic used by others in an argument or decision-making activity. [CCTP02-10A] (A) Communicates challenges to assumptions or weak logic used by others effectively. [CCTP02-20A] (A) Identifies more substantive arguments to be used as part of an argument or decision-making process. [CCTP02-30A]	Reports or technical papers to which they have contributed. [CCTP02-E10]
3	Develops robust arguments when responding to critical thinking analyses. [CCTP03]	(A) Uses critical thinking techniques to support others in constructing arguments. [CCTP03-10A] (A) Develops robust arguments when responding to critical thinking analyses. [CCTP03-20A]	Reports reviewing assumptions, approaches, arguments, conclusions, and decisions made. [CCTP03-E10] Documents which demonstrate development of complex arguments. [CCTP03-E20]
4	Uses a range of different critical thinking approaches to challenge conclusions of others. [CCTP04]	(K) Lists differing critical thinking techniques available. [CCTP04-10K] (K) Lists potential sources of cognitive bias in an argument or report. [CCTP04-20K] (A) Uses awareness of cognitive bias issues to challenge others regarding the way an argument was examined, or analyzed or conclusions drawn. [CCTP04-30A] (A) Uses critical thinking to challenge others regarding the way an argument was examined, or analyzed or conclusions drawn. [CCTP04-40A] (P) Communicates challenges to others effectively. [CCTP04-50P]	Reports or technical papers highlighting awareness of potential cognitive bias. [CCTP04-E10]

5	Guides new or supervised practitioners in system thinking in order to develop their knowledge, abilities, skills, or associated behaviors. [CCTP05]	(P) Guides new or supervised practitioners in executing activities that form part of this competency. [CCTP05-10P] (A) Trains individuals to an "Awareness" level in this competency area. [CCTP05-20A]	Organizational Breakdown Structure showing their responsibility for technical supervision in this area. [CCTP05-E10] On-the-job training objectives/guidance etc. [CCTP05-E20] Coaching or mentoring assignment records. [CCTP05-E30] Records highlighting their impact on another individual in terms of improvement or professional development in this competency. [CCTP05-E40]
6	Maintains and enhances own competence in this area through Continual Professional Development (CPD) activities. [CCTP06]	(A) Identifies potential development needs in this area, identifying opportunities to address these through continual professional development activities. [CCTP06-10A] (A) Performs continual professional development activities to maintain and enhance their competency in this area. [CCTP06-20A] (A) Records continual professional development activities undertaken including learning or insights gained. [CCTP06-30A]	Records of Continual Professional Development (CPD) performed and learning outcomes. [CCTP06-E10]

Lead Practitioner – Core: Critical Thinking			
ID	**Indicators of Competence *(in addition to those at Practitioner level)***	**Relevant knowledge, experience, and/or behaviors**	**Possible examples of objective evidence of personal contribution to activities performed, or professional behaviors applied**
1	Creates enterprise-level policies, procedures, guidance, and best practice for critical thinking, including associated tools. [CCTL01]	(A) Analyzes enterprise need for critical thinking policies, processes, tools, or guidance. [CCTL01-10A]	Records showing their role in embedding critical thinking into enterprise policies (e.g. guidance introduced at enterprise level, enterprise-level review minutes). [CCTL01-E10]
		(A) Creates enterprise policies, procedures or guidance for critical thinking activities. [CCTL01-20A]	Procedures they have written. [CCTL01-E20]
		(A) Selects and acquires appropriate tools supporting critical thinking activities. [CCTL01-30A]	Records of support for tool introduction. [CCTL01-E30]
2	Uses a range of techniques and viewpoints to critically evaluate assumptions, approaches, arguments, conclusions, and decisions made across the enterprise. [CCTL02]	(A) Uses a range of critical thinking techniques to assess a range of projects across the enterprise. [CCTL02-10A]	Analysis or decision-making reports and other documents highlighting own critical thinking. [CCTL02-E10]
		(A) Assesses the critical thinking used in order to test assumptions, approaches, arguments, conclusions, and decisions across the enterprise. [CCTL02-20A]	Organizational communications resulting from critical thinking. [CCTL02-E20]
		(A) Identifies potential errors or issues in approach or conclusions made. [CCTL02-30A]	
		(P) Communicates conclusions made based on a deep understanding of critical thinking. [CCTL02-40P]	
3	Identifies alternative approaches to an existing approach to problem solving, to address flawed thinking and its results. [CCTL03]	(A) Identifies possible alternative thinking approaches to an existing approach to problem solving, to address flawed thinking and its results. [CCTL03-10A]	Records showing alternative approaches to problem resolution. [CCTL03-E10]
		(P) Communicates new or improved approaches, leading to changed thinking on the affected projects. [CCTL03-20P]	Records showing critical thinking advice provided in this area. [CCTL03-E20]
4	Judges impact of weak, incomplete or flawed arguments, conclusions, and decisions made across the enterprise. [CCTL04]	(A) Assesses the impact of weak, incomplete or flawed arguments, conclusions, and decisions made across the enterprise. [CCTL04-10A]	Independent of the application of critical thinking in the review or assessment of ideas, proposals, decisions, and their underlying justification. [CCTL04-E10]
		(P) Communicates assessment of critical thinking ideas, proposals, decisions and their underlying justification. [CCTL04-20P]	Organizational communications resulting from critical thinking. [CCTL04-E20]

5	Produces logical and clear explanations in support of the resolution of intricate or difficult situations across the enterprise. [CCTL05]	(A) Assesses intricate or difficult situations across the enterprise using a critical thinking approach. [CCTL05-10A] (A) Produces logical and clear explanations in support of the resolution of these situations across the enterprise. [CCTL05-20A]	Analysis or decision-making reports and other documents highlighting own critical thinking. [CCTL05-E10] Analysis of problems which show the application of own critical thinking. [CCTL05-E20]
6	Assesses uncertainty in situational assessment made across the enterprise, recommending approaches to address the impact of this. [CCTL06]	(A) Uses critical thinking techniques to highlight uncertainty in an analysis or decision performed by others. [CCTL06-10A] (A) Communicates solutions which reduce or overcome the uncertainty to affected stakeholders. [CCTL06-20A]	Analysis or decision-making reports and other documents highlighting own contribution in this area. [CCTL06-E10] Organizational communications resulting from critical thinking. [CCTL06-E20]
7	Uses own experiences to inform the critical examination of novel scenarios or domains across the enterprise. [CCTL07]	(A) Uses own experience to inform the critical analysis of novel scenarios or domains. [CCTL07-10A] (A) Uses balanced judgement to highlight areas requiring deeper critical review as a result of incorrect critical thinking. [CCTL07-20A] (A) Identifies and engages individuals across the enterprise in order to bring independence or different perspective to assessments. [CCTL07-30A]	Analysis of problems which show the application of own critical thinking. [CCTL07-E10] Analysis or decision-making reports highlighting flawed decisions through critical thinking approaches. [CCTL07-E20] Reports highlighting active engagement with others in assessment activities. [CCTL07-E30]
8	Judges aspects of decision-making which require deeper critical review on behalf of the enterprise. [CCTL08]	(A) Reviews the quality of analyses or decisions made by others, in order to highlight areas requiring more detailed critical analysis. [CCTL08-10A] (A) Uses their own experience and judgement to highlight areas requiring more detailed critical analysis, resulting in an improved analysis or outcome. [CCTL08-20A]	Analysis of problems that show the application of own critical thinking. [CCTL08-E10]
9	Develops own critical thinking approaches through a regular analysis of both personal experiences and the experiences of others across the enterprise. [CCTL09]	(A) Develops own critical thinking skills through analysis of personal experiences or the experiences of others. [CCTL09-10A]	Self-development records that demonstrate learning from personal experiences or those of others. [CCTL09-E10] Extracts from own coaching or mentoring activities. [CCTL09-E20]
10	Coaches or mentors practitioners across the enterprise in critical thinking in order to develop their knowledge, abilities, skills, or associated behaviors. [CCTL10]	(P) Coaches or mentors practitioners across the enterprise in competency-related techniques, recommending development activities. [CCTL10-10P] (A) Develops or authorizes enterprise training materials in this competency area. [CCTL10-20A]	Coaching or mentoring assignment records. [CCTL10-E10] Records of formal training courses, workshops, seminars, and authored training material supported by successful post-training evaluation data. [CCTL10-E20]

(*Continued*)

ID	Indicators of Competence *(in addition to those at Practitioner level)*	Relevant knowledge, experience, and/or behaviors	Possible examples of objective evidence of personal contribution to activities performed, or professional behaviors applied
		(A) Provides enterprise workshops/seminars or training in this competency area. [CCTL10-30A]	Listing as an approved organizational trainer for this competency area. [CCTL10-E30]
11	Promotes the introduction and use of novel techniques and ideas in critical thinking across the enterprise, to improve enterprise competence in this area. [CCTL11]	(A) Analyzes different approaches across different domains through research. [CCTL11-10A]	Research records. [CCTL11-E10]
		(A) Defines novel approaches that could potentially improve the SE discipline within the enterprise. [CCTL11-20A]	Published papers in refereed journals/company literature. [CCTL11-E20]
		(P) Fosters awareness of these novel techniques within the enterprise. [CCTL11-30P]	Records showing support for enabling systems supporting the new techniques or ideas. [CCTL11-E30]
		(P) Collaborates with enterprise stakeholders to facilitate the introduction of techniques new to the enterprise. [CCTL11-40P]	Published papers (or similar) at enterprise level. [CCTL11-E40]
		(A) Monitors new techniques after their introduction to determine their effectiveness. [CCTL11-50A]	Records of improvements made against a recognized process improvement model in this area. [CCTL11-E50]
		(A) Adapts approach to reflect actual enterprise performance improvements. [CCTL11-60A]	
12	Develops expertise in this competency area through specialist Continual Professional Development (CPD) activities. [CCTL12]	(A) Identifies own needs for further professional development in order to increase competence beyond practitioner level. [CCTL12-10A]	Records of Continual Professional Development (CPD) performed and learning outcomes. [CCTL12-E10]
		(A) Performs professional development activities in order to move own competence toward expert level. [CCTL12-20A]	
		(A) Records continual professional development activities undertaken including learning or insights gained. [CCTL12-30A]	

NOTES

In addition to items above, enterprise-level or independent 3rd-Party-generated evidence may be used to amplify other evidence presented and may include:

a. Formally recognized by senior management in current organization as an expert in this competency area
b. Evidence of role as Product/System Design Authority or Technical Authority on a complex project with responsibilities in this area or where skills within this competency area were used
c. Recognized as an authorizing signatory on behalf of enterprise for formal documentation in this competency area (e.g. policies, processes, and deliverables)
d. Formal commendation or award within own enterprise for contribution or item of work successfully performed, which required proficiency in this competency area
e. Customer, Supplier, or other external project-specific key Stakeholder accolades for specific work performed in this competency area
f. Independently assessed or accredited work in this competency area (e.g. for independent publication or use)
g. Formal organizational HR records positively highlighting any specific professional competencies or behaviors identified (if applicable) plus any of the evidence indicators listed at Expert level below

Expert – Core: Critical Thinking			
ID	**Indicators of Competence** ***(in addition to those at Lead Practitioner level)***	**Relevant knowledge, experience, and/or behaviors**	**Possible examples of objective evidence of personal contribution to activities performed, or professional behaviors applied**
1	Communicates own knowledge and experience in critical thinking in order to improve best practice beyond the enterprise boundary. [CCTE01]	(A) Produces papers, seminars, or presentations outside own enterprise for publication in order to share own ideas and improve industry best practices in this competence area. [CCTE01-10A]	Published papers or books etc. on new technique in refereed journals/company literature. [CCTE01-E10]
		(P) Fosters incorporation of own ideas into industry best practices in this area. [CCTE01-20P]	Published papers in refereed journals or internal literature proposing new practices in this competence area (or presentations, tutorials, etc.). [CCTE01-E20]
		(P) Develops guidance materials identifying new (or updating existing) best practice in this competence area. [CCTE01-30P]	Own proposals adopted as industry best practices in this competence area. [CCTE01-E30]
2	Advises organizations beyond the enterprise boundary on the suitability of their approach to critical thinking activities. [CCTE02]	(A) Advises on the suitability of critical thinking applied. [CCTE02-10A]	Records demonstrating internal or external advisory or consultative role in critical thinking. [CCTE02-E10]
		(A) Advises on critical thinking strategies. [CCTE02-20A]	
3	Advises organizations beyond the enterprise boundary on complex or sensitive assumptions, approaches, arguments, conclusions, and decisions. [CCTE03]	(A) Advises stakeholders beyond the enterprise on their critical thinking handling of complex or sensitive issues. [CCTE03-10A]	Records showing critical thinking advice provided in this area together with evidence that the issues advised on were by their nature either complex or sensitive. [CCTE03-E10]
		(P) Conducts sensitive negotiations on a highly complex critical thinking issue, making limited use of specialized, technical terminology. [CCTE03-20P]	Records showing role as arbiter of the most appropriate approach to take. [CCTE03-E20]
		(A) Uses a critical thinking approach to complex issue resolution including balanced, rational arguments on way forward. [CCTE03-30A]	
4	Advises organizations beyond the enterprise boundary on the resolution of weak, incomplete, or flawed approaches impacting arguments, conclusions, and decisions made. [CCTE04]	(A) Advises stakeholders beyond the enterprise on resolution of weak, incomplete, or flawed approaches impacting arguments, conclusions, and decisions. [CCTE04-10A]	Records showing critical thinking advice provided in this area. [CCTE04-E10]
		(P) Conducts sensitive negotiations on resolution of weak, incomplete, or flawed approaches impacting arguments, conclusions and decisions. [CCTE04-20P]	Records showing use of critical thinking during negotiations. [CCTE04-E20]

(*Continued*)

ID	Indicators of Competence *(in addition to those at Lead Practitioner level)*	Relevant knowledge, experience, and/or behaviors	Possible examples of objective evidence of personal contribution to activities performed, or professional behaviors applied
		(A) Analyzes weak, incomplete, or flawed approaches impacting arguments, conclusions, and decisions using critical thinking. [CCTE04-30A]	
5	Develops own critical thinking expertise through regular review and analysis of critical thinking successes and failures documented beyond the enterprise boundary. [CCTE05]	(A) Reviews externally published reports or articles on critical thinking successes and failures. [CCTE05-10A]	Self-development records which demonstrate learning from personal experiences or those of others. [CCTE05-E10]
		(A) Analyzes articles for learning points to improve own critical thinking approaches or viewpoints. [CCTE05-20A]	Extracts from own coaching or mentoring activities. [CCTE05-E20]
6	Champions the introduction of novel techniques and ideas in critical thinking, beyond the enterprise boundary, in order to develop the wider Systems Engineering community in this competency. [CCTE06]	(A) Analyzes different approaches across different domains through research. [CCTE06-10A]	Records of activities promoting research and need to adopt novel technique or ideas. [CCTE06-E10]
		(A) Produces reports for the wider SE community on the effectiveness of new techniques after their introduction. [CCTE06-20A]	Records of improvements made to process and appraisal against a recognized process improvement model. [CCTE06-E20]
		(P) Collaborates with those introducing novel techniques within the wider SE community. [CCTE06-30P]	Research records. [CCTE06-E30]
		(A) Defines novel approaches that could potentially improve the wider SE discipline. [CCTE06-40A]	Published papers in refereed journals/company literature. [CCTE06-E40]
		(P) Fosters awareness of these novel techniques within the wider SE community. [CCTE06-50P]	Records showing introduction of enabling systems supporting the new techniques or ideas. [CCTE06-E50]
7	Coaches individuals beyond the enterprise boundary in critical thinking in order to further develop their knowledge, abilities, skills, or associated behaviors. [CCTE07]	(P) Coaches or mentors individuals beyond the enterprise boundary, in competency-related techniques, recommending development activities. [CCTE07-10P]	Coaching or mentoring assignment records. [CCTE07-E10]
		(A) Develops or authorizes training materials in this competency area, which are subsequently successfully delivered beyond the enterprise boundary. [CCTE07-20A]	Records of formal training courses, workshops, seminars, and authored training material supported by successful post-training evaluation data. [CCTE07-E20]

		(A) Provides workshops/seminars or training in this competency area for practitioners or lead practitioners beyond the enterprise boundary (e.g. conferences and open training days). [CCTE07-30A]	Records of Training/workshops/seminars created supported by successful post-training evaluation data. [CCTE07-E30]
8	Maintains expertise in this competency area through specialist Continual Professional Development (CPD) activities. [CCTE08]	(A) Reviews research, new ideas, and state of the art to identify relevant new areas requiring personal development in order to maintain expertise in this competency area. [CCTE08-10A]	Records of documents reviewed and insights gained as part of own research into this competency area. [CCTE08-E10]
		(A) Performs identified specialist professional development activities in order to maintain or further develop competence at expert level. [CCTE08-20A]	Records of Continual Professional Development (CPD) performed and learning outcomes. [CCTE08-E20]
		(A) Records continual professional development activities undertaken including learning or insights gained. [CCTE08-30A]	

NOTES	In addition to items above, enterprise-level or independent 3rd-Party-generated evidence may be used to amplify other evidence presented and may include: a. Formally recognized by a reputable external organization as an expert in this competency area b. Evidence of role as independent assessor or reviewer on project outside own organization where skills in this competency area were used c. Evidence of invitation(s) from wider community for contribution of systems engineering expertise in this area (e.g. industry conference panel, government advisory board etc. cross-industry working groups, partnerships, accredited advanced university courses or research, or as part of professional institute) d. Formal commendation beyond the enterprise (e.g. by INCOSE or other recognized authority) for work performed in this competency area e. Independently assessed or accredited work product in this competency area (e.g. for independent publication or use) f. Accolades of expertise in this area from recognized industry leaders

Competency area – Core: Systems Modeling and Analysis
Description Modeling is a physical, mathematical, or logical representation of a system entity, phenomenon, or process. System analysis provides a rigorous set of data and information to aid technical understanding and decision-making across the life cycle. A key part of systems analysis is modeling.
Why it matters Modeling, analysis, and simulation can provide early, cost effective, indications of function, and performance, thereby driving the solution design, enabling risk mitigation and supporting the verification and validation of a solution. Modeling and simulation also allow the exploration of scenarios outside the normal operating parameters of the system.
Possible contributory types of evidence Any combination of the types of evidence may be acceptable (depending on how the Framework is tailored and used). The evidence items identified at each level indicate example work products only. Contributions to work products will generally differ at each proficiency level.
Learning and development The INCOSE Professional Development Portal provides example guidance on how to gain an initial awareness of a competency area and options for developing further competence thereafter.

Awareness – Core: Systems Modeling and Analysis		
ID	**Indicators of Competence**	**Relevant knowledge sub-indicators**
1	Explains why system representations are required and the benefits they can bring to developments. [CSMA01]	(K) Explains how modeling and system analysis allows early understanding of the system. [CSMA01-10K] (K) Describes the impact of complexity and cost of implementation on models. [CSMA01-20K] (K) Explains why there is a need to perform trials and "what ifs" in developments. [CSMA01-30K] (K) Explains the concepts of virtual systems and demonstrators. [CSMA01-40K] (K) Explains why modeling requires Interactions, interfaces, boundaries, and flow diagrams. [CSMA01-50K]
2	Explains the scope and limitations of models and simulations, including definition, implementation, and analysis. [CSMA02]	(K) Describes key interfaces and links between systems modeling and analysis and stakeholders in the wider enterprise outside engineering (e.g. infrastructure management, human resource management, quality management, knowledge management, portfolio management, and life cycle model management). [CSMA02-10K] (K) Explains why models are abstractions. [CSMA02-20K] (K) Explains why all models and simulations contain assumptions and approximations and the concept of "garbage in, garbage out". [CSMA02-30K] (K) Explains the difference between real-time and iterative simulations. [CSMA02-40K] (K) Explains why models can be hierarchical. [CSMA02-50K] (K) Explains why models and simulations need to be validated to an appropriate level. [CSMA02-60K] (K) Explains why the adage "all models are wrong; some models are useful" is a good summary of models. [CSMA02-70K]
3	Explains different types of modeling and simulation approaches. [CSMA03]	(K) Lists different types of modeling and simulation, e.g. live, virtual, constructive. [CSMA03-10K]
4	Explains how the purpose of modeling and simulation affect the approach taken. [CSMA04]	(K) Describes the different uses of models and how these might affect the way the model is designed, built or the data from it used. [CSMA04-10K]

5	Explains why functional analysis and modeling is important in Systems Engineering. [CSMA05]	(K) Describes the purpose and key activities supported by functional modeling and analysis. [CSMA05-10K] (K) Describes the difference between "Functional" and "Non-functional" requirements and their impact on modeling and analysis. [CSMA05-20K] (K) Lists the key benefits of developing a functional architecture. [CSMA05-30K] (K) Explains why there is a need to establish an agreed system and model boundary. [CSMA05-40K]
6	Explains the relevance of outputs from systems modeling and analysis, and how these relate to overall system development. [CSMA06]	(K) Describes the key features of typical artifacts from functional modeling, such as Behavior Diagrams, Context Diagrams, Control Flow Diagrams, Data Flow Diagrams, Data Dictionaries, detailed specifications, functional hierarchy, diagram functional matrix (e.g. N2 diagram), and functional flow block diagram. [CSMA06-10K] (K) Describes how systems modeling and analysis might identify missing requirements or cause derived requirements to be developed. [CSMA06-20K] (K) Explains how systems modeling and analysis can helps identify poorly written and unrealistic requirements. [CSMA06-30K]
7	Explains the difference between modeling and simulation. [CSMA07]	(K) Describes the key features of typical artifacts from functional modeling, such as Behavior Diagrams, Context Diagrams, Control Flow Diagrams, Data Flow Diagrams, Data Dictionaries, detailed specifications, functional hierarchy, diagram functional matrix (e.g. N2 diagram), and functional flow block diagram [CSMA07-10K] (K) Explains how systems modeling and analysis can helps identify poorly written and unrealistic requirements. [CSMA07-20K]
8	Describes a variety of system analysis techniques that can be used to derive information about a system. [CSMA08]	(K) Lists various types of system analysis used in support of systems engineering, together with their primary purpose (e.g. life cycle cost, availability, effectiveness, electromagnetic compatibility, environmental, interoperability, logistics, manufacturing, mass properties, reliability, availability, maintainability, resilience, safety, security, training needs, usability, human systems integration, value engineering). [CSMA08-10K] (K) Explains how differing forms of analysis might be used to benefit different aspects of the system. [CSMA08-20K] (K) Explains how some analysis activities require a design to be completed, but others may be used before a design exists to derive or improve requirements. [CSMA08-30K]
9	Explains why the benefits of modeling can only be realized if choices made in defining the model are correct. [CSMA09]	(K) Explains why it is necessary to use the right choice of model and/or simulation tool, e.g. exploratory/fitted, specific/general, numerical/analytical, deterministic/stochastic, discrete/continuous, and quantitative/qualitative. [CSMA09-10K] (K) Explains why there is a trade between cost vs. value when selecting a model. [CSMA09-20K] (K) Explains the importance of ensuring the integrity of the model interface to the system. [CSMA09-30K] (K) Explains how modeling is affected by the criticality of the system element being modeled or simulated. [CSMA09-40K] (K) Explains how modeling is affected by the criticality of the results of the system element being modeled or simulated. [CSMA09-50K]
10	Explains why models and simulations have a limit of valid use, and the risks of using models and simulations outside those limits. [CSMA10]	(K) Describes limits of model validity and how these may impact model selection and usage. [CSMA10-10K] (K) Explains how model validity can relate to the number of iterations required. [CSMA10-20K] (K) Explains how model validity relates to the assumptions and approximations made when developing the model. [CSMA10-30K]

Supervised Practitioner – Core: Systems Modeling and Analysis			
ID	**Indicators of Competence** ***(in addition to those at Awareness level)***	**Relevant knowledge, experience, and/or behaviors**	**Possible examples of objective evidence of personal contribution to activities performed, or professional behaviors applied**
1	Uses modeling and simulation tools and techniques to represent a system or system element. [CSMS01]	(K) Explains the relationship between the system Functional Architecture – hierarchy of decomposed functions – and the model. [CSMS01-10K] (K) Explains the relationship of the model to the basic sub-functions arising from Functional Decomposition. [CSMS01-20K] (K) Explains how models support the development and definition of interfaces. [CSMS01-30K] (K) Describes key elements of functional models. [CSMS01-40K] (A) Identifies an appropriate model or simulation type to use, defining scope, assumptions, model strengths and weaknesses, validation approach, and how outputs could be used in support of a project. [CSMS01-50A] (A) Uses a model or simulation in support of a project to benefit the overall development. [CSMS01-60A]	Contributions to model and/or a simulation in support of development of a system or system element. [CSMS01-E10] Modeling or analysis model/tool work products. [CSMS01-E20] Contributions to analysis model developed or used or Work products from such a Model. [CSMS01-E30]
2	Analyzes outcomes of modeling and analysis and uses this to improve understanding of a system. [CSMS02]	(A) Uses modeling artifacts (e.g. diagrams, simulation, or other results) in their work. [CSMS02-10A]	Contributions to analysis model developed or used or Work products from such a Functional Model. [CSMS02-E10]
3	Analyzes risks or limits of a model or simulation. [CSMS03]	(A) Identifies potential limits to the accuracy or scope of a selected model. [CSMS03-10A] (A) Identifies assumptions associated with use of a selected model or simulation. [CSMS03-20A] (A) Identifies mechanisms in order to reduce risks associated with use of a selected model (e.g. increasing number of iterations). [CSMS03-30A]	Contributions to a model or work products associated from a model. [CSMS03-E10]
4	Uses systems modeling and analysis tools and techniques to verify a model or simulation. [CSMS04]	(A) Uses tools and techniques in order to verify a model. [CSMS04-10A]	Records of verification performed on a model or simulation. [CSMS04-E10]

5	Prepares inputs used in support of model development activities. [CSMS05]	(A) Prepares inputs used for model development activities. [CSMS05-10A]	Contributions to development of a model. [CSMS05-E10]
6	Uses different types of models for different reasons. [CSMS06]	(A) Uses different types of models for different purposes. [CSMS06-10A]	Contributions to different models. [CSMS06-E10]
7	Uses system analysis techniques to derive information about the real system. [CSMS07]	(A) Uses data produced by a model in order to derive new system information. [CSMS07-10A]	Analysis performed on a model during its development to derive information. [CSMS07-E10]
8	Develops own understanding of this competency area through Continual Professional Development (CPD). [CSMS08]	(A) Identifies potential gaps in own knowledge or development needs in this area, identifying opportunities to address these through continual professional development activities. [CSMS08-10A] (A) Performs continual professional development activities to improve their knowledge and understanding in this area. [CSMS08-20A] (A) Records continual professional development activities undertaken including learning or insights gained. [CSMS08-30A]	Records of Continual Professional Development (CPD) performed and learning outcomes. [CSMS08-E10]

Practitioner – Core: Systems Modeling and Analysis			
ID	**Indicators of Competence** ***(in addition to those at Supervised Practitioner level)***	**Relevant knowledge, experience, and/or behaviors**	**Possible examples of objective evidence of personal contribution to activities performed, or professional behaviors applied**
1	Identifies project-specific modeling or analysis needs that need to be addressed when performing modeling on a project. [CSMP01]	(A) Identifies project-specific modeling or analysis needs that need to be incorporated into SE planning. [CSMP01-10A]	System modeling analysis plans. [CSMP01-E10]
		(A) Identifies trade-offs (e.g. cost vs benefit/ value), which may need to be made when selecting modeling approach on a project. [CSMP01-20A]	System modeling analysis trade-offs performed. [CSMP01-E20]
		(A) Prepares project-specific modeling and analysis activities for SE planning purposes, using appropriate processes and procedures. [CSMP01-30A]	
		(A) Prepares project-specific modeling or analysis task estimates in support of SE planning. [CSMP01-40A]	
		(P) Liaises throughout with stakeholders to gain approval, updating strategy as necessary. [CSMP01-50P]	
2	Creates a governing process, plan, and associated tools for systems modeling, and analysis in order to monitor and control systems modeling and analysis activities on a system or system element. [CSMP02]	(K) Identifies the steps necessary to define a process and appropriate techniques to be adopted for system modeling and analysis. [CSMP02-10K]	SE system modeling or analysis planning or process work products developed. [CSMP02-E10]
		(K) Describes key elements of a successful system modeling and analysis process and how these activities relate to different stages of a system life cycle. [CSMP02-20K]	Certificates or proficiency evaluations in system modeling or management tools. [CSMP02-E20]
		(A) Uses system modeling and analysis tools or uses a methodology. [CSMP02-30A]	SE system interface definition or management tool guidance developed. [CSMP02-E30]
		(A) Uses appropriate standards and governing processes in their system modeling and analysis planning, tailoring as appropriate. [CSMP02-40A]	
		(A) Selects system modeling and analysis methods for each architectural interface. [CSMP02-50A]	
		(A) Creates system modeling and analysis plans and processes for use on a project. [CSMP02-60A]	
		(P) Liaises throughout with stakeholders to gain approval, updating plans as necessary. [CSMP02-70P]	
3	Determines key parameters or constraints, which scope or limit the modeling and analysis activities. [CSMP03]	(A) Determines the context in which a system of interest will operate (the super system) and uses this to contextualize modeling activities. [CSMP03-10A]	SE system modeling or analysis planning or process work products developed. [CSMP03-E10]

		(A) Identifies interfaces and interactions with a super system and uses these to characterize modeling and transition. [CSMP03-20A]	Certificates or proficiency evaluations in system modeling or management tools. [CSMP03-E20]
		(A) Identifies effects of the system map on the super system and vice versa and uses these to tailor modeling and analysis strategy. [CSMP03-30A]	SE system interface definition or management tool guidance developed. [CSMP03-E30]
		(A) Creates a modeling and analysis plan or strategy. [CSMP03-40A]	Records showing a strategy selection for Functional analysis or modeling including rationale for its selection. [CSMP03-E40]
		(A) Identifies choice of model(s). [CSMP03-50A]	
		(A) Identifies the flexibility of available models. [CSMP03-60A]	
		(A) Identifies the limitations of available models and simulations. [CSMP03-70A]	
		(A) Records the rationale for choice of strategy (e.g. alternatives and criteria) on a project. [CSMP03-80A]	
4	Uses a governing process and appropriate tools to manage and control their own system modeling and analysis activities. [CSMP04]	(A) Complies with defined plans, processes, and associated tools for Systems modeling and analysis activities on a project. [CSMP04-10A]	Records of challenges made to others' plans or work products in order to address project issues in this area. [CSMP04-E10]
		(A) Prepares systems modeling and analysis data and reports in support of wider system measurement, monitoring, and control. [CSMP04-20A]	Process or tool guidance. [CSMP04-E20]
		(P) Guides and actively coordinates the interpretation of systems modeling and analysis plans or processes in order to complete activities successfully. [CSMP04-30P]	Records of interpretations authorized in order to overcome challenges. [CSMP04-E30]
		(P) Recognizes situations where deviation from published plans and processes or clarification from others is appropriate in order to overcome complex challenges. [CSMP04-40P]	
		(P) Recognizes situations where existing strategy, process, plan, or tools require formal change, liaising with stakeholders to gain approval as necessary. [CSMP04-50P]	
5	Analyzes a system, determining the representation of the system or system element, collaborating with model stakeholders as required. [CSMP05]	(P) Analyzes a system (or major system element) on a project/program. [CSMP05-10P]	Records from a System (or system element) Analysis activity they led on a project/program. [CSMP05-E10]
		(A) Identifies derived requirements as a result of modeling system behavior. [CSMP05-20A]	Records of system behavior modeling performed to derive requirements. [CSMP05-E20]
		(A) Identifies low level functional requirements remain solution free within the model. [CSMP05-30A]	Low level solution-independent functional requirements derived. [CSMP05-E30]
		(A) Maintains traceability between decomposed functionality and system requirements. [CSMP05-40A]	Records of traceability information created or maintained between decomposed functionality and system requirements. [CSMP05-E40]

(Continued)

ID	Indicators of Competence *(in addition to those at Supervised Practitioner level)*	Relevant knowledge, experience, and/or behaviors	Possible examples of objective evidence of personal contribution to activities performed, or professional behaviors applied
		(A) Monitors allocation of functions to components in system architecture. [CSMP05-50A] (A) Uses existing Functional Analysis models to support analysis. [CSMP05-60A]	Records of Functions allocated to components in system architecture. [CSMP05-E50] Functional Analysis model reused or adapted for reuse. [CSMP05-E60]
6	Selects appropriate tools and techniques for system modeling and analysis. [CSMP06]	(A) Identifies appropriate tools and techniques for each aspect of the required Functional Analysis. [CSMP06-10A]	Records of approved tools listing created or edited, or records of tools used as a result of an existing list. [CSMP06-E10] Records of process documents authored. [CSMP06-E20]
7	Defines appropriate representations of a system or system element. [CSMP07]	(A) Identifies systems' constituent elements. [CSMP07-10A] (A) Identifies the appropriate models and simulation tools, e.g. exploratory/fitted, specific/general, numerical/analytical, deterministic/stochastic, discrete/continuous, and quantitative/qualitative. [CSMP07-20A] (A) Uses existing models and simulations when appropriate. [CSMP07-30A] (A) Creates interfaces and translates interface data appropriately. [CSMP07-40A] (A) Identifies appropriate model considering both cost and value. [CSMP07-50A] (A) Identifies model areas where there is a criticality of the system element being modeled or simulated. [CSMP07-60A] (A) Identifies model results where there is a criticality. [CSMP07-70A]	Records of model selections made including rationale. [CSMP07-E10]
8	Uses appropriate representations and analysis techniques to derive information about a real system. [CSMP08]	(A) Uses appropriate techniques to validate systems or system elements. [CSMP08-10A] (A) Develops multifunctional/multilevel models of (sub)systems by linking models together. [CSMP08-20A] (A) Maintains a linkage of models to "real" systems. [CSMP08-30A] (A) Selects appropriate systems analysis techniques. [CSMP08-40A] (A) Uses systems analysis techniques to derive information. [CSMP08-50A] (A) Uses information from systems analysis techniques to improve system understanding or design. [CSMP08-60A]	Records of model validation performed comparing theoretical versus actual performance. [CSMP08-E10] Use of appropriate validated model. [CSMP08-E20] Work products documenting different systems analysis techniques used. [CSMP08-E30] Records of model analysis performed comparing theoretical versus actual performance. [CSMP08-E40]

9	Ensures the content of models that are produced within a project are controlled and coordinated. [CSMP09]	(K) Explains why there is a need to manage models during a project development. [CSMP09-10K]	Records showing management of model content. [CSMP09-E10]
		(A) Identifies the activities required in order to manage a developing model. [CSMP09-20A]	Records of interfaces defined between model(s) and other tools for data coordination. [CSMP09-E20]
		(A) Defines data exchanges required between models and with other tools in order to coordinate importing, exporting and baselining of model data sets (e.g. from Requirements management tool to architecting tool). [CSMP09-30A]	
10	Uses systems modeling and analysis tools and techniques to validate a model or simulation. [CSMP10]	(A) Selects appropriate systems modeling and analysis tools and techniques to validate a model or simulation. [CSMP10-10A]	Model validation records. [CSMP10-E10]
		(A) Uses selected tools and techniques to perform a formal validation of a selected model, gaining approval from stakeholders. [CSMP10-20A]	Stakeholder model review or approval records. [CSMP10-E20]
11	Guides new or supervised practitioners in modeling and systems analysis to operation in order to develop their knowledge, abilities, skills, or associated behaviors. [CSMP11]	(P) Guides new or supervised practitioners in executing activities that form part of this competency. [CSMP11-10P]	Organizational Breakdown Structure showing their responsibility for technical supervision in this area. [CSMP11-E10]
		(A) Trains individuals to an "Awareness" level in this competency area. [CSMP11-20A]	Job training objectives/guidance etc. suggested or authorized for others. [CSMP11-E20]
			Coaching or mentoring assignment records. [CSMP11-E30]
			Records highlighting their impact on another individual in terms of improvement or professional development in this competency. [CSMP11-E40]
12	Maintains and enhances own competence in this area through Continual Professional Development (CPD) activities. [CSMP12]	(A) Identifies potential development needs in this area, identifying opportunities to address these through continual professional development activities. [CSMP12-10A]	Records of Continual Professional Development (CPD) performed and learning outcomes. [CSMP12-E10]
		(A) Performs continual professional development activities to maintain and enhance their competency in this area. [CSMP12-20A]	
		(A) Records continual professional development activities undertaken including learning or insights gained. [CSMP12-30A]	

Lead Practitioner – Core: Systems Modeling and Analysis			
ID	**Indicators of Competence** ***(in addition to those at Practitioner level)***	**Relevant knowledge, experience, and/or behaviors**	**Possible examples of objective evidence of personal contribution to activities performed, or professional behaviors applied**
1	Creates enterprise-level policies, procedures, guidance, and best practice for systems modeling and analysis definition and management, including associated tools. [CSML01]	(A) Analyzes enterprise need for system modeling and analysis policies, processes, tools, or guidance. [CSML01-10A]	Records showing their role in embedding systems modeling and analysis into enterprise policies (e.g. guidance introduced at enterprise level, enterprise -level review minutes). [CSML01-E10]
		(A) Creates enterprise policies, procedures or guidance for system modeling and analysis activities. [CSML01-20A]	Procedures they have written. [CSML01-E20]
		(A) Selects and acquires appropriate tools supporting modeling and analysis activities. [CSML01-30A]	Records of support for tool introduction. [CSML01-E30]
2	Judges the correctness of tailoring of enterprise-level modeling and analysis processes to meet the needs of a project, on behalf of the enterprise. [CSML02]	(A) Identifies and acquires appropriate tools supporting system modeling and analysis activities. [CSML02-10A]	Records demonstrating role as internal or external reviewer or consultant in the relevant areas. [CSML02-E10]
		(P) Communicates constructive feedback on systems modeling and analysis process. [CSML02-20P]	
3	Advises stakeholders across the enterprise, on systems modeling and analysis. [CSML03]	(A) Advises on modeling and systems analysis approaches to multiple projects across the organization. [CSML03-10A]	Records demonstrating advice provided in this area. [CSML03-E10]
4	Coordinates modeling or analysis activities across the enterprise in order to determine appropriate representations or analysis of complex system or system elements. [CSML04]	(A) Coordinates modeling or analysis activities across the enterprise in order to determine appropriate representations or analysis of complex system or system elements. [CSML04-10A]	Records highlighting personal contribution performed. [CSML04-E10]
		(P) Communicates on way forward to affected stakeholders. [CSML04-20P]	
5	Adapts approaches used to accommodate complex or challenging aspects of a system of interest being modeled or analyze on projects across the enterprise. [CSML05]	(A) Analyzes potential approaches to complex or challenging systems of interest requiring modeling or analysis on projects across the enterprise. [CSML05-10A]	Records of complex or challenging systems modeling or analysis performed. [CSML05-E10]
		(A) Adapts approach used to accommodate specific context or circumstances of complex or challenging system modeling or analysis activities. [CSML05-20A]	Records of advice provided to others in support of their analysis of complex or challenging systems. [CSML05-E20]

6	Assesses the outputs of systems modeling and analysis across the enterprise to ensure that the results can be used for the intended purpose. [CSML06]	(A) Reviews modeling and analysis across the enterprise. [CSML06-10A]	Records demonstrating role as internal or external reviewer or consultant in the relevant areas. [CSML06-E10]
		(A) Judges the modeling and analysis techniques performed. [CSML06-20A]	
7	Advises stakeholders across the enterprise on selection of appropriate modeling or analysis approach across the enterprise. [CSML07]	(A) Uses understanding of strengths and weaknesses of various modeling and analysis techniques when advising. [CSML07-10A]	Records of advice provided. [CSML07-E10]
		(A) Advises on the selection of a system modeling or analysis technique, based upon an assessment of its appropriateness to the purpose compared to other potentially applicable tools or techniques. [CSML07-20A]	
8	Coordinates the integration and combination of different models and analyses for a system or system element across the enterprise. [CSML08]	(A) Coordinates the integration or combination of different models or analyses. [CSML08-10A]	Records highlighting personal contribution performed. [CSML08-E10]
9	Coaches or mentors practitioners across the enterprise in systems modeling and analysis in order to develop their knowledge, abilities, skills, or associated behaviors. [CSML09]	(P) Coaches or mentors practitioners across the enterprise in competency-related techniques, recommending development activities. [CSML09-10P]	Coaching or mentoring assignment records. [CSML09-E10]
		(A) Develops or authorizes enterprise training materials in this competency area. [CSML09-20A]	Records of formal training courses, workshops, seminars, and authored training material supported by successful post-training evaluation data. [CSML09-E20]
		(A) Provides enterprise workshops/seminars or training in this competency area. [CSML09-30A]	Listing as an approved organizational trainer for this competency area. [CSML09-E30]
10	Promotes the introduction and use of novel techniques and ideas in Systems Modeling and Analysis across the enterprise, to improve enterprise competence in this area. [CSML10]	(A) Analyzes different approaches across different domains through research. [CSML10-10A]	Research records. [CSML10-E10]
		(A) Defines novel approaches that could potentially improve the SE discipline within the enterprise. [CSML10-20A]	Published papers in refereed journals/company literature. [CSML10-E20]
		(P) Fosters awareness of these novel techniques within the enterprise. [CSML10-30P]	Records showing introduction of enabling systems supporting the new techniques or ideas. [CSML10-E30]
		(P) Collaborates with enterprise stakeholders to facilitate the introduction of techniques new to the enterprise. [CSML10-40P]	Published papers (or similar) at enterprise level. [CSML10-E40]

(*Continued*)

ID	Indicators of Competence *(in addition to those at Practitioner level)*	Relevant knowledge, experience, and/or behaviors	Possible examples of objective evidence of personal contribution to activities performed, or professional behaviors applied
		(A) Monitors new techniques after their introduction to determine their effectiveness. [CSML10-50A] (A) Adapts approach to reflect actual enterprise performance improvements. [CSML10-60A]	Records of improvements made against a recognized process improvement model. [CSML10-E50]
11	Develops expertise in this competency area through specialist Continual Professional Development (CPD) activities. [CSML11]	(A) Identifies own needs for further professional development in order to increase competence beyond practitioner level. [CSML11-10A] (A) Performs professional development activities in order to move own competence toward expert level. [CSML11-20A] (A) Records continual professional development activities undertaken including learning or insights gained. [CSML11-30A]	Records of Continual Professional Development (CPD) performed and learning outcomes. [CSML11-E10]

NOTES	In addition to items above, enterprise-level or independent 3rd-Party-generated evidence may be used to amplify other evidence presented and may include: a. Formally recognized by senior management in current organization as an expert in this competency area b. Evidence of role as Product/System Design Authority or Technical Authority on a complex project with responsibilities in this area or where skills within this competency area were used c. Recognized as an authorizing signatory on behalf of enterprise for formal documentation in this competency area (e.g. policies, processes, and deliverables) d. Formal commendation or award within own enterprise for contribution or item of work successfully performed, which required proficiency in this competency area e. Customer, Supplier, or other external project-specific key Stakeholder accolades for specific work performed in this competency area f. Independently assessed or accredited work in this competency area (e.g. for independent publication or use) g. Formal organizational HR records positively highlighting any specific professional competencies or behaviors identified (if applicable) plus any of the evidence indicators listed at Expert level below

Expert – Core: Systems Modeling and Analysis			
ID	**Indicators of Competence *(in addition to those at Lead Practitioner level)***	**Relevant knowledge, experience, and/or behaviors**	**Possible examples of objective evidence of personal contribution to activities performed, or professional behaviors applied**
1	Communicates own knowledge and experience in Systems Modeling and Analysis in order to improve best practice beyond the enterprise boundary. [CSME01]	(A) Produces papers, seminars, or presentations outside own enterprise for publication in order to share own ideas and improve industry best practices in this competence area. [CSME01-10A]	Published papers or books etc. on new technique in refereed journals/company literature. [CSME01-E10]
		(P) Fosters incorporation of own ideas into industry best practices in this area. [CSME01-20P]	Published papers in refereed journals or internal literature proposing new practices in this competence area (or presentations, tutorials, etc.). [CSME01-E20]
		(P) Develops guidance materials identifying new (or updating existing) best practice in this competence area. [CSME01-30P]	Own proposals adopted as industry best practices in this competence area. [CSME01-E30]
2	Advises organizations beyond the enterprise on the appropriateness of their selected approaches in any given level of complexity and novelty. [CSME02]	(A) Advises external stakeholders on the usage of particular work products with specific analysis techniques. [CSME02-10A]	Technical document detailing consistent, validated performance. [CSME02-E10]
		(A) Uses knowledge of strengths and weaknesses of available modeling and analysis techniques to provide guidance. [CSME02-20A]	Project/program plan or document authored. [CSME02-E20]
		(A) Produces proposals identifying scenarios for validation of simulation. [CSME02-30A]	
3	Advises organizations beyond the enterprise boundary on the modeling and analysis of complex or novel systems, or system elements. [CSME03]	(A) Advises stakeholders beyond the enterprise boundary on their strategy and approach for modeling and analysis of complex or novel system or system elements (e.g. How the model will be used? What will be modeled? How will the results influence the design?). [CSME03-10A]	Records of advice provided on external engineering or modeling plans. [CSME03-E10]
		(A) Selects appropriate type of model. [CSME03-20A]	Published work. [CSME03-E20]
		(A) Communicates model information to stakeholders beyond the enterprise, effectively, and confidently. [CSME03-30A]	

(Continued)

ID	Indicators of Competence *(in addition to those at Lead Practitioner level)*	Relevant knowledge, experience, and/or behaviors	Possible examples of objective evidence of personal contribution to activities performed, or professional behaviors applied
4	Advises organizations beyond the enterprise boundary on the model or analysis validation issues and risks. [CSME04]	(A) Reviews systems modeling or analysis strategies. [CSME04-10A]	Records demonstrating internal or external advisory or consultative role in modeling and analysis. [CSME04-E10]
		(A) Advises on systems modeling or analysis strategies of changes required. [CSME04-20A]	Records of advice provided on the adoption/rejection of modeling and analysis solutions. [CSME04-E20]
		(A) Produces guidance and recommendations or best practice for external stakeholders. [CSME04-30A]	Records of advice provided on the suitability of modeling and analysis techniques used on projects beyond the enterprise boundary. [CSME04-E30]
5	Advises organizations beyond the enterprise boundary on the suitability of their approach to systems modeling and analysis. [CSME05]	(A) Reviews the systems modeling or analysis approach of external stakeholders. [CSME05-10A]	Records demonstrating internal or external advisory or consultative role in modeling and analysis. [CSME05-E10]
		(A) Advises stakeholders on the suitability of their approaches, recommending changes. [CSME05-20A]	Records of advice provided on adoption/rejection of modeling and analysis solutions. [CSME05-E20]
		(A) Communicates modeling and analysis guidance and recommendations effectively to key external stakeholders. [CSME05-30A]	Records of advice provided on the suitability of modeling and analysis techniques used on projects beyond the enterprise boundary. [CSME05-E30]
6	Advises organizations beyond the enterprise boundary on complex or sensitive systems modeling and analysis issues. [CSME06]	(A) Advises stakeholders beyond the enterprise boundary on complex or sensitive issues relating to modeling and analysis. [CSME06-10A]	Records demonstrating internal or external advisory or consultative role in systems modeling and analysis. [CSME06-E10]
			Records of advice provided together with evidence that the issues advised on were by their nature either complex or sensitive. [CSME06-E20]
7	Champions the introduction of novel techniques and ideas in Systems modeling and analysis, beyond the enterprise boundary, in order to develop the wider Systems Engineering community in this competency. [CSME07]	(A) Analyzes different approaches across different domains through research. [CSME07-10A]	Records of activities promoting research and need to adopt novel technique or ideas. [CSME07-E10]
		(A) Produces reports for the wider SE community on the effectiveness of new techniques after their introduction. [CSME07-20A]	Records of improvements made to process and appraisal against a recognized process improvement model. [CSME07-E20]

		(P) Collaborates with those introducing novel techniques within the wider SE community. [CSME07-30P]	Research records. [CSME07-E30]
		(A) Defines novel approaches that could potentially improve the wider SE discipline. [CSME07-40A]	Published papers in refereed journals/company literature. [CSME07-E40]
		(P) Fosters awareness of these novel techniques within the wider SE community. [CSME07-50P]	Records showing introduction of enabling systems supporting the new techniques or ideas. [CSME07-E50]
8	Coaches individuals beyond the enterprise boundary in systems modeling and analysis, in order to further develop their knowledge, abilities, skills, or associated behaviors. [CSME08]	(P) Coaches or mentors individuals beyond the enterprise boundary, in competency-related techniques, recommending development activities. [CSME08-10P]	Coaching or mentoring assignment records. [CSME08-E10]
		(A) Develops or authorizes training materials in this competency area, which are subsequently successfully delivered beyond the enterprise boundary. [CSME08-20A]	Records of formal training courses, workshops, seminars performed, or authored training material supported by successful post-training evaluation data. [CSME08-E20]
		(A) Provides workshops/seminars or training in this competency area for practitioners or lead practitioners beyond the enterprise boundary (e.g. conferences and open training days). [CSME08-30A]	Records of training/workshops/seminars created supported by successful post-training evaluation data. [CSME08-E30]
9	Maintains expertise in this competency area through specialist Continual Professional Development (CPD) activities. [CSME09]	(A) Reviews research, new ideas, and state of the art to identify relevant new areas requiring personal development in order to maintain expertise in this competency area. [CSME09-10A]	Records of documents reviewed and insights gained as part of own research into this competency area. [CSME09-E10]
		(A) Performs identified specialist professional development activities in order to maintain or further develop competence at expert level. [CSME09-20A]	Records of Continual Professional Development (CPD) performed and learning outcomes. [CSME09-E20]
		(A) Records continual professional development activities undertaken including learning or insights gained. [CSME09-30A]	

NOTES	In addition to items above, enterprise-level or independent 3rd-Party-generated evidence may be used to amplify other evidence presented and may include: a. Formally recognized by a reputable external organization as an expert in this competency area b. Evidence of role as independent assessor or reviewer on project outside own organization where skills in this competency area were used c. Evidence of invitation(s) from wider community for contribution of systems engineering expertise in this area (e.g. industry conference panel, government advisory board etc. cross-industry working groups, partnerships, accredited advanced university courses or research, or as part of professional institute) d. Formal commendation beyond the enterprise (e.g. by INCOSE or other recognized authority) for work performed in this competency area e. Independently assessed or accredited work product in this competency area (e.g. for independent publication or use) f. Accolades of expertise in this area from recognized industry leaders

Competency area – Professional: Communications
Description The dynamic process of transmitting or exchanging information using various principles such as verbal, speech, body-language, signals, behavior, writing, audio, video, graphics, and language. Communication includes all interactions between individuals, individuals and groups, or between different groups.
Why it matters Communication plays a fundamental role in all facets of business within an organization, in order to transfer information between individuals and groups to develop a common understanding and build and maintain relationships and other intangible benefits. Ineffective communication has been identified as the root cause of problems on projects.
Possible contributory types of evidence Any combination of the types of evidence may be acceptable (depending on how the Framework is tailored and used). The evidence items identified at each level indicate example work products only. Contributions to work products will generally differ at each proficiency level.
Learning and development The INCOSE Professional Development Portal provides example guidance on how to gain an initial awareness of a competency area and options for developing further competence thereafter.

Awareness – Professional: Communications		
ID	**Indicators of Competence**	**Relevant knowledge sub-indicators**
1	Explains communications in terms of the sender, the receiver, and the message and why these three parameters are central to the success of any team communication. [PCCA01]	(K) Lists examples of different interpersonal communications mechanisms in terms of sender, receiver, and the message. [PCCA01-10K] (K) Explains why sender, receiver, and message content are important. [PCCA01-20K] (K) Explains why successful communications is not solely due to the information, but includes how it is encoded, decoded, and perceived emotionally. [PCCA01-30K]
2	Explains why there is a need for clear and concise communications. [PCCA02]	(K) Identifies the main barriers to effective interpersonal communication. [PCCA02-10K] (K) Explains how communication can influence understanding and encourage compliance with management direction. [PCCA02-20K] (K) Describes key communications interfaces and links between the engineering function and stakeholders in the wider enterprise outside engineering (e.g. infrastructure management, human resource management, quality management, knowledge management, portfolio management, and life cycle model management). [PCCA02-30K] (K) Explains the role of feedback in certain communications. [PCCA02-40K]
3	Describes the role communications has in developing positive relationships. [PCCA03]	(K) Explains how social awareness and social facility impact on the effectiveness of communications. [PCCA03-10K] (K) Explains how organizational climate can affect or enable communications. [PCCA03-20K]

4	Explains why employing the appropriate means for communications is essential. [PCCA04]	(K) Lists different forms of interpersonal communication and the differing circumstances in which they can be effective (or ineffective). [PCCA04-10K] (K) Describes potential barriers to successful communications. [PCCA04-20K] (K) Explains how cultural differences effect communications. [PCCA04-30K]
5	Explains why openness and transparency in communications matters. [PCCA05]	(K) Describes the nature and significance of nonverbal communication cues and clusters (e.g. body language, paralanguage, and impression management). [PCCA05-10K] (K) Describes the role of trust in facilitating communications. [PCCA05-20K]
6	Explains why systems engineers need to listen to stakeholders' point of view. [PCCA06]	(K) Describes the role of differing questioning techniques (open, closed, probe, reflective, multiple, leading, and hypothetical) and when they might be deployed. [PCCA06-10K] (K) Explains why differing stakeholders may hold different views about the system need, requirements, and appropriate solutions. [PCCA06-20K] (K) Explains how sharing differing stakeholder viewpoints can improve the quality of a solution. [PCCA06-30K]

Supervised Practitioner – Professional: Communications			
ID	**Indicators of Competence** *(in addition to those at Awareness level)*	**Relevant knowledge, experience, and/or behaviors**	**Possible examples of objective evidence of personal contribution to activities performed, or professional behaviors applied**
1	Follows guidance received (e.g. from mentors) when using communications skills to plan and control their own communications activities. [PCCS01]	(A) Follows guidance received when using communications skills to plan and control their own communications activities. [PCCS01-10A]	Communications plans and tools they have used in their role. [PCCS01-E10]
2	Uses appropriate communications techniques to ensure a shared understanding of information with peers. [PCCS02]	(P) Uses appropriate communications techniques to ensure a shared understanding of information delivered to peers. [PCCS02-10P]	Records of different communications methods used for differing peer-level communications relationships. [PCCS02-E10]
		(P) Uses appropriate communications techniques to ensure a common understanding of information received from peers. [PCCS02-20P]	Records showing improvements to peer-level communications through changes have made. [PCCS02-E20]
		(P) Recognizes differing levels of effectiveness of communications with different peers and how these can be improved over time. [PCCS02-30P]	Records showing improved effectiveness through changes peer-level communications approaches. [PCCS02-E30]
3	Fosters positive relationships through effective communications. [PCCS03]	(P) Recognizes that the quality (or positively) of different stakeholder relationships manifests itself through communications. [PCCS03-10P]	Records showing improved personal stakeholder relationships over time and can illustrate how this was achieved. [PCCS03-E10]
		(P) Uses appropriate communications techniques to ensure a positive relationships with peers. [PCCS03-20P]	Records showing improved stakeholder relationship(s) through improved personal stakeholder communications. [PCCS03-E20]
		(P) Recognizes different levels of effectiveness with stakeholder relationships and how these can be improved over time. [PCCS03-30P]	Records showing improved effectiveness through changes peer-level communications approaches. [PCCS03-E30]
4	Uses appropriate communications techniques to interact with others, depending on the nature of the relationship. [PCCS04]	(P) Recognizes why different communications strategies may have different levels of effectiveness. [PCCS04-10P]	Records of distinct communications selected for different stakeholders and rationale. [PCCS04-E10]
		(A) Selects appropriate communications mechanisms for different peer-level relationships in their role, depending on the nature of the relationship. [PCCS04-20A]	Records demonstrating a modified communications mechanism used for one or more stakeholders in order to improve interactions with them. [PCCS04-E20]
		(P) Uses appropriate communications strategies to communicate with different stakeholders, depending on the nature of the relationship. [PCCS04-30P]	Records demonstrating improved stakeholder communication strategies with different stakeholders, depending on the nature of the relationship. [PCCS04-E30]

5	Fosters trust through openness and transparency in communication. [PCCS05]	(A) Identifies different personal stakeholder relationships with differing levels of trust or openness with rationale as to why this might have occurred. [PCCS05-10A]	Records demonstrating personal stakeholder relationships with differing perceived levels of trust or openness. [PCCS05-E10]
		(P) Fosters trust in one or more relationships over time. [PCCS05-20P]	Records illustrating how personal stakeholder relationship trust or openness has improved over time through communications approaches. [PCCS05-E20]
		(P) Fosters improved trust in one or more relationships over time. [PCCS05-30P]	Records demonstrating improved stakeholder relationship over time. [PCCS05-E30]
6	Uses active listening techniques to clarify understanding of information or views. [PCCS06]	(P) Recognizes techniques associated with "active listening." [PCCS06-10P]	Records demonstrating use of "active listening" techniques. [PCCS06-E10]
		(P) Uses some active listening techniques to clarify information or views. [PCCS06-20P]	Records demonstrating clarification of stakeholder points of view through use of active listening. [PCCS06-E20]
7	Develops own understanding of this competency area through Continual Professional Development (CPD). [PCCS07]	(A) Identifies potential gaps in own knowledge or development needs in this area, identifying opportunities to address these through continual professional development activities. [PCCS07-10A]	Records of Continual Professional Development (CPD) performed and learning outcomes. [PCCS07-E10]
		(A) Performs continual professional development activities to improve their knowledge and understanding in this area. [PCCS07-20A]	
		(A) Records continual professional development activities undertaken including learning or insights gained. [PCCS07-30A]	

Practitioner – Professional: Communications			
ID	**Indicators of Competence** ***(in addition to those at Supervised Practitioner level)***	**Relevant knowledge, experience, and/or behaviors**	**Possible examples of objective evidence of personal contribution to activities performed, or professional behaviors applied**
1	Uses a governing communications plan and appropriate tools to control communications. [PCCP01]	(A) Follows a governing communications plan and using appropriate tools to control project communications. [PCCP01-10A]	Communications plan(s) they have been required to use for differing stakeholder groups. [PCCP01-E10]
		(P) Recognizes situations where deviation from published communications plans and processes or clarification from others is appropriate in order to overcome complex communications challenges. [PCCP01-20P]	Published communications plans and processes have been deviated from in order to overcome communication challenges. [PCCP01-E20]
		(P) Recognizes situations where existing communications strategy, process, plan or tools require formal change, liaising with stakeholders to gain approval as necessary. [PCCP01-30P]	Published communications strategies, processes, plans, or tools have been deviated from in order to overcome communication challenges. [PCCP01-E30]
2	Uses appropriate communications techniques to ensure a shared understanding of information with all project stakeholders. [PCCP02]	(P) Uses appropriate communications techniques to ensure a shared understanding of information delivered to project stakeholders. [PCCP02-10P]	Personal communications demonstrating differing strategies used for differing different stakeholder groups. [PCCP02-E10]
		(P) Uses appropriate communications techniques to ensure a common understanding of information received from project stakeholders. [PCCP02-20P]	Personal communications demonstrating differing effectiveness levels for differing different stakeholder groups and changes over time to improve effectiveness where appropriate. [PCCP02-E20]
		(P) Recognizes differing levels of effectiveness of communications with different stakeholders and how these can be improved over time. [PCCP02-30P]	Records demonstrating improved effectiveness of differing stakeholder communications through an improvement in communication approach. [PCCP02-E30]
3	Uses appropriate communications techniques to ensure positive relationships are maintained. [PCCP03]	(A) Identifies differing quality (or positivity) of stakeholder relationships as a result of examining communications with them. [PCCP03-10A]	Personal communications demonstrating the link between the method used and the quality of stakeholder relationships. [PCCP03-E10]
		(P) Fosters improvements to poor stakeholder relationships over time through improved communications. [PCCP03-20P]	Related personal communications demonstrating how a relationship with particular stakeholder was improved over time through appropriate communications strategy. [PCCP03-E20]
		(P) Fosters maintenance of good stakeholder relationships through effective communications. [PCCP03-30P]	Records (e.g. from Human Resources department) indicating good communications skills in this area. [PCCP03-E30]
4	Uses appropriate communications techniques to express alternate points of view in a diplomatic manner using the appropriate means of communication. [PCCP04]	(P) Uses the concept of "diplomatic language" in relation to communications with stakeholders listing the communications techniques involved. [PCCP04-10P]	Personal communications demonstrating the successful diplomatic presentation of alternate or challenging stakeholder viewpoints. [PCCP04-E10]

		(P) Uses alternate viewpoints diplomatically in order to achieve a positive outcome. [PCCP04-20P]	Alternate viewpoints have been used diplomatically in order to achieve a positive outcome. [PCCP04-E20]
		(P) Uses appropriate diplomatic language to communicate alternate stakeholder viewpoints with a positive outcome. [PCCP04-30P]	Diplomatic language has been used appropriately to communicate alternate stakeholder viewpoints with positive outcomes. [PCCP04-E30]
5	Fosters a communicating culture by finding appropriate language and communication styles, augmenting where necessary to avoid misunderstanding. [PCCP05]	(P) Recognizes why alternate language and communication styles (augmented where necessary) were adopted for communications, with a positive outcome. [PCCP05-10P]	Personal communications demonstrating the deliberate adoption of alternative language or communication styles for differing audiences. [PCCP05-E10]
		(A) Identifies situations where alternate language and communication styles require adoption in order to communicate with a positive outcome. [PCCP05-20A]	Records (e.g. from Human Resources department) indicating good communications skills in this area. [PCCP05-E20]
		(P) Uses alternate language and communication styles for communications, with a positive outcome. [PCCP05-30P]	Alternate language and communication styles have been used for communications with a positive outcome. [PCCP05-E30]
6	Uses appropriate communications techniques to express own thoughts effectively and convincingly in order to reinforce the content of the message. [PCCP06]	(P) Recognizes reinforced message content used to express effective communications. [PCCP06-10P]	Personal communication(s) and associated feedback demonstrating that the communicated message was delivered effectively and convincingly. [PCCP06-E10]
			Records (e.g. from Human Resources department) indicating good communications skills in this area. [PCCP06-E20]
7	Uses full range of active listening techniques to clarify information or views. [PCCP07]	(A) Identifies project situations where active listening helps to confirm an idea and improve understanding. [PCCP07-10A]	Personal communication(s) demonstrating improved understanding through active listening approach. [PCCP07-E10]
		(P) Uses full range of active listening techniques to clarify information or views. [PCCP07-20P]	Records (e.g. from Human Resources department) indicating good communications skills in this area. [PCCP07-E20]
8	Uses appropriate feedback techniques to verify success of communications. [PCCP08]	(A) Identifies project situations where feedback helps confirm that a communication was correctly understood. [PCCP08-10A]	Personal communications which requested feedback to confirm understanding improved understanding. [PCCP08-E10]
		(A) Elicits feedback to confirm that a communication was correctly understood by all recipients. [PCCP08-20A]	Records (e.g. from Human Resources department) indicating good communications skills in this area. [PCCP08-E20]
		(P) Adapts communications techniques where feedback (or lack of feedback) necessitates in order to improve future communications. [PCCP08-30P]	Records showing how they adapted communications techniques as a result of feedback, which resulted in improved communications. [PCCP08-E30]

(Continued)

ID	Indicators of Competence *(in addition to those at Supervised Practitioner level)*	Relevant knowledge, experience, and/or behaviors	Possible examples of objective evidence of personal contribution to activities performed, or professional behaviors applied
9	Guides new or supervised Systems Engineering practitioners in Communications techniques in order to develop their knowledge, abilities, skills, or associated behaviors. [PCCP09]	(P) Guides new or supervised practitioners in executing activities, which form part of this competency. [PCCP09-10P]	Organizational Breakdown Structure showing their responsibility for technical supervision in this area. [PCCP09-E10]
		(A) Trains individuals to an "Awareness" level in this competency area. [PCCP09-20A]	On-the-job training objectives/guidance. [PCCP09-E20] Coaching or mentoring assignment records. [PCCP09-E30] Records highlighting their impact on another individual in terms of improvement or professional development in this competency. [PCCP09-E40]
10	Maintains and enhances own competence in this area through Continual Professional Development (CPD) activities. [PCCP10]	(A) Identifies potential development needs in this area, identifying opportunities to address these through continual professional development activities. [PCCP10-10A] (A) Performs continual professional development activities to maintain and enhance their competency in this area. [PCCP10-20A] (A) Records continual professional development activities undertaken including learning or insights gained. [PCCP10-30A]	Records of Continual Professional Development (CPD) performed and learning outcomes. [PCCP10-E10]

Lead Practitioner – Professional: Communications			
ID	**Indicators of Competence** ***(in addition to those at Practitioner level)***	**Relevant knowledge, experience, and/or behaviors**	**Possible examples of objective evidence of personal contribution to activities performed, or professional behaviors applied**
1	Creates enterprise-level policies, procedures, guidance, and best practice for systems engineering communications, including associated tools. [PCCL01]	(A) Analyzes enterprise need for systems engineering communications policies, processes, tools, or guidance. [PCCL01-10A]	Records showing their role in embedding systems engineering communications into enterprise policies (e.g. guidance introduced at enterprise level, enterprise -level review minutes). [PCCL01-E10]
		(A) Creates enterprise policies, procedures or guidance for systems engineering communications activities. [PCCL01-20A]	Procedures they have written. [PCCL01-E20]
		(A) Selects and acquires appropriate tools supporting systems engineering communications activities. [PCCL01-30A]	Records of support for tool introduction. [PCCL01-E30]
2	Uses best practice communications techniques to improve the effectiveness of Systems Engineering activities across the enterprise. [PCCL02]	(A) Develops own state-of-the-art communications skills in order to communicate more effectively with stakeholders across the enterprise. [PCCL02-10A]	Records demonstrating personal use of various communication(s) techniques and associated feedback indicating effectiveness. [PCCL02-E10]
		(P) Adapts best practice techniques to communicate more effectively with different stakeholders or stakeholder groups across the enterprise. [PCCL02-20P]	Records demonstrating personal communications to a variety of diverse stakeholders across the enterprise. [PCCL02-E20]
			Records (e.g. from Human Resources department) indicating excellent communications skills, highlighting this area. [PCCL02-E30]
3	Maintains positive relationships across the enterprise through effective communications in challenging situations, adapting as necessary to achieve communications clarity or to improve the relationship. [PCCL03]	(A) Identifies challenging situations requiring communications in order to positively impact stakeholder relationships. [PCCL03-10A]	Records of personal communication(s) in challenging situations, with positive outcomes. [PCCL03-E10]
		(A) Identifies the need for different communications styles in challenging situations. [PCCL03-20A]	Records demonstrating adaptation of personal communication(s) styles to reflect differing situations, with positive outcomes. [PCCL03-E20]
		(P) Uses differing communications styles to improve clarity for different audiences in challenging circumstances. [PCCL03-30P]	Records (e.g. from Human Resources department) indicating excellent communications skills, highlighting this area. [PCCL03-E30]

ID	Indicators of Competence *(in addition to those at Practitioner level)*	Relevant knowledge, experience, and/or behaviors	Possible examples of objective evidence of personal contribution to activities performed, or professional behaviors applied
4	Uses effective communications techniques to convince stakeholders across the enterprise to reach consensus in challenging situations. [PCCL04]	(A) Identifies challenging situations where effective communications are required in order to persuade and convince stakeholders to reach consensus. [PCCL04-10A] (P) Uses effective communications to persuade and convince stakeholders to reach consensus in challenging situations. [PCCL04-20P]	Records demonstrating personal communication(s), which persuaded stakeholders to reach consensus in a challenging situation. [PCCL04-E10] Records (e.g. from Human Resources department) indicating excellent communications skills, highlighting this area. [PCCL04-E20]
5	Uses a proactive style, building consensus among stakeholders across the enterprise using techniques supporting the verbal messages (e.g. nonverbal communication). [PCCL05]	(A) Identifies situations requiring a proactive approach to build consensus among stakeholders communications. [PCCL05-10A] (A) Identifies situations requiring supplementary communications techniques (e.g. nonverbal communications) to build consensus among stakeholders. [PCCL05-20A] (P) Uses a proactive approach to build consensus among stakeholders communications. [PCCL05-30P] (P) Uses supplementary techniques (e.g. nonverbal communications) to build consensus among stakeholders. [PCCL05-40P]	Records of support for novel idea or technique. [PCCL05-E10] Records (e.g. from Human Resources department) indicating excellent communications skills, highlighting this area. [PCCL05-E20]
6	Adapts communications techniques or expresses ideas differently to improve effectiveness of communications to stakeholders across the enterprise, by changing language, content, or style. [PCCL06]	(A) Identifies situations requiring reformulation of existing communications to improve understanding. [PCCL06-10A] (A) Adapts existing communications, reformulating or feeding back to originators in order to successfully clarify and improve understanding. [PCCL06-20A] (A) Identifies situations requiring direct feedback to stakeholders or team. [PCCL06-30A]	Records demonstrating personal communication(s) to key stakeholders where their own communications were reformulated and clarified to improve understanding. [PCCL06-E10] Records (e.g. from Human Resources department) indicating excellent communications skills, highlighting this area. [PCCL06-E20]
7	Reviews ongoing communications across the enterprise, anticipating and mitigating potential problems. [PCCL07]	(P) Uses supplementary communications techniques (e.g. nonverbal communications) to build consensus among stakeholders. [PCCL07-10P] (A) Determines actions to mitigate potential communications problems. [PCCL07-20A] (P) Acts proactively to mitigate identified communications problems. [PCCL07-30P]	Records (e.g. from Human Resources department) indicating excellent communications skills, highlighting this area. [PCCL07-E10]

		(P) Uses supplementary communications techniques (e.g. nonverbal communications) to build consensus among stakeholders. [PCCL07-40P]	
8	Fosters the wider enterprise vision, communicating it successfully across the enterprise. [PCCL08]	(A) Proposes contributions to the overall vision for the enterprise. [PCCL08-10A]	Records demonstrating personal contribution to enterprise communications vision. [PCCL08-E10]
		(P) Communicates the vision across the enterprise. [PCCL08-20P]	Records demonstrating activities associated with communicating the enterprise vision to others. [PCCL08-E20]
		(P) Acts to explain the rationale for the enterprise vision. [PCCL08-30P]	
9	Coaches or mentors practitioners across the enterprise, or those new to this competency are in order to develop their knowledge, abilities, skills, or associated behaviors. [PCCL09]	(P) Coaches or mentors practitioners across the enterprise in competency-related techniques, recommending development activities. [PCCL09-10P]	Coaching or mentoring assignment records. [PCCL09-E10]
		(A) Develops or authorizes enterprise training materials in this competency area. [PCCL09-20A]	Records of formal training courses, workshops, seminars, and authored training material supported by successful post-training evaluation data. [PCCL09-E20]
		(A) Provides enterprise workshops/seminars or training in this competency area. [PCCL09-30A]	Listing as an approved organizational trainer for this competency area. [PCCL09-E30]
10	Develops expertise in this competency area through specialist Continual Professional Development (CPD) activities. [PCCL10]	(A) Identifies own needs for further professional development in order to increase competence beyond practitioner level. [PCCL10-10A]	Records of Continual Professional Development (CPD) performed and learning outcome. [PCCL10-E10]
		(A) Performs professional development activities in order to move own competence toward expert level. [PCCL10-20A]	
		(A) Records continual professional development activities undertaken including learning or insights gained. [PCCL10-30A]	

NOTES	In addition to items above, enterprise-level or independent 3rd-Party-generated evidence may be used to amplify other evidence presented and may include: a. Formally recognized by senior management in current organization as an expert in this competency area b. Evidence of role as Product/System Design Authority or Technical Authority on a complex project with responsibilities in this area or where skills within this competency area were used c. Recognized as an authorizing signatory on behalf of enterprise for formal documentation in this competency area (e.g. policies, processes, and deliverables) d. Formal commendation or award within own enterprise for contribution or item of work successfully performed, which required proficiency in this competency area e. Customer, Supplier, or other external project-specific key Stakeholder accolades for specific work performed in this competency area f. Independently assessed or accredited work in this competency area (e.g. for independent publication or use) g. Formal organizational HR records positively highlighting any specific professional competencies or behaviors identified (if applicable) plus any of the evidence indicators listed at Expert level below

Expert – Professional: Communications			
ID	**Indicators of Competence *(in addition to those at Lead Practitioner level)***	**Relevant knowledge, experience, and/or behaviors**	**Possible examples of objective evidence of personal contribution to activities performed, or professional behaviors applied**
1	Communicates own knowledge and experience in Communications Techniques in order to improve best practice beyond the enterprise boundary. [PCCE01]	(A) Produces papers, seminars, or presentations outside own enterprise for publication in order to share own ideas and improve industry best practices in this competence area. [PCCE01-10A] (P) Fosters incorporation of own ideas into industry best practices in this area. [PCCE01-20P] (P) Develops guidance materials identifying new (or updating existing) best practice in this competence area. [PCCE01-30P]	Papers or books published on new technique in refereed journals/company literature. [PCCE01-E10] Papers published in refereed journals or internal literature proposing new practices in this competence area (or presentations, tutorials, etc.). [PCCE01-E20] Records of own proposals adopted as industry best practices in this competence area. [PCCE01-E30]
2	Advises organizations beyond the enterprise boundary on the suitability of their approach to communications. [PCCE02]	(P) Advises key stakeholders on how communications strategies, plans, processes, or tools may differ from organization to organization. [PCCE02-10P] (A) Advises beyond the enterprise boundary on the suitability of current or proposed communications strategies, plans, processes, or tools. [PCCE02-20A]	Records of advice provided on the suitability of communications strategies, plans, processes, or tools. [PCCE02-E10]
3	Fosters a collaborative learning, listening atmosphere among key stakeholders beyond the enterprise boundary. [PCCE03]	(A) Identifies situations requiring a collaborative learning and listening atmosphere between stakeholders beyond the enterprise boundary. [PCCE03-10A] (P) Fosters a collaborative atmosphere with beyond the enterprise boundary with measurable success. [PCCE03-20P]	Records demonstrating personal contribution to the development of a collaborative relationship with key stakeholders. [PCCE03-E10]
4	Advises organizations beyond the enterprise boundary on complex or sensitive communications-related matters affecting Systems Engineering. [PCCE04]	(A) Identifies complex or sensitive matters needing sensitive communication of advice. [PCCE04-10A] (A) Conducts communications on complex or sensitive issues with sensitivity. [PCCE04-20A] (P) Conducts sensitive negotiations communicating about a highly complex system but making limited use of specialized, technical terminology. [PCCE04-30P] (A) Uses a holistic approach to complex issue resolution including balanced, rational arguments on way forward. [PCCE04-40A] (A) Communicates complex or sensitive advice successfully to beyond the enterprise boundary. [PCCE04-50A]	Records of advice provided on complex or sensitive matters affecting Systems Engineering together with evidence that the issues advised on were by their nature either complex or sensitive. [PCCE04-E10]
5	Champions the introduction of novel techniques and ideas in "communications," beyond the enterprise boundary, in order to develop the wider Systems Engineering community in this competency. [PCCE05]	(A) Analyzes different approaches across different domains through research. [PCCE05-10A]	Records of activities promoting research and need to adopt novel technique or ideas. [PCCE05-E10]

		(A) Produces reports for the wider SE community on the effectiveness of new techniques after their introduction. [PCCE05-20A]	Records of improvements made to process and appraisal against a recognized process improvement model. [PCCE05-E20]
		(P) Collaborates with those introducing novel techniques within the wider SE community. [PCCE05-30P]	Research records. [PCCE05-E30]
		(A) Defines novel approaches that could potentially improve the wider SE discipline. [PCCE05-40A]	Published papers in refereed journals/company literature. [PCCE05-E40]
		(P) Fosters awareness of these novel techniques within the wider SE community. [PCCE05-50P]	Records showing introduction of enabling systems supporting the new techniques or ideas. [PCCE05-E50]
6	Coaches individuals beyond the enterprise boundary in Communications techniques, in order to further develop their knowledge, abilities, skills, or associated behaviors. [PCCE06]	(P) Coaches or mentors individuals beyond the enterprise boundary, in competency-related techniques, recommending development activities. [PCCE06-10P]	Coaching or mentoring assignment records. [PCCE06-E10]
		(A) Develops or authorizes training materials in this competency area, which are subsequently successfully delivered beyond the enterprise boundary. [PCCE06-20A]	Records of formal training courses, workshops, seminars, and authored training material supported by successful post-training evaluation data. [PCCE06-E20]
		(A) Provides workshops/seminars or training in this competency area for practitioners or lead practitioners beyond the enterprise boundary (e.g. conferences and open training days). [PCCE06-30A]	Records of Training/workshops/seminars created supported by successful post-training evaluation data. [PCCE06-E30]
7	Maintains expertise in this competency area through specialist Continual Professional Development (CPD) activities. [PCCE07]	(A) Reviews research, new ideas, and state of the art to identify relevant new areas requiring personal development in order to maintain expertise in this competency area. [PCCE07-10A]	Records of documents reviewed and insights gained as part of own research into this competency area. [PCCE07-E10]
		(A) Performs identified specialist professional development activities in order to maintain or further develop competence at expert level. [PCCE07-20A]	Records of Continual Professional Development (CPD) performed and learning outcomes. [PCCE07-E20]
		(A) Records continual professional development activities undertaken including learning or insights gained. [PCCE07-30A]	

NOTES	In addition to items above, enterprise-level or independent 3rd-Party-generated evidence may be used to amplify other evidence presented and may include: a. Formally recognized by a reputable external organization as an expert in this competency area b. Evidence of role as independent assessor or reviewer on project outside own organization where skills in this competency area were used c. Evidence of invitation(s) from wider community for contribution of systems engineering expertise in this area (e.g. industry conference panel, government advisory board etc. cross-industry working groups, partnerships, accredited advanced university courses or research, or as part of professional institute) d. Formal commendation beyond the enterprise (e.g. by INCOSE or other recognized authority) for work performed in this competency area e. Independently assessed or accredited work product in this competency area (e.g. for independent publication or use) f. Accolades of expertise in this area from recognized industry leaders

Competency area – Professional: Ethics and Professionalism
Description Professional ethics encompass the personal, organizational, and corporate standards of behavior expected of systems engineers. Professional ethics also encompasses the use of specialist knowledge and skills by systems engineers when providing a service to the public. Overall, competence in ethics and professionalism can be summarized by a personal commitment to professional standards, recognizing obligations to society, the profession and the environment.
Why it matters Systems engineers are routinely trusted to apply their skills, make judgments, and to reach unbiased, informed and potentially significant decisions because of their specialized knowledge and skills. It is important that the professional systems engineer always acts ethically, in order to maintain trust, ensure professional standards are upheld, and that their wider obligations to society, and the environment are met.
Possible contributory types of evidence Any combination of the types of evidence may be acceptable (depending on how the Framework is tailored and used). The evidence items identified at each level indicate example work products only. Contributions to work products will generally differ at each proficiency level.
Learning and development The INCOSE Professional Development Portal provides example guidance on how to gain an initial awareness of a competency area and options for developing further competence thereafter.

Awareness – Professional: Ethics and Professionalism		
ID	**Indicators of Competence**	**Relevant knowledge sub-indicators**
1	Explains why Systems Engineering has a social significance. [PEPA01]	(K) Explains why the social significant of Systems Engineering relates to ethics and professionalism. [PEPA01-10K] (K) Explains why Systems Engineering has a social significance. [PEPA01-20K]
2	Describes applicable codes of conduct for professional systems engineers including institutional or company codes of conduct. [PEPA02]	(K) Explains the rational for publishing a company code of ethics. [PEPA02-10K] (K) Explains the code of ethics of their employer or a professional organization to which they belong. [PEPA02-20K]
3	Lists typical safety standards and requirements. [PEPA03]	(K) Explains the rationale for ensuring independence in safety assessment and auditing. [PEPA03-10K] (K) Describes the safety process and standards of their employer or a professional organization to which they belong. [PEPA03-20K] (K) Explains the difference and similarities between personal safety (e.g. health and safety at work) and product safety. [PEPA03-30K] (K) Explains the ethics and professionalism that may be required to achieve personal safety (e.g. health and safety at work) and product safety. [PEPA03-40K]
4	Explains why security has become increasingly important general requirement in the development of systems and provides examples. [PEPA04]	(K) Describes how different security perspectives (e.g. national, organizational, and product, individual) can potentially impact system development. [PEPA04-10K]

		(K) Explains the rationale for ensuring independence in security assessment and auditing. [PEPA04-20K] (K) Explains how ethical and professionalism may need to be considered when ensuring security concerns are addressed. [PEPA04-30K]
5	Explains why there is a need to undertake engineering activities in a way that contributes to sustainable, environmentally sound development and the relationship these have with the economic sustainability of a system. [PEPA05]	(K) Explains the term "sustainable" and why there is a need to undertake engineering activities in a way that contributes to sustainable development. [PEPA05-10K] (K) Explains the term "environmental sustainability" and why there is a need to undertake all engineering activities in a way that considers their environmental impact. [PEPA05-20K] (K) Explains the relationship between sustainability, the environment and economic viability. [PEPA05-30K] (K) Describes their company's policy relating to sustainable development. [PEPA05-40K] (K) Describes their company's policies relating to environment. [PEPA05-50K]
6	Explains why there is a need to undertake engineering activities in a way that considers diversity, equality, and inclusivity, and provides examples. [PEPA06]	(K) Explains why there is a need to undertake engineering activities in a way that considers diversity, equality, and inclusivity. [PEPA06-10K] (K) Describes their company's policy on diversity, equality and inclusivity. [PEPA06-20K]
7	Explains why it is necessary to develop, plan, carry out, and record Continued Professional Development (CPD) in order to maintain and enhance competence in own area of practice. [PEPA07]	(K) Explains why it is necessary to undertake Continued Professional Development (CPD) in order to maintain and enhance competence in own area of practice. [PEPA07-10K] (K) Describes their organization's approach to professional development. [PEPA07-20K] (K) Explains why it is necessary to develop, plan, carry out, and record Continued Professional Development (CPD). [PEPA07-30K]
8	Explains why Systems Engineering has a relationship to ethics and professionalism. [PEPA08]	(K) Explains why Systems Engineering has a social significance. [PEPA08-10K] (K) Explains why the social significant of Systems Engineering relates to ethics and professionalism. [PEPA08-20K] (K) Explains why systems engineers have a responsibility to act in an ethical manner. [PEPA08-30K] (K) Explains potential consequences of systems engineers acting in an ethical manner. [PEPA08-40K] (K) Explains potential consequences of systems engineers not acting in an ethical manner. [PEPA08-50K]

Supervised Practitioner – Professional: Ethics and Professionalism			
ID	**Indicators of Competence *(in addition to those at Awareness level)***	**Relevant knowledge, experience, and/or behaviors**	**Possible examples of objective evidence of personal contribution to activities performed, or professional behaviors applied**
1	Complies with applicable codes of professional conduct within the enterprise. [PEPS01]	(A) Follows guidance received when using professional or ethical codes to plan and control their own professional activities. [PEPS01-10A]	Records of professional or ethical codes used to control activities. [PEPS01-E10]
		(A) Complies with applicable codes of professional conduct within the enterprise, seeking guidance if required. [PEPS01-20A]	Records of compliance with applicable codes of professional conduct within the enterprise. [PEPS01-E20]
		(A) Communicates any breaches of code of conduct as required. [PEPS01-30A]	Records of communication for any breaches of code of conduct. [PEPS01-E30]
		(P) Uses ethical judgement to challenge issues and situations. [PEPS01-40P]	Records of using ethical judgement to challenge issues and situations. [PEPS01-E40]
2	Follows safe systems principles at work, by interpreting relevant health, safety, and welfare processes, legislation, and standards seeking guidance if required. [PEPS02]	(K) Defines the key concepts of applying safe systems of work. [PEPS02-10K]	Records of actions to apply safe systems of work or guidance sought. [PEPS02-E10]
		(K) Lists potential health, safety and welfare hazards that might be present at work. [PEPS02-20K]	Records of actions to maintain safe systems of work or guidance sought. [PEPS02-E20]
		(A) Complies with relevant safety, health and employment frameworks and legislation. [PEPS02-30A]	
		(P) Acts to maintain safe systems of work, seeking guidance if required. [PEPS02-40P]	
3	Follows systems security principles at work, by interpreting relevant security processes, legislation, and standards seeking guidance if required. [PEPS03]	(K) Lists potential areas where security risks or issues might exist within the organization, the project and for individuals. [PEPS03-10K]	Records demonstrating acceptance of or compliance with security requirements at work. [PEPS03-E10]
		(K) Identifies relevant legislation covering these areas. [PEPS03-20K]	Records of actions taken to ensure or maintain security of own project work or guidance sought. [PEPS03-E20]
		(A) Complies with relevant security frameworks and legislation. [PEPS03-30A]	
		(P) Acts to ensure or maintain security in their own activities at work, seeking guidance if required. [PEPS03-40P]	

4	Acts to ensure their own activities are performed in a way that contributes to sustainable development. [PEPS04]	(K) Explains the merits of providing a visible example for others regarding sustainable development. [PEPS04-10K] (P) Acts to take on personal responsibility for ensuring their own activities are performed in a way that contributes to sustainable development. [PEPS04-20P]	Records of taking on personal responsibility for ensuring their own activities are performed in a way that contributes to sustainable development. [PEPS04-E10]
5	Acts to ensure their own activities are conducted in a way that reduces their environmental impact. [PEPS05]	(K) Explains the potential areas where their system development activities impact the environment. [PEPS05-10K] (P) Acts to take on personal responsibility for ensuring their own activities reduce impact on the environment. [PEPS05-20P]	Records of actions performed in a way that considered and reduced environmental impact. [PEPS05-E10]
6	Acts to take on personal responsibility for ensuring their own activities consider diversity, equality, and inclusivity. [PEPS06]	(K) Explains the principles of diversity, equality, and inclusivity. [PEPS06-10K] (P) Acts to take on personal responsibility for ensuring their own activities consider diversity, equality, and inclusivity. [PEPS06-20P]	Records of activities performed, which considered diversity, equality, and inclusivity. [PEPS06-E10]
7	Proposes changes to the project or organization which maintain and enhance the quality of the environment and community and meet financial objectives. [PEPS07]	(K) Explains how creativity and innovation may be important when considering changes to the organization which maintain and enhance the quality of the environment and community, and meet financial objectives. [PEPS07-10K] (K) Explains the importance of enhancing the quality of the environment and community while also meeting financial objectives. [PEPS07-20K] (A) Proposes changes to the organization, which maintain and enhance the quality of the environment and community and meet financial objectives. [PEPS07-30A]	Records of organizational changes proposed to maintain or enhance the quality of the environment and community and meet financial objectives. [PEPS07-E10]
8	Maintains personal continual development records and plans. [PEPS08]	(K) Explains the importance of continual personal development and record keeping. [PEPS08-10K] (K) Explains the link between personal development and awareness of community standards and expectation. [PEPS08-20K] (A) Maintains a plan for personal professional development. [PEPS08-30A] (A) Maintains a record of professional development activities. [PEPS08-40A]	Records of personal professional development plans. [PEPS08-E10] Records of professional development activities. [PEPS08-E20]

(Continued)

ID	Indicators of Competence *(in addition to those at Awareness level)*	Relevant knowledge, experience, and/or behaviors	Possible examples of objective evidence of personal contribution to activities performed, or professional behaviors applied
9	Acts with integrity when fulfilling own responsibilities. [PEPS09]	(K) Describes everyday issues and situations where integrity may be demonstrated. [PEPS09-10K] (K) Describes situations or difficult issues that would challenge integrity. [PEPS09-20K] (P) Acts with integrity when fulfilling own responsibilities, seeking guidance when appropriate. [PEPS09-30P]	Records of actions taken with integrity when fulfilling own responsibilities, seeking guidance when appropriate. [PEPS09-E10]
10	Acts ethically when fulfilling own responsibilities. [PEPS10]	(K) Describes situation demonstrating the exercise of responsibilities in an ethical manner. [PEPS10-10K] (K) Describes situations where guidance may be needed to exercise responsibilities in an ethical manner. [PEPS10-20K] (P) Acts ethically when fulfilling own responsibilities, seeking guidance when appropriate. [PEPS10-30P]	Records of actions taken ethically when fulfilling own responsibilities, seeking guidance when appropriate. [PEPS10-E10]
11	Develops own understanding of this competency area through Continual Professional Development (CPD). [PEPS11]	(A) Identifies potential gaps in own knowledge or development needs in this area, identifying opportunities to address these through continual professional development activities. [PEPS11-10A] (A) Performs continual professional development activities to improve their knowledge and understanding in this area. [PEPS11-20A] (A) Records continual professional development activities undertaken including learning or insights gained. [PEPS11-30A]	Records of Continual Professional Development (CPD) performed and learning outcomes. [PEPS11-E10]

Practitioner – Professional: Ethics and Professionalism

ID	Indicators of Competence *(in addition to those at Supervised Practitioner level)*	Relevant knowledge, experience, and/or behaviors	Possible examples of objective evidence of personal contribution to activities performed, or professional behaviors applied
1	Follows governing ethics and professionalism guidance, adapting as required to address new situations if required. [PEPP01]	(A) Follows governing guidance to address ethical, professionalism, or career development matters. [PEPP01-10A]	Records of ethics guidance they have been required to use for differing stakeholder groups. [PEPP01-E10]
		(P) Recognizes situations where deviation from published guidance or clarification from others is appropriate in order to overcome complex ethical or professionalism challenges. [PEPP01-20P]	Records of the use of differing professional development tools or methods. [PEPP01-E20]
		(P) Recognizes situations where existing guidance requires formal change, liaising with stakeholders to gain approval as necessary. [PEPP01-30P]	Records of recognizing situations where existing guidance required formal change and liaising with stakeholders to gain approval. [PEPP01-E30]
2	Acts to ensure safe systems are used at work, by interpreting relevant health, safety, and welfare legislation and standards. [PEPP02]	(A) Identifies potential hazards and work-related health, safety, and welfare risks. [PEPP02-10A]	Records of the hazard identification and risk management work products discussed, and the names of enterprise leaders aware of the individual's involvement. [PEPP02-E10]
		(A) Performs risk management to eliminate or mitigate health, safety, and welfare risks n. [PEPP02-20A]	Records of compliance with health, safety, and welfare requirements. [PEPP02-E20]
		(P) Fosters compliance with relevant safety, health, and employment frameworks and legislation at work. [PEPP02-30P]	Records of identifying hazards and work-related health, safety, and welfare risks. [PEPP02-E30]
		(P) Acts promptly to address potential issues in health, safety, or welfare within their scope of responsibility. [PEPP02-40P]	Records of performing risk management to eliminate or mitigate health, safety, and welfare risks. [PEPP02-E40]
			Records of actions to address potential issues in health, safety, or welfare within their scope of responsibility. [PEPP02-E50]
3	Acts to promote consideration and elimination of security issues or threats across project activities. [PEPP03]	(A) Identifies potential security risks on their project. [PEPP03-10A]	Records of the security risk management work products, and the names of enterprise leaders aware of the individual's involvement. [PEPP03-E10]
		(A) Performs risk management to eliminate or mitigate security risks on their project. [PEPP03-20A]	Records showing acceptance of or compliance with security requirements. [PEPP03-E20]
		(A) Fosters compliance with relevant security frameworks and legislation on their project. [PEPP03-30A]	Records of performing risk management to identify, eliminate, or mitigate security risks. [PEPP03-E30]
		(P) Acts promptly to address potential security threats or issues within their scope of responsibility. [PEPP03-40P]	
4	Ensures compliance with relevant workplace social and employment legislation and regulatory framework across the project. [PEPP04]	(A) Ensures project activities comply with all relevant employment regulatory frameworks and legislation. [PEPP04-10A]	Records of work products created to satisfy employment or other social regulatory frameworks or legislation discussed. [PEPP04-E10]

(Continued)

ID	Indicators of Competence *(in addition to those at Supervised Practitioner level)*	Relevant knowledge, experience, and/or behaviors	Possible examples of objective evidence of personal contribution to activities performed, or professional behaviors applied
		(A) Ensures project activities comply with all relevant social regulatory frameworks and legislation. [PEPP04-20A]	Records of work products created in support of diversity, equality and/or inclusivity. [PEPP04-E20]
		(A) Ensures project activities comply with all relevant diversity, equality, and inclusivity regulatory frameworks and legislation. [PEPP04-30A]	
5	Fosters a sustainable development, taking personal responsibility to promote this area in project activities. [PEPP05]	(A) Acts to promote the need for sustainability on their project. [PEPP05-10A]	Records of own work to promote sustainability on their project. [PEPP05-E10]
		(A) Prepares information highlighting potential areas where sustainable engineering could be improved. [PEPP05-20A]	Records of the work products created which contribute to sustainable development within their organization or domain. [PEPP05-E20]
		(A) Maintains awareness of sustainable development needs in their project. [PEPP05-30A]	Records of maintaining a sustainable development in their project or beyond. [PEPP05-E30]
6	Fosters an environmentally sound approach to project activities, taking personal responsibility to promote environmental and community considerations in project activities. [PEPP06]	(A) Acts to promote environmental awareness on their project. [PEPP06-10A]	Records of own work to promote environmental awareness on their project. [PEPP06-E10]
		(A) Maintains awareness of environmental issues in their project. [PEPP06-20A]	Records of the work products created which contribute to sustainable development within their organization or domain. [PEPP06-E20]
		(A) Prepares information highlighting potential areas where environmental considerations could be addressed on their project, or their treatment improved. [PEPP06-30A]	Records of maintaining a sustainable development in their project or beyond. [PEPP06-E30]
		(A) Identifies changes which enhance the quality of the project environment and community while still meeting financial objectives. [PEPP06-40A]	Records of the work products created, which resulted in project, or organization changes, which maintained and enhanced the quality of the environment and community. [PEPP06-E40]
		(P) Acts creatively and innovatively when performing own activities to improve quality of the environment and community. [PEPP06-50P]	Records of actions taken creatively and innovatively when performing own activities to improve quality of the environment and community. [PEPP06-E50]
		(A) Identifies wider enterprise-level challenges that could improve quality of the environment and community, documenting as necessary. [PEPP06-60A]	Records of identifying wider enterprise-level challenges, which could improve quality of the environment and community, documenting. [PEPP06-E60]
7	Acts to address own professional development needs in order to maintain and enhance professional competence in own area of practice, evaluating outcomes against any plans made. [PEPP07]	(A) Identifies own personal development requirements in order to maintain and enhance competence in own area of practice area. [PEPP07-10A]	Provides a copy of the professional development records kept by the individual. [PEPP07-E10]

		(A) Ensures own personal development needs are addressed in a timely manner. [PEPP07-20A] (A) Records professional development activities performed. [PEPP07-30A] (A) Reviews the outcome of professional development activities against plans made, making adjustment to future plans accordingly. [PEPP07-40A]	Copy of the of the professional development plans developed by the individual. [PEPP07-E20] Records of an evaluation performed against the individual's professional development plan. [PEPP07-E30] Records of reviewing the outcome of professional development activities against plans made and making adjustment to future plans accordingly. [PEPP07-E40]
8	Acts to ensure all members of the project/ team operate with integrity and in an ethical manner. [PEPP08]	(P) Acts to overcome personal challenges of others to integrity or ethics in the workplace. [PEPP08-10P] (P) Acts to overcome identified team challenges to integrity or ethics in the workplace. [PEPP08-20P]	Records of relevant work products (agenda, minutes, or reports) created to resolve the integrity/ethical issue, and the names of enterprise leaders aware of the individual's involvement. [PEPP08-E10] Records of actions to overcome identified team challenges to integrity or ethics in the workplace. [PEPP08-E20]
9	Acts in an ethical manner when fulfilling their own responsibilities, without support of guidance. [PEPP09]	(P) Acts to overcome personal challenges to integrity or ethics in the workplace. [PEPP09-10P] (P) Acts ethically when fulfilling own responsibilities as part of a team, without seeking guidance. [PEPP09-20P] (P) Acts with integrity when fulfilling own responsibilities as part of a team, without seeking guidance. [PEPP09-30P]	Records of relevant work products (agenda, minutes, or reports) created in the process of exercising the practitioner's responsibility without support or guidance, and the names of enterprise leaders aware of the individual's involvement. [PEPP09-E10] Records of acting ethically when fulfilling own responsibilities as part of a team, without seeking guidance. [PEPP09-E20] Records of acting with integrity when fulfilling own responsibilities as part of a team, without seeking guidance. [PEPP09-E30]
10	Guides new or supervised practitioners in matters relating to ethics and professionalism, including career development planning, in order to develop their knowledge, abilities, skills, or associated behaviors. [PEPP10]	(P) Guides new or supervised practitioners in executing activities that form part of this competency. [PEPP10-10P] (A) Trains individuals to an "Awareness" level in this competency area. [PEPP10-20A]	Organizational Breakdown Structure showing their responsibility for technical supervision in this area. [PEPP10-E10] Records of on-the-job training objectives/guidance etc. [PEPP10-E20] Coaching or mentoring assignment records. [PEPP10-E30] Records highlighting their impact on another individual in terms of improvement or professional development in this competency. [PEPP10-E40]
11	Maintains and enhances own competence in this area through Continual Professional Development (CPD) activities. [PEPP11]	(A) Identifies potential development needs in this area, identifying opportunities to address these through continual professional development activities. [PEPP11-10A] (A) Performs continual professional development activities to maintain and enhance their competency in this area. [PEPP11-20A] (A) Records continual professional development activities undertaken including learning or insights gained. [PEPP11-30A]	Records of Continual Professional Development (CPD) performed and learning outcomes. [PEPP11-E10]

Lead Practitioner – Professional: Ethics and Professionalism			
ID	**Indicators of Competence** ***(in addition to those at Practitioner level)***	**Relevant knowledge, experience, and/or behaviors**	**Possible examples of objective evidence of personal contribution to activities performed, or professional behaviors applied**
1	Promotes best practice ethics and professionalism across the enterprise. [PEPL01]	(A) Develops own understanding of state-of-the-art ethics and professionalism best practice. [PEPL01-10A] (P) Acts in situations where there is a challenge to integrity or professional ethics in the enterprise, with successful outcome. [PEPL01-20P] (P) Champions the ethics of Systems Engineering activities across the enterprise, providing advice and judgments. [PEPL01-30P] (P) Champions the integrity of Systems Engineering activities across the enterprise, providing advice and judgments. [PEPL01-40P] (P) Acts to improve the professionalism of Systems Engineering activities across the enterprise. [PEPL01-50P]	Records of introducing or promoting ethics or professionalism best practices. [PEPL01-E10] Records (e.g. from projects or HR) of dealings demonstrating ethical approach to activities. [PEPL01-E20] Records (e.g. from HR) confirming professional approach at all times in working environment. [PEPL01-E30] Articles, papers or training material authored to support ethics or professionalism. [PEPL01-E40] Records of recognition as a Role Model for Ethics by the community, outside employer organization. [PEPL01-E50] Records demonstrating the championing of ethics or integrity issues. [PEPL01-E60] Records of ethical or professional arguments or judgments made. [PEPL01-E70]
2	Judges compliance with relevant workplace social and employment legislation and regulatory framework on behalf of the enterprise. [PEPL02]	(A) Reviews activities across the enterprise ensuring compliance with all relevant employment regulatory frameworks and legislation. [PEPL02-10A] (A) Reviews activities across the enterprise ensuring compliance with all relevant social regulatory frameworks and legislation. [PEPL02-20A] (A) Reviews activities across the enterprise ensuring compliance with all relevant diversity, equality and inclusivity regulatory frameworks and legislation. [PEPL02-30A] (A) Uses knowledge of health, safety and welfare legislation to review safe systems at work. [PEPL02-40A] (A) Judges compliance with requirements, hazard identification, and risk management systems and safety culture, recording changes as necessary. [PEPL02-50A] (P) Acts to ensure safety systems at work are updated when deficiencies are detected. [PEPL02-60P]	Records of work products reviewed or recommendations made to ensure employment or social regulatory frameworks or legislation discussed. [PEPL02-E10] Records of work products reviewed or recommendations made to ensure diversity, equality, and/or inclusivity. [PEPL02-E20] Project records, agendas, organization charts, minutes, or reports demonstrating application of knowledge of health, safety and welfare legislation at organization level. [PEPL02-E30] Records of judging compliance with requirements, hazard identification and risk management systems and safety culture at organization level. [PEPL02-E40] Records of actions to foster or improve organizational safety culture. [PEPL02-E50] Records of actions to ensure safety systems at work are updated where deficiencies are detected. [PEPL02-E60]

3	[PEPL03] - ITEM DELETED		
4	Judges the security of systems across the organization, including compliance with requirements, security risk management, and security awareness culture, on behalf of the enterprise. [PEPL04]	(A) Uses knowledge of security considerations, frameworks, and relevant legislation to review the security of systems at work. [PEPL04-10A]	Project records, agendas, organization charts, minutes, or reports demonstrating application of knowledge of security processes, frameworks and legislation at organization level. [PEPL04-E10]
		(A) Judges compliance with requirements, security threat identification, security risk management, and security culture, recording changes as necessary. [PEPL04-20A]	Records of judging compliance with security requirements, security risk identification, and security risk management systems at organization level. [PEPL04-E20]
		(P) Acts to ensure security of systems is updated when deficiencies are detected. [PEPL04-30P]	Records of actions to foster or improve organizational security culture. [PEPL04-E30]
			Records of actions to ensure safety systems at work are updated where deficiencies are detected. [PEPL04-E40]
5	Promotes the goal of performing engineering activities in a sustainable manner across the enterprise. [PEPL05]	(P) Fosters information highlighting where sustainable engineering could be improved across the wider enterprise. [PEPL05-10P]	Project records, agendas, organization charts, minutes, or reports demonstrating personally ensuring that engineering activities across the enterprise are performed in a sustainable way. [PEPL05-E10]
		(P) Acts to contribute to sustainable development across the enterprise. [PEPL05-20P]	Documents of leading sustainability analysis of engineering activities across the enterprise. [PEPL05-E20]
6	Promotes the goal of performing engineering activities in an environmentally sound manner across the enterprise. [PEPL06]	(A) Acts to promote environmental awareness across the enterprise. [PEPL06-10A]	Project records, agendas, organization charts, minutes, or reports demonstrating personal contribution environmental awareness across the enterprise. [PEPL06-E10]
		(A) Maintains awareness of environmental issues across the enterprise. [PEPL06-20A]	Documents of leading environmental analysis of engineering activities across the enterprise. [PEPL06-E20]
7	Judges continual professional development planning activities at enterprise level to ensure they maintain and enhance organizational and individual competencies. [PEPL07]	(A) Reviews enterprise continual professional development planning activities to ensure they maintain and enhance organizational and individual competencies. [PEPL07-10A]	Project records, agendas, organization charts, minutes, or reports demonstrating personally reviewing enterprise continual development planning activities and evaluating outcomes against any plans made. [PEPL07-E10]
		(A) Evaluates changes made to continual professional development planning activities following review, to confirm impact of improvements made. [PEPL07-20A]	Documents of ensuring an organization maintains and enhances organization and individual competencies. [PEPL07-E20]
8	Coaches or mentors practitioners across the enterprise in matters relating to ethics and professionalism, including career development planning, in order to develop their knowledge, abilities, skills, or associated behaviors. [PEPL08]	(P) Coaches or mentors practitioners across the enterprise in competency-related techniques, recommending development activities. [PEPL08-10P]	Coaching or mentoring assignment records. [PEPL08-E10]
		(A) Develops or authorizes enterprise training materials in this competency area. [PEPL08-20A]	Records of formal training courses, workshops, seminars, and authored training material supported by successful post-training evaluation data. [PEPL08-E20]

(Continued)

ID	Indicators of Competence *(in addition to those at Practitioner level)*	Relevant knowledge, experience, and/or behaviors	Possible examples of objective evidence of personal contribution to activities performed, or professional behaviors applied
		(A) Provides enterprise workshops/seminars or training in this competency area. [PEPL08-30A]	Listing as an approved organizational trainer for this competency area. [PEPL08-E30]
9	Promotes the introduction and use of novel techniques and ideas in ethics and professionalism across the enterprise, to improve enterprise competence in this area. [PEPL09]	(A) Analyzes different approaches across different domains through research. [PEPL09-10A]	Research records. [PEPL09-E10]
		(A) Defines novel approaches that could potentially improve the SE discipline within the enterprise. [PEPL09-20A]	Published papers in refereed journals/company literature. [PEPL09-E20]
		(P) Fosters awareness of these novel techniques within the enterprise. [PEPL09-30P]	Records showing introduction of enabling systems supporting the new techniques or ideas. [PEPL09-E30]
		(P) Collaborates with enterprise stakeholders to facilitate the introduction of techniques new to the enterprise. [PEPL09-40P]	Published papers (or similar) at enterprise level. [PEPL09-E40]
		(A) Monitors new techniques after their introduction to determine their effectiveness. [PEPL09-50A]	Records of improvements made against a recognized process improvement model. [PEPL09-E50]
		(A) Adapts approach to reflect actual enterprise performance improvements. [PEPL09-60A]	Records of adapting approach to reflect actual enterprise performance improvements. [PEPL09-E60]
10	Develops expertise in this competency area through specialist Continual Professional Development (CPD) activities. [PEPL10]	(A) Identifies own needs for further professional development in order to increase competence beyond practitioner level. [PEPL10-10A]	Records of Continual Professional Development (CPD) performed and learning outcomes. [PEPL10-E10]
		(A) Performs professional development activities in order to move own competence toward expert level. [PEPL10-20A]	
		(A) Records continual professional development activities undertaken including learning or insights gained. [PEPL10-30A]	

NOTES	In addition to items above, enterprise-level or independent 3rd-Party-generated evidence may be used to amplify other evidence presented and may include:
	a. Formally recognized by senior management in current organization as an expert in this competency area
	b. Evidence of role as Product/System Design Authority or Technical Authority on a complex project with responsibilities in this area or where skills within this competency area were used
	c. Recognized as an authorizing signatory on behalf of enterprise for formal documentation in this competency area (e.g. policies, processes, and deliverables)
	d. Formal commendation or award within own enterprise for contribution or item of work successfully performed, which required proficiency in this competency area
	e. Customer, Supplier, or other external project-specific key Stakeholder accolades for specific work performed in this competency area
	f. Independently assessed or accredited work in this competency area (e.g. for independent publication or use)
	g. Formal organizational HR records positively highlighting any specific professional competencies or behaviors identified (if applicable) plus any of the evidence indicators listed at Expert level below

Expert – Professional: Ethics and Professionalism			
ID	**Indicators of Competence** ***(in addition to those at Lead Practitioner level)***	**Relevant knowledge, experience, and/or behaviors**	**Possible examples of objective evidence of personal contribution to activities performed, or professional behaviors applied**
1	Communicates own knowledge and experience in ethics and professionalism in order to improve best practice beyond the enterprise boundary. [PEPE01]	(A) Produces papers, seminars, or presentations outside own enterprise for publication in order to share own ideas and improve industry best practices in this competence area. [PEPE01-10A]	Published papers or books etc. on new technique in refereed journals/company literature. [PEPE01-E10]
		(P) Fosters incorporation of own ideas into industry best practices in this area. [PEPE01-20P]	Published papers in refereed journals or internal literature proposing new practices in this competence area (or presentations, tutorials, etc.). [PEPE01-E20]
		(P) Develops guidance materials identifying new (or updating existing) best practice in this competence area. [PEPE01-30P]	Records of own proposals adopted as industry best practices in this competence area. [PEPE01-E30]
2	Persuades legislative and regulatory framework stakeholders beyond the enterprise to follow a particular path for in support of improving professionalism and ethics within Systems Engineering. [PEPE02]	(P) Fosters a particular legislative or regulatory path for professionalism and ethics within Systems Engineering. [PEPE02-10P]	Records, agendas, organization charts, minutes, or reports demonstrating influencing the direction of relevant legislative and regulatory frameworks in support of improving professionalism and ethics within Systems Engineering. [PEPE02-E10]
		(P) Persuades stakeholders beyond enterprise boundary to take a particular path on the application of ethics and professionalism to Systems Engineering. [PEPE02-20P]	Records of ideas assimilated into international standards or referenced in others' books, papers, and peer-reviewed articles. [PEPE02-E20]
		(P) Persuades stakeholders beyond enterprise boundary to change existing thinking on the application of ethics and professionalism to Systems Engineering. [PEPE02-30P]	
3	Persuades stakeholders beyond the enterprise boundary to improve health, safety, and welfare issues, systems, or safety culture in their activities. [PEPE03]	(P) Analyzes health, safety, or welfare areas beyond the enterprise boundary. [PEPE03-10P]	Records or reviews of health, safety, and welfare issues beyond enterprise boundary. [PEPE03-E10]
		(P) Persuades stakeholders beyond the enterprise boundary to change their existing thinking on health, safety, or welfare issues. [PEPE03-20P]	Records indicating original and changed stakeholder thinking as a resolute of personal intervention. [PEPE03-E20]
		(P) Fosters an improvement to health, safety, welfare, or safety culture beyond the enterprise boundary. [PEPE03-30P]	Records of activities performed to change stakeholder thinking beyond the enterprise boundary on health, safety, and welfare issues. [PEPE03-E30]
4	Persuades stakeholders beyond the enterprise boundary to address security issues, systems, or security culture in their activities. [PEPE04]	(P) Analyzes security areas beyond the enterprise boundary. [PEPE04-10P]	Records of analysis performed. [PEPE04-E10]

(Continued)

ID	Indicators of Competence *(in addition to those at Lead Practitioner level)*	Relevant knowledge, experience, and/or behaviors	Possible examples of objective evidence of personal contribution to activities performed, or professional behaviors applied
		(P) Persuades stakeholders beyond the enterprise boundary to change their existing thinking on security issues. [PEPE04-20P]	Records indicating original and changed stakeholder thinking to security as a resolute of personal intervention. [PEPE04-E20]
		(P) Fosters an improvement to security approach or culture beyond the enterprise boundary. [PEPE04-30P]	Records of activities performed to change stakeholder thinking beyond the enterprise boundary on security-related issues. [PEPE04-E30]
5	Persuades stakeholders beyond the enterprise boundary to address relevant employment and social regulatory compliance issues within their activities. [PEPE05]	(A) Analyzes relevant employment and social regulatory areas in organizations beyond the enterprise boundary. [PEPE05-10A]	Records of analysis performed. [PEPE05-E10]
		(P) Persuades stakeholders beyond the enterprise boundary to address issues concerning compliance with all relevant employment regulatory frameworks and legislation. [PEPE05-20P]	Records indicating original and changed stakeholder thinking as a resolute of personal intervention. [PEPE05-E20]
		(P) Persuades stakeholders beyond the enterprise boundary to address issues concerning compliance with all relevant social regulatory frameworks and legislation. [PEPE05-30P]	
		(P) Persuades stakeholders beyond the enterprise boundary to address issues concerning compliance with all relevant diversity, equality, and inclusivity regulatory frameworks and legislation. [PEPE05-40P]	
6	Champions the development of a sustainable and environmentally sound approach to systems engineering beyond the enterprise boundary. [PEPE06]	(P) Champions causes beyond the enterprise boundary, which promote or improve sustainability of systems, leading to a successful outcome. [PEPE06-10P]	Records of causes or activities supported personally, work performed, or successes achieved. [PEPE06-E10]
		(P) Champions causes beyond the enterprise boundary, which promote or improve environmental approach or characteristics of systems, leading to a successful outcome. [PEPE06-20P]	Records of recognition as a Role Model in their support of sustainable development by the community, outside employer organization. [PEPE06-E20]
			Records of recognition as a Role Model in their support of environmentally sound developments by the community, outside employer organization. [PEPE06-E30]
7	Advises organizations beyond the enterprise boundary on the suitability of their approach to ethics and professionalism. [PEPE07]	(A) Analyzes different approaches across different domains through research. [PEPE07-10A]	Project records, agendas, organization charts, minutes, or reports demonstrating personally advising on development planning activities both within and beyond the enterprise boundary, to ensure they maintain and enhance organizational and individual competencies, evaluating outcomes against any plans made. [PEPE07-E10]

		(A) Produces reports for the wider SE community on the effectiveness of new techniques after their introduction. [PEPE07-20A]	Records of advice provided on the suitability of ethics strategies, plans, processes, or tools. [PEPE07-E20]
		(P) Collaborates with those introducing novel techniques within the wider SE community. [PEPE07-30P]	Records of advice provided on the suitability of professionalism strategies, plans, processes, or tools. [PEPE07-E30]
		(A) Defines novel approaches that could potentially improve the wider SE discipline. [PEPE07-40A]	Documents of ensuring an organization maintains and enhances organization and individual competencies. [PEPE07-E40]
		(P) Fosters awareness of these novel techniques within the wider SE community. [PEPE07-50P]	
8	Champions an ethical and professional culture beyond the enterprise boundary. [PEPE08]	(P) Champions causes beyond the enterprise boundary where there is a challenge to integrity or professional ethics in the enterprise, leading to a successful outcome. [PEPE08-10P]	Records of relevant work products (agenda, minutes, or reports) created in the process of exercising the practitioner's responsibility without support or guidance, and the names of enterprise leaders aware of the individual's involvement. [PEPE08-E10]
		(P) Champions the ethics of Systems Engineering activities beyond the enterprise boundary, providing advice and judgments. [PEPE08-20P]	Records of recognition as a Role Model for Ethics by the community, outside employer organization. [PEPE08-E20]
		(P) Champions integrity of Systems Engineering activities beyond the enterprise boundary, providing advice and judgments. [PEPE08-30P]	Documents of being recognized as a role model for ethics and integrity both within and beyond the enterprise. [PEPE08-E30]
9	Champions the introduction of novel techniques and ideas in Systems Engineering ethics and professionalism, beyond the enterprise boundary, in order to develop the wider Systems Engineering community in these competencies. [PEPE09]	(A) Analyzes different approaches across different domains through research. [PEPE09-10A]	Records of activities promoting research and need to adopt novel technique or ideas. [PEPE09-E10]
		(A) Produces reports for the wider SE community on the effectiveness of new techniques after their introduction. [PEPE09-20A]	Records of improvements made to process and appraisal against a recognized process improvement model. [PEPE09-E20]
		(P) Collaborates with those introducing novel techniques within the wider SE community. [PEPE09-30P]	Research records. [PEPE09-E30]
		(A) Defines novel approaches that could potentially improve the wider SE discipline. [PEPE09-40A]	Published papers in refereed journals/company literature. [PEPE09-E40]
		(P) Fosters awareness of these novel techniques within the wider SE community. [PEPE09-50P]	Records showing introduction of enabling systems supporting the new techniques or ideas. [PEPE09-E50]

(*Continued*)

ID	Indicators of Competence *(in addition to those at Lead Practitioner level)*	Relevant knowledge, experience, and/or behaviors	Possible examples of objective evidence of personal contribution to activities performed, or professional behaviors applied
10	Coaches individuals beyond the enterprise boundary in ethics and professionalism, including career development planning in order to further develop their knowledge, abilities, skills, or associated behaviors. [PEPE10]	(P) Coaches or mentors individuals beyond the enterprise boundary, in competency-related techniques, recommending development activities. [PEPE10-10P]	Coaching or mentoring assignment records. [PEPE10-E10]
		(A) Develops or authorizes training materials in this competency area, which are subsequently successfully delivered beyond the enterprise boundary. [PEPE10-20A]	Records of formal training courses, workshops, seminars, and authored training material supported by successful post-training evaluation data. [PEPE10-E20]
		(A) Provides workshops/seminars or training in this competency area for practitioners or lead practitioners beyond the enterprise boundary (e.g. conferences and open training days). [PEPE10-30A]	Records of Training/workshops/seminars created supported by successful post-training evaluation data. [PEPE10-E30]
11	Maintains expertise in this competency area through specialist Continual Professional Development (CPD) activities. [PEPE11]	(A) Reviews research, new ideas, and state of the art to identify relevant new areas requiring personal development in order to maintain expertise in this competency area. [PEPE11-10A]	Records of documents reviewed and insights gained as part of own research into this competency area. [PEPE11-E10]
		(A) Performs identified specialist professional development activities in order to maintain or further develop competence at expert level. [PEPE11-20A]	Records of Continual Professional Development (CPD) performed and learning outcomes. [PEPE11-E20]
		(A) Records continual professional development activities undertaken including learning or insights gained. [PEPE11-30A]	

NOTES	In addition to items above, enterprise-level or independent 3rd-Party-generated evidence may be used to amplify other evidence presented and may include: a. Formally recognized by a reputable external organization as an expert in this competency area b. Evidence of role as independent assessor or reviewer on project outside own organization where skills in this competency area were used c. Evidence of invitation(s) from wider community for contribution of systems engineering expertise in this area (e.g. industry conference panel, government advisory board etc. cross-industry working groups, partnerships, accredited advanced university courses or research, or as part of professional institute) d. Formal commendation beyond the enterprise (e.g. by INCOSE or other recognized authority) for work performed in this competency area e. Independently assessed or accredited work product in this competency area (e.g. for independent publication or use) f. Accolades of expertise in this area from recognized industry leaders

Competency area – Professional: Technical Leadership
Description Systems Engineering technical leadership is the combination of the application of technical knowledge and experience in Systems Engineering with appropriate professional competencies. This encompasses an understanding of customer need, problem solving, creativity and innovation skills, communications, team building, relationship management, operational oversight and accountability skills coupled with core Systems Engineering competency and engineering instinct.
Why it matters The complexity of modern system designs, the severity of their constraints and the need to succeed in a high tempo, high-stakes environment where competitive advantage matters, demands the highest levels of technical excellence and integrity throughout the life cycle. Systems Engineering technical leadership helps teams meet these challenges.
Possible contributory types of evidence Any combination of the types of evidence may be acceptable (depending on how the Framework is tailored and used). The evidence items identified at each level indicate example work products only. Contributions to work products will generally differ at each proficiency level.
Learning and development The INCOSE Professional Development Portal provides example guidance on how to gain an initial awareness of a competency area and options for developing further competence thereafter.

Awareness – Professional: Technical Leadership		
ID	**Indicators of Competence**	**Relevant knowledge sub-indicators**
1	Explains the role of technical leadership within Systems Engineering. [PTLA01]	(K) Describes how a systems engineer often "leads" through influence, not necessarily by organizational position or power. [PTLA01-10K] (K) Lists some unique characteristics of a Systems Engineering leader, such as systems thinking skills and experience across the life cycle of the product or service. [PTLA01-20K] (K) Defines Systems Engineering leadership as the combination of the application of technical knowledge and experience in Systems Engineering with appropriate professional competencies. [PTLA01-30K]
2	Defines the terms "vision," "strategy," and "goal" terms explaining why each is important in leadership. [PTLA02]	(K) Explains that the vision is the "what," the strategy is the high level "how," the goals are the measurable aspects of the strategy – the milestones. [PTLA02-10K] (K) Explains why it is the job of the leader to set the vision for the organization and to use the strategy and goals to keep the team on track. [PTLA02-20K]
3	Explains why understanding the strategy is central to Systems Engineering leadership. [PTLA03]	(K) Describes key interfaces and links between engineering technical leaders and stakeholders in the wider enterprise outside engineering (e.g. infrastructure management, human resource management, quality management, knowledge management, portfolio management, and life cycle model management). [PTLA03-10K] (K) Explains how the Systems Engineering strategy might differ from the strategy for other engineering domains – more of a collaborative approach as opposed to a highly technical strategy. [PTLA03-20K] (K) Explains the relationship between the Systems Engineering leader and the program/project leader on a project/program. [PTLA03-30K]

(Continued)

ID	Indicators of Competence	Relevant knowledge sub-indicators
4	Explains why fostering collaboration is central to Systems Engineering. [PTLA04]	(K) Explains that the Systems Engineering leader often "leads from the middle," i.e. may not be organizationally the leader of the overall effort but needs to demonstrate leadership through collaboration. [PTLA04-10K] (K) Explains that Systems Engineering cannot be performed in a vacuum. [PTLA04-20K] (K) Describes the importance of establishing relationships with stakeholders. [PTLA04-30K] (K) Lists at least one example of the impacts of poor collaboration, such as an incorrectly designed or built product. [PTLA04-40K] (K) Describes the relationship between collaboration and leadership. [PTLA04-50K]
5	Explains why the art of communications is central to Systems Engineering including the impact of poor communications. [PTLA05]	(K) Explains why that different modes of communication may be needed depending upon situation. [PTLA05-10K] (K) Lists examples of types of Systems Engineering information that could be easily misunderstood if not properly communicated, such as written versus verbal requirements. [PTLA05-20K] (K) Explains how poor collaboration impacts on the quality of leadership provided. [PTLA05-30K] (K) Explains why listening to external stakeholders (e.g. customers) and other stakeholders is important and understands the meaning of "active listening." [PTLA05-40K]
6	Explains how technical analysis, problem-solving techniques, and established best practices can be used to improve the excellence of Systems Engineering solutions. [PTLA06]	(K) Describes a method for structured problem solving, such as a formal trade study, and why it is needed for complex problems. [PTLA06-10K] (K) Defines areas where problem-solving techniques can be used to improve Systems Engineering solutions. [PTLA06-20K] (K) Defines the relationship between the use of best practices and improved Systems Engineering solutions. [PTLA06-30K]
7	Explains how creativity, ingenuity, experimentation, and accidents or errors, often lead to technological and engineering successes and advances and provides examples. [PTLA07]	(K) Lists examples of how creativity, experimentation, or accidents have led to successes and advances, such the incubator case study in the SE Handbook. [PTLA07-10K] (K) Explains the importance of asking open-ended questions and conducting brainstorming sessions in identifying creative ideas. [PTLA07-20K]
8	Explains how different sciences impact the technology domain and the engineering discipline. [PTLA08]	(K) Lists examples of how different sciences impact the technology domain, such as cybersecurity or artificial intelligence. [PTLA08-10K] (K) Explains how increased complexity impacts the importance of the role of the Systems Engineering leader. [PTLA08-20K] (K) Explains how model-based Systems Engineering impacts the practice of Systems Engineering. [PTLA08-30K]
9	Explains how complexity impacts the role of the engineering leader. [PTLA09]	(K) Explains the concept of complexity and its potential impact on the engineering of a system. [PTLA09-10K] (K) Explains how increased complexity impacts the importance of the role of the Systems Engineering leader. [PTLA09-20K] (K) Identifies mechanisms (e.g. Model based Systems Engineering), which can be used to simplify or assist in the leadership role for a complex system. [PTLA09-30K]

Supervised Practitioner – Professional: Technical Leadership			
ID	**Indicators of Competence** ***(in addition to those at Awareness level)***	**Relevant knowledge, experience, and/or behaviors**	**Possible examples of objective evidence of personal contribution to activities performed, or professional behaviors applied**
1	Follows guidance received (e.g. from mentors), to plan and control their own technical leadership activities or approaches. [PTLS01]	(A) Follows guidance received (e.g. from mentors) to plan and control their own technical leadership activities or approaches. [PTLS01-10A]	Documents of applicable technical leadership plans and tools to control their own activities. [PTLS01-E10]
2	Acts to gain trust in their Systems Engineering leadership activities. [PTLS02]	(K) Explains how earning trust as a technical leader is related to professional and technical competency. [PTLS02-10K] (K) Defines the term "trusted advisor" and illustrates how it relates to a Systems Engineering leadership role. [PTLS02-20K] (P) Uses professional and technical competencies successfully in a leadership role to gain trust, taking guidance when appropriate to improve performance. [PTLS02-30P]	Records using professional and technical competencies successfully in a leadership role to gain trust, taking guidance when appropriate to improve performance. [PTLS02-E10]
3	Complies with a project, or wider, vision in performing Systems Engineering leadership activities. [PTLS03]	(K) Explains how a vision can impact leadership on both the project and in the wider enterprise. [PTLS03-10K] (A) Uses the vision statement to guide their project leadership activities. [PTLS03-20A]	Records of using the vision statement to guide their project leadership activities. [PTLS03-E10]
4	Uses team and project to guide direction, thinking strategically, holistically, and systemically when performing own Systems Engineering leadership activities. [PTLS04]	(K) Lists the team and project goals. [PTLS04-10K] (K) Explains how team or project goals support the vision. [PTLS04-20K] (K) Explains how using systems thinking skills can support achieving leadership goals. [PTLS04-30K] (A) Uses systems thinking techniques to guide their leadership activities. [PTLS04-40A]	Records of using systems thinking techniques to guide their leadership activities. [PTLS04-E40]
5	Recognizes constructive criticism from others following guidance to improve their SE leadership. [PTLS05]	(P) Recognizes the rationale for constructive criticism of their leadership activities from others. [PTLS05-10P]	Records showing use of constructive criticism and supporting guidance from others (e.g. mentors) to improve their SE leadership. [PTLS05-E10]

(Continued)

ID	Indicators of Competence *(in addition to those at Awareness level)*	Relevant knowledge, experience, and/or behaviors	Possible examples of objective evidence of personal contribution to activities performed, or professional behaviors applied
		(A) Uses constructive criticism and supporting guidance from others (e.g. mentors) to improve their SE leadership. [PTLS05-20A]	
6	Uses appropriate mechanisms to offer constructive criticism to others on the team. [PTLS06]	(K) Explains best practice mechanisms for delivering "constructive criticism" to others, in the context of a SE leadership role. [PTLS06-10K] (P) Acts to self-improve by taking on board constructive criticism. [PTLS06-20P] (P) Uses appropriate communications techniques to offer constructive criticism to others on the team. [PTLS06-30P]	Records of acting to self-improve by taking on board constructive criticism. [PTLS06-E10] Records of using appropriate communications techniques to offer constructive criticism to others on the team. [PTLS06-E20]
7	Elicits viewpoints from others when developing solutions as part of their Systems Engineering leadership role. [PTLS07]	(K) Explains how actively seeking input from all team members can result in a better solution and appropriate ways this can be done. [PTLS07-10K] (A) Elicits viewpoints from others using appropriate communications techniques as part of their leadership role. [PTLS07-20A] (A) Uses viewpoints obtained from others to improve solutions formed as part of their leadership role. [PTLS07-30A]	Records of correspondence to solicit input from other team members for their activities. [PTLS07-E10] Records of using viewpoints obtained from others to improve solutions formed as part of their leadership role. [PTLS07-E20]
8	Uses appropriate communications mechanisms to reinforce their Systems Engineering leadership activities. [PTLS08]	(K) Explains why it is important for those in a technical leadership role to understand the audience needs and familiarity with the technical vocabulary prior to engaging. [PTLS08-10K] (P) Uses appropriate (e.g. verbal, written, and presentation) mode of communication when communicating ideas to reinforce technical leadership actions or activities. [PTLS08-20P]	Records of technical communications made in support of their systems engineering leadership actions. [PTLS08-E10]
9	Acts creatively and innovatively in their SE leadership activities. [PTLS09]	(K) Describes appropriate techniques for simple informal and complex formal problem solving (e.g. Plus/minus or a weighted criteria trade study.). [PTLS09-10K] (P) Acts with creativity and innovation in their leadership activities. [PTLS09-20P]	Records of documentation of innovative approaches to support their own work activities, which have been recognized by others. [PTLS09-E10] Records of documentation of structured decision-making tools to support problem solving. [PTLS09-E20]

		(P) Recognizes uncertainty in their technical leadership activities, but continues to lead activities effectively even when outside comfort zone. [PTLS09-30P] (P) Acts with willingness when addressing challenging leadership problems at the socio-technical interface. [PTLS09-40P]	Records of acting with willingness when addressing challenging leadership problems at the socio-technical interface. [PTLS09-E30]
10	Identifies concepts and ideas in sciences, technologies, or engineering disciplines beyond their own discipline, applying them to benefit their own Systems Engineering leadership activities on a project. [PTLS10]	(K) Explains how the use of external concepts and ideas can result in a better solution. [PTLS10-10K] (A) Uses concepts and ideas from beyond their own discipline to support leadership activities. [PTLS10-20A]	Records of using concepts and ideas from beyond their own discipline to support leadership activities. [PTLS10-E10]
11	Develops own understanding of this competency area through Continual Professional Development (CPD). [PTLS11]	(A) Identifies potential gaps in own knowledge or development needs in this area, identifying opportunities to address these through continual professional development activities. [PTLS11-10A] (A) Performs continual professional development activities to improve their knowledge and understanding in this area. [PTLS11-20A] (A) Records continual professional development activities undertaken including learning or insights gained. [PTLS11-30A]	Records of Continual Professional Development (CPD) performed and learning outcomes. [PTLS11-E10]

Practitioner – Professional: Technical Leadership			
ID	**Indicators of Competence** ***(in addition to those at Supervised Practitioner level)***	**Relevant knowledge, experience, and/or behaviors**	**Possible examples of objective evidence of personal contribution to activities performed, or professional behaviors applied**
1	Follows guidance received to develop their own technical leadership skills, using leadership techniques and tools as instructed. [PTLP01]	(A) Follows guidance received to develop their own technical leadership skills, using leadership techniques and tools as instructed. [PTLP01-10A]	Records of leadership guidance they have been required to use for differing stakeholder groups. [PTLP01-E10]
		(P) Recognizes situations where deviation from received guidance or clarification from others is appropriate in order to overcome complex technical leadership challenges. [PTLP01-20P]	Records of recognizing situations where deviation from received guidance or clarification from others is appropriate in order to overcome complex technical leadership challenges. [PTLP01-E20]
		(P) Recognizes situations where existing guidance requires formal change, liaising with stakeholders to gain approval as necessary. [PTLP01-30P]	Records of recognizing situations where existing guidance requires formal change, liaising with stakeholders to gain approval as necessary. [PTLP01-E30]
2	Acts with integrity in their leadership activities, being trusted by their team. [PTLP02]	(K) Explains what it means to lead with integrity. [PTLP02-10K]	Records of acting with integrity in a leadership role. [PTLP02-E10]
		(K) Describes how professional and technical competencies support integrity in a successful leader. [PTLP02-20K]	Records of using professional and technical competencies successfully in a leadership role to improve levels of trust. [PTLP02-E20]
		(P) Acts with integrity in a leadership role. [PTLP02-30P]	
		(P) Uses professional and technical competencies successfully in a leadership role to improve levels of trust. [PTLP02-40P]	
3	Guides and actively coordinates Systems Engineering activities across a team, combining appropriate professional and technical competencies, with demonstrable success. [PTLP03]	(K) Describes the technical and professional competencies needed for team to be successful. [PTLP03-10K]	Records of leading Systems Engineering activities for a team, with demonstrable success. [PTLP03-E10]
		(K) Defines the idea of an "Integrated Product Team" and how this operates. [PTLP03-20K]	
		(K) Explains how diversity on a team can provide constructive disruption, resulting in better solutions. [PTLP03-30K]	
		(P) Guides and actively coordinates Systems Engineering activities across a team, with demonstrable success. [PTLP03-40P]	

4	Develops technical vision for a project team, influencing and integrating the viewpoints of others in order to gain acceptance. [PTLP04]	(K) Explains the importance of listening to and understanding team member viewpoints, prior to trying to influence them. [PTLP04-10K] (A) Uses the viewpoints of others to create a project technical vision, integrating or adapting as required. [PTLP04-20A] (A) Prepares a project technical or product vision for use on a project, aligned with wider project or organizational vision statements. [PTLP04-30A] (P) Fosters acceptance of a technical or product vision by the project team. [PTLP04-40P]	Records of using the viewpoints of others to create a project technical vision, integrating or adapting as required. [PTLP04-E10] Records of technical project report or product model. [PTLP04-E20] Records of fostering acceptance of a technical or product vision by the project team. [PTLP04-E30]
5	Identifies a leadership strategy to support of project goals, changing as necessary, to ensure success. [PTLP05]	(K) Explains balancing cost, schedule, and product performance goals to create a balanced solution. [PTLP05-10K] (A) Develops leadership strategy to meet project goals, without compromising system quality. [PTLP05-20A] (A) Adapts leadership strategy if required, to accommodate new situations, without compromising system quality. [PTLP05-30A]	Records of developing leadership strategy to meet project goals, without compromising system quality. [PTLP05-E10] Records of adapting leadership strategy if required to accommodate new situations, without compromising system quality. [PTLP05-E20]
6	Recognizes constructive criticism from others within the enterprise following guidance to improve their SE leadership. [PTLP06]	(P) Recognizes accurate constructive criticism of their leadership activities from others with a professional demeanor at all times. [PTLP06-10P] (P) Recognizes inaccurate or unfair criticism of their leadership from others, challenging its accuracy while maintaining professional demeanor at all times. [PTLP06-20P] (A) Uses constructive criticism to improve their SE leadership. [PTLP06-30A]	Records of recognizing inaccurate or unfair criticism of their leadership from others, challenging its accuracy while maintaining professional demeanor at all times. [PTLP06-E10] Records of using constructive criticism to improve their SE leadership. [PTLP06-E20]
7	Uses appropriate communications techniques to offer constructive criticism to others on the team. [PTLP07]	(P) Uses appropriate communications techniques to offer constructive criticism to others on the team. [PTLP07-10P] (P) Acts to improve others on the team following receipt of constructive criticism. [PTLP07-20P]	Records of acting to improve others on the team following receipt of constructive criticism. [PTLP07-E10]

(Continued)

ID	Indicators of Competence *(in addition to those at Supervised Practitioner level)*	Relevant knowledge, experience, and/or behaviors	Possible examples of objective evidence of personal contribution to activities performed, or professional behaviors applied
8	Fosters a collaborative approach in their Systems Engineering leadership activities. [PTLP08]	(P) Fosters a collaborative approach with other leaders, such as Program/Project Lead. [PTLP08-10P] (P) Fosters a collaborative approach with other disciplines or functions, such as software engineering. [PTLP08-20P]	Records of fostering a collaborative approach with other disciplines or functions, such as software engineering. [PTLP08-E10]
9	Fosters the empowerment of team members, by supporting, facilitating, promoting, giving ownership, and supporting them in their endeavors. [PTLP09]	(K) Explains the difference between accountability (meet the goal) and responsibility (perform the assigned task) for team members. [PTLP09-10K] (P) Acts to empower team members using transformational leadership techniques (development, building trust, etc. as opposed to managing task assignments). [PTLP09-20P]	Records of role definitions reflecting delegation and empowerment of own role. [PTLP09-E10]
10	Uses best practice communications techniques in their leadership activities, in order to express their ideas clearly and effectively. [PTLP10]	(K) Describes different techniques for communication to team (understands when to use a written instruction versus a discussion). [PTLP10-10K] (P) Identifies best practices communications techniques appropriate to different types of technical information in a leadership role. [PTLP10-20P] (P) Uses appropriate best practices communications techniques to convey differing types of technical information in a leadership role. [PTLP10-30P]	Records of identifying best practices communications techniques appropriate to different types of technical information in a leadership role. [PTLP10-E10] Records of using appropriate best practices communications techniques to convey differing types of technical information in a leadership role. [PTLP10-E20]
11	Develops strategies for leadership activities or the resolution of team issues, using creativity and innovation. [PTLP11]	(A) Uses a range of problem-solving techniques to resolve different team or project technical issues. [PTLP11-10A] (P) Acts imaginatively, creatively and innovatively when resolving technical issues in their leadership role. [PTLP11-20P] (P) Fosters the use of novel methods (e.g. rapid prototyping, model-based Systems Engineering) on the team to support problem resolution in their leadership role. [PTLP11-30P]	Records of creative or innovative solutions developed. [PTLP11-E10] Record of structured decision document (e.g. formal trade study). [PTLP11-E20] Records of root cause analysis documentation (e.g. fishbone diagram, to understand a team or project issue). [PTLP11-E30]

12	Guides new or supervised practitioners in matters relating to technical leadership in Systems Engineering, in order to develop their knowledge, abilities, skills, or associated behaviors. [PTLP12]	(P) Guides new or supervised practitioners in executing activities that form part of this competency. [PTLP12-10P]	Organizational Breakdown Structure showing their responsibility for technical supervision in this area. [PTLP12-E10]
		(A) Trains individuals to an "Awareness" level in this competency area. [PTLP12-20A]	Records of on-the-job training objectives/guidance etc. [PTLP12-E20] Coaching or mentoring assignment records. [PTLP12-E30] Records highlighting their impact on another individual in terms of improvement or professional development in this competency. [PTLP12-E40]
13	Maintains and enhances own competence in this area through Continual Professional Development (CPD) activities. [PTLP13]	(A) Identifies potential development needs in this area, identifying opportunities to address these through continual professional development activities. [PTLP13-10A] (A) Performs continual professional development activities to maintain and enhance their competency in this area. [PTLP13-20A] (A) Records continual professional development activities undertaken including learning or insights gained. [PTLP13-30A]	Records of Continual Professional Development (CPD) performed and learning outcomes. [PTLP13-E10]

Lead Practitioner – Professional: Technical Leadership			
ID	**Indicators of Competence** ***(in addition to those at Practitioner level)***	**Relevant knowledge, experience, and/or behaviors**	**Possible examples of objective evidence of personal contribution to activities performed, or professional behaviors applied**
1	Uses best practice technical leadership techniques to guide, influence, and gain trust from systems engineering stakeholders across the enterprise. [PTLL01]	(A) Develops own state-of-the-art technical leadership skills. [PTLL01-10A]	Records showing positive feedback from stakeholders (e.g. customers, peers) regarding approach to activities such as problem solving or difficult decisions. [PTLL01-E10]
		(P) Uses best practice techniques to guide or influence different stakeholders or stakeholder groups across the enterprise. [PTLL01-20P]	Document demonstrating an ethical issue highlighted for acknowledgement or resolution at enterprise level. [PTLL01-E20]
		(P) Advises others on the use of best practice techniques to guide or influence different stakeholders or stakeholder groups across the enterprise. [PTLL01-30P]	Records of advising why Systems Engineering activities often require professional and technical competencies to be combined. [PTLL01-E30]
		(P) Uses best practice technical leadership skills to earn trust from others across the enterprise. [PTLL01-40P]	Records of advising why specific required professional and technical competencies are valuable at enterprise level. [PTLL01-E40]
		(P) Acts with integrity at all times, raising ethical issues to enterprise level as required. [PTLL01-50P]	Documents containing proof of customer acceptance regarding key technical systems documents, such as a system specification or Systems Engineering Management Plan. [PTLL01-E50]
		(P) Advises systems engineering stakeholders across the enterprise on how best to combine technical and professional leadership skills. [PTLL01-60P]	Document indicating Systems Engineering Integration Team (SEIT) or similar leadership role. [PTLL01-E60]
2	Reacts professionally and positively to constructive criticism received from others across the enterprise. [PTLL02]	(P) Reacts positively to accurate constructive criticism of their leadership activities from others beyond the enterprise, with professional demeanor. [PTLL02-10P]	Records of constructive criticism accepted, such as making a change in leadership style for a specific situation. [PTLL02-E10]
		(P) Challenges the accuracy of inaccurate or unfair criticism while maintaining professional demeanor at all times. [PTLL02-20P]	Records of offering or challenging constructive criticism, such as during a work product review or regarding a dry run of a customer presentation. [PTLL02-E20]
		(A) Uses constructive criticism to improve their SE leadership. [PTLL02-30A]	Document from stakeholders, which highlights positive reaction to constructive criticism. [PTLL02-E30]
3	Uses appropriate communications techniques to offer constructive criticism to others across the enterprise. [PTLL03]	(P) Uses appropriate communications techniques to offer constructive criticism to others across the enterprise. [PTLL03-10P]	Record of constructive criticism accepted, such as making a change in leadership style for a specific situation. [PTLL03-E10]
		(P) Acts to support others across the enterprise, in improving following receipt of constructive criticism. [PTLL03-20P]	Record of support offered to others in improving following receipt of constructive criticism. [PTLL03-E20]
4	Fosters stakeholder collaboration across the enterprise, sharing ideas and knowledge, and establishing mutual trust. [PTLL04]	(K) Explains why it can be more difficult to regain trust once it is lost, with examples. [PTLL04-10K]	Records of collaboration that extend beyond own team to other parts of the enterprise, such as a documented process that crosses organizational boundaries. [PTLL04-E10]

		(A) Identifies possible approaches to fostering and maintaining collaboration. [PTLL04-20A]	Records of demonstration of establishing trust with a stakeholder in another part of the enterprise. [PTLL04-E20]
		(P) Fosters collaboration across the enterprise, maintaining momentum even when the situation is difficult. [PTLL04-30P]	Records of advising why it is more difficult it is to regain trust once it is lost. [PTLL04-E30]
		(P) Acts to establish or regain trust with a key stakeholder in another part of the enterprise. [PTLL04-40P]	Records of acting to establish or regain trust with a key stakeholder in another part of the enterprise. [PTLL04-E40]
		(P) Fosters collaboration with a key project stakeholder elsewhere in the enterprise. [PTLL04-50P]	Records fostering collaboration with a key project stakeholder elsewhere in the enterprise. [PTLL04-E50]
5	Fosters the empowerment of individuals across the enterprise, by supporting, facilitating, promoting, giving ownership, and supporting them in their endeavors. [PTLL05]	(P) Selects distinct roles and responsibilities for Systems Engineering team working on a program and how they collaborate to be successful. [PTLL05-10P]	Records showing roles/responsibilities or team charter to show how team members are empowered. [PTLL05-E10]
		(P) Acts to empower individual team members, for example through delegation of authority to others for specific decisions. [PTLL05-20P]	Records of identification of individual team members as decision authority within the organization to demonstrate empowerment. [PTLL05-E20]
			Document demonstrating delegation of authority (e.g. for specific decisions). [PTLL05-E30]
6	Acts with creativity and innovation, applying problem-solving techniques to develop strategies or resolve complex project or enterprise technical leadership issues. [PTLL06]	(K) Explains how emerging behaviors in complex problems plays a role in the need for innovative solutions. [PTLL06-10K]	Documents highlighting complex problem solved, highlighting the creative or innovative solutions used or developed. [PTLL06-E10]
		(A) Uses recognized creative problem-solving techniques (e.g. the "Six Thinking Hats" or "Negative Brainstorming") to identify solutions to complex problems. [PTLL06-20A]	Documents highlighting structured problem-solving technique used to solve a complex problem, such as formal trade study. [PTLL06-E20]
		(P) Acts with creativity and innovation in the resolution of complex technical leadership issues. [PTLL06-30P]	Records of acting with creativity and innovation in the resolution of complex technical leadership issues. [PTLL06-E30]
7	Coaches or mentors practitioners across the enterprise in technical and leadership issues in order to develop their knowledge, abilities, skills, or associated behaviors. [PTLL07]	(P) Coaches or mentors practitioners across the enterprise in competency-related techniques, recommending development activities. [PTLL07-10P]	Coaching or mentoring assignment records. [PTLL07-E10]
		(A) Develops or authorizes enterprise training materials in this competency area. [PTLL07-20A]	Records of formal training courses, workshops, seminars, and authored training material supported by successful post-training evaluation data. [PTLL07-E20]
		(A) Provides enterprise workshops/seminars or training in this competency area. [PTLL07-30A]	Listing as an approved organizational trainer for this competency area. [PTLL07-E30]

(Continued)

ID	Indicators of Competence *(in addition to those at Practitioner level)*	Relevant knowledge, experience, and/or behaviors	Possible examples of objective evidence of personal contribution to activities performed, or professional behaviors applied
8	Promotes the introduction and use of novel techniques and ideas in SE technical leadership across the enterprise, to improve enterprise competence in this area. [PTLL08]	(A) Analyzes different approaches across different domains through research. [PTLL08-10A]	Research records. [PTLL08-E10]
		(A) Defines novel approaches that could potentially improve the SE discipline within the enterprise. [PTLL08-20A]	Published papers in refereed journals/company literature. [PTLL08-E20]
		(P) Fosters awareness of these novel techniques within the enterprise. [PTLL08-30P]	Records showing introduction of enabling systems supporting the new techniques or ideas. [PTLL08-E30]
		(P) Collaborates with enterprise stakeholders to facilitate the introduction of techniques new to the enterprise. [PTLL08-40P]	Published papers (or similar) at enterprise level. [PTLL08-E40]
		(A) Monitors new techniques after their introduction to determine their effectiveness. [PTLL08-50A]	Records of improvements made against a recognized process improvement model. [PTLL08-E50]
		(A) Adapts approach to reflect actual enterprise performance improvements. [PTLL08-60A]	Records of adapting approach to reflect actual enterprise performance improvements. [PTLL08-E60]
9	Develops expertise in this competency area through specialist Continual Professional Development (CPD) activities. [PTLL09]	(A) Identifies own needs for further professional development in order to increase competence beyond practitioner level. [PTLL09-10A]	Records of Continual Professional Development (CPD) performed and learning outcomes. [PTLL09-E10]
		(A) Performs professional development activities in order to move own competence toward expert level. [PTLL09-20A]	
		(A) Records continual professional development activities undertaken including learning or insights gained. [PTLL09-30A]	

NOTES	In addition to items above, enterprise-level or independent 3rd-Party-generated evidence may be used to amplify other evidence presented and may include: a. Formally recognized by senior management in current organization as an expert in this competency area b. Evidence of role as Product/System Design Authority or Technical Authority on a complex project with responsibilities in this area or where skills within this competency area were used c. Recognized as an authorizing signatory on behalf of enterprise for formal documentation in this competency area (e.g. policies, processes, and deliverables) d. Formal commendation or award within own enterprise for contribution or item of work successfully performed, which required proficiency in this competency area e. Customer, Supplier, or other external project-specific key Stakeholder accolades for specific work performed in this competency area f. Independently assessed or accredited work in this competency area (e.g. for independent publication or use) g. Formal organizational HR records positively highlighting any specific professional competencies or behaviors identified (if applicable) plus any of the evidence indicators listed at Expert level below

Expert – Professional: Technical Leadership			
ID	**Indicators of Competence** ***(in addition to those at Lead Practitioner level)***	**Relevant knowledge, experience, and/or behaviors**	**Possible examples of objective evidence of personal contribution to activities performed, or professional behaviors applied**
1	Communicates own knowledge and experience in technical leadership in order to improve best practice beyond the enterprise boundary. [PTLE01]	(A) Produces papers, seminars, or presentations outside own enterprise for publication in order to share own ideas and improve industry best practices in this competence area. [PTLE01-10A]	Published papers or books etc. on new technique in refereed journals/company literature. [PTLE01-E10]
		(P) Fosters incorporation of own ideas into industry best practices in this area. [PTLE01-20P]	Published papers in refereed journals or internal literature proposing new practices in this competence area (or presentations, tutorials, etc.). [PTLE01-E20]
		(P) Develops guidance materials identifying new (or updating existing) best practice in this competence area. [PTLE01-30P]	Records of own proposals adopted as industry best practices in this competence area. [PTLE01-E30]
2	Advises organizations beyond the enterprise boundary on the suitability of their approach to technical leadership issues. [PTLE02]	(A) Identifies technical leadership issues in organizations beyond the enterprise boundary. [PTLE02-10A]	Records highlighting successful influence of stakeholders beyond the enterprise. [PTLE02-E10]
		(P) Persuades key stakeholders to address identified technical leadership issues earning trust from others. [PTLE02-20P]	Records highlighting an ethical issue raised beyond enterprise level for acknowledgement or resolution. [PTLE02-E20]
		(P) Fosters support from stakeholders beyond enterprise boundary in resolving identified issues. [PTLE02-30P]	Records indicating changed stakeholder thinking from personal intervention. [PTLE02-E30]
		(P) Persuades stakeholders beyond the enterprise to change their position on technical leadership issues. [PTLE02-40P]	Records of acting to earn trust from leaders beyond the enterprise boundary (e.g. customers, or from technical leaders outside the organization). [PTLE02-E40]
		(P) Acts to earn trust from leaders beyond the enterprise boundary (e.g. customers, or from technical leaders outside the organization). [PTLE02-50P]	Records of acting with integrity at all times, raising ethical issues and challenging those beyond the enterprise boundary where appropriate. [PTLE02-E50]
		(P) Acts with integrity at all times, challenging the ethics of proposed approaches if appropriate. [PTLE02-60P]	
3	Guides and actively coordinates the progress of Systems Engineering activities beyond the enterprise boundary, combining appropriate professional competencies with technical knowledge and experience. [PTLE03]	(A) Identifies appropriate professional and technical competencies in a leadership activity in order to achieve a successful result beyond the enterprise boundary. [PTLE03-10A]	Records highlighting activities supporting SE leadership in external organizations (e.g. provision of independent expertise in reviews or decision-making, panels, presentations, paper reviews, external bodies). [PTLE03-E10]
		(P) Uses appropriate technical leadership skills to convey technical knowledge outside the enterprise at a peer level such as an industry forum. [PTLE03-20P]	Records highlighting technical knowledge shared outside the enterprise at a peer level (e.g. industry forum.). [PTLE03-E20]

ID	Indicators of Competence *(in addition to those at Lead Practitioner level)*	Relevant knowledge, experience, and/or behaviors	Possible examples of objective evidence of personal contribution to activities performed, or professional behaviors applied
4	Guides and actively coordinates the progress of collaborative activities beyond the enterprise boundary, establishing mutual trust. [PTLE04]	(P) Advises why there is a need for a strong foundation in both technical domain knowledge and management skills. [PTLE04-10P]	Records highlighting collaborative leadership activities outside the enterprise. [PTLE04-E10]
		(A) Develops joint work products or processes outside the enterprise by persuading disparate or unwilling groups to collaborate. [PTLE04-20A]	Records of joint work products or processes from groups that do not normally collaborate (e.g. customers or competitors). [PTLE04-E20]
		(P) Acts to ensure team capabilities are leveraged to yield synergistic benefits. [PTLE04-30P]	Records highlighting work products where team collaboration yielded synergistic benefits. [PTLE04-E30]
5	Fosters empowerment of others beyond the enterprise boundary. [PTLE05]	(P) Fosters empowerment of others outside the enterprise, such as team members on an industry working group. [PTLE05-10P]	Records of development of future engineers (e.g. at high school, Science, Technology, Engineering, and Mathematics (STEM) initiative). [PTLE05-E10]
		(P) Fosters engagement of those not directly involved in Systems Engineering to see the value of a Systems Engineering approach. [PTLE05-20P]	Records of fostering engagement of those not directly involved in Systems Engineering to see the value of a Systems Engineering approach. [PTLE05-E20]
		(P) Fosters development of future engineers to develop interest in Systems Engineering. [PTLE05-30P]	Records of fostering development of future engineers to develop interest in Systems Engineering. [PTLE05-E30]
6	Advises organizations beyond the enterprise boundary on complex or sensitive team leadership problems or issues, applying creativity and innovation to ensure successful delivery. [PTLE06]	(P) Fosters the work of a team or individual to resolve a complex technical issue based on previous experience. [PTLE06-10P]	Records of fostering the work of a team or individual to resolve a complex technical issue based on previous experience. [PTLE06-E10]
		(P) Uses creativity or innovation in guiding the team to resolve a complex or sensitive issue. [PTLE06-20P]	Records of using creativity or innovation in guiding the team to resolve a complex or sensitive issue. [PTLE06-E20]
7	Uses their extended network and influencing skills to gain collaborative agreement with key stakeholders beyond the enterprise boundary in order to progress project or their own enterprise needs. [PTLE07]	(A) Uses own extended network, such as fellow Systems Engineering leaders in other parts of the enterprise, to gain agreement with stakeholders. [PTLE07-10A]	Records of using extended network contacts to help agreement between stakeholders. [PTLE07-E10]
		(A) Uses influencing skills to gain agreement with stakeholders where Systems Engineering leader had no official authority. [PTLE07-20A]	Documents indicating role in fostering stakeholder agreement despite having no direct authority. [PTLE07-E20]
8	Champions the introduction of novel techniques and ideas in Systems Engineering technical leadership, beyond the enterprise boundary, in order to develop the wider Systems Engineering community in this competency. [PTLE08]	(A) Analyzes different approaches across different domains through research. [PTLE08-10A]	Records of activities promoting research and need to adopt novel technique or ideas. [PTLE08-E10]

		(A) Produces reports for the wider SE community on the effectiveness of new techniques after their introduction. [PTLE08-20A]	Records of improvements made to process and appraisal against a recognized process improvement model. [PTLE08-E20]
		(P) Collaborates with those introducing novel techniques within the wider SE community. [PTLE08-30P]	Research records. [PTLE08-E30]
		(A) Defines novel approaches that could potentially improve the wider SE discipline. [PTLE08-40A]	Published papers in refereed journals/company literature. [PTLE08-E40]
		(P) Fosters awareness of these novel techniques within the wider SE community. [PTLE08-50P]	Records showing introduction of enabling systems supporting the new techniques or ideas. [PTLE08-E50]
9	Coaches individuals beyond the enterprise boundary, in technical leadership techniques in order to further develop their knowledge, abilities, skills, or associated behaviors. [PTLE09]	(P) Coaches or mentors individuals beyond the enterprise boundary, in competency-related techniques, recommending development activities. [PTLE09-10P]	Coaching or mentoring assignment records. [PTLE09-E10]
		(A) Develops or authorizes training materials in this competency area, which are subsequently successfully delivered beyond the enterprise boundary. [PTLE09-20A]	Records of formal training courses, workshops, seminars, and authored training material supported by successful post-training evaluation data. [PTLE09-E20]
		(A) Provides workshops/seminars or training in this competency area for practitioners or lead practitioners beyond the enterprise boundary (e.g. conferences and open training days). [PTLE09-30A]	Records of Training/workshops/seminars created supported by successful post-training evaluation data. [PTLE09-E30]
10	Maintains expertise in this competency area through specialist Continual Professional Development (CPD) activities. [PTLE10]	(A) Reviews research, new ideas, and state of the art to identify relevant new areas requiring personal development in order to maintain expertise in this competency area. [PTLE10-10A]	Records of documents reviewed and insights gained as part of own research into this competency area. [PTLE10-E10]
		(A) Performs identified specialist professional development activities in order to maintain or further develop competence at expert level. [PTLE10-20A]	Records of Continual Professional Development (CPD) performed and learning outcomes. [PTLE10-E20]
		(A) Records continual professional development activities undertaken including learning or insights gained. [PTLE10-30A]	

NOTES	In addition to items above, enterprise-level or independent 3rd-Party-generated evidence may be used to amplify other evidence presented and may include: a. Formally recognized by a reputable external organization as an expert in this competency area b. Evidence of role as independent assessor or reviewer on project outside own organization where skills in this competency area were used c. Evidence of invitation(s) from wider community for contribution of systems engineering expertise in this area (e.g. industry conference panel, government advisory board etc. cross-industry working groups, partnerships, accredited advanced university courses or research, or as part of professional institute) d. Formal commendation beyond the enterprise (e.g. by INCOSE or other recognized authority) for work performed in this competency area e. Independently assessed or accredited work product in this competency area (e.g. for independent publication or use) f. Accolades of expertise in this area from recognized industry leaders

Competency area – Professional: Negotiation
Description Negotiation is a dialogue between two or more parties intended to reach a beneficial outcome over one or more issues where differences exist with respect to at least one of these issues. This beneficial outcome can be for all parties involved, or just for one or some of them. Negotiation aims to resolve points of difference, to gain advantage for an individual or collective, or to craft outcomes to satisfy various interests. It is often conducted by putting forward a position and making small concessions to achieve an agreement.
Why it matters Systems Engineers are the "glue" that hold elements of a complex system development together. To achieve success, they need to involve themselves in many aspects of a project, interacting with different types of stakeholders and organizations. This necessitates resolution of many different types of issue in order to gain agreement between differing groups of stakeholders. Good negotiation skills are central to this activity.
Possible contributory types of evidence Any combination of the types of evidence may be acceptable (depending on how the Framework is tailored and used). The evidence items identified at each level indicate example work products only. Contributions to work products will generally differ at each proficiency level.
Learning and development The INCOSE Professional Development Portal provides example guidance on how to gain an initial awareness of a competency area and options for developing further competence thereafter.

Awareness – Professional: Negotiation		
ID	**Indicators of Competence**	**Relevant knowledge sub-indicators**
1	Explains key terminology associated with negotiation. [PNEA01]	(K) Describes different types, scenarios, and styles of negotiation and explains the high-level differences between them. [PNEA01-10K] (K) Describes examples of each style of negotiation. [PNEA01-20K] (K) Identifies different types of negotiating skill. [PNEA01-30K]
2	Describes situations where it may be necessary to negotiate and why. [PNEA02]	(K) Explains the key drivers to undertake negotiation. [PNEA02-10K]
3	Explains how different stakeholders hold different positions and bargaining power. [PNEA03]	(K) Explains different types of bargaining power in negotiations. [PNEA03-10K] (K) Describes key interfaces and links between engineering negotiations and stakeholders in the wider enterprise outside engineering (e.g. infrastructure management, human resource management, quality management, knowledge management, portfolio management, and life cycle model management). [PNEA03-20K] (K) Describes examples of situations where stakeholders may hold different bargaining power. [PNEA03-30K]
4	Identifies situations which do or do not require negotiation, to support negotiating strategies. [PNEA04]	(K) Describes general topics which may be relevant in determining when and when not to negotiate. [PNEA04-10K]

Supervised Practitioner – Professional: Negotiation			
ID	**Indicators of Competence *(in addition to those at Awareness level)***	**Relevant knowledge, experience, and/or behaviors**	**Possible examples of objective evidence of personal contribution to activities performed, or professional behaviors applied**
1	Develops good working level relationships with counterparts by negotiating to resolve routine issues. [PNES01]	(A) Identifies possible routine issues which may require negotiation with working level counterparts. [PNES01-10A] (P) Acts to address issues to improve working level relationships with counterparts. [PNES01-20P]	Records of acting to establish good working level relationships with counterparts to resolve routine issues. [PNES01-E10]
2	Collates data from a range of sources through research and analysis to provide useful input to a negotiation team. [PNES02]	(K) Describes how different types of information may be useful to the team in particular negotiations. [PNES02-10K] (K) Describes sources of negotiation information and how this information may be researched. [PNES02-20K] (K) Describes what type of analysis may be performed on raw information to improve its usefulness. [PNES02-30K] (A) Prepares information used by the team during a negotiation process. [PNES02-40A]	Records of preparing negotiation information used by the team during a negotiation process. [PNES02-E10]
3	Identifies stakeholders with different bargaining power on a project. [PNES03]	(K) Identifies stakeholders are on a particular project who might be subject of a negotiation activity. [PNES03-10K] (A) Identifies different positions and bargaining power of stakeholders on a particular project. [PNES03-20A]	Records of preparing negotiation information used by the team during a negotiation process. [PNES03-E10]
4	Describes key stakeholders' negotiation positions of these stakeholders. [PNES04]	(K) Describes how different situations and individual requirements of a key stakeholder might be of interest in a particular negotiation. [PNES04-10K] (A) Identifies stakeholder "negotiating positions" in a particular negotiation. [PNES04-20A]	Records of preparing negotiation information used by the team during a negotiation process. [PNES04-E10]
5	Prepares inputs to the review of a negotiation, covering the broad implications and unintended consequences of a negotiation decision. [PNES05]	(K) Describes the benefits of reviewing the immediate results, broad implications, and unintended consequences of a negotiation decision. [PNES05-10K] (A) Prepares inputs to the review of a negotiation, covering the broad implications and unintended consequences of a negotiation decision. [PNES05-20A]	Records of preparing negotiation information used by the team during a negotiation process. [PNES05-E10]

(Continued)

ID	Indicators of Competence *(in addition to those at Awareness level)*	Relevant knowledge, experience, and/or behaviors	Possible examples of objective evidence of personal contribution to activities performed, or professional behaviors applied
6	Maintains own confidence in the face of objections during negotiations. [PNES06]	(K) Explains why it is important to retain self-confidence in the face of objections during negotiation. [PNES06-10K] (P) Acts with confidence in the face of objections. [PNES06-20P]	3rd-party reflections on negotiation performance (e.g. individual performance records). [PNES06-E10]
7	Develops own understanding of this competency area through Continual Professional Development (CPD). [PNES07]	(A) Identifies potential gaps in own knowledge or development needs in this area, identifying opportunities to address these through continual professional development activities. [PNES07-10A] (A) Performs continual professional development activities to improve their knowledge and understanding in this area. [PNES07-20A] (A) Records continual professional development activities undertaken including learning or insights gained. [PNES07-30A]	Records of Continual Professional Development (CPD) performed and learning outcomes. [PNES07-E10]

Practitioner – Professional: Negotiation			
ID	**Indicators of Competence** ***(in addition to those at Supervised Practitioner level)***	**Relevant knowledge, experience, and/or behaviors**	**Possible examples of objective evidence of personal contribution to activities performed, or professional behaviors applied**
1	Follows established best practice strategies for negotiation in terms of preparation, approach, strategy, tactics, and style. [PNEP01]	(K) Describes established best practice for negotiation in terms of preparation, approach, strategy, tactics, and style. [PNEP01-10K] (A) Uses using best practice approaches with internal and external project stakeholders. [PNEP01-20A]	Reports or references from recognized project stakeholders confirming the role and contribution to a particular negotiation. [PNEP01-E10]
2	Negotiates successfully with internal and external project stakeholders. [PNEP02]	(A) Identifies differing styles of negotiating parties. [PNEP02-10A] (A) Adapts their negotiating style to the particular parties involved. [PNEP02-20A]	Records from project records, agendas, organization charts, minutes, or reports. [PNEP02-E10] Reports or references from recognized project stakeholders confirming the role and contribution to a particular negotiation. [PNEP02-E20] Records of adapting their negotiating style to the particular parties involved. [PNEP02-E30]
3	Acts to ensure buy-in and gain trust with internal stakeholders prior to and during negotiations. [PNEP03]	(P) Acts to gain trust and buy-in from internal stakeholders prior to and during negotiations. [PNEP03-10P]	Records from project records, agendas, organization charts, minutes, or reports. [PNEP03-E10] Reports or references from recognized project stakeholders confirming the role and contribution to a particular negotiation. [PNEP03-E20]
4	Communicates negotiation developments to internal stakeholders in order to manage expectations while keeping all parties informed. [PNEP04]	(A) Maintains regular communications in order to keep all parties informed of developments. [PNEP04-10A] (P) Communicates negotiation developments to internal stakeholders in order to manage expectations while keeping all parties informed. [PNEP04-20P]	Records from project records, agendas, organization charts, minutes, or reports. [PNEP04-E10] Reports or references from recognized project stakeholders confirming the role and contribution to a particular negotiation. [PNEP04-E20]
5	Analyzes data from a range of sources to make robust fact-based statements during negotiations, to make available choices clear and simple to stakeholders. [PNEP05]	(K) Describes the benefit analysis of data from a range of sources and making robust fact-based statements during negotiations, to make available choices clear and simple to stakeholders. [PNEP05-10K] (A) Analyzes data from a range of sources to make robust fact-based statements during negotiations. [PNEP05-20A] (A) Records available choices to stakeholders in a clear and simple way during a negotiation. [PNEP05-30A]	Reports or references from recognized project stakeholders confirming the role and contribution to a particular negotiation. [PNEP05-E10] Records of documenting available choices to stakeholders in a clear and simple way during a negotiation. [PNEP05-E20]

(Continued)

ID	Indicators of Competence *(in addition to those at Supervised Practitioner level)*	Relevant knowledge, experience, and/or behaviors	Possible examples of objective evidence of personal contribution to activities performed, or professional behaviors applied
6	Reacts positively when handling objections or points of view expressed by others challenging these views without damaging stakeholder relationship. [PNEP06]	(P) Reacts positively when receiving objections or points of view from other stakeholders on the project. [PNEP06-10P]	Records from project records, agendas, organization charts, minutes, or reports. [PNEP06-E10]
		(A) Uses negotiating skills to challenge the points of view expressed by others on the project. [PNEP06-20A]	Reports or references from recognized project stakeholders confirming the role and contribution to a particular negotiation. [PNEP06-E20]
		(P) Acts to overcome objections or points of view expressed by others without damaging stakeholder relationships. [PNEP06-30P]	Records from project records, agendas, organization charts, minutes, or reports. [PNEP06-E30]
			Reports or references from recognized project stakeholders confirming the role and contribution to a particular negotiation. [PNEP06-E40]
7	Reviews the immediate results, broad implications, and unintended consequences of a negotiation decision to ensure decision is sound. [PNEP07]	(A) Reviews the immediate results, broad implications, and unintended consequences of a negotiation decision to ensure decision is sound. [PNEP07-10A]	Records from project records, agendas, organization charts, minutes, or reports. [PNEP07-E10]
8	Acts with political awareness when negotiating with key decision-makers. [PNEP08]	(K) Describes the benefit of political awareness regarding key decision-makers. [PNEP08-10K]	Reports or references from recognized project stakeholders confirming the role and contribution to a particular negotiation. [PNEP08-E10]
		(A) Uses political awareness when negotiating with key decision-makers. [PNEP08-20A]	
9	Acts to gain credibility and gains trust and respect of all parties to negotiations. [PNEP09]	(P) Acts to establish credibility and gain trust and respect of all parties to negotiations. [PNEP09-10P]	Records from project records, agendas, organization charts, minutes, or reports. [PNEP09-E10]
			Reports or references from recognized project stakeholders confirming the role and contribution to a particular negotiation. [PNEP09-E20]
10	Guides new or supervised practitioners in negotiation techniques, in order to develop their knowledge, abilities, skills, or associated behaviors. [PNEP10]	(P) Guides new or supervised practitioners in executing activities that form part of this competency. [PNEP10-10P]	Organizational Breakdown Structure showing their responsibility for technical supervision in this area. [PNEP10-E10]
		(A) Trains individuals to an "Awareness" level in this competency area. [PNEP10-20A]	Records of on-the-job training objectives/guidance etc. [PNEP10-E20]

			Assignment as coach or mentor. [PNEP10-E30] Records highlighting their impact on another individual in terms of improvement or professional development in this competency. [PNEP10-E40]
11	Maintains and enhances own competence in this area through Continual Professional Development (CPD) activities. [PNEP11]	(A) Identifies potential development needs in this area, identifying opportunities to address these through continual professional development activities. [PNEP11-10A] (A) Performs continual professional development activities to maintain and enhance their competency in this area. [PNEP11-20A] (A) Records continual professional development activities undertaken including learning or insights gained. [PNEP11-30A]	Records of Continual Professional Development (CPD) performed and learning outcomes. [PNEP11-E10]

Lead Practitioner – Professional: Negotiation

ID	Indicators of Competence *(in addition to those at Practitioner level)*	Relevant knowledge, experience, and/or behaviors	Possible examples of objective evidence of personal contribution to activities performed, or professional behaviors applied
1	Promotes best practice negotiation techniques across the enterprise to improve the effectiveness of systems engineering negotiations. [PNEL01]	(A) Develops own state-of-the-art negotiations skills. [PNEL01-10A]	Documents demonstrating introducing or controlling negotiation best practices. [PNEL01-E10]
		(P) Advises others on best practice when preparing for negotiations. [PNEL01-20P]	Articles, papers or training materials authored to support negotiation. [PNEL01-E20]
		(P) Advises others on best practice when negotiating, with success. [PNEL01-30P]	Records from negotiations showing advice provided and used successfully. [PNEL01-E30]
2	Judges the suitability of the planned approach or strategy for negotiations affecting Systems Engineering across the enterprise. [PNEL02]	(A) Assesses the suitability of negotiation strategy and meetings. [PNEL02-10A]	Records demonstrating a personal contribution to the assessment and improvement of negotiation strategy and meetings. [PNEL02-E10]
3	Guides and actively coordinates the direction of negotiation teams across the enterprise, accepting accountability for final negotiation outcomes whether successful or not. [PNEL03]	(P) Guides and actively coordinates the activities of negotiation teams to demonstrable success in closing negotiations. [PNEL03-10P]	Documents from enterprise or external program records. [PNEL03-E10]
		(P) Challenges the appropriateness of negotiating stance or reasoning, with notable success. [PNEL03-20P]	Records of leading negotiating teams. [PNEL03-E20]
		(P) Acts to accept accountability for final negotiation outcomes whether successful or not. [PNEL03-30P]	Records of where the negotiation teams they have led, have had demonstrable success in closing negotiation. [PNEL03-E30]
4	Adapts personal positions and style quickly if circumstances change favorably and unfavorably. [PNEL04]	(A) Adapts negotiating technical stance if circumstances change favorably or unfavorably. [PNEL04-10A]	Records from project records, agendas, organization charts, minutes, or reports. [PNEL04-E10]
		(A) Adapts personal negotiation style quickly to reflect changing circumstance. [PNEL04-20A]	Reports or references from recognized project stakeholders confirming the role and contribution to a particular negotiation. [PNEL04-E20]
5	Acts on behalf of the wider enterprise during tough, challenging negotiating situations with both external and internal stakeholders. [PNEL05]	(P) Acts with sensitivity and control in tough, challenging situations with both external and internal stakeholders. [PNEL05-10P]	Records from project records, agendas, organization charts, minutes, or reports. [PNEL05-E10]
6	Acts on behalf of the wider enterprise to gain credibility and gains trust and respect of all parties during difficult negotiations. [PNEL06]	(P) Acts to establish credibility and gain trust and respect of all parties to negotiations. [PNEL06-10P]	Records from project records, agendas, organization charts, minutes, or reports. [PNEL06-E10]
			Reports or references from recognized project stakeholders confirming the role and contribution to a particular negotiation. [PNEL06-E20]

7	Acts positively when handling objections or points of view expressed by senior enterprise stakeholders challenging views without damaging stakeholder relationship and persuading them to change their mind. [PNEL07]	(P) Acts positively when receiving objections or points of view from other stakeholders across the enterprise. [PNEL07-10P]	Records from project records, agendas, organization charts, minutes, or reports. [PNEL07-E10]
		(A) Uses negotiating skills to challenge the points of view expressed by others across the enterprise. [PNEL07-20A]	Reports or references from recognized project stakeholders confirming the role and contribution to a particular negotiation. [PNEL07-E20]
		(A) Assesses objections or points of view expressed by others across the enterprise, without damaging stakeholder relationships. [PNEL07-30A]	Records of assessing objections or points of view expressed by others across the enterprise, without damaging stakeholder relationships. [PNEL07-E30]
		(P) Persuades enterprise stakeholders to modify viewpoint successfully. [PNEL07-40P]	Records of persuading enterprise stakeholders to modify viewpoint successfully. [PNEL07-E40]
8	Persuades third-party decision-makers to move toward wider enterprise goals, using good political awareness. [PNEL08]	(A) Determines best enterprise outcomes from third-party decisions. [PNEL08-10A]	Records from project records, agendas, organization charts, minutes, or reports. [PNEL08-E10]
		(P) Persuades third-party decision-makers through good political awareness. [PNEL08-20P]	Reports or references from recognized project stakeholders confirming the role and contribution to a particular negotiation. [PNEL08-E20]
9	Acts to accept accountability for final negotiation outcomes on behalf of the enterprise, whether successful or not. [PNEL09]	(P) Acts to accept accountability for final negotiation outcomes for both successful outcomes and otherwise. [PNEL09-10P]	Documents from enterprise or external program records. [PNEL09-E10]
		(A) Analyzes results of negotiation and uses the analysis to improve future strategies. [PNEL09-20A]	
10	Coaches or mentors practitioners across the enterprise in negotiation techniques in order to develop their knowledge, abilities, skills, or associated behaviors. [PNEL10]	(P) Coaches or mentors practitioners across the enterprise in competency-related techniques, recommending development activities. [PNEL10-10P]	Coaching or mentoring assignment records. [PNEL10-E10]
		(A) Develops or authorizes enterprise training materials in this competency area. [PNEL10-20A]	Records of formal training courses, workshops, seminars, and authored training material supported by successful post-training evaluation data. [PNEL10-E20]
		(A) Provides enterprise workshops/seminars or training in this competency area. [PNEL10-30A]	Listing as an approved organizational trainer for this competency area. [PNEL10-E30]
11	Promotes the introduction and use of novel techniques and ideas in negotiation across the enterprise, to improve enterprise competence in this area. [PNEL11]	(A) Analyzes different approaches across different domains through research. [PNEL11-10A]	Research records. [PNEL11-E10]

(*Continued*)

ID	Indicators of Competence *(in addition to those at Practitioner level)*	Relevant knowledge, experience, and/or behaviors	Possible examples of objective evidence of personal contribution to activities performed, or professional behaviors applied
		(A) Defines novel approaches that could potentially improve the SE discipline within the enterprise. [PNEL11-20A]	Published papers in refereed journals/company literature. [PNEL11-E20]
		(P) Fosters awareness of these novel techniques within the enterprise. [PNEL11-30P]	Records showing introduction of enabling systems supporting the new techniques or ideas. [PNEL11-E30]
		(P) Collaborates with enterprise stakeholders to facilitate the introduction of techniques new to the enterprise. [PNEL11-40P]	Published papers (or similar) at enterprise level. [PNEL11-E40]
		(A) Monitors new techniques after their introduction to determine their effectiveness. [PNEL11-50A]	Records of improvements made against a recognized process improvement model. [PNEL11-E50]
		(A) Adapts approach to reflect actual enterprise performance improvements. [PNEL11-60A]	Records of adapting approach to reflect actual enterprise performance improvements. [PNEL11-E60]
12	Develops expertise in this competency area through specialist Continual Professional Development (CPD) activities. [PNEL12]	(A) Identifies own needs for further professional development in order to increase competence beyond practitioner level. [PNEL12-10A]	Records of Continual Professional Development (CPD) performed and learning outcomes. [PNEL12-E10]
		(A) Performs professional development activities in order to move own competence toward expert level. [PNEL12-20A]	
		(A) Records continual professional development activities undertaken including learning or insights gained. [PNEL12-30A]	

NOTES	In addition to items above, enterprise-level or independent 3rd-Party-generated evidence may be used to amplify other evidence presented and may include: a. Formally recognized by senior management in current organization as an expert in this competency area b. Evidence of role as Product/System Design Authority or Technical Authority on a complex project with responsibilities in this area or where skills within this competency area were used c. Recognized as an authorizing signatory on behalf of enterprise for formal documentation in this competency area (e.g. policies, processes, and deliverables) d. Formal commendation or award within own enterprise for contribution or item of work successfully performed, which required proficiency in this competency area e. Customer, Supplier, or other external project-specific key Stakeholder accolades for specific work performed in this competency area f. Independently assessed or accredited work in this competency area (e.g. for independent publication or use) g. Formal organizational HR records positively highlighting any specific professional competencies or behaviors identified (if applicable) plus any of the evidence indicators listed at Expert level below

Expert – Professional: Negotiation			
ID	**Indicators of Competence *(in addition to those at Lead Practitioner level)***	**Relevant knowledge, experience, and/or behaviors**	**Possible examples of objective evidence of personal contribution to activities performed, or professional behaviors applied**
1	Communicates own knowledge and experience in negotiation skills in order to improve best practice beyond the enterprise boundary. [PNEE01]	(A) Produces papers, seminars, or presentations outside own enterprise for publication in order to share own ideas and improve industry best practices in this competence area. [PNEE01-10A]	Published papers or books etc. on new technique in refereed journals/company literature. [PNEE01-E10]
		(P) Fosters incorporation of own ideas into industry best practices in this area. [PNEE01-20P]	Published papers in refereed journals or internal literature proposing new practices in this competence area (or presentations, tutorials, etc.). [PNEE01-E20]
		(P) Develops guidance materials identifying new (or updating existing) best practice in this competence area. [PNEE01-30P]	Records of own proposals adopted as industry best practices in this competence area. [PNEE01-E30]
2	Influences stakeholders beyond the enterprise boundary in support of negotiations activities affecting Systems Engineering. [PNEE02]	(P) Persuades stakeholders beyond the enterprise boundary on negotiating activities affecting Systems Engineering. [PNEE02-10P]	Records of ideas assimilated into international standards or referenced in others' books, papers, and peer-reviewed articles. [PNEE02-E10]
		(A) Identifies changes to existing stakeholder thinking in negotiation activities affecting Systems Engineering. [PNEE02-20A]	Records demonstrating a personal contribution to negotiation strategic changes. [PNEE02-E20]
3	Guides and actively coordinates the direction of negotiations beyond the enterprise boundary, on complex or strategic decisions. [PNEE03]	(P) Persuades key strategic decision-makers extending beyond own enterprise, on negotiation activities affecting Systems Engineering. [PNEE03-10P]	Documents from enterprise or external program records. [PNEE03-E10]
		(P) Challenges the appropriateness of negotiating stance or reasoning of stakeholders beyond the enterprise boundary, with notable success. [PNEE03-20P]	Documents from enterprise or external program records. [PNEE03-E20]
		(P) Guides and actively coordinates the direction of negotiations on complex strategic issues beyond of the enterprise. [PNEE03-30P]	
4	Advises organizations beyond the enterprise boundary on the suitability of their negotiating strategies. [PNEE04]	(A) Advises organizations beyond the enterprise boundary on current or proposed negotiating strategies or approaches. [PNEE04-10A]	Documents from enterprise or external program records. [PNEE04-E10]
		(A) Advises negotiators beyond the enterprise boundary on strategies designed to change the views of, or persuade key or challenging stakeholders, with demonstrable results. [PNEE04-20A]	Documents from enterprise or external program records. [PNEE04-E20]

(Continued)

ID	Indicators of Competence *(in addition to those at Lead Practitioner level)*	Relevant knowledge, experience, and/or behaviors	Possible examples of objective evidence of personal contribution to activities performed, or professional behaviors applied
		(A) Advises organizations beyond the enterprise boundary on appropriate negotiating positions and negotiating styles. [PNEE04-30A]	Records of advising stakeholders beyond enterprise boundary on negotiating positions and styles. [PNEE04-E30]
		(A) Advises key stakeholders beyond enterprise boundary on their negotiating positions and styles. [PNEE04-40A]	
5	Champions the introduction of novel techniques and ideas in negotiation techniques, beyond the enterprise boundary, in order to develop the wider Systems Engineering community in this competency. [PNEE05]	(A) Analyzes different approaches across different domains through research. [PNEE05-10A]	Records of activities promoting research and need to adopt novel technique or ideas. [PNEE05-E10]
		(A) Produces reports for the wider SE community on the effectiveness of new techniques after their introduction. [PNEE05-20A]	Records of improvements made to process and appraisal against a recognized process improvement model. [PNEE05-E20]
		(P) Collaborates with those introducing novel techniques within the wider SE community. [PNEE05-30P]	Research records. [PNEE05-E30]
		(A) Defines novel approaches that could potentially improve the wider SE discipline. [PNEE05-40A]	Published papers in refereed journals/company literature. [PNEE05-E40]
		(P) Fosters awareness of these novel techniques within the wider SE community. [PNEE05-50P]	Records showing introduction of enabling systems supporting the new techniques or ideas. [PNEE05-E50]
6	Coaches individuals beyond the enterprise boundary, in negotiation techniques in order to further develop their knowledge, abilities, skills, or associated behaviors. [PNEE06]	(P) Coaches or mentors individuals beyond the enterprise boundary, in competency-related techniques, recommending development activities. [PNEE06-10P]	Coaching or mentoring assignment records. [PNEE06-E10]
		(A) Develops or authorizes training materials in this competency area, which are subsequently successfully delivered beyond the enterprise boundary. [PNEE06-20A]	Records of formal training courses, workshops, seminars, and authored training material supported by successful post-training evaluation data. [PNEE06-E20]
		(A) Provides workshops/seminars or training in this competency area for practitioners or lead practitioners beyond the enterprise boundary (e.g. conferences and open training days). [PNEE06-30A]	Records of Training/workshops/seminars created supported by successful post-training evaluation data. [PNEE06-E30]

7	Maintains expertise in this competency area through specialist Continual Professional Development (CPD) activities. [PNEE07]	(A) Reviews research, new ideas, and state of the art to identify relevant new areas requiring personal development in order to maintain expertise in this competency area. [PNEE07-10A]	Records of documents reviewed and insights gained as part of own research into this competency area. [PNEE07-E10]
		(A) Performs identified specialist professional development activities in order to maintain or further develop competence at expert level. [PNEE07-20A]	Records of Continual Professional Development (CPD) performed and learning outcomes. [PNEE07-E20]
		(A) Records continual professional development activities undertaken including learning or insights gained. [PNEE07-30A]	

NOTES	In addition to items above, enterprise-level or independent 3rd-Party-generated evidence may be used to amplify other evidence presented and may include: a. Formally recognized by a reputable external organization as an expert in this competency area b. Evidence of role as independent assessor or reviewer on project outside own organization where skills in this competency area were used c. Evidence of invitation(s) from wider community for contribution of systems engineering expertise in this area (e.g. industry conference panel, government advisory board etc. cross-industry working groups, partnerships, accredited advanced university courses or research, or as part of professional institute) d. Formal commendation beyond the enterprise (e.g. by INCOSE or other recognized authority) for work performed in this competency area e. Independently assessed or accredited work product in this competency area (e.g. for independent publication or use) f. Accolades of expertise in this area from recognized industry leaders

Competency area – Professional: Team Dynamics
Description Team dynamics are the unconscious, psychological forces that influence the direction of a team's behavior and performance. Team dynamics are created by the nature of the team's work, the personalities within the team, their working relationships with other people, and the environment in which the team works.
Why it matters Team dynamics can be good – for example, when they improve overall team performance and/or get the best out of individual team members. They can also be bad – for example, when they cause unproductive conflict, demotivation, and prevent the team from achieving its goals.
Possible contributory types of evidence Any combination of the types of evidence may be acceptable (depending on how the Framework is tailored and used). The evidence items identified at each level indicate example work products only. Contributions to work products will generally differ at each proficiency level.
Learning and development The INCOSE Professional Development Portal provides example guidance on how to gain an initial awareness of a competency area and options for developing further competence thereafter.

Awareness – Professional: Team Dynamics		
ID	**Indicators of Competence**	**Relevant knowledge sub-indicators**
1	Lists different types of team and the role of each team within the project or organization. [PTDA01]	(K) Defines the difference between a team and a group. [PTDA01-10K] (K) Explains the value of working in teams vs groups. [PTDA01-20K]
2	Explains the different stages of team development and how they affect team dynamics and performance. [PTDA02]	(K) Defines a common team dynamics model (e.g. Tuckman: Forming, Storming, Norming, Performing, and "mourning") or Peck (Pseudo-community, Chaos, Emptiness, true Community) including how the different stages of a group might impact effectiveness of the group. [PTDA02-10K] (K) Explains why team performance is impacted by the stages of development. [PTDA02-20K]
3	Explains the positive and negative features of cooperation and competition within teams. [PTDA03]	(K) Explains positive features of cooperation and competition within teams. [PTDA03-10K] (K) Describes links between engineering and the wider enterprise outside engineering (e.g. infrastructure management, human resource management, quality management, knowledge management, portfolio management, and life cycle model management), which may impact team dynamics. [PTDA03-20K]
4	Explains how the effectiveness of communications affects team dynamics. [PTDA04]	(K) Defines what "effective communication" is and illustrates, with examples, how it is important the receiver understands the message the sender is sending. [PTDA04-10K] (K) Explains how the characteristics of the sender, the receiver, and the message can affect the impact of team communications. [PTDA04-20K]

		(K) Defines the value of effective communication on team performance. [PTDA04-30K] (K) Explains how active listening can improve communication. [PTDA04-40K]
5	Explains the differing nature of disagreement, conflict, and criticism in teams and core strategies for resolving conflict. [PTDA05]	(K) Lists potential sources of conflict at work (e.g. personality, working style, miscommunications or misunderstandings, resource availability, lack of individual or team support, poor customer service, poorly organized workplace, poor management, discrimination or harassment, contract terms for individuals). [PTDA05-10K] (K) Defines well-established mechanisms to resolve conflict (e.g. informal discussions, mediation, arbitration, conciliation, promoting a culture of openness). [PTDA05-20K] (K) Lists the steps of a conflict resolution process. [PTDA05-30K]
6	Explains why team building can help form effective teams, what it involves, and its key challenges. [PTDA06]	(K) Explains why team building is an effective way to improve the dynamics of a team. [PTDA06-10K] (K) Explains how team building activities helps to build trust and how trust affects team dynamics. [PTDA06-20K]
7	Identifies different types of team-building activities, their aims, and provides examples. [PTDA07]	(K) Identifies different team building activities. [PTDA07-10K] (K) Explains how team building activities can support team development and when to use them. [PTDA07-20K]

Supervised Practitioner – Professional: Team Dynamics			
ID	**Indicators of Competence *(in addition to those at Awareness level)***	**Relevant knowledge, experience, and/or behaviors**	**Possible examples of objective evidence of personal contribution to activities performed, or professional behaviors applied**
1	Identifies when and when not to identify own positions, roles, and responsibilities within different teams within the project, or organization. [PTDS01]	(K) Explains how roles are formed and their role in current team. [PTDS01-10K] (A) Describes how own role interfaces with other roles in the organization. [PTDS01-20A]	Records showing team organization and/or team roles and interfaces. [PTDS01-E10]
2	Uses team dynamics to improve their effectiveness in performing team goals. [PTDS02]	(A) Uses team dynamics to improve their effectiveness in performing team goals. [PTDS02-10A]	Records showing use of team dynamics to improve own effectiveness. [PTDS02-E10]
3	Identifies the stage (e.g. forming, Storming, and Norming) at which each of the teams within which they participate is operating and provides rationale. [PTDS03]	(A) Identifies the stage (e.g. Forming, Storming, Norming, and Performing) of their current team and provides rationale. [PTDS03-10A]	Records of having participated in a new, developing, or performing team. [PTDS03-E10]
4	Explains the building blocks of successful team performance and why they affect performance. [PTDS04]	(K) Explains the impact of team dynamics on team performance, covering areas such as effective communication, common purpose, accepted leadership, effective processes, and solid relationships. [PTDS04-10K] (K) Explains how building blocks of successful team performance can potentially (or actually) effected team effectiveness. [PTDS04-20K] (A) Identifies use of team building blocks on a project. [PTDS04-30A]	Records showing team dynamics usage on project. [PTDS04-E10]
5	Explains how team goals, communication, and interpersonal actions are affected by competitive behaviors. [PTDS05]	(K) Identifies current team goals. [PTDS05-10K] (K) Identifies situations where communication, or interpersonal actions may be affected by competitive behaviors. [PTDS05-20K] (A) Describes situations where communication, or interpersonal actions may have been affected by competitive behaviors. [PTDS05-30A]	Records supporting examples of competitive behavior on a project. [PTDS05-E10]
6	Identifies competitive behaviors within a team and their potential cause (e.g. cultural, personal, and organizational reasons). [PTDS06]	(K) Describes competitive behaviors, their potential causes, and how these affected team dynamics. [PTDS06-10K] (P) Acts to overcome competitive behaviors within their team. [PTDS06-20P]	Records of having experienced and overcome competitive behaviors within their team or project. [PTDS06-E10]

7	Describes different potential types of team conflict and the differing techniques available to resolve them. [PTDS07]	(K) Identifies how team conflict situations can be successfully resolved (or averted). [PTDS07-10K] (A) Identifies areas where team conflict has existed and been addressed. [PTDS07-20A]	Records of team conflict and their resolution in their team or project. [PTDS07-E10]
8	Explains how team dynamics affect decision-making. [PTDS08]	(K) Explains how team relationships can cause barriers which affect decision-making (e.g. information barriers, cognitive bias, individual personalities, cultural or organizational norms, attitude to risk or uncertainty, and competence of decision-maker). [PTDS08-10K] (K) Explains how team dynamics can affect problem solving (e.g. confirmation bias, mental set, functional fixedness, unnecessary constraints, and irrelevant information). [PTDS08-20K] (A) Describes situations they have experienced where barriers or other team dynamics issues may have impacted decision-making or its effectiveness. [PTDS08-30A]	Records supporting examples of team dynamics influencing decision-making on a project. [PTDS08-E10]
9	Develops own understanding of this competency area through Continual Professional Development (CPD). [PTDS09]	(A) Identifies potential gaps in own knowledge or development needs in this area, identifying opportunities to address these through continual professional development activities. [PTDS09-10A] (A) Performs continual professional development activities to improve their knowledge and understanding in this area. [PTDS09-20A] (A) Records continual professional development activities undertaken including learning or insights gained. [PTDS09-30A]	Records of Continual Professional Development (CPD) performed and learning outcomes. [PTDS09-E10]

Practitioner – Professional: Team Dynamics			
ID	**Indicators of Competence** ***(in addition to those at Supervised Practitioner level)***	**Relevant knowledge, experience, and/or behaviors**	**Possible examples of objective evidence of personal contribution to activities performed, or professional behaviors applied**
1	Acts collaboratively with other teams to accomplish interdependent project or organizational goals. [PTDP01]	(P) Collaborates (e.g. with other teams) to accomplish interdependent project or organizational goals. [PTDP01-10P]	Records demonstrating team collaboration toward independent goals. [PTDP01-E10]
2	Recognizes the dynamic of their team and applies best practice to improve this as necessary. [PTDP02]	(K) Describes potential differing dynamic of teams and how this can affect the project (positively or negatively). [PTDP02-10K]	
		(A) Identifies the dynamic of their team. [PTDP02-20A]	Records demonstrating activities designed to identify team dynamics. [PTDP02-E20]
		(A) Uses best practice team dynamics techniques to improve the dynamic of a project team. [PTDP02-30A]	Records demonstrating activities designed to improve team dynamics. [PTDP02-E30]
3	Fosters a common understanding of an assignment in line with organizational intent within their team. [PTDP03]	(K) Describes how different team members may have a different understanding of an assignment to each other or to organizational intent and how this can be addressed. [PTDP03-10K]	
		(P) Fosters a common understanding of an assignment in line with organizational intent within their team. [PTDP03-20P]	Records demonstrating activities designed to improve team dynamics. [PTDP03-E20]
4	Fosters cooperation and pride within the team through strategies focused on group goals, communication, and interpersonal actions. [PTDP04]	(P) Fosters cooperation and pride within the team through strategies focused on group goals, communication, and interpersonal actions. [PTDP04-10P]	Records demonstrating activities designed to improve team pride based upon group goals, communication, and interpersonal actions. [PTDP04-E10]
5	Identifies negative behaviors within the team, challenging these to create positive outcomes. [PTDP05]	(A) Identifies negative behaviors within the team, challenging these to create positive outcomes. [PTDP05-10A]	Records of behavioral challenge addressed. [PTDP05-E10]
		(A) Uses team dynamics techniques to ensure a positive outcome when resolving negative behaviors. [PTDP05-20A]	Records of using team dynamics techniques to ensure a positive outcome when resolving negative behaviors. [PTDP05-E20]
6	Uses communications skills to offer constructive feedback to improve team performance, managing emotions as an important aspect of team's communications. [PTDP06]	(K) Explains why the use of constructive feedback helps to improve team performance. [PTDP06-10K]	
		(P) Communicates feedback constructively in order to improve team performance, and the resultant outcome. [PTDP06-20P]	Records of a communications challenge addressed. [PTDP06-E20]

7	Recognizes conflict in a team in order to resolve it. [PTDP07]	(P) Recognizes conflict within their team and acts to resolve. [PTDP07-10P] (P) Acts to resolve conflict within their team to ensure a positive outcome. [PTDP07-20P]	Records of a conflict challenge within their team that was addressed. [PTDP07-E10] Records of the resolution of a conflict challenge within their team with a positive outcome. [PTDP07-E20]
8	Fosters an open team dynamic within the team so that all team members can express their opinions and feelings. [PTDP08]	(P) Fosters an open expression of opinions in order to improve team dynamics. [PTDP08-10P] (A) Uses openness to share team ideas in order to create alternative improved solutions. [PTDP08-20A]	Records demonstrating that opinions and feelings were actively taken into consideration. [PTDP08-E10] Reports or references from recognized enterprise leaders confirming contribution to strategy regarding the application of team dynamics to Systems Engineering. [PTDP08-E20]
9	Uses best practice team dynamics techniques to obtain team consensus when making decisions. [PTDP09]	(A) Uses best practice team dynamics techniques to obtain team consensus when making decisions, including the decision situation, and the techniques used to gain team consensus. [PTDP09-10A]	Records of gaining team consensus as part of decision-making process. [PTDP09-E10]
10	Uses team-building activities to improve team dynamics. [PTDP10]	(A) Uses best practice team-building techniques to improve team dynamics, the techniques used, and the improvements achieved. [PTDP10-10A]	Records of team-building techniques used and measured successes. [PTDP10-E10]
11	Guides new or supervised practitioners in negotiation techniques, in order to develop their knowledge, abilities, skills, or associated behaviors. [PTDP11]	(P) Guides new or supervised practitioners in executing activities that form part of this competency. [PTDP11-10P] (A) Trains individuals to an "Awareness" level in this competency area. [PTDP11-20A]	Organizational Breakdown Structure (OBS) showing their responsibility for technical supervision in this area. [PTDP11-E10] Records of on-the-job training objectives/guidance. [PTDP11-E20] Coaching or mentoring assignment records. [PTDP11-E30] Records highlighting their impact on another individual in terms of improvement or professional development in this competency. [PTDP11-E40]
12	Maintains and enhances own competence in this area through Continual Professional Development (CPD) activities. [PTDP12]	(A) Identifies potential development needs in this area, identifying opportunities to address these through continual professional development activities. [PTDP12-10A] (A) Performs continual professional development activities to maintain and enhance their competency in this area. [PTDP12-20A] (A) Records continual professional development activities undertaken including learning or insights gained. [PTDP12-30A]	Records of Continual Professional Development (CPD) performed and learning outcomes. [PTDP12-E10]

Lead Practitioner – Professional: Team Dynamics			
ID	**Indicators of Competence *(in addition to those at Practitioner level)***	**Relevant knowledge, experience, and/or behaviors**	**Possible examples of objective evidence of personal contribution to activities performed, or professional behaviors applied**
1	Uses best practice team dynamics techniques to improve the effectiveness of Systems Engineering activities across the enterprise. [PTDL01]	(A) Develops own state-of-the-art team dynamics skills in order to improve understanding of where their application would be most effective. [PTDL01-10A] (P) Uses best practice team dynamics techniques to improve interactions with stakeholders or stakeholder groups across the enterprise. [PTDL01-20P]	Records demonstrating personal use of various team dynamics(s) techniques and associated feedback indicating effectiveness. [PTDL01-E10] Records demonstrating use of team dynamics successfully to a variety of diverse stakeholders across the enterprise. [PTDL01-E20] Records (e.g. from Human Resources department) indicating excellent team dynamics skills. [PTDL01-E30]
2	Judges the dynamic of teams across the enterprise, advising where improvement is necessary. [PTDL02]	(A) Reviews team structures across the enterprise with regard to their team dynamics. [PTDL02-10A] (A) Assesses the performance of teams, making recommendations as necessary. [PTDL02-20A]	Records demonstrating a personal contribution to the assessment and improvement of team dynamics. [PTDL02-E10] Records of having assessed the performance of teams and making recommendations as necessary. [PTDL02-E20]
3	Advises stakeholders across the enterprise, on the selection of measurable group goals, communication, or interpersonal actions designed to improve team performance. [PTDL03]	(A) Advises given with regard to selection of measurable goals, communication, or interpersonal actions designed to improve team performance. [PTDL03-10A] (A) Defines measurable team or group-related goals, communication, or interpersonal actions across the enterprise, which are designed to improve team performance. [PTDL03-20A]	Records demonstrating a personal contribution to improvements to measurable goals, communication, or interpersonal actions. [PTDL03-E10] Records of having defined measurable team or group-related goals, communication or interpersonal actions across the enterprise, which were designed to improve team performance. [PTDL03-E20]
4	Challenges negative behaviors of key enterprise stakeholders, with measurable success. [PTDL04]	(P) Acts to highlight negative behavior from key stakeholders. [PTDL04-10P] (P) Uses team dynamics skills to address or mitigate causes of stakeholder negative behaviors. [PTDL04-20P]	Records demonstrating a personal contribution to improving a stakeholders behaviors. [PTDL04-E10] Records of actions to ensure positive outcomes when addressing negative stakeholder behaviors. [PTDL04-E20]
5	Advises stakeholders across the enterprise on different best practice team dynamics techniques across the enterprise depending on the situation and decision required. [PTDL05]	(A) Advises on selection of appropriate team-centric techniques in order to improve dynamics and effectiveness across the enterprise. [PTDL05-10A] (A) Advises on selection of appropriate team-centric decision techniques designed to improve team decision-making across the enterprise. [PTDL05-20A]	Records demonstrating personal suggestions for improvement of team dynamics. [PTDL05-E10] Records demonstrating personal suggestions for improvement of team decision-making. [PTDL05-E20]

6	Fosters communication across the wider enterprise building trust through the application of team dynamics techniques. [PTDL06]	(A) Selects appropriate team dynamics techniques to improve communication between enterprise teams based upon the situation. [PTDL06-10A]	Records demonstrating personal suggestions for improvement of team communications. [PTDL06-E10]
		(P) Fosters communication between enterprise teams in order to improve wider dynamics and effectiveness. [PTDL06-20P]	Records demonstrating personal actions designed to improve trust within a team environment. [PTDL06-E20]
		(A) Uses team dynamics techniques to build trust and improvement across the enterprise. [PTDL06-30A]	Records of having used team dynamics techniques to build trust and improvement across the enterprise. [PTDL06-E30]
7	Influences key stakeholders across the enterprise to follow a revised path to improve a project or enterprise team dynamics. [PTDL07]	(P) Identifies stakeholder viewpoints, which require changing to improve wider project or enterprise team dynamics. [PTDL07-10P]	Records demonstrating how advice provided influenced key stakeholders and improved a team dynamics issue (e.g. conflict). [PTDL07-E10]
		(P) Persuades key stakeholders to follow a new path resulting in positive improvements for the team. [PTDL07-20P]	
8	Uses different types of team-building activities depending on the team context, to improve team dynamics across the enterprise. [PTDL08]	(A) Uses different types of team-building techniques to improve dynamics of different teams across the wider enterprise. [PTDL08-10A]	Records of team building activities. [PTDL08-E10]
9	Coaches or mentors practitioners across the enterprise in team dynamics techniques in order to develop their knowledge, abilities, skills, or associated behaviors. [PTDL09]	(P) Coaches or mentors practitioners across the enterprise in competency-related techniques, recommending development activities. [PTDL09-10P]	Coaching or mentoring assignment records. [PTDL09-E10]
		(A) Develops or authorizes enterprise training materials in this competency area. [PTDL09-20A]	Records of formal training courses, workshops, seminars, and authored training material supported by successful post-training evaluation data. [PTDL09-E20]
		(A) Provides enterprise workshops/seminars or training in this competency area. [PTDL09-30A]	Listing as an approved organizational trainer for this competency area. [PTDL09-E30]
10	Promotes the introduction and use of novel techniques and ideas in team dynamics across the enterprise, to improve enterprise competence in this area. [PTDL10]	(A) Analyzes different approaches across different domains through research. [PTDL10-10A]	Research records. [PTDL10-E10]
		(A) Defines novel approaches that could potentially improve the SE discipline within the enterprise. [PTDL10-20A]	Papers published in refereed journals/company literature. [PTDL10-E20]
		(P) Fosters awareness of these novel techniques within the enterprise. [PTDL10-30P]	Records showing introduction of enabling systems supporting the new techniques or ideas. [PTDL10-E30]

(Continued)

ID	Indicators of Competence *(in addition to those at Practitioner level)*	Relevant knowledge, experience, and/or behaviors	Possible examples of objective evidence of personal contribution to activities performed, or professional behaviors applied
		(P) Collaborates with enterprise stakeholders to facilitate the introduction of techniques new to the enterprise. [PTDL10-40P]	Papers (or similar) published at the enterprise level. [PTDL10-E40]
		(A) Monitors new techniques after their introduction to determine their effectiveness. [PTDL10-50A]	Records of improvements made against a recognized process improvement model. [PTDL10-E50]
		(A) Adapts approach to reflect actual enterprise performance improvements. [PTDL10-60A]	Records of adapting approach to reflect actual enterprise performance improvements. [PTDL10-E60]
11	Develops expertise in this competency area through specialist Continual Professional Development (CPD) activities. [PTDL11]	(A) Identifies own needs for further professional development in order to increase competence beyond practitioner level. [PTDL11-10A]	Records of Continual Professional Development (CPD) performed and learning outcomes. [PTDL11-E10]
		(A) Performs professional development activities in order to move own competence toward expert level. [PTDL11-20A]	
		(A) Records continual professional development activities undertaken including learning or insights gained. [PTDL11-30A]	

NOTES	In addition to items above, enterprise-level or independent 3rd-Party-generated evidence may be used to amplify other evidence presented and may include:
	a. Formally recognized by senior management in current organization as an expert in this competency area b. Evidence of role as Product/System Design Authority or Technical Authority on a complex project with responsibilities in this area or where skills within this competency area were used c. Recognized as an authorizing signatory on behalf of enterprise for formal documentation in this competency area (e.g. policies, processes, and deliverables) d. Formal commendation or award within own enterprise for contribution or item of work successfully performed, which required proficiency in this competency area e. Customer, Supplier, or other external project-specific key Stakeholder accolades for specific work performed in this competency area f. Independently assessed or accredited work in this competency area (e.g. for independent publication or use) g. Formal organizational HR records positively highlighting any specific professional competencies or behaviors identified (if applicable) plus any of the evidence indicators listed at Expert level below

Expert – Professional: Team Dynamics			
ID	**Indicators of Competence *(in addition to those at Lead Practitioner level)***	**Relevant knowledge, experience, and/or behaviors**	**Possible examples of objective evidence of personal contribution to activities performed, or professional behaviors applied**
1	Communicates own knowledge and experience in negotiation skills in order to improve best practice beyond the enterprise boundary. [PTDE01]	(A) Produces papers, seminars, or presentations outside own enterprise for publication in order to share own ideas and improve industry best practices in this competence area. [PTDE01-10A]	Papers or books published on new technique in refereed journals/company literature. [PTDE01-E10]
		(P) Fosters incorporation of own ideas into industry best practices in this area. [PTDE01-20P]	Papers published in refereed journals or internal literature proposing new practices in this competence area (or presentations, tutorials). [PTDE01-E20]
		(P) Develops guidance materials identifying new (or updating existing) best practice in this competence area. [PTDE01-30P]	Records of own proposals adopted as industry best practices in this competence area. [PTDE01-E30]
2	Advises organizations beyond the enterprise boundary on the suitability of their approach to team dynamics. [PTDE02]	(A) Advises external organizations on current or proposed strategies, plans, and tools used in support to team dynamics. [PTDE02-10A]	Records of advice requested and subsequently provided to external organizations. [PTDE02-E10]
		(A) Advises external organizations on strategies for improving team dynamics on complex or challenging teams. [PTDE02-20A]	Records of projects, agendas, organization charts, minutes, or reports demonstrating where they have advised on the dynamics within a team or a group. [PTDE02-E20]
3	Advises organizations beyond the enterprise boundary on the selection and interpretation of goals used to challenge, measure, and assess team performance. [PTDE03]	(A) Advises on selection and interpretation of goals used to challenge, measure, and assess team performance. [PTDE03-10A]	Records of projects, agendas, organization charts, minutes, or reports demonstrating where they have advised on goals to challenge or measure a team or a group. [PTDE03-E10]
4	Advises organizations beyond the enterprise boundary on how team members can be rewarded to act cooperatively. [PTDE04]	(A) Advises on strategies for rewarding team members to act cooperatively. [PTDE04-10A]	Records of projects, agendas, organization charts, minutes, or reports demonstrating where they have advised on reward strategies for a team or a group. [PTDE04-E10]
5	Challenges negative behaviors of beyond the enterprise boundary, with measurable success. [PTDE05]	(P) Challenges negative behaviors of beyond the enterprise boundary. [PTDE05-10P]	Records demonstrating a personal contribution to improving the behavior of one or more stakeholders beyond the enterprise boundary. [PTDE05-E10]
		(P) Acts to ensure positive outcomes when addressing negative stakeholder behaviors beyond the enterprise boundary. [PTDE05-20P]	Records of actions to ensure positive outcomes when addressing negative stakeholder behaviors beyond the enterprise boundary. [PTDE05-E20]

(Continued)

ID	Indicators of Competence *(in addition to those at Lead Practitioner level)*	Relevant knowledge, experience, and/or behaviors	Possible examples of objective evidence of personal contribution to activities performed, or professional behaviors applied
6	Influences key stakeholders beyond the enterprise boundary to follow a revised path to improve team dynamics across or beyond the enterprise. [PTDE06]	(P) Identifies viewpoints in stakeholders beyond the enterprise boundary, which require changing to improve wider enterprise team dynamics. [PTDE06-10P] (P) Persuades key stakeholders beyond the enterprise boundary to follow a new path resulting in positive improvements for the wider team. [PTDE06-20P]	Records demonstrating how advice provided influenced key stakeholders and improved a team dynamics issue (e.g. conflict) beyond the enterprise boundary. [PTDE06-E10]
7	Champions the introduction of novel techniques and ideas in team dynamics, beyond the enterprise boundary, in order to develop the wider Systems Engineering community in this competency. [PTDE07]	(A) Analyzes different approaches across different domains through research. [PTDE07-10A] (A) Produces reports for the wider SE community on the effectiveness of new techniques after their introduction. [PTDE07-20A] (P) Collaborates with those introducing novel techniques within the wider SE community. [PTDE07-30P] (A) Defines novel approaches that could potentially improve the wider SE discipline. [PTDE07-40A] (P) Fosters awareness of these novel techniques within the wider SE community. [PTDE07-50P]	Records of activities promoting research and need to adopt novel technique or ideas. [PTDE07-E10] Records of improvements made to process and appraisal against a recognized process improvement model. [PTDE07-E20] Research records. [PTDE07-E30] Published papers in refereed journals/company literature. [PTDE07-E40] Records showing introduction of enabling systems supporting the new techniques or ideas. [PTDE07-E50]
8	Coaches individuals beyond the enterprise boundary, in team dynamics in order to further develop their knowledge, abilities, skills, or associated behaviors. [PTDE08]	(P) Coaches or mentors individuals beyond the enterprise boundary, in competency-related techniques, recommending development activities. [PTDE08-10P] (A) Develops or authorizes training materials in this competency area, which are subsequently successfully delivered beyond the enterprise boundary. [PTDE08-20A] (A) Provides workshops/seminars or training in this competency area for practitioners or lead practitioners beyond the enterprise boundary (e.g. conferences and open training days). [PTDE08-30A]	Coaching or mentoring assignment records. [PTDE08-E10] Records of formal training courses, workshops, seminars, and authored training material supported by successful post-training evaluation data. [PTDE08-E20] Records of training/workshops/seminars created supported by successful post-training evaluation data. [PTDE08-E30]

9	Maintains expertise in this competency area through specialist Continual Professional Development (CPD) activities. [PTDE09]	(A) Reviews research, new ideas, and state of the art to identify relevant new areas requiring personal development in order to maintain expertise in this competency area. [PTDE09-10A] (A) Performs identified specialist professional development activities in order to maintain or further develop competence at expert level. [PTDE09-20A] (A) Records continual professional development activities undertaken including learning or insights gained. [PTDE09-30A]	Records of documents reviewed and insights gained as part of own research into this competency area. [PTDE09-E10] Records of Continual Professional Development (CPD) performed and learning outcomes. [PTDE09-E20]

NOTES	In addition to items above, enterprise-level or independent 3rd-Party-generated evidence may be used to amplify other evidence presented and may include: a. Formally recognized by a reputable external organization as an expert in this competency area b. Evidence of role as independent assessor or reviewer on project outside own organization where skills in this competency area were used c. Evidence of invitation(s) from wider community for contribution of systems engineering expertise in this area (e.g. industry conference panel, government advisory board etc. cross-industry working groups, partnerships, accredited advanced university courses or research, or as part of professional institute) d. Formal commendation beyond the enterprise (e.g. by INCOSE or other recognized authority) for work performed in this competency area e. Independently assessed or accredited work product in this competency area (e.g. for independent publication or use) f. Accolades of expertise in this area from recognized industry leaders

Competency area – Professional: Facilitation
Description The act of helping others to deal with a process, solve a problem, or reach a goal without getting directly involved. The goal is set by the individuals or groups, not by the facilitator.
Why it matters Modern systems engineers must perform successfully in environments where accountability expectations are increasing, but where the use of direct authority may not achieve the desired results. Numerous sources indicate that an alternative form of leadership can address these seemingly contradictory conditions. This form of leadership has been named "facilitative leadership," and is the ability to lead without controlling while making it easier for everyone in the organization to achieve agreed- upon goals.
Possible contributory types of evidence Any combination of the types of evidence may be acceptable (depending on how the Framework is tailored and used). The evidence items identified at each level indicate example work products only. Contributions to work products will generally differ at each proficiency level.
Learning and development The INCOSE Professional Development Portal provides example guidance on how to gain an initial awareness of a competency area and options for developing further competence thereafter.

Awareness – Professional: Facilitation		
ID	**Indicators of Competence**	**Relevant knowledge sub-indicators**
1	Explains the concept of "facilitation," summarizing its key characteristics and techniques. [PFAA01]	(K) Defines the role of the facilitator is to support individuals within a group in understanding a set of common objectives by moving collaboratively through a structured process. This can help individuals or the group as a whole. [PFAA01-10K] (K) Defines key differences between a trainer/training and a facilitator/facilitation. (e.g. Trainer is hierarchical relationship vs facilitator, which is more collaborative; Training generally a linear path, driven by a training lesson plan, whereas facilitation is more adaptable and flexible, driven by the group. Training aims generally for an immediate improvement; facilitation is generally a longer-term strategy; Training is about learning how to apply content rather than facilitation which is more about providing a thinking space for group and for communication of ideas across the group). [PFAA01-20K] (K) Defines the key characteristics of a good facilitator (e.g. catalyst, conductor and coach). [PFAA01-30K] (K) Defines essential facilitation skills (e.g. planning and organization, communication – especially verbal, adaptability, conflict resolution, creating an inclusive environment, understanding Group dynamics and group management, Empathy, Active listening, Consensus-building, time management, Gauging the energy level of a room, remaining neutral, and Recording outcomes). [PFAA01-40K]
2	Explains how facilitation can help individuals and groups to achieve their goals. [PFAA02]	(K) Explains why having a skilled facilitator can help through effectively managing a group or an individual journey through one or more facilitated sessions. [PFAA02-10K] (K) Describes key interfaces and links between engineering and stakeholders in the wider enterprise outside engineering (e.g. infrastructure management, human resource management, quality management, knowledge management, portfolio management, and life cycle model management), which can help in facilitation activities. [PFAA02-20K] (K) Defines the ideas of divergent and convergent thinking and how this can be used to facilitate group discussions. [PFAA02-30K]

		(K) Defines the techniques that might be used to ensure an inclusive environment is created in a group. [PFAA02-40K] (K) Describes common verbal tools and how a facilitator might use them to support facilitation (e.g. active listening, probing, paraphrasing, redirecting questions or comments, bridging or referring back, shifting perspective, summarizing, positive reinforcement, and ensuring quieter members are included). [PFAA02-50K] (K) Explains how a facilitator might control an individual or group session (e.g. setting context and ground rules, encouraging participation, staying neutral, maintaining track of time, maintaining a focused and participative environment, adapting the plan/process as necessary, and recording results). [PFAA02-60K]
3	Describes why the effectiveness of facilitation can differ during different stages of group formation. [PFAA03]	(K) Explains the basic concepts of group dynamics and group management and how group dynamics can impact facilitation (e.g. understanding the participants, reading emotions, empathy, and maintaining inclusivity). [PFAA03-10K] (K) Defines a common team development model (e.g. Tuckman: Forming, Storming, Norming, Performing, and adjourning) or Peck (Pseudo-community, Chaos, Emptiness, true Community) including how the different stages of a group might impact effectiveness of the group. [PFAA03-20K]
4	Describes how different facilitation skills can help resolve different forms of conflict and dissent in a group to mitigate their impact. [PFAA04]	(K) Explains key categories of workplace group conflict (e.g. intrapersonal, interpersonal, intragroup, and intergroup). [PFAA04-10K] (K) Lists potential sources of conflict at work (e.g. personality, working style, miscommunications or misunderstandings, resource availability, lack of individual or team support, poor customer service, poorly organized workplace, poor management, discrimination or harassment, and contract terms for individuals). [PFAA04-20K] (K) Defines differing potential ways to prevent conflict (e.g. job role change, workplace changes – change layout, training (e.g. equality, diversity, and conflict management), and putting a conflict resolution policy in place). [PFAA04-30K] (K) Defines techniques managers might use to address a situation (e.g. a conflict risk assessment, dealing with the problem – i.e. not ignoring it, implementing an open-door policy, promoting different opinions and working (or life) styles, being supportive, listening actively, addressing the problem, not the person). [PFAA04-40K] (K) Defines well-established mechanisms to resolve conflict (e.g. informal discussions, mediation, arbitration, conciliation, and promoting a culture of openness). [PFAA04-50K]
5	Describes how facilitation skills supplement different approaches to problem solving and the patterns of thinking associated with each. [PFAA05]	(K) Defines key characteristics of several different possible approaches to problem solving (e.g. Abstraction, analogy, Brainstorming, Divide and conquer, Hypothesis testing, Lateral thinking, Means-end analysis, morphologic, proof, reduction, research, root cause analysis, and trial-and-error). [PFAA05-10K] (K) Defines key characteristics of a number of distinct established problem-solving methods (e.g. Eight Disciplines (8D). GROW, OODA loop (observe, orient, decide, and act), PDCA (plan–do–check–act), Root cause analysis, RPR problem diagnosis (rapid problem resolution) TRIZ, A3 problem solving, System dynamics, Hive mind, Lean Six-Sigma). [PFAA05-20K] (K) Explains the impact of common thinking barriers to problem solving (e.g. confirmation bias, mental set, functional fixedness, unnecessary constraints, irrelevant information, framing effect, bandwagon effect, information bias, anchoring bias, availability bias, hindsight bias, recently effect, and self-serving bias.). [PFAA05-30K] (K) Explains how the agenda and process for a typical facilitated session to resolve a problem might be planned, organized, and communicated to participants. [PFAA05-40K]

Supervised Practitioner – Professional: Facilitation			
ID	**Indicators of Competence** ***(in addition to those at Awareness level)***	**Relevant knowledge, experience, and/or behaviors**	**Possible examples of objective evidence of personal contribution to activities performed, or professional behaviors applied**
1	Acts as a neutral servant of a group performing a facilitated task. [PFAS01]	(K) Explains why the role of a facilitator requires neutrality. [PFAS01-10K] (K) Explains why the role of a facilitator is in effect a "servant" of a facilitation group, rather than its "master." [PFAS01-20K] (P) Uses facilitation skills to act as a neutral servant of a group performing a facilitated task. [PFAS01-30P]	Records of discussions they have co-facilitated or supported. [PFAS01-E10]
2	Identifies members of a group in order to perform a facilitated task. [PFAS02]	(K) Explains why a facilitator needs to understand the dynamics of a group they are working with in order to be effective. [PFAS02-10K] (A) Coordinates a group in order to perform a facilitated task. [PFAS02-20A]	Records of a group they created to address a facilitated task. [PFAS02-E10]
3	Identifies rules of conduct between individuals within a facilitated group, with guidance. [PFAS03]	(K) Describes typical "ground rules" for a facilitation session. [PFAS03-10K] (A) Defines rules of conduct between individuals within a facilitated group, with guidance. [PFAS03-20A]	Ground rules from group discussions facilitated. [PFAS03-E10]
4	Acts as impartial observer focused on facilitated group activities, with guidance. [PFAS04]	(K) Explains why the role of a facilitator is an "impartial observe" of the facilitation group. [PFAS04-10K] (P) Uses facilitation skills to act as impartial observer focused on facilitated group activities, with guidance. [PFAS04-20P]	Records of a group discussion they co-facilitated or supported. [PFAS04-E10]
5	Acts in support of facilitation of a group problem-solving session. [PFAS05]	(K) Explains how problem solving can be supported by group facilitation techniques. [PFAS05-10K] (A) Performs activities facilitating a group problem-solving session. [PFAS05-20A]	Records of contributions to a group problem-solving session. [PFAS05-E10]
6	Conducts a small, facilitated group problem-solving session on their team. [PFAS06]	(K) Describes the role and key responsibilities of a facilitator during a group problem-solving session. [PFAS06-10K] (A) Conducts a small, facilitated group problem-solving session within a team. [PFAS06-20A]	Records of a group discussion they co-facilitated or supported. [PFAS06-E10]
7	Develops own understanding of this competency area through Continual Professional Development (CPD). [PFAS07]	(A) Identifies potential gaps in own knowledge or development needs in this area, identifying opportunities to address these through continual professional development activities. [PFAS07-10A] (A) Performs continual professional development activities to improve their knowledge and understanding in this area. [PFAS07-20A] (A) Records continual professional development activities undertaken including learning or insights gained. [PFAS07-30A]	Records of Continual Professional Development (CPD) performed and learning outcomes. [PFAS07-E10]

Practitioner – Professional: Facilitation			
ID	**Indicators of Competence *(in addition to those at Supervised Practitioner level)***	**Relevant knowledge, experience, and/or behaviors**	**Possible examples of objective evidence of personal contribution to activities performed, or professional behaviors applied**
1	Identifies rules of conduct between individuals within a facilitated group. [PFAP01]	(K) Explains why there may be potential challenges within a facilitation session due to personality differences within a group to be facilitated. [PFAP01-10K] (A) Creates rules of conduct between individuals within a facilitated group. [PFAP01-20A]	Records of ground rules used for a group discussion they have facilitated. [PFAP01-E10]
2	Guides a facilitated group problem-solving session. [PFAP02]	(K) Describes the role and key responsibilities of a facilitator during a group problem solving session. [PFAP02-10K] (P) Guides and actively coordinates a facilitated group problem-solving session. [PFAP02-20P]	Records of co-facilitating or supporting a group discussion. [PFAP02-E10]
3	Acts to ensure own views and feelings remain hidden, when facilitating group activities on a project. [PFAP03]	(P) Acts with a neutral stance during facilitation activity despite holding personal views supporting a particular viewpoint. [PFAP03-10P]	Records of facilitating a group. [PFAP03-E10]
4	Acts to facilitate self-improvement of the performance of a group. [PFAP04]	(P) Acts to facilitate self-improvement of the performance of a group as a whole. [PFAP04-10P]	Records of facilitating a group with measurable improvement outcome. [PFAP04-E10]
5	Acts to engage individuals to improve performance of a group. [PFAP05]	(A) Identifies underproductive individuals within a group. [PFAP05-10A] (P) Acts to engage individuals within a group in situations where they are uncommunicative or working without reference to or interaction with the rest of the group. [PFAP05-20P]	Records of facilitating a group with measurable improvement outcome. [PFAP05-E10] Records of engaging with individuals to improve their group engagement. [PFAP05-E20]
6	Acts to protect individuals and their ideas from attack within a facilitated group. [PFAP06]	(P) Acts to give individuals within a group the freedom to express a particular viewpoint, even if this creates conflict (e.g. if was believed to be incorrect or otherwise wrong). [PFAP06-10P]	Records from a group they personally facilitated. [PFAP06-E10]
7	Adapts strategy if a facilitated group requires a change of direction. [PFAP07]	(A) Adapts strategy in order to improve the effectiveness of the session as a whole. [PFAP07-10A]	Records from a group they personally facilitated. [PFAP07-E10]
8	Guides and actively coordinates a facilitated group problem-solving session on the project. [PFAP08]	(P) Guides and actively coordinates a facilitated group problem-solving session on the project. [PFAP08-10P]	Records from a group they personally facilitated. [PFAP08-E10]

(*Continued*)

ID	Indicators of Competence *(in addition to those at Supervised Practitioner level)*	Relevant knowledge, experience, and/or behaviors	Possible examples of objective evidence of personal contribution to activities performed, or professional behaviors applied
9	Guides new or supervised practitioners in facilitation techniques, in order to develop their knowledge, abilities, skills, or associated behaviors. [PFAP09]	(P) Guides new or supervised practitioners in executing activities that form part of this competency. [PFAP09-10P]	Organizational Breakdown Structure showing their responsibility for technical supervision in this area. [PFAP09-E10]
		(A) Trains individuals to an "Awareness" level in this competency area. [PFAP09-20A]	Records of on-the-job training objectives/guidance. [PFAP09-E20] Coaching or mentoring assignment records. [PFAP09-E30] Records highlighting their impact on another individual in terms of improvement or professional development in this competency. [PFAP09-E40]
10	Maintains and enhances own competence in this area through Continual Professional Development (CPD) activities. [PFAP10]	(A) Identifies potential development needs in this area, identifying opportunities to address these through continual professional development activities. [PFAP10-10A] (A) Performs continual professional development activities to maintain and enhance their competency in this area. [PFAP10-20A] (A) Records continual professional development activities undertaken including learning or insights gained. [PFAP10-30A]	Records of Continual Professional Development (CPD) performed and learning outcomes. [PFAP10-E10]

Lead Practitioner – Professional: Facilitation			
ID	**Indicators of Competence** ***(in addition to those at Practitioner level)***	**Relevant knowledge, experience, and/or behaviors**	**Possible examples of objective evidence of personal contribution to activities performed, or professional behaviors applied**
1	Uses best practice facilitation techniques to improve the effectiveness of Systems Engineering activities across the enterprise. [PFAL01]	(A) Develops own state-of-the-art facilitation skills in order to improve understanding of where their application would be most effective. [PFAL01-10A]	Records demonstrating personal use of various facilitation techniques and associated feedback indicating effectiveness. [PFAL01-E10]
		(P) Uses best practice facilitation techniques to improve interactions with stakeholders or stakeholder groups across the enterprise. [PFAL01-20P]	Records demonstrating use of facilitation successfully to a variety of diverse stakeholders across the enterprise. [PFAL01-E20]
			Records (e.g. from Human Resources department) indicating excellent facilitation skills. [PFAL01-E30]
2	Creates a plan for a facilitated enterprise-level activity, defining methods to be used, coordinating the logistics of the meeting arrangements. [PFAL02]	(A) Creates a plan documenting the approach and methods for a series of facilitation workshops or a complex workshop with multiple key stakeholders. [PFAL02-10A]	Records of planning in support of facilitation workshops. [PFAL02-E10]
		(A) Defines workshop logistics and ensures facilities are in place for an enterprise-level workshop with multiple key stakeholders. [PFAL02-20A]	Records of logistics required in support of facilitation workshops. [PFAL02-E20]
3	Selects the most appropriate style of facilitation based upon enterprise-level facilitated group maturity. [PFAL03]	(K) Explains the differing approaches available for facilitation of different types of facilitation, and the rationale for selecting a particular approach. [PFAL03-10K]	Records of differing approaches used at different facilitation workshops, providing appropriate justification for selecting the approaches used. [PFAL03-E10]
		(A) Selects the most appropriate style of facilitation based upon group maturity. [PFAL03-20A]	
4	Uses facilitation skills to ensure that an enterprise-level facilitated group clarifies its goals. [PFAL04]	(K) Explains how facilitation workshop goals often require clarification and possible approaches that can be used to clarify these goals. [PFAL04-10K]	Records where facilitation workshop goals were clarified, including before and afterward. [PFAL04-E10]
		(P) Uses facilitation skills to ensure that the group clarifies its goals. [PFAL04-20P]	
5	Uses facilitation skills to facilitate an enterprise-level facilitated group toward achieving its objectives. [PFAL05]	(K) Explains how facilitation can be used to achieve goals and possible approaches that can be used to facilitate a group. [PFAL05-10K]	Records where facilitation workshop objectives were achieved. [PFAL05-E10]
		(P) Uses facilitation skills to facilitate an enterprise-level group toward achieving its objectives. [PFAL05-20P]	

(Continued)

ID	Indicators of Competence *(in addition to those at Practitioner level)*	Relevant knowledge, experience, and/or behaviors	Possible examples of objective evidence of personal contribution to activities performed, or professional behaviors applied
6	Acts as a referee in times of conflict, disagreement, or tension within an enterprise-level facilitated group. [PFAL06]	(K) Explains why facilitation workshops may produce conflict or disagreement between participants (e.g. possible causes of disagreement or conflict) and possible approaches that can be used to reduce, remove, or prevent conflict from occurring. [PFAL06-10K]	Records where facilitation workshop conflict occurred and successfully resolved. [PFAL06-E10]
		(K) Explains why facilitation workshops at enterprise level may surface different conflicts or disagreements between participants compared with a group on a project team. [PFAL06-20K]	
		(P) Acts as a referee in times of conflict, disagreement, or tension within an enterprise-level facilitated group. [PFAL06-30P]	
7	Fosters systematic patterns of thinking during the facilitated enterprise-level group problem-solving process. [PFAL07]	(K) Explains how systematic thinking patterns can help support facilitated group problem solving. [PFAL07-10K]	Records where group systems thinking patterns were changed during a facilitation workshop with successful results. [PFAL07-E10]
		(P) Fosters systematic patterns of thinking during the facilitated group problem solving process. [PFAL07-20P]	
8	Coaches or mentors practitioners across the enterprise in facilitation techniques in order to develop their knowledge, abilities, skills, or associated behaviors. [PFAL08]	(P) Coaches or mentors practitioners across the enterprise in competency-related techniques, recommending development activities. [PFAL08-10P]	Coaching or mentoring assignment records. [PFAL08-E10]
		(A) Develops or authorizes enterprise training materials in this competency area. [PFAL08-20A]	Records of formal training courses, workshops, seminars, and authored training material supported by successful post-training evaluation data. [PFAL08-E20]
		(A) Provides enterprise workshops/seminars or training in this competency area. [PFAL08-30A]	Listing as an approved organizational trainer for this competency area. [PFAL08-E30]
9	Promotes the introduction and use of novel techniques and ideas in facilitation across the enterprise, to improve enterprise competence in this area. [PFAL09]	(A) Analyzes different approaches across different domains through research. [PFAL09-10A]	Research records. [PFAL09-E10]
		(A) Defines novel approaches that could potentially improve the SE discipline within the enterprise. [PFAL09-20A]	Papers published in refereed journals/company literature. [PFAL09-E20]
		(P) Fosters awareness of these novel techniques within the enterprise. [PFAL09-30P]	Records showing introduction of enabling systems supporting the new techniques or ideas. [PFAL09-E30]
		(P) Collaborates with enterprise stakeholders to facilitate the introduction of techniques new to the enterprise. [PFAL09-40P]	Papers (or similar) published at enterprise level. [PFAL09-E40]

		(A) Monitors new techniques after their introduction to determine their effectiveness. [PFAL09-50A]	Records of improvements made against a recognized process improvement model. [PFAL09-E50]
		(A) Adapts approach to reflect actual enterprise performance improvements. [PFAL09-60A]	Records of adapting approach to reflect actual enterprise performance improvements. [PFAL09-E60]
10	Develops expertise in this competency area through specialist Continual Professional Development (CPD) activities. [PFAL10]	(A) Identifies own needs for further professional development in order to increase competence beyond practitioner level. [PFAL10-10A]	Records of Continual Professional Development (CPD) performed and learning outcomes. [PFAL10-E10]
		(A) Performs professional development activities in order to move own competence toward expert level. [PFAL10-20A]	
		(A) Records continual professional development activities undertaken including learning or insights gained. [PFAL10-30A]	

NOTES In addition to items above, enterprise-level or independent 3rd-Party-generated evidence may be used to amplify other evidence presented and may include:

a. Formally recognized by senior management in current organization as an expert in this competency area
b. Evidence of role as Product/System Design Authority or Technical Authority on a complex project with responsibilities in this area or where skills within this competency area were used
c. Recognized as an authorizing signatory on behalf of enterprise for formal documentation in this competency area (e.g. policies, processes, and deliverables)
d. Formal commendation or award within own enterprise for contribution or item of work successfully performed, which required proficiency in this competency area
e. Customer, Supplier, or other external project-specific key Stakeholder accolades for specific work performed in this competency area
f. Independently assessed or accredited work in this competency area (e.g. for independent publication or use)
g. Formal organizational HR records positively highlighting any specific professional competencies or behaviors identified (if applicable) plus any of the evidence indicators listed at Expert level below

Expert – Professional: Facilitation			
ID	**Indicators of Competence *(in addition to those at Lead Practitioner level)***	**Relevant knowledge, experience, and/or behaviors**	**Possible examples of objective evidence of personal contribution to activities performed, or professional behaviors applied**
1	Communicates own knowledge and experience in facilitation skills in order to improve best practice beyond the enterprise boundary. [PFAE01]	(A) Produces papers, seminars, or presentations outside own enterprise for publication in order to share own ideas and improve industry best practices in this competence area. [PFAE01-10A]	Papers or books published on new technique in refereed journals/company literature. [PFAE01-E10]
		(P) Fosters incorporation of own ideas into industry best practices in this area. [PFAE01-20P]	Papers published in refereed journals or internal literature proposing new practices in this competence area (or presentations, tutorials, etc.). [PFAE01-E20]
		(P) Develops guidance materials identifying new (or updating existing) best practice in this competence area. [PFAE01-30P]	Records of own proposals adopted as industry best practices in this competence area. [PFAE01-E30]
2	Advises organizations beyond the enterprise boundary on the suitability of their approach to facilitation. [PFAE02]	(A) Advises key stakeholders beyond the enterprise facilitation on plans, methods, and logistics to be used for facilitated sessions. [PFAE02-10A]	Records of projects, agendas, organization charts, minutes, or reports demonstrating where they have advised key stakeholders both within and beyond the enterprise boundary when facilitation a group or a problem. [PFAE02-E10]
3	Persuades key stakeholders beyond the enterprise boundary to support facilitated group activities. [PFAE03]	(P) Persuades stakeholders beyond the enterprise boundary to support a facilitated group session. [PFAE03-10P]	Records, agendas, organization charts, minutes, or reports demonstrating influencing the direction of relevant legislative and regulatory frameworks in support of facilitation within Systems Engineering. [PFAE03-E10]
		(P) Acts to change external stakeholder thinking regarding facilitated activities affecting Systems Engineering. [PFAE03-20P]	Records of ideas assimilated into international standards or referenced in others' books, papers, and peer-reviewed articles. [PFAE03-E20]
4	Reviews the suitability of facilitation programs affecting Systems Engineering beyond the enterprise boundary. [PFAE04]	(A) Reviews the suitability of facilitation programs affecting Systems Engineering beyond the enterprise. [PFAE04-10A]	Review records demonstrating a personal contribution to the assessment and improvement of facilitation programs. [PFAE04-E10]
5	Identifies alternative ways of working to reinforce collaboration within the context of a facilitated group with membership beyond the enterprise boundary. [PFAE05]	(A) Identifies alternative ways of working beyond the enterprise boundary to reinforce collaboration within the context of the facilitated group. [PFAE05-10A]	Review records demonstrating a personal contribution to the assessment and improvement of facilitation programs. [PFAE05-E10]
6	Advises organizations beyond the enterprise boundary on complex or sensitive matters, conflict disagreement, or tension affecting facilitated group. [PFAE06]	(A) Advises on complex or sensitive matters, conflict disagreement, or tension affecting facilitated group, arbitrating as necessary. [PFAE06-10A]	Records from enterprise facilitation records. [PFAE06-E10]

		(P) Arbitrates successfully on complex or sensitive matter within a facilitation sessions. [PFAE06-20P]	Records of successfully arbitrating on complex or sensitive matter within a facilitation sessions. [PFAE06-E20]
7	Acts to anticipate and mitigate potential problems in facilitation of a group extending beyond the enterprise boundary. [PFAE07]	(A) Identifies possible challenges within a planned facilitation session through advance planning and communications with stakeholders. [PFAE07-10A]	Records demonstrating a personal contribution to the assessment and improvement of facilitation programs. [PFAE07-E10]
		(P) Acts to address predicted facilitation challenges, leading to a successful facilitated session. [PFAE07-20P]	Records address predicted facilitation challenges, leading to a successful facilitated session. [PFAE07-E20]
8	Fosters open communications, which surface prevailing mental models and challenge a facilitation group extending beyond the enterprise boundary to build a shared vision. [PFAE08]	(A) Analyzes mental models of other facilitation group members to surface their mental models and help establish a shared vision. [PFAE08-10A]	Records where facilitation workshop mental models were shared, resulting in a successful workshop outcome and a shared vision. [PFAE08-E10]
		(P) Fosters open communications, which surface prevailing mental models and challenge a facilitation group to build a shared vision. [PFAE08-20P]	Records of fostering open communications, which surfaced prevailing mental models and challenged a facilitation group to build a shared vision. [PFAE08-E20]
9	Champions the introduction of novel techniques and ideas in facilitation, beyond the enterprise boundary, in order to develop the wider Systems Engineering community in this competency. [PFAE09]	(A) Analyzes different approaches across different domains through research. [PFAE09-10A]	Records of activities promoting research and need to adopt novel technique or ideas. [PFAE09-E10]
		(A) Produces reports for the wider SE community on the effectiveness of new techniques after their introduction. [PFAE09-20A]	Records of improvements made to process and appraisal against a recognized process improvement model. [PFAE09-E20]
		(P) Collaborates with those introducing novel techniques within the wider SE community. [PFAE09-30P]	Research records. [PFAE09-E30]
		(A) Defines novel approaches that could potentially improve the wider SE discipline. [PFAE09-40A]	Published papers in refereed journals/company literature. [PFAE09-E40]
		(P) Fosters awareness of these novel techniques within the wider SE community. [PFAE09-50P]	Records showing introduction of enabling systems supporting the new techniques or ideas. [PFAE09-E50]
10	Coaches individuals beyond the enterprise boundary, in facilitation techniques in order to further develop their knowledge, abilities, skills, or associated behaviors. [PFAE10]	(P) Coaches or mentors individuals beyond the enterprise boundary, in competency-related techniques, recommending development activities. [PFAE10-10P]	Coaching or mentoring assignment records. [PFAE10-E10]
		(A) Develops or authorizes training materials in this competency area, which are subsequently successfully delivered beyond the enterprise boundary. [PFAE10-20A]	Records of formal training courses, workshops, seminars, and authored training material supported by successful post-training evaluation data. [PFAE10-E20]

(Continued)

ID	Indicators of Competence *(in addition to those at Lead Practitioner level)*	Relevant knowledge, experience, and/or behaviors	Possible examples of objective evidence of personal contribution to activities performed, or professional behaviors applied
		(A) Provides workshops/seminars or training in this competency area for practitioners or lead practitioners beyond the enterprise boundary (e.g. conferences and open training days). [PFAE10-30A]	Records of Training/workshops/seminars created supported by successful post-training evaluation data. [PFAE10-E30]
11	Maintains expertise in this competency area through specialist Continual Professional Development (CPD) activities. [PFAE11]	(A) Reviews research, new ideas, and state of the art to identify relevant new areas requiring personal development in order to maintain expertise in this competency area. [PFAE11-10A]	Records of documents reviewed and insights gained as part of own research into this competency area. [PFAE11-E10]
		(A) Performs identified specialist professional development activities in order to maintain or further develop competence at expert level. [PFAE11-20A]	Records of Continual Professional Development (CPD) performed and learning outcomes. [PFAE11-E20]
		(A) Records continual professional development activities undertaken including learning or insights gained. [PFAE11-30A]	

NOTES

In addition to items above, enterprise-level or independent 3rd-Party-generated evidence may be used to amplify other evidence presented and may include:

a. Formally recognized by a reputable external organization as an expert in this competency area
b. Evidence of role as independent assessor or reviewer on project outside own organization where skills in this competency area were used
c. Evidence of invitation(s) from wider community for contribution of systems engineering expertise in this area (e.g. industry conference panel, government advisory board etc. cross-industry working groups, partnerships, accredited advanced university courses or research, or as part of professional institute)
d. Formal commendation beyond the enterprise (e.g. by INCOSE or other recognized authority) for work performed in this competency area
e. Independently assessed or accredited work product in this competency area (e.g. for independent publication or use)
f. Accolades of expertise in this area from recognized industry leaders

Competency area – Professional: Emotional Intelligence
Description Emotional intelligence is the ability to monitor one's own and others' feelings, to discriminate among them, and to use this information to guide thinking and action. This is usually broken down into four distinct but related proposed abilities: perceiving, using, understanding, and managing emotions.
Why it matters Emotional intelligence is regularly cited as a critical competency for effective leadership and team performance in organizations. It influences the success with which individuals in organizations interact with colleagues, the approaches they use to manage conflict and stress, and their overall job performance. As Systems Engineering involves interacting with many diverse stakeholders, emotional intelligence is critical to its success.
Possible contributory types of evidence Any combination of the types of evidence may be acceptable (depending on how the Framework is tailored and used). The evidence items identified at each level indicate example work products only. Contributions to work products will generally differ at each proficiency level.
Learning and development The INCOSE Professional Development Portal provides example guidance on how to gain an initial awareness of a competency area and options for developing further competence thereafter.

Awareness – Professional: Emotional Intelligence		
ID	**Indicators of Competence**	**Relevant knowledge sub-indicators**
1	Explains why the perception of emotion is important including differentiating one's own emotions from those of others. [PEIA01]	(K) Describes what emotions are and how to differentiate between own emotions and others' emotions. [PEIA01-10K] (K) Explains how to perceive the emotions of others (e.g. tone and volume of voice, facial expressions, body language, and eye contact). [PEIA01-20K] (K) Describes indicators of own emotions (e.g. heart racing or "jitters" indicates anxiety, a clenched jaw, and tenseness indicate anger). [PEIA01-30K]
2	Explains how emotions can be used to facilitate thinking such as reasoning, problem solving, and interpersonal communication and explains why this is important. [PEIA02]	(K) Explains how emotions affect people and the importance of recognizing the emotions of others. [PEIA02-10K] (K) Explains how emotions affect thinking processes (e.g. anger or anxiety may result in hasty decision-making). [PEIA02-20K]
3	Explains why it is important to be able to understand and analyze emotions. [PEIA03]	(K) Describes someone displaying emotion (e.g. fear or distrust) and how that might make them react (e.g. less willing to negotiate). [PEIA03-10K] (K) Describes the language and meaning of emotions and relationships between emotions. [PEIA03-20K]
4	Explains why managing and regulating emotions in both oneself and in others is important. [PEIA04]	(K) Explains how emotions impact others (e.g. expressing happiness makes others want to stay engaged in the conversation, anger could make others defensive and unwilling to continue communication). [PEIA04-10K]

Supervised Practitioner – Professional: Emotional Intelligence			
ID	**Indicators of Competence** ***(in addition to those at Awareness level)***	**Relevant knowledge, experience, and/or behaviors**	**Possible examples of objective evidence of personal contribution to activities performed, or professional behaviors applied**
1	Identifies emotions in one's physical states, feelings, and thoughts. [PEIS01]	(K) Explains "triggers" that may drive a change in an emotional state. [PEIS01-10K] (A) Identifies particular personal "triggers" that they are sensitive to and acts accordingly to control emotions. [PEIS01-20A] (A) Uses emotional intelligence for self-awareness, identifying what they are capable of and when they require help from others. [PEIS01-30A]	Records of identifying particular personal "triggers" that they are sensitive to and acts accordingly to control emotions. [PEIS01-E10] Records of using emotional intelligence for self-awareness, identifying what they are capable of and when they require help from others. [PEIS01-E20]
2	Uses emotional intelligence techniques to identify the emotions of others via verbal and nonverbal cues. [PEIS02]	(A) Identifies non-verbal cues of another person's emotional state. [PEIS02-10A]	Records of identifying nonverbal cues of another person's emotional state. [PEIS02-E10]
3	Explains the language used to label emotions. [PEIS03]	(K) Defines examples of emotions or emotional models (e.g. Robert Plutchik: joy, trust, fear, surprise, sadness, disgust, anger, anticipation, or the 22 emotions identified by Abraham Hicks). [PEIS03-10K] (A) Identifies examples of emotional states they have experienced in different project settings. [PEIS03-20A]	Personal records of states recorded. [PEIS03-E10]
4	Develops own understanding of this competency area through Continual Professional Development (CPD). [PEIS04]	(A) Identifies potential gaps in own knowledge or development needs in this area, identifying opportunities to address these through continual professional development activities. [PEIS04-10A] (A) Performs continual professional development activities to improve their knowledge and understanding in this area. [PEIS04-20A] (A) Records continual professional development activities undertaken including learning or insights gained. [PEIS04-30A]	Records of Continual Professional Development (CPD) performed and learning outcomes. [PEIS04-E10]

Practitioner – Professional: Emotional Intelligence

ID	Indicators of Competence *(in addition to those at Supervised Practitioner level)*	Relevant knowledge, experience, and/or behaviors	Possible examples of objective evidence of personal contribution to activities performed, or professional behaviors applied
1	Uses Emotional Intelligence techniques to interpret meanings and origins of emotions and acts accordingly. [PEIP01]	(K) Explains how that emotions originate in the limbic system and happen before the area of reason and logic in the brain. [PEIP01-10K]	Records of using Emotional Intelligence techniques to interpret meanings and origins of emotions and acts accordingly. [PEIP01-E10]
		(A) Uses Emotional Intelligence techniques to interpret meanings and origins of emotions and acts accordingly. [PEIP01-20A]	
2	Uses Emotional Intelligence techniques to identify needs related to emotional feelings. [PEIP02]	(A) Analyzes own emotional state and responds accordingly. [PEIP02-10A]	Records of intentional management of emotional reaction (e.g. pausing a discussion to be resumed later when emotions are calmer, acknowledging perceived disappointment, and addressing corrective actions). [PEIP02-E10]
		(A) Identifies their own personal values and how they might leverage these for positive result in stressful situations. [PEIP02-20A]	Records of identifying their own personal values and how they might leverage these for positive result in stressful situations. [PEIP02-E20]
3	Uses emotional intelligence techniques to monitor their own emotions in relation to others. [PEIP03]	(A) Identifies own emotions, and how they could disrupt work-related situations. [PEIP03-10A]	Records of journal entries identifying own emotions. [PEIP03-E10]
		(P) Uses emotional intelligence techniques to monitor their own monitor emotions in relation to others. [PEIP03-20P]	Records of formal requests of reflections and feedback from others. [PEIP03-E20]
		(P) Acts empathically by responding appropriately to the emotions of others. [PEIP03-30P]	Records of acting empathically by responding appropriately to the emotions of others. [PEIP03-E30]
4	Acts to capitalize fully upon changing moods in order to best fit the task at hand. [PEIP04]	(P) Acts to capitalize fully upon changing moods in order to best fit the task at hand. [PEIP04-10P]	Records of intentional management of emotional reaction (e.g. pausing a discussion to be resumed later when emotions are calmer, acknowledging perceived disappointment, and addressing corrective actions). [PEIP04-E10]
		(P) Acts to control own emotions such as maintains calm during disagreements, controls overreactions, avoids undermining behavior such as self-pity or panic. [PEIP04-20P]	Records of actions to control own emotions such as maintains calm during disagreements, controls overreactions, avoids undermining behavior such as self-pity or panic. [PEIP04-E20]
		(P) Acts to manage conflict occurring due to anger or anxiety. [PEIP04-30P]	Records of actions to manage conflict occurring due to anger or anxiety. [PEIP04-E30]
		(A) Identifies common ground in the face of strong emotions. [PEIP04-40A]	Records of identifying common ground in the face of strong emotions. [PEIP04-E40]
		(P) Uses negotiating skills to persuade others in the face of strong emotions. [PEIP04-50P]	Records of using negotiating skills to persuade others in the face of strong emotions. [PEIP04-E50]

(Continued)

ID	Indicators of Competence *(in addition to those at Supervised Practitioner level)*	Relevant knowledge, experience, and/or behaviors	Possible examples of objective evidence of personal contribution to activities performed, or professional behaviors applied
5	Acts to remain open to feelings, both those that are pleasant and those that are unpleasant. [PEIP05]	(K) Explains how breathing, physical activity, physical posture, and meditation can help manage emotions. [PEIP05-10K] (A) Identifies responses to own emotions and their potential impact on a situation. [PEIP05-20A]	Records of intentional management of emotional reaction (e.g. pausing a discussion to be resumed later when emotions are calmer, acknowledging perceived disappointment, and addressing corrective actions). [PEIP05-E10]
6	Acts to control own emotion by preventing, reducing, enhancing, or modifying an emotional response. [PEIP06]	(K) Explains why self-management is a key component of emotional intelligence, with examples. [PEIP06-10K] (P) Acts to control own emotion by preventing, reducing, enhancing, or modifying an emotional response. [PEIP06-20P]	Records of a situation where they identified a potentially disruptive own emotion and were able to manage it. [PEIP06-E10] Records from HR highlighting control or good management of own emotions. [PEIP06-E20]
7	Guides new or supervised practitioners in emotional intelligence techniques, in order to develop their knowledge, abilities, skills, or associated behaviors. [PEIP07]	(P) Guides new or supervised practitioners in executing activities that form part of this competency. [PEIP07-10P] (A) Trains individuals to an "Awareness" level in this competency area. [PEIP07-20A]	Organizational Breakdown Structure showing their responsibility for technical supervision in this area. [PEIP07-E10] Records of on-the-job training objectives/guidance etc. [PEIP07-E20] Coaching or mentoring assignment records. [PEIP07-E30] Records highlighting their impact on another individual in terms of improvement or professional development in this competency. [PEIP07-E40]
8	Maintains and enhances own competence in this area through Continual Professional Development (CPD) activities. [PEIP08]	(A) Identifies potential development needs in this area, identifying opportunities to address these through continual professional development activities. [PEIP08-10A] (A) Performs continual professional development activities to maintain and enhance their competency in this area. [PEIP08-20A] (A) Records continual professional development activities undertaken including learning or insights gained. [PEIP08-30A]	Records of Continual Professional Development (CPD) performed and learning outcomes. [PEIP08-E10]

Lead Practitioner – Professional: Emotional Intelligence			
ID	**Indicators of Competence** ***(in addition to those at Practitioner level)***	**Relevant knowledge, experience, and/or behaviors**	**Possible examples of objective evidence of personal contribution to activities performed, or professional behaviors applied**
1	Uses best practice emotional intelligence techniques to improve the effectiveness of Systems Engineering activities across the enterprise. [PEIL01]	(A) Develops own state-of-the-art emotional intelligence skills in order to improve understanding of where their application would be most effective. [PEIL01-10A] (P) Uses best practice emotional intelligence techniques to improve interactions with stakeholders or stakeholder groups across the enterprise. [PEIL01-20P]	Records demonstrating personal use of various emotional intelligence techniques and associated feedback indicating effectiveness. [PEIL01-E10] Records demonstrating use of emotional intelligence successfully to a variety of diverse stakeholders across the enterprise. [PEIL01-E20] Records (e.g. from Human Resources department) indicating excellent emotional intelligence skills. [PEIL01-E30]
2	Guides others across the enterprise in controlling their own emotional responses. [PEIL02]	(A) Collaborates with others to identify potential situations where emotional responses may be challenged. [PEIL02-10A] (A) Collaborates with others to highlight situations where previous emotional responses need improvement. [PEIL02-20A] (P) Acts with empathy toward others when discussing others' emotional responses. [PEIL02-30P] (P) Fosters a climate of where emotional responses are improved over time. [PEIL02-40P]	Records of a situation where they identified a potentially disruptive emotion in others and were able to manage it. [PEIL02-E10] Records from HR highlighting empathy to others, particularly in difficult or stressful situations. [PEIL02-E20] Records of an experience of fully appreciating a perspective very different from their own and how that changed their response. [PEIL02-E30]
3	Uses emotional intelligence techniques in tough, challenging situations with both external and internal stakeholders, with demonstrable results. [PEIL03]	(K) Explains why social competence is made up of social awareness and relationship management. [PEIL03-10K] (P) Adapts behavior to take account of social and relationship management factors. [PEIL03-20P] (P) Acts to encourage time/space to manage emotions, then proposes solutions based upon mutual benefit. [PEIL03-30P] (A) Identifies own emotion and its potential impact on the situation and determines whether to leverage the emotion or detach from it. [PEIL03-40A] (P) Uses emotional intelligence techniques reflectively, detached from an emotion, depending on its perceived utility in complex, challenging situations. [PEIL03-50P]	Records of a situation where they identified potentially disruptive emotions in self or others and were able to manage it. [PEIL03-E10] Records of proposition of solutions appropriate social skills to manage challenging situations. [PEIL03-E20] Records showing a situation where they identified a potentially disruptive own emotion and chose to detach from it, or identified a positive emotion and were able to capitalize on it. [PEIL03-E30]

(Continued)

ID	Indicators of Competence *(in addition to those at Practitioner level)*	Relevant knowledge, experience, and/or behaviors	Possible examples of objective evidence of personal contribution to activities performed, or professional behaviors applied
4	Uses emotional intelligence to influence key stakeholders within the enterprise. [PEIL04]	(A) Identifies strategies for managing expected emotional reactions to information from key stakeholders. [PEIL04-10A]	Records of a situation where an expected emotional reaction was managed to positive result (e.g. happiness after a successful event was parlayed into additional funding support for follow-on work). [PEIL04-E10]
		(P) Identifies cultural differences in emotional expression and responds accordingly. [PEIL04-20P]	Records of an instance of recognizing culturally different emotional reactions and responding appropriately. [PEIL04-E20]
		(P) Acts to leverage "mood contagion" where a leader's emotional style drives others' emotions and behaviors. [PEIL04-30P]	Records of actions to leverage "mood contagion" where a leader's emotional style drives others' emotions and behaviors. [PEIL04-E30]
		(P) Acts to manage emotion in teams, bringing them deliberately to the surface, understanding their impact on the team and developing a path to successful team dynamics. [PEIL04-40P]	Records of guiding the management of emotion in teams, bringing emotions deliberately to the surface, understanding their impact on the team and developing a path to successful team dynamics. [PEIL04-E40]
5	Coaches or mentors practitioners across the enterprise in emotional intelligence techniques in order to develop their knowledge, abilities, skills, or associated behaviors. [PEIL05]	(P) Coaches or mentors practitioners across the enterprise in competency-related techniques, recommending development activities. [PEIL05-10P]	Coaching or mentoring assignment records. [PEIL05-E10]
		(A) Develops or authorizes enterprise training materials in this competency area. [PEIL05-20A]	Records of formal training courses, workshops, seminars, and authored training material supported by successful post-training evaluation data. [PEIL05-E20]
		(A) Provides enterprise workshops/seminars or training in this competency area. [PEIL05-30A]	Listing as an approved organizational trainer for this competency area. [PEIL05-E30]
6	Promotes the introduction and use of novel techniques and ideas in Emotional Intelligence techniques across the enterprise, to improve enterprise competence in this area. [PEIL06]	(A) Analyzes different approaches across different domains through research. [PEIL06-10A]	Research records. [PEIL06-E10]
		(A) Defines novel approaches that could potentially improve the SE discipline within the enterprise. [PEIL06-20A]	Published papers in refereed journals/company literature. [PEIL06-E20]
		(P) Fosters awareness of these novel techniques within the enterprise. [PEIL06-30P]	Records showing introduction of enabling systems supporting the new techniques or ideas. [PEIL06-E30]
		(P) Collaborates with enterprise stakeholders to facilitate the introduction of techniques new to the enterprise. [PEIL06-40P]	Published papers (or similar) at enterprise level. [PEIL06-E40]

		(A) Monitors new techniques after their introduction to determine their effectiveness. [PEIL06-50A] (A) Adapts approach to reflect actual enterprise performance improvements. [PEIL06-60A]	Records of improvements made against a recognized process improvement model. [PEIL06-E50] Records of adapting approach to reflect actual enterprise performance improvements. [PEIL06-E60]
7	Develops expertise in this competency area through specialist Continual Professional Development (CPD) activities. [PEIL07]	(A) Identifies own needs for further professional development in order to increase competence beyond practitioner level. [PEIL07-10A] (A) Performs professional development activities in order to move own competence toward expert level. [PEIL07-20A] (A) Records continual professional development activities undertaken including learning or insights gained. [PEIL07-30A]	Records of Continual Professional Development (CPD) performed and learning outcomes. [PEIL07-E10]

NOTES	In addition to items above, enterprise-level or independent 3rd-Party-generated evidence may be used to amplify other evidence presented and may include: a. Formally recognized by senior management in current organization as an expert in this competency area b. Evidence of role as Product/System Design Authority or Technical Authority on a complex project with responsibilities in this area or where skills within this competency area were used c. Recognized as an authorizing signatory on behalf of enterprise for formal documentation in this competency area (e.g. policies, processes, and deliverables) d. Formal commendation or award within own enterprise for contribution or item of work successfully performed, which required proficiency in this competency area e. Customer, Supplier, or other external project-specific key Stakeholder accolades for specific work performed in this competency area f. Independently assessed or accredited work in this competency area (e.g. for independent publication or use) g. Formal organizational HR records positively highlighting any specific professional competencies or behaviors identified (if applicable) plus any of the evidence indicators listed at Expert level below

Expert – Professional: Emotional Intelligence			
ID	**Indicators of Competence** ***(in addition to those at Lead Practitioner level)***	**Relevant knowledge, experience, and/or behaviors**	**Possible examples of objective evidence of personal contribution to activities performed, or professional behaviors applied**
1	Communicates own knowledge and experience in emotional intelligence in order to improve best practice beyond the enterprise boundary. [PEIE01]	(A) Produces papers, seminars, or presentations outside own enterprise for publication in order to share own ideas and improve industry best practices in this competence area. [PEIE01-10A]	Published papers or books etc. on new technique in refereed journals/company literature. [PEIE01-E10]
		(P) Fosters incorporation of own ideas into industry best practices in this area. [PEIE01-20P]	Published papers in refereed journals or internal literature proposing new practices in this competence area (or presentations, tutorials, etc.). [PEIE01-E20]
		(P) Develops guidance materials identifying new (or updating existing) best practice in this competence area. [PEIE01-30P]	Records of own proposals adopted as industry best practices in this competence area. [PEIE01-E30]
2	Uses emotional intelligence to influence beyond the enterprise boundary. [PEIE02]	(P) Persuades stakeholders beyond the enterprise boundary to adopt a different path, through the application of emotional intelligence. [PEIE02-10P]	Records of ideas assimilated into international standards or referenced in others' books, papers, and peer-reviewed articles. [PEIE02-E10]
		(A) Identifies changes to existing stakeholder leveraging emotional intelligence. [PEIE02-20A]	Records demonstrating a personal contribution to the application of emotional intelligence. [PEIE02-E20]
3	Advises organizations beyond the enterprise boundary on the suitability of their approach to emotional intelligence awareness and its utilization. [PEIE03]	(A) Reviews strategies of organizations beyond the enterprise for the usage of emotional intelligence within systems engineering activities. [PEIE03-10A]	Records of reviews of external organization emotional intelligence programs. [PEIE03-E10]
		(A) Advises beyond the enterprise boundary on differing strategies for improving the usage of emotional intelligence techniques within systems engineering activities. [PEIE03-20A]	Records of advice provided in this area. [PEIE03-E20]
4	Advises beyond the enterprise boundary on complex or sensitive emotionally charged issues. [PEIE04]	(P) Advises external organizations in the strategies to be adopted to deal with emotionally charged situations, leading to positive outcomes. [PEIE04-10P]	Records of employing emotional intelligence when dealing with difficult situation(s) in external organizations, with a positive outcome. [PEIE04-E10]
		(P) Advises individuals beyond the enterprise boundary in the strategies to be adopted to deal with emotionally charged situations, leading to positive outcomes. [PEIE04-20P]	Records of employing emotional intelligence to support individuals beyond the boundary, dealing with difficult situation(s), with a positive outcome. [PEIE04-E20]
5	Champions the introduction of novel techniques and ideas in the application of emotional intelligence, beyond the enterprise boundary, in order to develop the wider Systems Engineering community in this competency. [PEIE05]	(A) Analyzes different approaches across different domains through research. [PEIE05-10A]	Records of activities promoting research and need to adopt novel technique or ideas. [PEIE05-E10]

		(A) Produces reports for the wider SE community on the effectiveness of new techniques after their introduction. [PEIE05-20A]	Records of improvements made to process and appraisal against a recognized process improvement model. [PEIE05-E20]
		(P) Collaborates with those introducing novel techniques within the wider SE community. [PEIE05-30P]	Research records. [PEIE05-E30]
		(A) Defines novel approaches that could potentially improve the wider SE discipline. [PEIE05-40A]	Published papers in refereed journals/company literature. [PEIE05-E40]
		(P) Fosters awareness of these novel techniques within the wider SE community. [PEIE05-50P]	Records showing introduction of enabling systems supporting the new techniques or ideas. [PEIE05-E50]
6	Coaches individuals beyond the enterprise boundary, in emotional intelligence techniques in order to further develop their knowledge, abilities, skills, or associated behaviors. [PEIE06]	(P) Coaches or mentors individuals beyond the enterprise boundary, in competency-related techniques, recommending development activities. [PEIE06-10P]	Coaching or mentoring assignment records. [PEIE06-E10]
		(A) Develops or authorizes training materials in this competency area, which are subsequently successfully delivered beyond the enterprise boundary. [PEIE06-20A]	Records of formal training courses, workshops, seminars, and authored training material supported by successful post-training evaluation data. [PEIE06-E20]
		(A) Provides workshops/seminars or training in this competency area for practitioners or lead practitioners beyond the enterprise boundary (e.g. conferences and open training days). [PEIE06-30A]	Records of Training/workshops/seminars created supported by successful post-training evaluation data. [PEIE06-E30]
7	Maintains expertise in this competency area through specialist Continual Professional Development (CPD) activities. [PEIE07]	(A) Reviews research, new ideas, and state of the art to identify relevant new areas requiring personal development in order to maintain expertise in this competency area. [PEIE07-10A]	Records of documents reviewed and insights gained as part of own research into this competency area. [PEIE07-E10]
		(A) Performs identified specialist professional development activities in order to maintain or further develop competence at expert level. [PEIE07-20A]	Records of Continual Professional Development (CPD) performed and learning outcomes. [PEIE07-E20]
		(A) Records continual professional development activities undertaken including learning or insights gained. [PEIE07-30A]	

NOTES	In addition to items above, enterprise-level or independent 3rd-Party-generated evidence may be used to amplify other evidence presented and may include: a. Formally recognized by a reputable external organization as an expert in this competency area b. Evidence of role as independent assessor or reviewer on project outside own organization where skills in this competency area were used c. Evidence of invitation(s) from wider community for contribution of systems engineering expertise in this area (e.g. industry conference panel, government advisory board etc. cross-industry working groups, partnerships, accredited advanced university courses or research, or as part of professional institute) d. Formal commendation beyond the enterprise (e.g. by INCOSE or other recognized authority) for work performed in this competency area e. Independently assessed or accredited work product in this competency area (e.g. for independent publication or use) f. Accolades of expertise in this area from recognized industry leaders

Competency area – Professional: Coaching and Mentoring
Description Coaching and mentoring are development approaches based on the use of one-to-one conversations to enhance an individual's skills, knowledge, or work performance. Coaching is a non-directive form of development aiming to produce optimal performance and improvement at work. It focuses on specific skills and goals, although may impact an individual's personal attributes. The process typically lasts for a defined period. Mentoring is a relationship where a more experienced colleague shares their greater knowledge to support development of a less experienced member of staff. It uses many of the techniques associated with coaching. One key distinction is that mentoring relationships tend to be longer term than coaching arrangements.
Why it matters Coaching and mentoring play an important role in the development of Systems Engineering professionals, providing targeted development and guidance, organizational and cultural insights. They represent learning opportunities for both parties, encouraging sharing and learning across generations and/or between roles. In addition, an organization may benefit through greater retention of staff, improved skills and productivity, improved communication, etc.
Possible contributory types of evidence Any combination of the types of evidence may be acceptable (depending on how the Framework is tailored and used). The evidence items identified at each level indicate example work products only. Contributions to work products will generally differ at each proficiency level.
Learning and development The INCOSE Professional Development Portal provides example guidance on how to gain an initial awareness of a competency area and options for developing further competence thereafter.

Awareness – Professional: Coaching and Mentoring		
ID	**Indicators of Competence**	**Relevant knowledge sub-indicators**
1	Describes key characteristics and personal attributes of coach and mentor roles, and how both approaches help to develop individual potential. [PMEA01]	(K) Explains the high-level differences between coaching and mentoring and how each activity can provide a mutually beneficial partnership. [PMEA01-10K] (K) Lists key coach/mentor attributes such as: depth and breadth of experience, good listening skills, specific coaching/mentoring training/experience. [PMEA01-20K] (K) Lists key mentee/trainee attributes such as: an openness to discussing their skills/experience/environment, an openness to advice, an openness to new perspectives, a want to improve specific skills. [PMEA01-30K]
2	Explains how those undergoing coaching and mentoring need to act in order to benefit from the activity. [PMEA02]	(K) Describes the advantages of considering career goals and objective before undergoing coaching and mentoring. [PMEA02-10K] (K) Describes the advantages of preparing to discuss goals and objectives in advance of coaching and mentoring sessions. [PMEA02-20K] (K) Describes key interfaces and links between engineering and stakeholders in the wider enterprise outside engineering (e.g. infrastructure management, human resource management, quality management, knowledge management, portfolio management, and life cycle model management) which could help coaching and mentoring activities. [PMEA02-30K]

3	Explains why listening to an individual's goals and objectives is important. [PMEA03]	(K) Describes the benefits of ensuring that coaches and mentors listen to goals before discussion or making suggestions. [PMEA03-10K]
4	Lists enterprise goals and describes the influence mentoring may have on meeting those goals. [PMEA04]	(K) Explains why it is necessary to clearly understand and accommodate the enterprise goal and objectives. [PMEA04-10K] (K) Explains why it is necessary to realistically review individual goals and objectives in the context of the enterprise goals and objectives. [PMEA04-20K]
5	Explains why taking a comprehensive approach to assess an individual's challenge is important. [PMEA05]	(K) Describes the benefits of assessing the effort and likelihood of an individual's ability to achieve their goals and objectives. [PMEA05-10K] (K) Describes the benefits of planning using a sequence of smaller more achieve sub-goals and objectives over larger objectives. [PMEA05-20K]
6	Describes the design and operation of the enterprise's coaching and mentoring program. [PMEA06]	(K) Explains who/how/when/where/why of the enterprise coaching and mentoring program. [PMEA06-10K] (K) Explains the role of coach and mentor within the organization or enterprise. [PMEA06-20K] (K) Lists typical mentoring and coaching opportunities in the enterprise. [PMEA06-30K]

Supervised Practitioner – Professional: Coaching and Mentoring			
ID	**Indicators of Competence** ***(in addition to those at Awareness level)***	**Relevant knowledge, experience, and/or behaviors**	**Possible examples of objective evidence of personal contribution to activities performed, or professional behaviors applied**
1	Identifies areas of own skills, knowledge or experience which could be improved. [PMES01]	(K) Explains the concept of self-assessment and how this can be used to determine areas that could benefit from coaching and mentoring. [PMES01-10K] (A) Identifies areas within own skills, knowledge, or performance that could benefit from coaching or mentoring. [PMES01-20A] (A) Identifies the links between areas that could benefit from coaching and mentoring and individual goals and objectives. [PMES01-30A]	Records of development self-assessment needs as part of an enterprise coaching and mentoring program. [PMES01-E10] Personal Development Plan(s). [PMES01-E20]
2	Identifies personal challenges through various perspectives. [PMES02]	(K) Describes differing perspectives on personal challenges and explains why they assist understanding. [PMES02-10K] (A) Identifies personal challenges through various perspectives. [PMES02-20A]	Records of discussion of own personal challenges as part of an enterprise coaching and mentoring program. [PMES02-E10] Personal Development Plan(s) documents personal challenges. [PMES02-E20]
3	Prepares information supporting the development of others within the team. [PMES03]	(K) Describes issues and challenges related to assisting the development of others in the team. [PMES03-10K] (A) Prepares documentation supporting the development of others within the team. [PMES03-20A]	Records of the development of development plans for others as part of an enterprise coaching and mentoring program. [PMES03-E10]
4	Develops own understanding of this competency area through Continual Professional Development (CPD). [PMES04]	(A) Identifies potential gaps in own knowledge or development needs in this area, identifying opportunities to address these through continual professional development activities. [PMES04-10A] (A) Performs continual professional development activities to improve their knowledge and understanding in this area. [PMES04-20A] (A) Records continual professional development activities undertaken including learning or insights gained. [PMES04-30A]	Records of Continual Professional Development (CPD) performed and learning outcomes. [PMES04-E10]

Practitioner – Professional: Coaching and Mentoring			
ID	**Indicators of Competence** ***(in addition to those at Supervised Practitioner level)***	**Relevant knowledge, experience, and/or behaviors**	**Possible examples of objective evidence of personal contribution to activities performed, or professional behaviors applied**
1	Coaches (or mentors) others on the project as part of an enterprise coaching and mentoring program. [PMEP01]	(P) Coaches (or mentors) others as part of an enterprise coaching and mentoring program. [PMEP01-10P]	Coaching or mentoring assignment records. [PMEP01-E10] Profile may be found in an individual's CV or LinkedIn. [PMEP01-E20]
2	Creates career development goals and objectives with individuals. [PMEP02]	(A) Creates alternative development goals and objectives for discussion. [PMEP02-10A] (A) Creates agreed individual development goals and objectives. [PMEP02-20A]	Coaching or mentoring assignment records. [PMEP02-E10] Profile may be found in an individual's CV or LinkedIn profile. [PMEP02-E20]
3	Develops individual career development paths based on development goals and objectives. [PMEP03]	(A) Creates individual career development paths based upon goals and objectives. [PMEP03-10A] (P) Fosters the progress of individuals along their agreed career development paths. [PMEP03-20P]	Coaching or mentoring assignment records. [PMEP03-E10] Profile may be found in an individual's CV or LinkedIn profile. [PMEP03-E20]
4	Uses available coaching and mentoring opportunities to develop individuals within the enterprise. [PMEP04]	(A) Identifies enterprise opportunities, which can be exploited to support the development of individuals through coaching and/or mentoring. [PMEP04-10A]	Coaching or mentoring assignment records. [PMEP04-E10] Profile may be found in an individual's CV or LinkedIn profile. [PMEP04-E20]
5	Develops individuals within their team by supporting them in solving their individual challenges. [PMEP05]	(P) Coaches or mentors individuals in their team to overcome different challenges. [PMEP05-10P] (A) Selects appropriate coaching or mentoring support mechanism for individuals, depending on the individual challenges involved. [PMEP05-20A]	Coaching or mentoring assignment records. [PMEP05-E10] Profile may be found in an individual's CV or LinkedIn profile. [PMEP05-E20]

(Continued)

ID	Indicators of Competence *(in addition to those at Supervised Practitioner level)*	Relevant knowledge, experience, and/or behaviors	Possible examples of objective evidence of personal contribution to activities performed, or professional behaviors applied
6	Guides new or supervised practitioners in coaching and mentoring techniques, in order to develop their knowledge, abilities, skills, or associated behaviors. [PMEP06]	(P) Guides new or supervised practitioners in executing activities that form part of this competency. [PMEP06-10P] (A) Trains individuals to an "Awareness" level in this competency area. [PMEP06-20A]	Record of Organizational Breakdown Structure showing their responsibility for technical supervision in this area. [PMEP06-E10] Record of on-the-job training objectives/guidance. [PMEP06-E20] Coaching or mentoring assignment records. [PMEP06-E30] Records highlighting their impact on another individual in terms of improvement or professional development in this competency. [PMEP06-E40]
7	Maintains and enhances own competence in this area through Continual Professional Development (CPD) activities. [PMEP07]	(A) Identifies potential development needs in this area, identifying opportunities to address these through continual professional development activities. [PMEP07-10A] (A) Performs continual professional development activities to maintain and enhance their competency in this area. [PMEP07-20A] (A) Records continual professional development activities undertaken including learning or insights gained. [PMEP07-30A]	Records of Continual Professional Development (CPD) performed and learning outcomes. [PMEP07-E10]

Lead Practitioner – Professional: Coaching and Mentoring			
ID	**Indicators of Competence *(in addition to those at Practitioner level)***	**Relevant knowledge, experience, and/or behaviors**	**Possible examples of objective evidence of personal contribution to activities performed, or professional behaviors applied**
1	Promotes the use of best practice coaching and mentoring techniques to improve the effectiveness of Systems Engineering activities across the enterprise. [PMEL01]	(A) Develops own state-of-the-art skills in support of coaching and mentoring activities. [PMEL01-10A]	Records of introducing or controlling coaching and mentoring best practices. [PMEL01-E10]
		(A) Uses best practice techniques when coaching and mentoring others across the enterprise. [PMEL01-20A]	Articles, papers or training material authored to support coaching and mentoring. [PMEL01-E20]
		(P) Advises others across the enterprise on appropriate approaches to use to ensure effectiveness of specific coaching or mentoring activities. [PMEL01-30P]	Records of advice provided in support of coaching or mentoring. [PMEL01-E30]
2	Judges the suitability of planned coaching and mentoring programs affecting Systems Engineering within the enterprise. [PMEL02]	(A) Assesses the suitability of coaching and mentoring programs affecting Systems Engineering within the enterprise. [PMEL02-10A]	Records demonstrating a personal contribution to the assessment and improvement of coaching and mentoring program. [PMEL02-E10]
3	Defines the direction of enterprise coaching and mentoring program development. [PMEL03]	(A) Defines specific career development paths available for individuals across the enterprise. [PMEL03-10A]	Records of leading introduction of coaching and mentoring program. [PMEL03-E10]
		(P) Acts to maintain enterprise coaching or mentoring programs changing them as necessary to reflect best practice. [PMEL03-20P]	Records of enterprise coaching and mentoring programs change. [PMEL03-E20]
4	Guides and actively coordinates the implementation of an enterprise-level coaching and mentoring program. [PMEL04]	(A) Identifies potential issues that might be encountered during the execution of an enterprise-level coaching and mentoring program. [PMEL04-10A]	Organization charts demonstrating position as coach and mentor. [PMEL04-E10]
		(P) Guides and actively coordinates the direction of an enterprise Systems Engineering coaching and mentoring program. [PMEL04-20P]	Materials demonstrating specific contribution to enterprise coaching and mentoring. [PMEL04-E20]
		(P) Acts to overcome issues while supervising an enterprise coaching and mentoring program. [PMEL04-30P]	Profile includes recognition of this nature referenced in an individual's CV or LinkedIn. [PMEL04-E30]
5	Assesses career development path activities for individuals across the enterprise, providing regular feedback. [PMEL05]	(A) Assesses career development paths for individuals within the enterprise, providing feedback which led to improvements. [PMEL05-10A]	Records demonstrating a personal contribution to the coaching and mentoring of individuals. [PMEL05-E10]
		(P) Communicates feedback to others during coaching/mentoring sessions on career path choices. [PMEL05-20P]	Coaching or mentoring assignment records. [PMEL05-E20]
6	Advises stakeholders across the enterprise on individual coaching and mentoring issues with demonstrable success. [PMEL06]	(A) Advises on individual coaching and mentoring issues with demonstrable success. [PMEL06-10A]	Coaching or mentoring assignment records. [PMEL06-E10]
7	Coaches or mentors practitioners across the enterprise in coaching and mentoring techniques in order to develop their knowledge, abilities, skills, or associated behaviors. [PMEL07]	(P) Coaches or mentors practitioners across the enterprise in competency-related techniques, recommending development activities. [PMEL07-10P]	Coaching or mentoring assignment records. [PMEL07-E10]

(Continued)

ID	Indicators of Competence *(in addition to those at Practitioner level)*	Relevant knowledge, experience, and/or behaviors	Possible examples of objective evidence of personal contribution to activities performed, or professional behaviors applied
		(A) Develops or authorizes enterprise training materials in this competency area. [PMEL07-20A]	Records of formal training courses, workshops, seminars, and authored training material supported by successful post-training evaluation data. [PMEL07-E20]
		(A) Provides enterprise workshops/seminars or training in this competency area. [PMEL07-30A]	Listing as an approved organizational trainer for this competency area. [PMEL07-E30]
8	Promotes the introduction and use of novel techniques and ideas in Coaching and Mentoring across the enterprise, to improve enterprise competence in this area. [PMEL08]	(A) Analyzes different approaches across different domains through research. [PMEL08-10A]	Research records. [PMEL08-E10]
		(A) Defines novel approaches that could potentially improve the SE discipline within the enterprise. [PMEL08-20A]	Papers published in refereed journals/company literature. [PMEL08-E20]
		(P) Fosters awareness of these novel techniques within the enterprise. [PMEL08-30P]	Records showing introduction of enabling systems supporting the new techniques or ideas. [PMEL08-E30]
		(P) Collaborates with enterprise stakeholders to facilitate the introduction of techniques new to the enterprise. [PMEL08-40P]	Papers (or similar) published at enterprise level. [PMEL08-E40]
		(A) Monitors new techniques after their introduction to determine their effectiveness. [PMEL08-50A]	Records of improvements made against a recognized process improvement model. [PMEL08-E50]
		(A) Adapts approach to reflect actual enterprise performance improvements. [PMEL08-60A]	
9	Develops expertise in this competency area through specialist Continual Professional Development (CPD) activities. [PMEL09]	(A) Identifies own needs for further professional development in order to increase competence beyond practitioner level. [PMEL09-10A]	Records of Continual Professional Development (CPD) performed and learning outcomes. [PMEL09-E10]
		(A) Performs professional development activities in order to move own competence toward expert level. [PMEL09-20A]	
		(A) Records continual professional development activities undertaken including learning or insights gained. [PMEL09-30A]	

NOTES In addition to items above, enterprise-level or independent 3rd-Party-generated evidence may be used to amplify other evidence presented and may include:

a. Formally recognized by senior management in current organization as an expert in this competency area
b. Evidence of role as Product/System Design Authority or Technical Authority on a complex project with responsibilities in this area or where skills within this competency area were used
c. Recognized as an authorizing signatory on behalf of enterprise for formal documentation in this competency area (e.g. policies, processes, and deliverables)
d. Formal commendation or award within own enterprise for contribution or item of work successfully performed, which required proficiency in this competency area
e. Customer, Supplier, or other external project-specific key Stakeholder accolades for specific work performed in this competency area
f. Independently assessed or accredited work in this competency area (e.g. for independent publication or use)
g. Formal organizational HR records positively highlighting any specific professional competencies or behaviors identified (if applicable) plus any of the evidence indicators listed at Expert level below

Expert – Professional: Coaching and Mentoring			
ID	**Indicators of Competence** ***(in addition to those at Lead Practitioner level)***	**Relevant knowledge, experience, and/or behaviors**	**Possible examples of objective evidence of personal contribution to activities performed, or professional behaviors applied**
1	Communicates own knowledge and experience in coaching and mentoring skills in order to improve best practice beyond the enterprise boundary. [PMEE01]	(A) Produces papers, seminars, or presentations outside own enterprise for publication in order to share own ideas and improve industry best practices in this competence area. [PMEE01-10A]	Papers or books published on new technique in refereed journals/company literature. [PMEE01-E10]
		(P) Fosters incorporation of own ideas into industry best practices in this area. [PMEE01-20P]	Papers published in refereed journals or internal literature proposing new practices in this competence area (or presentations, tutorials, etc.). [PMEE01-E20]
		(P) Develops guidance materials identifying new (or updating existing) best practice in this competence area. [PMEE01-30P]	Records of own proposals adopted as industry best practices in this competence area. [PMEE01-E30]
2	Persuades key stakeholders beyond the enterprise boundary to follow a particular path for coaching and mentoring activities affecting Systems Engineering. [PMEE02]	(P) Persuades key stakeholders beyond the enterprise boundary to follow a particular path for coaching and mentoring activities affecting Systems Engineering. [PMEE02-10P]	Records of ideas assimilated into international standards or referenced in others' books, papers, and peer-reviewed articles. [PMEE02-E10]
		(P) Acts to change external stakeholder thinking in coaching and mentoring activities affecting Systems Engineering. [PMEE02-20P]	Records demonstrating a personal contribution to coaching and mentoring strategic changes. [PMEE02-E20]
3	Advises organizations beyond the enterprise boundary on the suitability of their approach to coaching and mentoring. [PMEE03]	(A) Advises others on differing strategies for coaching and mentoring programs affecting Systems Engineering. [PMEE03-10A]	Enterprise coaching and mentoring program records. [PMEE03-E10]
		(A) Advises stakeholders beyond the enterprise on strategies for coaching and mentoring, gaining acceptance. [PMEE03-20A]	
4	Advises organizations beyond the enterprise boundary on the development of coaching and mentoring programs. [PMEE04]	(A) Advises on development on coaching and mentoring activities affecting Systems Engineering beyond the enterprise boundary. [PMEE04-10A]	Coaching or mentoring assignment records. [PMEE04-E10]
5	Assesses the effectiveness of a mentoring program for an organization beyond the enterprise boundary, providing regular feedback. [PMEE05]	(A) Assesses the effectiveness of a mentoring program for an organization beyond the enterprise boundary. [PMEE05-10A]	Coaching or mentoring assignment records. [PMEE05-E10]
		(A) Communicates feedback regularly on the mentoring program, resulting in recorded improvements. [PMEE05-20A]	

(Continued)

ID	Indicators of Competence *(in addition to those at Lead Practitioner level)*	Relevant knowledge, experience, and/or behaviors	Possible examples of objective evidence of personal contribution to activities performed, or professional behaviors applied
6	Advises organizations beyond the enterprise boundary on complex or challenging coaching and mentoring issues. [PMEE06]	(A) Assesses complex or challenging coaching and mentoring issues on behalf of external organizations or stakeholders. [PMEE06-10A]	Records of analysis made. [PMEE06-E10]
		(P) Communicates recommendations effectively on the challenging issues, resulting in improvements. [PMEE06-20P]	Coaching or mentoring assignment records. [PMEE06-E20]
7	Champions the introduction of novel techniques and ideas in coaching and mentoring, beyond the enterprise boundary, in order to develop the wider Systems Engineering community in this competency. [PMEE07]	(A) Analyzes different approaches across different domains through research. [PMEE07-10A]	Records of activities promoting research and need to adopt novel technique or ideas. [PMEE07-E10]
		(A) Produces reports for the wider SE community on the effectiveness of new techniques after their introduction. [PMEE07-20A]	Records of improvements made to process and appraisal against a recognized process improvement model. [PMEE07-E20]
		(P) Collaborates with those introducing novel techniques within the wider SE community. [PMEE07-30P]	Research records. [PMEE07-E30]
		(A) Defines novel approaches that could potentially improve the wider SE discipline. [PMEE07-40A]	Published papers in refereed journals/company literature. [PMEE07-E40]
		(P) Fosters awareness of these novel techniques within the wider SE community. [PMEE07-50P]	Records showing introduction of enabling systems supporting the new techniques or ideas. [PMEE07-E50]
8	Coaches individuals beyond the enterprise boundary, in coaching and mentoring techniques in order to further develop their knowledge, abilities, skills, or associated behaviors. [PMEE08]	(P) Coaches or mentors individuals beyond the enterprise boundary, in competency-related techniques, recommending development activities. [PMEE08-10P]	Coaching or mentoring assignment records. [PMEE08-E10]
		(A) Develops or authorizes training materials in this competency area, which are subsequently successfully delivered beyond the enterprise boundary. [PMEE08-20A]	Records of formal training courses, workshops, seminars, and authored training material supported by successful post-training evaluation data. [PMEE08-E20]
		(A) Provides workshops/seminars or training in this competency area for practitioners or lead practitioners beyond the enterprise boundary (e.g. conferences and open training days). [PMEE08-30A]	Records of Training/workshops/seminars created supported by successful post-training evaluation data. [PMEE08-E30]

9	Maintains expertise in this competency area through specialist Continual Professional Development (CPD) activities. [PMEE09]	(A) Reviews research, new ideas, and state of the art to identify relevant new areas requiring personal development in order to maintain expertise in this competency area. [PMEE09-10A]	Records of documents reviewed and insights gained as part of own research into this competency area. [PMEE09-E10]
		(A) Performs identified specialist professional development activities in order to maintain or further develop competence at expert level. [PMEE09-20A]	Records of Continual Professional Development (CPD) performed and learning outcomes. [PMEE09-E20]
		(A) Records continual professional development activities undertaken including learning or insights gained. [PMEE09-30A]	

NOTES In addition to items above, enterprise-level or independent 3rd-Party-generated evidence may be used to amplify other evidence presented and may include:

a. Formally recognized by a reputable external organization as an expert in this competency area
b. Evidence of role as independent assessor or reviewer on project outside own organization where skills in this competency area were used
c. Evidence of invitation(s) from wider community for contribution of systems engineering expertise in this area (e.g. industry conference panel, government advisory board etc. cross-industry working groups, partnerships, accredited advanced university courses or research, or as part of professional institute)
d. Formal commendation beyond the enterprise (e.g. by INCOSE or other recognized authority) for work performed in this competency area
e. Independently assessed or accredited work product in this competency area (e.g. for independent publication or use)
f. Accolades of expertise in this area from recognized industry leaders

Competency area – Technical: Requirements Definition
Description To analyze the stakeholder needs and expectations to establish the requirements for a system.
Why it matters The requirements of a system describe the problem to be solved (its purpose, how it performs, how it is to be used, maintained and disposed of, and what the expectations of the stakeholders are).
Possible contributory types of evidence Any combination of the types of evidence may be acceptable (depending on how the Framework is tailored and used). The evidence items identified at each level indicate example work products only. Contributions to work products will generally differ at each proficiency level.
Learning and development The INCOSE Professional Development Portal provides example guidance on how to gain an initial awareness of a competency area and options for developing further competence thereafter.

Awareness – Technical: Requirements Definition		
ID	**Indicators of Competence**	**Relevant knowledge sub-indicators**
1	Describes what a requirement is, the purpose of requirements, and why requirements are important. [TRDA01]	(K) Describes what a "requirement" is. [TRDA01-10K]
		(K) Describes the difference between a "want" and a "need" in terms of requirements. [TRDA01-20K]
		(K) Describes why requirements exist (e.g. commercial, common understanding). [TRDA01-30K]
		(K) Describes how poorly formed requirements can potentially impact a development (e.g. its cost, schedule, quality, or its ultimate success as a system). [TRDA01-40K]
2	Describes different types of requirements and constraints that may be placed on a system. [TRDA02]	(K) Describes with examples, what a functional requirement is. [TRDA02-10K]
		(K) Describes with examples, what a "performance" requirement is. [TRDA02-20K]
		(K) Lists different categories of constraint, providing examples of each (e.g. constraints (physical, realization, integration, verification, validation, production, maintenance, and disposal). [TRDA02-30K]
		(K) Lists categories of "interface" requirements, providing examples of each. [TRDA02-40K]
		(K) Describes with examples, what a "behavioral" requirement is. [TRDA02-50K]
		(K) Describes what an "operational" requirement is and how operational conditions result in requirements. [TRDA02-60K]
		(K) Describes with examples, "packaging," "transportation," "handling," and "storage" requirements. [TRDA02-70K]

		(K) Lists categories of regulatory requirement, providing examples of each (e.g. safety, health, security, and environmental). [TRDA02-80K] (K) Describes how the needs of the wider "enterprise" may result in additional requirements being placed on a system. [TRDA02-90K] (K) Describes how an obligation in a legal contract can define a requirement. [TRDA02-100K]
3	Explains why there is a need for good quality requirements. [TRDA03]	(K) Explains how good quality requirements help us understand what the customer wants or needs. [TRDA03-10K] (K) Explains the impact of the following quality requirements attributes: verifiable, unambiguous, complete, singular, achievable, concise, consistent, Necessary, implementation independent (aka "design neutral"). [TRDA03-20K] (K) Explains how good quality requirements help reduce risk/uncertainty. [TRDA03-30K] (K) Explains why bad quality requirements make the job more difficult due to ambiguity, confusion, lack of clarity, conflict, etc. [TRDA03-40K] (K) Explains why establishing a mechanism for confirming how a requirement has been fulfilled in a solution is an integral part of defining the requirement (rather than waiting for the solution to be developed). [TRDA03-50K] (K) Explains how even good quality requirements may need to be considered as part of the wider super-system to be fully understood. [TRDA03-60K]
4	Identifies major stakeholders and their needs. [TRDA04]	(K) Lists internal and external stakeholders or stakeholder groups: Customer/Sponsor, User, Designer, Maintainer, Legislative or Regulatory bodies (verify system is compliant with laws), Suppliers, Business, Sub-contractor supplier, Manufacturer, Testers, Trainers, Local Community, Political/Social organizations, Etc. [TRDA04-10K] (K) Explains how understanding stakeholder and market need can help ensure that program objectives are being met and/or re-baselined if required. [TRDA04-20K]
5	Explains why managing requirements throughout the life cycle is important. [TRDA05]	(K) Explains why a change to a requirement needs to be controlled formally. [TRDA05-10K] (K) Explains how managing requirements through the life cycle ensures that what is delivered meets an agreed set of requirements. [TRDA05-20K] (K) Explains how managing requirements helps to maintain integrity through the life cycle. [TRDA05-30K] (K) Explains how managing requirements helps reduce risk caused by unassessed requirements changes. [TRDA05-40K] (K) Explains why there is a need to manage all types of requirements. [TRDA05-50K] (K) Explains why requirements beyond system function and performance also need to be managed (e.g. cost, schedule, quality, delivery, standards/certification, and packaging quantities). [TRDA05-60K]
6	Describes the relationship between requirements, testing, and acceptance. [TRDA06]	(K) Explains why we need to verify that we have met all requirements the user specified (e.g. verification of user's need) and why not doing this could cause problems. [TRDA06-10K] (K) Explains how requirements verification ensures we have built the system we have specified. [TRDA06-20K] (K) Explains how requirements validation helps ensure that the needs of the stakeholder(s) have been met. [TRDA06-30K] (K) Explains the relationship between verification, validation and acceptance. [TRDA06-40K]

Supervised Practitioner – Technical: Requirements Definition			
ID	**Indicators of Competence** ***(in addition to those at Awareness level)***	**Relevant knowledge, experience, and/or behaviors**	**Possible examples of objective evidence of personal contribution to activities performed, or professional behaviors applied**
1	Uses a governing process using appropriate tools to manage and control their own requirements definition activities. [TRDS01]	(A) Uses a traceable approach to requirements definition. [TRDS01-10A] (K) Describes key elements of a requirements definition process (i.e. the process to transform the stakeholder, user-oriented view of desired capabilities into a technical view of a solution that meets the operational needs of the user). [TRDS01-20K]	Requirements Definitions document, or Requirement Diagram in a model-based Systems Engineering tool (e.g. SysML diagram). [TRDS01-E10] Organizational document or self-developed document describing a set processes for managing requirements as part of the system engineering life cycle. [TRDS01-E20] Requirements traceability matrix. [TRDS01-E30]
2	Identifies examples of internal and external project stakeholders highlighting their sphere of influence. [TRDS02]	(A) Analyzes the system to determine all its direct stakeholders including their importance or influence on the required system design, development, or operation. [TRDS02-10A] (A) Analyzes the system to determine all its indirect stakeholders. (e.g. stakeholders may include business engineering and procurement; project management and control; production management; product assurance, quality, and safety; operations, maintenance, and servicing). [TRDS02-20A]	Stakeholder list, map, diagram, or matrix. [TRDS02-E10] Records (e.g. meeting minutes) showing participation in activity to determine relevant stakeholders. [TRDS02-E20]
3	Elicits requirements from stakeholders under guidance, in order to understand their need and ensuring requirement validity. [TRDS03]	(A) Uses recognized methods to elicit requirements (e.g. interviews, focus groups, the Delphi method, and soft systems methodology). [TRDS03-10A] (A) Uses Quality Function Deployment (QFD), questionnaires, workshops or similar approaches. [TRDS03-20A]	Records (e.g. meeting minutes) showing participation in use case generation, scenario development, simulation, or requirements questionnaires. [TRDS03-E10] Completed QFD documents, questionnaires, or workshop notes. [TRDS03-E20]
4	Describes the characteristics of good quality requirements and provides examples. [TRDS04]	(K) Explains the impact of key individual requirement characteristics such as necessary, implementation independent (aka "design neutral"), unambiguous, complete, singular, achievable, verifiable, and conforming. [TRDS04-10K] (K) Explains the impact of the characteristics of a set of requirements as a whole such as complete, consistent, feasible/affordable and bounded. [TRDS04-20K] (A) Analyzes the quality characteristics of existing requirements. [TRDS04-30A] (A) Defines requirements possessing all key quality characteristics. [TRDS04-40A] (A) Analyzes requirements as part of the wider super-system. [TRDS04-50A]	Sample requirements statements improved or created. [TRDS04-E10] Working papers showing an evolution of requirement statement(s) from poorly to properly written. [TRDS04-E20] Sample requirements listing with proposed verification and validation approaches. [TRDS04-E30]

		(A) Defines approaches to be used for confirming that a solution meets requirements and needs as part of the requirement definition activity (rather than later in the development). [TRDS04-60A]	
5	Describes different mechanisms used to gather requirements. [TRDS05]	(A) Uses recognized elicitation methods in order to establish requirements (e.g. workshops, brainstorming, seminar, prototyping, demonstrations, standards review, interviews, focus groups, the Delphi method, and soft systems methodology). [TRDS05-10A] (K) Explains what bias and sampling is and how this can influence requirement elicitation. [TRDS05-20K] (A) Ensures completeness, following up when an incomplete set of requirements is detected. [TRDS05-30A]	Records (e.g. meeting minutes) showing participation in workshops, brainstorming, seminar, prototyping, demonstrations, standards review or similar sessions in order to establish or clarify a set of requirements. [TRDS05-E10] Requirements document or model showing traceability to various source documents (e.g. formal documents, informal documents, conversations with users, and reviews with other stakeholders). [TRDS05-E20] Use Case Diagrams, Concepts of Operation document, initial Requirement lists. [TRDS05-E30]
6	Defines acceptance criteria for requirements, under guidance. [TRDS06]	(A) Defines acceptance criteria in order to define how user needs will be fulfilled. [TRDS06-10A] (A) Ensures requirements are testable, defined, and have acceptance criteria, in order to ensure fulfillment of contract and justify payment. [TRDS06-20A] (A) Identifies appropriate method of acceptance when determining acceptance criteria for requirements (e.g. test, analysis, analogy or similarity, demonstration, inspection, simulation, and sampling). [TRDS06-30A]	Requirements acceptance/verification matrix. [TRDS06-E10] Acceptance criteria they have defined for requirements. [TRDS06-E20]
7	Explains why there may be potential requirement conflicts within a requirement set. [TRDS07]	(K) Explains how regulations may conflict with user needs. [TRDS07-10K] (K) Explains how cost requirements may conflict with performance requirements. [TRDS07-20K] (K) Explains how technology may conflict with performance requirements. [TRDS07-30K] (A) Analyzes conflicts in a requirements set. [TRDS07-40A]	Results of analysis of a requirements set with conflicting requirements indicated. [TRDS07-E10] Analysis of alternatives indicating what is feasible given current constraints and how requirements could be re-written and thus be satisfied. [TRDS07-E20]
8	Explains how requirements affect design and vice versa. [TRDS08]	(K) Explains the relationship between requirements and design (e.g. Requirements specify what the system needs to do, within what constraints, while Design defines how a set of requirements may be implemented). [TRDS08-10K] (A) Analyzes impact on design of different requirements. [TRDS08-20A]	Document or model showing requirements allocated to design (behavior diagrams, state machine diagrams, etc. traced back to requirements). [TRDS08-E10]

(Continued)

ID	Indicators of Competence *(in addition to those at Awareness level)*	Relevant knowledge, experience, and/or behaviors	Possible examples of objective evidence of personal contribution to activities performed, or professional behaviors applied
		(A) Analyzes impact on requirements of different design approaches. [TRDS08-30A]	
9	Defines (or maintains) requirements traceability information. [TRDS09]	(K) Explains how traceability ensures control of the system development. [TRDS09-10K]	Traceability matrix. [TRDS09-E10]
		(K) Explains what a requirements baseline is and how it is used. [TRDS09-20K]	Model view showing requirements traceability (e.g. in SysML). [TRDS09-E20]
		(A) Identifies source references for requirements (e.g. though a document and use case). [TRDS09-30A]	Model view showing requirements traceability (e.g. in SysML). [TRDS09-E30]
		(A) Maintains traceability matrix as requirements are verified. [TRDS09-40A]	Requirements baseline listing. [TRDS09-E40]
		(A) Maintains traceability backwards to requirement sources, forward to design element(s) satisfying the requirement, and to test event(s) that confirm the requirement was met. [TRDS09-50A]	
		(A) Ensures accountability and full allocation of requirements. [TRDS09-60A]	Analysis of progress toward completing a requirements Traceability matrix or similar (sometimes referred to as Requirements Burn Down). [TRDS09-E60]
		(A) Ensures consistency of traceability. [TRDS09-70A]	Records of requirements and traceability evolution as new requirements are imposed or identified (e.g. iterative baselines of a model or traceability matrix). [TRDS09-E70]
10	Reviews developed requirements. [TRDS10]	(K) Identifies different types of reviews used to assess requirements quality. [TRDS10-10K]	Records of requirements review attended. [TRDS10-E10]
		(A) Reviews technical requirements for a system. [TRDS10-20A]	
11	Develops own understanding of this competency area through Continual Professional Development (CPD). [TRDS11]	(A) Identifies potential gaps in own knowledge or development needs in this area, identifying opportunities to address these through continual professional development activities. [TRDS11-10A]	Records of Continual Professional Development (CPD) performed and learning outcomes. [TRDS11-E10]
		(A) Performs continual professional development activities to improve their knowledge and understanding in this area. [TRDS11-20A]	
		(A) Records continual professional development activities undertaken including learning or insights gained. [TRDS11-30A]	

Practitioner – Technical: Requirements Definition			
ID	**Indicators of Competence *(in addition to those at Supervised Practitioner level)***	**Relevant knowledge, experience, and/or behaviors**	**Possible examples of objective evidence of personal contribution to activities performed, or professional behaviors applied**
1	Creates a strategy for requirements definition on a project to support SE project and wider enterprise needs. [TRDP01]	(A) Identifies project-specific system requirements management constraints which need to be incorporated into SE planning. [TRDP01-10A]	System Engineering Management Plan (or similar planning document). [TRDP01-E10]
		(A) Prepares project-specific system requirements management inputs for SE planning purposes, using appropriate processes and procedures. [TRDP01-20A]	System requirements management strategy work products. [TRDP01-E20]
		(A) Prepares project-specific system requirements management task estimates in support of SE planning. [TRDP01-30A]	Task estimates for requirements management efforts. [TRDP01-E30]
		(P) Collaborates throughout with stakeholders to gain approval, updating strategy as necessary. [TRDP01-40P]	
2	Creates a governing process, plan, and associated tools for Requirements Definition, which reflect project and business strategy. [TRDP02]	(K) Identifies the steps necessary to define a process and appropriate techniques to be adopted for requirements definition. [TRDP02-10K]	Systems Engineering requirements definition planning or process work products they have developed. [TRDP02-E10]
		(K) Describes key elements of a successful requirements definition process and how these activities relate to different stages of a system life cycle. [TRDP02-20K]	Certificates or proficiency evaluations in requirements definition and management tools (e.g. DOORS, Rational, or SysML). [TRDP02-E20]
		(A) Uses requirements definition tools or uses a methodology. [TRDP02-30A]	Systems Engineering requirements definition guidance developed. [TRDP02-E30]
		(A) Uses appropriate standards and governing processes in their requirements definition planning, tailoring as appropriate. [TRDP02-40A]	Systems Engineering requirements management guidance developed. [TRDP02-E40]
		(A) Selects requirements definition methods for the development. [TRDP02-50A]	Systems Engineering Management Plan. [TRDP02-E50]
		(A) Creates requirements definition plans and processes for use on a project. [TRDP02-60A]	
		(P) Liaises throughout with stakeholders to gain approval, updating plans as necessary. [TRDP02-70P]	

(Continued)

ID	Indicators of Competence *(in addition to those at Supervised Practitioner level)*	Relevant knowledge, experience, and/or behaviors	Possible examples of objective evidence of personal contribution to activities performed, or professional behaviors applied
3	Uses plans and processes for requirements definition, interpreting, evolving, or seeking guidance where appropriate. [TRDP03]	(A) Complies with defined plans, processes and associated tools for requirements definition activities on a project. [TRDP03-10A]	SE requirements definition plans or work products they have interpreted or challenged to address project issues. [TRDP03-E10]
		(A) Prepares requirements definition data and reports in support of wider system measurement, monitoring, and control. [TRDP03-20A]	Process or tool guidance. [TRDP03-E20]
		(P) Guides and actively coordinates the interpretation of requirements definition plans or processes in order to complete activities successfully. [TRDP03-30P]	
		(P) Recognizes situations where deviation from published plans and processes or clarification from others is appropriate in order to overcome complex challenges. [TRDP03-40P]	
		(P) Recognizes situations where existing strategy, process, plan or tools require formal change, liaising with stakeholders to gain approval as necessary. [TRDP03-50P]	
4	Elicits requirements from stakeholders ensuring their validity, to understand their need. [TRDP04]	(A) Develops Use Cases, scenarios, simulation, questionnaires. [TRDP04-10A]	Completed Use Cases, scenarios, simulations, questionnaires, or similar. [TRDP04-E10]
		(A) Uses Quality Function Deployment (QFD), questionnaires, workshops, or similar approaches. [TRDP04-20A]	Records (e.g. meeting minutes) of leading stakeholder requirements elicitation activities. [TRDP04-E20]
		(P) Challenges appropriateness of requirements in a rational way to improve quality of requirements. [TRDP04-30P]	Records (e.g. meeting minutes) of challenges made to stakeholder requirements, which produced changes. [TRDP04-E30]
5	Develops good quality, consistent requirements. [TRDP05]	(A) Develops requirement statements that fulfill quality attributes such as verifiable, unambiguous, complete, singular, achievable, concise, consistent, necessary, implementation independent (aka "design neutral"). [TRDP05-10A]	Requirements document or model. [TRDP05-E10]
		(A) Reviews the overall requirements set for its suitability and completeness to ensure appropriate quality level has been achieved. [TRDP05-20A]	Documentation showing an evolution of requirement statement(s) from poorly to properly written. [TRDP05-E20]

6	Determines derived requirements. [TRDP06]	(A) Determines requirements for lower-level architecture elements by breaking-down higher-level requirements into individual testable requirements applicable at the lower level(s). [TRDP06-10A] (A) Analyzes beyond the boundary of the system of interest to determine interface requirements. [TRDP06-20A] (A) Reviews statutory and regulatory documents to derive relevant requirements. [TRDP06-30A]	Requirements document showing derived requirements. [TRDP06-E10] Requirements table or model output showing derived requirements. [TRDP06-E20]
7	Creates a system to support requirements management and traceability. [TRDP07]	(A) Identifies business and project set-up requirements for requirements management and traceability. [TRDP07-10A] (A) Creates a matrix or database ready for requirements traceability capture. [TRDP07-20A]	List of constraints or project set-up requirements for traceability. [TRDP07-E10] Initialized Traceability matrix or database (e.g. using a specialist requirements database or a model-based systems engineering (MBSE) tool). [TRDP07-E20]
8	Determines acceptance criteria for requirements. [TRDP08]	(A) Determines appropriate acceptance criteria required in order to ensure fulfillment of contract and justify payment. [TRDP08-10A] (A) Determines appropriate testing methods/ approach to be used to document a requirement has been met by the solution. [TRDP08-20A]	Requirements acceptance/verification matrix. [TRDP08-E10] Validation or acceptance criteria for individual requirements. [TRDP08-E20] Acceptance criteria for individual requirements categorized by method (e.g. Inspection, Analysis, Demonstration, and Test). [TRDP08-E30]
9	Negotiates agreement in requirement conflicts within a requirement set. [TRDP09]	(A) Identifies regulations in conflict with conflict with user desires. [TRDP09-10A] (A) Identifies cost requirements in conflict with performance requirements. [TRDP09-20A] (A) Identifies technology requirements in conflict with performance requirements. [TRDP09-30A] (A) Identifies potential conflicting stakeholder requirements within the requirement set. [TRDP09-40A] (P) Negotiates recommended options with stakeholders to address conflicting requirements. [TRDP09-50P]	Results of analysis of a requirements set with conflicting requirements indicated. [TRDP09-E10] Document identifying suggested methods of resolution with stakeholders. [TRDP09-E20] Records (e.g. meeting minutes) of stakeholders meetings to discuss and address requirements conflicts. [TRDP09-E30]
10	Analyzes the impact of changes to requirements on the solution and program. [TRDP10]	(A) Performs impact analysis based upon a requirement change. [TRDP10-10A] (A) Performs impact analysis based upon a design change (e.g. item obsolescence.). [TRDP10-20A]	Model or document showing impact analysis of design variations on requirements. [TRDP10-E10] Model or document showing relationship of new or changed requirements on existing design. [TRDP10-E20]

(Continued)

ID	Indicators of Competence *(in addition to those at Supervised Practitioner level)*	Relevant knowledge, experience, and/or behaviors	Possible examples of objective evidence of personal contribution to activities performed, or professional behaviors applied
11	Maintains requirements traceability information to ensure source(s) and test records are correctly linked over the life cycle. [TRDP11]	(A) Ensures existence of source references for requirements (e.g. through a document and use case). [TRDP11-10A]	Traceability matrix. [TRDP11-E10]
		(A) Performs traceability linking system (products or service) requirements to source(s). [TRDP11-20A]	Analysis of progress toward meeting requirements using Traceability matrix or similar (sometimes referred to as Requirements Burn Down). [TRDP11-E20]
		(A) Ensures traceability backwards to requirement sources, forward to design element(s) satisfying the requirement, and to test event(s) that confirm the requirement was met. [TRDP11-30A]	Record of requirements and traceability evolution as new requirements are imposed or identified (e.g. iterative baselines of a model or traceability matrix). [TRDP11-E30]
		(A) Ensures traceability matrix is updated as requirements are verified. [TRDP11-40A]	
12	Guides new or supervised practitioners in Systems Engineering Requirements Definition to develop their knowledge, abilities, skills, or associated behaviors. [TRDP12]	(P) Guides new or supervised practitioners in executing activities that form part of this competency. [TRDP12-10P]	Organizational Breakdown Structure showing their responsibility for technical supervision in this area. [TRDP12-E10]
		(A) Trains individuals to an "Awareness" level in this competency area. [TRDP12-20A]	Job training objectives/guidance etc. suggested or authorized for others. [TRDP12-E20]
			Coaching or mentoring assignment records. [TRDP12-E30]
			Records highlighting their impact on another individual in terms of improvement or professional development in this competency. [TRDP12-E40]
13	Maintains and enhances own competence in this area through Continual Professional Development (CPD) activities. [TRDP13]	(A) Identifies potential development needs in this area, identifying opportunities to address these through continual professional development activities. [TRDP13-10A]	Records of Continual Professional Development (CPD) performed and learning outcomes. [TRDP13-E10]
		(A) Performs continual professional development activities to maintain and enhance their competency in this area. [TRDP13-20A]	
		(A) Records continual professional development activities undertaken including learning or insights gained. [TRDP13-30A]	

Lead Practitioner – Technical: Requirements Definition			
ID	**Indicators of Competence** ***(in addition to those at Practitioner level)***	**Relevant knowledge, experience, and/or behaviors**	**Possible examples of objective evidence of personal contribution to activities performed, or professional behaviors applied**
1	Creates enterprise-level policies, procedures, guidance, and best practice for requirements elicitation and management, including associated tools. [TRDL01]	(A) Analyzes enterprise need for requirements definition policies, processes, tools, or guidance. [TRDL01-10A]	Records showing their role in embedding requirements definition into enterprise policies (e.g. guidance introduced at enterprise level and enterprise-level review minutes). [TRDL01-E10]
		(A) Creates enterprise policies, procedures, or guidance for requirements definition activities. [TRDL01-20A]	Procedures they have written. [TRDL01-E20]
		(A) Selects and acquires appropriate tools supporting requirements definition activities. [TRDL01-30A]	Records of support for tool introduction. [TRDL01-E30]
2	Judges the tailoring of enterprise-level requirements elicitation and management processes to meet the needs of a project. [TRDL02]	(A) Evaluates the enterprise process against the business and external stakeholder (e.g. customer) needs in order to tailor processes to enable project success. [TRDL02-10A]	Documented tailored process. [TRDL02-E10]
		(A) Provides constructive feedback on Requirements Management Plans. [TRDL02-20A]	Requirement Management Plan (or similar feedback and revised plan. [TRDL02-E20]
3	Advises on complex or challenging requirements from across the enterprise to ensure completeness and suitability. [TRDL03]	(A) Reviews complex or challenging requirements for their suitability and quality across the enterprise. [TRDL03-10A]	Review comments. [TRDL03-E10]
		(A) Provides advice on projects across the enterprise on complex or challenging requirements that has led to changes being implemented. [TRDL03-20A]	Records of advice provided on requirements suitability. [TRDL03-E20]
			System Requirements Review, System Functional Review, or similar documentation. [TRDL03-E30]
4	Defines strategies for requirements resolution in situations across the enterprise where stakeholders (or their requirements) demand unusual or sensitive treatment. [TRDL04]	(A) Defines strategies for resolving requirements in unusual or sensitive stakeholder situations across the enterprise. [TRDL04-10A]	Review comments and responses. [TRDL04-E10]
		(P) Collaborates with stakeholders to implement identified strategies in a rational way resulting in beneficial changes. [TRDL04-20P]	Records of advice provided on documenting successes in changing stakeholder approach. [TRDL04-E20]

(Continued)

ID	Indicators of Competence *(in addition to those at Practitioner level)*	Relevant knowledge, experience, and/or behaviors	Possible examples of objective evidence of personal contribution to activities performed, or professional behaviors applied
5	Persuades key stakeholders across the enterprise to address identified enterprise-level requirements elicitation and management issues to reduce enterprise-level risk. [TRDL05]	(A) Identifies and engages with key stakeholders across the enterprise. [TRDL05-10A]	Minutes of meetings. [TRDL05-E10]
		(P) Fosters agreement between key stakeholders at enterprise level to resolve requirements definition issues, by promoting a holistic viewpoint. [TRDL05-20P]	Minutes of key stakeholder review meetings and actions taken. [TRDL05-E20]
		(P) Persuades key stakeholders to resolve conflict and identify optimal design options. [TRDL05-30P]	Requirements Review documentation. [TRDL05-E30]
6	Coaches or mentors practitioners across the enterprise in Systems Engineering Requirements Definition in order to develop their knowledge, abilities, skills, or associated behaviors. [TRDL06]	(P) Coaches or mentors practitioners across the enterprise in competency-related techniques, recommending development activities. [TRDL06-10P]	Coaching or mentoring assignment records. [TRDL06-E10]
		(A) Develops or authorizes enterprise training materials in this competency area. [TRDL06-20A]	Records of formal training courses, workshops, seminars, and authored training material supported by successful post-training evaluation data. [TRDL06-E20]
		(A) Provides enterprise workshops/seminars or training in this competency area. [TRDL06-30A]	Listing as an approved organizational trainer for this competency area. [TRDL06-E30]
7	Promotes the introduction and use of novel techniques and ideas in Requirements Definition across the enterprise, to improve enterprise competence in this area. [TRDL07]	(A) Analyzes different approaches across different domains through research. [TRDL07-10A]	Research records. [TRDL07-E10]
		(A) Defines novel approaches that could potentially improve the SE discipline within the enterprise. [TRDL07-20A]	Published papers in refereed journals/company literature. [TRDL07-E20]
		(P) Fosters awareness of these novel techniques within the enterprise. [TRDL07-30P]	Records showing introduction of enabling systems supporting the new techniques or ideas. [TRDL07-E30]
		(P) Collaborates with enterprise stakeholders to facilitate the introduction of techniques new to the enterprise. [TRDL07-40P]	Published papers (or similar) at enterprise level. [TRDL07-E40]

		(A) Monitors new techniques after their introduction to determine their effectiveness. [TRDL07-50A] (A) Adapts approach to reflect actual enterprise performance improvements. [TRDL07-60A]	Records of improvements made against a recognized process improvement model in this area. [TRDL07-E50]
8	Develops expertise in this competency area through specialist Continual Professional Development (CPD) activities. [TRDL08]	(A) Identifies own needs for further professional development in order to increase competence beyond practitioner level. [TRDL08-10A] (A) Performs professional development activities in order to move own competence toward expert level. [TRDL08-20A] (A) Records continual professional development activities undertaken including learning or insights gained. [TRDL08-30A]	Records of Continual Professional Development (CPD) performed and learning outcomes. [TRDL08-E10]

NOTES	In addition to items above, enterprise-level or independent 3rd-Party-generated evidence may be used to amplify other evidence presented and may include: a. Formally recognized by senior management in current organization as an expert in this competency area b. Evidence of role as Product/System Design Authority or Technical Authority on a complex project with responsibilities in this area or where skills within this competency area were used c. Recognized as an authorizing signatory on behalf of enterprise for formal documentation in this competency area (e.g. policies, processes, and deliverables) d. Formal commendation or award within own enterprise for contribution or item of work successfully performed, which required proficiency in this competency area e. Customer, Supplier, or other external project-specific key Stakeholder accolades for specific work performed in this competency area f. Independently assessed or accredited work in this competency area (e.g. for independent publication or use) g. Formal organizational HR records positively highlighting any specific professional competencies or behaviors identified (if applicable) plus any of the evidence indicators listed at Expert level below

Expert – Technical: Requirements Definition			
ID	**Indicators of Competence *(in addition to those at Lead Practitioner level)***	**Relevant knowledge, experience, and/or behaviors**	**Possible examples of objective evidence of personal contribution to activities performed, or professional behaviors applied**
1	Communicates own knowledge and experience in Systems Engineering requirements definition in order to promote best practice beyond the enterprise boundary. [TRDE01]	(A) Produces papers, seminars, or presentations outside own enterprise for publication in order to share own ideas and improve industry best practices in this competence area. [TRDE01-10A]	Published papers or books etc. on new technique in refereed journals/company literature. [TRDE01-E10]
		(P) Fosters incorporation of own ideas into industry best practices in this area. [TRDE01-20P]	Published papers in refereed journals or internal literature proposing new practices in this competence area (or presentations, tutorials, etc.). [TRDE01-E20]
		(P) Develops guidance materials identifying new (or updating existing) best practice in this competence area. [TRDE01-30P]	Proposals adopted as industry best practice. [TRDE01-E30]
2	Persuades key stakeholders beyond the enterprise boundary to address identified requirements definition issues to reduce project risk. [TRDE02]	(A) Identifies and engages with external stakeholders to elicit needed requirements. [TRDE02-10A]	Minutes of meetings. [TRDE02-E10]
		(P) Acts to change external stakeholder thinking in requirements definition activities affecting project risk. [TRDE02-20P]	Revised requirements. [TRDE02-E20]
3	Advises organizations beyond the enterprise boundary on the suitability of their approach to requirements definition. [TRDE03]	(A) Advises stakeholders beyond enterprise boundary regarding their Requirements Definition activities. [TRDE03-10A]	Records of membership on oversight committee with relevant terms of reference. [TRDE03-E10]
		(P) Acts with sensitivity on highly complex system issues making limited use of specialized, technical terminology and taking account of external stakeholders' (e.g. customer's) background and knowledge. [TRDE03-20P]	Participates in an oversight committee or similar body that deals with approval of such plans. [TRDE03-E20]
		(P) Communicates using a holistic approach to complex Requirements Definition resolution including balanced, rational arguments on way forward. [TRDE03-30P]	Review comments and revised Requirements Management plan. [TRDE03-E30]
		(A) Advises on Systems Engineering Requirements Definition decisions on behalf of stakeholders beyond enterprise boundary, arbitrating if required. [TRDE03-40A]	Records indicating changed stakeholder thinking from personal intervention. [TRDE03-E40]
		(P) Persuades stakeholders beyond the enterprise boundary to accept difficult project-specific Requirements Definition recommendations or actions. [TRDE03-50P]	

4	Advises organizations beyond the enterprise boundary on the handling of complex or sensitive Systems Engineering requirements definition issues. [TRDE04]	(P) Coaches or mentors external stakeholders (e.g. customers) on their requirements related issues. [TRDE04-10P]	Records of advice provided on requirements issues. [TRDE04-E10]
		(P) Conducts sensitive negotiations on requirements associated with a highly complex system making limited use of specialized, technical terminology. [TRDE04-20P]	Records from negotiations demonstrating awareness of customer's background and knowledge. [TRDE04-E20]
		(A) Uses a holistic approach to complex issue resolution including balanced, rational arguments on way forward. [TRDE04-30A]	Stakeholder approval of system requirements. [TRDE04-E30]
5	Champions the introduction of novel techniques and ideas in the requirements definition, beyond the enterprise boundary, in order to develop the wider Systems Engineering community in this competency. [TRDE05]	(A) Analyzes different approaches across different domains through research. [TRDE05-10A]	Records of activities promoting research and need to adopt novel technique or ideas. [TRDE05-E10]
		(A) Produces reports for the wider SE community on the effectiveness of new techniques after their introduction. [TRDE05-20A]	Records of improvements made to process and appraisal against a recognized process improvement model. [TRDE05-E20]
		(P) Collaborates with those introducing novel techniques within the wider SE community. [TRDE05-30P]	Research records. [TRDE05-E30]
		(A) Defines novel approaches that could potentially improve the wider SE discipline. [TRDE05-40A]	Published papers in refereed journals/company literature. [TRDE05-E40]
		(P) Fosters awareness of these novel techniques within the wider SE community. [TRDE05-50P]	Records showing introduction of enabling systems supporting the new techniques or ideas. [TRDE05-E50]
6	Coaches individuals beyond the enterprise boundary, in requirements definition in order to further develop their knowledge, abilities, skills, or associated behaviors. [TRDE06]	(P) Coaches or mentors individuals beyond the enterprise boundary, in competency-related techniques, recommending development activities. [TRDE06-10P]	Coaching or mentoring assignment records. [TRDE06-E10]
		(A) Develops or authorizes training materials in this competency area, which are subsequently successfully delivered beyond the enterprise boundary. [TRDE06-20A]	Records of formal training courses, workshops, seminars, and authored training material supported by successful post-training evaluation data. [TRDE06-E20]
		(A) Provides workshops/seminars or training in this competency area for practitioners or lead practitioners beyond the enterprise boundary (e.g. conferences and open training days). [TRDE06-30A]	Records of Training/workshops/seminars created supported by successful post-training evaluation data. [TRDE06-E30]

(Continued)

ID	Indicators of Competence *(in addition to those at Lead Practitioner level)*	Relevant knowledge, experience, and/or behaviors	Possible examples of objective evidence of personal contribution to activities performed, or professional behaviors applied
7	Maintains expertise in this competency area through specialist Continual Professional Development (CPD) activities. [TRDE07]	(A) Reviews research, new ideas, and state of the art to identify relevant new areas requiring personal development in order to maintain expertise in this competency area. [TRDE07-10A] (A) Performs identified specialist professional development activities in order to maintain or further develop competence at expert level. [TRDE07-20A] (A) Records continual professional development activities undertaken including learning or insights gained. [TRDE07-30A]	Records of documents reviewed and insights gained as part of own research into this competency area. [TRDE07-E10] Records of Continual Professional Development (CPD) performed and learning outcomes. [TRDE07-E20]

NOTES	In addition to items above, enterprise-level or independent 3rd-Party-generated evidence may be used to amplify other evidence presented and may include: a. Formally recognized by a reputable external organization as an expert in this competency area b. Evidence of role as independent assessor or reviewer on project outside own organization where skills in this competency area were used c. Evidence of invitation(s) from wider community for contribution of systems engineering expertise in this area (e.g. industry conference panel, government advisory board etc. cross-industry working groups, partnerships, accredited advanced university courses or research, or as part of professional institute) d. Formal commendation beyond the enterprise (e.g. by INCOSE or other recognized authority) for work performed in this competency area e. Independently assessed or accredited work product in this competency area (e.g. for independent publication or use) f. Accolades of expertise in this area from recognized industry leaders

Competency area – Technical: Systems Architecting
Description The definition of the system structure, interfaces, and associated derived requirements to produce a solution that can be implemented to enable a balanced and optimum result that considers all stakeholder requirements (business, technical….). This includes the early generation of potential system concepts that meet a set of needs and demonstration that one or more credible, feasible options exist.
Why it matters Effective architectural design enables systems to be partitioned into realizable system elements, which can be brought together to meet the requirements. Failure to explore alternative conceptual options as part of architectural analysis may result in a non-optimal system. There may be no viable option (e.g. technology not available).
Possible contributory types of evidence Any combination of the types of evidence may be acceptable (depending on how the Framework is tailored and used). The evidence items identified at each level indicate example work products only. Contributions to work products will generally differ at each proficiency level.
Learning and development The INCOSE Professional Development Portal provides example guidance on how to gain an initial awareness of a competency area and options for developing further competence thereafter.

Awareness – Technical: System Architecting		
ID	**Indicators of Competence**	**Relevant knowledge sub-indicators**
1	Describes the principles of architectural design and its role within the life cycle. [TSAA01]	(K) Defines the boundary of a system, identifies major interfaces to the system, and generates a functional analysis, which can be used to assess multiple physical design options. [TSAA01-10K] (K) Explains the purpose of architectural design within the overall system life cycle. [TSAA01-20K] (K) Describes key elements of an architecture definition process (e.g. prepare for architecture definition, develop architecture viewpoints, develop models and views of candidate architectures, relate the architecture to design, assess architecture candidates, and managed the selected architecture). [TSAA01-30K] (K) Describes the importance of architectural design (e.g. common vehicle for communication between stakeholders, allows quality attributes such as performance to be modeled, etc.) and understands the criteria for good design. [TSAA01-40K] (K) Describes architecture in terms of a decomposition of a system into its components, their interrelationships and the constraints that apply. [TSAA01-50K]
2	Describes different types of architecture and provides examples. [TSAA02]	(K) Explains why there is not a 'one size fits all' approach to architectural design. Can name at least two architecture frameworks (e.g. Zachman, DoDAF, MoDAF, TOGAF). [TSAA02-10K] (K) Explains how a system can be abstracted into a structured functional and physical representation (e.g. a complex weapon system and an information technology (IT) system). [TSAA02-20K] (K) Lists types of architecture and their key characteristics such as functional, physical, logical, and operational. [TSAA02-30K]

ID	Indicators of Competence	Relevant knowledge sub-indicators
3	Explains why architectural decisions can constrain and limit future use and evolution and provides examples. [TSAA03]	(K) Describes examples of limitations and constraints in an architecture. [TSAA03-10K]
4	Explains why there is a need to explore alternative and innovative ways of satisfying the requirements. [TSAA04]	(K) Explains why the initial solution concept is often not the best option. [TSAA04-10K] (K) Explains how Architectures facilitate development and assessment of alternative design solutions. [TSAA04-20K] (K) Explains the idea of a solution being "good enough" from a cost-benefit perspective when generating a solution (e.g. why an "80% correct solution" might be sufficient if the extra 20% costs the majority of the available project budget). [TSAA04-30K] (K) Explains why adaptation of an existing solutions is often not the best option. [TSAA04-40K] (K) Defines terms such as design-neutral, select, trade-off, solution space. [TSAA04-50K] (K) Explains why there is a need to avoid cognitive bias or decision traps. [TSAA04-60K] (K) Explains how creative thinking techniques and formal design methodologies aid in exploring solution space. [TSAA04-70K]
5	Explains why alternative discipline technologies can be used to satisfy the same requirement and provides examples. [TSAA05]	(K) Explains how different technologies can often achieve the same effect, but in a different way (e.g. combustion engine VS battery power). [TSAA05-10K] (K) Explains how different technologies – e.g. software vs. hardware – can provide solutions. [TSAA05-20K]
6	Describes the process and key artifacts of functional analysis. [TSAA06]	(K) Describes how a system architecture can be formally represented through the relationships between a set of interrelated architectural entities (e.g. functions, function flows, interfaces, resource flow items, information/data elements, physical elements, containers, nodes, links, and communication resources). [TSAA06-10K] (K) Describes how these architectural entities may possess architectural characteristics (e.g. dimensions, environmental resilience, availability, robustness, learnability, execution efficiency, and mission effectiveness). [TSAA06-20K] (K) Explains with examples, types of diagrammatic notations that can be used to capture or describe system behavior. [TSAA06-30K]
7	Explains why there is a need for functional models of the system. [TSAA07]	(K) Explains why a functional model allows thorough and robust understanding of the system performance requirements. [TSAA07-10K] (K) Explains how functional models enable the assessment and comparison of multiple physical design models to determine a best option. [TSAA07-20K]
8	Explains how outputs from functional analysis relate to the overall system design and provides examples. [TSAA08]	(K) Explains how a functional model often helps identify gaps in system requirements. [TSAA08-10K] (K) Explains how functional models enable the assessment and comparison of multiple physical design models to determine a best option. [TSAA08-20K]

Supervised Practitioner – Technical: System Architecting			
ID	**Indicators of Competence** ***(in addition to those at Awareness level)***	**Relevant knowledge, experience, and/or behaviors**	**Possible examples of objective evidence of personal contribution to activities performed, or professional behaviors applied**
1	Uses a governing process using appropriate tools to manage and control their own system architectural design activities. [TSAS01]	(K) Describes key elements of an architectural design process (e.g. define scope of architecture, define purpose, determine appropriate tool & language, and establish or leverage enterprise guidance to ensure consistency and coherence of model). [TSAS01-10K]	System architecture diagrams traceable to requirements; Functional Model; System model (e.g. in SysML). [TSAS01-E10]
		(A) Uses a traceable approach to architectural design. [TSAS01-20A]	Processes used for managing a system architecture (e.g. Model Configuration Management Plan). [TSAS01-E20]
2	Uses analysis techniques or principles used to support an architectural design process. [TSAS02]	(K) Describes a systems architectural design, identifying its key architectural features and why they are present. [TSAS02-10K]	Records (e.g. meeting minutes) showing participation in architectural design review. [TSAS02-E10]
		(K) Describes concepts of abstraction and the benefits of controlling complexity. [TSAS02-20K]	Architecture design methodology or modeling methodology document used. [TSAS02-E20]
		(K) Explains the differences between types of architectures. [TSAS02-30K]	Architecture standard(s) document used. [TSAS02-E30]
		(K) Describes a set of architectural design principles. [TSAS02-40K]	
		(A) Uses architectural design techniques to support systems architecting work on a project. [TSAS02-50A]	
		(K) Describes the advantages of a formal approach. [TSAS02-60K]	
3	Develops multiple different architectural solutions (or parts thereof) meeting the same set of requirements to highlight different options available. [TSAS03]	(K) Explains the idea of having multiple "views" of a system model. [TSAS03-10K]	Analysis-of-alternatives study. [TSAS03-E10]
		(A) Performs architecture trade-offs in terms of finding an acceptable balance between constraints such as performance, cost, and time parameters. [TSAS03-20A]	System architecture diagrams traceable to requirements; Functional Model; System model (e.g. in SysML). [TSAS03-E20]
		(A) Develops alternative architectural designs from a set of requirements. [TSAS03-30A]	

(*Continued*)

ID	Indicators of Competence *(in addition to those at Awareness level)*	Relevant knowledge, experience, and/or behaviors	Possible examples of objective evidence of personal contribution to activities performed, or professional behaviors applied
		(A) Identifies different architectural design considerations when following different approaches to architectural design. [TSAS03-40A] (A) Uses differing approaches required for different architectural design considerations. [TSAS03-50A]	
4	Produces traceability information linking differing architectural design solutions to requirements. [TSAS04]	(A) Produces traceability information linking differing architectural design solutions to requirements. [TSAS04-10A] (A) Identifies areas where architectural design solution(s) fail to meet requirements or go beyond identified requirements. [TSAS04-20A]	System architecture diagrams traceable to requirements; Functional Model; System model (e.g. in SysML). [TSAS04-E10]
5	Uses different techniques to develop architectural solutions. [TSAS05]	(K) Describes a range of different creativity techniques such as; brainstorming, lateral thinking, TRIZ, highlighting their strengths, and weaknesses. [TSAS05-10K] (K) Explains why there is a need for research & data collection to generate concepts. [TSAS05-20K] (K) Explains set based design principles. [TSAS05-30K]	Multiple functional behavior diagrams and design diagrams, traceable to requirements. [TSAS05-E10]
6	Compares the characteristics of different concepts to determine their strengths and weaknesses. [TSAS06]	(K) Explains how different concepts can exist against the same requirement. [TSAS06-10K] (K) Explains how to identify appropriate selection criteria. [TSAS06-20K] (A) Analyzes the feasibility of possible solutions. [TSAS06-30A]	Feasibility studies, trade studies, Quality Function Deployment (QFD), or similar need analysis documentation. [TSAS06-E10] Documented analysis approach, selection criteria, and weighting factors. [TSAS06-E20]
7	Prepares a functional analysis using appropriate tools and techniques to characterize a system. [TSAS07]	(A) Uses tools and techniques used in support of functional analysis. [TSAS07-10A] (A) Uses physics based or stochastic models to simulate performance against requirements. [TSAS07-20A] (A) Prepares executable system models. [TSAS07-30A]	Decision Management Plan, or similar document defining approaches and tools for analyzing solutions. [TSAS07-E10] Results of analysis models or simulations. [TSAS07-E20]

8	Prepares architectural design work products (or parts thereof) traceable to the requirements. [TSAS08]	(A) Prepares architectural design documentation. [TSAS08-10A] (A) Prepares multiple "views" of a system model. [TSAS08-20A] (A) Prepares decision management or trade-off documentation as part of an architectural solution selection activity. [TSAS08-30A]	System architecture diagrams traceable to requirements; Functional Model; System model (e.g. in SysML). [TSAS08-E10] System architecture views (e.g. for an Architectural framework such as DoDAF/MoDAF OV-1, SV-4). [TSAS08-E20]
9	Develops own understanding of this competency area through Continual Professional Development (CPD). [TSAS09]	(A) Identifies potential gaps in own knowledge or development needs in this area, identifying opportunities to address these through continual professional development activities. [TSAS09-10A] (A) Performs continual professional development activities to improve their knowledge and understanding in this area. [TSAS09-20A] (A) Records continual professional development activities undertaken including learning or insights gained. [TSAS09-30A]	Records of Continual Professional Development (CPD) performed and learning outcomes. [TSAS09-E10]

Practitioner – Technical: System Architecting			
ID	**Indicators of Competence** ***(in addition to those at Supervised Practitioner level)***	**Relevant knowledge, experience, and/or behaviors**	**Possible examples of objective evidence of personal contribution to activities performed, or professional behaviors applied**
1	Creates a strategy for system architecting on a project to support SE project and wider enterprise needs. [TSAP01]	(A) Identifies project-specific system architecting needs that need to be incorporated into SE planning. [TSAP01-10A]	SE system architecting strategy work products. [TSAP01-E10]
		(A) Prepares project-specific system architecting inputs for SE planning purposes, using appropriate processes, and procedures. [TSAP01-20A]	Task estimates for system architecting efforts. [TSAP01-E20]
		(A) Prepares project-specific system architecting task estimates in support of SE planning. [TSAP01-30A]	
		(P) Collaborates throughout with stakeholders to gain approval, updating strategy as necessary. [TSAP01-40P]	
2	Creates a governing process, plan, and associated tools for systems architecting, which reflect project and business strategy. [TSAP02]	(K) Identifies the steps necessary to define a process and appropriate techniques to be adopted for the system architecting of system elements. [TSAP02-10K]	SE system architecting planning work products they have developed. [TSAP02-E10]
		(K) Describes key elements of a successful systems architecting process and how systems architecting activities relate to different stages of a system life cycle. [TSAP02-20K]	System specific architecting approach they have developed. [TSAP02-E20]
		(A) Selects an appropriate system architecting tools or uses a methodology. [TSAP02-30A]	
		(A) Uses appropriate standards and governing processes in their system architecting planning, tailoring as appropriate. [TSAP02-40A]	
		(A) Selects system architecting method for each system element. [TSAP02-50A]	
		(A) Creates system architecting plans and processes for use on a project. [TSAP02-60A]	
		(P) Collaborates throughout with stakeholders to gain approval, updating plans as necessary. [TSAP02-70P]	

3	Uses plans and processes for system architecting, interpreting, evolving, or seeking guidance where appropriate. [TSAP03]	(A) Follows defined plans, processes and associated tools to perform system architecting on a project. [TSAP03-10A]	System architecting plans or work products interpreted to address project issues. [TSAP03-E10]
		(P) Guides and actively coordinates others interpretation of system architecting plans or processes in order to complete activities system architecting activities successfully. [TSAP03-20P]	Process or tool guidance. [TSAP03-E20]
		(A) Prepares System architecting data and reports in support of wider system measurement, monitoring, and control. [TSAP03-30A]	
		(P) Recognizes situations where deviation from published plans and processes or clarification from others is appropriate in order to overcome complex system architecting challenges. [TSAP03-40P]	
		(P) Recognizes situations where existing strategy, process, plan or tools require formal change, liaising with stakeholders to gain approval as necessary. [TSAP03-50P]	
4	Creates alternative architectural designs traceable to the requirements to demonstrate different approaches to the solution. [TSAP04]	(K) Explains how concepts such as Horizon scanning and technology watching can assist in developing a strategy for architecting. [TSAP04-10K]	Architectural design documentation (conceptual, functional, logical, and physical). [TSAP04-E10]
		(A) Develops alternative architectural design solutions from a set of requirements. [TSAP04-20A]	Certificate of proficiency with SysML or similar system modeling languages. [TSAP04-E20]
		(A) Uses architectural frameworks in assisting consistency and re-usability of architectural design. [TSAP04-30A]	Architectural impact analysis from cost-benefit documentation. [TSAP04-E30]
		(A) Identifies the merits or consideration in different architectural design solutions. [TSAP04-40A]	Architectural impact analysis of user interviews. [TSAP04-E40]
		(A) Uses an architectural design tool, methodology, or modeling language. [TSAP04-50A]	Identified new technologies. [TSAP04-E50]
		(A) Uses different architectural approaches to establish preferred solution approaches of different stakeholders. [TSAP04-60A]	Identified new ideas. [TSAP04-E60]
		(A) Develops new ideas as part of alternative concept development ("solutioneering"). [TSAP04-70A]	
		(A) Identifies new technologies. [TSAP04-80A]	

(Continued)

ID	Indicators of Competence *(in addition to those at Supervised Practitioner level)*	Relevant knowledge, experience, and/or behaviors	Possible examples of objective evidence of personal contribution to activities performed, or professional behaviors applied
5	Analyzes options and concepts in order to demonstrate that credible, feasible options exist. [TSAP05]	(K) Describes the purpose and potential challenges of reviewing different architectural design solutions. [TSAP05-10K]	Trade study showing alternatives and the solution selected. [TSAP05-E10]
		(A) Performs architecture trade-offs in terms of finding an acceptable balance between constraints such as performance, cost, and time parameters. [TSAP05-20A]	Architectural design document. [TSAP05-E20]
		(A) Selects preferred options from those available, listing advantages and disadvantages. [TSAP05-30A]	Reports/minutes of brainstorming sessions. [TSAP05-E30]
6	Uses appropriate analysis techniques to ensure different viewpoints are considered. [TSAP06]	(K) Lists and describes key characteristics of different analysis techniques (e.g. cost analysis, technical risk analysis effectiveness analysis, or other recognized formal analysis techniques). [TSAP06-10K]	Results of a cost analysis, technical risk analysis effectiveness analysis or other recognized formal analysis techniques. [TSAP06-E10]
		(A) Uses techniques for analyzing the effectiveness of a particular architectural solution and selecting the most appropriate solution. [TSAP06-20A]	Minutes of meetings, reports, design documents. [TSAP06-E20]
		(K) Describes the advantages and limitations of the use of architectural design tools in relation to at least one tool. [TSAP06-30K]	
7	Elicits derived discipline specific architectural constraints from specialists to support partitioning and decomposition. [TSAP07]	(A) Identifies areas where discipline implementation or technology constraints dictate partitioning of functionality (e.g. software, hardware, human factors, packaging, and safety). [TSAP07-10A]	Solution partitioning documentation. [TSAP07-E10]
		(A) Develops architectural partitioning between discipline technologies. [TSAP07-20A]	
		(A) Elicits partitioning requirements from discipline technologies such as software, hardware, human factors, packaging, and safety [TSAP07-30A]	
8	Uses the results of system analysis activities to inform system architectural design. [TSAP08]	(A) Analyzes potential options against selection criteria. [TSAP08-10A]	Trade study reports/conclusions. [TSAP08-E10]
		(A) Selects credible solutions using criteria. [TSAP08-20A]	QFD analysis diagram. [TSAP08-E20]
		(A) Analyzes selection in qualitative and quantitative terms, justifying as required. [TSAP08-30A]	Cost-benefit/effectiveness analysis report. [TSAP08-E30]

		(A) Uses trade-off studies to assist in addressing solution challenges (e.g. feasibility, risk, cost, schedule, technology requirements, human factors, and "-ilities"). [TSAP08-40A] (K) Describes the characteristics of Quality Function Deployment (QFD) and how it can be used to support solution selection. [TSAP08-50K] (A) Uses cost-benefit/effectiveness analysis techniques to inform the architectural design activity. [TSAP08-60A]	
9	Identifies the strengths and weaknesses of relevant technologies in the context of the requirement and provides examples. [TSAP09]	(A) Prepares inputs to trade studies, feasibility analysis, QFD, & creativity techniques. [TSAP09-10A] (A) Identifies strengths and weaknesses of a concept against its requirements. [TSAP09-20A] (K) Describes different technologies that can be used to implement solutions (hardware, software, people, processes, etc.). [TSAP09-30K]	Records of reports/papers drawing conclusions of trade studies, feasibility analysis, QFD, & creativity techniques. [TSAP09-E10]
10	Monitors key aspects of the evolving design solution in order to adjust architecture, if appropriate. [TSAP10]	(A) Reviews the impact of inserting new design components on overall system design. [TSAP10-10A] (A) Adapts an existing system design to incorporate new design components. [TSAP10-20A] (A) Analyzes the impact of evolving requirements in the system architecture. [TSAP10-30A]	Configuration managed architecture model. [TSAP10-E10] Changes requirements and associated impact analysis on solution. [TSAP10-E20]
11	Guides new or supervised practitioners in Systems Architecting to develop their knowledge, abilities, skills, or associated behaviors. [TSAP11]	(P) Guides new or supervised practitioners in executing activities that form part of this competency. [TSAP11-10P] (A) Trains individuals to an "Awareness" level in this competency area. [TSAP11-20A]	Organizational Breakdown Structure showing their responsibility for technical supervision in this area. [TSAP11-E10] Records of on-the-job training objectives/guidance etc. [TSAP11-E20] Coaching or mentoring assignment records. [TSAP11-E30] Records highlighting their impact on another individual in terms of improvement or professional development in this competency. [TSAP11-E40]
12	Maintains and enhances own competence in this area through Continual Professional Development (CPD) activities. [TSAP12]	(A) Identifies potential development needs in this area, identifying opportunities to address these through continual professional development activities. [TSAP12-10A] (A) Performs continual professional development activities to maintain and enhance their competency in this area. [TSAP12-20A] (A) Records continual professional development activities undertaken including learning or insights gained. [TSAP12-30A]	Records of Continual Professional Development (CPD) performed and learning outcomes. [TSAP12-E10]

Lead Practitioner – Technical: System Architecting			
ID	**Indicators of Competence** ***(in addition to those at Practitioner level)***	**Relevant knowledge, experience, and/or behaviors**	**Possible examples of objective evidence of personal contribution to activities performed, or professional behaviors applied**
1	Creates enterprise-level policies, procedures, guidance, and best practice for system architectural design including associated tools. [TSAL01]	(A) Analyzes enterprise need for system architecting policies, processes, tools, or guidance. [TSAL01-10A]	Records showing their role in embedding systems architecting into enterprise policies (e.g. guidance introduced at enterprise level and enterprise-level review minutes). [TSAL01-E10]
		(A) Creates enterprise policies, procedures, or guidance for system architecting activities. [TSAL01-20A]	Procedures they have written. [TSAL01-E20]
		(A) Selects and acquires appropriate tools supporting system architecting activities. [TSAL01-30A]	Records of support for tool introduction. [TSAL01-E30]
2	Assesses the tailoring of enterprise-level system architectural design processes to meet the needs of a project. [TSAL02]	(A) Evaluates the enterprise process against the business and external stakeholder (e.g. customer) needs in order to tailor processes to enable project success. [TSAL02-10A]	Records of process architectural tailoring. [TSAL02-E10]
		(A) Produces constructive feedback on architectural design process for projects across the enterprise. [TSAL02-20A]	Design process feedback. [TSAL02-E20]
3	Advises stakeholders across the enterprise on selection of architectural design and functional analysis techniques to ensure effectiveness and efficiency of approach. [TSAL03]	(K) Describes a full range of architectural design techniques and their potential applicability for different types of system or complexity. [TSAL03-10K]	Extracts demonstrating architectural design tools or techniques used (e.g. Solution abstraction; Clustering; Interface minimization; and Layering). [TSAL03-E10]
		(A) Uses a full range of architectural design techniques for a range of systems based upon complexity. [TSAL03-20A]	
4	Judges the suitability of architectural solutions across the enterprise in areas of complex or challenging technical requirements or needs. [TSAL04]	(A) Reviews architectural solutions across the enterprise for the approach taken to address complex or challenging technical areas. [TSAL04-10A]	Records of a review process for complex architectural area in which they have been involved. [TSAL04-E10]
		(A) Advises on appropriateness of architecture solution proposed for these areas. [TSAL04-20A]	Records of advice or comments provided on a complex architectural solution. [TSAL04-E20]
			Architectural design approvals, Preliminary Design Review (PDR), Solution Functional Review (SFR) documentation. [TSAL04-E30]
5	Assesses system architectures across the enterprise, to determine whether they meet the overall needs of individual projects. [TSAL05]	(A) Assesses architectural design documentation (e.g. paperwork, modeling) to determine its adequacy, correctness, completeness, coherence, or consistency. [TSAL05-10A]	Records of advice or assessment comments provided on an architectural design. [TSAL05-E10]

		(A) Produces architecture design assessment information. [TSAL05-20A]	Architectural design approvals, Preliminary Design Review (PDR), Solution Functional Review (SFR) documentation. [TSAL05-E20]
6	Persuades key stakeholders across the enterprise to address identified enterprise-level Systems Engineering architectural design issues to reduce project cost, schedule, or technical risk. [TSAL06]	(A) Identifies and engages with key stakeholders. [TSAL06-10A]	Minutes of meetings. [TSAL06-E10]
		(P) Fosters agreement between key stakeholders at enterprise level to resolve architectural design issues, by promoting a holistic viewpoint. [TSAL06-20P]	Minutes of key stakeholder review meetings and actions taken. [TSAL06-E20]
		(P) Persuades key stakeholders to resolve conflict and identify optimal design options. [TSAL06-30P]	Preliminary Design Review (PDR), Solution Functional Review (SFR) documentation, or Architectural review documentation. [TSAL06-E30]
7	Coaches or mentors practitioners across the enterprise in Systems Architecting in order to develop their knowledge, abilities, skills, or associated behaviors. [TSAL07]	(P) Coaches or mentors practitioners across the enterprise in competency-related techniques, recommending development activities. [TSAL07-10P]	Coaching or mentoring assignment records. [TSAL07-E10]
		(A) Develops or authorizes enterprise training materials in this competency area. [TSAL07-20A]	Records of formal training courses, workshops, seminars, and authored training material supported by successful post-training evaluation data. [TSAL07-E20]
		(A) Provides enterprise workshops/seminars or training in this competency area. [TSAL07-30A]	Listing as an approved organizational trainer for this competency area. [TSAL07-E30]
8	Promotes the introduction and use of novel techniques and ideas in Systems Architecting across the enterprise, to improve enterprise competence in this area. [TSAL08]	(A) Analyzes different approaches across different domains through research. [TSAL08-10A]	Research records. [TSAL08-E10]
		(A) Defines novel approaches that could potentially improve the SE discipline within the enterprise. [TSAL08-20A]	Published papers in refereed journals/company literature. [TSAL08-E20]
		(P) Fosters awareness of these novel techniques within the enterprise. [TSAL08-30P]	Records showing introduction of enabling systems supporting the new techniques or ideas. [TSAL08-E30]
		(P) Collaborates with enterprise stakeholders to facilitate the introduction of techniques new to the enterprise. [TSAL08-40P]	Published papers (or similar) at enterprise level. [TSAL08-E40]
		(A) Monitors new techniques after their introduction to determine their effectiveness. [TSAL08-50A]	Records of improvements made against a recognized process improvement model in this area. [TSAL08-E50]
		(A) Adapts approach to reflect actual enterprise performance improvements. [TSAL08-60A]	

(Continued)

ID	Indicators of Competence *(in addition to those at Practitioner level)*	Relevant knowledge, experience, and/or behaviors	Possible examples of objective evidence of personal contribution to activities performed, or professional behaviors applied
9	Develops expertise in this competency area through specialist Continual Professional Development (CPD) activities. [TSAL09]	(A) Identifies own needs for further professional development in order to increase competence beyond practitioner level. [TSAL09-10A] (A) Performs professional development activities in order to move own competence toward expert level. [TSAL09-20A] (A) Records continual professional development activities undertaken including learning or insights gained. [TSAL09-30A]	Records of Continual Professional Development (CPD) performed and learning outcomes. [TSAL09-E10]

NOTES	In addition to items above, enterprise-level or independent 3rd-Party-generated evidence may be used to amplify other evidence presented and may include: a. Formally recognized by senior management in current organization as an expert in this competency area b. Evidence of role as Product/System Design Authority or Technical Authority on a complex project with responsibilities in this area or where skills within this competency area were used c. Recognized as an authorizing signatory on behalf of enterprise for formal documentation in this competency area (e.g. policies, processes, and deliverables) d. Formal commendation or award within own enterprise for contribution or item of work successfully performed, which required proficiency in this competency area e. Customer, Supplier, or other external project-specific key Stakeholder accolades for specific work performed in this competency area f. Independently assessed or accredited work in this competency area (e.g. for independent publication or use) g. Formal organizational HR records positively highlighting any specific professional competencies or behaviors identified (if applicable) plus any of the evidence indicators listed at Expert level below

Expert – Technical: System Architecting			
ID	**Indicators of Competence** ***(in addition to those at Lead Practitioner level)***	**Relevant knowledge, experience, and/or behaviors**	**Possible examples of objective evidence of personal contribution to activities performed, or professional behaviors applied**
1	Communicates own knowledge and experience in Systems Architecting in order to promote best practice beyond the enterprise boundary. [TSAE01]	(A) Produces papers, seminars, or presentations outside own enterprise for publication in order to share own ideas and improve industry best practices in this competence area. [TSAE01-10A]	Published papers or books etc. on new technique in refereed journals/company literature. [TSAE01-E10]
		(P) Fosters incorporation of own ideas into industry best practices in this area. [TSAE01-20P]	Published papers in refereed journals or internal literature proposing new practices in this competence area (or presentations, tutorials, etc.). [TSAE01-E20]
		(P) Develops guidance materials identifying new (or updating existing) best practice in this competence area. [TSAE01-30P]	Proposals adopted as industry best practice. [TSAE01-E30]
2	Persuades key stakeholders beyond the enterprise boundary in order to facilitate the system architectural design. [TSAE02]	(A) Identifies and engages with external stakeholders to facilitate architectural design. [TSAE02-10A]	Minutes of meetings. [TSAE02-E10]
		(P) Acts to change external stakeholder thinking in the area of system architecture and design. [TSAE02-20P]	Revised architectural design. [TSAE02-E20]
3	Advises organizations beyond the enterprise boundary on the suitability of their approach to system architectural design. [TSAE03]	(A) Assesses system architectural designs as independent reviewer, on behalf of an external organization. [TSAE03-10A]	Records of membership on oversight committee with relevant terms of reference. [TSAE03-E10]
		(P) Acts with sensitivity on highly complex system architecting issues making limited use of specialized, technical terminology and taking account of external stakeholders' (e.g. customer's) background and knowledge. [TSAE03-20P]	Participates in a review committee or similar body that deals with approval of such plans. [TSAE03-E20]
		(P) Communicates using a holistic approach to complex system architecting issue resolution including balanced, rational arguments on way forward. [TSAE03-30P]	Review comments or revised document. [TSAE03-E30]
		(A) Selects balanced Systems Engineering system architecting solution from options available, on behalf of stakeholders beyond enterprise boundary. [TSAE03-40A]	Records indicating changed stakeholder thinking from personal intervention. [TSAE03-E40]
		(P) Persuades stakeholders beyond the enterprise boundary to accept difficult project-specific system architecting recommendations or actions. [TSAE03-50P]	

(Continued)

ID	Indicators of Competence *(in addition to those at Lead Practitioner level)*	Relevant knowledge, experience, and/or behaviors	Possible examples of objective evidence of personal contribution to activities performed, or professional behaviors applied
4	Advises organizations beyond the enterprise boundary on improving their handling of complex or sensitive Systems Architecting issues. [TSAE04]	(A) Advises external stakeholders (e.g. customers) on their system architecture requirements and issues. [TSAE04-10A]	Records of advice to customers on architecture issues. [TSAE04-E10]
		(P) Conducts sensitive negotiations on a highly complex system architecture making limited use of specialized, technical terminology. [TSAE04-20P]	Records from negotiations demonstrating awareness of customer's background and knowledge. [TSAE04-E20]
		(A) Uses a holistic approach to complex issue resolution including balanced, rational arguments on way forward. [TSAE04-30A]	Stakeholder approval of architectural design. [TSAE04-E30]
5	Advises organizations beyond the enterprise boundary on improving their concept generation activities. [TSAE05]	(A) Produces guidance in concepts generation for external stakeholders. [TSAE05-10A]	Concept document and associated guidance provided. [TSAE05-E10]
6	Champions the introduction of novel techniques and ideas in systems architecting, beyond the enterprise boundary, in order to develop the wider Systems Engineering community in this competency. [TSAE06]	(A) Analyzes different approaches across different domains through research. [TSAE06-10A]	Records of activities promoting research and need to adopt novel technique or ideas. [TSAE06-E10]
		(A) Defines novel approaches that could potentially improve the wider SE discipline. [TSAE06-20A]	Records of improvements made to process and appraisal against a recognized process improvement model. [TSAE06-E20]
		(P) Fosters awareness of these novel techniques within the wider SE community. [TSAE06-30P]	Research records. [TSAE06-E30]
		(P) Collaborates with those introducing novel techniques within the wider SE community. [TSAE06-40P]	Published papers in refereed journals/company literature. [TSAE06-E40]
		(A) Produces reports for the wider SE community on the effectiveness of new techniques after their introduction. [TSAE06-50A]	Records showing introduction of enabling systems supporting the new techniques or ideas. [TSAE06-E50]
7	Coaches individuals beyond the enterprise boundary in Systems Architecting, in order to further develop their knowledge, abilities, skills, or associated behaviors. [TSAE07]	(P) Coaches or mentors individuals beyond the enterprise boundary, in competency-related techniques, recommending development activities. [TSAE07-10P]	Coaching or mentoring assignment records. [TSAE07-E10]

		(A) Develops or authorizes training materials in this competency area, which are subsequently successfully delivered beyond the enterprise boundary. [TSAE07-20A]	Records of formal training courses, workshops, seminars, and authored training material supported by successful post-training evaluation data. [TSAE07-E20]
		(A) Provides workshops/seminars or training in this competency area for practitioners or lead practitioners beyond the enterprise boundary (e.g. conferences and open training days). [TSAE07-30A]	Records of Training/workshops/seminars created supported by successful post-training evaluation data. [TSAE07-E30]
8	Maintains expertise in this competency area through specialist Continual Professional Development (CPD) activities. [TSAE08]	(A) Reviews research, new ideas, and state of the art to identify relevant new areas requiring personal development in order to maintain expertise in this competency area. [TSAE08-10A]	Records of documents reviewed and insights gained as part of own research into this competency area. [TSAE08-E10]
		(A) Performs identified specialist professional development activities in order to maintain or further develop competence at expert level. [TSAE08-20A]	Records of Continual Professional Development (CPD) performed and learning outcomes. [TSAE08-E20]
		(A) Records continual professional development activities undertaken including learning or insights gained. [TSAE08-30A]	

NOTES	In addition to items above, enterprise-level or independent 3rd-Party-generated evidence may be used to amplify other evidence presented and may include: a. Formally recognized by a reputable external organization as an expert in this competency area b. Evidence of role as independent assessor or reviewer on project outside own organization where skills in this competency area were used c. Evidence of invitation(s) from wider community for contribution of systems engineering expertise in this area (e.g. industry conference panel, government advisory board etc. cross-industry working groups, partnerships, accredited advanced university courses or research, or as part of professional institute) d. Formal commendation beyond the enterprise (e.g. by INCOSE or other recognized authority) for work performed in this competency area e. Independently assessed or accredited work product in this competency area (e.g. for independent publication or use) f. Accolades of expertise in this area from recognized industry leaders

Competency area – Technical: Design for…
Description Ensuring that the requirements of all life cycle stages are addressed at the correct point in the system design. During the design process consideration should be given to the design attributes such as manufacturability, testability, reliability, maintainability, affordability, safety, security, human factors, environmental impacts, robustness and resilience, flexibility, interoperability, capability growth, disposal, cost, and natural variations. Includes the need to design for robustness. A robust system is tolerant of misuse, out of spec scenarios, component failure, environmental stress, and evolving needs.
Why it matters Failure to design for these attributes at the correct point in the development life cycle may result in the attributes never being achieved or achieved at escalated cost. In particular, a robust system provides greater availability during operation.
Possible contributory types of evidence Any combination of the types of evidence may be acceptable (depending on how the Framework is tailored and used). The evidence items identified at each level indicate example work products only. Contributions to work products will generally differ at each proficiency level.
Learning and development The INCOSE Professional Development Portal provides example guidance on how to gain an initial awareness of a competency area and options for developing further competence thereafter.

Awareness – Technical: Design for…		
ID	**Indicators of Competence**	**Relevant knowledge sub-indicators**
1	Explains why there is a need to accommodate the requirements of all life cycle stages when determining a solution. [TDFA01]	(K) Identifies 'Design for…' attributes of a system within their domain. [TDFA01-10K] (K) Identifies from later parts of the life cycle those activities for which 'Design for…' expertise would be beneficial during the design phase, such as testability, manufacturability, and maintainability. [TDFA01-20K] (K) Describes the advantages of considering such design attributes early on to mitigate against increased costs further downstream. [TDFA01-30K] (K) Explains why whole life cycle cost is important. [TDFA01-40K] (K) Explains why there is a need for design trade-offs to address potential conflicts in "design for…" requirements. Describes the importance of legislation. [TDFA01-50K] (K) Identifies how legislation can impact different specialties (e.g. safety, security, reliability, and environmental impact). [TDFA01-60K]
2	Identifies design attributes and explains why attributes must be balanced using trade-off studies. [TDFA02]	(K) Describes examples of design attributes associated with a design specialty and how to trade-off them against other system requirements. [TDFA02-10K] (K) Explains how analysis resulting from tradeoff studies enables the selection of the most effective alternative or candidate design. [TDFA02-20K]

3	Identifies different design specialties and describes their role and key activities. [TDFA03]	(K) Describes an example of a design specialty – something that may impose additional functional or nonfunctional requirements on the system design. [TDFA03-10K]
		(K) Describes multiple examples of design specialties (e.g. manufacturability, reliability, maintainability, affordability, safety, security, human system integration, training, environmental impact analysis, resilience, interoperability, affordability, and supportability). [TDFA03-20K]
4	Explains why it is important to integrate design specialties into the solution and how this can be a potential source of conflict with requirements. [TDFA04]	(K) Explains the meaning of the term integration of specialties. [TDFA04-10K] (K) Explains the types of conflict that may occur between specialties – for example a requirement for reliability may increase cost or decrease ability to implement modular design. [TDFA04-20K] (K) Identifies the practical implementation of considering design specialties early in system design, with regard to resources. [TDFA04-30K] (K) Describes how customer satisfaction can impact choices made when addressing the needs of different specialties. [TDFA04-40K]
5	Explains how design specialties can affect the cost of ownership and provides examples. [TDFA05]	(K) Explains how different implementation levels of some specialties (such as availability and reliability) may affect the cost of ownership (total costs of delivered solution). [TDFA05-10K] (K) Identifies typical analytical techniques used to calculate whole life costs or cost of ownership. [TDFA05-20K]
6	Explains how the design, throughout the life cycle, affects the robustness of the solution. [TDFA06]	(K) Explains the relationship between design and life cycle. [TDFA06-10K] (K) Explains why there may be many drivers in determining the necessary level of robustness for a system. [TDFA06-20K] (K) Explains the meaning of "robustness" and why robustness has to be designed in (e.g. the effects of intentional or unintentional misuse). [TDFA06-30K] (K) Explains how robustness affects reliability. [TDFA06-40K] (K) Explains why human factors are likely to play a part in the ultimate robustness of a system (both explicitly in a system containing humans and through human involvement in the Systems Engineering process). [TDFA06-50K]
7	Describes the relationship between reliability, availability, maintainability, and safety. [TDFA07]	(K) Explains the mathematical relationships between reliability, availability, and maintainability characteristics (e.g. Mean Time To Repair). [TDFA07-10K] (K) Explains the idea of "design integrity" and how this relates to safety, reliability, availability, and maintainability. [TDFA07-20K] (K) Explains how requirements for safety, reliability, availability, and maintainability requirements can be interrelated. [TDFA07-30K]

Supervised Practitioner – Technical: Design For…			
ID	**Indicators of Competence** ***(in addition to those at Awareness level)***	**Relevant knowledge, experience, and/or behaviors**	**Possible examples of objective evidence of personal contribution to activities performed, or professional behaviors applied**
1	Uses a governing process using appropriate tools and techniques to manage and control their own specialty engineering activities, interpreting, evolving, or seeking guidance where appropriate. [TDFS01]	(K) Explains why that there are areas of expertise that need greater depth of knowledge. [TDFS01-10K]	Records (e.g. meeting minutes) showing participation with an interdisciplinary team including subject matter experts for specialties. [TDFS01-E10]
		(K) Explains why advances in certain specialties may give market advantage. [TDFS01-20K]	System requirements matrices or models showing incorporation of specialty requirements. [TDFS01-E20]
		(K) Explains why early involvement of specialties may reduce cost and timescales by avoiding later problems. [TDFS01-30K]	Organizational document or self-developed document describing a set processes for managing design specialties such as reliability, cybersecurity, and interoperability. [TDFS01-E30]
		(A) Ensures specialty requirements are incorporated into system documentation. [TDFS01-40A]	
		(K) Explains why specialties may be key design drivers, such as safety or security. [TDFS01-50K]	
2	Explains the concept of design attributes, explaining how they influence the design. [TDFS02]	(K) Describes the benefits of a multidisciplinary system design team and Identifies "design for…" practitioners both generically and with reference to specific domains. [TDFS02-10K]	System design documents or models showing incorporation of specialty requirements in system design. [TDFS02-E10]
		(K) Explains why there is a need to tailor a team for different systems. [TDFS02-20K]	Records (e.g. meeting minutes) showing participation in peer-review of designs. [TDFS02-E20]
		(K) Explains how design attributes from different specialties and how they may influence a design. [TDFS02-30K]	
		(A) Identifies the impact of design attributes from different specialties on a design. [TDFS02-40A]	
3	Selects design attributes in order to balance differing specialty engineering needs. [TDFS03]	(A) Identifies generic "design for…" attributes and those specific to their domain. [TDFS03-10A]	Records (e.g. meeting minutes) showing participation in workshops for developing "design for…" design attributes within a system development. [TDFS03-E10]
		(A) Identifies why applicable attributes may change depending on context. [TDFS03-20A]	
4	Reviews design attributes with specialists to ensure they are addressed. [TDFS04]	(A) Identifies required specialists and ensures their expertise is included in design considerations. [TDFS04-10A]	Records (e.g. meeting minutes) showing participation with an interdisciplinary team including subject matter experts for specialties. [TDFS04-E10]
5	Identifies conflicting demands from differing design specialties and records these in trade studies in order to compare alternative solutions. [TDFS05]	(K) Explains the purpose of trade-offs and why they may be required to address inconsistencies between cost, schedule, performance, safety, and other specialties. [TDFS05-10K]	Records of activities that addressed requirement inconsistencies or conflicts. [TDFS05-E10]

		(A) Communicates with specialists explaining the need for compromise. [TDFS05-20A] (K) Identifies potential motivations and viewpoints of specialists. [TDFS05-30K] (A) Analyzes inputs from specialists, to ensure they meet the needs of the system development. [TDFS05-40A]	Project/Program documents capturing agreed trade-offs. [TDFS05-E20] Records (e.g. meeting minutes) showing participation with an interdisciplinary team including subject matter experts for specialties. [TDFS05-E30]
6	Elicits operational environment characteristics from specialty engineers in support of specialty engineering activities. [TDFS06]	(K) Explains how an operational environment impacts on the implementation of design specialties. [TDFS06-10K]	Records (e.g. meeting minutes) showing participation in specialty assessment activities (e.g. assessments of system safety requirements in the operational environment). [TDFS06-E10]
7	Compares specialty characteristics for proposed solutions and records these in trade studies in order to compare differing solutions. [TDFS07]	(A) Compares techniques for analyzing system specialties (e.g. brainstorming, checklists, Failure Mode and Effects Analysis (FMEA), fault tree analysis (FTA), Monte Carlo simulation, Bayesian statistics, and Bayes nets). [TDFS07-10A]	Records from formal analysis of specialties performed. [TDFS07-E10]
8	Describes how design integrity affects their project and provides examples. [TDFS08]	(A) Identifies how reliability and maintainability affect availability on their project. [TDFS08-10A] (A) Identifies how reliability, maintainability or availability could affect affects safety on their project. [TDFS08-20A] (A) Identifies how interoperability could affect system security on their project. [TDFS08-30A]	Document showing availability, reliability, and maintainability calculations. [TDFS08-E10] Documentation identifying conflicts between reliability and safety requirements. [TDFS08-E20] Documentation identifying conflicts between interoperability and system security requirements. [TDFS08-E30]
9	Identifies constraints placed on the system because of the needs of design specialties. [TDFS09]	(A) Identifies and documents new requirements based upon design specialties. [TDFS09-10A]	Requirements documents or model reflecting the needs of the specialty. [TDFS09-E10]
10	Uses specialty engineering techniques and tools to ensure delivery of designs meeting specialty needs. [TDFS10]	(A) Uses specialty engineering techniques. [TDFS10-10A] (A) Analyzes outputs from tools/techniques to deliver designs meeting specialty needs. [TDFS10-20A]	Outputs or results from a variety of tools and techniques to address proper design. [TDFS10-E10]
11	Develops own understanding of this competency area through Continual Professional Development (CPD). [TDFS11]	(A) Identifies potential gaps in own knowledge or development needs in this area, identifying opportunities to address these through continual professional development activities. [TDFS11-10A] (A) Performs continual professional development activities to improve their knowledge and understanding in this area. [TDFS11-20A] (A) Records continual professional development activities undertaken including learning or insights gained. [TDFS11-30A]	Records of Continual Professional Development (CPD) performed and learning outcomes. [TDFS11-E10]

Practitioner – Technical: Design for…			
ID	**Indicators of Competence *(in addition to those at Supervised Practitioner level)***	**Relevant knowledge, experience, and/or behaviors**	**Possible examples of objective evidence of personal contribution to activities performed, or professional behaviors applied**
1	Creates a strategy for "Designing for…" specialties on a project to support SE project and wider enterprise needs. [TDFP01]	(A) Identifies project-specific Specialty engineering needs in support of Systems Engineering which need to be incorporated into SE planning. [TDFP01-10A]	SE Specialty engineering strategy work products. [TDFP01-E10]
		(A) Prepares project-specific Specialty engineering inputs in support of Systems Engineering planning, using appropriate processes and procedures. [TDFP01-20A]	Task estimates for specific specialty engineering efforts. [TDFP01-E20]
		(A) Coordinates the production of project-specific Specialty engineering task estimates in support of SE planning. [TDFP01-30A]	Documented modifications to strategy based upon review by Specialty Engineers. [TDFP01-E30]
		(P) Liaises throughout with Specialty engineers to gain approval of the strategy, updating as necessary. [TDFP01-40P]	
2	Creates a governing process, plan, and associated tools for Specialty engineering, which reflect project and business strategy. [TDFP02]	(K) Identifies the steps necessary to define a process and appropriate techniques to be adopted to design for Specialty engineering in system elements. [TDFP02-10K]	Specialty engineering planning work products they have developed. [TDFP02-E10]
		(K) Describes key elements of a successful process which integrates Specialty engineering needs and how different Specialty engineering disciplines relate to different stages of a system life cycle. [TDFP02-20K]	Records detailing selection of relevant specialties for inclusion and associated activities. [TDFP02-E20]
		(A) Uses outputs from Specialty engineering tools or uses a methodology to ensure a system has been designed to accommodate the specialty. [TDFP02-30A]	Inputs provided to systems engineering management plan. [TDFP02-E30]
		(A) Uses appropriate standards and governing processes to ensure that Specialty engineering needs are incorporated in wider SE planning, tailoring as appropriate. [TDFP02-40A]	
		(A) Selects appropriate Specialty engineering methods for each system element, liaising with Specialist Engineers as necessary. [TDFP02-50A]	
		(A) Creates plans and processes for use on a project when ensure specialty engineering needs are accommodated on the project. [TDFP02-60A]	
		(P) Liaises throughout with stakeholders to gain approval, updating plans as necessary. [TDFP02-70P]	

3	Uses governing plans and processes to ensure Specialty engineering is accommodated in the evolving design, interpreting, evolving, or seeking guidance where appropriate. [TDFP03]	(A) Uses defined plans, processes, and associated tools to ensure Specialty engineering needs are addressed on a project. [TDFP03-10A]	Specialty engineering plans or work products they have interpreted or challenged to address project issues. [TDFP03-E10]
		(P) Guides and actively coordinates the interpretation of plans or processes in order to ensures that Specialty engineering needs are addressed on the project. [TDFP03-20P]	Specialty engineering Processes or tools guidance. [TDFP03-E20]
		(A) Prepares specialty engineering data and reports in support of wider system measurement, monitoring and control. [TDFP03-30A]	Process and appropriate techniques adopted for specialty engineering. [TDFP03-E30]
		(P) Recognizes situations where deviation from published plans and processes or clarification from others is appropriate in order to overcome complex Specialty engineering challenges. [TDFP03-40P]	
		(P) Recognizes situations where existing Specialty engineering strategy, process, plan or tools require formal change, liaising with Specialty engineering stakeholders to gain approval as necessary. [TDFP03-50P]	
4	Selects design attributes throughout the design process balancing these to support specialty engineering needs. [TDFP04]	(A) Identifies "design for…" design attributes within a system development. [TDFP04-10A]	Relevant section of Systems Engineering Management Plan or Systems Requirement Document. [TDFP04-E10]
		(A) Identifies the interrelationship between design attributes and how they affect each other. [TDFP04-20A]	Relevant section of Systems Engineering Management Plan or other project/program plans. [TDFP04-E20]
		(A) Compares and analyzes attributes to create a balanced design. [TDFP04-30A]	Design notes and reports. [TDFP04-E30]
		(A) Identifies "dependencies" between attributes. [TDFP04-40A]	Design decision logs. [TDFP04-E40]
5	Identifies the appropriate specialists to ensure it addresses design attributes effectively and at the correct time. [TDFP05]	(A) Defines the members of a system design team at the appropriate phase in the life cycle. [TDFP05-10A]	Document or model showing abstraction of system in terms needed by specialists. [TDFP05-E10]
		(A) Communicates needs of the system to the specialists to enable the requirements of the "design for…" attributes to be addressed. [TDFP05-20A]	Requirements document showing appropriate translation of specialists requirements into system requirements. [TDFP05-E20]
		(A) Identifies system requirements by translating (as necessary) specialist requirements. [TDFP05-30A]	Organizational chart or task planning document showing involvement of appropriate specialists. [TDFP05-E30]
		(A) Identifies how a design may be changed by considering "design for" attributes. [TDFP05-40A]	

(Continued)

ID	Indicators of Competence *(in addition to those at Supervised Practitioner level)*	Relevant knowledge, experience, and/or behaviors	Possible examples of objective evidence of personal contribution to activities performed, or professional behaviors applied
6	Analyzes demands from differing design specialties highlighted in trade studies, resolving identified conflicts as necessary. [TDFP06]	(A) Develops a trade-off study. [TDFP06-10A]	Demonstrate use of trade-off tools. [TDFP06-E10]
		(P) Recognizes the implications of conflicting demands to meeting requirements and acts accordingly to remove conflict. [TDFP06-20P]	Records of activities to resolve issues. [TDFP06-E20]
		(K) Explains the purpose of trade-offs and why they may be required to address inconsistencies between cost, schedule, performance, safety, and other specialties. [TDFP06-30K]	Completed trade studies. [TDFP06-E30]
		(P) Collaborates with specialists to understand their viewpoints. [TDFP06-40P]	
		(P) Communicates with specialists explaining the need for compromise. [TDFP06-50P]	
		(A) Analyzes inputs from specialists, ensuring they meet the needs of the system development. [TDFP06-60A]	
7	Produces a sensitivity analysis on specialty engineering trade-off criteria. [TDFP07]	(A) Uses a sensitivity analysis (or similar) technique to assess the impact on the overall system of specialty design decisions or changes. [TDFP07-10A]	Results of sensitivity analysis performed. [TDFP07-E10]
		(A) Produces rigorous assessments of design trade-offs to determine best implementation of the relevant engineering specialties. [TDFP07-20A]	
8	Uses appropriate techniques to characterize the operational environment in order to support specialty engineering activities. [TDFP08]	(A) Identifies the operational environment for a project and identifies which specialties require consideration during the design. [TDFP08-10A]	Documentation describing the operational environment, which identifies the subject matter experts (SMEs) to include in the design team. [TDFP08-E10]
		(A) Selects appropriate techniques in order to ensure system design characteristics are fully considered. [TDFP08-20A]	Independently assessed documentation selecting the appropriate techniques for ensuring system design characteristics are considered. [TDFP08-E20]
		(K) Explains why a particular technique is appropriate. [TDFP08-30K]	Independently assessed scenarios for determination of "design for" specialties. [TDFP08-E30]
		(A) Selects scenarios which contribute to the determination of required "design for" specialties. [TDFP08-40A]	
9	Uses appropriate techniques and trade studies to determine and characterize specialty characteristics of proposed solutions. [TDFP09]	(A) Identifies the specialties to be considered as part of the design, given the operational environment. [TDFP09-10A]	Documentation showing analysis or alternative solutions. [TDFP09-E10]

		(P) Collaborates with specialties. [TDFP09-20P] (A) Ensures selected solutions address key specialty engineering design parameters. [TDFP09-30A]	Technical reports. [TDFP09-E20] Records of collaboration with a Project/program specialist. [TDFP09-E30] Whole life-cost model/total operational costs analysis. [TDFP09-E40]
10	Ensures specialty engineering experts and specialty engineering activities are fully integrated into Systems Engineering development activities. [TDFP10]	(A) Maintains interfaces between specialist and the rest of the team. [TDFP10-10A] (P) Acts to ensure integration of specialty subject matter experts into the project team. [TDFP10-20P] (P) Collaborates with specialties to address identified specialty needs and activities. [TDFP10-30P]	Organizational charts, meeting records. [TDFP10-E10] Records (e.g. meeting minutes) showing discussion and issue resolution with Specialty Engineers. [TDFP10-E20]
11	Identifies constraints on a system which reflect the needs of different design specialties. [TDFP11]	(K) Explains how different implementation levels of some specialties (such as availability, and reliability) may affect the cost of ownership (total costs of delivered solution). [TDFP11-10K] (A) Identifies different design constraints. [TDFP11-20A] (A) Identifies limits imposed on system development and their impact. [TDFP11-30A]	Documented system constraints. [TDFP11-E10] Total cost of ownership calculations. [TDFP11-E20]
12	Guides new or supervised practitioners in the area of specialty engineering to develop their knowledge, abilities, skills, or associated behaviors. [TDFP12]	(P) Guides new or supervised practitioners in executing activities that form part of this competency. [TDFP12-10P] (A) Trains individuals to an "Awareness" level in this competency area. [TDFP12-20A]	Organizational Breakdown Structure showing their responsibility for technical supervision in this area. [TDFP12-E10] Records of on-the-job training objectives/guidance etc. [TDFP12-E20] Coaching or mentoring assignment records. [TDFP12-E30] Records highlighting their impact on another individual in terms of improvement or professional development in this competency. [TDFP12-E40]
13	Maintains and enhances own competence in this area through Continual Professional Development (CPD) activities. [TDFP13]	(A) Identifies potential development needs in this area, identifying opportunities to address these through continual professional development activities. [TDFP13-10A] (A) Performs continual professional development activities to maintain and enhance their competency in this area. [TDFP13-20A] (A) Records continual professional development activities undertaken including learning or insights gained. [TDFP13-30A]	Records of Continual Professional Development (CPD) performed and learning outcomes. [TDFP13-E10]

Lead Practitioner – Technical: Design for…

ID	Indicators of Competence *(in addition to those at Practitioner level)*	Relevant knowledge, experience, and/or behaviors	Possible examples of objective evidence of personal contribution to activities performed, or professional behaviors applied
1	Creates enterprise-level policies, procedures, guidance, and best practice relating to specialty engineering including associated tools. [TDFL01]	(A) Analyzes enterprise need for design specialty policies, processes, tools, or guidance. [TDFL01-10A]	Records showing their role in embedding specialty engineering into enterprise policies (e.g. guidance introduced at enterprise level and enterprise-level review minutes). [TDFL01-E10]
		(A) Creates enterprise policies, procedures or guidance for specialty engineering activities. [TDFL01-20A]	Procedures they have written. [TDFL01-E20]
		(A) Selects and acquires appropriate tools supporting specialty engineering activities. [TDFL01-30A]	Records of support for tool introduction. [TDFL01-E30]
2	Judges the tailoring of enterprise-level system specialty engineering processes to meet the needs of a project. [TDFL02]	(A) Evaluates the enterprise process against the business and external stakeholder (e.g. customer) needs in order to tailor processes to enable project success. [TDFL02-10A]	Documented tailored process. [TDFL02-E10]
		(A) Produces constructive feedback on system specialty engineering process. [TDFL02-20A]	Specialty engineering process feedback. [TDFL02-E20]
3	Judges the strategy to be adopted on projects across the enterprise to ensure required specialty engineering characteristics are met. [TDFL03]	(A) Analyzes specialties to ensure they are fully integrated into the system development in a coherent and timely way and address the relevant issues. [TDFL03-10A]	Review comments or meeting minutes. [TDFL03-E10]
		(A) Analyzes the different implementation levels of specialties (such as availability and reliability) and their effect on cost of ownership (total costs of delivered solution). [TDFL03-20A]	Total Life Cycle Cost (TLCC) models. [TDFL03-E20]
		(A) Analyzes the combined effect of the specialties on the cost of ownership and the system development. [TDFL03-30A]	Design review meeting minutes. [TDFL03-E30]
		(A) Identifies and resolves conflicts between specialty engineering activities. [TDFL03-40A]	Records showing membership of an oversight committee. [TDFL03-E40]
		(A) Advises enterprise-level committees (or similar) on approval of plans. [TDFL03-50A]	Review comments. [TDFL03-E50]
		(A) Reviews engineering plans as part of an approval process. [TDFL03-60A]	

4	Judges the adequacy of sensitivity analysis made for specialty engineering criteria across the enterprise. [TDFL04]	(A) Reviews sensitivity analyses performed against specialty engineering criteria, from across the enterprise, for their adequacy. [TDFL04-10A]	Results of sensitivity analysis reviewed. [TDFL04-E10]
		(A) Analyzes specialty engineering trade-offs, from across the enterprise, for their adequacy. [TDFL04-20A]	Results of trade-off analyses reviewed. [TDFL04-E20]
5	Judges the suitability of plans across the enterprise, for the incorporation of all life cycle design attributes at the correct point within the design process. [TDFL05]	(A) Defines the attributes of a specialty engineering strategy in the context of the project/ program /domain / business. [TDFL05-10A]	Specialty engineering strategies that proved successful. [TDFL05-E10]
		(A) Identifies risks and mitigation techniques appropriate to the situation. [TDFL05-20A]	Records of risk mitigations performed. [TDFL05-E20]
6	Judges selected solutions from across the enterprise, against key specialty engineering design parameters to support the decision-making process. [TDFL06]	(A) Reviews design selections against specialty engineering requirements. [TDFL06-10A]	Review comments. [TDFL06-E10]
		(A) Advises on the suitability of specialty design requirements leading to changes. [TDFL06-20A]	PDR Documentation. [TDFL06-E20]
7	Persuades key stakeholders to address identified enterprise-level specialty-related design issues to reduce project cost, schedule, or technical risk. [TDFL07]	(A) Identifies and engages with key stakeholders. [TDFL07-10A]	Minutes of meetings. [TDFL07-E10]
		(P) Fosters agreement between key stakeholders at enterprise level to resolve specialty design issues, by promoting a holistic viewpoint. [TDFL07-20P]	Minutes of key stakeholder review meetings and actions taken. [TDFL07-E20]
8	Coaches or mentors practitioners across the enterprise in "designing for…" specialties, in order to develop their knowledge, abilities, skills, or associated behaviors. [TDFL08]	(P) Coaches or mentors practitioners across the enterprise in competency-related techniques, recommending development activities. [TDFL08-10P]	Coaching or mentoring assignment records. [TDFL08-E10]
		(A) Develops or authorizes enterprise training materials in this competency area. [TDFL08-20A]	Records of formal training courses, workshops, seminars, and authored training material supported by successful post-training evaluation data. [TDFL08-E20]
		(A) Provides enterprise workshops/seminars or training in this competency area. [TDFL08-30A]	Listing as an approved organizational trainer for this competency area. [TDFL08-E30]

(Continued)

ID	Indicators of Competence *(in addition to those at Practitioner level)*	Relevant knowledge, experience, and/or behaviors	Possible examples of objective evidence of personal contribution to activities performed, or professional behaviors applied
9	Promotes the introduction and use of novel techniques and ideas in "designing for…" specialties across the enterprise, to improve enterprise competence in this area. [TDFL09]	(A) Analyzes different approaches across different domains through research. [TDFL09-10A]	Research records. [TDFL09-E10]
		(A) Defines novel approaches that could potentially improve the SE discipline within the enterprise. [TDFL09-20A]	Published papers in refereed journals/company literature. [TDFL09-E20]
		(P) Fosters awareness of these novel techniques within the enterprise. [TDFL09-30P]	Enabling systems introduced in support of new techniques or ideas. [TDFL09-E30]
		(P) Collaborates with enterprise stakeholders to facilitate the introduction of techniques new to the enterprise. [TDFL09-40P]	Published papers (or similar) at enterprise level. [TDFL09-E40]
		(A) Monitors new techniques after their introduction to determine their effectiveness. [TDFL09-50A]	Records of improvements made against a recognized process improvement model in this area. [TDFL09-E50]
		(A) Adapts approach to reflect actual enterprise performance improvements. [TDFL09-60A]	
10	Develops expertise in this competency area through specialist Continual Professional Development (CPD) activities. [TDFL10]	(A) Identifies own needs for further professional development in order to increase competence beyond practitioner level. [TDFL10-10A]	Records of Continual Professional Development (CPD) performed and learning outcomes. [TDFL10-E10]
		(A) Performs professional development activities in order to move own competence toward expert level. [TDFL10-20A]	
		(A) Records continual professional development activities undertaken including learning or insights gained. [TDFL10-30A]	

NOTES	In addition to items above, enterprise-level or independent 3rd-Party-generated evidence may be used to amplify other evidence presented and may include: a. Formally recognized by senior management in current organization as an expert in this competency area b. Evidence of role as Product/System Design Authority or Technical Authority on a complex project with responsibilities in this area or where skills within this competency area were used c. Recognized as an authorizing signatory on behalf of enterprise for formal documentation in this competency area (e.g. policies, processes, and deliverables) d. Formal commendation or award within own enterprise for contribution or item of work successfully performed, which required proficiency in this competency area e. Customer, Supplier, or other external project-specific key Stakeholder accolades for specific work performed in this competency area f. Independently assessed or accredited work in this competency area (e.g. for independent publication or use) g. Formal organizational HR records positively highlighting any specific professional competencies or behaviors identified (if applicable) plus any of the evidence indicators listed at Expert level below

Expert – Technical: Design for…			
ID	**Indicators of Competence *(in addition to those at Lead Practitioner level)***	**Relevant knowledge, experience, and/or behaviors**	**Possible examples of objective evidence of personal contribution to activities performed, or professional behaviors applied**
1	Communicates own knowledge and experience in specialty engineering, in order to promote best practice beyond the enterprise boundary. [TDFE01]	(A) Produces papers, seminars, or presentations outside own enterprise for publication in order to share own ideas and improve industry best practices in this competence area. [TDFE01-10A]	Published papers or books etc. on new technique in refereed journals/company literature. [TDFE01-E10]
		(P) Fosters incorporation of own ideas into industry best practices in this area. [TDFE01-20P]	Published papers in refereed journals or internal literature proposing new practices in this competence area (or presentations, tutorials, etc.). [TDFE01-E20]
		(P) Develops guidance materials identifying new (or updating existing) best practice in this competence area. [TDFE01-30P]	Proposals adopted as industry best practice. [TDFE01-E30]
2	Persuades key stakeholders beyond the enterprise boundary to accept recommendations in support of specialty engineering activities. [TDFE02]	(A) Identifies and engages with external stakeholders to facilitate specialty design considerations. [TDFE02-10A]	Minutes of meetings. [TDFE02-E10]
		(P) Acts to change external stakeholder thinking regarding specialty engineering activities. [TDFE02-20P]	Documentation of specialty engineering design aspects. [TDFE02-E20]
3	Advises organizations beyond the enterprise boundary on the suitability of their approach to specialty engineering including its organization and integration across their enterprise. [TDFE03]	(A) Assesses the system impact of specialty engineering activities as independent reviewer, on behalf of an external organization. [TDFE03-10A]	Records of membership on oversight committee with relevant terms of reference. [TDFE03-E10]
		(A) Advises on the integration and organization of specialty engineering activities across their enterprise. [TDFE03-20A]	Participates in a review committee or similar body that deals with approval of specialty engineering plans. [TDFE03-E20]
		(A) Selects balanced specialty engineering solutions from options available, on behalf of stakeholders beyond enterprise boundary. [TDFE03-30A]	Review comments or revised document. [TDFE03-E30]
		(P) Persuades stakeholders beyond the enterprise boundary to accept difficult project-specific specialty engineering recommendations or actions. [TDFE03-40P]	Records indicating changed stakeholder thinking from personal intervention. [TDFE03-E40]
4	Advises organizations beyond the enterprise boundary on complex or sensitive specialty engineering planning issues. [TDFE04]	(A) Advises stakeholders beyond the boundary on their specialty design requirements and issues (e.g. cost of ownership or allocation of technical margins). [TDFE04-10A]	Records of advice provided on special design consideration issues. [TDFE04-E10]

(Continued)

ID	Indicators of Competence *(in addition to those at Lead Practitioner level)*	Relevant knowledge, experience, and/or behaviors	Possible examples of objective evidence of personal contribution to activities performed, or professional behaviors applied
		(P) Acts with sensitivity on highly complex system specialty engineering issues making limited use of specialized, technical terminology and taking account of external stakeholders' (e.g. customer's) background and knowledge. [TDFE04-20P]	Records from negotiations taking account of customer's background and knowledge, for example in minutes of meetings, position papers, and emails. [TDFE04-E20]
		(P) Communicates using a holistic approach to complex system specialty engineering issue resolution including balanced, rational arguments on way forward. [TDFE04-30P]	Stakeholder approval of system design including all required specialties. [TDFE04-E30]
		(P) Acts with sensitivity during negotiations on a highly complex system making limited use of specialized, technical terminology. [TDFE04-40P]	
		(A) Uses a holistic approach to complex issue resolution including balanced, rational arguments on way forward. [TDFE04-50A]	
5	Coordinates activities beyond the enterprise boundary which cover multiple specialties. [TDFE05]	(P) Coaches or mentors specialty subject matter experts beyond enterprise boundary (e.g. safety SMEs, reliability SMEs, and System Security Engineers) in their integration of efforts in order to ensure potential conflict or rework is minimized. [TDFE05-10P]	Records of coordination meetings. [TDFE05-E10]
			Records of advice provided to subject matter experts. [TDFE05-E20]
6	Identifies conflicts involving specialty engineering issues that extend beyond the enterprise boundary in order to enable the project to progress. [TDFE06]	(P) Acts to resolve conflict across multiple design specialties. [TDFE06-10P]	Minutes of meetings; revised design requirements with conflicts resolved. [TDFE06-E10]
7	Advises organizations beyond the enterprise boundary on how evolving needs impacts specialty engineering. [TDFE07]	(A) Advises stakeholders beyond the boundary how evolving needs may impact on the system. [TDFE07-10A]	Documentation covering evolving needs and their impact on the system. [TDFE07-E10]
		(A) Analyzes evolving system needs with regard to one or more design specialties (e.g. a new cybersecurity threat, and increased reliability requirement). [TDFE07-20A]	Records of advice provided. [TDFE07-E20]
8	Champions the introduction of novel techniques and ideas in the "Design for…" area, beyond the enterprise boundary, in order to develop the wider Systems Engineering community in this competency. [TDFE08]	(A) Analyzes different approaches across different domains through research. [TDFE08-10A]	Records of activities promoting research and need to adopt novel technique or ideas. [TDFE08-E10]

		(A) Defines novel approaches that could potentially improve the wider SE discipline. [TDFE08-20A]	Records of improvements made to process and appraisal against a recognized process improvement model. [TDFE08-E20]
		(P) Fosters awareness of these novel techniques within the wider SE community. [TDFE08-30P]	Research records. [TDFE08-E30]
		(P) Collaborates with those introducing novel techniques within the wider SE community. [TDFE08-40P]	Published papers in refereed journals/company literature. [TDFE08-E40]
		(A) Produces reports for the wider SE community on the effectiveness of new techniques after their introduction. [TDFE08-50A]	Records showing introduction of enabling systems supporting the new techniques or ideas. [TDFE08-E50]
9	Coaches individuals beyond the enterprise boundary in designing for specialty engineering, in order to further develop their knowledge, abilities, skills, or associated behaviors. [TDFE09]	(P) Coaches or mentors individuals beyond the enterprise boundary, in competency-related techniques, recommending development activities. [TDFE09-10P]	Coaching or mentoring assignment records. [TDFE09-E10]
		(A) Develops or authorizes training materials in this competency area, which are subsequently successfully delivered beyond the enterprise boundary. [TDFE09-20A]	Records of formal training courses, workshops, seminars, and authored training material supported by successful post-training evaluation data. [TDFE09-E20]
		(A) Provides workshops/seminars or training in this competency area for practitioners or lead practitioners beyond the enterprise boundary (e.g. conferences and open training days). [TDFE09-30A]	Records of Training/workshops/seminars created supported by successful post-training evaluation data. [TDFE09-E30]
10	Maintains expertise in this competency area through specialist Continual Professional Development (CPD) activities. [TDFE10]	(A) Reviews research, new ideas, and state of the art to identify relevant new areas requiring personal development in order to maintain expertise in this competency area. [TDFE10-10A]	Records of documents reviewed and insights gained as part of own research into this competency area. [TDFE10-E10]
		(A) Performs identified specialist professional development activities in order to maintain or further develop competence at expert level. [TDFE10-20A]	Records of Continual Professional Development (CPD) performed and learning outcomes. [TDFE10-E20]
		(A) Records continual professional development activities undertaken including learning or insights gained. [TDFE10-30A]	

NOTES	In addition to items above, enterprise-level or independent 3rd-Party-generated evidence may be used to amplify other evidence presented and may include: a. Formally recognized by a reputable external organization as an expert in this competency area b. Evidence of role as independent assessor or reviewer on project outside own organization where skills in this competency area were used c. Evidence of invitation(s) from wider community for contribution of systems engineering expertise in this area (e.g. industry conference panel, government advisory board etc. cross-industry working groups, partnerships, accredited advanced university courses or research, or as part of professional institute) d. Formal commendation beyond the enterprise (e.g. by INCOSE or other recognized authority) for work performed in this competency area e. Independently assessed or accredited work product in this competency area (e.g. for independent publication or use) f. Accolades of expertise in this area from recognized industry leaders

Competency area – Technical: Integration
Description Systems Integration is the logical process for assembling a set of system elements and aggregates into the realized system, product, or service that satisfies system requirements, architecture, and design. Systems integration focuses on the testing of interfaces, data flows, and control mechanisms, checking that realized elements and aggregates perform as predicted by their design and architectural solution, since it may not always be practicable or cost-effective to confirm these lower-level aspects at higher levels of system integration.
Why it matters Systems Integration should be planned so that system elements are brought together in a logical sequence to avoid wasted effort. Systematic and incremental integration makes it easier to find, isolate, diagnose, and correct problems. A system or system element that has not been integrated systematically cannot be relied on to meet its requirements.
Possible contributory types of evidence Any combination of the types of evidence may be acceptable (depending on how the Framework is tailored and used). The evidence items identified at each level indicate example work products only. Contributions to work products will generally differ at each proficiency level.
Learning and development The INCOSE Professional Development Portal provides example guidance on how to gain an initial awareness of a competency area and options for developing further competence thereafter.

Awareness – Technical: Integration		
ID	**Indicators of Competence**	**Relevant knowledge sub-indicators**
1	Explains why integration is important and how it confirms the system design, architecture, and interfaces. [TINA01]	(K) Describes the process of building up the system from smaller elements into larger and larger aggregations, or perhaps in related collections of elements, verifying along the way to ensure system elements work as designed. [TINA01-10K] (K) Explains why, if you put the entire system together and find it does not work as intended, it is difficult to identify clearly what elements may not be performing properly. [TINA01-20K] (K) Explains the potential impact of a system being a "system of systems" on the integration process. [TINA01-30K]
2	Explains why it is important to integrate the system in a logical sequence. [TINA02]	(K) Explains how integration is conducted using a progressive, logical process of assembling system elements, building an element, evaluating it then assembling several elements at the next level (system build). [TINA02-10K] (K) Describes alternative integration sequences (top down, bottom up, middle out, etc.) and how they may be assessed in order to define the most appropriate sequence in terms of overall cost and risk. [TINA02-20K] (K) Describes with examples how, if integration is performed in the wrong sequence, re-work and extra cost may be incurred (dependency on suppliers, development, new technology, obsolescence, etc.). [TINA02-30K]

3	Explains why planning and management of systems integration is necessary. [TINA03]	(K) Explains why planning for integration should occur at the beginning of the project/program. [TINA03-10K] (K) Explains how a failure to plan could result in a delay to integration; procedures may not be written; the sequences may not have been defined and the environment may not be available. [TINA03-20K] (K) Explains why the integration sequence should be documented. [TINA03-30K] (K) Explains how test requirements may influence the design. [TINA03-40K] (K) Describes key elements of an integration plan (e.g. identification of resources, equipment and that develop test requirements (influence the design). [TINA03-50K]
4	Explains the relationship between integration and verification. [TINA04]	(K) Explains how testing system element performance during integration can help establish the verification of some system requirements. [TINA04-10K] (K) Explains how documenting system element testing can provide evidence in support of verification and potentially even acceptance. [TINA04-20K]

Supervised Practitioner – Technical: Integration			
ID	**Indicators of Competence *(in addition to those at Awareness level)***	**Relevant knowledge, experience, and/or behaviors**	**Possible examples of objective evidence of personal contribution to activities performed, or professional behaviors applied**
1	Uses a governing process using appropriate tools to manage and control their own integration activities. [TINS01]	(A) Reviews element information (e.g. tests passed and certificate of conformity received for commercial-off-the-shelf (COTS) products) in order to confirm readiness for integration. [TINS01-10A] (A) Reviews test environment (test equipment, tools, procedures, sequence, etc.) to confirm its readiness for integration. [TINS01-20A] (A) Follows an integration or test procedure and identify non conformances against the plan. [TINS01-30A]	Documented results of system element integration (e.g. checklists and verification matrices). [TINS01-E10] Documented results of system integration tests. [TINS01-E20]
2	Prepares inputs to integration plans based upon governing standards and processes including identification of method and timing for each activity to meet project requirements. [TINS02]	(A) Uses governing process and tools to create information in support of integration sequence planning activities, including the environment and the integration approach. [TINS02-10A] (A) Ensures integration plans include measures of successful integration, such as how to verify conformance. [TINS02-20A] (A) Ensures integration plans include all required enterprise or statutory/regulatory process requirements. [TINS02-30A]	Integration plan. [TINS02-E10] Schedule of integration activities. [TINS02-E20]
3	Prepares plans which address integration for system elements (or noncomplex systems) in order to define or scope that activity. [TINS03]	(A) Prepares inputs to integration plan generation documenting the integration sequence, the environment, and approach. [TINS03-10A] (A) Prepares inputs to integration plan generation documenting measures of successful integration and how to verify conformance. [TINS03-20A]	Detailed integration plan including specifics on environment, required equipment and tools, procedures, and verification measures. [TINS03-E10]
4	Records the causes of simple faults typically found during integration activities in order to communicate with stakeholders. [TINS04]	(A) Identifies where results differ from those expected. [TINS04-10A] (A) Records faults appropriately (process, tools used, method). [TINS04-20A] (A) Analyzes simple faults in a logical manner and initiate the corrective action process. [TINS04-30A]	Documentation of faults, e.g. Trouble Report, Nonconformance Report, Fault Log. [TINS04-E10] Documentation of fault resolution, e.g. Corrective Action, Engineering Investigation. [TINS04-E20]

		(A) Performs corrective actions as authorized. [TINS04-40A] (A) Records corrective actions taken and close the outstanding fault log. [TINS04-50A]	
5	Collates evidence during integration in support of downstream test and acceptance activities. [TINS05]	(K) Explains methods for documenting conformance and how it contributes to verification of the system. [TINS05-10K] (A) Collates evidence from various sources during integration in support of downstream test and acceptance activities. [TINS05-20A]	Documented results of system element integration (e.g. checklists). [TINS05-E10]
6	Identifies an integration environment to facilitate system integration activities. [TINS06]	(A) Identifies required tools, equipment, personnel, etc. required to perform integration. [TINS06-10A] (A) Prepares work products that characterize an integration environment designed to facilitate system integration activities. [TINS06-20A]	Detailed integration plan including specifics on environment, required equipment and tools, procedures, verification measures, etc. [TINS06-E10]
7	Develops own understanding of this competency area through Continual Professional Development (CPD). [TINS07]	(A) Identifies potential gaps in own knowledge or development needs in this area, identifying opportunities to address these through continual professional development activities. [TINS07-10A] (A) Performs continual professional development activities to improve their knowledge and understanding in this area. [TINS07-20A] (A) Records continual professional development activities undertaken including learning or insights gained. [TINS07-30A]	Records of Continual Professional Development (CPD) performed and learning outcomes. [TINS07-E10]

Practitioner – Technical: Integration			
ID	**Indicators of Competence** ***(in addition to those at Supervised Practitioner level)***	**Relevant knowledge, experience, and/or behaviors**	**Possible examples of objective evidence of personal contribution to activities performed, or professional behaviors applied**
1	Creates a strategy for system integration on a project to support SE project and wider enterprise needs. [TINP01]	(A) Identifies project-specific system integration needs that need to be incorporated into SE planning. [TINP01-10A] (A) Identifies the impact on system integration of wider "systems of systems" integration issues. [TINP01-20A] (A) Prepares project-specific system integration inputs for SE planning purposes, using appropriate processes and procedures. [TINP01-30A] (A) Prepares project-specific system integration task estimates in support of SE planning. [TINP01-40A] (P) Liaises throughout with stakeholders to gain approval, updating strategy as necessary. [TINP01-50P]	System integration strategy work products. [TINP01-E10]
2	Creates a governing process, plan, and associated tools for systems integration, which reflect project and business strategy. [TINP02]	(K) Identifies the steps necessary to define a process and appropriate techniques to be adopted for system integration. [TINP02-10K] (K) Describes key elements of a successful integration process and how systems integration activities relate to different stages of a system life cycle. [TINP02-20K] (A) Uses integration tools or uses a methodology. [TINP02-30A] (A) Uses appropriate standards and governing processes in their system integration planning, tailoring as appropriate. [TINP02-40A] (A) Selects system integration methods for each system element. [TINP02-50A] (A) Creates system integration plans and processes for use on a project. [TINP02-60A] (P) Liaises throughout with stakeholders to gain approval, updating plans as necessary. [TINP02-70P]	SE system integration planning or process work products they have developed. [TINP02-E10] Certificates or proficiency evaluations in system integration tools. [TINP02-E20] SE system integration tool guidance they have developed. [TINP02-E30] Records of a range of meetings with different stakeholders, which they attended including any actions taken and completed. [TINP02-E40]
3	Uses governing plans and processes to plan and execute system integration activities, interpreting, evolving, or seeking guidance where appropriate. [TINP03]	(A) Follows defined plans, processes and associated tools to perform system Integration on a project. [TINP03-10A]	System Integration plans or work products. [TINP03-E10]

		(P) Guides and actively coordinates the interpretation of system Integration plans or processes in order to complete activities system Integration activities successfully. [TINP03-20P] (A) Prepares system integration data in support of wider system monitoring and control activities. [TINP03-30A] (P) Recognizes situations where deviation from published plans and processes or clarification from others is appropriate in order to overcome complex system Integration challenges. [TINP03-40P] (P) Recognizes situations where existing system Integration strategy, process, plan, or tools require formal change, liaising with stakeholders to gain approval as necessary. [TINP03-50P]	Process or tool guidance. [TINP03-E20]
4	Performs rectification of faults found during integration activities. [TINP04]	(A) Identifies where results differ from those expected. [TINP04-10A] (A) Records faults appropriately (process, tools used, method). [TINP04-20A] (A) Analyzes faults in a logical manner and contributes to corrective actions. [TINP04-30A] (A) Records corrective actions taken and communicates this. [TINP04-40A] (P) Communicates corrective actions taken to all stakeholders. [TINP04-50P] (A) Performs corrective actions. [TINP04-60A] (A) Identifies consequences of corrective actions ensuring they are fully resolved. [TINP04-70A] (A) Maintains the outstanding fault log when all outstanding actions have been completed. [TINP04-80A]	Fault log, Trouble Report, or similar test record. [TINP04-E10] Corrective actions. [TINP04-E20] Minutes of fault analysis meetings. [TINP04-E30]
5	Prepares evidence obtained during integration in support of downstream test and acceptance activities. [TINP05]	(A) Identifies the precise evidence required for external stakeholder (e.g. customer) acceptance and certification. [TINP05-10A] (A) Ensures all activities necessary in order to produce required evidence are encompassed in integration plans. [TINP05-20A]	Traceability matrices. [TINP05-E10] Compliance/verification matrices. [TINP05-E20]

(*Continued*)

ID	Indicators of Competence *(in addition to those at Supervised Practitioner level)*	Relevant knowledge, experience, and/or behaviors	Possible examples of objective evidence of personal contribution to activities performed, or professional behaviors applied
			Test reports. [TINP05-E30] Certification data package. [TINP05-E40] Acceptance data package. [TINP05-E50]
6	Guides and actively coordinates integration activities for a system. [TINP06]	(A) Ensures systems integration activity is successfully performed by an integration team. [TINP06-10A] (K) Explains key aspects of performing integration activities (e.g. entry/exit criteria, analysis of results, and reviews). [TINP06-20K] (A) Identifies integration issues and overcomes them (e.g. problems with schedule and lateness of equipment). [TINP06-30A]	Compliance/verification matrices. [TINP06-E10] Test reports. [TINP06-E20] Integration measures showing actual performance against plan. [TINP06-E30] Minutes of test readiness review including relevant action log. [TINP06-E40] Certification data package. [TINP06-E50]
7	Identifies a suitable integration environment. [TINP07]	(A) Identifies the facilities to be used. Consideration should be given to the size of the area, furniture required, power requirements, the IT requirements, and the security of the facility. [TINP07-10A] (A) Identifies external test facilities that may be used. [TINP07-20A] (A) Identifies any bespoke tools and equipment that are required for integration, e.g. simulators and emulators. [TINP07-30A] (A) Identifies resources and skills required. [TINP07-40A]	Integration plans. [TINP07-E10] Procurement of equipment. [TINP07-E20] Records demonstrating arranging use of specialized/external facilities. [TINP07-E30]
8	Creates detailed integration procedures. [TINP08]	(A) Creates integration procedures that relate directly to the requirements (design and system). [TINP08-10A] (A) Creates clear, concise instructions for the activities to be performed, pre-requisites, the expected outcome, and action in case of a failure. [TINP08-20A]	Approved integration procedures. [TINP08-E10]
9	Guides new or supervised practitioners in Systems integration to develop their knowledge, abilities, skills, or associated behaviors. [TINP09]	(P) Guides new or supervised practitioners in executing activities that form part of this competency. [TINP09-10P]	Organizational Breakdown Structure showing their responsibility for technical supervision in this area. [TINP09-E10]

		(A) Trains individuals to an "Awareness" level in this competency area. [TINP09-20A]	On-the-job training records. [TINP09-E20] Coaching or mentoring assignment records. [TINP09-E30] Records highlighting their impact on another individual in terms of improvement or professional development in this competency. [TINP09-E40]
10	Maintains and enhances own competence in this area through Continual Professional Development (CPD) activities. [TINP10]	(A) Identifies potential development needs in this area, identifying opportunities to address these through continual professional development activities. [TINP10-10A] (A) Performs continual professional development activities to maintain and enhance their competency in this area. [TINP10-20A] (A) Records continual professional development activities undertaken including learning or insights gained. [TINP10-30A]	Records of Continual Professional Development (CPD) performed and learning outcomes. [TINP10-E10]

Lead Practitioner – Technical: Integration			
ID	**Indicators of Competence** ***(in addition to those at Practitioner level)***	**Relevant knowledge, experience, and/or behaviors**	**Possible examples of objective evidence of personal contribution to activities performed, or professional behaviors applied**
1	Creates enterprise-level policies, procedures, guidance, and best practice for integration, including associated tools for a project. [TINL01]	(A) Analyzes enterprise need for integration policies, processes, tools, or guidance. [TINL01-10A]	Records showing their role in embedding system integration into enterprise policies (e.g. guidance introduced at enterprise level and enterprise-level review minutes). [TINL01-E10]
		(A) Creates governing policies, procedures, or guidance for integration activities. [TINL01-20A]	Procedures they have written. [TINL01-E20]
		(A) Selects and acquires appropriate tools supporting integration activities. [TINL01-30A]	Records of support for tool introduction. [TINL01-E30]
2	Judges the tailoring of enterprise-level integration processes to meet the needs of a project. [TINL02]	(A) Evaluates the enterprise process against the business and external stakeholder (e.g. customer) needs in order to tailor processes to enable project success. [TINL02-10A]	Documented tailored process. [TINL02-E10]
		(A) Produces constructive feedback on integration process across the enterprise. [TINL02-20A]	Integration process feedback. [TINL02-E20]
3	Judges the suitability of integration plans from projects across the enterprise, to ensure project success. [TINL03]	(K) Describes the attributes of a successful integration strategy in the context of the project/program /domain/ business. [TINL03-10K]	Integration strategies that proved successful. [TINL03-E10]
		(A) Evaluates the attributes of integration plans in the context of the project/ program /domain/business to enable project success. [TINL03-20A]	Risk mitigations performed. [TINL03-E20]
		(A) Describes risks and mitigation techniques applied in the context of the project/program /domain/business. [TINL03-30A]	Risk mitigations performed. [TINL03-E30]
4	Judges detailed integration procedures from projects across the enterprise, to ensure project success. [TINL04]	(A) Reviews and comments on integration plans and procedures. [TINL04-10A]	Review comments. [TINL04-E10]
5	Judges integration evidence generated by projects across the enterprise, to ensure adequacy of information. [TINL05]	(A) Reviews and comments on integration conformance documentation. [TINL05-10A]	Records of a review process in which they have been involved. [TINL05-E10]
		(A) Describes occasions where they have provided advice on integration conformance documentation that has led to changes being implemented. [TINL05-20A]	Records of recommendations and resulting changes. [TINL05-E20]

6	Guides and actively coordinates integration activities on complex systems or across multiple projects from projects across the enterprise. [TINL06]	(A) Describes typical approaches to complex integration activities. [TINL06-10A]	Integration plans. [TINL06-E10]
		(A) Identifies integration for system elements within a wider "system of systems" integration activity. [TINL06-20A]	Integration plan and records of conformance. [TINL06-E20]
		(A) Defines detailed integration sequences and the readiness criteria for each complex system element. [TINL06-30A]	System of system integration plans. [TINL06-E30]
		(A) Produces an integration schedule showing dependencies of each activity (critical path analysis). [TINL06-40A]	Organizational responsibility matrix. [TINL06-E40]
		(A) Defines the integration environment required, including outsourcing of qualification tests as required. [TINL06-50A]	Minutes of meetings. [TINL06-E50]
		(P) Guides and actively coordinates the integration of complex systems or projects. [TINL06-60P]	
7	Persuades key stakeholders to address identified enterprise-level system integration issues to reduce project cost, schedule, or technical risk. [TINL07]	(A) Identifies and engages with key stakeholders. [TINL07-10A]	Minutes of meetings. [TINL07-E10]
		(P) Persuades stakeholders to address enterprise level issues using a holistic viewpoint. [TINL07-20P]	
8	Coaches or mentors practitioners across the enterprise in systems Integration in order to develop their knowledge, abilities, skills, or associated behaviors. [TINL08]	(P) Coaches or mentors practitioners across the enterprise in competency-related techniques, recommending development activities. [TINL08-10P]	Coaching or mentoring assignment records. [TINL08-E10]
		(A) Develops or authorizes enterprise training materials in this competency area. [TINL08-20A]	Records of formal training courses, workshops, seminars, and authored training material supported by successful post-training evaluation data. [TINL08-E20]
		(A) Provides enterprise workshops/seminars or training in this competency area. [TINL08-30A]	Listing as an approved organizational trainer for this competency area. [TINL08-E30]

(Continued)

ID	Indicators of Competence *(in addition to those at Practitioner level)*	Relevant knowledge, experience, and/or behaviors	Possible examples of objective evidence of personal contribution to activities performed, or professional behaviors applied
9	Promotes the introduction and use of novel techniques and ideas in systems integration across the enterprise, to improve enterprise competence in this area. [TINL09]	(A) Analyzes different approaches across different domains through research. [TINL09-10A]	Research records. [TINL09-E10]
		(A) Defines novel approaches that could potentially improve the SE discipline within the enterprise. [TINL09-20A]	Published papers in refereed journals/company literature. [TINL09-E20]
		(P) Fosters awareness of these novel techniques within the enterprise. [TINL09-30P]	Records showing introduction of enabling systems supporting the new techniques or ideas. [TINL09-E30]
		(P) Collaborates with enterprise stakeholders to facilitate the introduction of techniques new to the enterprise. [TINL09-40P]	Published papers (or similar) at enterprise level. [TINL09-E40]
		(A) Monitors new techniques after their introduction to determine their effectiveness. [TINL09-50A]	Records of improvements made against a recognized process improvement model in this area. [TINL09-E50]
		(A) Adapts approach to reflect actual enterprise performance improvements. [TINL09-60A]	
10	Develops expertise in this competency area through specialist Continual Professional Development (CPD) activities. [TINL10]	(A) Identifies own needs for further professional development in order to increase competence beyond practitioner level. [TINL10-10A]	Records of Continual Professional Development (CPD) performed and learning outcomes. [TINL10-E10]
		(A) Performs professional development activities in order to move own competence toward expert level. [TINL10-20A]	
		(A) Records continual professional development activities undertaken including learning or insights gained. [TINL10-30A]	

NOTES In addition to items above, enterprise-level or independent 3rd-Party-generated evidence may be used to amplify other evidence presented and may include:

a. Formally recognized by senior management in current organization as an expert in this competency area
b. Evidence of role as Product/System Design Authority or Technical Authority on a complex project with responsibilities in this area or where skills within this competency area were used
c. Recognized as an authorizing signatory on behalf of enterprise for formal documentation in this competency area (e.g. policies, processes, and deliverables)
d. Formal commendation or award within own enterprise for contribution or item of work successfully performed, which required proficiency in this competency area
e. Customer, Supplier, or other external project-specific key Stakeholder accolades for specific work performed in this competency area
f. Independently assessed or accredited work in this competency area (e.g. for independent publication or use)
g. Formal organizational HR records positively highlighting any specific professional competencies or behaviors identified (if applicable) plus any of the evidence indicators listed at Expert level below

Expert – Technical: Integration			
ID	**Indicators of Competence** ***(in addition to those at Lead Practitioner level)***	**Relevant knowledge, experience, and/or behaviors**	**Possible examples of objective evidence of personal contribution to activities performed, or professional behaviors applied**
1	Communicates own knowledge and experience in Systems Integration in order to promote best practice beyond the enterprise boundary. [TINE01]	(A) Produces papers, seminars, or presentations outside own enterprise for publication in order to share own ideas and improve industry best practices in this competence area. [TINE01-10A]	Published papers or books etc. on new technique in refereed journals/company literature. [TINE01-E10]
		(P) Fosters incorporation of own ideas into industry best practices in this area. [TINE01-20P]	Published papers in refereed journals or internal literature proposing new practices in this competence area (or presentations, tutorials, etc.). [TINE01-E20]
		(P) Develops guidance materials identifying new (or updating existing) best practice in this competence area. [TINE01-30P]	Proposals adopted as industry best practice. [TINE01-E30]
2	Persuades key stakeholders beyond the enterprise boundary to accept recommendation associated with integration activities. [TINE02]	(A) Identifies and engages with external stakeholders to facilitate integration. [TINE02-10A]	Minutes of meetings. [TINE02-E10]
		(P) Acts to change external stakeholder thinking regarding systems integration activities. [TINE02-20P]	Revised Integration plans, procedures or conformance documentation. [TINE02-E20]
3	Advises organizations beyond the enterprise boundary on the suitability of their approach to integration to support enterprise needs. [TINE03]	(A) Advises stakeholders beyond enterprise boundary regarding their Systems Integration activities. [TINE03-10A]	Records of membership on oversight committee with relevant terms of reference. [TINE03-E10]
		(P) Acts with sensitivity on highly complex Systems Integration issues making limited use of specialized, technical terminology and taking account of external stakeholders' (e.g. customer's) background and knowledge. [TINE03-20P]	Records of participation in an oversight committee or similar body that deals with approval of such plans. [TINE03-E20]
		(P) Communicates using a holistic approach to complex Systems Integration issue resolution including balanced, rational arguments on way forward. [TINE03-30P]	Review comments made or comments implemented. [TINE03-E30]
		(A) Advises on Systems Integration decisions on behalf of stakeholders beyond enterprise boundary, arbitrating if required. [TINE03-40A]	Advice provided together with evidence that the issues advised on were by their nature either complex or sensitive. [TINE03-E40]
		(P) Persuades stakeholders beyond the enterprise boundary to accept difficult project-specific Systems Integration recommendations or actions. [TINE03-50P]	
4	Advises organizations beyond the enterprise boundary on evidence generated during integration to support enterprise needs. [TINE04]	(A) Reviews and advises on project/program integration results/documentation from across the organization. [TINE04-10A]	Review comments. [TINE04-E10]

(Continued)

ID	Indicators of Competence *(in addition to those at Lead Practitioner level)*	Relevant knowledge, experience, and/or behaviors	Possible examples of objective evidence of personal contribution to activities performed, or professional behaviors applied
5	Advises organizations beyond the enterprise boundary on complex or sensitive integration-related issues to support enterprise needs. [TINE05]	(A) Advises external stakeholders (e.g. customers) on their integration requirements and issues. [TINE05-10A]	Records of advice provided together with evidence that the issues advised on were by their nature either complex or sensitive. [TINE05-E10]
		(P) Conducts sensitive negotiations regarding integration of a highly complex system making limited use of specialized, technical terminology. [TINE05-20P]	Records from negotiations demonstrating awareness of customer's background and knowledge. [TINE05-E20]
		(A) Uses a holistic approach to complex issue resolution including balanced, rational arguments on way forward. [TINE05-30A]	Stakeholder approval of integration plans and procedures. [TINE05-E30]
6	Champions the introduction of novel techniques and ideas in systems integration, beyond the enterprise boundary, in order to develop the wider Systems Engineering community in this competency. [TINE06]	(A) Analyzes different approaches across different domains through research. [TINE06-10A]	Records of activities promoting research and need to adopt novel technique or ideas. [TINE06-E10]
		(A) Defines novel approaches that could potentially improve the wider SE discipline. [TINE06-20A]	Records of improvements made to process and appraisal against a recognized process improvement model. [TINE06-E20]
		(P) Fosters awareness of these novel techniques within the wider SE community. [TINE06-30P]	Research records. [TINE06-E30]
		(P) Collaborates with those introducing novel techniques within the wider SE community. [TINE06-40P]	Published papers in refereed journals/company literature. [TINE06-E40]
		(A) Produces reports for the wider SE community on the effectiveness of new techniques after their introduction. [TINE06-50A]	Records showing introduction of enabling systems supporting the new techniques or ideas. [TINE06-E50]
7	Coaches individuals beyond the enterprise boundary in Systems Integration, in order to further develop their knowledge, abilities, skills, or associated behaviors. [TINE07]	(P) Coaches or mentors individuals beyond the enterprise boundary, in competency-related techniques, recommending development activities. [TINE07-10P]	Coaching or mentoring assignment records. [TINE07-E10]
		(A) Develops or authorizes training materials in this competency area, which are subsequently successfully delivered beyond the enterprise boundary. [TINE07-20A]	Records of formal training courses, workshops, seminars, and authored training material supported by successful post-training evaluation data. [TINE07-E20]
		(A) Provides workshops/seminars or training in this competency area for practitioners or lead practitioners beyond the enterprise boundary (e.g. conferences and open training days). [TINE07-30A]	Records of Training/workshops/seminars created supported by successful post-training evaluation data. [TINE07-E30]

8	Maintains expertise in this competency area through specialist Continual Professional Development (CPD) activities. [TINE08]	(A) Reviews research, new ideas, and state of the art to identify relevant new areas requiring personal development in order to maintain expertise in this competency area. [TINE08-10A] (A) Performs identified specialist professional development activities in order to maintain or further develop competence at expert level. [TINE08-20A] (A) Records continual professional development activities undertaken including learning or insights gained. [TINE08-30A]	Records of documents reviewed and insights gained as part of own research into this competency area. [TINE08-E10] Records of Continual Professional Development (CPD) performed and learning outcomes. [TINE08-E20]

NOTES	In addition to items above, enterprise-level or independent 3rd-Party-generated evidence may be used to amplify other evidence presented and may include: a. Formally recognized by a reputable external organization as an expert in this competency area b. Evidence of role as independent assessor or reviewer on project outside own organization where skills in this competency area were used c. Evidence of invitation(s) from wider community for contribution of systems engineering expertise in this area (e.g. industry conference panel, government advisory board etc. cross-industry working groups, partnerships, accredited advanced university courses or research, or as part of professional institute) d. Formal commendation beyond the enterprise (e.g. by INCOSE or other recognized authority) for work performed in this competency area e. Independently assessed or accredited work product in this competency area (e.g. for independent publication or use) f. Accolades of expertise in this area from recognized industry leaders

Competency area – Technical: Interfaces
Description Interfaces occur where system elements interact, for example human, mechanical, electrical, thermal, and data. Interface Management comprises the identification, definition, and control of interactions across system or system element boundaries.
Why it matters Poor interface definition and management can result in incompatible system elements (either internal to the system or between the system and its environment), which may ultimately result in system failure or project overrun.
Possible contributory types of evidence Any combination of the types of evidence may be acceptable (depending on how the Framework is tailored and used). The evidence items identified at each level indicate example work products only. Contributions to work products will generally differ at each proficiency level.
Learning and development The INCOSE Professional Development Portal provides example guidance on how to gain an initial awareness of a competency area and options for developing further competence thereafter.

Awareness – Technical: Interfaces		
ID	**Indicators of Competence**	**Relevant knowledge sub-indicators**
1	Defines key concepts within interface definition and management. [TIFA01]	(K) Describes what an interface is. [TIFA01-10K] (K) Defines different strategies for the successful identification of interfaces. [TIFA01-20K] (K) Describes interface stakeholders. [TIFA01-30K] (K) Describes the importance of ensuring each interface is "owned" by someone and the effects of failing to have this in place. [TIFA01-40K] (K) Explains how interfaces are usually defined. [TIFA01-50K] (K) Describes different strategies for defining the content of interfaces and managing thereafter and when each strategy may be applicable. [TIFA01-60K]
2	Explains how interface definition and management affects the integrity of the system solution. [TIFA02]	(K) Explains why control of the management of interface development is necessary. [TIFA02-10K] (K) Describes the potential impact on the system of failure to define interfaces properly. [TIFA02-20K] (K) Describes the potential impact on the system of failure to manage interfaces properly. [TIFA02-30K] (K) Explains how and why internal and external interfaces may evolve and subsequently be managed differently. [TIFA02-40K] (K) Explains how different stakeholder types may affect the definition or management of interfaces. [TIFA02-50K] (K) Describes the importance of configuration management when managing interfaces. [TIFA02-60K]
3	Identifies possible sources of complexity in interface definition and management. [TIFA03]	(K) Describes different type of interface (functional and physical) covering different domains. [TIFA03-10K] (K) Describes possible sources of interface complexity. [TIFA03-20K]
4	Explains how different sources of complexity affect interface definition and management. [TIFA04]	(K) Explains how different interfaces (functional and physical) across different domains are managed differently. [TIFA04-10K] (K) Explains how different sources of interface complexity affect the management of systems interfaces. [TIFA04-20K]

Supervised Practitioner – Technical: Interfaces			
ID	**Indicators of Competence *(in addition to those at Awareness level)***	**Relevant knowledge, experience, and/or behaviors**	**Possible examples of objective evidence of personal contribution to activities performed, or professional behaviors applied**
1	Uses a governing process to manage and control their own interface management activities. [TIFS01]	(A) Uses an interface management procedure/interface management plan to ensure interface coherence across a system model (e.g. Interface ID, ownership, interface control document/specification, change, or CM processes). [TIFS01-10A]	Interface management procedures. [TIFS01-E10]
2	Identifies the properties of simple interfaces in order to define them. [TIFS02]	(K) Explains how there are multiple aspects/characteristics to even simple interfaces. [TIFS02-10K]	Interface diagram. [TIFS02-E10]
		(A) Identifies simple interfaces for a system. [TIFS02-20A]	Interface definition/description document (IDD). [TIFS02-E20]
		(A) Defines different parameters of a simple interface (both functional and physical). [TIFS02-30A]	Interface Control document (ICD). [TIFS02-E30]
3	Explains the potential consequences of changes on system interfaces to coordinate and control ongoing development. [TIFS03]	(K) Explains how a change at one end of the interface can impact the other end of the interface. [TIFS03-10K]	Records of interfaces and the changes made to them (functionally or physically). [TIFS03-E10]
		(K) Explains how system performance may be affected by a change to an interface. [TIFS03-20K]	
		(A) Identifies the impact on an interface as a result of changes elsewhere in a system. [TIFS03-30A]	
4	Maintains technical parameters associated with an interface to ensure continued stability of definition. [TIFS04]	(K) Lists typical technical parameters requiring maintenance on an interface in order to ensure stability during its development or evolution. [TIFS04-10K]	Records of parameters maintained. [TIFS04-E10]
		(A) Maintains technical parameters for an interface during its development or evolution. [TIFS04-20A]	
5	Develops own understanding of this competency area through Continual Professional Development (CPD). [TIFS05]	(A) Identifies potential gaps in own knowledge or development needs in this area, identifying opportunities to address these through continual professional development activities. [TIFS05-10A]	Records of Continual Professional Development (CPD) performed and learning outcomes. [TIFS05-E10]
		(A) Performs continual professional development activities to improve their knowledge and understanding in this area. [TIFS05-20A]	
		(A) Records continual professional development activities undertaken including learning or insights gained. [TIFS05-30A]	

(*Continued*)

Practitioner – Technical: Interfaces			
ID	**Indicators of Competence** ***(in addition to those at Supervised Practitioner level)***	**Relevant knowledge, experience, and/or behaviors**	**Possible examples of objective evidence of personal contribution to activities performed, or professional behaviors applied**
1	Creates a strategy for interface definition and management on a project to support SE project and wider enterprise needs. [TIFP01]	(A) Identifies project-specific interface definition and management needs that need to be incorporated into SE planning. [TIFP01-10A] (A) Prepares project-specific interface definition and management inputs for SE planning purposes, using appropriate processes and procedures. [TIFP01-20A] (A) Prepares project-specific interface definition and management task estimates in support of SE planning. [TIFP01-30A] (P) Liaises throughout with stakeholders to gain approval, updating strategy as necessary. [TIFP01-40P]	SE interface definition and management strategy work products. [TIFP01-E10]
2	Creates a governing process, plan, and associated tools for interface definition and management, which reflect project and business strategy. [TIFP02]	(K) Identifies the steps necessary to define a process and appropriate techniques to be adopted for systems interface definition and management. [TIFP02-10K] (K) Describes key elements of a successful systems interface definition and management process and how these activities relate to different stages of a system life cycle. [TIFP02-20K] (A) Uses systems interface definition and management tools or uses a methodology. [TIFP02-30A] (A) Uses appropriate standards and governing processes in their systems interface definition and management planning, tailoring as appropriate. [TIFP02-40A] (A) Selects systems interface definition and management methods for each architectural interface. [TIFP02-50A] (A) Creates systems interface definition and management plans and processes for use on a project. [TIFP02-60A] (P) Liaises throughout with stakeholders to gain approval, updating plans as necessary. [TIFP02-70P]	SE system interface definition planning or process work products. [TIFP02-E10] Certificates or proficiency evaluations in system interface definition or management tools. [TIFP02-E20] SE system interface definition or management tool guidance. [TIFP02-E30]

3	Uses interface management techniques and governing processes, to manage and control their own interface management activities. [TIFP03]	(A) Complies with defined plans, processes and associated tools to perform system Interfaces on a project. [TIFP03-10A]	System Interfaces plans or work products. [TIFP03-E10]
		(P) Guides and actively coordinates the interpretation of system Interfaces plans or processes in order to complete activities system Interfaces activities successfully. [TIFP03-20P]	Interface definition Process or tool guidance. [TIFP03-E20]
		(A) Prepares interface definition and management data and reports in support of wider system measurement, monitoring, and control. [TIFP03-30A]	
		(P) Recognizes situations where deviation from published plans and processes or clarification from others is appropriate in order to overcome complex system Interfaces challenges. [TIFP03-40P]	
		(P) Recognizes situations where the existing system interfaces strategy, process, plan, or tools require formal change, liaising with stakeholders to gain approval as necessary. [TIFP03-50P]	
4	Maintains interfaces over time to ensure continued coherence and alignment with project need. [TIFP04]	(A) Elicits system element interfaces for a project using interface definition techniques. [TIFP04-10A]	Interface definition/description document (IDD). [TIFP04-E10]
		(A) Defines the scope and content of system element interfaces. [TIFP04-20A]	Interface Control Document (ICD). [TIFP04-E20]
		(A) Maintains system element interfaces as the project evolves, updating project work products as required. [TIFP04-30A]	
5	Explains the effect of complexity on interface definition and management. [TIFP05]	(A) Identifies different sources of complexity affect the management of interfaces on a project. [TIFP05-10A]	System Interface Management plans addressing areas such as time zones, culture, language, perspectives, domains, legislation, differing suppliers, different contract types, varying standards, and novel technologies. [TIFP05-E10]
		(A) Uses different approaches to address sources of interface complexity on a project. [TIFP05-20A]	

(Continued)

ID	Indicators of Competence *(in addition to those at Supervised Practitioner level)*	Relevant knowledge, experience, and/or behaviors	Possible examples of objective evidence of personal contribution to activities performed, or professional behaviors applied
6	Negotiates interfaces between interface stakeholders to facilitate system development. [TIFP06]	(K) Explains how different sources of complexity have affects the management of interfaces on a project. [TIFP06-10K] (P) Negotiates agreement of interfaces in situations where conflict exists. [TIFP06-20P] (A) Identifies conflicts in the definition of interfaces, resolving as required. [TIFP06-30A]	Revisions of Interface definition/description document (IDD). [TIFP06-E10] Updates to an Interface Control Document (ICD). [TIFP06-E20] Records (e.g. meeting minutes) showing participation in interface negotiation and revision. [TIFP06-E30]
7	Identifies impact on interface definitions as a result of wider changes. [TIFP07]	(A) Identifies impact of a change at one end of the interface on the other end of the interface. [TIFP07-10A] (A) Evaluates impact of interface changes on system performance parameters and communicates these to stakeholders. [TIFP07-20A] (A) Identifies impacts of interface changes on other aspects of a system and communicates these to stakeholders. [TIFP07-30A]	Revisions of Interface definition/description document (IDD). [TIFP07-E10] Revisions of Interface Control document (ICD). [TIFP07-E20]
8	Guides new or supervised practitioners in Systems Engineering interface management in order to develop their knowledge, abilities, skills, or associated behaviors. [TIFP08]	(P) Guides new or supervised practitioners in executing activities that form part of this competency. [TIFP08-10P] (A) Trains individuals to an "Awareness" level in this competency area. [TIFP08-20A]	Organizational Breakdown Structure showing their responsibility for technical supervision in this area. [TIFP08-E10] On-the-job training records. [TIFP08-E20] Coaching or mentoring assignment records. [TIFP08-E30] Records highlighting their impact on another individual in terms of improvement or professional development in this competency. [TIFP08-E40]
9	Maintains and enhances own competence in this area through Continual Professional Development (CPD) activities. [TIFP09]	(A) Identifies potential development needs in this area, identifying opportunities to address these through continual professional development activities. [TIFP09-10A] (A) Performs continual professional development activities to maintain and enhance their competency in this area. [TIFP09-20A] (A) Records continual professional development activities undertaken including learning or insights gained. [TIFP09-30A]	Records of Continual Professional Development (CPD) performed and learning outcomes. [TIFP09-E10]

Lead Practitioner – Technical: Interfaces			
ID	**Indicators of Competence** ***(in addition to those at Practitioner level)***	**Relevant knowledge, experience, and/or behaviors**	**Possible examples of objective evidence of personal contribution to activities performed, or professional behaviors applied**
1	Creates enterprise- level policies, procedures, guidance, and best practice for interface definition and management, including associated tools. [TIFL01]	(A) Analyzes enterprise need for interface management policies, processes, tools, or guidance. [TIFL01-10A]	Records showing their role in embedding interface definition and management into enterprise policies (e.g. guidance introduced at enterprise level and enterprise -level review minutes). [TIFL01-E10]
		(A) Creates enterprise policies, procedures or guidance for interface management activities. [TIFL01-20A]	Procedures they have written. [TIFL01-E20]
		(A) Selects and acquires appropriate tools supporting interface management activities. [TIFL01-30A]	Records of support for tool introduction. [TIFL01-E30]
		(A) Defines a full range of Interface Management techniques for a range of systems. [TIFL01-40A]	
2	Judges the tailoring of enterprise-level interface definition and management processes to meet the needs of a project. [TIFL02]	(A) Evaluates the enterprise process against the business and external stakeholder (e.g. customer) needs in order to tailor processes to enable project success. [TIFL02-10A]	Records of tailoring to process in this area. [TIFL02-E10]
		(A) Provides constructive feedback on interface management process. [TIFL02-20A]	Interface management process feedback. [TIFL02-E20]
3	Judges the suitability and completeness of interfaces and associated management practices used on projects across the enterprise. [TIFL03]	(A) Reviews interface documentation. [TIFL03-10A]	Records of a review process in which they have been involved. [TIFL03-E10]
		(A) Provides advice on interface documentation that has led to changes being implemented. [TIFL03-20A]	Records of advice provided on interface documentation. [TIFL03-E20]
		(A) Identifies improper interface definitions (e.g. discontinuities and unidentified elements). [TIFL03-30A]	
4	Identifies conflicts in the definition or management of interfaces requiring resolution on projects across the enterprise. [TIFL04]	(A) Identifies and engages with key stakeholders. [TIFL04-10A]	Minutes of meetings. [TIFL04-E10]
		(A) Proposes solutions to enterprise level issues with a holistic viewpoint. [TIFL04-20A]	

(*Continued*)

ID	Indicators of Competence *(in addition to those at Practitioner level)*	Relevant knowledge, experience, and/or behaviors	Possible examples of objective evidence of personal contribution to activities performed, or professional behaviors applied
5	Acts to arbitrate when there are conflicts in the definition of interfaces or their management on projects across the enterprise. [TIFL05]	(A) Proposes different strategies in order to resolve management of complex interfaces on a project. [TIFL05-10A]	Interface documents and review comments. [TIFL05-E10]
		(A) Proposes different strategies in order to resolve conflict resulting from management or definition of project interfaces. [TIFL05-20A]	Interface management plans and review comments. [TIFL05-E20]
		(P) Advises stakeholders at enterprise level to resolve conflicts arising from interface management strategies or interface definitions. [TIFL05-30P]	Interface management strategies and comments they made showing issues. [TIFL05-E30]
6	Coaches or mentors practitioners across the enterprise in Systems Engineering interface management in order to develop their knowledge, abilities, skills, or associated behaviors. [TIFL06]	(P) Coaches or mentors practitioners across the enterprise in competency-related techniques, recommending development activities. [TIFL06-10P]	Coaching or mentoring assignment records. [TIFL06-E10]
		(A) Develops or authorizes enterprise training materials in this competency area. [TIFL06-20A]	Records of formal training courses, workshops, seminars, and authored training material supported by successful post-training evaluation data. [TIFL06-E20]
		(A) Provides enterprise workshops/seminars or training in this competency area. [TIFL06-30A]	Listing as an approved organizational trainer for this competency area. [TIFL06-E30]
7	Promotes the introduction and use of novel techniques and ideas in interface management across the enterprise, to improve enterprise competence in this area. [TIFL07]	(A) Analyzes different approaches across different domains through research. [TIFL07-10A]	Research records. [TIFL07-E10]
		(A) Defines novel approaches that could potentially improve the SE discipline within the enterprise. [TIFL07-20A]	Published papers in refereed journals/company literature. [TIFL07-E20]
		(P) Fosters awareness of these novel techniques within the enterprise. [TIFL07-30P]	Enabling systems introduced in support of new techniques or ideas. [TIFL07-E30]
		(P) Collaborates with enterprise stakeholders to facilitate the introduction of techniques new to the enterprise. [TIFL07-40P]	Published papers (or similar) at enterprise level. [TIFL07-E40]
		(A) Monitors new techniques after their introduction to determine their effectiveness. [TIFL07-50A]	Records of improvements made against a recognized process improvement model in this area. [TIFL07-E50]
		(A) Adapts approach to reflect actual enterprise performance improvements. [TIFL07-60A]	

8	Develops expertise in this competency area through specialist Continual Professional Development (CPD) activities. [TIFL08]	(A) Identifies own needs for further professional development in order to increase competence beyond practitioner level. [TIFL08-10A] (A) Performs professional development activities in order to move own competence toward expert level. [TIFL08-20A] (A) Records continual professional development activities undertaken including learning or insights gained. [TIFL08-30A]	Records of Continual Professional Development (CPD) performed and learning outcomes. [TIFL08-E10]

NOTES	In addition to items above, enterprise-level or independent 3rd-Party-generated evidence may be used to amplify other evidence presented and may include: a. Formally recognized by senior management in current organization as an expert in this competency area b. Evidence of role as Product/System Design Authority or Technical Authority on a complex project with responsibilities in this area or where skills within this competency area were used c. Recognized as an authorizing signatory on behalf of enterprise for formal documentation in this competency area (e.g. policies, processes, and deliverables) d. Formal commendation or award within own enterprise for contribution or item of work successfully performed, which required proficiency in this competency area e. Customer, Supplier, or other external project-specific key Stakeholder accolades for specific work performed in this competency area f. Independently assessed or accredited work in this competency area (e.g. for independent publication or use) g. Formal organizational HR records positively highlighting any specific professional competencies or behaviors identified (if applicable) plus any of the evidence indicators listed at Expert level below

Expert – Technical: Interfaces			
ID	**Indicators of Competence** ***(in addition to those at Lead Practitioner level)***	**Relevant knowledge, experience, and/or behaviors**	**Possible examples of objective evidence of personal contribution to activities performed, or professional behaviors applied**
1	Communicates own knowledge and experience in Systems Engineering interface definition and management in order to improve best practice beyond the enterprise boundary. [TIFE01]	(A) Produces papers, seminars, or presentations outside own enterprise for publication in order to share own ideas and improve industry best practices in this competence area. [TIFE01-10A]	Published papers or books etc. on new technique in refereed journals/company literature. [TIFE01-E10]
		(P) Fosters incorporation of own ideas into industry best practices in this area. [TIFE01-20P]	Published papers in refereed journals or internal literature proposing new practices in this competence area (or presentations, tutorials, etc.). [TIFE01-E20]
		(P) Develops guidance materials identifying new (or updating existing) best practice in this competence area. [TIFE01-30P]	Own proposals adopted as industry best practices in this competence area. [TIFE01-E30]
2	Influences key stakeholders beyond the enterprise boundary in interface definition and management to support enterprise needs. [TIFE02]	(A) Identifies and engages with external stakeholders to facilitate interface definition and management. [TIFE02-10A]	Minutes of meetings. [TIFE02-E10]
			Revised interface management plans or procedures. [TIFE02-E20]
3	Advises organizations beyond the enterprise boundary on the suitability of their approach to Systems Engineering Interface Management and Control. [TIFE03]	(A) Provides independent advice to stakeholders beyond enterprise boundary regarding their Interface Management activities. [TIFE03-10A]	Records of advice provided. [TIFE03-E10]
		(P) Advises on highly complex Interface Management issues making limited use of specialized, technical terminology and taking account of external stakeholders' (e.g. customer's) background and knowledge. [TIFE03-20P]	
		(P) Communicates using a holistic approach to complex Interface Management issue resolution including balanced, rational arguments on way forward. [TIFE03-30P]	
		(P) Arbitrates on Systems Engineering Interface Management decisions on behalf of stakeholders beyond enterprise boundary. [TIFE03-40P]	
		(A) Persuades stakeholders beyond the enterprise boundary to accept difficult project-specific Interface Management recommendations or actions. [TIFE03-50A]	

4	Advises organizations beyond the enterprise boundary on their handling of complex or sensitive Systems Engineering Interface management issues. [TIFE04]	(A) Advises external stakeholders (e.g. customers) on their interface requirements and issues. [TIFE04-10A]	Records of advice provided on interface management issues. [TIFE04-E10]
		(P) Acts with sensitivity on highly complex system interface issues making limited use of specialized, technical terminology and taking into account external stakeholders' (e.g. customer's) background and knowledge. [TIFE04-20P]	Records from negotiations demonstrating awareness of customer's background and knowledge. [TIFE04-E20]
		(P) Fosters a holistic approach to complex issue resolution including balanced, rational arguments on way forward. [TIFE04-30P]	Stakeholder approval of interface documentation. [TIFE04-E30]
5	Champions the introduction of novel techniques and ideas in interface management, beyond the enterprise boundary, in order to develop the wider Systems Engineering community in this competency. [TIFE05]	(A) Analyzes different approaches across different domains through research. [TIFE05-10A]	Records of activities promoting research and need to adopt novel technique or ideas. [TIFE05-E10]
		(A) Produces reports for the wider SE community on the effectiveness of new techniques after their introduction. [TIFE05-20A]	Records of improvements made to process and appraisal against a recognized process improvement model. [TIFE05-E20]
		(P) Collaborates with those introducing novel techniques within the wider SE community. [TIFE05-30P]	Research records. [TIFE05-E30]
		(A) Defines novel approaches that could potentially improve the wider SE discipline. [TIFE05-40A]	Published papers in refereed journals/company literature. [TIFE05-E40]
		(P) Fosters awareness of these novel techniques within the wider SE community. [TIFE05-50P]	Records showing introduction of enabling systems supporting the new techniques or ideas. [TIFE05-E50]
6	Coaches individuals beyond the enterprise boundary in Interface Management, in order to further develop their knowledge, abilities, skills, or associated behaviors. [TIFE06]	(P) Coaches or mentors individuals beyond the enterprise boundary, in competency-related techniques, recommending development activities. [TIFE06-10P]	Coaching or mentoring assignment records. [TIFE06-E10]
		(A) Develops or authorizes training materials in this competency area, which are subsequently successfully delivered beyond the enterprise boundary. [TIFE06-20A]	Records of formal training courses, workshops, seminars, and authored training material supported by successful post-training evaluation data. [TIFE06-E20]
		(A) Provides workshops/seminars or training in this competency area for practitioners or lead practitioners beyond the enterprise boundary (e.g. conferences and open training days). [TIFE06-30A]	Records of Training/workshops/seminars created supported by successful post-training evaluation data. [TIFE06-E30]

ID	Indicators of Competence *(in addition to those at Lead Practitioner level)*	Relevant knowledge, experience, and/or behaviors	Possible examples of objective evidence of personal contribution to activities performed, or professional behaviors applied
7	Maintains expertise in this competency area through specialist Continual Professional Development (CPD) activities. [TIFE07]	(A) Reviews research, new ideas, and state of the art to identify relevant new areas requiring personal development in order to maintain expertise in this competency area. [TIFE07-10A] (A) Performs identified specialist professional development activities in order to maintain or further develop competence at expert level. [TIFE07-20A] (A) Records continual professional development activities undertaken including learning or insights gained. [TIFE07-30A]	Records of documents reviewed and insights gained as part of own research into this competency area. [TIFE07-E10] Records of Continual Professional Development (CPD) performed and learning outcomes. [TIFE07-E20]

NOTES

In addition to items above, enterprise-level or independent 3rd-Party-generated evidence may be used to amplify other evidence presented and may include:

a. Formally recognized by a reputable external organization as an expert in this competency area
b. Evidence of role as independent assessor or reviewer on project outside own organization where skills in this competency area were used
c. Evidence of invitation(s) from wider community for contribution of systems engineering expertise in this area (e.g. industry conference panel, government advisory board etc. cross-industry working groups, partnerships, accredited advanced university courses or research, or as part of professional institute)
d. Formal commendation beyond the enterprise (e.g. by INCOSE or other recognized authority) for work performed in this competency area
e. Independently assessed or accredited work product in this competency area (e.g. for independent publication or use)
f. Accolades of expertise in this area from recognized industry leaders

Competency area – Technical: Verification
Description Verification is the formal process of obtaining objective evidence that a system or system element, product or service fulfils its specified requirements and characteristics. Verification includes formal testing of the system against the system requirements; including qualification against the super system environment (e.g. electro-magnetic compatibility, thermal, vibration, humidity, and fungus growth). Put simply, it answers the question "Did we build the system right?"
Why it matters System verification should be planned so that system elements are tested in a logical sequence to avoid wasted effort. Systematic and incremental verification makes it easier to find, isolate, diagnose, and correct problems. A system or system element that has not been verified cannot be relied on to meet its requirements. Systems Verification is an essential pre-requisite to customer acceptance and certification.
Possible contributory types of evidence Any combination of the types of evidence may be acceptable (depending on how the Framework is tailored and used). The evidence items identified at each level indicate example work products only. Contributions to work products will generally differ at each proficiency level.
Learning and development The INCOSE Professional Development Portal provides example guidance on how to gain an initial awareness of a competency area and options for developing further competence thereafter.

Awareness – Technical: Verification		
ID	**Indicators of Competence**	**Relevant knowledge sub-indicators**
1	Explains what verification is, the purpose of verification, and why verification against the system requirements is important. [TVEA01]	(K) Describes why the system should be verified against the requirements (i.e. the system requirements and not external stakeholder or customer requirements) in order to ensure that the specified design requirements are fulfilled by the system. [TVEA01-10K]
2	Explains why there is a need to verify the system in a logical sequence. [TVEA02]	(K) Describes why Verification should be conducted using a progressive, logical process. [TVEA02-10K] (K) Explains why alternative verification sequences may be assessed in order to define the most appropriate sequence in terms of overall cost and risk (this means that the sequence should not necessarily be based on a success assumed process). [TVEA02-20K] (K) Explains why, if verification is performed in the wrong sequence re-work and extra cost may be incurred (dependency on suppliers, development, new technology, obsolescence, etc.). [TVEA02-30K]
3	Explains why planning for system verification is necessary. [TVEA03]	(K) Explains why planning for verification should occur at the beginning of the project/program. [TVEA03-10K] (K) Describes key interfaces and links between verification and stakeholders in the wider enterprise outside engineering (e.g. infrastructure management, human resource management, quality management, knowledge management, portfolio management, and life cycle model management). [TVEA03-20K] (K) Explains why the verification sequence should be documented. [TVEA03-30K]

(Continued)

ID	Indicators of Competence	Relevant knowledge sub-indicators
		(K) Explains why there is a need to identify the resources, equipment, and develop test requirements (influence the design). [TVEA03-40K]
4	Explains how traceability can be used to establish whether a system meets requirements. [TVEA04]	(K) Explains how tracing the system requirements to a specific verification action provides evidence the requirement was met. [TVEA04-10K] (K) Explains why, if all system requirements are traced to at least one verification action, the completed verification event documents that the system meets all requirements. [TVEA04-20K]
5	Describes the relationship between verification, validation, qualification, certification, and acceptance. [TVEA05]	(K) Explains how a system may be verified against the requirements but may not be accepted (validated) by an external stakeholder as fit for purpose (i.e. “built the system right” but “not built the right system”). [TVEA05-10K] (K) Explains how verification evidence may or may not be used in support of acceptance. [TVEA05-20K] (K) Explains the key characteristics of qualification and certification and how they relate to verification. [TVEA05-30K]

Supervised Practitioner – Technical: Verification			
ID	**Indicators of Competence** ***(in addition to those at Awareness level)***	**Relevant knowledge, experience, and/or behaviors**	**Possible examples of objective evidence of personal contribution to activities performed, or professional behaviors applied**
1	Complies with a governing process and appropriate tools to plan and control their own verification activities. [TVES01]	(A) Reviews element information (e.g. tests passed, certificate of conformity received for COTS products) in order to confirm readiness for verification. [TVES01-10A]	System verification test planning documents including schedules and resources required. [TVES01-E10]
		(A) Reviews test environment (test equipment, tools, procedures, sequence, etc.) to confirm its readiness for verification. [TVES01-20A]	System verification test procedures. [TVES01-E20]
		(A) Follows a verification test procedure and identifies non conformances against the plan. [TVES01-30A]	System verification test reports. [TVES01-E30]
2	Prepares inputs to verification plans. [TVES02]	(K) Describes the key elements of a system verification plan. [TVES02-10K]	System verification test plan. [TVES02-E10]
		(A) Prepares a simple verification procedure. [TVES02-20A]	System verification test procedure. [TVES02-E20]
		(A) Identifies method and timing for each verification activity. [TVES02-30A]	
3	Prepares verification plans for smaller projects. [TVES03]	(K) Describes key elements of a system verification plan. [TVES03-10K]	System verification test plan including schedules and resources required. [TVES03-E10]
		(A) Prepares a simple verification plan for a project or project element. [TVES03-20A]	
4	Performs verification testing as part of system verification activities. [TVES04]	(A) Reviews inputs for verification activities ensuring they meet requirements. [TVES04-10A]	Records from Test Readiness Review (TRR). [TVES04-E10]
		(A) Performs verification activities following defined procedures, recording results. [TVES04-20A]	Verification data and/or test reports. [TVES04-E20]
		(A) Prepares verification work products. [TVES04-30A]	
		(A) Reviews verification work products, highlighting anomalies and supporting resolution of any failures. [TVES04-40A]	
5	Identifies simple faults found during verification through diagnosis and consequential corrective actions. [TVES05]	(K) Explains how results may differ from those expected and how this is handled. [TVES05-10K]	Verification test results. [TVES05-E10]
		(A) Records faults appropriately (process, tools used, method). [TVES05-20A]	Fault log, Trouble Report, or similar test record. [TVES05-E20]
		(A) Analyzes simple faults in a logical manner and initiate the corrective action process. [TVES05-30A]	

(Continued)

ID	Indicators of Competence *(in addition to those at Awareness level)*	Relevant knowledge, experience, and/or behaviors	Possible examples of objective evidence of personal contribution to activities performed, or professional behaviors applied
		(A) Records corrective action taken. [TVES05-40A] (A) Maintains the outstanding fault log when all outstanding actions have been completed. [TVES05-50A]	
6	Collates evidence in support of verification, qualification, certification, and acceptance. [TVES06]	(K) Describes different types of information that should be documented as part of verification evidence. [TVES06-10K]	Verification data and/or test reports. [TVES06-E10]
		(K) Describes how verification evidence may also be used in support of qualification, certification, validation or acceptance. [TVES06-20K]	Test records. [TVES06-E20]
		(A) Collates evidence in support of verification, qualification certification. [TVES06-30A]	Fault log, trouble report, or similar verification test record. [TVES06-E30]
		(A) Identifies exceptional and unexpected verification issues, resolving as required. [TVES06-40A]	Records of issue resolution. [TVES06-E40]
		(A) Uses different methods to verify system requirements. [TVES06-50A]	
7	Reviews verification evidence to establish whether a system meets requirements. [TVES07]	(A) Ensures test results confirm system requirements are met. [TVES07-10A]	Requirements acceptance/verification matrix. [TVES07-E10]
		(A) Ensures test results confirm system requirements are met. [TVES07-20A]	Requirements acceptance/verification matrix. [TVES07-E20]
8	Selects a verification environment to ensure requirements can be fully verified. [TVES08]	(K) Lists areas to consider when arranging for the availability of verification environment assets and why. [TVES08-10K]	Detailed verification test planning documents. [TVES08-E10]
		(A) Selects tools, equipment or facilities needed to execute system verification. [TVES08-20A]	
9	Develops own understanding of this competency area through Continual Professional Development (CPD). [TVES09]	(A) Identifies potential gaps in own knowledge or development needs in this area, identifying opportunities to address these through continual professional development activities. [TVES09-10A]	Records of Continual Professional Development (CPD) performed and learning outcomes. [TVES09-E10]
		(A) Performs continual professional development activities to improve their knowledge and understanding in this area. [TVES09-20A]	
		(A) Records continual professional development activities undertaken including learning or insights gained. [TVES09-30A]	

Practitioner – Technical: Verification			
ID	**Indicators of Competence *(in addition to those at Supervised Practitioner level)***	**Relevant knowledge, experience, and/or behaviors**	**Possible examples of objective evidence of personal contribution to activities performed, or professional behaviors applied**
1	Creates a strategy for system verification on a project to support SE project and wider enterprise needs. [TVEP01]	(A) Identifies project-specific system verification needs that need to be incorporated into SE planning. [TVEP01-10A]	System verification strategy or test plan, including schedules and resources required. [TVEP01-E10]
		(A) Prepares project-specific system verification inputs for SE planning purposes, using appropriate processes and procedures. [TVEP01-20A]	
		(A) Prepares project-specific system verification task estimates in support of SE planning. [TVEP01-30A]	
		(P) Liaises throughout with stakeholders to gain approval, updating strategy as necessary. [TVEP01-40P]	
2	Creates a governing process, plan, and associated tools for systems verification, which reflect project and business strategy. [TVEP02]	(K) Identifies the steps necessary to define a process and appropriate techniques to be adopted for the verification of system elements. [TVEP02-10K]	System verification planning work products. [TVEP02-E10]
		(K) Describes key elements of a successful systems verification process and how systems verification activities relate to different stages of a system life cycle. [TVEP02-20K]	Planning documents (e.g. systems engineering management plan and system test plan), other project/program plan or organizational process. [TVEP02-E20]
		(A) Uses system verification tools or uses a methodology. [TVEP02-30A]	System verification test procedures. [TVEP02-E30]
		(A) Uses appropriate standards and governing processes in their system verification planning, tailoring as appropriate. [TVEP02-40A]	Requirements acceptance/verification matrix. [TVEP02-E40]
		(A) Selects system verification method for each system element. [TVEP02-50A]	
		(A) Creates system verification plans and processes for use on a project. [TVEP02-60A]	
		(P) Liaises with stakeholders throughout development of the plan, to gain approval, updating as necessary. [TVEP02-70P]	
3	Uses governing plans and processes for System verification, interpreting, evolving, or seeking guidance where appropriate. [TVEP03]	(A) Follows defined plans, processes and associated tools to perform system Verification on a project. [TVEP03-10A]	System Verification plans or work products they have interpreted or challenged to address project issues. [TVEP03-E10]
		(P) Guides and actively coordinates the interpretation of system verification plans or processes in order to complete activities system Verification activities successfully. [TVEP03-20P]	System verification process or tool guidance. [TVEP03-E20]

(*Continued*)

ID	Indicators of Competence *(in addition to those at Supervised Practitioner level)*	Relevant knowledge, experience, and/or behaviors	Possible examples of objective evidence of personal contribution to activities performed, or professional behaviors applied
		(A) Prepares verification data and reports in support of wider system measurement, monitoring and control. [TVEP03-30A] (P) Recognizes situations where deviation from published plans and processes or clarification from others is appropriate in order to overcome complex system Verification challenges. [TVEP03-40P] (P) Recognizes situations where existing system Verification strategy, process, plan or tools require formal change, liaising with stakeholders to gain approval as necessary. [TVEP03-50P]	
4	Prepares verification plans for systems or projects. [TVEP04]	(K) Explains the different verification methods and how/when/why to select the most appropriate method (test, analysis, inspection, similarity, and comparison). [TVEP04-10K] (A) Records the appropriate degree of verification evidence (for example, safety critical software requires a greater degree of verification than non-safety critical). [TVEP04-20A] (A) Develops a verification test plan containing objectives, conditions, priorities, schedules and responsibilities, tools, facilities, procedures and standards to be applied, and the success criteria to be applied, etc. [TVEP04-30A] (A) Reviews the depth of testing required for verification and plans accordingly. [TVEP04-40A] (A) Defines detailed verification test sequences and the readiness criteria for each system element. [TVEP04-50A] (A) Creates a verification test schedule showing dependencies of each activity (critical path analysis). [TVEP04-60A] (A) Defines the verification test environment required, including outsourcing of qualification tests as required. [TVEP04-70A] (A) Identifies the evidence required for external stakeholder (e.g. customer) acceptance and certification and ensure production of evidence is in integration and verification plans. [TVEP04-80A] (A) Identifies the relevant certification authorities and the process by which certification must be obtained. [TVEP04-90A]	Requirements acceptance/verification matrix. [TVEP04-E10] System verification test procedures. [TVEP04-E20] Compliance/verification matrices. [TVEP04-E30]
5	Reviews project-level system verification plans. [TVEP05]	(A) Reviews verification plans for a system for its suitability for the project. [TVEP05-10A] (P) Collaborates with project stakeholders to improve verification plans as necessary to ensure success. [TVEP05-20P]	Records showing review and approval of verification plan. [TVEP05-E10]

		(A) Identifies required changes and monitors plan updates until they have been appropriately implemented. [TVEP05-30A]	
6	Reviews verification results, diagnosing complex faults found during verification activities. [TVEP06]	(A) Analyzes test execution and test results to confirm validity of test and results collected. [TVEP06-10A]	Fault log, Trouble Report, or similar test record. [TVEP06-E10]
		(A) Uses statistical or modeling techniques to demonstrate sufficient and necessary verification activities have taken place. [TVEP06-20A]	Record of corrective actions or fault resolution. [TVEP06-E20]
		(A) Identifies where results differ from those expected. [TVEP06-30A]	Minutes of fault analysis meetings. [TVEP06-E30]
		(A) Analyzes complex faults in a logical manner to determine corrective actions. [TVEP06-40A]	
		(A) Identifies consequences of corrective actions (re-planning, re-test, etc.). [TVEP06-50A]	
		(A) Monitors corrective actions to completion, ensuring closure. [TVEP06-60A]	
		(A) Reviews test evidence to ensure suitability for use in proving compliance. [TVEP06-70A]	
7	Prepares evidence obtained during verification testing to support system verification or downstream qualification, certification, and acceptance activities. [TVEP07]	(A) Identifies the precise evidence required in support of verification, qualification, certification, or acceptance testing. [TVEP07-10A]	Requirements acceptance/verification matrix. [TVEP07-E10]
		(A) Uses a systematic method for classifying the results of verification test reports. [TVEP07-20A]	Compliance/verification matrices. [TVEP07-E20]
		(A) Prepares inputs to formal verification test reviews and associated planning activities. [TVEP07-30A]	Test reports. [TVEP07-E30]
		(A) Uses an appropriate verification test results classification method, for example: pass, mild deficiency, annoyance, and catastrophic. [TVEP07-40A]	Inputs to Certification data package. [TVEP07-E40]
			Inputs to Qualification data package. [TVEP07-E50]
			Inputs to Acceptance data package. [TVEP07-E60]
			Minutes of test data reviews. [TVEP07-E70]
8	Monitors the traceability of verification requirements and tests to system requirements and vice versa. [TVEP08]	(A) Monitors system and verification test requirements in order to obtain forward- and backward-traceability ensuring integrity is maintained. [TVEP08-10A]	Requirements acceptance/verification matrix. [TVEP08-E10]

(*Continued*)

ID	Indicators of Competence *(in addition to those at Supervised Practitioner level)*	Relevant knowledge, experience, and/or behaviors	Possible examples of objective evidence of personal contribution to activities performed, or professional behaviors applied
9	Identifies a suitable verification environment. [TVEP09]	(A) Defines requirements for system verification facilities (e.g. size of the area, furniture required, power requirements, the IT requirements, and the security of the facility). [TVEP09-10A]	Verification plans. [TVEP09-E10]
		(A) Defines requirements for external test facilities. [TVEP09-20A]	Procurement of equipment. [TVEP09-E20]
		(A) Defines the requirements for bespoke tools and equipment needed in support of verification, e.g. simulators and emulators. [TVEP09-30A]	Arrangement of external facilities or required certification authority activities. [TVEP09-E30]
		(A) Determines resources and skills required. [TVEP09-40A]	
10	Creates detailed verification procedures. [TVEP10]	(A) Identifies appropriate techniques for requirements verification, for example: requirements analysis, exploration of requirements adequacy and completion, assessment of prototypes, stimulations, simulations, models, scenarios, and mock-ups. [TVEP10-10A]	Approved verification procedures. [TVEP10-E10]
		(A) Develops verification tests that will confirm system requirements have been met. [TVEP10-20A]	Requirements acceptance/verification matrix. [TVEP10-E20]
		(A) Analyzes verification test cases linking them to requirements. [TVEP10-30A]	Requirements acceptance/verification matrix. [TVEP10-E30]
		(A) Creates clear, concise verification procedures detailing the activities to be performed, pre-requisites, the expected outcome and action in case of a failure. [TVEP10-40A]	Test Reports. [TVEP10-E40]
		(A) Identifies pass/fail criteria for verification tests, maintaining the link to the appropriate system requirement (while understanding that the two need to be developed together). [TVEP10-50A]	Minutes of meetings determining agreement of verification strategies. [TVEP10-E50]
11	Performs system verification activities. [TVEP11]	(P) Performs Verification testing activities, including appropriate reviews e.g. test readiness review. [TVEP11-10P]	Verification measures showing actual performance against plan. [TVEP11-E10]
		(A) Reviews Verification evidence to ensure full compliance against requirement. [TVEP11-20A]	Minutes of test readiness review including relevant action log. [TVEP11-E20]
		(A) Identifies problems encountered during Verification (e.g. problems with schedule and lateness of equipment), resolving as required. [TVEP11-30A]	

12	Prepares evidence obtained during verification testing to support downstream verification testing, integration, or validation activities. [TVEP12]	(A) Identifies the precise evidence required in support of downstream verification, integration, or validation activities. [TVEP12-10A]	Minutes of verification reviews. [TVEP12-E10]
		(A) Uses a systematic method for classifying the results of verification test reports. [TVEP12-20A]	Verification test documentation or final report. [TVEP12-E20]
		(A) Prepares inputs to downstream test reviews (e.g. test readiness reviews) and associated planning activities. [TVEP12-30A]	Requirements verification matrix. [TVEP12-E30]
		(A) Uses an appropriate verification test results classification method, for example: pass, mild deficiency, annoyance, and catastrophic. [TVEP12-40A]	Compliance matrix. [TVEP12-E40]
			Verification Test reports. [TVEP12-E50]
			Data packages for other test activities (e.g. verification, validation, qualification, and Certification). [TVEP12-E60]
13	Guides new or supervised practitioners in Systems verification in order to develop their knowledge, abilities, skills, or associated behaviors. [TVEP13]	(P) Guides new or supervised practitioners in executing activities that form part of this competency. [TVEP13-10P]	Organizational Breakdown Structure showing their responsibility for technical supervision in this area. [TVEP13-E10]
		(A) Trains individuals to an "Awareness" level in this competency area. [TVEP13-20A]	On-the-job training records. [TVEP13-E20]
			Coaching or mentoring assignment records. [TVEP13-E30]
			Records highlighting their impact on individual in terms of improvement or professional development in this competency. [TVEP13-E40]
14	Maintains and enhances own competence in this area through Continual Professional Development (CPD) activities. [TVEP14]	(A) Identifies potential development needs in this area, identifying opportunities to address these through continual professional development activities. [TVEP14-10A]	Records of Continual Professional Development (CPD) performed and learning outcomes. [TVEP14-E10]
		(A) Performs continual professional development activities to maintain and enhance their competency in this area. [TVEP14-20A]	
		(A) Records continual professional development activities undertaken including learning or insights gained. [TVEP14-30A]	

Lead Practitioner – Technical: Verification			
ID	**Indicators of Competence** *(in addition to those at Practitioner level)*	**Relevant knowledge, experience, and/or behaviors**	**Possible examples of objective evidence of personal contribution to activities performed, or professional behaviors applied**
1	Creates enterprise-level policies, procedures, guidance, and best practice for verification, including associated tools. [TVEL01]	(A) Analyzes enterprise need for verification policies, processes, tools, or guidance. [TVEL01-10A]	Records showing their role in embedding system verification into enterprise policies (e.g. guidance introduced at enterprise level and enterprise-level review minutes). [TVEL01-E10]
		(A) Creates enterprise policies, procedures, or guidance for verification activities. [TVEL01-20A]	Procedures they have written. [TVEL01-E20]
		(A) Selects and acquires appropriate tools supporting verification activities. [TVEL01-30A]	Records of support for tool introduction. [TVEL01-E30]
2	Judges the tailoring of enterprise-level verification processes to meet the needs of a project. [TVEL02]	(A) Evaluates the enterprise process against the business and external stakeholder (e.g. customer) needs in order to tailor processes to enable project success. [TVEL02-10A]	Documented tailored process. [TVEL02-E10]
		(P) Communicates constructive feedback on verification process. [TVEL02-20P]	Verification process feedback. [TVEL02-E20]
3	Judges the suitability of verification plans, from multiple projects, on behalf of the enterprise. [TVEL03]	(A) Evaluates verification strategies from multiple project/programs for consistency, correctness, and to ensure enterprise risk is minimized. [TVEL03-10A]	Verification strategies that proved successful. [TVEL03-E10]
		(A) Evaluates verification risks and proposes mitigation strategies to reduce project or enterprise risk. [TVEL03-20A]	Risk mitigations performed. [TVEL03-E20]
4	Advises on verification approaches on complex or challenging systems or projects across the enterprise. [TVEL04]	(A) Reviews verification plans for highly complex systems or projects, on behalf of the enterprise. [TVEL04-10A]	Records showing review and approval of verification plans. [TVEL04-E10]
		(P) Challenges verification plans as necessary to ensure success, liaising with stakeholders if required. [TVEL04-20P]	
		(P) Collaborates with enterprise stakeholders to improve verification plans as necessary to ensure success. [TVEL04-30P]	
		(A) Authorizes verification plans for highly complex systems or project on behalf of enterprise. [TVEL04-40A]	
5	Judges detailed verification procedures from multiple projects, on behalf of the enterprise. [TVEL05]	(A) Reviews and comments on project/program verification procedures from across the organization. [TVEL05-10A]	Records of a review process in which they have been involved. [TVEL05-E10]
		(A) Judges the adequacy of verification procedures identifying required changes. [TVEL05-20A]	Records of advice provided in this area. [TVEL05-E20]
			Test Readiness Review (TRR) Documentation. [TVEL05-E30]

6	Judges verification evidence generated from multiple projects on behalf of the enterprise. [TVEL06]	(A) Reviews and comments on verification evidence from multiple projects on behalf of the enterprise. [TVEL06-10A]	Records of a review process in which they have been involved. [TVEL06-E10]
		(A) Advises stakeholders across the enterprise on verification evidence, leading to changes being implemented. [TVEL06-20A]	Records of advice provided in this area. [TVEL06-E20]
7	Guides and actively coordinates verification activities for complex systems or projects across the enterprise. [TVEL07]	(A) Describes typical approaches to complex verification activities. [TVEL07-10A]	Verification plans. [TVEL07-E10]
		(P) Guides and actively coordinates verification activities for complex systems or projects. [TVEL07-20P]	Verification measures showing actual performance against plan. [TVEL07-E20]
8	Coaches or mentors practitioners across the enterprise in systems verification in order to develop their knowledge, abilities, skills, or associated behaviors. [TVEL08]	(P) Coaches or mentors practitioners across the enterprise in competency-related techniques, recommending development activities. [TVEL08-10P]	Coaching or mentoring assignment records. [TVEL08-E10]
		(A) Develops or authorizes enterprise training materials in this competency area. [TVEL08-20A]	Records of formal training courses, workshops, seminars, and authored training material supported by successful post-training evaluation data. [TVEL08-E20]
		(A) Provides enterprise workshops/seminars or training in this competency area. [TVEL08-30A]	Listing as an approved organizational trainer for this competency area. [TVEL08-E30]
9	Promotes the introduction and use of novel techniques and ideas in verification across the enterprise, to improve enterprise competence in this area. [TVEL09]	(A) Analyzes different approaches across different domains through research. [TVEL09-10A]	Research records. [TVEL09-E10]
		(A) Defines novel approaches that could potentially improve the SE discipline within the enterprise. [TVEL09-20A]	Published papers in refereed journals/company literature. [TVEL09-E20]
		(P) Fosters awareness of these novel techniques within the enterprise. [TVEL09-30P]	Enabling systems introduced in support of new techniques or ideas. [TVEL09-E30]
		(P) Collaborates with enterprise stakeholders to facilitate the introduction of techniques new to the enterprise. [TVEL09-40P]	Published papers (or similar) at enterprise level. [TVEL09-E40]
		(A) Monitors new techniques after their introduction to determine their effectiveness. [TVEL09-50A]	Records of improvements made against a recognized process improvement model in this area. [TVEL09-E50]
		(A) Adapts approach to reflect actual enterprise performance improvements. [TVEL09-60A]	

(Continued)

ID	Indicators of Competence *(in addition to those at Practitioner level)*	Relevant knowledge, experience, and/or behaviors	Possible examples of objective evidence of personal contribution to activities performed, or professional behaviors applied
10	Develops expertise in this competency area through specialist Continual Professional Development (CPD) activities. [TVEL10]	(A) Identifies own needs for further professional development in order to increase competence beyond practitioner level. [TVEL10-10A] (A) Performs professional development activities in order to move own competence toward expert level. [TVEL10-20A] (A) Records continual professional development activities undertaken including learning or insights gained. [TVEL10-30A]	Records of Continual Professional Development (CPD) performed and learning outcomes. [TVEL10-E10]

NOTES

In addition to items above, enterprise-level or independent 3rd-Party-generated evidence may be used to amplify other evidence presented and may include:

a. Formally recognized by senior management in current organization as an expert in this competency area
b. Evidence of role as Product/System Design Authority or Technical Authority on a complex project with responsibilities in this area or where skills within this competency area were used
c. Recognized as an authorizing signatory on behalf of enterprise for formal documentation in this competency area (e.g. policies, processes, and deliverables)
d. Formal commendation or award within own enterprise for contribution or item of work successfully performed, which required proficiency in this competency area
e. Customer, Supplier, or other external project-specific key Stakeholder accolades for specific work performed in this competency area
f. Independently assessed or accredited work in this competency area (e.g. for independent publication or use)
g. Formal organizational HR records positively highlighting any specific professional competencies or behaviors identified (if applicable) plus any of the evidence indicators listed at Expert level below

Expert – Technical: Verification			
ID	**Indicators of Competence** ***(in addition to those at Lead Practitioner level)***	**Relevant knowledge, experience, and/or behaviors**	**Possible examples of objective evidence of personal contribution to activities performed, or professional behaviors applied**
1	Communicates own knowledge and experience in Systems Engineering verification in order to improve best practice beyond the enterprise boundary. [TVEE01]	(A) Produces papers, seminars, or presentations outside own enterprise for publication in order to share own ideas and improve industry best practices in this competence area. [TVEE01-10A]	Published papers or books etc. on new technique in refereed journals/company literature. [TVEE01-E10]
		(P) Fosters incorporation of own ideas into industry best practices in this area. [TVEE01-20P]	Published papers in refereed journals or internal literature proposing new practices in this competence area (or presentations, tutorials, etc.). [TVEE01-E20]
		(P) Develops guidance materials identifying new (or updating existing) best practice in this competence area. [TVEE01-30P]	Proposals adopted as industry best practice. [TVEE01-E30]
2	Advises organizations beyond the enterprise boundary on the suitability of their approach to Systems Engineering verification. [TVEE02]	(A) Advises on the suitability of external verification plans. [TVEE02-10A]	Revised verification plans or procedures. [TVEE02-E10]
		(A) Advises on the suitability of external verification strategies. [TVEE02-20A]	Records of membership on oversight committee with relevant terms of reference. [TVEE02-E20]
		(P) Acts as independent reviewer within an enterprise review body, in order to approve such plans. [TVEE02-30P]	
3	Advises organizations beyond the enterprise boundary on their Systems Engineering Verification plans or practices on complex systems or projects. [TVEE03]	(A) Advises on verification strategies that has led to changes being implemented. [TVEE03-10A]	Review comments or revised document. [TVEE03-E10]
		(A) Advises external stakeholders (e.g. customers) on their verification requirements and issues. [TVEE03-20A]	Records of advice provided on Verification issues. [TVEE03-E20]
		(P) Conducts sensitive negotiations regarding verification of a highly complex system making limited use of specialized, technical terminology. [TVEE03-30P]	Records from negotiations demonstrating awareness of customer's background and knowledge. [TVEE03-E30]
		(A) Uses a holistic approach to complex issue resolution including balanced, rational arguments on way forward. [TVEE03-40A]	Stakeholder approval of verification plans or procedures. [TVEE03-E40]

(*Continued*)

ID	Indicators of Competence *(in addition to those at Lead Practitioner level)*	Relevant knowledge, experience, and/or behaviors	Possible examples of objective evidence of personal contribution to activities performed, or professional behaviors applied
4	Advises organizations beyond the enterprise boundary on complex or sensitive verification-related issues. [TVEE04]	(A) Advises on external verification plans. [TVEE04-10A]	Records of membership on oversight committee with relevant terms of reference. [TVEE04-E10]
		(A) Advises on verification strategies. [TVEE04-20A]	Review comments or revised document. [TVEE04-E20]
		(A) Advises external organizations on their verification requirements and issues. [TVEE04-30A]	Records of advice provided on verification issues. [TVEE04-E30]
		(P) Conducts sensitive negotiations regarding verification of a highly complex system making limited use of specialized, technical terminology. [TVEE04-40P]	Records from negotiations demonstrating awareness of customer's background and knowledge. [TVEE04-E40]
		(A) Uses a holistic approach to complex issue resolution including balanced, rational arguments on way forward. [TVEE04-50A]	Stakeholder approval of transition plans or procedures. [TVEE04-E50]
5	Champions the introduction of novel techniques and ideas in systems verification, beyond the enterprise boundary, in order to develop the wider Systems Engineering community in this competency. [TVEE05]	(A) Analyzes different approaches across different domains through research. [TVEE05-10A]	Records of activities promoting research and need to adopt novel technique or ideas. [TVEE05-E10]
		(A) Produces reports for the wider SE community on the effectiveness of new techniques after their introduction. [TVEE05-20A]	Records of improvements made to process and appraisal against a recognized process improvement model. [TVEE05-E20]
		(P) Collaborates with those introducing novel techniques within the wider SE community. [TVEE05-30P]	Research records. [TVEE05-E30]
		(A) Defines novel approaches that could potentially improve the wider SE discipline. [TVEE05-40A]	Published papers in refereed journals/company literature. [TVEE05-E40]
		(P) Fosters awareness of these novel techniques within the wider SE community. [TVEE05-50P]	Records showing introduction of enabling systems supporting the new techniques or ideas. [TVEE05-E50]
6	Coaches individuals beyond the enterprise boundary in Systems Verification, in order to further develop their knowledge, abilities, skills, or associated behaviors. [TVEE06]	(P) Coaches or mentors individuals beyond the enterprise boundary, in competency-related techniques, recommending development activities. [TVEE06-10P]	Coaching or mentoring assignment records. [TVEE06-E10]

		(A) Develops or authorizes training materials in this competency area, which are subsequently successfully delivered beyond the enterprise boundary. [TVEE06-20A]	Records of formal training courses, workshops, seminars, and authored training material supported by successful post-training evaluation data. [TVEE06-E20]
		(A) Provides workshops/seminars or training in this competency area for practitioners or lead practitioners beyond the enterprise boundary (e.g. conferences and open training days). [TVEE06-30A]	Records of Training/workshops/seminars created supported by successful post-training evaluation data. [TVEE06-E30]
7	Maintains expertise in this competency area through specialist Continual Professional Development (CPD) activities. [TVEE07]	(A) Reviews research, new ideas, and state of the art to identify relevant new areas requiring personal development in order to maintain expertise in this competency area. [TVEE07-10A]	Records of documents reviewed and insights gained as part of own research into this competency area. [TVEE07-E10]
		(A) Performs identified specialist professional development activities in order to maintain or further develop competence at expert level. [TVEE07-20A]	Records of Continual Professional Development (CPD) performed and learning outcomes. [TVEE07-E20]
		(A) Records continual professional development activities undertaken including learning or insights gained. [TVEE07-30A]	

NOTES	In addition to items above, enterprise-level or independent 3rd-Party-generated evidence may be used to amplify other evidence presented and may include: a. Formally recognized by a reputable external organization as an expert in this competency area b. Evidence of role as independent assessor or reviewer on project outside own organization where skills in this competency area were used c. Evidence of invitation(s) from wider community for contribution of systems engineering expertise in this area (e.g. industry conference panel, government advisory board etc. cross-industry working groups, partnerships, accredited advanced university courses or research, or as part of professional institute) d. Formal commendation beyond the enterprise (e.g. by INCOSE or other recognized authority) for work performed in this competency area e. Independently assessed or accredited work product in this competency area (e.g. for independent publication or use) f. Accolades of expertise in this area from recognized industry leaders

Competency area – Technical: Validation
Description The purpose of validation is to provide objective evidence that the system, product or service when in use, fulfills its business or mission objectives, and stakeholder requirements, achieving its intended use in its intended operational environment. Put simply, validation checks that the needs of the customer/end user have been met and answers the question "Did we build the right system?"
Why it matters Validation is used to check that the system meets the needs of the customer/end user. Failure to satisfy the customer will impact future business. Validation provides some important inputs to future system development.
Possible contributory types of evidence Any combination of the types of evidence may be acceptable (depending on how the Framework is tailored and used). The evidence items identified at each level indicate example work products only. Contributions to work products will generally differ at each proficiency level.
Learning and development The INCOSE Professional Development Portal provides example guidance on how to gain an initial awareness of a competency area and options for developing further competence thereafter.

Awareness – Technical: Validation		
ID	**Indicators of Competence**	**Relevant knowledge sub-indicators**
1	Explains what validation is, the purpose of validation, and why validation is important. [TVAA01]	(K) Explains how validation comprises 'product' validation i.e. the product satisfies user needs in operation and "requirements" validation, i.e. set of system requirements meets the user needs. [TVAA01-10K] (K) Explains how validation helps to reduce the risk of system failure to an acceptable level. [TVAA01-20K] (K) Defines the difference between verification activities, which address whether a system has been built correctly in accordance with the system requirements, and validation, which addresses whether the correct system has been built against the user needs. [TVAA01-30K] (K) Explains why validation activities should be undertaken by someone different from the people who designed and built the system. [TVAA01-40K]
2	Explains why there is a need for early planning for validation. [TVAA02]	(K) Explains why there is a need for early validation planning. [TVAA02-10K] (K) Describes key interfaces and links between validation and stakeholders in the wider enterprise outside engineering (e.g. infrastructure management, human resource management, quality management, knowledge management, portfolio management, and life cycle model management). [TVAA02-20K] (K) Describes the reasons why every user need should have an associated validation activity. [TVAA02-30K] (K) Explains why there is a need to plan for the validation of the system in the correct operational environment wherever practicable (or through simulated environments where that is impracticable). [TVAA02-40K]
3	Describes the relationship between validation, verification, qualification, certification, and acceptance. [TVAA03]	(K) Explains how a system may be verified against the requirements but may not then be accepted (validated) by an external stakeholder (e.g. customers) as fit for purpose. "May have built the system right but it may not be the right system." [TVAA03-10K] (K) Explains why Verification evidence may support acceptance, validation evidence does support acceptance. [TVAA03-20K] (K) Explains why Qualification and Certification typically involves third party assessors. [TVAA03-30K]
4	Describes the relationship between traceability and validation. [TVAA04]	(K) Explains why a validation procedure should be defined to the satisfaction of the user, such that when the validation activities are successfully completed, the user will be satisfied the system meets their needs and will accept the delivery of the system during transition. [TVAA04-10K]

Supervised Practitioner – Technical: Validation			
ID	**Indicators of Competence** ***(in addition to those at Awareness level)***	**Relevant knowledge, experience, and/or behaviors**	**Possible examples of objective evidence of personal contribution to activities performed, or professional behaviors applied**
1	Complies with a governing process and appropriate tools to plan and control their own validation activities. [TVAS01]	(A) Reviews element information (e.g. tests passed and certificate of conformity received for COTS products) in order to confirm readiness for validation. [TVAS01-10A] (A) Reviews test environment (test equipment, tools, procedures, sequence, etc.) to confirm its readiness for validation. [TVAS01-20A] (A) Follows a validation test procedure and identifies non conformances against the plan. [TVAS01-30A]	System validation test planning documents including schedules and resources required. [TVAS01-E10] System validation test procedures. [TVAS01-E20] System validation test reports. [TVAS01-E30]
2	Prepares inputs to validation plans. [TVAS02]	(K) Explains why validation is "customer" or "end user" focused and thus needs to be executed and documented from the perspective of the customer or end user. [TVAS02-10K] (K) Explains how a validation "customer" or "end user" could be external or internal, depending on the nature of the enterprise. [TVAS02-20K] (K) Describes the key elements of a system validation plan. [TVAS02-30K] (A) Identifies method and timing for each validation activity. [TVAS02-40A] (A) Prepares a simple validation procedure. [TVAS02-50A]	Records of review of validation plans with customer. [TVAS02-E10] System validation test plan. [TVAS02-E20] System validation test procedure. [TVAS02-E30]
3	Prepares validation plans for smaller projects. [TVAS03]	(K) Describes key elements of a system validation or acceptance plan. [TVAS03-10K] (A) Prepares a simple validation, acceptance or qualification plan for a project. [TVAS03-20A]	System Validation test plan including schedules and resources required. [TVAS03-E10]
4	Performs validation testing as part of system validation or system acceptance. [TVAS04]	(A) Reviews inputs for validation activities ensuring they meet requirements. [TVAS04-10A] (A) Performs validation activities following defined procedures, recording results. [TVAS04-20A] (A) Prepares validation work products. [TVAS04-30A] (A) Reviews validation work products, highlighting anomalies and supporting resolution of any failures. [TVAS04-40A]	Records from Test Readiness Review (TRR). [TVAS04-E10] Validation data and/or test reports. [TVAS04-E20]
5	Identifies simple faults found during validation through diagnosis and consequential corrective actions. [TVAS05]	(K) Explains how validation results may differ from verification results and how this may be handled. [TVAS05-10K]	Validation test results. [TVAS05-E10]

(Continued)

ID	Indicators of Competence *(in addition to those at Awareness level)*	Relevant knowledge, experience, and/or behaviors	Possible examples of objective evidence of personal contribution to activities performed, or professional behaviors applied
		(A) Records faults appropriately (process, tools used, method) so that details can be communicated to Customer. [TVAS05-20A] (A) Analyzes simple validation faults in a logical manner to diagnose fault. [TVAS05-30A] (A) Records corrective actions required so that they can be communicated to Customer representative. [TVAS05-40A] (A) Reviews fault log to ensure outstanding actions have been addressed. [TVAS05-50A]	Fault log, Trouble Report, or similar test record. [TVAS05-E20]
6	Collates evidence in support of validation, qualification, certification, and acceptance. [TVAS06]	(K) Describes different types of information that should be documented as part of validation evidence. [TVAS06-10K] (K) Describes how validation evidence may utilize test evidence collected in earlier test phases (e.g. verification). [TVAS06-20K] (A) Collates evidence in support of validation or acceptance. [TVAS06-30A] (A) Identifies exceptional and unexpected validation issues, resolving as required. [TVAS06-40A] (A) Uses different methods to validate operational system performance. [TVAS06-50A]	Validation data and/or test reports. [TVAS06-E10] Test records. [TVAS06-E20] Fault log, trouble report, or similar validation test record. [TVAS06-E30] Records of issue resolution. [TVAS06-E40]
7	Reviews validation evidence to establish whether a system will meet the operational need. [TVAS07]	(A) Ensures validation test results confirm system will meet operational need when in service. [TVAS07-10A]	Requirements acceptance/verification matrix. [TVAS07-E10] Validation test result analysis records. [TVAS07-E20]
8	Selects a validation environment to ensure requirements can be fully validated. [TVAS08]	(K) Lists areas to consider when arranging for the availability of validation environment assets and why. [TVAS08-10K] (A) Selects tools, equipment or facilities needed to execute system validation. [TVAS08-20A]	Detailed validation test planning documents. [TVAS08-E10] Records of validation discussions with Customer. [TVAS08-E20]
9	Develops own understanding of this competency area through Continual Professional Development (CPD). [TVAS09]	(A) Identifies potential gaps in own knowledge or development needs in this area, identifying opportunities to address these through continual professional development activities. [TVAS09-10A] (A) Performs continual professional development activities to improve their knowledge and understanding in this area. [TVAS09-20A] (A) Records continual professional development activities undertaken including learning or insights gained. [TVAS09-30A]	Records of Continual Professional Development (CPD) performed and learning outcomes. [TVAS09-E10]

Practitioner – Technical: Validation			
ID	**Indicators of Competence** ***(in addition to those at Supervised Practitioner level)***	**Relevant knowledge, experience, and/or behaviors**	**Possible examples of objective evidence of personal contribution to activities performed, or professional behaviors applied**
1	Creates a strategy for system validation on a project to support SE project and wider enterprise needs. [TVAP01]	(A) Identifies project-specific system Validation needs that need to be incorporated into SE planning. [TVAP01-10A]	System validation strategy or test plan, including schedules and resources required. [TVAP01-E10]
		(A) Prepares project-specific system Validation inputs for SE planning purposes, using appropriate processes and procedures. [TVAP01-20A]	
		(A) Prepares project-specific system Validation task estimates in support of SE planning. [TVAP01-30A]	
		(P) Liaises throughout with stakeholders to gain approval, updating strategy as necessary. [TVAP01-40P]	
2	Creates a governing process, plan, and associated tools for system validation, which reflect project and business strategy. [TVAP02]	(K) Identifies the steps necessary to define a process and appropriate techniques to be adopted for the Validation of system elements. [TVAP02-10K]	System validation test plans including schedules and resources required. [TVAP02-E10]
		(K) Describes key elements of a successful systems validation process and how systems validation activities relate to different stages of a system life cycle. [TVAP02-20K]	Planning documents (e.g. systems engineering management plan, system test plan, other project/ program plan), or organizational process. [TVAP02-E20]
		(A) Uses system Validation tools or uses a methodology. [TVAP02-30A]	System validation or acceptance test procedures. [TVAP02-E30]
		(A) Uses appropriate standards and governing processes in their system Validation planning, tailoring as appropriate. [TVAP02-40A]	Requirements acceptance/verification matrix. [TVAP02-E40]
		(A) Selects system Validation method for each system element. [TVAP02-50A]	
		(A) Creates system Validation plans and processes for use on a project. [TVAP02-60A]	
		(P) Liaises with stakeholders throughout development of the plan, to gain approval, updating as necessary. [TVAP02-70P]	
3	Uses governing plans and processes for System validation, interpreting, evolving, or seeking guidance where appropriate. [TVAP03]	(A) Follows defined plans, processes and associated tools to perform system Validation on a project. [TVAP03-10A]	System Validation plans or work products they have interpreted or challenged to address project issues. [TVAP03-E10]

(Continued)

ID	Indicators of Competence *(in addition to those at Supervised Practitioner level)*	Relevant knowledge, experience, and/or behaviors	Possible examples of objective evidence of personal contribution to activities performed, or professional behaviors applied
		(P) Guides and actively coordinates the interpretation of system validation plans or processes in order to complete activities system Validation activities successfully. [TVAP03-20P] (A) Prepares Validation data and reports in support of wider system measurement, monitoring and control. [TVAP03-30A] (P) Recognizes situations where deviation from published plans and processes or clarification from others is appropriate in order to overcome complex system Validation challenges. [TVAP03-40P] (P) Recognizes situations where existing system Validation strategy, process, plan or tools require formal change, liaising with stakeholders to gain approval as necessary. [TVAP03-50P]	System Validation process or tool guidance. [TVAP03-E20]
4	Communicates using the terminology of the customer while focusing on customer need. [TVAP04]	(A) Elicits external stakeholder (e.g. customer) needs and produces the associated validation test requirements. [TVAP04-10A] (A) Uses validation plans as the basis for test scripts that external stakeholders (e.g. customers) understand. [TVAP04-20A] (A) Identifies the language or terminology of external stakeholders (e.g. customers) or end users. [TVAP04-30A] (P) Communicates validation using the language or terminology of external stakeholders (e.g. customers) or end users. [TVAP04-40P] (A) Uses external stakeholder (e.g. customer) language in validation test scripts and associated validation reports. [TVAP04-50A]	System Validation test plans including schedules and resources required. [TVAP04-E10] Documented validation requirements. [TVAP04-E20] Validation test documentation or final report. [TVAP04-E30] Validation cross reference matrix. [TVAP04-E40]
5	Prepares validation plans for systems or projects. [TVAP05]	(K) Explains the different validation methods and how/when/why to select the most appropriate method (test, analysis, inspection, similarity, comparison, etc.). [TVAP05-10K] (A) Records the appropriate degree of validation evidence (for example, safety critical software requires a greater degree of validation than non-safety critical). [TVAP05-20A] (A) Develops a validation test plan containing objectives, conditions, priorities, schedules and responsibilities, tools, facilities, procedures and standards to be applied, and the success criteria to be applied. [TVAP05-30A]	Validation cross reference matrix. [TVAP05-E10] System Validation test plans including schedules and resources required. [TVAP05-E20]

		(A) Reviews the depth of testing required for validation and plans accordingly. [TVAP05-40A] (A) Defines detailed validation test sequences and the readiness criteria for each system element. [TVAP05-50A] (A) Creates a validation test schedule showing dependencies of each activity (critical path analysis). [TVAP05-60A] (A) Defines the validation test environment required, including outsourcing of qualification tests as required. [TVAP05-70A] (A) Defines validation records that need to be created and kept. [TVAP05-80A]	
6	Reviews project-level system validation plans. [TVAP06]	(A) Reviews validation plans for a system for its suitability for the project. [TVAP06-10A] (P) Collaborates with project stakeholders to improve validation plans as necessary to ensure success. [TVAP06-20P] (A) Identifies required changes and monitors plan updates until they have been appropriately implemented. [TVAP06-30A]	Records showing review and approval of validation plan. [TVAP06-E10]
7	Reviews validation results, diagnosing complex faults found during validation activities. [TVAP07]	(A) Analyzes test execution and test results to confirm validity of test and results collected. [TVAP07-10A] (A) Uses statistical or modeling techniques to demonstrate sufficient and necessary verification activities have taken place. [TVAP07-20A] (A) Identifies where results differ from those expected. [TVAP07-30A] (A) Analyzes complex faults in a logical manner to determine corrective actions. [TVAP07-40A] (A) Identifies consequences of corrective actions (re-planning, re-test etc.). [TVAP07-50A] (A) Monitors corrective actions to completion, ensuring closure. [TVAP07-60A] (A) Reviews test evidence to ensure suitability for use in proving compliance. [TVAP07-70A]	Fault log, Trouble Report, or similar test record. [TVAP07-E10] Record of corrective actions or fault resolution. [TVAP07-E20] Minutes of fault analysis meetings. [TVAP07-E30]
8	Identifies a suitable validation environment. [TVAP08]	(A) Defines requirements for system validation facilities. (e.g. size of the area, furniture required, power requirements, the IT requirements, and the security of the facility). [TVAP08-10A]	System Validation test plans including schedules and resources required. [TVAP08-E10]

(Continued)

ID	Indicators of Competence *(in addition to those at Supervised Practitioner level)*	Relevant knowledge, experience, and/or behaviors	Possible examples of objective evidence of personal contribution to activities performed, or professional behaviors applied
		(A) Defines requirements for external test facilities. [TVAP08-20A] (A) Defines the requirements for bespoke tools and equipment needed in support of validation, e.g. simulators and emulators. [TVAP08-30A] (A) Determines resources and skills required. [TVAP08-40A]	Records showing equipment procurement specifications or requirements. [TVAP08-E20] Records of external facilities or required certification authority activities. [TVAP08-E30]
9	Creates detailed validation procedures. [TVAP09]	(A) Defines techniques for requirements validation, for example: requirements analysis, exploration of requirements adequacy and completion, assessment of prototypes, stimulations, simulations, models, scenarios, and mock-ups. [TVAP09-10A] (A) Develops validation scenarios that are agreed with users in order to confirm needs have been met. [TVAP09-20A] (A) Analyzes validation test cases linking them to requirements. [TVAP09-30A] (A) Creates clear, concise verification procedures detailing the activities to be performed, pre-requisites, the expected outcome and action in case of a failure. [TVAP09-40A] (A) Identifies pass/fail criteria for validation tests, maintaining the link to the appropriate user requirement (while understanding that the two need to be developed together). [TVAP09-50A] (A) Develops robustness tests in support of validation test cases (e.g. covering doing things wrong, using the system in the wrong way, doing nothing, doing too little, and doing too much). [TVAP09-60A]	Approved Validation scenarios. [TVAP09-E10] Approved Validation procedures. [TVAP09-E20] Validation test documentation or final report. [TVAP09-E30] Requirements acceptance/Validation matrix. [TVAP09-E40]
10	Performs system validation activities. [TVAP10]	(P) Performs Validation testing activities, including appropriate reviews e.g. test readiness review. [TVAP10-10P] (A) Creates a validation test organization. [TVAP10-20A] (A) Identifies passed and failed items and taken corrective action to make the failed items conform to requirements. [TVAP10-30A] (A) Prepares a procedure for identifying unambiguously the inspection and test status of system components being validated including provision for quarantine status. [TVAP10-40A] (A) Identifies and uses validation test tools. [TVAP10-50A]	Organizational structures. [TVAP10-E10] Test documentation, e.g. metrics. [TVAP10-E20]

11	Prepares evidence obtained during validation testing to support certification and acceptance activities. [TVAP11]	(A) Identifies the precise evidence required in support of external stakeholders (e.g. customer) acceptance. [TVAP11-10A]	Minutes of customer acceptance reviews. [TVAP11-E10]
		(A) Uses a systematic method for classifying the results of validation test reports. [TVAP11-20A]	Validation test documentation or final report. [TVAP11-E20]
		(A) Prepares inputs to external stakeholder (e.g. customer) acceptance reviews and associated planning activities. [TVAP11-30A]	Requirements acceptance/Validation matrix. [TVAP11-E30]
		(A) Uses an appropriate validation test results classification method, for example: pass, mild deficiency, annoyance, and catastrophic. [TVAP11-40A]	Compliance matrix. [TVAP11-E40]
			Test reports. [TVAP11-E50]
			Data packages from other test activities (e.g. verification, qualification and Certification). [TVAP11-E60]
12	Monitors the traceability of validation requirements and tests to system requirements and vice versa. [TVAP12]	(A) Monitors external stakeholder (e.g. customer) requirements and validation test requirements in order to obtain forward- and backward-traceability ensuring integrity is maintained. [TVAP12-10A]	Validation cross reference matrix. [TVAP12-E10]
13	Guides new or supervised practitioners in System Validation in order to develop their knowledge, abilities, skills, or associated behaviors. [TVAP13]	(P) Guides new or supervised practitioners in executing activities that form part of this competency. [TVAP13-10P]	Organizational Breakdown Structure showing their responsibility for technical supervision in this area. [TVAP13-E10]
		(A) Trains individuals to an "Awareness" level in this competency area. [TVAP13-20A]	On-the-job training records. [TVAP13-E20]
			Coaching or mentoring assignment records. [TVAP13-E30]
			Records highlighting their impact on individual in terms of improvement or professional development in this competency. [TVAP13-E40]
14	Maintains and enhances own competence in this area through Continual Professional Development (CPD) activities. [TVAP14]	(A) Identifies potential development needs in this area, identifying opportunities to address these through continual professional development activities. [TVAP14-10A]	Records of Continual Professional Development (CPD) performed and learning outcomes. [TVAP14-E10]
		(A) Performs continual professional development activities to maintain and enhance their competency in this area. [TVAP14-20A]	
		(A) Records continual professional development activities undertaken including learning or insights gained. [TVAP14-30A]	

Lead Practitioner – Technical: Validation			
ID	**Indicators of Competence** ***(in addition to those at Practitioner level)***	**Relevant knowledge, experience, and/or behaviors**	**Possible examples of objective evidence of personal contribution to activities performed, or professional behaviors applied**
1	Creates enterprise-level policies, procedures, guidance, and best practice for validation, including associated tools. [TVAL01]	(A) Analyzes enterprise need for validation policies, processes, tools, or guidance. [TVAL01-10A]	Records showing their role in embedding system validation into enterprise policies (e.g. guidance introduced at enterprise level and enterprise -level review minutes). [TVAL01-E10]
		(A) Creates enterprise policies, procedures or guidance for validation activities. [TVAL01-20A]	Procedures they have written. [TVAL01-E20]
		(A) Selects and acquires appropriate tools supporting validation activities. [TVAL01-30A]	Records of support for tool introduction. [TVAL01-E30]
2	Judges the tailoring of enterprise-level validation processes to meet the needs of a project. [TVAL02]	(A) Evaluates enterprise process against the business and external stakeholder (e.g. customer) needs in order to tailor processes to enable project success. [TVAL02-10A]	Documented tailored process. [TVAL02-E10]
		(P) Communicates constructive feedback on validation process. [TVAL02-20P]	Validation process feedback. [TVAL02-E20]
3	Judges the suitability of validation plans from multiple projects, on behalf of the enterprise. [TVAL03]	(A) Evaluates validation strategies from multiple project/ programs for consistency, correctness and to ensure enterprise risk is minimized. [TVAL03-10A]	Validation strategies that proved successful. [TVAL03-E10]
		(A) Evaluates validation risks and proposes mitigation strategies to reduce project or enterprise risk. [TVAL03-20A]	Risk mitigations performed. [TVAL03-E20]
4	Advises on validation approaches on complex or challenging systems or projects across the enterprise. [TVAL04]	(A) Reviews validation plans for highly complex systems or projects, on behalf of the enterprise. [TVAL04-10A]	Records showing review and approval of validation plans. [TVAL04-E10]
		(P) Challenges validation plans as necessary to ensure success, liaising with stakeholders if required. [TVAL04-20P]	
		(P) Collaborates with enterprise stakeholders to improve validation plans as necessary to ensure success. [TVAL04-30P]	
		(A) Authorizes validation plans for highly complex systems or project on behalf of enterprise. [TVAL04-40A]	
5	Judges detailed validation procedures from multiple projects, on behalf of the enterprise. [TVAL05]	(A) Reviews and comments on project/program validation procedures from across the organization. [TVAL05-10A]	Records of a review process in which they have been involved. [TVAL05-E10]
		(A) Judges the adequacy of validation procedures identifying required changes. [TVAL05-20A]	Records of advice provided. [TVAL05-E20]
			Test Readiness Review (TRR) Documentation. [TVAL05-E30]

6	Judges validation evidence generated from multiple projects on behalf of the enterprise. [TVAL06]	(A) Reviews and comments on validation evidence from multiple projects on behalf of the enterprise. [TVAL06-10A]	Records of a review process in which they have been involved. [TVAL06-E10]
		(A) Advises stakeholders across the enterprise on Validation evidence, leading to changes being implemented. [TVAL06-20A]	Records of advice provided. [TVAL06-E20]
7	Guides and actively coordinates validation activities on complex systems or projects across the enterprise. [TVAL07]	(A) Describes typical approaches to complex validation activities. [TVAL07-10A]	Validation plans. [TVAL07-E10]
		(P) Guides and actively coordinates validation activities for complex systems or projects. [TVAL07-20P]	Validation measures showing actual performance against plan. [TVAL07-E20]
8	Coaches or mentors practitioners across the enterprise in systems validation in order to develop their knowledge, abilities, skills, or associated behaviors. [TVAL08]	(P) Coaches or mentors practitioners across the enterprise in competency-related techniques, recommending development activities. [TVAL08-10P]	Coaching or mentoring assignment records. [TVAL08-E10]
		(A) Develops or authorizes enterprise training materials in this competency area. [TVAL08-20A]	Records of formal training courses, workshops, seminars, and authored training material supported by successful post-training evaluation data. [TVAL08-E20]
		(A) Provides enterprise workshops/seminars or training in this competency area. [TVAL08-30A]	Listing as an approved organizational trainer for this competency area. [TVAL08-E30]
9	Promotes the introduction and use of novel techniques and ideas in Validation across the enterprise, to improve enterprise competence in this area. [TVAL09]	(A) Analyzes different approaches across different domains through research. [TVAL09-10A]	Research records. [TVAL09-E10]
		(A) Defines novel approaches that could potentially improve the SE discipline within the enterprise. [TVAL09-20A]	Published papers in refereed journals/company literature. [TVAL09-E20]
		(P) Fosters awareness of these novel techniques within the enterprise. [TVAL09-30P]	Enabling systems introduced in support of new techniques or ideas. [TVAL09-E30]
		(P) Collaborates with enterprise stakeholders to facilitate the introduction of techniques new to the enterprise. [TVAL09-40P]	Published papers (or similar) at enterprise level. [TVAL09-E40]
		(A) Monitors new techniques after their introduction to determine their effectiveness. [TVAL09-50A]	Records of improvements made against a recognized process improvement model in this area. [TVAL09-E50]
		(A) Adapts approach to reflect actual enterprise performance improvements. [TVAL09-60A]	

(Continued)

ID	Indicators of Competence *(in addition to those at Practitioner level)*	Relevant knowledge, experience, and/or behaviors	Possible examples of objective evidence of personal contribution to activities performed, or professional behaviors applied
10	Develops expertise in this competency area through specialist Continual Professional Development (CPD) activities. [TVAL10]	(A) Identifies own needs for further professional development in order to increase competence beyond practitioner level. [TVAL10-10A] (A) Performs professional development activities in order to move own competence toward expert level. [TVAL10-20A] (A) Records continual professional development activities undertaken including learning or insights gained. [TVAL10-30A]	Records of Continual Professional Development (CPD) performed and learning outcomes. [TVAL10-E10]

NOTES	In addition to items above, enterprise-level or independent 3rd-Party-generated evidence may be used to amplify other evidence presented and may include: a. Formally recognized by senior management in current organization as an expert in this competency area b. Evidence of role as Product/System Design Authority or Technical Authority on a complex project with responsibilities in this area or where skills within this competency area were used c. Recognized as an authorizing signatory on behalf of enterprise for formal documentation in this competency area (e.g. policies, processes, and deliverables) d. Formal commendation or award within own enterprise for contribution or item of work successfully performed, which required proficiency in this competency area e. Customer, Supplier, or other external project-specific key Stakeholder accolades for specific work performed in this competency area f. Independently assessed or accredited work in this competency area (e.g. for independent publication or use) g. Formal organizational HR records positively highlighting any specific professional competencies or behaviors identified (if applicable) plus any of the evidence indicators listed at Expert level below

Expert – Technical: Validation			
ID	**Indicators of Competence** ***(in addition to those at Lead Practitioner level)***	**Relevant knowledge, experience, and/or behaviors**	**Possible examples of objective evidence of personal contribution to activities performed, or professional behaviors applied**
1	Communicates own knowledge and experience in Systems Engineering Validation in order to improve best practice beyond the enterprise boundary. [TVAE01]	(A) Produces papers, seminars, or presentations outside own enterprise for publication in order to share own ideas and improve industry best practices in this competence area. [TVAE01-10A]	Published papers or books etc. on new technique in refereed journals/company literature. [TVAE01-E10]
		(P) Fosters incorporation of own ideas into industry best practices in this area. [TVAE01-20P]	Published papers in refereed journals or internal literature proposing new practices in this competence area (or presentations, tutorials, etc.). [TVAE01-E20]
		(P) Develops guidance materials identifying new (or updating existing) best practice in this competence area. [TVAE01-30P]	Proposals adopted as industry best practice. [TVAE01-E30]
2	Advises organizations beyond the enterprise boundary on the suitability of their approach to Systems Engineering validation. [TVAE02]	(P) Acts as independent reviewer within an enterprise review body, in order to approve such plans. [TVAE02-10P]	Records of membership on oversight committee with relevant terms of reference. [TVAE02-E10]
		(A) Reviews and advises on Validation strategies that has led to changes being implemented. [TVAE02-20A]	Review comments or revised document. [TVAE02-E20]
3	Advises organizations beyond the enterprise boundary on their handling of complex or sensitive Systems Engineering validation issues. [TVAE03]	(A) Advises external stakeholders (e.g. customers) on their Validation requirements and issues. [TVAE03-10A]	Records of advice provided on Validation issues. [TVAE03-E10]
		(P) Conducts sensitive negotiations regarding validation of a highly complex system making limited use of specialized, technical terminology. [TVAE03-20P]	Records from negotiations demonstrating awareness of customer's background and knowledge. [TVAE03-E20]
		(A) Uses a holistic approach to complex issue resolution including balanced, rational arguments on way forward. [TVAE03-30A]	Stakeholder approval of Validation plans or procedures. [TVAE03-E30]
4	Advises organizations beyond the enterprise boundary on complex or sensitive validation-related issues. [TVAE04]	(A) Advises on external Validation plans. [TVAE04-10A]	Records of membership on oversight committee with relevant terms of reference. [TVAE04-E10]
		(A) Advises on Validation strategies. [TVAE04-20A]	Review comments or revised document. [TVAE04-E20]
		(A) Advises external organizations on their Validation requirements and issues. [TVAE04-30A]	Records of advice provided on Validation issues. [TVAE04-E30]

(Continued)

ID	Indicators of Competence *(in addition to those at Lead Practitioner level)*	Relevant knowledge, experience, and/or behaviors	Possible examples of objective evidence of personal contribution to activities performed, or professional behaviors applied
		(P) Conducts sensitive negotiations regarding validation of a highly complex system making limited use of specialized, technical terminology. [TVAE04-40P]	Records from negotiations demonstrating awareness of customer's background and knowledge. [TVAE04-E40]
		(A) Uses a holistic approach to complex issue resolution including balanced, rational arguments on way forward. [TVAE04-50A]	Stakeholder approval of transition plans or procedures. [TVAE04-E50]
5	Champions the introduction of novel techniques and ideas in system validation, beyond the enterprise boundary, in order to develop the wider Systems Engineering community in this competency. [TVAE05]	(A) Analyzes different approaches across different domains through research. [TVAE05-10A]	Records of activities promoting research and need to adopt novel technique or ideas. [TVAE05-E10]
		(A) Produces reports for the wider SE community on the effectiveness of new techniques after their introduction. [TVAE05-20A]	Records of improvements made to process and appraisal against a recognized process improvement model. [TVAE05-E20]
		(P) Collaborates with those introducing novel techniques within the wider SE community. [TVAE05-30P]	Research records. [TVAE05-E30]
		(A) Defines novel approaches that could potentially improve the wider SE discipline. [TVAE05-40A]	Published papers in refereed journals/company literature. [TVAE05-E40]
		(P) Fosters awareness of these novel techniques within the wider SE community. [TVAE05-50P]	Records showing introduction of enabling systems supporting the new techniques or ideas. [TVAE05-E50]
6	Coaches individuals beyond the enterprise boundary in Systems Validation, in order to further develop their knowledge, abilities, skills, or associated behaviors. [TVAE06]	(P) Coaches or mentors individuals beyond the enterprise boundary, in competency-related techniques, recommending development activities. [TVAE06-10P]	Coaching or mentoring assignment records. [TVAE06-E10]
		(A) Develops or authorizes training materials in this competency area, which are subsequently successfully delivered beyond the enterprise boundary. [TVAE06-20A]	Records of formal training courses, workshops, seminars, and authored training material supported by successful post-training evaluation data. [TVAE06-E20]
		(A) Provides workshops/seminars or training in this competency area for practitioners or lead practitioners beyond the enterprise boundary (e.g. conferences and open training days). [TVAE06-30A]	Records of Training/workshops/seminars created supported by successful post-training evaluation data. [TVAE06-E30]

7	Maintains expertise in this competency area through specialist Continual Professional Development (CPD) activities. [TVAE07]	(A) Reviews research, new ideas, and state of the art to identify relevant new areas requiring personal development in order to maintain expertise in this competency area. [TVAE07-10A]	Records of documents reviewed and insights gained as part of own research into this competency area. [TVAE07-E10]
		(A) Performs identified specialist professional development activities in order to maintain or further develop competence at expert level. [TVAE07-20A]	Records of Continual Professional Development (CPD) performed and learning outcomes. [TVAE07-E20]
		(A) Records continual professional development activities undertaken including learning or insights gained. [TVAE07-30A]	

NOTES	In addition to items above, enterprise-level or independent 3rd-Party-generated evidence may be used to amplify other evidence presented and may include: a. Formally recognized by a reputable external organization as an expert in this competency area b. Evidence of role as independent assessor or reviewer on project outside own organization where skills in this competency area were used c. Evidence of invitation(s) from wider community for contribution of systems engineering expertise in this area (e.g. industry conference panel, government advisory board etc. cross-industry working groups, partnerships, accredited advanced university courses or research, or as part of professional institute) d. Formal commendation beyond the enterprise (e.g. by INCOSE or other recognized authority) for work performed in this competency area e. Independently assessed or accredited work product in this competency area (e.g. for independent publication or use) f. Accolades of expertise in this area from recognized industry leaders

Competency area – Technical: Transition
Description Transition is the integration of a verified system, product or service into its operational environment including the wider ("Super") system of which it forms a part. Transition is performed in accordance with stakeholder agreements and includes support activities and provision of relevant enabling systems (e.g. production and volume manufacturing, site preparation, support and logistics systems, operator training). Transition is used at each level in the system structure.
Why it matters Incorrectly transitioning the system into operation can lead to misuse, failure to perform, and customer or end-user dissatisfaction. Failure to plan for transition to operation may result in a system that is delayed into service or market with a consequential impact on the customer or business. Failure to satisfy the customer will impact future business.
Possible contributory types of evidence Any combination of the types of evidence may be acceptable (depending on how the Framework is tailored and used). The evidence items identified at each level indicate example work products only. Contributions to work products will generally differ at each proficiency level.
Learning and development The INCOSE Professional Development Portal provides example guidance on how to gain an initial awareness of a competency area and options for developing further competence thereafter.

Awareness – Technical: Transition		
ID	**Indicators of Competence**	**Relevant knowledge sub-indicators**
1	Explains why there is a need to carry out transition to operation. [TTRA01]	(K) Explains the purpose of transition (i.e. to establish a capability for a system to provide services specified by stakeholder requirements in the operational environment). [TTRA01-10K] (K) Explains the relationship between transition, completion of development, production, and readiness for use. [TTRA01-20K] (K) Explains the benefits of formally controlling transition activities. [TTRA01-30K]
2	Explains key activities and work products required for transition to operation. [TTRA02]	(K) Explains the key activities that ensure the system is ready for installation, delivery and use. [TTRA02-10K] (K) Explains how maintenance procedures relate to transition and service. [TTRA02-20K] (K) Describes different ways users and maintainers may be trained. [TTRA02-30K] (K) Lists key work products associated with transition (e.g. guides, manuals, demonstrations, and instructions). [TTRA02-40K] (K) Identifies requirements associated with packaging, storage, export control, and similar transitional areas. [TTRA02-50K]

Supervised Practitioner – Technical: Transition			
ID	**Indicators of Competence** ***(in addition to those at Awareness level)***	**Relevant knowledge, experience, and/or behaviors**	**Possible examples of objective evidence of personal contribution to activities performed, or professional behaviors applied**
1	Uses a governing process and appropriate tools to plan and control their own Transition activities. [TTRS01]	(A) Uses governing process and tools to create transition enabling products. [TTRS01-10A]	Transition plan. [TTRS01-E10]
		(A) Uses governing process and tools to perform shipping and storage activities. [TTRS01-20A]	Transition enabling products. [TTRS01-E20]
		(A) Uses governing process and tools to prepare sites where end products will be stored, installed, used, maintained, or serviced. [TTRS01-30A]	Packing and shipping records. [TTRS01-E30]
		(A) Uses governing process and tools to ensure delivery of the system at the correct location and time. [TTRS01-40A]	Commissioning plans or records. [TTRS01-E40]
		(A) Uses governing process and tools to perform system commissioning activities. [TTRS01-50A]	Maintainer training records. [TTRS01-E50]
		(A) Uses governing process and tools in support of the implementation of a service level agreement. [TTRS01-60A]	End-user training records. [TTRS01-E60]
		(A) Uses governing process and tools to perform user or maintainer training. [TTRS01-70A]	Utilization, support, or maintenance activity records. [TTRS01-E70]
		(A) Uses a governing process and appropriate tools to perform utilization and support-related activities. [TTRS01-80A]	
2	Performs transitioning of a system into production or operation in order to meet the requirements of a plan. [TTRS02]	(K) Describes a sequence of events, which characterize transition to operation of a simple system. [TTRS02-10K]	Transition plan. [TTRS02-E10]
		(A) Performs tasks forming part of the plan for transitioning a system into operation or production. [TTRS02-20A]	Transition enabling products. [TTRS02-E20]
3	Describes the system's contribution to the wider system (super-system) of which it forms a part. [TTRS03]	(K) Describes the system's contribution to the wider system (super-system) of which it forms a part. [TTRS03-10K]	Transition execution records (packing & shipping, commissioning, training…). [TTRS03-E10]
		(K) Describes what to supply for the transition of the system into the next level up. [TTRS03-20K]	
		(A) Performs tasks forming part of transitioning of a system into operation within a wider (super) system. [TTRS03-30A]	
4	Develops own understanding of this competency area through Continual Professional Development (CPD). [TTRS04]	(A) Identifies potential gaps in own knowledge or development needs in this area, identifying opportunities to address these through continual professional development activities. [TTRS04-10A]	Records of Continual Professional Development (CPD) performed and learning outcomes. [TTRS04-E10]
		(A) Performs continual professional development activities to improve their knowledge and understanding in this area. [TTRS04-20A]	
		(A) Records continual professional development activities undertaken including learning or insights gained. [TTRS04-30A]	

Practitioner – Technical: Transition			
ID	**Indicators of Competence** ***(in addition to those at Supervised Practitioner level)***	**Relevant knowledge, experience, and/or behaviors**	**Possible examples of objective evidence of personal contribution to activities performed, or professional behaviors applied**
1	Creates a strategy for transitioning a system, which supports project and wider enterprise needs. [TTRP01]	(A) Identifies project-specific system transition needs that need to be incorporated into SE planning. [TTRP01-10A] (A) Prepares project-specific system transition inputs for SE planning purposes, using appropriate processes and procedures. [TTRP01-20A] (A) Prepares project-specific system transition task estimates in support of SE planning. [TTRP01-30A] (P) Liaises throughout with stakeholders to gain approval, updating strategy as necessary. [TTRP01-40P]	System transition strategy work product. [TTRP01-E10]
2	Creates a governing process, plan, and associated tools for systems transition activities that reflect project and business strategy. [TTRP02]	(K) Identifies the steps necessary to define a process and appropriate techniques to be adopted for system transition management. [TTRP02-10K] (K) Describes key elements of a successful system transition management process and how these activities relate to different stages of a system life cycle. [TTRP02-20K] (A) Uses system transition management tools or uses a methodology. [TTRP02-30A] (A) Uses appropriate standards and governing processes in their system transition management planning, tailoring as appropriate. [TTRP02-40A] (A) Selects system transition management methods for each architectural interface. [TTRP02-50A] (A) Creates system transition management plans and processes for use on a project. [TTRP02-60A] (P) Liaises with stakeholders throughout system transition to gain approval, updating plans as necessary. [TTRP02-70P]	Transition plan or work product. [TTRP02-E10] Certificates or proficiency evaluations in system transition management tools. [TTRP02-E20] SE system transition management tool guidance. [TTRP02-E30]
3	Uses governing plans and processes to plan and execute system transition activities, interpreting, evolving, or seeking guidance where appropriate. [TTRP03]	(A) Follows defined plans, processes, and associated tools to perform system transition on a project. [TTRP03-10A]	Transition plan or work product. [TTRP03-E10]

		(P) Guides and actively coordinates the interpretation of system transition plans or processes in order to complete activities system transition activities successfully. [TTRP03-20P] (A) Prepares system transition data in support of wider system monitoring and control activities. [TTRP03-30A] (P) Recognizes situations where deviation from published plans and processes or clarification from others is appropriate in order to overcome complex system transition challenges. [TTRP03-40P] (P) Recognizes situations where existing system transition strategy, process, plan, or tools require formal change, liaising with stakeholders to gain approval as necessary. [TTRP03-50P]	Transition process or tool guidance. [TTRP03-E20]
4	Performs a system transition to production and operation taking into consideration its contribution to the wider (super) system. [TTRP04]	(A) Guides and actively coordinates activities that transition a system from development into production or operation. [TTRP04-10A]	Project/program Transition to Operation plans and readiness review meeting minutes. [TTRP04-E10]
		(A) Records transition activities. [TTRP04-20A]	Project/program activities and their contribution to the success of the transition. [TTRP04-E20]
		(A) Analyzes anomalies and issues resolving them as necessary. [TTRP04-30A]	
		(A) Communicates anomalies or concerns found during transition, which relate to the wider (super) system. [TTRP04-40A]	
5	Communicates transition activities using user terminology to ensure clear communications. [TTRP05]	(K) Explains how the operator view of a system differs from the developer view. [TTRP05-10K]	Manuals, guides etc. for user consumption. [TTRP05-E10]
		(A) Performs successful transition with a custom. [TTRP05-20A]	Project/program Transition plans, readiness review meeting minutes, or records. [TTRP05-E20]
		(P) Communicates with the user using their own terminology. [TTRP05-30P]	
6	Acts to ensure system transition addresses export control and licensing obligations. [TTRP06]	(A) Ensures system meets export control legislation. [TTRP06-10A]	Export regulation analysis, authorization records or paperwork produced. [TTRP06-E10]
		(A) Ensures system meets 3rd-party licensing obligations (e.g. software licensing export and usage restrictions). [TTRP06-20A]	License agreement records, or records of agreements made. [TTRP06-E20]

(Continued)

ID	Indicators of Competence *(in addition to those at Supervised Practitioner level)*	Relevant knowledge, experience, and/or behaviors	Possible examples of objective evidence of personal contribution to activities performed, or professional behaviors applied
7	Acts to ensure system transition activities gain customer approval. [TTRP07]	(A) Performs a transition to operation activity. [TTRP07-10A]	Project/program Transition plans, readiness review meeting minutes, or records. [TTRP07-E10]
8	Guides new or supervised practitioners in System transition to operation in order to develop their knowledge, abilities, skills, or associated behaviors. [TTRP08]	(P) Guides new or supervised practitioners in executing activities that form part of this competency. [TTRP08-10P] (A) Trains individuals to an "Awareness" level in this competency area. [TTRP08-20A]	Organizational Breakdown Structure showing their responsibility for technical supervision in this area. [TTRP08-E10] On-the-job training records. [TTRP08-E20] Coaching or mentoring assignment records. [TTRP08-E30] Records highlighting their impact on another individual in terms of improvement or professional development in this competency. [TTRP08-E40]
9	Maintains and enhances own competence in this area through Continual Professional Development (CPD) activities. [TTRP09]	(A) Identifies potential development needs in this area, identifying opportunities to address these through continual professional development activities. [TTRP09-10A] (A) Performs continual professional development activities to maintain and enhance their competency in this area. [TTRP09-20A] (A) Records continual professional development activities undertaken including learning or insights gained. [TTRP09-30A]	Records of Continual Professional Development (CPD) performed and learning outcomes. [TTRP09-E10]

Lead Practitioner – Technical: Transition			
ID	**Indicators of Competence *(in addition to those at Practitioner level)***	**Relevant knowledge, experience, and/or behaviors**	**Possible examples of objective evidence of personal contribution to activities performed, or professional behaviors applied**
1	Creates enterprise-level policies, procedures, guidance, and best practice for system transition, including associated tools. [TTRL01]	(A) Analyzes enterprise need for transition policies, processes, tools, or guidance. [TTRL01-10A]	Records showing their role in embedding system transition into enterprise policies (e.g. guidance introduced at enterprise level, enterprise -level review minutes). [TTRL01-E10]
		(A) Creates enterprise policies, procedures or guidance for transition activities. [TTRL01-20A]	Procedures they have written. [TTRL01-E20]
		(A) Selects and acquires appropriate tools supporting transition activities. [TTRL01-30A]	Records of support for tool introduction. [TTRL01-E30]
2	Judges the tailoring of enterprise-level transition processes to meet the needs of a project. [TTRL02]	(A) Evaluates the enterprise process against the business and external stakeholder (e.g. customer) needs in order to tailor processes to enable project success. [TTRL02-10A]	Documented tailored process. [TTRL02-E10]
		(P) Communicates constructive feedback on transition process. [TTRL02-20P]	Transition process feedback. [TTRL02-E20]
3	Judges the adequacy of transition approaches and procedures on complex or challenging systems or projects across the enterprise. [TTRL03]	(A) Reviews and comments on transition approaches or procedures for complex or challenging systems. [TTRL03-10A]	Records of a review process in which they have been involved. [TTRL03-E10]
		(A) Judges the adequacy of transition procedures identifying required changes. [TTRL03-20A]	Records of advice provided. [TTRL03-E20]
4	Judges the suitability of transition plans to ensure wider enterprise needs are addressed. [TTRL04]	(A) Evaluates the attributes of a transition strategy in the context of the project/program /domain/business to ensure a successful outcome. [TTRL04-10A]	Transition strategies that proved successful. [TTRL04-E10]
		(A) Evaluates transition risks and proposes mitigation strategies to reduce project or enterprise risk. [TTRL04-20A]	Records of risk mitigation activities. [TTRL04-E20]
5	Persuades key stakeholders to address identified enterprise-level transition issues to reduce project and business risk. [TTRL05]	(A) Identifies and engages with key stakeholders. [TTRL05-10A]	Minutes of meetings. [TTRL05-E10]
		(P) Fosters agreement between key stakeholders at enterprise level to resolve transition issues, by promoting a holistic viewpoint. [TTRL05-20P]	Minutes of key stakeholder review meetings and actions taken. [TTRL05-E20]
		(P) Persuades key stakeholders to address identified enterprise-level transition issues. [TTRL05-30P]	Records indicating changed stakeholder thinking from personal intervention. [TTRL05-E30]

ID	Indicators of Competence *(in addition to those at Practitioner level)*	Relevant knowledge, experience, and/or behaviors	Possible examples of objective evidence of personal contribution to activities performed, or professional behaviors applied
6	Guides the transition of a complex system or projects across the enterprise, into service. [TTRL06]	(A) Describes critical considerations for transition of complex systems (e.g. thorough understanding of operational environment, strong relationship with receiving external stakeholder). [TTRL06-10A]	Transition Plan or other project/program engineering plans. [TTRL06-E10]
		(P) Guides and actively coordinates transition to operation for highly complex systems, e.g. adverse conditions, highly political, multinational, very large scale, replacing legacy systems, and technically complex. [TTRL06-20P]	Transition Completion Reports. [TTRL06-E20]
			System being used successfully by end users. [TTRL06-E30]
7	Coaches or mentors practitioners across the enterprise in systems transition in order to develop their knowledge, abilities, skills, or associated behaviors. [TTRL07]	(P) Coaches or mentors practitioners across the enterprise in competency-related techniques, recommending development activities. [TTRL07-10P]	Coaching or mentoring assignment records. [TTRL07-E10]
		(A) Develops or authorizes enterprise training materials in this competency area. [TTRL07-20A]	Records of formal training courses, workshops, seminars, and authored training material supported by successful post-training evaluation data. [TTRL07-E20]
		(A) Provides enterprise workshops/seminars or training in this competency area. [TTRL07-30A]	Listing as an approved organizational trainer for this competency area. [TTRL07-E30]
8	Promotes the introduction and use of novel techniques and ideas in transition management across the enterprise, to improve enterprise competence in this area. [TTRL08]	(A) Analyzes different approaches across different domains through research. [TTRL08-10A]	Research records. [TTRL08-E10]
		(A) Defines novel approaches that could potentially improve the SE discipline within the enterprise. [TTRL08-20A]	Published papers in refereed journals/company literature. [TTRL08-E20]
		(P) Fosters awareness of these novel techniques within the enterprise. [TTRL08-30P]	Records showing introduction of enabling systems supporting the new techniques or ideas. [TTRL08-E30]
		(P) Collaborates with enterprise stakeholders to facilitate the introduction of techniques new to the enterprise. [TTRL08-40P]	Published papers (or similar) at enterprise level. [TTRL08-E40]

		(A) Monitors new techniques after their introduction to determine their effectiveness. [TTRL08-50A] (A) Adapts approach to reflect actual enterprise performance improvements. [TTRL08-60A]	Records of improvements made against a recognized process improvement model in this area. [TTRL08-E50]
9	Develops expertise in this competency area through specialist Continual Professional Development (CPD) activities. [TTRL09]	(A) Identifies own needs for further professional development in order to increase competence beyond practitioner level. [TTRL09-10A] (A) Performs professional development activities in order to move own competence toward expert level. [TTRL09-20A] (A) Records continual professional development activities undertaken including learning or insights gained. [TTRL09-30A]	Records of Continual Professional Development (CPD) performed and learning outcomes. [TTRL09-E10]

NOTES	In addition to items above, enterprise-level or independent 3rd-Party-generated evidence may be used to amplify other evidence presented and may include: a. Formally recognized by senior management in current organization as an expert in this competency area b. Evidence of role as Product/System Design Authority or Technical Authority on a complex project with responsibilities in this area or where skills within this competency area were used c. Recognized as an authorizing signatory on behalf of enterprise for formal documentation in this competency area (e.g. policies, processes, and deliverables) d. Formal commendation or award within own enterprise for contribution or item of work successfully performed, which required proficiency in this competency area e. Customer, Supplier, or other external project-specific key Stakeholder accolades for specific work performed in this competency area f. Independently assessed or accredited work in this competency area (e.g. for independent publication or use) g. Formal organizational HR records positively highlighting any specific professional competencies or behaviors identified (if applicable) plus any of the evidence indicators listed at Expert level below

Expert – Technical: Transition			
ID	**Indicators of Competence** ***(in addition to those at Lead Practitioner level)***	**Relevant knowledge, experience, and/or behaviors**	**Possible examples of objective evidence of personal contribution to activities performed, or professional behaviors applied**
1	Communicates own knowledge and experience in Systems transition in order to improve best practice beyond the enterprise boundary. [TTRE01]	(A) Produces papers, seminars, or presentations outside own enterprise for publication in order to share own ideas and improve industry best practices in this competence area. [TTRE01-10A]	Published papers or books etc. on new technique in refereed journals/company literature. [TTRE01-E10]
		(P) Fosters incorporation of own ideas into industry best practices in this area. [TTRE01-20P]	Published papers in refereed journals or internal literature proposing new practices in this competence area (or presentations, tutorials, etc.). [TTRE01-E20]
		(P) Develops guidance materials identifying new (or updating existing) best practice in this competence area. [TTRE01-30P]	Own proposals adopted as industry best practices in this competence area. [TTRE01-E30]
2	Advises organizations beyond the enterprise boundary on the suitability of their approach to Systems Engineering Transition activities. [TTRE02]	(A) Advises on external transition plans. [TTRE02-10A]	Records of membership on oversight committee with relevant terms of reference. [TTRE02-E10]
		(A) Advises on transition strategies. [TTRE02-20A]	Review comments or revised document. [TTRE02-E20]
3	Advises organizations beyond the enterprise boundary on the handling of complex or sensitive Systems transition issues. [TTRE03]	(A) Advises external stakeholders on their transition requirements and issues. [TTRE03-10A]	Records of advice provided on transition issues. [TTRE03-E10]
		(P) Conducts sensitive negotiations regarding transition of a highly complex system making limited use of specialized, technical terminology. [TTRE03-20P]	Records from negotiations demonstrating awareness of customer's background and knowledge. [TTRE03-E20]
		(A) Uses a holistic approach to complex issue resolution including balanced, rational arguments on way forward. [TTRE03-30A]	Stakeholder approval of transition plans or procedures. [TTRE03-E30]
4	Champions the introduction of novel techniques and ideas in system transition, beyond the enterprise boundary, in order to develop the wider Systems Engineering community in this competency. [TTRE04]	(A) Analyzes different approaches across different domains through research. [TTRE04-10A]	Records of activities promoting research and need to adopt novel technique or ideas. [TTRE04-E10]
		(A) Produces reports for the wider SE community on the effectiveness of new techniques after their introduction. [TTRE04-20A]	Records of improvements made to process and appraisal against a recognized process improvement model. [TTRE04-E20]

		(P) Collaborates with those introducing novel techniques within the wider SE community. [TTRE04-30P]	Research records. [TTRE04-E30]
		(A) Defines novel approaches that could potentially improve the wider SE discipline. [TTRE04-40A]	Published papers in refereed journals/company literature. [TTRE04-E40]
		(P) Fosters awareness of these novel techniques within the wider SE community. [TTRE04-50P]	Records showing introduction of enabling systems supporting the new techniques or ideas. [TTRE04-E50]
5	Coaches individuals beyond the enterprise boundary in Systems Transition, in order to further develop their knowledge, abilities, skills, or associated behaviors. [TTRE05]	(P) Coaches or mentors individuals beyond the enterprise boundary, in competency-related techniques, recommending development activities. [TTRE05-10P]	Coaching or mentoring assignment records. [TTRE05-E10]
		(A) Develops or authorizes training materials in this competency area, which are subsequently successfully delivered beyond the enterprise boundary. [TTRE05-20A]	Records of formal training courses, workshops, seminars, and authored training material supported by successful post-training evaluation data. [TTRE05-E20]
		(A) Provides workshops/seminars or training in this competency area for practitioners or lead practitioners beyond the enterprise boundary (e.g. conferences and open training days). [TTRE05-30A]	Records of Training/workshops/seminars created supported by successful post-training evaluation data. [TTRE05-E30]
6	Maintains expertise in this competency area through specialist Continual Professional Development (CPD) activities. [TTRE06]	(A) Reviews research, new ideas, and state of the art to identify relevant new areas requiring personal development in order to maintain expertise in this competency area. [TTRE06-10A]	Records of documents reviewed and insights gained as part of own research into this competency area. [TTRE06-E10]
		(A) Performs identified specialist professional development activities in order to maintain or further develop competence at expert level. [TTRE06-20A]	Records of Continual Professional Development (CPD) performed and learning outcomes. [TTRE06-E20]
		(A) Records continual professional development activities undertaken including learning or insights gained. [TTRE06-30A]	

NOTES	In addition to items above, enterprise-level or independent 3rd-Party-generated evidence may be used to amplify other evidence presented and may include: a. Formally recognized by a reputable external organization as an expert in this competency area b. Evidence of role as independent assessor or reviewer on project outside own organization where skills in this competency area were used c. Evidence of invitation(s) from wider community for contribution of systems engineering expertise in this area (e.g. industry conference panel, government advisory board etc. cross-industry working groups, partnerships, accredited advanced university courses or research, or as part of professional institute) d. Formal commendation beyond the enterprise (e.g. by INCOSE or other recognized authority) for work performed in this competency area e. Independently assessed or accredited work product in this competency area (e.g. for independent publication or use) f. Accolades of expertise in this area from recognized industry leaders

Competency area – Technical: Utilization and Support
Description Utilization is the stage of development when a system, product, or service is used (operated) in its intended environment to deliver its intended capabilities. The support stage of a system life cycle encompasses the activities required to sustain operation of the system, product or service over time, such as maintaining the system to continue or extend its operational life, address performance issues, evolving needs, obsolescence and technology upgrades, and changes. Support entails monitoring system performance, addressing system failures and performance issues and updating the system to accommodate evolving needs and technology.
Why it matters The Utilization and support stages of a system, product or service typically account for the largest portion of the total life cycle cost. Proactive and systematic responses to operational issues contribute significantly to user satisfaction and operational cost management.
Possible contributory types of evidence Any combination of the types of evidence may be acceptable (depending on how the Framework is tailored and used). The evidence items identified at each level indicate example work products only. Contributions to work products will generally differ at each proficiency level.
Learning and development The INCOSE Professional Development Portal provides example guidance on how to gain an initial awareness of a competency area and options for developing further competence thereafter.

Awareness – Technical: Utilization and Support		
ID	**Indicators of Competence**	**Relevant knowledge sub-indicators**
1	Explains why a system needs to be supported during operation. [TUSA01]	(K) Explains why no system is flawless, and that it will inevitably require some repair or upgrading. [TUSA01-10K] (K) Explains how, as end users operate the system, unanticipated system behaviors or performance issues may occur that require addressing. [TUSA01-20K]
2	Describes the difference between preventive and corrective maintenance. [TUSA02]	(K) Describes key interfaces and links between utilization and support and stakeholders in the wider enterprise outside engineering (e.g. infrastructure management, human resource management, quality management, knowledge management, portfolio management, and life cycle model management). [TUSA02-10K] (K) Describes the pre-emptive nature of "Preventive maintenance." [TUSA02-20K]
3	Explains why it is necessary to address failures, parts obsolescence, and evolving user requirements during system operation. [TUSA03]	(K) Explains how common or repeated failures may indicated a need for a system design change. [TUSA03-10K] (K) Explains how unavailability of parts (particularly COTS elements) affects system repairs and how this may be addressed. [TUSA03-20K] (K) Explains how changing user environments and use needs can render a system ineffective or possibly useless if not addressed. [TUSA03-30K]
4	Lists the different levels of repair capability and describes the characteristics of each. [TUSA04]	(K) Describes different levels of repair (such as on-site (at the system location), Intermediate or depot locations, and original equipment manufacturer (OEM) and the key characteristics of each. [TUSA04-10K]

5	Explains the impact of operations and support on specialty engineering areas. [TUSA05]	(K) Explains how affordability (e.g. cost-effectiveness and life cycle cost) considerations can affect the design or implementation of system utilization and support. [TUSA05-10K] (K) Explains how cybersecurity considerations can affect the design or implementation of system utilization and support. [TUSA05-20K] (K) Explains how environmental considerations can affect the design or implementation of system utilization and support. [TUSA05-30K] (K) Explains how interoperability considerations can affect the design or implementation of system utilization and support. [TUSA05-40K] (K) Explains how reliability, availability, and maintainability considerations can affect the design or implementation of system utilization and support. [TUSA05-50K]

Supervised Practitioner – Technical: Utilization and Support			
ID	**Indicators of Competence** ***(in addition to those at Awareness level)***	**Relevant knowledge, experience, and/or behaviors**	**Possible examples of objective evidence of personal contribution to activities performed, or professional behaviors applied**
1	Uses a governing process and appropriate tools to plan and control their own operations and support activities. [TUSS01]	(K) Explains the primary elements of system support (monitoring performance, addressing failures, identifying, and enabling proper maintenance, proactively managing design changes…). [TUSS01-10K] (A) Identifies key elements of an operational support plan (e.g. maintenance planning, supply support, storage and transportation, support equipment, training, system performance monitoring…). [TUSS01-20A] (A) Uses governing process and tools to create operations and support. [TUSS01-30A] (A) Uses governing process and tools to perform operations and support activities enabling products. [TUSS01-40A] (A) Uses a governing process and appropriate tools to utilization and support-related activities. [TUSS01-50A]	Operational support plan. [TUSS01-E10] Maintenance plans or other support-enabling products. [TUSS01-E20] Records from In-Service Review (ISR) or support or maintenance activities. [TUSS01-E30]
2	Identifies operational data in order to assess system performance. [TUSS02]	(K) Describes how the capture of meaningful performance data, may involve challenges and how they may be resolved. [TUSS02-10K] (A) Records operational data in order to support assessment of system performance. [TUSS02-20A]	Operational performance data collection plans. [TUSS02-E10] Operational performance data. [TUSS02-E20]
3	Reviews system failures or performance issues, proposing design changes to rectify such failures. [TUSS03]	(A) Identifies performance data indicating a need to consider a design change. [TUSS03-10A] (A) Reviews operational data in order to assess system performance. [TUSS03-20A] (A) Reviews system failures to identify potential design enhancements. [TUSS03-30A] (A) Prepares inputs to change proposals to address improvements following review of a system failure. [TUSS03-40A]	System failure reporting data and analysis reports. [TUSS03-E10] Change proposal documentation (e.g. Engineering Change Proposal (ECP), or similar). [TUSS03-E20]
4	Performs rectification of system failures or performance issues. [TUSS04]	(A) Identifies potential causes of system failure or performance issues. [TUSS04-10A] (A) Prepares input to design changes to rectify such failures. [TUSS04-20A] (A) Reviews proposed design changes to ensure that they fully rectify identified failures, without side-effects. [TUSS04-30A]	Failure review reports or records of performance (or similar) review meetings. [TUSS04-E10] Failure root cause analysis documentation. [TUSS04-E20] Detailed change proposal analysis. [TUSS04-E30]

5	Reviews the feasibility and impact of evolving user need on operations, maintenance, and support. [TUSS05]	(A) Reviews evolving user need, including activities assessing the feasibility of updates in response to the changed need. [TUSS05-10A]	Change proposal documentation (e.g. ECP or similar). [TUSS05-E10]
		(A) Reviews evolving user need, including activities assessing the impact on operations, support or maintenance of changing user need. [TUSS05-20A]	Detailed change proposal analysis. [TUSS05-E20]
		(A) Prepares inputs to change proposals to document evolving user need. [TUSS05-30A]	
6	Prepares inputs to concept studies to document the impact or feasibility of new technologies or possible system updates. [TUSS06]	(K) Describes factors to be considered when assessing the potential technology change (e.g. cost, complexity, impact to the rest of the system & verified system requirements, impact to system support elements, timeline required to implement…). [TUSS06-10K]	Detailed change proposal analysis. [TUSS06-E10]
		(A) Reviews new technology to benefit a system (e.g. new battery technology, larger data storage capacity for same size/weight/power/cost, and quieter or less heat-producing system element). [TUSS06-20A]	Return on Investment report. [TUSS06-E20]
		(A) Reviews new technology for its impact on system operation, maintenance, and support. [TUSS06-30A]	
7	Prepares inputs to obsolescence studies to identify obsolescent components and suitable replacements. [TUSS07]	(K) Describes the concept of Line Replaceable Unit (LRU). [TUSS07-10K]	Obsolescence studies. [TUSS07-E10]
		(A) Performs research to determine potential for obsolescence of parts. [TUSS07-20A]	
		(A) Identifies obsolescence replacements (e.g. identify key component requirements, research other sources, assess suitability against requirements, and solicit peer-review to ensure all relevant factors have been considered). [TUSS07-30A]	
		(A) Identifies impact of obsolescence on system utilization and support activities. [TUSS07-40A]	
8	Prepares updates to technical data (e.g. procedures, guidelines, checklists, and training materials) to ensure operations and maintenance activities and data are current. [TUSS08]	(K) Describes options for technical data review (e.g. allowing user feedback, formal review meetings with subject matter experts, and informal or formal audits). [TUSS08-10K]	Documentation from technical data review (e.g. configuration audit, training review…). [TUSS08-E10]
		(K) Explains why it is important to have formal and rigorous configuration control of technical data for operational systems throughout the life cycle. [TUSS08-20K]	Technical data package, showing management of changes. [TUSS08-E20]

(*Continued*)

ID	Indicators of Competence *(in addition to those at Awareness level)*	Relevant knowledge, experience, and/or behaviors	Possible examples of objective evidence of personal contribution to activities performed, or professional behaviors applied
		(A) Identifies required updates to technical data work products (e.g. procedures, guidelines, checklists, training, and maintenance materials) to ensure they are current. [TUSS08-30A] (A) Prepares updates to technical data work products to reflect changes identifies to maintain their currency. [TUSS08-40A]	
9	Identifies potential changes to system operational environment or external interfaces. [TUSS09]	(A) Identifies potential changes to system operational environment or external interfaces. [TUSS09-10A]	System in-service operational analysis. [TUSS09-E10]
10	Develops own understanding of this competency area through Continual Professional Development (CPD). [TUSS10]	(A) Identifies potential gaps in own knowledge or development needs in this area, identifying opportunities to address these through continual professional development activities. [TUSS10-10A] (A) Performs continual professional development activities to improve their knowledge and understanding in this area. [TUSS10-20A] (A) Records continual professional development activities undertaken including learning or insights gained. [TUSS10-30A]	Records of Continual Professional Development (CPD) performed and learning outcomes. [TUSS10-E10]

Practitioner – Technical: Utilization and Support			
ID	**Indicators of Competence** ***(in addition to those at Supervised Practitioner level)***	**Relevant knowledge, experience, and/or behaviors**	**Possible examples of objective evidence of personal contribution to activities performed, or professional behaviors applied**
1	Creates a strategy for system utilization and support, which reflects wider project and business strategies. [TUSP01]	(A) Identifies project-specific system utilization and support needs that need to be incorporated into SE planning. [TUSP01-10A]	Operational support plan. [TUSP01-E10]
		(A) Prepares project-specific system utilization and support task estimates in support of SE planning. [TUSP01-20A]	Maintenance plans or other support-enabling products. [TUSP01-E20]
		(A) Prepares project-specific system utilization and support inputs for SE planning purposes, using appropriate processes and procedures. [TUSP01-30A]	
		(P) Liaises throughout with stakeholders to gain approval, updating strategy as necessary. [TUSP01-40P]	
2	Creates a governing process, plan, and associated tools for system utilization and support, which reflect wider project and business plans. [TUSP02]	(K) Identifies the steps necessary to define a process and appropriate techniques to be adopted for the operations and support of system elements. [TUSP02-10K]	Operational support plan. [TUSP02-E10]
		(K) Describes key elements of a successful systems operations and support process and how these activities relate to different stages of a system life cycle. [TUSP02-20K]	Maintenance plans or other support-enabling products. [TUSP02-E20]
		(A) Uses system operations and support tools or uses a methodology. [TUSP02-30A]	Inputs to planning documents (e.g. Systems Engineering Management Plan (SEMP)), other project/program plan, or organizational process. [TUSP02-E30]
		(A) Uses appropriate standards and governing processes in their system operations and support planning, tailoring as appropriate. [TUSP02-40A]	
		(A) Selects system operations and support method for each system element. [TUSP02-50A]	
		(A) Creates system operations and support plans and processes for use on a project. [TUSP02-60A]	
		(P) Liaises with stakeholders throughout development of the plan, to gain approval, updating as necessary. [TUSP02-70P]	

(Continued)

ID	Indicators of Competence *(in addition to those at Supervised Practitioner level)*	Relevant knowledge, experience, and/or behaviors	Possible examples of objective evidence of personal contribution to activities performed, or professional behaviors applied
3	Uses governing plans and processes for System Utilization and support, interpreting, evolving, or seeking guidance where appropriate. [TUSP03]	(A) Follows defined plans, processes and associated tools to perform system operations and maintenance on a project. [TUSP03-10A]	System Operations and Maintenance plans or work products they have interpreted or challenged to address project issues. [TUSP03-E10]
		(P) Guides and actively coordinates the interpretation of system Operations and Maintenance plans or processes in order to complete activities system operations and Maintenance activities successfully. [TUSP03-20P]	System Operations and Maintenance process or tool guidance. [TUSP03-E20]
		(A) Prepares Operations and Maintenance data and reports in support of wider system measurement, monitoring, and control. [TUSP03-30A]	
		(P) Recognizes situations where deviation from published plans and processes or clarification from others is appropriate in order to overcome complex system Operations and Maintenance challenges. [TUSP03-40P]	
		(P) Recognizes situations where existing system operations and maintenance strategy, process, plan or tools require formal change, liaising with stakeholders to gain approval as necessary. [TUSP03-50P]	
4	Guides and actively coordinates in-service support activities for a system. [TUSP04]	(A) Defines critical capabilities of an in-service support activity (e.g. help desk, tiered support, response teams, supply management…). [TUSP04-10A]	System support SOPs. [TUSP04-E10]
		(P) Guides and actively coordinates in-service support activities. [TUSP04-20P]	Organizational Breakdown Structure. [TUSP04-E20]
5	Identifies data to be collected in order to assess system operational performance. [TUSP05]	(A) Identifies relevant measurable performance data. [TUSP05-10A]	Data collection plans. [TUSP05-E10]
		(A) Identifies methods by which performance data can be collected. [TUSP05-20A]	Knowledge of data collection tools or systems. [TUSP05-E20]
6	Reviews system failures or performance issues in order to initiate design change proposals rectifying these failures. [TUSP06]	(A) Analyzes performance data to identify issues. [TUSP06-10A]	Data analysis reports. [TUSP06-E10]
		(A) Analyzes system failures to identify root cause and recommend corrections. [TUSP06-20A]	Root cause analyses of failures. [TUSP06-E20]
		(A) Develops engineering change proposals to address system performance issues. [TUSP06-30A]	Engineering Change Proposals. [TUSP06-E30]

7	Identifies system elements approaching obsolescence and conducts studies to identify suitable replacements. [TUSP07]	(K) Describes methods for identifying potential obsolescence of system elements. [TUSP07-10K]	Obsolescence reports, Diminishing Manufacturing Sources and Material Shortages (DMSMS) reports. [TUSP07-E10]
		(A) Uses system requirements to identify suitable replacement elements. [TUSP07-20A]	
		(A) Ensures disposal impact is considered as part of the obsolescence activity. [TUSP07-30A]	
8	Maintains system elements and associated documentation following their replacement due to obsolescence. [TUSP08]	(A) Ensures periodic checks are performed on system or elements to identify potential obsolescence. [TUSP08-10A]	Records of system activities performed in support of obsolescence management. [TUSP08-E10]
		(A) Identifies all system or element changes required (e.g. to components or to documentation) to implement updates required as a result of potential obsolescence. [TUSP08-20A]	
		(A) Performs system or element update(s) to address potential obsolescence. [TUSP08-30A]	
		(A) Ensures all affected documentation is maintained to reflect impact of changes made. [TUSP08-40A]	
9	Monitors the effectiveness of system support or operations. [TUSP09]	(A) Identifies system performance monitoring tools to evaluate system operation (e.g. performance based logistic support). [TUSP09-10A]	Sample data from system monitoring tools. [TUSP09-E10]
		(A) Identifies advanced software maintenance capabilities (e.g. automatic software upgrades) to support system. [TUSP09-20A]	Standard Operating Procedures defined for executing software monitoring of systems. [TUSP09-E20]
		(A) Monitors system support or operations for their effectiveness using specialist support tools. [TUSP09-30A]	
10	Reviews the timing of technology upgrade implementations in order to improve the cost-benefit ratio of an upgraded design solution. [TUSP10]	(A) Evaluates engineering change proposals to assess impacts and determine optimal approach. [TUSP10-10A]	Change proposal review notes and results. [TUSP10-E10]
		(A) Monitors overall system performance and advises on appropriate time for major upgrades to address performance issues or significant requirement changes. [TUSP10-20A]	White paper or proposal documenting need for system upgrade. [TUSP10-E20]
		(A) Evaluates wider impact of technology upgrade on planned operations and maintenance activities. [TUSP10-30A]	

(*Continued*)

ID	Indicators of Competence *(in addition to those at Supervised Practitioner level)*	Relevant knowledge, experience, and/or behaviors	Possible examples of objective evidence of personal contribution to activities performed, or professional behaviors applied
11	Reviews potential changes to the system operational environment or external interfaces. [TUSP11]	(A) Identifies possible changes that could cause impacts to system operation (e.g. new operational requirement, new system threat, and change to external interface). [TUSP11-10A] (A) Identifies possible changes on system maintenance (e.g. new maintenance requirement, new maintenance impact, and change to maintenance plan). [TUSP11-20A] (K) Describes methods for monitoring potential changes (e.g. user group meetings, cross-system integrated project teams (IPTs)). [TUSP11-30K]	Engineering Change Proposals. [TUSP11-E10] Records of participation in user group meetings or cross-system IPTs. [TUSP11-E20]
12	Reviews technical support data (e.g. procedures, guidelines, checklists, training, and maintenance materials) to ensure it is current. [TUSP12]	(A) Develops and executes plans for periodic in-service reviews. [TUSP12-10A] (A) Prepares information used for in service reviews (e.g. updates to training material, manuals, and checklists). [TUSP12-20A]	In Service Review meeting documentation. [TUSP12-E10] Updated training material, manuals, and checklists. [TUSP12-E20]
13	Guides new or supervised practitioners in System operation, support, and maintenance, in order to develop their knowledge, abilities, skills, or associated behaviors. [TUSP13]	(P) Guides new or supervised practitioners in executing activities that form part of this competency. [TUSP13-10P] (A) Trains individuals to an “Awareness” level in this competency area. [TUSP13-20A]	Organizational Breakdown Structure showing their responsibility for technical supervision in this area. [TUSP13-E10] On-the-job training records. [TUSP13-E20] Coaching or mentoring assignment records. [TUSP13-E30] Records highlighting their impact on another individual in terms of improvement or professional development in this competency. [TUSP13-E40]
14	Maintains and enhances own competence in this area through Continual Professional Development (CPD) activities. [TUSP14]	(A) Identifies potential development needs in this area, identifying opportunities to address these through continual professional development activities. [TUSP14-10A] (A) Performs continual professional development activities to maintain and enhance their competency in this area. [TUSP14-20A] (A) Records continual professional development activities undertaken including learning or insights gained. [TUSP14-30A]	Records of Continual Professional Development (CPD) performed and learning outcomes. [TUSP14-E10]

Lead Practitioner – Technical: Utilization and Support			
ID	**Indicators of Competence *(in addition to those at Practitioner level)***	**Relevant knowledge, experience, and/or behaviors**	**Possible examples of objective evidence of personal contribution to activities performed, or professional behaviors applied**
1	Creates enterprise-level policies, procedures, guidance, and best practice for utilization and support, including associated tools. [TUSL01]	(A) Analyzes enterprise need for utilization and support policies, processes, tools, or guidance. [TUSL01-10A]	Records showing their role in embedding system utilization and support into enterprise policies (e.g. guidance introduced at enterprise level and enterprise -level review minutes). [TUSL01-E10]
		(A) Produces enterprise-level policies, procedures, or guidance on utilization and support. [TUSL01-20A]	Procedures they have written. [TUSL01-E20]
		(A) Selects and acquires appropriate tools supporting utilization and support. [TUSL01-30A]	Records of support for tool introduction. [TUSL01-E30]
2	Judges the tailoring of enterprise-level utilization and support processes to meet the needs of a project. [TUSL02]	(A) Evaluates the enterprise operations and support processes against the business and external stakeholder (e.g. customer) needs in order to tailor processes to enable project success. [TUSL02-10A]	Documented tailored process. [TUSL02-E10]
		(P) Communicates constructive feedback on utilization and support processes. [TUSL02-20P]	Utilization and support process feedback. [TUSL02-E20]
3	Advises across the enterprise on the application of advanced practices to improve the effectiveness of project-level system support or operations. [TUSL03]	(A) Identifies opportunities to implement improvements to performance monitoring approaches used across the enterprise. [TUSL03-10A]	Knowledge of or sample data from system monitoring tools. [TUSL03-E10]
		(A) Identifies opportunities to implement new or advanced software maintenance capabilities (e.g. automatic software upgrades) to improve support system effectiveness. [TUSL03-20A]	Standard Operating Procedures defined for executing software upgrades. [TUSL03-E20]
		(P) Persuades key stakeholders across the enterprise to adopt new practices in operations and support. [TUSL03-30P]	Records indicating changed stakeholder thinking from personal intervention. [TUSL03-E30]
4	Advises across the enterprise on technology upgrade implementations in order to improve the cost-benefit ratio of an upgraded design solution. [TUSL04]	(A) Evaluates engineering change proposals across the enterprise to assess impact and determine optimal approach. [TUSL04-10A]	Change or implementation assessment/review notes. [TUSL04-E10]
		(A) Advises across the enterprise on appropriate implementation strategies to address performance issues or significant requirement changes. [TUSL04-20A]	White paper or proposals documenting strategies for system upgrade from enterprise perspective. [TUSL04-E20]
		(A) Evaluates the wider impact of technology upgrade on planned activities across the enterprise. [TUSL04-30A]	

(Continued)

ID	Indicators of Competence *(in addition to those at Practitioner level)*	Relevant knowledge, experience, and/or behaviors	Possible examples of objective evidence of personal contribution to activities performed, or professional behaviors applied
5	Persuades key stakeholders across the enterprise to address identified operation, maintenance, and support issues to reduce project or wider enterprise risk. [TUSL05]	(A) Identifies and engages with key stakeholders across the enterprise. [TUSL05-10A]	Minutes of meetings. [TUSL05-E10]
		(P) Fosters agreement between key stakeholders at enterprise level to resolve maintenance and support issues, by promoting a holistic viewpoint. [TUSL05-20P]	Records indicating changed stakeholder thinking from personal intervention. [TUSL05-E20]
		(P) Persuades key stakeholders to address identified enterprise-level utilization and support issues. [TUSL05-30P]	
6	Coaches or mentors practitioners across the enterprise in systems utilization and support in order to develop their knowledge, abilities, skills, or associated behaviors. [TUSL06]	(P) Coaches or mentors practitioners across the enterprise in competency-related techniques, recommending development activities. [TUSL06-10P]	Coaching or mentoring assignment records. [TUSL06-E10]
		(A) Develops or authorizes enterprise training materials in this competency area. [TUSL06-20A]	Records of formal training courses, workshops, seminars, and authored training material supported by successful post-training evaluation data. [TUSL06-E20]
		(A) Provides enterprise workshops/seminars or training in this competency area. [TUSL06-30A]	Listing as an approved organizational trainer for this competency area. [TUSL06-E30]
7	Promotes the introduction and use of novel techniques and ideas in systems operations and Support across the enterprise, to improve enterprise competence in this area. [TUSL07]	(A) Analyzes different approaches across different domains through research. [TUSL07-10A]	Research records. [TUSL07-E10]
		(A) Defines novel approaches that could potentially improve the SE discipline within the enterprise. [TUSL07-20A]	Published papers in refereed journals/company literature. [TUSL07-E20]
		(P) Fosters awareness of these novel techniques within the enterprise. [TUSL07-30P]	Enabling systems introduced in support of new techniques or ideas. [TUSL07-E30]
		(P) Collaborates with enterprise stakeholders to facilitate the introduction of techniques new to the enterprise. [TUSL07-40P]	Published papers (or similar) at enterprise level. [TUSL07-E40]
		(A) Monitors new techniques after their introduction to determine their effectiveness. [TUSL07-50A]	Records of improvements made against a recognized process improvement model in this area. [TUSL07-E50]
		(A) Adapts approach to reflect actual enterprise performance improvements. [TUSL07-60A]	

8	Develops expertise in this competency area through specialist Continual Professional Development (CPD) activities. [TUSL08]	(A) Identifies own needs for further professional development in order to increase competence beyond practitioner level. [TUSL08-10A] (A) Performs professional development activities in order to move own competence toward expert level. [TUSL08-20A] (A) Records continual professional development activities undertaken including learning or insights gained. [TUSL08-30A]	Records of Continual Professional Development (CPD) performed and learning outcomes. [TUSL08-E10]

NOTES	In addition to items above, enterprise-level or independent 3rd-Party-generated evidence may be used to amplify other evidence presented and may include: a. Formally recognized by senior management in current organization as an expert in this competency area b. Evidence of role as Product/System Design Authority or Technical Authority on a complex project with responsibilities in this area or where skills within this competency area were used c. Recognized as an authorizing signatory on behalf of enterprise for formal documentation in this competency area (e.g. policies, processes, and deliverables) d. Formal commendation or award within own enterprise for contribution or item of work successfully performed, which required proficiency in this competency area e. Customer, Supplier, or other external project-specific key Stakeholder accolades for specific work performed in this competency area f. Independently assessed or accredited work in this competency area (e.g. for independent publication or use) g. Formal organizational HR records positively highlighting any specific professional competencies or behaviors identified (if applicable) plus any of the evidence indicators listed at Expert level below

Expert – Technical: Utilization and Support			
ID	**Indicators of Competence** ***(in addition to those at Lead Practitioner level)***	**Relevant knowledge, experience, and/or behaviors**	**Possible examples of objective evidence of personal contribution to activities performed, or professional behaviors applied**
1	Communicates own knowledge and experience in systems utilization and support in order to improve best practice beyond the enterprise boundary. [TUSE01]	(A) Produces papers, seminars, or presentations outside own enterprise for publication in order to share own ideas and improve industry best practices in this competence area. [TUSE01-10A]	Published papers or books etc. on new technique in refereed journals/company literature. [TUSE01-E10]
		(P) Fosters incorporation of own ideas into industry best practices in this area. [TUSE01-20P]	Published papers in refereed journals or internal literature proposing new practices in this competence area (or presentations, tutorials, etc.). [TUSE01-E20]
		(P) Develops guidance materials identifying new (or updating existing) best practice in this competence area. [TUSE01-30P]	Proposals adopted as industry best practice. [TUSE01-E30]
2	Advises organizations beyond the enterprise boundary on the suitability of their approach to system utilization and support. [TUSE02]	(P) Acts as independent reviewer within an enterprise review body, in order to approve operations and support plans. [TUSE02-10P]	Records of membership on oversight committee with relevant terms of reference. [TUSE02-E10]
		(A) Reviews and advises on utilization and support strategies that has led to changes being implemented. [TUSE02-20A]	Review comments or revised document. [TUSE02-E20]
3	Advises organizations beyond the enterprise boundary on the handling of complex or sensitive operations, maintenance, and support-related issues. [TUSE03]	(A) Advises external stakeholders (e.g. customers) on their operations and support requirements and issues. [TUSE03-10A]	Records of advice provided on utilization and support-related issues. [TUSE03-E10]
		(P) Acts with sensitivity on highly complex system operations or support issues making limited use of specialized, technical terminology and taking account of external stakeholders' (e.g. customer's) background and knowledge. [TUSE03-20P]	Records from negotiations demonstrating awareness of customer's background and knowledge. [TUSE03-E20]
		(A) Uses a holistic approach to complex issue resolution including balanced, rational arguments on way forward. [TUSE03-30A]	Stakeholder approval of utilization and support plans or procedures. [TUSE03-E30]
4	Champions the introduction of novel techniques and ideas in systems utilization and support, beyond the enterprise boundary, in order to develop the wider Systems Engineering community in this competency. [TUSE04]	(A) Analyzes different approaches across different domains through research. [TUSE04-10A]	Records of activities promoting research and need to adopt novel technique or ideas. [TUSE04-E10]
		(A) Produces reports for the wider SE community on the effectiveness of new techniques after their introduction. [TUSE04-20A]	Records of improvements made to process and appraisal against a recognized process improvement model. [TUSE04-E20]

		(P) Collaborates with those introducing novel techniques within the wider SE community. [TUSE04-30P]	Research records. [TUSE04-E30]
		(A) Defines novel approaches that could potentially improve the wider SE discipline. [TUSE04-40A]	Published papers in refereed journals/company literature. [TUSE04-E40]
		(P) Fosters awareness of these novel techniques within the wider SE community. [TUSE04-50P]	Records showing introduction of enabling systems supporting the new techniques or ideas. [TUSE04-E50]
5	Coaches individuals beyond the enterprise boundary in System Utilization and support, in order to further develop their knowledge, abilities, skills, or associated behaviors. [TUSE05]	(P) Coaches or mentors individuals beyond the enterprise boundary, in competency-related techniques, recommending development activities. [TUSE05-10P]	Coaching or mentoring assignment records. [TUSE05-E10]
		(A) Develops or authorizes training materials in this competency area, which are subsequently successfully delivered beyond the enterprise boundary. [TUSE05-20A]	Records of formal training courses, workshops, seminars, and authored training material supported by successful post-training evaluation data. [TUSE05-E20]
		(A) Provides workshops/seminars or training in this competency area for practitioners or lead practitioners beyond the enterprise boundary (e.g. conferences and open training days). [TUSE05-30A]	Records of Training/workshops/seminars created supported by successful post-training evaluation data. [TUSE05-E30]
6	Maintains expertise in this competency area through specialist Continual Professional Development (CPD) activities. [TUSE06]	(A) Reviews research, new ideas, and state of the art to identify relevant new areas requiring personal development in order to maintain expertise in this competency area. [TUSE06-10A]	Records of documents reviewed and insights gained as part of own research into this competency area. [TUSE06-E10]
		(A) Performs identified specialist professional development activities in order to maintain or further develop competence at expert level. [TUSE06-20A]	Records of Continual Professional Development (CPD) performed and learning outcomes. [TUSE06-E20]
		(A) Records continual professional development activities undertaken including learning or insights gained. [TUSE06-30A]	

NOTES

In addition to items above, enterprise-level or independent 3rd-Party-generated evidence may be used to amplify other evidence presented and may include:

a. Formally recognized by a reputable external organization as an expert in this competency area
b. Evidence of role as independent assessor or reviewer on project outside own organization where skills in this competency area were used
c. Evidence of invitation(s) from wider community for contribution of systems engineering expertise in this area (e.g. industry conference panel, government advisory board etc. cross-industry working groups, partnerships, accredited advanced university courses or research, or as part of professional institute)
d. Formal commendation beyond the enterprise (e.g. by INCOSE or other recognized authority) for work performed in this competency area
e. Independently assessed or accredited work product in this competency area (e.g. for independent publication or use)
f. Accolades of expertise in this area from recognized industry leaders

Competency area – Technical: Retirement
Description The retirement stage of a system is the final stage of a system life cycle, where the existence of a system or product is ended for a specific use, through appropriate and often controlled handling or recycling. System engineering activities in this stage are primarily focus on ensuring that disposal requirements are addressed. They could also concern preparing for the next generation of a system.
Why it matters Retirement is an essential part of a system overall concept as experience has shown that failure to plan for retirement (e.g. disposal, recycling, and reuse) early can be both costly and time-consuming. Indeed, many countries have changed their laws to insist that the developer is now responsible for ensuring the proper end-of-life disposal of all system components.
Possible contributory types of evidence Any combination of the types of evidence may be acceptable (depending on how the Framework is tailored and used). The evidence items identified at each level indicate example work products only. Contributions to work products will generally differ at each proficiency level.
Learning and development The INCOSE Professional Development Portal provides example guidance on how to gain an initial awareness of a competency area and options for developing further competence thereafter.

Awareness – Technical: Retirement		
ID	**Indicators of Competence**	**Relevant knowledge sub-indicators**
1	Explains why the needs of system retirement need to be considered, even as part of the original system design concept. [TREA01]	(K) Explains the need for retirement requirements. [TREA01-10K] (K) Explains how, retirement needs may influence a system concept or design. [TREA01-20K] (K) Identifies key activities and considerations associated with retirement. [TREA01-30K]
2	Identifies areas requiring special consideration when determining retirement requirements across each of the life cycle stages. [TREA02]	(K) Identifies typical areas which may have complex retirement needs in each of the different life cycle stages. [TREA02-10K]
3	Explains how evolving user requirements could affect retirement. [TREA03]	(K) Explains how evolving user requirements could bring about design changes that could impact retirement plans, e.g. addition of a hazardous material, or fielding the system to a new operational area with different disposal requirements. [TREA03-10K]

Supervised Practitioner – Technical: Retirement			
ID	**Indicators of Competence** ***(in addition to those at Awareness level)***	**Relevant knowledge, experience, and/or behaviors**	**Possible examples of objective evidence of personal contribution to activities performed, or professional behaviors applied**
1	Uses a governing process and appropriate tools to plan and control their retirement activities. [TRES01]	(K) Explains the primary elements of system retirement. [TRES01-10K]	Retirement support plan. [TRES01-E10]
		(A) Identifies key elements of a retirement plan. [TRES01-20A]	Detailed retirement processes. [TRES01-E20]
		(A) Uses governing process and tools to create retirement work products. [TRES01-30A]	Records of retirement activities. [TRES01-E30]
		(A) Uses governing process and tools to perform retirement activities. [TRES01-40A]	
2	Identifies required retirement requirements or design changes in order to address system retirement needs. [TRES02]	(A) Identifies required retirement requirements in order to assess system disposal needs. [TRES02-10A]	Documented retirement requirements or needs. [TRES02-E10]
		(A) Identifies potential item retirement issues, proposing design changes to mitigate or eliminate those issues. [TRES02-20A]	Documentation of issue resolution. [TRES02-E20]
3	Reviews the feasibility and impact of evolving user need on system retirement. [TRES03]	(A) Reviews evolving user need, including activities assessing the impact on retirement of changing user need. [TRES03-10A]	Records of changes to user need, including their impact on current and future stages of the system life cycle. [TRES03-E10]
		(A) Prepares inputs to change proposals to document evolving user need. [TRES03-20A]	Detailed change proposal analysis. [TRES03-E20]
4	Prepares updates to technical data (e.g. procedures, guidelines, checklists, and training) to ensure retirement activities and data are current. [TRES04]	(K) Describes options for technical data review (e.g. allowing user feedback, formal review meetings with subject matter experts, and informal or formal audits). [TRES04-10K]	Records from In-Service Review (ISR), retirement reviews or audits. [TRES04-E10]
		(K) Explains why it is important to have formal and rigorous configuration control of technical data for systems right up until retirement. [TRES04-20K]	Technical data package, showing management of changes. [TRES04-E20]
		(A) Identifies required updates to retirement work products (e.g. procedures, guidelines, checklists, training, and maintenance materials) to ensure they are current. [TRES04-30A]	Updated retirement support plan. [TRES04-E30]
		(A) Prepares updates to technical data work products to reflect changes identifies to maintain their currency. [TRES04-40A]	

(Continued)

ID	Indicators of Competence *(in addition to those at Awareness level)*	Relevant knowledge, experience, and/or behaviors	Possible examples of objective evidence of personal contribution to activities performed, or professional behaviors applied
5	Prepares inputs to obsolescence studies to identify impact on retirement of obsolescent components and their replacements. [TRES05]	(A) Performs research to determine impact on disposal of obsolescent parts or their replacements. [TRES05-10A] (A) Identifies impact of obsolescence on system retirement activities. [TRES05-20A]	Detailed change proposal analysis. [TRES05-E10] Updated retirement support plan. [TRES05-E20]
6	Identifies potential changes to system retirement environment or external interfaces as a result of system evolution or other technology change. [TRES06]	(A) Identifies potential changes to system retirement environment or external interfaces as a result of system evolution or other technology change. [TRES06-10A]	Environmental or retirement impact assessment records. [TRES06-E10]
7	Identifies potential changes to system retirement process or interfaces as a result of changes to interfacing systems or usage changes. [TRES07]	(A) Identifies the impact on the retirement process of changes to an existing system environment or interfaces. [TRES07-10A]	Environmental or retirement impact assessment records. [TRES07-E10]
8	Develops own understanding of this competency area through Continual Professional Development (CPD). [TRES08]	(A) Identifies potential gaps in own knowledge or development needs in this area, identifying opportunities to address these through continual professional development activities. [TRES08-10A] (A) Performs continual professional development activities to improve their knowledge and understanding in this area. [TRES08-20A] (A) Records continual professional development activities undertaken including learning or insights gained. [TRES08-30A]	Records of Continual Professional Development (CPD) performed and learning outcomes. [TRES08-E10]

Practitioner – Technical: Retirement			
ID	**Indicators of Competence** ***(in addition to those at Supervised Practitioner level)***	**Relevant knowledge, experience, and/or behaviors**	**Possible examples of objective evidence of personal contribution to activities performed, or professional behaviors applied**
1	Creates a strategy for system retirement which reflects wider project and business strategies. [TREP01]	(A) Identifies project-specific system retirement needs that need to be incorporated into SE planning. [TREP01-10A]	System retirement strategy work products. [TREP01-E10]
		(A) Prepares project-specific system retirement task estimates in support of SE planning. [TREP01-20A]	System retirement strategy work products. [TREP01-E20]
		(A) Prepares project-specific system retirement inputs for SE planning purposes, using appropriate processes and procedures. [TREP01-30A]	
		(P) Liaises throughout with stakeholders to gain approval, updating strategy as necessary. [TREP01-40P]	
2	Uses governing plans and processes for system retirement, interpreting, evolving, or seeking guidance where appropriate. [TREP02]	(K) Identifies the steps necessary to define a process and appropriate techniques to be adopted for the retirement of system elements. [TREP02-10K]	System retirement work products. [TREP02-E10]
		(K) Describes key elements of a successful retirement process and how these activities relate to different stages of a system life cycle. [TREP02-20K]	Inputs to planning documents, systems engineering management plans, other project/program plan, or organizational processes. [TREP02-E20]
		(A) Uses system retirement tools or uses a methodology. [TREP02-30A]	
		(A) Uses appropriate standards and governing processes in their system retirement planning, tailoring as appropriate. [TREP02-40A]	
		(A) Selects system retirement methods for each system element. [TREP02-50A]	
		(A) Creates system retirement plans and processes for use on a project. [TREP02-60A]	
		(P) Liaises with stakeholders throughout development of the plan, to gain approval, updating as necessary. [TREP02-70P]	
3	Complies with governing plans and processes for system retirement, interpreting, evolving, or seeking guidance where appropriate. [TREP03]	(A) Follows defined plans, processes, and associated tools to perform system retirement on a project. [TREP03-10A]	System retirement plans or work products they have interpreted or challenged to address project issues. [TREP03-E10]

(Continued)

ID	Indicators of Competence *(in addition to those at Supervised Practitioner level)*	Relevant knowledge, experience, and/or behaviors	Possible examples of objective evidence of personal contribution to activities performed, or professional behaviors applied
		(P) Guides and actively coordinates the interpretation of system retirement plans or processes in order to complete activities system retirement activities successfully. [TREP03-20P]	System retirement process or tool guidance. [TREP03-E20]
		(A) Prepares retirement data and reports in support of wider system measurement, monitoring and control. [TREP03-30A]	
		(P) Recognizes situations where deviation from published plans and processes or clarification from others is appropriate in order to overcome complex system retirement challenges. [TREP03-40P]	
		(P) Recognizes situations where existing system disposal strategy, process, plan or tools require formal change, liaising with stakeholders to gain approval as necessary. [TREP03-50P]	
4	Guides and actively coordinates the retirement of a system at end-of-life. [TREP04]	(A) Defines key considerations for system retirement (e.g. retirement or removal procedures, management of hazardous materials, disposal or re-use of system components…). [TREP04-10A]	System Retirement Plan. [TREP04-E10]
		(P) Guides and actively coordinates the retirement of a system at end-of life (e.g. disposal, removal activities). [TREP04-20P]	Systems retirement documentation. [TREP04-E20]
5	Determines the data to be collected in order to assess system retirement performance. [TREP05]	(A) Identifies relevant measurable retirement data. [TREP05-10A]	Data collection plans. [TREP05-E10]
		(A) Identifies methods by which retirement data can be collected. [TREP05-20A]	Knowledge of data collection tools or systems. [TREP05-E20]
6	Monitors the implementation of changes to system retirement environment or external interfaces. [TREP06]	(A) Identifies possible changes that could cause impacts to system retirement (e.g. new retirement requirement, new retirement threat, change to external interface, and change to retirement plan). [TREP06-10A]	Engineering Change Proposals. [TREP06-E10]
		(A) Identifies possible changes on system retirement (e.g. new retirement requirement, new retirement impact, and change to retirement plan). [TREP06-20A]	Participation in user group meetings or cross-system IPTs. [TREP06-E20]

7	Reviews retirement technical support data (e.g. procedures, guidelines, checklists, training, and materials) to ensure it is current. [TREP07]	(A) Develops and executes plans for periodic retirement reviews. [TREP07-10A]	Retirement review meeting documentation. [TREP07-E10]
		(A) Prepares information used for retirement reviews (e.g. updates to training material, manuals, and checklists). [TREP07-20A]	Updated training material, manuals, and checklists. [TREP07-E20]
		(A) Prepares information used for retirement reviews (e.g. updates to procedures, materials, manuals, and checklists). [TREP07-30A]	Updated retirement materials, manuals, and checklists. [TREP07-E30]
8	Guides new or supervised practitioners in system retirement in order to develop their knowledge, abilities, skills, or associated behaviors. [TREP08]	(P) Guides new or supervised practitioners in executing activities that form part of this competency. [TREP08-10P]	Organizational Breakdown Structure showing their responsibility for technical supervision in this area. [TREP08-E10]
		(A) Trains individuals to an "Awareness" level in this competency area. [TREP08-20A]	On-the-job training records. [TREP08-E20]
			Coaching or mentoring assignment records. [TREP08-E30]
			Records highlighting their impact on another individual in terms of improvement or professional development in this competency. [TREP08-E40]
9	Maintains and enhances own competence in this area through Continual Professional Development (CPD) activities. [TREP09]	(A) Identifies potential development needs in this area, identifying opportunities to address these through continual professional development activities. [TREP09-10A]	Records of Continual Professional Development (CPD) performed and learning outcomes. [TREP09-E10]
		(A) Performs continual professional development activities to maintain and enhance their competency in this area. [TREP09-20A]	
		(A) Records continual professional development activities undertaken including learning or insights gained. [TREP09-30A]	

Lead Practitioner – Technical: Retirement			
ID	**Indicators of Competence** ***(in addition to those at Practitioner level)***	**Relevant knowledge, experience, and/or behaviors**	**Possible examples of objective evidence of personal contribution to activities performed, or professional behaviors applied**
1	Creates enterprise-level policies, procedures, guidance, and best practice for retirement, including associated tools. [TREL01]	(A) Analyzes enterprise need for system retirement policies, processes, and tools. [TREL01-10A]	Records showing their role in embedding system retirement into enterprise policies (e.g. guidance introduced at enterprise level and enterprise-level review minutes). [TREL01-E10]
		(A) Produces enterprise-level policies, procedures, and guidance on system retirement. [TREL01-20A]	Procedures they have written. [TREL01-E20]
		(A) Identifies and acquires appropriate tools supporting system retirement. [TREL01-30A]	Records of support for tool introduction. [TREL01-E30]
2	Judges the tailoring of enterprise-level retirement processes to meet the needs of a project. [TREL02]	(A) Evaluates the enterprise retirement process against the business and external stakeholders (e.g. customer) needs in order to tailor processes to enable project success. [TREL02-10A]	Documented tailored process. [TREL02-E10]
		(P) Communicates constructive feedback on retirement process. [TREL02-20P]	Disposal/Retirement process feedback. [TREL02-E20]
3	Advises across the enterprise on the application of advanced practices to improve the effectiveness of project-level retirement activities. [TREL03]	(A) Identifies opportunities to implement improvements to retirement used across the enterprise. [TREL03-10A]	Records from novel retirement tool evaluations. [TREL03-E10]
		(P) Persuades key stakeholders across the enterprise to adopt new practices in retirement. [TREL03-20P]	Records indicating changed stakeholder thinking from personal intervention. [TREL03-E20]
4	Persuades key stakeholders across the enterprise to address identified retirement issues to reduce project and business risk. [TREL04]	(A) Identifies and engages with key stakeholders. [TREL04-10A]	Minutes of meetings. [TREL04-E10]
		(P) Fosters agreement between key stakeholders at enterprise level to resolve disposal issues, by promoting a holistic viewpoint. [TREL04-20P]	Records indicating changed stakeholder thinking from personal intervention. [TREL04-E20]
		(P) Persuades key stakeholders to address identified enterprise-level disposal issues. [TREL04-30P]	
5	Coaches or mentors practitioners across the enterprise in system retirement in order to develop their knowledge, abilities, skills, or associated behaviors. [TREL05]	(P) Coaches or mentors practitioners across the enterprise in competency-related techniques, recommending development activities. [TREL05-10P]	Coaching or mentoring assignment records. [TREL05-E10]
		(A) Develops or authorizes enterprise training materials in this competency area. [TREL05-20A]	Records of formal training courses, workshops, seminars, and authored training material supported by successful post-training evaluation data. [TREL05-E20]

		(A) Provides enterprise workshops/seminars or training in this competency area. [TREL05-30A]	Listing as an approved organizational trainer for this competency area. [TREL05-E30]
6	Promotes the introduction and use of novel techniques and ideas in system retirement across the enterprise, to improve enterprise competence in this area. [TREL06]	(A) Analyzes different approaches across different domains through research. [TREL06-10A]	Research records. [TREL06-E10]
		(A) Defines novel approaches that could potentially improve the SE discipline within the enterprise. [TREL06-20A]	Published papers in refereed journals/company literature. [TREL06-E20]
		(P) Fosters awareness of these novel techniques within the enterprise. [TREL06-30P]	Enabling systems introduced in support of new techniques or ideas. [TREL06-E30]
		(P) Collaborates with enterprise stakeholders to facilitate the introduction of techniques new to the enterprise. [TREL06-40P]	Published papers (or similar) at enterprise level. [TREL06-E40]
		(A) Monitors new techniques after their introduction to determine their effectiveness. [TREL06-50A]	Records of improvements made against a recognized process improvement model in this area. [TREL06-E50]
		(A) Adapts approach to reflect actual enterprise performance improvements. [TREL06-60A]	
7	Develops expertise in this competency area through specialist Continual Professional Development (CPD) activities. [TREL07]	(A) Identifies own needs for further professional development in order to increase competence beyond practitioner level. [TREL07-10A]	Records of Continual Professional Development (CPD) performed and learning outcomes. [TREL07-E10]
		(A) Performs professional development activities in order to move own competence toward expert level. [TREL07-20A]	
		(A) Records continual professional development activities undertaken including learning or insights gained. [TREL07-30A]	

NOTES

In addition to items above, enterprise-level or independent 3rd-Party-generated evidence may be used to amplify other evidence presented and may include:

a. Formally recognized by senior management in current organization as an expert in this competency area
b. Evidence of role as Product/System Design Authority or Technical Authority on a complex project with responsibilities in this area or where skills within this competency area were used
c. Recognized as an authorizing signatory on behalf of enterprise for formal documentation in this competency area (e.g. policies, processes, and deliverables)
d. Formal commendation or award within own enterprise for contribution or item of work successfully performed, which required proficiency in this competency area
e. Customer, Supplier, or other external project-specific key Stakeholder accolades for specific work performed in this competency area
f. Independently assessed or accredited work in this competency area (e.g. for independent publication or use)
g. Formal organizational HR records positively highlighting any specific professional competencies or behaviors identified (if applicable) plus any of the evidence indicators listed at Expert level below

Expert – Technical: Retirement			
ID	**Indicators of Competence** ***(in addition to those at Lead Practitioner level)***	**Relevant knowledge, experience, and/or behaviors**	**Possible examples of objective evidence of personal contribution to activities performed, or professional behaviors applied**
1	Communicates own knowledge and experience in System retirement, in order to improve best practice beyond the enterprise boundary. [TREE01]	(A) Produces papers, seminars, or presentations outside own enterprise for publication in order to share own ideas and improve industry best practices in this competence area. [TREE01-10A]	Published papers or books etc. on new technique in refereed journals/company literature. [TREE01-E10]
		(P) Fosters incorporation of own ideas into industry best practices in this area. [TREE01-20P]	Published papers in refereed journals or internal literature proposing new practices in this competence area (or presentations, tutorials, etc.). [TREE01-E20]
		(P) Develops guidance materials identifying new (or updating existing) best practice in this competence area. [TREE01-30P]	Proposals adopted as industry best practice. [TREE01-E30]
2	Advises organizations beyond the enterprise boundary on the suitability of their approach to system retirement. [TREE02]	(P) Acts as independent reviewer within an enterprise review body, in order to approve disposal plans. [TREE02-10P]	Records of membership on oversight committee with relevant terms of reference. [TREE02-E10]
		(A) Reviews and advises on disposal strategies that has led to changes being implemented. [TREE02-20A]	Review comments or revised document. [TREE02-E20]
3	Advises organizations beyond the enterprise boundary on their handling of complex or sensitive retirement-related issues. [TREE03]	(A) Advises external stakeholders (e.g. customers) on their retirement requirements and issues. [TREE03-10A]	Records of advice provided on retirement issues. [TREE03-E10]
		(P) Acts with sensitivity on highly complex system retirement issues making limited use of specialized, technical terminology and taking into account external stakeholders' (e.g. customer's) background and knowledge. [TREE03-20P]	Records from negotiations demonstrating awareness of customer's background and knowledge. [TREE03-E20]
		(A) Uses a holistic approach to complex retirement issue resolution including balanced, rational arguments on way forward. [TREE03-30A]	Stakeholder approval of retirement plans or procedures. [TREE03-E30]
4	Champions the introduction of novel techniques and ideas in system retirement, beyond the enterprise boundary, in order to develop the wider Systems Engineering community in this competency. [TREE04]	(A) Analyzes different approaches across different domains through research. [TREE04-10A]	Records of activities promoting research and need to adopt novel technique or ideas. [TREE04-E10]
		(A) Produces reports for the wider SE community on the effectiveness of new techniques after their introduction. [TREE04-20A]	Records of improvements made to process and appraisal against a recognized process improvement model. [TREE04-E20]

		(P) Collaborates with those introducing novel techniques within the wider SE community. [TREE04-30P]	Research records. [TREE04-E30]
		(A) Defines novel approaches that could potentially improve the wider SE discipline. [TREE04-40A]	Published papers in refereed journals/company literature. [TREE04-E40]
		(P) Fosters awareness of these novel techniques within the wider SE community. [TREE04-50P]	Records showing introduction of enabling systems supporting the new techniques or ideas. [TREE04-E50]
5	Coaches individuals beyond the enterprise boundary, in System retirement in order to further develop their knowledge, abilities, skills, or associated behaviors. [TREE05]	(P) Coaches or mentors individuals beyond the enterprise boundary, in competency-related techniques, recommending development activities. [TREE05-10P]	Coaching or mentoring assignment records. [TREE05-E10]
		(A) Develops or authorizes training materials in this competency area, which are subsequently successfully delivered beyond the enterprise boundary. [TREE05-20A]	Records of formal training courses, workshops, seminars, and authored training material supported by successful post-training evaluation data. [TREE05-E20]
		(A) Provides workshops/seminars or training in this competency area for practitioners or lead practitioners beyond the enterprise boundary (e.g. conferences and open training days). [TREE05-30A]	Records of Training/workshops/seminars created supported by successful post-training evaluation data. [TREE05-E30]
6	Maintains expertise in this competency area through specialist Continual Professional Development (CPD) activities. [TREE06]	(A) Reviews research, new ideas, and state of the art to identify relevant new areas requiring personal development in order to maintain expertise in this competency area. [TREE06-10A]	Records of documents reviewed and insights gained as part of own research into this competency area. [TREE06-E10]
		(A) Performs identified specialist professional development activities in order to maintain or further develop competence at expert level. [TREE06-20A]	Records of Continual Professional Development (CPD) performed and learning outcomes. [TREE06-E20]
		(A) Records continual professional development activities undertaken including learning or insights gained. [TREE06-30A]	

NOTES	In addition to items above, enterprise-level or independent 3rd-Party-generated evidence may be used to amplify other evidence presented and may include: a. Formally recognized by a reputable external organization as an expert in this competency area b. Evidence of role as independent assessor or reviewer on project outside own organization where skills in this competency area were used c. Evidence of invitation(s) from wider community for contribution of systems engineering expertise in this area (e.g. industry conference panel, government advisory board etc. cross-industry working groups, partnerships, accredited advanced university courses or research, or as part of professional institute) d. Formal commendation beyond the enterprise (e.g. by INCOSE or other recognized authority) for work performed in this competency area e. Independently assessed or accredited work product in this competency area (e.g. for independent publication or use) f. Accolades of expertise in this area from recognized industry leaders

Competency area – Management: Planning
Description The purpose of planning is to produce, coordinate, and maintain effective and workable plans across multiple disciplines. Systems Engineering planning includes planning the way the engineering of the system will be performed and managed, tailoring generic engineering processes to address specific project context, technical activities, and identified risks. This includes estimating the effort, resources and timescales required to complete the project to the required quality level. Planning is performed in association with the Project Manager. Plans and estimates may need updating to reflect changes or to overcome unexpected issues encountered during the development process.
Why it matters It is important to identify the full scope and timing of all Systems Engineering activities and their associated resource needs and to link this with task effort and cost estimation through controlled planning. Alignment between Systems Engineering planning and estimation is vital to ensure that assumptions made when developing a plan, such as ways of working and process tailoring are taken into consideration. Failure to plan correctly will mean inadequate visibility of progress and is likely to cause ongoing problems with time, budget and quality.
Possible contributory types of evidence Any combination of the types of evidence may be acceptable (depending on how the Framework is tailored and used). The evidence items identified at each level indicate example work products only. Contributions to work products will generally differ at each proficiency level.
Learning and development The INCOSE Professional Development Portal provides example guidance on how to gain an initial awareness of a competency area and options for developing further competence thereafter.

Awareness – Management: Planning		
ID	**Indicators of Competence**	**Relevant knowledge sub-indicators**
1	Identifies key planning and estimating terms and acronyms and the relationships between them. [MPLA01]	(K) Lists common project planning stakeholders with rationale (e.g. individuals, groups, or organizations who affect or are affected by the project). [MPLA01-10K] (K) Defines common planning terms or acronyms (e.g. "organization," "responsibility," "authority," "milestone," "stage review," "gate review," "work package," "metrics," "Schedule Performance Index (SPI)," "Cost Performance Index (CPI)"). [MPLA01-20K] (K) Defines common estimating terms (e.g. "budgetary," "fixed price," "fixed effort," "not-to-exceed"). [MPLA01-30K] (K) Describes common estimating methods (e.g. "using historical data," "Delphi," "expert judgement," "by similarity," "metricated," or "size-based estimating"). [MPLA01-40K]
2	Explains why planning Systems Engineering activities is important and how planning interacts across disciplines and organizations. [MPLA02]	(K) Defines the purpose of Systems Engineering planning (e.g. to identify and organize the set of required activities to complete a project from a technical point of view). [MPLA02-10K]

		(K) Describes key interfaces and links between engineering planning and planning stakeholders in the wider enterprise outside engineering (e.g. infrastructure management, human resource management, quality management, knowledge management, portfolio management, and life cycle model management). [MPLA02-20K] (K) Explains how monitoring and control activities support planning (e.g. to provide an understanding of the project/program's progress so that appropriate corrective actions can be taken when progress deviates from the plan). [MPLA02-30K] (K) Describes the scope of information required for Systems Engineering management planning (e.g. sufficient to define the work, processes and activities, schedule the duration, and budget the allocated funds). [MPLA02-40K] (K) Identifies common information sources for SE planning (e.g. project/program requirements, statements of work, and estimates of effort and cost). [MPLA02-50K] (K) Explains how Systems Engineering maintains the balance between requirements, architecture, program management, scheduling and budget. [MPLA02-60K] (K) Explains why a failure to plan could have a negative impact (e.g. significantly increases risk and will probably lead to schedule and cost overrun). [MPLA02-70K] (K) Explains the relationship between system and project planning activities and the management plan (e.g. Systems Engineering Management Plan, SEMP). [MPLA02-80K]
3	Identifies key areas that need to be addressed in a project Systems Engineering plan. [MPLA03]	(K) Lists topics to be considered when constructing a system engineering management plan (e.g. from a technical execution point of view: team organization and areas of responsibility, technical environment and tools, technical decision process, requirements management process, architecture design process, configuration management process with references to the configuration management plan, description of program interfaces, data collection and management process, system implementation and support process, production support, and specialty engineering needs). [MPLA03-10K] (K) Explains how key areas affect Systems Engineering planning. [MPLA03-20K]
4	Explains the principles of Systems Engineering process tailoring including its benefits and potential issues. [MPLA04]	(K) Describes the concepts and principles of tailoring a process or tools, tailoring benefits, and how tailoring is typically managed organizationally. [MPLA04-10K] (K) Explains why different types of development may require different tailoring of processes and tools. [MPLA04-20K] (K) Explains why different projects or external stakeholders (e.g. customers) may require different tailoring of processes and tools. [MPLA04-30K] (K) Explains why tailoring too much or too little may both be detrimental to a project. [MPLA04-40K]

ID	Indicators of Competence	Relevant knowledge sub-indicators
5	Identifies key potential sources of change on a project and why the impact of such changes needs to be carefully assessed and planned. [MPLA05]	(K) Identifies common sources of change on a project (e.g. requirements creep, clarification, funding source changes, evolving priorities, agile development approach, and risk realization). [MPLA05-10K] (K) Explains how technical parameters identified during Systems Engineering Planning are used to monitor design margin as the design evolves. [MPLA05-20K] (K) Explains how a change may impact different areas of a project plan (e.g. project re-baselining, cost and schedule, re-estimation, re-planning, documenting decisions, staffing, technology, and process changes). [MPLA05-30K]
6	Explains the relationship between life cycle reviews and planning. [MPLA06]	(K) Describes life cycle decision gates and reviews across the full SE life cycle and the purpose of each. [MPLA06-10K] (K) Identifies common project "milestones" and explains how these milestones impact planning activities (e.g. milestones are specific points along the project timeline). [MPLA06-20K] (K) Explains why life cycle reviews may impact planning activities. [MPLA06-30K] (K) Identifies common reviews, which are not decision gates. [MPLA06-40K]

Supervised Practitioner – Management: Planning			
ID	**Indicators of Competence** ***(in addition to those at Awareness level)***	**Relevant knowledge, experience, and/or behaviors**	**Possible examples of objective evidence of personal contribution to activities performed, or professional behaviors applied**
1	Follows a defined governing process, using appropriate tools, to guide their Systems Engineering planning activities. [MPLS01]	(K) Identifies processes and tools used in support of planning activities. [MPLS01-10K]	Planning processes and tools they used on a given project. [MPLS01-E10]
		(A) Follows Systems Engineering processes, applying appropriate tools, in order to plan activities. [MPLS01-20A]	Planning outputs from Systems Engineering activities they performed or used on a project. [MPLS01-E20]
2	Explains the role of Systems Engineering planning and its relationship to wider project planning and management. [MPLS02]	(K) Describes the relationship between the Systems Engineering life cycle and the project/program life cycle (see Life Cycle Process Definition competency). [MPLS02-10K]	Contributions they made to a Systems Engineering management plan or relevant section of a project/program management plan. [MPLS02-E10]
		(K) Describes the interfaces between a Systems Engineering plan, other plans on the same project and other organizational plans. [MPLS02-20K]	Organizational chart demonstrating their personal responsibilities. [MPLS02-E20]
		(A) Describes how SE planning is coordinated with other activities across the wider project and wider organization on a given project. [MPLS02-30A]	Information or process(-es) they coordinated to generate a plan. [MPLS02-E30]
		(A) Describes scope of own responsibilities and how this scope aligns with the scope of other key project roles. [MPLS02-40A]	
3	Prepares information required in order to create an SE plan, in order to control the management of a system development. [MPLS03]	(K) Explains how a work breakdown structure is used to develop basic schedule milestones and reviews. [MPLS03-10K]	Systems Engineering Plan inputs. [MPLS03-E10]
		(K) Describes the relationship between staffing, resources,` and scheduling. [MPLS03-20K]	
		(K) Lists key attributes of typical development cycles and how these affect SE Plans. [MPLS03-30K]	
		(K) Lists key attributes of typical review decision gates and how these are captured in SE Plans. [MPLS03-40K]	
		(K) Lists parameters identified within Systems Engineering planning used in support of monitoring and control. [MPLS03-50K]	

(Continued)

ID	Indicators of Competence *(in addition to those at Awareness level)*	Relevant knowledge, experience, and/or behaviors	Possible examples of objective evidence of personal contribution to activities performed, or professional behaviors applied
		(K) Lists technical parameters identified within Systems Engineering planning which are used to monitor design margin. [MPLS03-60K] (A) Produces information used in a SE Plan, using appropriate tools. [MPLS03-70A]	
4	Prepares inputs to SE management plan. [MPLS04]	(A) Prepares (sections of) a plan for managing the engineering of a system. [MPLS04-10A]	Engineering Management Plans (or SEMP). [MPLS04-E10]
5	Explains how Systems Engineering estimates are compiled in order to scope the size of a development. [MPLS05]	(K) Explains how a Work Breakdown Structure (WBS) is used to support Systems Engineering estimating. [MPLS05-10K] (K) Explains how effort and resource estimates and are used to create Systems Engineering related tasking and resource estimates. [MPLS05-20K] (A) Collates effort or resource requirements in support of Systems Engineering estimates. [MPLS05-30A]	Effort or resource estimates (e.g. own effort or resource needs or those of others). [MPLS05-E10]
6	Prepares inputs to Systems Engineering work packages to support the scoping of Systems Engineering tasks. [MPLS06]	(K) Defines work packages and how are related to project pricing and staffing. [MPLS06-10K] (A) Collates effort and resource estimates taking into account inputs from others as necessary. [MPLS06-20A] (A) Coordinates and manages output content with others to avoid uncertainty, duplication or omission. [MPLS06-30A]	Systems Engineering Work Package inputs. [MPLS06-E10] Effort and resource estimates for Systems Engineering tasks. [MPLS06-E20]
7	Prepares inputs to Systems Engineering replanning activities in order to implement engineering changes. [MPLS07]	(K) Explains how Configuration Management activities (and Change Management in particular) interact with Planning to help maintain a product baseline. [MPLS07-10K] (K) Explains how "engineering change proposals" are used to effect changes to plans. [MPLS07-20K]	Inputs they updated in support of Systems Engineering changes. [MPLS07-E10] Effort and resource estimates that they updated in support of Systems Engineering changes. [MPLS07-E20]

		(K) Lists possible sources of future changes to existing plans. [MPLS07-30K] (A) Analyzes the potential impact of identified future changes on existing Systems Engineering plans. [MPLS07-40A] (A) Follows instructions to update existing Systems Engineering plans in order to take account of identified future changes. [MPLS07-50A]	Inputs they created to knowledge repository which document organizational improvements as a result of lessons learned. [MPLS07-E30]
8	Prepares updates to Systems Engineering plans to reflect authorized changes. [MPLS08]	(A) Maintains existing plans for Systems Engineering following guidance received. [MPLS08-10A] (A) Prepares information for stakeholder meetings including inputs, actions or minutes. [MPLS08-20A]	Systems Engineering management plan (or similar) they updated on a specific project. [MPLS08-E10] Systems Engineering estimates they updated on a specific project. [MPLS08-E20] Change review material they created. [MPLS08-E30]
9	Identifies development lessons learned performing SE planning to inform future projects. [MPLS09]	(A) Records personal lessons learned. [MPLS09-10A] (A) Communicates lessons learned to others. [MPLS09-20A]	Logbook, engineering notebook (or similar) showing personal lessons they learned. [MPLS09-E10] Memorandums written to document lessons learned or blogs kept in an electronic repository for others to see. [MPLS09-E20]
10	Develops own understanding of this competency area through Continual Professional Development (CPD). [MPLS10]	(A) Identifies potential gaps in own knowledge or development needs in this area, identifying opportunities to address these through continual professional development activities. [MPLS10-10A] (A) Performs continual professional development activities to improve their knowledge and understanding in this area. [MPLS10-20A] (A) Records continual professional development activities undertaken including learning or insights gained. [MPLS10-30A]	Records of Continual Professional Development (CPD) performed and learning outcomes. [MPLS10-E10]

Practitioner – Management: Planning			
ID	**Indicators of Competence** ***(in addition to those at Supervised Practitioner level)***	**Relevant knowledge, experience, and/or behaviors**	**Possible examples of objective evidence of personal contribution to activities performed, or professional behaviors applied**
1	Creates a strategy for performing Systems Engineering life cycle activities considering the wider project plan to ensure integration and coherence across the development. [MPLP01]	(A) Identifies constraints on Systems Engineering planning for a project. [MPLP01-10A]	Systems Engineering Management Plan (or similar). [MPLP01-E10]
		(A) Identifies potential life cycle approaches applicable to the project from those available within the organization. [MPLP01-20A]	List of factors and rationale influencing their tailoring approach on a specific project (e.g. customer program life cycle, complexity of the system, stability of requirements, milestone and delivery dates, technology insertion/readiness, standards, internal policy and process requirements, product life cycle, and availability of tools). [MPLP01-E20]
		(A) Identifies the needs of the project for Systems Engineering planning. [MPLP01-30A]	
		(A) Identifies needs of wider project or business including potential areas of conflict. [MPLP01-40A]	
		(A) Analyzes strategies for all technical activities to create a single coherent project strategy, negotiating as required with other stakeholders. [MPLP01-50A]	
		(A) Ensures identified Systems Engineering planning needs are incorporated into the wider project/program strategy or plan. [MPLP01-60A]	
2	Creates a governing Systems Engineering management plan, which reflect project and business strategy and all development constraints. [MPLP02]	(K) Identifies the steps necessary to define a plan for the management of Systems Engineering activities. [MPLP02-10K]	SE system planning or process work products they have developed. [MPLP02-E10]
		(K) Describes key elements of a successful Systems Engineering management plan and how these activities relate to different stages of a system life cycle. [MPLP02-20K]	Certificates or proficiency evaluations in system management tools. [MPLP02-E20]
		(A) Uses tools or a methodology to support Systems Engineering management planning activities. [MPLP02-30A]	SE system management tool guidance they have developed. [MPLP02-E30]

		(A) Uses appropriate standards and governing processes in their Systems Engineering management planning, tailoring as appropriate. [MPLP02-40A] (A) Selects appropriate management approaches to control each key activity (e.g. interfaces, developments, and procurements). [MPLP02-50A] (A) Creates a system engineering management plan and processes for use on a project (e.g. SEMP). [MPLP02-60A]	
3	Develops effort, resource, and schedule estimates to scope Systems Engineering life cycle activities. [MPLP03]	(A) Prepares project-specific Systems Engineering effort, schedule, and resource estimates taking account of identified constraints and the proposed Systems Engineering plan. [MPLP03-10A] (A) Identifies required effort and resource estimates, justifying as required. [MPLP03-20A]	Estimates prepared for an SE management plan they defined. [MPLP03-E10] Reports communicating the tracking of resource efforts and estimates they compiled. [MPLP03-E20]
4	Selects key design parameters required to track critical aspects of the design during development. [MPLP04]	(A) Selects key design parameters required to track critical aspects of the design during development. [MPLP04-10A]	List of design parameters selected. [MPLP04-E10]
5	Negotiates successfully with others to secure identified future Systems Engineering needs of a project. [MPLP05]	(K) Explains how project management and in particular Project Planning roles interact with Systems Engineering roles when planning a system, highlighting potential issues and challenges. [MPLP05-10K] (P) Negotiates needs successfully with project/program management lead/chief engineer. [MPLP05-20P]	Documents from project planning negotiations they have performed, highlighting issues encountered and subsequent outcomes. [MPLP05-E10] Minutes of review meetings (or similar) covering their resource needs negotiations. [MPLP05-E20]
6	Negotiates successfully with project management to secure identified future Systems Engineering needs of a project. [MPLP06]	(P) Negotiates successfully with project management to secure identified future Systems Engineering needs of a project. [MPLP06-10P]	Plans or reports addressing resource needs before and after negotiation. [MPLP06-E10]
7	Prepares updates required to Systems Engineering plan to address internal or external changes. [MPLP07]	(K) Describes potential sources of change to an existing system engineering management plan (e.g. internal, external, organizational, risk, technology, and resource). [MPLP07-10K] (A) Identifies the impact of change on Systems Engineering plans and related technical data items as a result of different types of change (e.g. internal and external). [MPLP07-20A]	Engineering change review material. [MPLP07-E10] Engineering change review material. [MPLP07-E10]

(Continued)

ID	Indicators of Competence *(in addition to those at Supervised Practitioner level)*	Relevant knowledge, experience, and/or behaviors	Possible examples of objective evidence of personal contribution to activities performed, or professional behaviors applied
		(A) Creates updates to Systems Engineering planning documentation to reflect approved engineering changes. [MPLP07-30A]	Plans or reports updates resulting from changes. [MPLP07-E30]
		(A) Identifies change management process utilized in executed programs and overall configuration management approach. [MPLP07-40A]	Planning information they updated as a result of changes. [MPLP07-E40]
8	Coordinates implementation of updates to Systems Engineering plan to address internal or external changes. [MPLP08]	(A) Ensures Systems Engineering re-planning needs are incorporated into the wider project/program re-planning activities. [MPLP08-10A]	Systems Engineering management plan (or similar) they updated on a specific project. [MPLP08-E10]
		(A) Maintains existing plans for Systems Engineering taking into account revised estimates of effort and cost, statements of work, requirements and constraints. [MPLP08-20A]	Systems Engineering estimates they updated on a specific project. [MPLP08-E20]
			Change review material they created in this area. [MPLP08-E30]
9	Guides new or supervised practitioners in Systems Engineering planning to develop their knowledge, abilities, skills, or associated behaviors. [MPLP09]	(P) Guides new or supervised practitioners in executing activities that form part of this competency. [MPLP09-10P]	Organizational Breakdown Structure showing their responsibility for technical supervision in this area. [MPLP09-E10]
		(A) Trains individuals to an "Awareness" level in this competency area. [MPLP09-20A]	Lessons, seminars, presentations made to educate junior engineers or train incoming engineers on project processes. [MPLP09-E20]
			Records highlighting their impact on another individual in terms of improvement or professional development in this competency. [MPLP09-E30]
10	Maintains and enhances own competence in this area through Continual Professional Development (CPD) activities. [MPLP10]	(A) Identifies potential development needs in this area, identifying opportunities to address these through continual professional development activities. [MPLP10-10A]	Records of Continual Professional Development (CPD) performed and learning outcomes. [MPLP10-E10]
		(A) Performs continual professional development activities to maintain and enhance their competency in this area. [MPLP10-20A]	
		(A) Records continual professional development activities undertaken including learning or insights gained. [MPLP10-30A]	

Lead Practitioner – Management: Planning			
ID	**Indicators of Competence** ***(in addition to those at Practitioner level)***	**Relevant knowledge, experience, and/or behaviors**	**Possible examples of objective evidence of personal contribution to activities performed, or professional behaviors applied**
1	Creates enterprise-level policies, procedures, guidance, and best practice for Systems Engineering planning, including associated tools, to improve organizational effectiveness. [MPLL01]	(A) Identifies the need for enterprise-level documents (e.g. policies, procedures, guidance, and best practice) or tools supporting Systems Engineering planning. [MPLL01-10A]	Systems Engineering planning policies, processes, best practice guidance they created, which have been adopted by the enterprise. [MPLL01-E10]
		(A) Produces enterprise-level document(s) or selects appropriate enterprise-level] tool(s) in support of Systems Engineering planning, taking into account current and future enterprise needs. [MPLL01-20A]	Organization charts or documents recognizing their authority in Systems Engineering planning at enterprise level. [MPLL01-E20]
		(P) Collaborates with enterprise-level stakeholders across the enterprise to gain approval, updating documents or tool requirements as necessary. [MPLL01-30P]	Organization charts or documents recognizing their authority for defining Systems Engineering planning documentation across the enterprise. [MPLL01-E30]
2	Judges tailoring of enterprise-level Systems Engineering planning processes to balance the needs of the project and business. [MPLL02]	(A) Reviews Systems Engineering plans against enterprise tailoring guidance. [MPLL02-10A]	Systems Engineering planning reviews they attended or review comments they recorded. [MPLL02-E10]
		(A) Identifies constructive feedback on Systems Engineering plan tailoring performed. [MPLL02-20A]	Tailored Systems Engineering plans they have authorized. [MPLL02-E20]
		(A) Authorizes tailoring within Systems Engineering plans on behalf of enterprise. [MPLL02-30A]	
3	Creates Systems Engineering plans integrating multiple diverse projects or a complex system to ensure coherence across the development. [MPLL03]	(A) Produces Systems Engineering plans for multiple diverse projects or a single complex project. [MPLL03-10A]	Multiple Systems Engineering Management Plans or other planning documents they have written. [MPLL03-E10]
		(A) Analyzes individual Systems Engineering plans to ensure they address specific project or enterprise context or need. [MPLL03-20A]	Lessons learned and distributed documented at enterprise level. [MPLL03-E20]
		(A) Adapts own Systems Engineering planning practices based upon lessons learned. [MPLL03-30A]	
		(A) Adapts enterprise Systems Engineering planning practices or documentation based upon lessons learned. [MPLL03-40A]	
4	Judges Systems Engineering plans from across the enterprise for their suitability in meeting both the needs of the project and the wider enterprise. [MPLL04]	(A) Reviews SE plans from across the enterprise against business and external stakeholder (e.g. customer) needs. [MPLL04-10A]	Multiple SEMPs or other planning documents they have reviewed. [MPLL04-E10]
		(P) Acts as independent reviewer of high-value or high-risk program Systems Engineering plans. [MPLL04-20P]	Review comments they have recorded on SE plans. [MPLL04-E20]

(Continued)

ID	Indicators of Competence *(in addition to those at Practitioner level)*	Relevant knowledge, experience, and/or behaviors	Possible examples of objective evidence of personal contribution to activities performed, or professional behaviors applied
		(A) Identifies constructive feedback on SE plans. [MPLL04-30A]	SE Plans or estimates they have authorized on behalf of the enterprise. [MPLL04-E30]
		(A) Authorizes project Systems Engineering plans on behalf of enterprise. [MPLL04-40A]	
5	Judges Systems Engineering effort, resource, and schedule estimates from across the enterprise for their quality. [MPLL05]	(A) Reviews SE estimates for effort, resource and schedule from across the enterprise against business and external stakeholder (e.g. customer) needs. [MPLL05-10A]	Multiple estimates or other costing documents they have reviewed on behalf of the enterprise. [MPLL05-E10]
		(P) Acts as independent reviewer of high-value or high-risk program Systems Engineering estimates. [MPLL05-20P]	Review comments they have recorded on SE estimates. [MPLL05-E20]
		(A) Identifies constructive feedback on SE estimates. [MPLL05-30A]	SE estimates they have authorized on behalf of the enterprise. [MPLL05-E30]
		(A) Authorizes project Systems Engineering estimates on behalf of enterprise. [MPLL05-40A]	
6	Reviews engineering changes from across the enterprise to establish the impact on both the project itself and the wider enterprise. [MPLL06]	(A) Evaluates updates to Systems Engineering plans as a result of change requests ensuring business and external stakeholder (e.g. customer) needs continue to be met and processes remain appropriately tailored to enable project success. [MPLL06-10A]	Reviews they attended. [MPLL06-E10]
		(P) Acts as independent reviewer in change reviews for Systems Engineering plans and estimates. [MPLL06-20P]	Review comments they recorded on changes. [MPLL06-E20]
		(A) Authorizes Systems Engineering change information on behalf of enterprise. [MPLL06-30A]	
7	Persuades key stakeholders to address identified enterprise-level Systems Engineering project planning issues to reduce project cost, schedule, or technical risk. [MPLL07]	(A) Identifies issues affecting wider enterprise from project plans and estimates. [MPLL07-10A]	Issues resolved due to their personal intervention. [MPLL07-E10]
		(P) Fosters support from enterprise stakeholders in resolving identified issues. [MPLL07-20P]	Enterprise expert panel/process improvement/core competency they have participated in. [MPLL07-E20]
		(P) Collaborates with enterprise stakeholders to implement or tailor Systems Engineering principles across the enterprise within the context of the programs at hand. [MPLL07-30P]	Enterprise expert panel/process improvement/core competency lead. [MPLL07-E30]
8	Coaches or mentors practitioners across the enterprise in Systems Engineering planning in order to develop their knowledge, abilities, skills, or associated behaviors. [MPLL08]	(P) Coaches or mentors practitioners across the enterprise in competency-related techniques, recommending development activities. [MPLL08-10P]	Coaching or mentoring assignment records. [MPLL08-E10]
		(A) Develops or authorizes enterprise training materials in this competency area. [MPLL08-20A]	Records of formal training courses, workshops, seminars, and authored training material supported by successful post-training evaluation data. [MPLL08-E20]

		(P) Provides enterprise workshops/seminars or training in this competency area. [MPLL08-30P]	Listing as an approved organizational trainer for this competency area. [MPLL08-E30]
9	Promotes the introduction and use of novel techniques and ideas in Systems Engineering planning across the enterprise, to improve enterprise competence in this area. [MPLL09]	(A) Analyzes different approaches across different domains through research. [MPLL09-10A]	Research records. [MPLL09-E10]
		(A) Defines novel approaches that could potentially improve the SE discipline within the enterprise. [MPLL09-20A]	Published papers in refereed journals/company literature. [MPLL09-E20]
		(P) Fosters awareness of these novel techniques within the enterprise. [MPLL09-30P]	Records showing introduction of enabling systems supporting the new techniques or ideas. [MPLL09-E30]
		(P) Collaborates with enterprise stakeholders to facilitate the introduction of techniques new to the enterprise. [MPLL09-40P]	Published papers (or similar) at enterprise level. [MPLL09-E40]
		(A) Monitors new techniques after their introduction to determine their effectiveness. [MPLL09-50A]	Records of improvements made against a recognized process improvement model in this area. [MPLL09-E50]
		(A) Adapts approach to reflect actual enterprise performance improvements. [MPLL09-60A]	
10	Develops expertise in this competency area through specialist Continual Professional Development (CPD) activities. [MPLL10]	(A) Identifies own needs for further professional development in order to increase competence beyond practitioner level. [MPLL10-10A]	Records of Continual Professional Development (CPD) performed and learning outcomes. [MPLL10-E10]
		(A) Performs professional development activities in order to move own competence toward expert level. [MPLL10-20A]	
		(A) Records continual professional development activities undertaken including learning or insights gained. [MPLL10-30A]	

NOTES	In addition to items above, enterprise-level or independent 3rd-Party-generated evidence may be used to amplify other evidence presented and may include: a. Formally recognized by senior management in current organization as an expert in this competency area b. Evidence of role as Product/System Design Authority or Technical Authority on a complex project with responsibilities in this area or where skills within this competency area were used c. Recognized as an authorizing signatory on behalf of enterprise for formal documentation in this competency area (e.g. policies, processes, and deliverables) d. Formal commendation or award within own enterprise for contribution or item of work successfully performed, which required proficiency in this competency area e. Customer, Supplier, or other external project-specific key Stakeholder accolades for specific work performed in this competency area f. Independently assessed or accredited work in this competency area (e.g. for independent publication or use) g. Formal organizational HR records positively highlighting any specific professional competencies or behaviors identified (if applicable) plus any of the evidence indicators listed at Expert level below

Expert – Management: Planning			
ID	**Indicators of Competence** ***(in addition to those at Lead Practitioner level)***	**Relevant knowledge, experience, and/or behaviors**	**Possible examples of objective evidence of personal contribution to activities performed, or professional behaviors applied**
1	Communicates own knowledge and experience in Systems Engineering planning, in order to promote best practice beyond the enterprise boundary. [MPLE01]	(A) Produces papers, seminars, or presentations outside own enterprise for publication in order to share own ideas and improve industry best practices in this competence area. [MPLE01-10A]	Published papers or books etc. on new technique in refereed journals/company literature. [MPLE01-E10]
		(P) Fosters incorporation of own ideas into industry best practices in this area. [MPLE01-20P]	Published papers in refereed journals or internal literature proposing new practices in this competence area (or presentations, tutorials, etc.). [MPLE01-E20]
		(P) Develops guidance materials identifying new (or updating existing) best practice in this competence area. [MPLE01-30P]	Own proposals adopted as industry best practices in this competence area. [MPLE01-E30]
2	Advises organizations beyond the enterprise boundary on the suitability of their approach to Systems Engineering planning and estimation. [MPLE02]	(A) Provides independent advice to stakeholders beyond enterprise boundary regarding their SE plans or estimates. [MPLE02-10A]	Records of advice provided. [MPLE02-E10]
		(P) Acts with sensitivity on highly complex Systems Engineering planning issues, making limited use of specialized, technical terminology and taking account of external stakeholders' (e.g. customer's) background and knowledge. [MPLE02-20P]	Records indicating changed stakeholder thinking from personal intervention. [MPLE02-E20]
		(P) Communicates using a holistic approach to complex issue resolution including balanced, rational arguments on way forward. [MPLE02-30P]	
		(P) Arbitrates on Systems Engineering planning decisions on behalf of stakeholders beyond enterprise boundary. [MPLE02-40P]	
		(A) Persuades stakeholders beyond the enterprise boundary to accept difficult project-specific SE planning recommendations or actions. [MPLE02-50A]	
3	Advises organizations beyond the enterprise boundary on the handling of complex or sensitive Systems Engineering planning issues. [MPLE03]	(A) Provides independent advice to SE stakeholders beyond the enterprise boundary on complex or sensitive SE planning issues. [MPLE03-10A]	Records of advice provided. [MPLE03-E10]

		(P) Arbitrates on complex or sensitive Systems Engineering planning issues beyond the enterprise boundary. [MPLE03-20P]	Records indicating changed stakeholder thinking from personal intervention. [MPLE03-E20]
		(P) Persuades stakeholders beyond the enterprise boundary to accept difficult project-specific SE planning recommendations or actions. [MPLE03-30P]	
4	Champions the introduction of novel techniques and ideas in Systems Engineering planning, beyond the enterprise boundary, in order to develop the wider Systems Engineering community in this competency. [MPLE04]	(A) Analyzes different approaches across different domains through research. [MPLE04-10A]	Records of activities promoting research and need to adopt novel technique or ideas. [MPLE04-E10]
		(A) Produces reports for the wider SE community on the effectiveness of new techniques after their introduction. [MPLE04-20A]	Records of improvements made to process and appraisal against a recognized process improvement model. [MPLE04-E20]
		(P) Collaborates with those introducing novel techniques within the wider SE community. [MPLE04-30P]	Research records. [MPLE04-E30]
		(A) Defines novel approaches that could potentially improve the wider SE discipline. [MPLE04-40A]	Published papers in refereed journals/company literature. [MPLE04-E40]
		(P) Fosters awareness of these novel techniques within the wider SE community. [MPLE04-50P]	Records showing introduction of enabling systems supporting the new techniques or ideas. [MPLE04-E50]
5	Coaches individuals beyond the enterprise boundary in Systems Engineering planning, in order to further develop their knowledge, abilities, skills, or associated behaviors. [MPLE05]	(P) Coaches or mentors individuals beyond the enterprise boundary, in competency-related techniques, recommending development activities. [MPLE05-10P]	Coaching or mentoring assignment records. [MPLE05-E10]
		(A) Develops or authorizes training materials in this competency area, which are subsequently successfully delivered beyond the enterprise boundary. [MPLE05-20A]	Records of formal training courses, workshops, seminars, and authored training material supported by successful post-training evaluation data. [MPLE05-E20]
		(P) Provides workshops/seminars or training in this competency area for practitioners or lead practitioners beyond the enterprise boundary (e.g. conferences and open training days). [MPLE05-30P]	Records of Training/workshops/seminars created supported by successful post-training evaluation data. [MPLE05-E30]

(*Continued*)

ID	Indicators of Competence *(in addition to those at Lead Practitioner level)*	Relevant knowledge, experience, and/or behaviors	Possible examples of objective evidence of personal contribution to activities performed, or professional behaviors applied
6	Maintains expertise in this competency area through specialist Continual Professional Development (CPD) activities. [MPLE06]	(A) Reviews research, new ideas, and state of the art to identify relevant new areas requiring personal development in order to maintain expertise in this competency area. [MPLE06-10A] (A) Performs identified specialist professional development activities in order to maintain or further develop competence at expert level. [MPLE06-20A] (A) Records continual professional development activities undertaken including learning or insights gained. [MPLE06-30A]	Records of documents reviewed and insights gained as part of own research into this competency area. [MPLE06-E10] Records of Continual Professional Development (CPD) performed and learning outcomes. [MPLE06-E20]

NOTES	In addition to items above, enterprise-level or independent 3rd-Party-generated evidence may be used to amplify other evidence presented and may include: a. Formally recognized by a reputable external organization as an expert in this competency area b. Evidence of role as independent assessor or reviewer on project outside own organization where skills in this competency area were used c. Evidence of invitation(s) from wider community for contribution of systems engineering expertise in this area (e.g. industry conference panel, government advisory board etc. cross-industry working groups, partnerships, accredited advanced university courses or research, or as part of professional institute) d. Formal commendation beyond the enterprise (e.g. by INCOSE or other recognized authority) for work performed in this competency area e. Independently assessed or accredited work product in this competency area (e.g. for independent publication or use) f. Accolades of expertise in this area from recognized industry leaders

Competency area – Management: Monitoring and Control
Description Monitoring and control assesses the project to see if the current plans are aligned and feasible; determines the status of a project, technical and process performance and directs execution to ensure that performance is according to plans and schedule, within project budgets, and satisfies technical objectives.
Why it matters Failure to adequately assess and monitor performance against the plan prevents visibility of progress and, in consequence, appropriate corrective actions may not be identified and/or taken when project performance deviates from that required.
Possible contributory types of evidence Any combination of the types of evidence may be acceptable (depending on how the Framework is tailored and used). The evidence items identified at each level indicate example work products only. Contributions to work products will generally differ at each proficiency level.
Learning and development The INCOSE Professional Development Portal provides example guidance on how to gain an initial awareness of a competency area and options for developing further competence thereafter.

Awareness – Management: Monitoring and Control		
ID	**Indicators of Competence**	**Relevant knowledge sub-indicators**
1	Explains why monitoring and controlling Systems Engineering activities is important. [MMCA01]	(K) Explains why monitoring and control is required to provide an understanding of the project/program's progress so that appropriate corrective actions can be taken when progress deviates from the plan. [MMCA01-10K] (K) Explains why failure to monitor and control significantly increases risk and will probably lead to schedule and cost overrun. [MMCA01-20K] (K) Describes the relationship between monitoring and control and the systems engineering management plan, the project/program management plan, and project/program estimates. [MMCA01-30K]
2	Explains how Systems Engineering monitoring and control fits within the wider execution and control of a project. [MMCA02]	(K) Explains how planning activities support monitoring and control (e.g. a project plan defines the required infrastructure, resources and parameters required to monitor and control the project; monitoring and control may require a plan to be updated). [MMCA02-10K] (K) Describes the relationship between the Systems Engineering monitoring and control and the project/program monitoring and control. [MMCA02-20K] (K) Describes key interfaces and links between engineering monitoring and control and stakeholders in the wider enterprise outside engineering (e.g. infrastructure management, human resource management, quality management, knowledge management, portfolio management, and life cycle model management). [MMCA02-30K] (K) Explains how Systems Engineering defines and monitors the performance metrics used to track progress. [MMCA02-40K]

(Continued)

ID	Indicators of Competence	Relevant knowledge sub-indicators
		(K) Explains how SE monitoring and control is coordinated with other activities across the wider project and wider organization. [MMCA02-50K] (K) Explains how SE monitoring and control links to organizational knowledge repository (e.g. lessons learned). [MMCA02-60K] (K) Explains the relationship between a Work Breakdown Structure and the monitoring and control activity. [MMCA02-70K]
3	Explains the purpose of reviews and decision gates and their relationship to the monitoring and control of Systems Engineering tasks. [MMCA03]	(K) Identifies different types of technical and non-technical reviews explaining their primary purpose, key outputs, and entry and exit criteria. [MMCA03-10K] (K) Identifies commonly used decision gates and describes their primary purpose, key outputs and entry and exit criteria. [MMCA03-20K] (K) Explains the difference between a milestone and a review and what each is used for. [MMCA03-30K] (K) Describes a typical life cycle stage review activity. [MMCA03-40K]
4	Explains how Systems Engineering metrics and measures contribute to monitoring and controlling Systems Engineering on a project. [MMCA04]	(K) Lists examples of good and bad measures and metrics, including rationale for why they are good or bad. [MMCA04-10K] (K) Defines a "measure of effectiveness," and how it relates to "measures of performance," "technical performance measures," and "key performance parameters." [MMCA04-20K] (K) Explains how metrics and measures contribute to monitoring and controlling Systems Engineering on a project. [MMCA04-30K] (K) Explains how to select and define measurable and actionable technical performance measures. [MMCA04-40K] (K) Explains how and when technical performance measures are tracked and how this links with other activities (e.g. design, (re-)planning). [MMCA04-50K] (K) Explains why there is a need to track performance against utilizing the selected performance parameters. [MMCA04-60K]
5	Explains how communications support the successful monitoring and control of a Systems Engineering project. [MMCA05]	(K) Explains the relationship between communications activities and Systems Engineering monitoring and control. [MMCA05-10K] (K) Explains how a common understanding of the current state of a program achieved through good monitoring and control helps entire success. [MMCA05-20K] (K) Explains how good communication of Systems Engineering status as a result of monitoring activities helps ensure convergence in program evolution. [MMCA05-30K]

Supervised Practitioner – Management: Monitoring and Control			
ID	**Indicators of Competence** ***(in addition to those at Awareness level)***	**Relevant knowledge, experience, and/or behaviors**	**Possible examples of objective evidence of personal contribution to activities performed, or professional behaviors applied**
1	Follows a defined governing process, using appropriate tools, to guide their Systems Engineering monitoring and control activities. [MMCS01]	(K) Identifies processes and tools used in support of monitoring and control activities. [MMCS01-10K]	Monitoring and control processes and tools they used on a given project. [MMCS01-E10]
		(A) Follows Systems Engineering processes, applying appropriate tools, in order to monitor and control activities. [MMCS01-20A]	Monitoring and control outputs from Systems Engineering activities they performed or used on a project. [MMCS01-E20]
2	Records technical data identified as requiring monitoring or control in plans to facilitate analysis. [MMCS02]	(A) Records technical data items in support of progress and performance monitoring over time. [MMCS02-10A]	Data to support technical, performance and programmatic monitoring. [MMCS02-E10]
		(A) Records design margins over time. [MMCS02-20A]	Design margin data. [MMCS02-E20]
		(A) Monitors progress against a project's baseline schedule over time, coordinating inputs from task owners. [MMCS02-30A]	Schedule and technical baseline monitoring data. [MMCS02-E30]
		(A) Prepares reports for monitored parameters. [MMCS02-40A]	Reports containing technical parameters. [MMCS02-E40]
		(A) Prepares review data in accordance with review requirements. [MMCS02-50A]	Review data inputs. [MMCS02-E50]
3	Monitors key data parameters against expectations to determine deviations. [MMCS03]	(A) Analyzes monitored data continually, highlighting trends early in order give sufficient time to act. [MMCS03-10A]	Technical performance measures assessments they have performed. [MMCS03-E10]
		(A) Monitors technical performance parameters through the design cycle changing as necessary if another element becomes key. [MMCS03-20A]	Technical parameters they have assessed for gate reviews and contract status reports. [MMCS03-E20]
			Project performance measures they have assessed. [MMCS03-E30]
			Technical trends and parameters they have analyzed. [MMCS03-E40]
4	Identifies potential corrective actions to control and correct deviations from expectations. [MMCS04]	(A) Communicates concerns, issues or trends promptly in order maximize time to take action. [MMCS04-10A]	Reports they generated which describe potential or actual corrective actions. [MMCS04-E10]
		(A) Identifies potential remedial actions when parameters show deviations from plan. [MMCS04-20A]	Records showing changes they made to monitored parameters as a result of corrective actions. [MMCS04-E20]

(Continued)

ID	Indicators of Competence *(in addition to those at Awareness level)*	Relevant knowledge, experience, and/or behaviors	Possible examples of objective evidence of personal contribution to activities performed, or professional behaviors applied
		(A) Identifies technical performance indicator updates as necessary to reflect changing design priorities. [MMCS04-30A] (A) Identifies tracked parameter updates, which will monitor and control corrective actions. [MMCS04-40A] (A) Coordinates updates across project to ensure success. [MMCS04-50A] (A) Uses information provided by others to implement corrective activities. [MMCS04-60A]	Change review material they created. [MMCS04-E30]
5	Monitors technical margins both horizontally and vertically through the project hierarchy and over time, to control and monitor design evolution. [MMCS05]	(A) Analyzes technical performance parameters throughout the design cycle updating as necessary if another element becomes key. [MMCS05-10A] (A) Monitors technical performance parameter allocations through hierarchy. [MMCS05-20A]	Data they have collected and analyzed relating to margin tracking. [MMCS05-E10]
6	Identifies development lessons learned performing monitoring and control to inform future projects. [MMCS06]	(A) Records personal lessons learned. [MMCS06-10A] (A) Communicates lessons learned to others. [MMCS06-20A]	Logbook, engineering notebook (or similar) showing personal lessons they learned. [MMCS06-E10] Memorandums written to document lessons learned or blogs kept in an electronic repository for others to see. [MMCS06-E20]
7	Develops own understanding of this competency area through Continual Professional Development (CPD). [MMCS07]	(A) Identifies potential gaps in own knowledge or development needs in this area, identifying opportunities to address these through continual professional development activities. [MMCS07-10A] (A) Performs continual professional development activities to improve their knowledge and understanding in this area. [MMCS07-20A] (A) Records continual professional development activities undertaken including learning or insights gained. [MMCS07-30A]	Records of Continual Professional Development (CPD) performed and learning outcomes. [MMCS07-E10]

Practitioner – Management: Monitoring and Control			
ID	**Indicators of Competence** ***(in addition to those at Supervised Practitioner level)***	**Relevant knowledge, experience, and/or behaviors**	**Possible examples of objective evidence of personal contribution to activities performed, or professional behaviors applied**
1	Creates a strategy for Monitoring and Control on a project to support SE project and wider enterprise needs. [MMCP01]	(A) Identifies project-specific System Monitoring and Control needs that need to be incorporated into SE planning. [MMCP01-10A] (A) Identifies project-specific system engineering technical risk and opportunity management needs that need to be incorporated into SE planning. [MMCP01-20A] (A) Prepares project-specific System Monitoring and Control inputs for SE planning purposes, using appropriate processes and procedures. [MMCP01-30A] (A) Prepares system engineering technical risk and opportunity management inputs for SE planning purposes, using appropriate processes and procedures. [MMCP01-40A] (A) Prepares project-specific System Monitoring and Control task estimates in support of SE planning. [MMCP01-50A] (A) Prepares project-specific System Monitoring and Control task estimates in support of SE planning. [MMCP01-60A]	SE System Monitoring and Control work products. [MMCP01-E10] SE System engineering technical risk and opportunity management work products. [MMCP01-E20]
2	Creates a governing process, plan, and associated tools for systems decision management, which reflect project and business strategy. [MMCP02]	(K) Identifies the steps necessary to define a process and appropriate techniques to be adopted for the monitoring and control of system elements. [MMCP02-10K] (K) Describes key elements of a successful systems monitoring and control process and how systems monitoring and control activities relate to different stages of a system life cycle. [MMCP02-20K] (A) Uses system monitoring and control tools or uses a methodology. [MMCP02-30A] (A) Uses appropriate standards and governing processes in their system monitoring and control planning, tailoring as appropriate. [MMCP02-40A] (A) Selects system monitoring and control method for each system element. [MMCP02-50A] (A) Creates system monitoring and control plans and processes for use on a project. [MMCP02-60A]	SE system monitoring and control planning work products they have developed. [MMCP02-E10] Planning documents such as a systems engineering management plan, other project/program plan or organizational process they have written. [MMCP02-E20] Monitoring and control plan/schedule. [MMCP02-E30] Monitoring and control matrix. [MMCP02-E40]

(Continued)

ID	Indicators of Competence *(in addition to those at Supervised Practitioner level)*	Relevant knowledge, experience, and/or behaviors	Possible examples of objective evidence of personal contribution to activities performed, or professional behaviors applied
3	Complies with governing plans and processes for system monitoring and control, interpreting, evolving, or seeking guidance where appropriate. [MMCP03]	(A) Follows defined plans, processes and associated tools to perform System Monitoring and Control on a project. [MMCP03-10A]	System monitoring and control plans or work products they have interpreted or challenged to address project issues. [MMCP03-E10]
		(P) Guides and actively coordinates the interpretation of System Monitoring and Control plans or processes in order to complete System Monitoring and Control activities successfully. [MMCP03-20P]	System monitoring and control tools or methodologies used (e.g. a requirements verification matrix). [MMCP03-E20]
		(P) Recognizes situations where deviation from published plans and processes or clarification from others is necessary to overcome complex challenges. [MMCP03-30P]	
		(P) Recognizes situations where a formal update is preferred, liaising with lead engineers stakeholders as necessary to implement successfully. [MMCP03-40P]	
4	Monitors Systems Engineering activities in order to determine and report progress against estimates and plans on a project. [MMCP04]	(A) Monitors Systems Engineering performance against estimates during overall project execution to determine progress. [MMCP04-10A]	Project progress reports and presentations. [MMCP04-E10]
		(A) Analyzes Systems Engineering performance to determine or predict deviations from plans. [MMCP04-20A]	Trend analyses and parameter tracking database/plot examples. [MMCP04-E20]
		(A) Communicates with other disciplines to ensure errant Systems Engineering activities are corrected as necessary during overall project execution. [MMCP04-30A]	
		(A) Communicates measurements and assessments as necessary during overall project execution. [MMCP04-40A]	
		(P) Collaborates with other disciplines regarding estimates and resources to maintain baselines and to effect desired outcomes. [MMCP04-50P]	
5	Monitors Systems Engineering activities by processing measurement data in order to determine deviations or trends against plans. [MMCP05]	(A) Monitors Systems Engineering activities by processing measurement data in order to determine deviations or trends against plans. [MMCP05-10A]	Project data tracking information that they generated and analyzed. [MMCP05-E10]
6	Analyzes measurement and assessment data to determine and implement necessary remedial corrective actions in order to control SE activities. [MMCP06]	(A) Defines appropriate project/program performance measures and sets threshold limits for expected values. [MMCP06-10A]	Project/program quantitative management plan. [MMCP06-E10]
		(A) Analyzes project/program performance measures, defining corrective actions and tracking to closure when actual values deviate from those expected. [MMCP06-20A]	Project/program performance measures (run charts, control charts, Pareto analysis, root cause analysis, etc.). [MMCP06-E20]

		(A) Uses historic, quantitative data from past projects/ programs to predict current project/program performance. [MMCP06-30A]	Records showing project/program tracking, e.g. updated Gantt chart and updated plans. [MMCP06-E30]
7	Prepares recommendations for updates to existing monitoring and control plans to address internal or external changes. [MMCP07]	(A) Identifies required Systems Engineering monitoring and control updates as a result of internal or extremal changes. [MMCP07-10A] (A) Prepares updates to existing Systems Engineering monitoring and control determining revised estimates of effort and cost, statements of work, requirements, and constraints as necessary. [MMCP07-20A]	Systems Engineering Monitoring and Control plan (or similar) they updated on a specific project. [MMCP07-E10] Systems Engineering monitoring and control estimates they updated on a specific project. [MMCP07-E20] Change review material they created in this area. [MMCP07-E30]
8	Reviews technical margins both horizontally and vertically through the project hierarchy to maintain overall required margins. [MMCP08]	(A) Identifies and reviews parameter and performance budgets. [MMCP08-10A] (P) Collaborates with engineering and program management to ensure suppliers to control and manage system element margins. [MMCP08-20P] (A) Records issues detected in margin management. [MMCP08-30A]	Collaborative agreements. [MMCP08-E10] Budget allocation tables with margin. [MMCP08-E20]
9	Guides new or supervised practitioners in Systems Engineering monitoring and control in order to develop their knowledge, abilities, skills, or associated behaviors. [MMCP09]	(P) Guides new or supervised practitioners in executing activities that form part of this competency. [MMCP09-10P] (A) Trains individuals to an "Awareness" level in this competency area. [MMCP09-20A]	Organizational Breakdown Structure showing their responsibility for technical supervision in this area. [MMCP09-E10] On-the-job training records. [MMCP09-E20] Coaching or mentoring assignment records. [MMCP09-E30] Records highlighting their impact on another individual in terms of improvement or professional development in this competency. [MMCP09-E40]
10	Maintains and enhances own competence in this area through Continual Professional Development (CPD) activities. [MMCP10]	(A) Identifies potential development needs in this area, identifying opportunities to address these through continual professional development activities. [MMCP10-10A] (A) Performs continual professional development activities to maintain and enhance their competency in this area. [MMCP10-20A] (A) Records continual professional development activities undertaken including learning or insights gained. [MMCP10-30A]	Records of Continual Professional Development (CPD) performed and learning outcomes. [MMCP10-E10]

(Continued)

Lead Practitioner – Management: Monitoring and Control			
ID	**Indicators of Competence** ***(in addition to those at Practitioner level)***	**Relevant knowledge, experience, and/or behaviors**	**Possible examples of objective evidence of personal contribution to activities performed, or professional behaviors applied**
1	Creates enterprise-level policies, procedures, guidance, and best practice for Systems Engineering monitoring and control, including associated tools to improve organizational effectiveness. [MMCL01]	(A) Identifies the need for enterprise-level documents (e.g. policies, procedures, guidance, and best practice) or tools supporting Systems Engineering planning. [MMCL01-10A]	Systems Engineering monitoring and control policies, processes, best practice guidance they created, which have been adopted by the enterprise. [MMCL01-E10]
		(A) Produces enterprise-level document(s) or selects appropriate enterprise-level] tool(s) in support of Systems Engineering planning, taking into account current and future enterprise needs. [MMCL01-20A]	Organization charts or documents recognizing their authority in Systems Engineering monitoring and control at enterprise level. [MMCL01-E20]
		(P) Collaborates throughout with enterprise-level stakeholders to gain approval, updating documents or tool requirements as necessary. [MMCL01-30P]	Organization charts or documents recognizing their authority for defining Systems Engineering monitoring and control documentation across the enterprise. [MMCL01-E30]
2	Assesses tailoring of enterprise-level Systems Engineering monitoring and control processes to balance the needs of the project and business. [MMCL02]	(A) Reviews Systems Engineering monitoring and control processes against enterprise tailoring guidance. [MMCL02-10A]	Development monitoring or control reviews they attended or review comments they recorded. [MMCL02-E10]
		(A) Identifies critical parameters, measurement methods, establishes control values, determines process for managing inconsistencies. [MMCL02-20A]	Measurement items or control strategies they have identified or judged. [MMCL02-E20]
		(A) Identifies constructive feedback on tailoring performed. [MMCL02-30A]	Tailored processes they have authorized. [MMCL02-E30]
		(A) Authorizes tailoring within Systems Engineering monitoring and control on behalf of enterprise. [MMCL02-40A]	
3	Assesses ongoing Systems Engineering projects at enterprise-level to ensure they are being monitored and controlled successfully. [MMCL03]	(A) Reviews projects or enterprise proposals during their development, acting as an independent internal expert. [MMCL03-10A]	Independent reviews or panels they have participated in. [MMCL03-E10]
		(A) Reviews SE monitoring and control information against business and external stakeholder (e.g. customer) needs. [MMCL03-20A]	Reviews and proposals they have supported at enterprise level. [MMCL03-E20]
		(P) Acts as independent reviewer of monitoring and control of high-value or high-risk program Systems Engineering. [MMCL03-30P]	

		(A) Identifies constructive feedback on SE monitoring and control strategies, processes, tools or approaches. [MMCL03-40A] (A) Identifies updates to project as required based upon review outcome. [MMCL03-50A] (A) Maintains enterprise Systems Engineering monitoring and control practices or documentation to reflect upon lessons learned. [MMCL03-60A] (A) Uses predictive measures to keep a project/program on track. [MMCL03-70A] (A) Authorizes project Systems Engineering monitoring and control documents on behalf of enterprise. [MMCL03-80A]	
4	Judges the suitability of management and trading of design technical margins to satisfy the needs of the project, on ongoing projects across the enterprise. [MMCL04]	(A) Reviews ongoing project technical margins both horizontally and vertically through the project hierarchy to ensure they meet the needs of the project and business. [MMCL04-10A] (A) Identifies design technical trades and margins updates as required based upon review outcome. [MMCL04-20A] (A) Authorizes project Systems Engineering margins or trade son behalf of enterprise. [MMCL04-30A]	Multiple monitoring and control documents they have authorized or reviewed. [MMCL04-E10] Review comments they have recorded on SE monitoring and control. [MMCL04-E20] SE monitoring and control work products they have authorized on behalf of the enterprise. [MMCL04-E30]
5	Assesses proposals for preventative or remedial actions when assessment indicates a trend toward deviation on multiple distinct projects across the enterprise or a complex project. [MMCL05]	(A) Reviews projects and proposals during their development. [MMCL05-10A] (A) Reviews SE monitoring and control information against business and customer needs. [MMCL05-20A] (P) Acts as independent reviewer of monitoring and control of high-value or high-risk program Systems Engineering. [MMCL05-30P] (A) Identifies constructive feedback on SE monitoring and control strategies, processes, tools or approaches. [MMCL05-40A] (A) Authorizes project Systems Engineering monitoring and control documents on behalf of enterprise. [MMCL05-50A] (A) Reviews large scale efforts that span the enterprise resources. [MMCL05-60A]	Records of panel participation/contract reviews/ review reports. [MMCL05-E10]

(Continued)

ID	Indicators of Competence *(in addition to those at Practitioner level)*	Relevant knowledge, experience, and/or behaviors	Possible examples of objective evidence of personal contribution to activities performed, or professional behaviors applied
6	Analyzes monitoring and control data from multiple diverse projects or a complex system to provide enterprise level coordination of SE. [MMCL06]	(A) Produces Systems Engineering monitoring and control work products for multiple diverse projects or a single complex project. [MMCL06-10A]	Multiple monitoring and control documents they have written or processes they have used. [MMCL06-E10]
		(A) Adapts monitoring and control approaches based upon project or enterprise context or need. [MMCL06-20A]	Project/program performance measures they have used successfully or unsuccessfully (and why). [MMCL06-E20]
		(A) Analyzes monitoring and control data inputs to determine enterprise or complex system trends. [MMCL06-30A]	Records showing tracked performance measures (e.g. defect containment by phase and Schedule performance index) they have created. [MMCL06-E30]
		(A) Identifies updates to project Systems Engineering approaches as required to improve technical and project coordination or performance across the enterprise. [MMCL06-40A]	Lessons they have learned and documented at enterprise level or personally in this area. [MMCL06-E40]
			Monitoring or other measurement status reports they created. [MMCL06-E50]
7	Persuades key stakeholders to address identified enterprise-level Systems Engineering monitoring and control issues to reduce project cost, schedule, or technical risk. [MMCL07]	(A) Identifies issues affecting wider enterprise from project monitoring and control data. [MMCL07-10A]	Issues resolved due to their personal intervention. [MMCL07-E10]
		(P) Fosters support from enterprise stakeholders in resolving identified issues. [MMCL07-20P]	Enterprise expert panel/process improvement/core competency in which they have participated. [MMCL07-E20]
		(P) Collaborates with enterprise stakeholders to implement or tailor Systems Engineering principles across the enterprise within the context of the programs at hand. [MMCL07-30P]	
8	Coaches or mentors practitioners across the enterprise in Systems Engineering monitoring and control in order to develop their knowledge, abilities, skills, or associated behaviors. [MMCL08]	(P) Coaches or mentors practitioners across the enterprise in competency-related techniques, recommending development activities. [MMCL08-10P]	Coaching or mentoring assignment records. [MMCL08-E10]
		(A) Develops or authorizes enterprise training materials in this competency area. [MMCL08-20A]	Records of formal training courses, workshops, seminars, and authored training material supported by successful post-training evaluation data. [MMCL08-E20]
		(A) Provides enterprise workshops/seminars or training in this competency area. [MMCL08-30A]	Listing as an approved organizational trainer for this competency area. [MMCL08-E30]

9	Promotes the introduction and use of novel techniques and ideas in monitoring and control of Systems Engineering, across the enterprise, to improve enterprise competence in this area. [MMCL09]	(A) Analyzes different approaches across different domains through research. [MMCL09-10A]	Research records. [MMCL09-E10]
		(A) Defines novel approaches that could potentially improve the SE discipline within the enterprise. [MMCL09-20A]	Published papers in refereed journals/company literature. [MMCL09-E20]
		(P) Fosters awareness of these novel techniques within the enterprise. [MMCL09-30P]	Records showing introduction of enabling systems supporting the new techniques or ideas. [MMCL09-E30]
		(P) Collaborates with enterprise stakeholders to facilitate the introduction of techniques new to the enterprise. [MMCL09-40P]	Published papers (or similar) at enterprise level. [MMCL09-E40]
		(A) Monitors new techniques after their introduction to determine their effectiveness. [MMCL09-50A]	Records of improvements made against a recognized process improvement model in this area. [MMCL09-E50]
		(A) Adapts approach to reflect actual enterprise performance improvements. [MMCL09-60A]	
10	Develops expertise in this competency area through specialist Continual Professional Development (CPD) activities. [MMCL10]	(A) Identifies own needs for further professional development in order to increase competence beyond practitioner level. [MMCL10-10A]	Records of Continual Professional Development (CPD) performed and learning outcomes. [MMCL10-E10]
		(A) Performs professional development activities in order to move own competence toward expert level. [MMCL10-20A]	
		(A) Records continual professional development activities undertaken including learning or insights gained. [MMCL10-30A]	

NOTES	In addition to items above, enterprise-level or independent 3rd-Party-generated evidence may be used to amplify other evidence presented and may include: a. Formally recognized by senior management in current organization as an expert in this competency area b. Evidence of role as Product/System Design Authority or Technical Authority on a complex project with responsibilities in this area or where skills within this competency area were used c. Recognized as an authorizing signatory on behalf of enterprise for formal documentation in this competency area (e.g. policies, processes, and deliverables) d. Formal commendation or award within own enterprise for contribution or item of work successfully performed, which required proficiency in this competency area e. Customer, Supplier, or other external project-specific key Stakeholder accolades for specific work performed in this competency area f. Independently assessed or accredited work in this competency area (e.g. for independent publication or use) g. Formal organizational HR records positively highlighting any specific professional competencies or behaviors identified (if applicable) plus any of the evidence indicators listed at Expert level below

Expert – Management: Monitoring and Control			
ID	**Indicators of Competence *(in addition to those at Lead Practitioner level)***	**Relevant knowledge, experience, and/or behaviors**	**Possible examples of objective evidence of personal contribution to activities performed, or professional behaviors applied**
1	Communicates own knowledge and experience in Systems Engineering monitoring and control in order to improve best practice beyond the enterprise boundary. [MMCE01]	(A) Produces papers, seminars, or presentations outside own enterprise for publication in order to share own ideas and improve industry best practices in this competence area. [MMCE01-10A]	Published papers or books etc. on new technique in refereed journals/company literature. [MMCE01-E10]
		(P) Fosters incorporation of own ideas into industry best practices in this area. [MMCE01-20P]	Published papers in refereed journals or internal literature proposing new practices in this competence area (or presentations, tutorials, etc.). [MMCE01-E20]
		(P) Develops guidance materials identifying new (or updating existing) best practice in this competence area. [MMCE01-30P]	Own proposals adopted as industry best practices in this competence area. [MMCE01-E30]
2	Advises organizations beyond the enterprise boundary on the handling of complex or sensitive Systems Engineering monitoring and control issues. [MMCE02]	(A) Provides independent advice to SE stakeholders beyond the enterprise boundary on complex or sensitive SE monitoring and control issues. [MMCE02-10A]	Records of advice provided. [MMCE02-E10]
		(P) Arbitrates on complex or sensitive Systems Engineering monitoring and control issues beyond the enterprise boundary. [MMCE02-20P]	Records indicating changed stakeholder thinking from personal intervention. [MMCE02-E20]
		(P) Persuades stakeholders beyond the enterprise boundary. to accept difficult project-specific SE monitoring and control recommendations or actions. [MMCE02-30P]	
3	Advises organizations beyond the enterprise boundary on the suitability of their approach to Systems Engineering monitoring and control. [MMCE03]	(A) Provides independent advice to stakeholders beyond enterprise boundary regarding their SE monitoring and control activities. [MMCE03-10A]	Records of advice provided. [MMCE03-E10]
		(P) Acts with sensitivity on highly complex system monitoring and control issues making limited use of specialized, technical terminology and taking account of external stakeholders' (e.g. customer's) background and knowledge. [MMCE03-20P]	Records indicating changed stakeholder thinking from personal intervention. [MMCE03-E20]

		(P) Communicates using a holistic approach to complex issue resolution including balanced, rational arguments on way forward. [MMCE03-30P]	
		(P) Arbitrates on Systems Engineering monitoring and control decisions on behalf of stakeholders beyond enterprise boundary. [MMCE03-40P]	
		(P) Persuades stakeholders beyond the enterprise boundary to accept difficult project-specific SE monitoring and control recommendations or actions. [MMCE03-50P]	
4	Champions the introduction of novel techniques and ideas in Systems Engineering monitoring and control, beyond the enterprise boundary, in order to develop the wider Systems Engineering community in this competency. [MMCE04]	(A) Analyzes different approaches across different domains through research. [MMCE04-10A]	Records of activities promoting research and need to adopt novel technique or ideas. [MMCE04-E10]
		(A) Produces reports for the wider SE community on the effectiveness of new techniques after their introduction. [MMCE04-20A]	Records of improvements made to process and appraisal against a recognized process improvement model. [MMCE04-E20]
		(P) Collaborates with those introducing novel techniques within the wider SE community. [MMCE04-30P]	Research records. [MMCE04-E30]
		(A) Defines novel approaches that could potentially improve the wider SE discipline. [MMCE04-40A]	Published papers in refereed journals/company literature. [MMCE04-E40]
		(P) Fosters awareness of these novel techniques within the wider SE community. [MMCE04-50P]	Records showing introduction of enabling systems supporting the new techniques or ideas. [MMCE04-E50]
5	Coaches individuals beyond the enterprise boundary in Systems Engineering monitoring and control, in order to further develop their knowledge, abilities, skills, or associated behaviors. [MMCE05]	(P) Coaches or mentors individuals beyond the enterprise boundary, in competency-related techniques, recommending development activities. [MMCE05-10P]	Coaching or mentoring assignment records. [MMCE05-E10]
		(A) Develops or authorizes training materials in this competency area, which are subsequently successfully delivered beyond the enterprise boundary. [MMCE05-20A]	Records of formal training courses, workshops, seminars, and authored training material supported by successful post-training evaluation data. [MMCE05-E20]
		(A) Provides workshops/seminars or training in this competency area for practitioners or lead practitioners beyond the enterprise boundary (e.g. conferences and open training days). [MMCE05-30A]	Records of Training/workshops/seminars created supported by successful post-training evaluation data. [MMCE05-E30]

(Continued)

ID	Indicators of Competence *(in addition to those at Lead Practitioner level)*	Relevant knowledge, experience, and/or behaviors	Possible examples of objective evidence of personal contribution to activities performed, or professional behaviors applied
6	Maintains expertise in this competency area through specialist Continual Professional Development (CPD) activities. [MMCE06]	(A) Reviews research, new ideas, and state of the art to identify relevant new areas requiring personal development in order to maintain expertise in this competency area. [MMCE06-10A] (A) Performs identified specialist professional development activities in order to maintain or further develop competence at expert level. [MMCE06-20A] (A) Records continual professional development activities undertaken including learning or insights gained. [MMCE06-30A]	Records of documents reviewed and insights gained as part of own research into this competency area. [MMCE06-E10] Records of Continual Professional Development (CPD) performed and learning outcomes. [MMCE06-E20]

NOTES	In addition to items above, enterprise-level or independent 3rd-Party-generated evidence may be used to amplify other evidence presented and may include: a. Formally recognized by a reputable external organization as an expert in this competency area b. Evidence of role as independent assessor or reviewer on project outside own organization where skills in this competency area were used c. Evidence of invitation(s) from wider community for contribution of systems engineering expertise in this area (e.g. industry conference panel, government advisory board etc. cross-industry working groups, partnerships, accredited advanced university courses or research, or as part of professional institute) d. Formal commendation beyond the enterprise (e.g. by INCOSE or other recognized authority) for work performed in this competency area e. Independently assessed or accredited work product in this competency area (e.g. for independent publication or use) f. Accolades of expertise in this area from recognized industry leaders

Competency area – Management: Risk and Opportunity Management
Description Risk is an uncertain event or condition that, if it occurs, has a positive or negative effect on project or enterprise objectives. The purpose of risk and opportunity management is to reduce potential risks to an acceptable level before they occur, or maximize the potential of any opportunity, throughout the life of the project. Risk and opportunity management is a continuous, forward-looking process that is applied to anticipate and avert risks that may adversely impact the project and can be considered both a project management and a Systems Engineering activity.
Why it matters Every new system (or existing system modification) has inherent risk but is also based upon the pursuit of an opportunity. Risk and opportunity are both present throughout the life cycle of systems and the primary objective of managing these areas as part of Systems Engineering activities is to balance the allocation of resources to achieve greatest risk mitigation (or opportunity benefits).
Possible contributory types of evidence
Any combination of the types of evidence may be acceptable (depending on how the Framework is tailored and used). The evidence items identified at each level indicate example work products only. Contributions to work products will generally differ at each proficiency level.
Learning and development The INCOSE Professional Development Portal provides example guidance on how to gain an initial awareness of a competency area and options for developing further competence thereafter.

Awareness – Management: Risk and Opportunity Management		
ID	**Indicators of Competence**	**Relevant knowledge sub-indicators**
1	Describes the distinction between risk, issue, and opportunity, and can provide examples of each. [MROA01]	(K) Describes the concept of risk (e.g. terms of likelihood of adverse future event) issues are problems that have already occurred. [MROA01-10K]
		(K) Describes the difference between a risk and an issue. [MROA01-20K]
		(K) Explains why "opportunities" also need to be monitored and how they can be used to gain capability. [MROA01-30K]
2	Identifies key factors associated with good risk management and why these factors are important. [MROA02]	(K) Identifies key factors associated with good risk management (e.g. planning, management, monitoring, analyzing, and reporting). [MROA02-10K]
		(K) Describes each of the key factors (e.g. planning, management, monitoring, analyzing, and reporting). [MROA02-20K]
		(K) Describes key interfaces and links between engineering risk and opportunity management and stakeholders in the wider enterprise outside engineering (e.g. infrastructure management, human resource management, quality management, knowledge management, portfolio management, and life cycle model management). [MROA02-30K]
		(K) Explains how Configuration control boards may be used to report risk management to stakeholders. [MROA02-40K]

(Continued)

ID	Indicators of Competence	Relevant knowledge sub-indicators
3	Identifies different classes of risk and can provide examples of each. [MROA03]	(K) Describes key characteristics of Technical risks. [MROA03-10K] (K) Describes key characteristics of Technical performance risks. [MROA03-20K] (K) Describes key characteristics of schedule/cost risks. [MROA03-30K] (K) Describes key characteristics of Programmatic risks. [MROA03-40K]
4	Identifies different types of risk treatment available and can provide examples of each. [MROA04]	(K) Describes key characteristics of different risk management treatments, such as acceptance, avoidance, mitigation, transfer, and ignore. [MROA04-10K]
5	Identifies different types of opportunity and can provide examples of each. [MROA05]	(K) Describes the key characteristics of schedule, cost, technical, programmatic opportunities. [MROA05-10K]
6	Describes a typical high-level process for risk and opportunity management. [MROA06]	(K) Identifies key components of a Risk management plan, and how tailoring might affect plan content. [MROA06-10K] (K) Identifies how tailoring can affect risk both positively and negatively. [MROA06-20K] (K) Describes different risk level taxonomies and typical guidance on assigning risk values. [MROA06-30K] (K) Explains why risk mitigation is integral to the project management plan, schedule, and budget. [MROA06-40K]
7	Explains how risk is typically assessed and can provide examples. [MROA07]	(K) Describes techniques for assessing the likelihood of a risk, including their strengths and weaknesses. [MROA07-10K] (K) Explains how risk impact is calculated. [MROA07-20K] (K) Explains why risk assessment and status checking needs to be conducted regularly. [MROA07-30K] (K) Describes the purpose and typical membership of a Risk Reviews Board. [MROA07-40K]

Supervised Practitioner – Management: Risk and Opportunity Management			
ID	**Indicators of Competence** ***(in addition to those at Awareness level)***	**Relevant knowledge, experience, and/or behaviors**	**Possible examples of objective evidence of personal contribution to activities performed, or professional behaviors applied**
1	Follows a governing process and appropriate tools to plan and control their own risk and opportunity management activities. [MROS01]	(A) Identifies governing risk management processes and tools used. [MROS01-10A]	Contributions to Risk management plans. [MROS01-E10]
		(A) Uses processes and tools to plan and control own risk activities. [MROS01-20A]	Outputs from risk and mitigation database tools they used on a project. [MROS01-E20]
		(K) Describes the stages of risk management on a project. [MROS01-30K]	Outputs from Risk management process reviews reports and activities in CCBs they supported. [MROS01-E30]
2	Identifies potential risks and opportunities on a project. [MROS02]	(A) Identifies and records risks or opportunities on a project. [MROS02-10A]	
		(A) Prepares information supporting the assessment or analysis of risks or opportunities identified on a project. [MROS02-20A]	Records of contributions to risk trend analysis reports. [MROS02-E20]
		(A) Prepares information inputs for risk review boards. [MROS02-30A]	
		(A) Prepares risk and opportunity reports based on collected project status information and interviews. [MROS02-40A]	
		(A) Prepares information identifying the treatment strategies for identified risks. [MROS02-50A]	
3	Identifies action plans to treat risks and opportunities on a project. [MROS03]	(A) Prepares inputs to the analysis of treatment strategies for identified risks and opportunities. [MROS03-10A]	Records of contributions to Risk Management data used to develop mitigation strategy reports. [MROS03-E10]
		(A) Prepares inputs in support of the deployment of mitigation strategies for identified risks. [MROS03-20A]	
		(A) Prepares inputs to the risk monitoring activities to risk closure, including technical and program parameters. [MROS03-30A]	
		(A) Prepares inputs in support of identification of treatment strategies for identified risks. [MROS03-40A]	

(Continued)

ID	Indicators of Competence *(in addition to those at Awareness level)*	Relevant knowledge, experience, and/or behaviors	Possible examples of objective evidence of personal contribution to activities performed, or professional behaviors applied
4	Develops own understanding of this competency area through Continual Professional Development (CPD). [MROS04]	(A) Identifies potential gaps in own knowledge or development needs in this area, identifying opportunities to address these through continual professional development activities. [MROS04-10A] (A) Performs continual professional development activities to improve their knowledge and understanding in this area. [MROS04-20A] (A) Records continual professional development activities undertaken including learning or insights gained. [MROS04-30A]	Records of Continual Professional Development (CPD) performed and learning outcomes. [MROS04-E10]

Practitioner – Management: Risk and Opportunity Management			
ID	**Indicators of Competence** ***(in addition to those at Supervised Practitioner level)***	**Relevant knowledge, experience, and/or behaviors**	**Possible examples of objective evidence of personal contribution to activities performed, or professional behaviors applied**
1	Creates a strategy for risk and opportunity management on a project to support SE project and wider enterprise needs. [MROP01]	(A) Identifies project-specific risk and opportunity management needs that need to be incorporated into SE planning. [MROP01-10A]	SE risk and opportunity management work products. [MROP01-E10]
		(A) Prepares project-specific risk and opportunity management inputs for SE planning purposes, using appropriate processes and procedures. [MROP01-20A]	Risk Management Plans authored. [MROP01-E20]
		(A) Prepares project-specific risk and opportunity management task estimates in support of SE planning. [MROP01-30A]	
2	Creates a governing process, plan, and associated tools for risk and opportunity management, which reflect project and business strategy. [MROP02]	(A) Creates risk management plans based on enterprise policies, external stakeholder (e.g. customer) needs, and industry best practices. [MROP02-10A]	Risk Management Plan (RMP) created. [MROP02-E10]
		(A) Uses risk and opportunity tools or uses a methodology. [MROP02-20A]	Reports from an enterprise tracking tool for risk and opportunity management. [MROP02-E20]
		(A) Uses appropriate standards and governing processes in their system risk and opportunity planning, tailoring as appropriate. [MROP02-30A]	Contributes to risk reports. [MROP02-E30]
		(A) Selects system risk and opportunity methods for each system element. [MROP02-40A]	Uses (or develops) a risk tracking tool (e.g. spreadsheet) to prepare risk data on a project. [MROP02-E40]
		(A) Creates system risk and opportunity plans and processes for use on a project. [MROP02-50A]	Uses risk management tool for risk tracking on a project. [MROP02-E50]
3	Develops a project risk and opportunity profile including context, probability, consequences, thresholds, priority, and risk action and status. [MROP03]	(A) Defines risk taxonomy for a project. [MROP03-10A]	Records showing a taxonomy produced for a project or system. [MROP03-E10]
		(A) Maintains project risk and opportunity profile including context, probability, consequences, thresholds, priority, and risk action and status. [MROP03-20A]	Risk Management Plan (RMP) maintained. [MROP03-E20]
		(A) Performs required initialization actions for a risk and opportunity tracking tool for project use. [MROP03-30A]	Reports generated and tools utilized. [MROP03-E30]

(Continued)

ID	Indicators of Competence *(in addition to those at Supervised Practitioner level)*	Relevant knowledge, experience, and/or behaviors	Possible examples of objective evidence of personal contribution to activities performed, or professional behaviors applied
4	Analyzes risks and opportunities for likelihood and consequence in order to determine magnitude and priority for treatment. [MROP04]	(A) Performs all stages of risk management; planning, risk identification, risk assessment, risk reduction strategies/fall back plan, implementation of chosen strategy (risk mitigation actions etc.), quantitative assessment, risk monitoring. [MROP04-10A]	Systems Engineering Risk register. [MROP04-E10]
		(A) Maintains a risk management activity. [MROP04-20A]	Risk management plan, reports, Change Control Board (CCB) inputs, mitigation effort tracking, continuous identification of potential new risks, and opportunities. [MROP04-E20]
		(A) Identifies different types of Systems Engineering risks, associated mitigation actions, subsequently recording actual outcomes. [MROP04-30A]	Minutes of risk review meetings. [MROP04-E30]
		(A) Ensures risks and uncertainties pertaining to requirements are fully addressed over time. [MROP04-40A]	Assumption analysis. [MROP04-E40]
			List of dependencies created. [MROP04-E50]
5	Monitors Systems Engineering risks and opportunities during project execution. [MROP05]	(P) Guides and actively coordinates risk review panels or risk control boards on a project. [MROP05-10P]	Chairs risk review panels or risk control boards on a project. [MROP05-E10]
		(A) Coordinates team inputs for updating risk and opportunity database. [MROP05-20A]	Records of outputs from a risk management tool used or created. [MROP05-E20]
		(A) Analyzes risks which occur but had not been identified to determine the cause, recording conclusions made. [MROP05-30A]	
		(A) Prepares risk reports. [MROP05-40A]	
6	Analyzes risks and opportunities effectively, considering alternative treatments and generating a plan of action when thresholds exceed certain levels. [MROP06]	(A) Conducts risk review boards. [MROP06-10A]	Risk reports. [MROP06-E10]
		(A) Uses appropriate risk and opportunity strategies to address different types of risk and opportunity effectively. [MROP06-20A]	Risk and opportunity treatments. [MROP06-E20]
		(A) Identifies and compares alternative risk treatments for risks. [MROP06-30A]	
		(A) Identifies and documents plan of action when risk thresholds are exceeded. [MROP06-40A]	
		(P) Guides and actively coordinates mitigation efforts against program plan and budget. [MROP06-50P]	

7	Communicates risk and opportunity status to affected stakeholders. [MROP07]	(A) Ensures stakeholders participate in risk review board. [MROP07-10A] (A) Creates and delivers risk review presentations. [MROP07-20A] (A) Prepares risk review reports and distributes to key stakeholders. [MROP07-30A]	Risk reports. [MROP07-E10]
8	Guides new or supervised practitioners in risk and opportunity Management techniques in order to develop their knowledge, abilities, skills, or associated behaviors. [MROP08]	(P) Guides new or supervised practitioners in executing activities that form part of this competency. [MROP08-10P] (A) Trains individuals to an "Awareness" level in this competency area. [MROP08-20A]	Organizational Breakdown Structure showing their responsibility for technical supervision in this area. [MROP08-E10] On-the-job training objectives/guidance etc. [MROP08-E20] Coaching or mentoring assignment records. [MROP08-E30] Records highlighting their impact on another individual in terms of improvement or professional development in this competency. [MROP08-E40]
9	Maintains and enhances own competence in this area through Continual Professional Development (CPD) activities. [MROP09]	(A) Identifies potential development needs in this area, identifying opportunities to address these through continual professional development activities. [MROP09-10A] (A) Performs continual professional development activities to maintain and enhance their competency in this area. [MROP09-20A] (A) Records continual professional development activities undertaken including learning or insights gained. [MROP09-30A]	Records of Continual Professional Development (CPD) performed and learning outcomes. [MROP09-E10]

Lead Practitioner – Management: Risk and Opportunity Management			
ID	**Indicators of Competence** ***(in addition to those at Practitioner level)***	**Relevant knowledge, experience, and/or behaviors**	**Possible examples of objective evidence of personal contribution to activities performed, or professional behaviors applied**
1	Creates enterprise-level policies, procedures, guidance, and best practice for Systems Engineering risk and opportunity management, including associated tools. [MROL01]	(A) Analyzes enterprise need for risk and opportunity management policies, processes, tools, or guidance. [MROL01-10A]	enterprise level reports and recommendations. [MROL01-E10]
		(A) Creates enterprise policies, procedures or guidance on risk and opportunity management. [MROL01-20A]	Part of enterprise SE resource pool and proposal or internal R&D reviews. [MROL01-E20]
		(A) Selects and acquires appropriate tools supporting risk and opportunity management. [MROL01-30A]	
2	Judges the tailoring of enterprise-level risk and opportunity management processes and associated work products to meet the needs of a project. [MROL02]	(A) Evaluates the enterprise risk and opportunity Management processes against the business and external stakeholder (e.g. customer) needs in order to tailor processes to enable project success. [MROL02-10A]	Part of enterprise SE resource pool and proposal or internal R&D reviews. [MROL02-E10]
		(A) Provides constructive feedback on risk and opportunity Management processes. [MROL02-20A]	Number of programs supported with successful results. [MROL02-E20]
		(A) Reviews high value, high risk or high opportunity risk and opportunity Management planning issues across the enterprise. [MROL02-30A]	Panel reports and white papers generated. [MROL02-E30]
3	Guides and actively coordinates Systems Engineering risk and opportunity management across multiple diverse projects or across a complex system, with proven success. [MROL03]	(A) Compares and contrasts the enterprise benefits of a range of techniques for performing Risk and opportunity Management. [MROL03-10A]	Records showing internal or external consultation in the relevant areas. [MROL03-E10]
		(A) Uses knowledge and experience of the application of different risk and opportunity management approaches with successful results. [MROL03-20A]	Reports or white papers generated on risk management. [MROL03-E20]
		(A) Uses Risk and opportunity policies and procedures across multiple projects for components of a complex system. [MROL03-30A]	
4	Produces an enterprise-level risk profile including context, probability, consequences, thresholds, priority, and risk action and status. [MROL04]	(A) Analyzes risk context, probability, consequences, thresholds, priority, and risk action and status multiple diverse projects across the enterprise. [MROL04-10A]	Risk profiling schema. [MROL04-E10]
		(A) Creates or maintains enterprise-wide risk profile, communicating with key stakeholders. [MROL04-20A]	Risk Management Plan. [MROL04-E20]
		(A) Develops enterprise risk profiling schema. [MROL04-30A]	

5	Judges on the treatment of risks and opportunities across multiple diverse projects or a complex project, with proven success. [MROL05]	(A) Analyzes proposed risk or opportunity treatments on multiple diverse projects (or a complex project) across the enterprise. [MROL05-10A]	Update to risk register showing changes to risks and/or mitigation actions. [MROL05-E10]
		(P) Persuades stakeholders to maintain or change existing project treatment where appropriate, with proven success. [MROL05-20P]	Minutes of risk review meetings. [MROL05-E20]
6	Persuades key enterprise stakeholders to address identified enterprise-level project risks and opportunities to reduce enterprise-level risks. [MROL06]	(P) Uses facilitation skills in project/program reviews to ensure Systems Engineering processes addresses enterprise-level Risk and opportunity management issues and vice versa. [MROL06-10P]	White papers and reports. [MROL06-E10]
		(A) Reviews high value, high risk, or high opportunity risk and opportunity Management issues across the enterprise. [MROL06-20A]	Records of enterprise-level meetings. [MROL06-E20]
		(P) Persuades stakeholders to address enterprise-level Risk and opportunity management issues. [MROL06-30P]	Records indicating changed stakeholder thinking from personal intervention. [MROL06-E30]
7	Coaches or mentors practitioners across the enterprise in Systems Engineering risk and opportunity management in order to develop their knowledge, abilities, skills, or associated behaviors. [MROL07]	(P) Coaches or mentors practitioners across the enterprise in competency-related techniques, recommending development activities. [MROL07-10P]	Coaching or mentoring assignment records. [MROL07-E10]
		(A) Develops or authorizes enterprise training materials in this competency area. [MROL07-20A]	Records of formal training courses, workshops, seminars, and authored training material supported by successful post-training evaluation data. [MROL07-E20]
		(A) Provides enterprise workshops/seminars or training in this competency area. [MROL07-30A]	Listing as an approved organizational trainer for this competency area. [MROL07-E30]
8	Promotes the introduction and use of novel techniques and ideas in Systems Engineering risk and opportunity management across the enterprise, to improve enterprise competence in this area. [MROL08]	(A) Analyzes different approaches across different domains through research. [MROL08-10A]	Research records. [MROL08-E10]
		(A) Defines novel approaches that could potentially improve the SE discipline within the enterprise. [MROL08-20A]	Published papers in refereed journals/company literature. [MROL08-E20]
		(P) Fosters awareness of these novel techniques within the enterprise. [MROL08-30P]	Records showing introduction of enabling systems supporting the new techniques or ideas. [MROL08-E30]

(Continued)

ID	Indicators of Competence *(in addition to those at Practitioner level)*	Relevant knowledge, experience, and/or behaviors	Possible examples of objective evidence of personal contribution to activities performed, or professional behaviors applied
		(P) Collaborates and advises enterprise stakeholders to facilitate the introduction of techniques new to the enterprise. [MROL08-40P] (A) Monitors new techniques after their introduction to determine their effectiveness. [MROL08-50A] (A) Adapts approach to reflect actual enterprise performance improvements. [MROL08-60A]	Published papers (or similar) at enterprise level. [MROL08-E40] Records of improvements made against a recognized process improvement model in this area. [MROL08-E50]
9	Develops expertise in this competency area through specialist Continual Professional Development (CPD) activities. [MROL09]	(A) Identifies own needs for further professional development in order to increase competence beyond practitioner level. [MROL09-10A] (A) Performs professional development activities in order to move own competence toward expert level. [MROL09-20A] (A) Records continual professional development activities undertaken including learning or insights gained. [MROL09-30A]	Records of Continual Professional Development (CPD) performed and learning outcomes. [MROL09-E10]

NOTES	In addition to items above, enterprise-level or independent 3rd-Party-generated evidence may be used to amplify other evidence presented and may include: a. Formally recognized by senior management in current organization as an expert in this competency area b. Evidence of role as Product/System Design Authority or Technical Authority on a complex project with responsibilities in this area or where skills within this competency area were used c. Recognized as an authorizing signatory on behalf of enterprise for formal documentation in this competency area (e.g. policies, processes, and deliverables) d. Formal commendation or award within own enterprise for contribution or item of work successfully performed, which required proficiency in this competency area e. Customer, Supplier, or other external project-specific key Stakeholder accolades for specific work performed in this competency area f. Independently assessed or accredited work in this competency area (e.g. for independent publication or use) g. Formal organizational HR records positively highlighting any specific professional competencies or behaviors identified (if applicable) plus any of the evidence indicators listed at Expert level below

Expert – Management: Risk and Opportunity Management			
ID	**Indicators of Competence** ***(in addition to those at Lead Practitioner level)***	**Relevant knowledge, experience, and/or behaviors**	**Possible examples of objective evidence of personal contribution to activities performed, or professional behaviors applied**
1	Communicates own knowledge and experience in Systems Engineering risk and opportunity management, in order to best practice beyond the enterprise boundary. [MROE01]	(A) Produces papers, seminars, or presentations outside own enterprise for publication in order to share own ideas and improve industry best practices in this competence area. [MROE01-10A]	Published papers or books etc. on new technique in refereed journals/company literature. [MROE01-E10]
		(P) Fosters incorporation of own ideas into industry best practices in this area. [MROE01-20P]	Published papers in refereed journals or internal literature proposing new practices in this competence area (or presentations, tutorials, etc.). [MROE01-E20]
		(P) Develops guidance materials identifying new (or updating existing) best practice in this competence area. [MROE01-30P]	Proposals adopted as industry best practice. [MROE01-E30]
2	Influences key beyond the enterprise boundary in support of risk and opportunity management. [MROE02]	(P) Acts as independent reviewer in industry panels and professional societies. [MROE02-10P]	Records of projects supported successfully. [MROE02-E10]
		(A) Provides evidence and arguments influencing external stakeholder Risk and opportunity management activities. [MROE02-20A]	Publications and presentations used in support of stakeholder arguments. [MROE02-E20]
		(P) Acts as independent external member of review panels for high risk and opportunity management external reviews or external stakeholder-led reviews. [MROE02-30P]	Review minutes showing contribution. [MROE02-E30]
3	Advises organizations beyond the enterprise boundary on the handling of complex or sensitive risk and opportunity issues. [MROE03]	(A) Advises external stakeholders (e.g. customers) on their risk and opportunity management issues. [MROE03-10A]	Meeting minutes or report outlining role as arbitrator. [MROE03-E10]
		(A) Conducts sensitive risk and opportunity management negotiations. [MROE03-20A]	
		(A) Conducts complex risk and opportunity management negotiations taking account of taking account of external stakeholders' (e.g. Customer's) background and knowledge background and knowledge. [MROE03-30A]	
		(P) Arbitrates on risk and opportunity decisions. [MROE03-40P]	

(Continued)

ID	Indicators of Competence *(in addition to those at Lead Practitioner level)*	Relevant knowledge, experience, and/or behaviors	Possible examples of objective evidence of personal contribution to activities performed, or professional behaviors applied
4	Advises organizations beyond the enterprise boundary on the suitability of their approach to risk and opportunity management. [MROE04]	(P) Acts as independent reviewer on approval of Risk and opportunity management plans. [MROE04-10P]	Records of membership on oversight committee with relevant terms of reference. [MROE04-E10]
		(A) Reviews and advises Risk and opportunity management strategies that has led to changes being implemented. [MROE04-20A]	Review comments or revised document. [MROE04-E20]
		(A) Advises stakeholders beyond the enterprise boundary to improve their risk and opportunity strategies. [MROE04-30A]	
5	Champions the introduction of novel techniques and ideas in Systems Engineering risk and opportunity management, beyond the enterprise boundary, in order to develop the wider Systems Engineering community in this competency. [MROE05]	(A) Analyzes different approaches across different domains through research. [MROE05-10A]	Records of activities promoting research and need to adopt novel technique or ideas. [MROE05-E10]
		(A) Produces reports for the wider SE community on the effectiveness of new techniques after their introduction. [MROE05-20A]	Records of improvements made to process and appraisal against a recognized process improvement model. [MROE05-E20]
		(P) Collaborates with those introducing novel techniques within the wider SE community. [MROE05-30P]	Research records. [MROE05-E30]
		(A) Defines novel approaches that could potentially improve the wider SE discipline. [MROE05-40A]	Published papers in refereed journals/company literature. [MROE05-E40]
		(P) Fosters awareness of these novel techniques within the wider SE community. [MROE05-50P]	Records showing introduction of enabling systems supporting the new techniques or ideas. [MROE05-E50]
6	Coaches individuals beyond the enterprise boundary in Systems Engineering risk and opportunity management, in order to further develop their knowledge, abilities, skills, or associated behaviors. [MROE06]	(P) Coaches or mentors individuals beyond the enterprise boundary, in competency-related techniques, recommending development activities. [MROE06-10P]	Coaching or mentoring assignment records. [MROE06-E10]
		(A) Develops or authorizes training materials in this competency area, which are subsequently successfully delivered beyond the enterprise boundary. [MROE06-20A]	Records of formal training courses, workshops, seminars, and authored training material supported by successful post-training evaluation data. [MROE06-E20]

		(A) Provides workshops/seminars or training in this competency area for practitioners or lead practitioners beyond the enterprise boundary (e.g. conferences and open training days). [MROE06-30A]	Records of Training/workshops/seminars created supported by successful post-training evaluation data. [MROE06-E30]
7	Maintains expertise in this competency area through specialist Continual Professional Development (CPD) activities. [MROE07]	(A) Reviews research, new ideas, and state of the art to identify relevant new areas requiring personal development in order to maintain expertise in this competency area. [MROE07-10A]	Records of documents reviewed and insights gained as part of own research into this competency area. [MROE07-E10]
		(A) Performs identified specialist professional development activities in order to maintain or further develop competence at expert level. [MROE07-20A]	Records of Continual Professional Development (CPD) performed and learning outcomes. [MROE07-E20]
		(A) Records continual professional development activities undertaken including learning or insights gained. [MROE07-30A]	

NOTES	In addition to items above, enterprise-level or independent 3rd-Party-generated evidence may be used to amplify other evidence presented and may include: a. Formally recognized by a reputable external organization as an expert in this competency area b. Evidence of role as independent assessor or reviewer on project outside own organization where skills in this competency area were used c. Evidence of invitation(s) from wider community for contribution of systems engineering expertise in this area (e.g. industry conference panel, government advisory board etc. cross-industry working groups, partnerships, accredited advanced university courses or research, or as part of professional institute) d. Formal commendation beyond the enterprise (e.g. by INCOSE or other recognized authority) for work performed in this competency area e. Independently assessed or accredited work product in this competency area (e.g. for independent publication or use) f. Accolades of expertise in this area from recognized industry leaders

Competency area – Management: Decision Management
Description Decision management provides a structured, analytical framework for objectively identifying, characterizing, and evaluating a set of alternatives for a decision at any point in the life cycle in order to select the most beneficial course of action.
Why it matters System development entails an array of interrelated decisions that require the holistic perspective of the Systems Engineering discipline. Decisions include selection of preferred solution at every level of the system, including technology option selection, architecture selection, make-or-buy decisions, strategy selection for maintenance, and disposal. While some low value decisions can be made and recorded simply, key decisions that might affect the long-term success and value delivery of the project need to be controlled using a formalized decision management process.
Possible contributory types of evidence Any combination of the types of evidence may be acceptable (depending on how the Framework is tailored and used). The evidence items identified at each level indicate example work products only. Contributions to work products will generally differ at each proficiency level.
Learning and development The INCOSE Professional Development Portal provides example guidance on how to gain an initial awareness of a competency area and options for developing further competence thereafter.

Awareness – Management: Decision Management		
ID	**Indicators of Competence**	**Relevant knowledge sub-indicators**
1	Identifies the Systems Engineering situations where a structured decision is, and is not, appropriate. [MDMA01]	(K) Explains why complex trades can be expensive and time consuming to conduct. [MDMA01-10K] (K) Explains why layered decision may be utilized to weed out non-promising cases. [MDMA01-20K] (K) Explains why trades that are not complex do not need the cost and time required for a structured decision. [MDMA01-30K]
2	Explains why there is a need to select a preferred solution. [MDMA02]	(K) Explains why a preferred solution needs to be selected and communicated to the development team to allow continuation to the next Systems Engineering process. [MDMA02-10K]
3	Describes the relevance of comparative techniques (e.g. trade studies, and make/buy) to assist decision processes. [MDMA03]	(K) Explains how formal processes may be used to enable the decision-making process and aid in arriving at a preferred solution. [MDMA03-10K] (K) Describes key interfaces and links between engineering decisions and stakeholders in the wider enterprise outside engineering (e.g. infrastructure management, human resource management, quality management, knowledge management, portfolio management, and life cycle model management). [MDMA03-20K] (K) Describes the difference between "musts" and "wants" I the decision-making activity. [MDMA03-30K]

4	Explains how to frame, tailor, and structure a decision including its objectives and measures and outlines the key characteristics of a structured decision-making approach. [MDMA04]	(K) Describes the pros and cons of various decision-making processes in various application environments. [MDMA04-10K]
		(K) Explains how key measures and their measurement may be tailored in support of decision-making. [MDMA04-20K]
		(K) Explains why the kinds of decisions to be made is a function of system life cycle. [MDMA04-30K]
5	Explains how uncertainty impacts on decision-making. [MDMA05]	(K) Explains why there is a need for statistical analysis in the decision-making process. [MDMA05-10K]
		(K) Explains why there is a need to apply differing levels of performance analysis downstream in the design cycle. [MDMA05-20K]
		(K) Explains how architecture selection assumptions relate to later physical implementation option alternatives. [MDMA05-30K]
6	Explains why there is a need for communication and accurate recording in all aspects of the decision-making process. [MDMA06]	(K) Explains why communicating the results of trade-offs and decisions to stakeholders improves outcome and provides feedback for what is really important. [MDMA06-10K]
		(K) Explains why providing a high-quality actionable report enhances the selection of approved choices and correct downstream implementation. [MDMA06-20K]
		(K) Explains the importance of archiving clear process, decisions and actionable results for future use and or tailoring on other projects. [MDMA06-30K]

Supervised Practitioner – Management: Decision Management			
ID	**Indicators of Competence** *(in addition to those at Awareness level)*	**Relevant knowledge, experience, and/or behaviors**	**Possible examples of objective evidence of personal contribution to activities performed, or professional behaviors applied**
1	Follows a governing process and appropriate tools to plan and control their own decision management activities. [MDMS01]	(K) Explains the importance of capturing baseline and explaining need for alternatives within that baseline description. [MDMS01-10K] (A) Describes decision process utilized in a project need to maintain timeline within project scope and schedule milestones. [MDMS01-20A] (K) Explains why there is a need to schedule appropriate resources. [MDMS01-30K] (A) Uses approved decision management processes and tools to control and manage own decision management activities. [MDMS01-40A]	Results of tailoring they performed on industry and enterprise processes to meet program objectives. [MDMS01-E10] Resource scheduling they performed within a trade study. [MDMS01-E20] Tailoring and process reports they created. [MDMS01-E30]
2	Identifies potential decision criteria and performance parameters for consideration. [MDMS02]	(A) Identifies potential decision criteria and performance parameters for consideration. [MDMS02-10A]	Decision activities showing criteria they identified for consideration. [MDMS02-E10]
3	Identifies tools and techniques for the decision process. [MDMS03]	(A) Describes the pros and cons of major decision management techniques used in various programs (e.g. statistical and deterministic). [MDMS03-10A] (K) Describes tools utilized at the architecture level for trade studies and to manage the statistical result management of these trades. [MDMS03-20K] (K) Explains how system complexity influences the selection of processes, techniques and tools, as well as the resources needed. [MDMS03-30K] (A) Prepares information supporting the selection of tools and techniques for the decision process. [MDMS03-40A]	Tailoring and process reports they created. [MDMS03-E10] Decision Management Plans generated. [MDMS03-E20]
4	Prepares information in support of decision trade studies. [MDMS04]	(A) Identifies information in support of decision trade studies, communicating this as required. [MDMS04-10A]	Decision information they created in support of decision trade studies. [MDMS04-E10]

5	Monitors the decision process to catalog actions taken and their supporting rationale. [MDMS05]	(K) Describes the importance of cataloguing results and creating clear, actionable reports that document process, decisions, resources, and results. [MDMS05-10K] (A) Records the decision-making process used. [MDMS05-20A] (A) Records decision actions made. [MDMS05-30A]	Activities performed in support of decision-making. [MDMS05-E10] Decision management reports and briefings or delivered. [MDMS05-E20] Decision management reports, which included baselines. [MDMS05-E30]
6	Develops own understanding of this competency area through Continual Professional Development (CPD). [MDMS06]	(A) Identifies potential gaps in own knowledge or development needs in this area, identifying opportunities to address these through continual professional development activities. [MDMS06-10A] (A) Performs continual professional development activities to improve their knowledge and understanding in this area. [MDMS06-20A] (A) Records continual professional development activities undertaken including learning or insights gained. [MDMS06-30A]	Records of Continual Professional Development (CPD) performed and learning outcomes. [MDMS06-E10]

Practitioner – Management: Decision Management			
ID	**Indicators of Competence** ***(in addition to those at Supervised Practitioner level)***	**Relevant knowledge, experience, and/or behaviors**	**Possible examples of objective evidence of personal contribution to activities performed, or professional behaviors applied**
1	Creates a strategy for decision management on a project to support SE project and wider enterprise needs. [MDMP01]	(A) Identifies project-specific decision management needs that need to be incorporated into SE planning. [MDMP01-10A] (A) Prepares project-specific decision management inputs for SE planning purposes, using appropriate processes and procedures. [MDMP01-20A] (A) Prepares project-specific decision management task estimates in support of SE planning. [MDMP01-30A] (P) Liaises throughout with stakeholders to gain approval, updating strategy as necessary. [MDMP01-40P]	SE decision management work products. [MDMP01-E10]
2	Creates a governing process, plan, and associated tools for systems decision management, which reflect project and business strategy. [MDMP02]	(K) Identifies the steps necessary to define a process and appropriate techniques to be adopted to establish decision management for system elements. [MDMP02-10K] (K) Describes key elements of a successful decision management process and how decision management activities relate to different stages of a system life cycle. [MDMP02-20K] (A) Uses decision management tools or uses a methodology. [MDMP02-30A] (A) Uses appropriate standards and governing processes in their decision management planning, tailoring as appropriate. [MDMP02-40A] (A) Selects decision management method appropriate to each decision. [MDMP02-50A] (A) Creates decision management plans and processes for use on a project. [MDMP02-60A]	SE system decision management planning work products they have developed. [MDMP02-E10] Planning documents (e.g. SEMP), other project/program plan or organizational processes they have written identifying how key decisions are to be made. [MDMP02-E20]
3	Complies with governing plans and processes for system decision management, interpreting, evolving, or seeking guidance where appropriate. [MDMP03]	(A) Follows defined plans, processes and associated tools to perform decision management on a project. [MDMP03-10A] (P) Guides and actively coordinates the interpretation of decision management plans or processes in order to complete decision management activities successfully. [MDMP03-20P]	Decision management plans or work products they have interpreted or challenged to address project issues. [MDMP03-E10] Extracts showing decision management tools or methodologies used. [MDMP03-E20]

		(P) Recognizes situations where deviation from published plans and processes or clarification from others is necessary to overcome complex challenges. [MDMP03-30P]	
4	Develops governing decision management plans, processes, and appropriate tools and uses these to control and monitor decision management activities. [MDMP04]	(K) Explains how decision timeline may impact project scope and schedule milestones. [MDMP04-10K]	Decision Management Plans. [MDMP04-E10]
		(K) Explains why decision may need scheduling of appropriate resources. [MDMP04-20K]	Tailoring Plans. [MDMP04-E20]
		(A) Performs decision management. [MDMP04-30A]	Reports and briefings. [MDMP04-E30]
		(A) Uses enterprise-level plans, processes, and appropriate tools to control and monitor decision management activities. [MDMP04-40A]	
		(A) Adapts (tailors) existing plans to meet project needs. [MDMP04-50A]	
5	Guides and actively coordinates ongoing decision management activities to ensure successful outcomes with decision management stakeholders. [MDMP05]	(A) Prepares reports to communicate decisions to all appropriate stakeholders. [MDMP05-10A]	Decision Management Plans. [MDMP05-E10]
		(A) Defines selection criteria as part of a decision-making process, e.g. technology requirements, off-the-shelf availability, competitive considerations, performance assessment, maintainability, capacity to evolve, standardization considerations, integration concerns, cost, and schedule. [MDMP05-20A]	Tailoring Plans. [MDMP05-E20]
			Reports and briefings. [MDMP05-E30]
6	Determines decision selection criteria, weightings of the criteria, and assess alternatives against selection criteria. [MDMP06]	(A) Creates weighting of selection criteria for a decision process. [MDMP06-10A]	Authored output from the decision-making process. [MDMP06-E10]
		(A) Performs a cost analysis as part of a decision-making process. [MDMP06-20A]	
		(A) Performs a risk and opportunity analysis as part of a decision-making process. [MDMP06-30A]	
		(K) Explains why differing tools and techniques are required to support different types of decision. [MDMP06-40K]	
		(K) Explains why differing tools and techniques are required to support different types of decision. [MDMP06-50K]	

(Continued)

ID	Indicators of Competence *(in addition to those at Supervised Practitioner level)*	Relevant knowledge, experience, and/or behaviors	Possible examples of objective evidence of personal contribution to activities performed, or professional behaviors applied
7	Selects appropriate tools and techniques for making different types of decision. [MDMP07]	(A) Selects the appropriate decision tool/technique for identifying a preferred solution (e.g. trade studies, make/buy, cost/benefit analysis, Quality Function Deployment (QFD), or other formal decision-making process). [MDMP07-10A]	Outputs from different decision tools used. [MDMP07-E10]
		(A) Selects appropriate tools and techniques for different types of decision. [MDMP07-20A]	Tailoring plans outlining tool selection. [MDMP07-E20]
		(A) Prepares a business case or report based on outputs of trade analysis. [MDMP07-30A]	
8	Prepares trade-off analyses and justifies the selection in terms that can be quantified and qualified. [MDMP08]	(P) Communicates trade alternatives to senior technical leaders and/or management stakeholders. [MDMP08-10P]	Authored output from trade analysis. [MDMP08-E10]
			Minutes of meeting describing decision made. [MDMP08-E20]
9	Assesses sensitivity of selection criteria through a sensitivity analysis, reporting as required. [MDMP09]	(A) Analyzes sensitivity of selection criteria, through a sensitivity analysis, reporting on this analysis. [MDMP09-10A]	Authored report of sensitivity analysis. [MDMP09-E10]
10	Guides new or supervised practitioners in Decision management techniques in order to develop their knowledge, abilities, skills, or associated behaviors. [MDMP10]	(P) Guides new or supervised practitioners in executing activities that form part of this competency. [MDMP10-10P]	Organizational Breakdown Structure showing their responsibility for technical supervision in this area. [MDMP10-E10]
		(A) Trains individuals to an "Awareness" level in this competency area. [MDMP10-20A]	On-the-job training objectives/guidance etc. [MDMP10-E20]
			Coaching or mentoring assignment records. [MDMP10-E30]
			Records highlighting their impact on another individual in terms of improvement or professional development in this competency. [MDMP10-E40]
11	Maintains and enhances own competence in this area through Continual Professional Development (CPD) activities. [MDMP11]	(A) Identifies potential development needs in this area, identifying opportunities to address these through continual professional development activities. [MDMP11-10A]	Records of Continual Professional Development (CPD) performed and learning outcomes. [MDMP11-E10]
		(A) Performs continual professional development activities to maintain and enhance their competency in this area. [MDMP11-20A]	
		(A) Records continual professional development activities undertaken including learning or insights gained. [MDMP11-30A]	

Lead Practitioner – Management: Decision Management			
ID	**Indicators of Competence *(in addition to those at Practitioner level)***	**Relevant knowledge, experience, and/or behaviors**	**Possible examples of objective evidence of personal contribution to activities performed, or professional behaviors applied**
1	Creates enterprise-level policies, procedures, guidance, and best practice for decision management and communication, including associated tools. [MDML01]	(A) Analyzes enterprise need for decision management policies, processes, tools, or guidance. [MDML01-10A]	Enterprise level reports and recommendations. [MDML01-E10]
		(A) Creates enterprise policies, procedures, or guidance on decision management. [MDML01-20A]	Part of enterprise SE resource pool and proposal or internal R&D reviews. [MDML01-E20]
		(A) Selects and acquires appropriate tools supporting decision management. [MDML01-30A]	
2	Judges the tailoring of enterprise-level decision management processes and associated work products to meet the needs of a project. [MDML02]	(A) Evaluates the enterprise decision management processes against the business and external stakeholder (e.g. customer) needs in order to tailor processes to enable project success. [MDML02-10A]	Details of independent internal R&D reviews performed. [MDML02-E10]
		(A) Provides constructive feedback on decision management processes. [MDML02-20A]	Details of programs supported with successful results. [MDML02-E20]
		(A) Reviews high value or high-risk program decision management planning and review issues across the enterprise. [MDML02-30A]	Enterprise level reports and recommendations. [MDML02-E30]
			Details of enterprise reports and white papers generated. [MDML02-E40]
3	Coordinates decision management and trade analysis using different techniques, across multiple diverse projects or across a complex system, with proven success. [MDML03]	(A) Compares and contrasts the benefits of a range of techniques for performing trade-offs across the enterprise. [MDML03-10A]	Internal or external consultant in the relevant areas. [MDML03-E10]
		(A) Uses knowledge and experience of the application of different techniques with successful results. [MDML03-20A]	
4	Persuades key stakeholders across the enterprise to address identified enterprise-level decision management issues. [MDML04]	(P) Uses facilitation skills at project/program process tailoring reviews to ensure the Systems Engineering process meets the needs of the project/program and vice versa. [MDML04-10P]	White papers and reports. [MDML04-E10]
		(A) Advises on high value, high risk or high opportunity enterprise-level management issues. [MDML04-20A]	Records indicating changed stakeholder thinking from personal intervention. [MDML04-E20]
		(A) Persuades stakeholders to adopt a different path regarding enterprise decision management issues. [MDML04-30A]	

(Continued)

ID	Indicators of Competence *(in addition to those at Practitioner level)*	Relevant knowledge, experience, and/or behaviors	Possible examples of objective evidence of personal contribution to activities performed, or professional behaviors applied
5	Negotiates complex trades on behalf of the enterprise. [MDML05]	(A) Proposes complex trade alternatives to relevant stakeholders enabling consensus on the preferred solution. [MDML05-10A]	Authored report of complex trade analysis. [MDML05-E10]
6	Judges decisions affecting solutions and the criteria for making the solution across the enterprise. [MDML06]	(A) Judges criteria used for making decisions and the decisions made by projects across the enterprise. [MDML06-10A]	Records of advice provided in the selection of preferred solutions. [MDML06-E10]
		(A) Analyzes the validity of decisions made by third party and their selection methods. [MDML06-20A]	Authored report outlining third party decision reviews. [MDML06-E20]
7	Coaches or mentors practitioners across the enterprise in Systems Engineering decision management in order to develop their knowledge, abilities, skills, or associated behaviors. [MDML07]	(P) Coaches or mentors practitioners across the enterprise in competency-related techniques, recommending development activities. [MDML07-10P]	Coaching or mentoring assignment records. [MDML07-E10]
		(A) Develops or authorizes enterprise training materials in this competency area. [MDML07-20A]	Records of formal training courses, workshops, seminars, and authored training material supported by successful post-training evaluation data. [MDML07-E20]
		(A) Provides enterprise workshops/seminars or training in this competency area. [MDML07-30A]	Listing as an approved organizational trainer for this competency area. [MDML07-E30]
8	Promotes the introduction and use of novel techniques and ideas in decision resolution and management across the enterprise, to improve enterprise competence in this area. [MDML08]	(A) Analyzes different approaches across different domains through research. [MDML08-10A]	Research records. [MDML08-E10]
		(A) Defines novel approaches that could potentially improve the SE discipline within the enterprise. [MDML08-20A]	Published papers in refereed journals/company literature. [MDML08-E20]
		(P) Fosters awareness of these novel techniques within the enterprise. [MDML08-30P]	Records showing introduction of enabling systems supporting the new techniques or ideas. [MDML08-E30]
		(P) Collaborates with enterprise stakeholders to facilitate the introduction of techniques new to the enterprise. [MDML08-40P]	Published papers (or similar) at enterprise level. [MDML08-E40]

		(A) Monitors new techniques after their introduction to determine their effectiveness. [MDML08-50A] (A) Adapts approach to reflect actual enterprise performance improvements. [MDML08-60A]	Records of improvements made against a recognized process improvement model in this area. [MDML08-E50]
9	Develops expertise in this competency area through specialist Continual Professional Development (CPD) activities. [MDML09]	(A) Identifies own needs for further professional development in order to increase competence beyond practitioner level. [MDML09-10A] (A) Performs professional development activities in order to move own competence toward expert level. [MDML09-20A] (A) Records continual professional development activities undertaken including learning or insights gained. [MDML09-30A]	Records of Continual Professional Development (CPD) performed and learning outcomes. [MDML09-E10]

NOTES	In addition to items above, enterprise-level or independent 3rd-Party-generated evidence may be used to amplify other evidence presented and may include: a. Formally recognized by senior management in current organization as an expert in this competency area b. Evidence of role as Product/System Design Authority or Technical Authority on a complex project with responsibilities in this area or where skills within this competency area were used c. Recognized as an authorizing signatory on behalf of enterprise for formal documentation in this competency area (e.g. policies, processes, and deliverables) d. Formal commendation or award within own enterprise for contribution or item of work successfully performed, which required proficiency in this competency area e. Customer, Supplier, or other external project-specific key Stakeholder accolades for specific work performed in this competency area f. Independently assessed or accredited work in this competency area (e.g. for independent publication or use) g. Formal organizational HR records positively highlighting any specific professional competencies or behaviors identified (if applicable) plus any of the evidence indicators listed at Expert level below

Expert – Management: Decision Management			
ID	**Indicators of Competence** ***(in addition to those at Lead Practitioner level)***	**Relevant knowledge, experience, and/or behaviors**	**Possible examples of objective evidence of personal contribution to activities performed, or professional behaviors applied**
1	Communicates own knowledge and experience in Systems Engineering decision management, in order to improve best practice beyond the enterprise boundary. [MDME01]	(A) Produces papers, seminars, or presentations outside own enterprise for publication in order to share own ideas and improve industry best practices in this competence area. [MDME01-10A] (P) Fosters incorporation of own ideas into industry best practices in this area. [MDME01-20P] (P) Develops guidance materials identifying new (or updating existing) best practice in this competence area. [MDME01-30P]	Published papers or books etc. on new technique in refereed journals/company literature. [MDME01-E10] Published papers in refereed journals or internal literature proposing new practices in this competence area (or presentations, tutorials, etc.). [MDME01-E20] Own proposals adopted as industry best practices in this competence area. [MDME01-E30]
2	Influences key decision stakeholders beyond the enterprise boundary. [MDME02]	(P) Acts as independent member of industry panels and professional societies. [MDME02-10P] (A) Provides evidence and arguments influencing external stakeholder decisions. [MDME02-20A] (P) Acts as independent external member of review panels for decision management external reviews or external stakeholder-led reviews. [MDME02-30P]	Number of projects supported successfully. [MDME02-E10] Publications and presentations used in support of stakeholder arguments. [MDME02-E20] Review minutes showing contribution. [MDME02-E30]
3	Advises organizations beyond the enterprise boundary on complex or sensitive decision management or trade-off issues. [MDME03]	(A) Advises external stakeholders (e.g. customers) on their decision management or trade-off issues. [MDME03-10A] (A) Conducts sensitive decision management or trade-off negotiations. [MDME03-20A] (A) Conducts complex decision management or trade-off negotiations taking into account external stakeholders' (e.g. customer's) background and knowledge. [MDME03-30A] (P) Arbitrates on marginal decisions. [MDME03-40P]	Meeting minutes or report outlining role as arbitrator. [MDME03-E10]
4	Advises organizations beyond the enterprise boundary on the suitability of their approach to decision management. [MDME04]	(P) Acts as independent external reviewer of decisions and decision criteria on external reviews or external stakeholder-led reviews. [MDME04-10P] (A) Advises on appropriate approach or strategy in order to reflect required decision type or program context. [MDME04-20A]	Records of independent reviewer role on a number of projects. [MDME04-E10] Publications and presentations used in support of stakeholder arguments. [MDME04-E20] Review minutes showing contribution. [MDME04-E30]

5	Identifies strategies for organizations beyond the enterprise boundary, in order to resolve their issues with complex system trade-offs. [MDME05]	(P) Communicates complex trade-off alternatives to relevant stakeholders to assist in reaching a consensus agreement on the preferred solution. [MDME05-10P]	Authored report of complex trade analysis. [MDME05-E10]
		(A) Advises projects/programs on decision-making to re-balance requirement allocation if any development activity is unable to meet its requirements. [MDME05-20A]	System trade studies. [MDME05-E20]
		(A) Advises on system trade-offs requiring design changes while maintaining original intent. [MDME05-30A]	Minutes of meetings. [MDME05-E30]
			Reviews. [MDME05-E40]
			Design documentation. [MDME05-E50]
6	Champions the introduction of novel techniques and ideas in Systems Engineering decision management, beyond the enterprise boundary, in order to develop the wider Systems Engineering community in this competency. [MDME06]	(A) Analyzes different approaches across different domains through research. [MDME06-10A]	Records of activities promoting research and need to adopt novel technique or ideas. [MDME06-E10]
		(A) Produces reports for the wider SE community on the effectiveness of new techniques after their introduction. [MDME06-20A]	Records of improvements made to process and appraisal against a recognized process improvement model. [MDME06-E20]
		(P) Collaborates with those introducing novel techniques within the wider SE community. [MDME06-30P]	Research records. [MDME06-E30]
		(A) Defines novel approaches that could potentially improve the wider SE discipline. [MDME06-40A]	Published papers in refereed journals/company literature. [MDME06-E40]
		(P) Fosters awareness of these novel techniques within the wider SE community. [MDME06-50P]	Records showing introduction of enabling systems supporting the new techniques or ideas. [MDME06-E50]
7	Coaches individuals beyond the enterprise boundary in Systems Engineering decision management, in order to further develop their knowledge, abilities, skills, or associated behaviors. [MDME07]	(P) Coaches or mentors individuals beyond the enterprise boundary, in competency-related techniques, recommending development activities. [MDME07-10P]	Coaching or mentoring assignment records. [MDME07-E10]
		(A) Develops or authorizes training materials in this competency area, which are subsequently successfully delivered beyond the enterprise boundary. [MDME07-20A]	Records of formal training courses, workshops, seminars, and authored training material supported by successful post-training evaluation data. [MDME07-E20]
		(A) Provides workshops/seminars or training in this competency area for practitioners or lead practitioners beyond the enterprise boundary (e.g. conferences and open training days). [MDME07-30A]	Records of Training/workshops/seminars created supported by successful post-training evaluation data. [MDME07-E30]

ID	Indicators of Competence *(in addition to those at Lead Practitioner level)*	Relevant knowledge, experience, and/or behaviors	Possible examples of objective evidence of personal contribution to activities performed, or professional behaviors applied
8	Maintains expertise in this competency area through specialist Continual Professional Development (CPD) activities. [MDME08]	(A) Reviews research, new ideas, and state of the art to identify relevant new areas requiring personal development in order to maintain expertise in this competency area. [MDME08-10A]	Records of documents reviewed and insights gained as part of own research into this competency area. [MDME08-E10]
		(A) Performs identified specialist professional development activities in order to maintain or further develop competence at expert level. [MDME08-20A]	Records of Continual Professional Development (CPD) performed and learning outcomes. [MDME08-E20]
		(A) Records continual professional development activities undertaken including learning or insights gained. [MDME08-30A]	

NOTES	In addition to items above, enterprise-level or independent 3rd-Party-generated evidence may be used to amplify other evidence presented and may include: a. Formally recognized by a reputable external organization as an expert in this competency area b. Evidence of role as independent assessor or reviewer on project outside own organization where skills in this competency area were used c. Evidence of invitation(s) from wider community for contribution of systems engineering expertise in this area (e.g. industry conference panel, government advisory board etc. cross-industry working groups, partnerships, accredited advanced university courses or research, or as part of professional institute) d. Formal commendation beyond the enterprise (e.g. by INCOSE or other recognized authority) for work performed in this competency area e. Independently assessed or accredited work product in this competency area (e.g. for independent publication or use) f. Accolades of expertise in this area from recognized industry leaders

Competency area – Management: Concurrent Engineering
Description Concurrent engineering is a work methodology based on the parallelization of tasks (i.e. performing tasks concurrently). It refers to an approach used in Systems Engineering in which functions of design and development engineering, manufacturing engineering, and other enterprise functions are integrated to reduce the elapsed time required to bring a new system, product, or service to market.
Why it matters Systems Engineering life cycles involve multiple, concurrent processes and activities, which must be coordinated to mitigate risk and prevent unnecessary work, paralysis, and a lack of convergence to an effective solution. Concurrency may be the only approach capable of meeting the customer schedule or gaining a competitive advantage. Performance can be constrained unnecessarily by allowing individual system elements to progress too quickly.
Possible contributory types of evidence Any combination of the types of evidence may be acceptable (depending on how the Framework is tailored and used). The evidence items identified at each level indicate example work products only. Contributions to work products will generally differ at each proficiency level.
Learning and development The INCOSE Professional Development Portal provides example guidance on how to gain an initial awareness of a competency area and options for developing further competence thereafter.

Awareness – Management: Concurrent Engineering		
ID	**Indicators of Competence**	**Relevant knowledge sub-indicators**
1	Explains how Systems Engineering life cycle processes and activities and the development of systems elements can be concurrent and provides examples. [MCEA01]	(K) Lists different aspects of systems that can be developed concurrently. [MCEA01-10K] (K) Explains the advantages of a Multidisciplinary team. [MCEA01-20K] (K) Explains how development can be managed so as to move forward over diverse range of disciplines. [MCEA01-30K] (K) Explains how development can be managed so as to move forward over diverse range of teams. [MCEA01-40K] (K) Explains the concept of concurrent design, with respect to the overall life cycle. [MCEA01-50K] (K) Describes key interfaces and concurrent development links to stakeholders in the wider enterprise outside engineering (e.g. infrastructure management, human resource management, quality management, knowledge management, portfolio management, and life cycle model management). [MCEA01-60K] (K) Explains the practical implementation of concurrency with regard to resources. [MCEA01-70K]
2	Describes the advantages and disadvantages of concurrent engineering. [MCEA02]	(K) Explains how concurrent development can reduce development time and how this can lead to reduce cost. [MCEA02-10K] (K) Explains why reduced development time and 'time to market' is important. [MCEA02-20K] (K) Explains why increased communications is important in concurrent development. [MCEA02-30K] (K) Explains how concurrent development may compromise design (e.g. reduced analysis time). [MCEA02-40K] (K) Explains how concurrent development may increase risk. [MCEA02-50K] (K) Explains strategies for reducing issues, such as increased management vigilance. [MCEA02-60K] (K) Explains why an efficient information control infrastructure is essential. [MCEA02-70K]

Supervised Practitioner – Management: Concurrent Engineering			
ID	**Indicators of Competence** ***(in addition to those at Awareness level)***	**Relevant knowledge, experience, and/or behaviors**	**Possible examples of objective evidence of personal contribution to activities performed, or professional behaviors applied**
1	Describes Systems Engineering life cycle processes in place on their project and how concurrency issues may impact its successful execution. [MCES01]	(K) Explains the concept of task interdependencies. [MCES01-10K] (K) Explains the challenges of Configuration control in concurrent development. [MCES01-20K] (K) Explains the challenges of Interface definition in concurrent development. [MCES01-30K] (K) Explains the challenges of communicating across multidisciplinary design team. [MCES01-40K] (A) Identifies concurrent engineering activities and associated challenges on their project. [MCES01-50A]	Records showing concurrent engineering tasks performed. [MCES01-E10] Records from baselined work products as part of milestones and associated tasks. [MCES01-E20] Project schedule showing tasks interdependencies. [MCES01-E30]
2	Coordinates concurrent engineering activities on a Systems Engineering project. [MCES02]	(A) Uses governing process and tools to plan and control concurrent development schedule. [MCES02-10A] (A) Identifies interdependencies during plan creation, resolving them as necessary. [MCES02-20A] (P) Acts to facilitate concurrent flow of information across team. [MCES02-30P]	Records showing contribution to the scheduling of multiple concurrent tasks. [MCES02-E10]
3	Prepares inputs to concurrency-related inputs to management plans for a Systems Engineering project. [MCES03]	(A) Coordinates information supporting management plan generation across multidisciplinary design team. [MCES03-10A] (A) Identifies interdependencies during plan creation, resolving them as necessary. [MCES03-20A] (A) Prepares information supporting performance parameters and their management. [MCES03-30A]	Records showing contribution to the planning and execution of concurrent engineering tasks across multiple teams. [MCES03-E10] Records showing contribution to planning of multiple interdependent tasks. [MCES03-E20] Records showing contributions to measurement of concurrent task performance. [MCES03-E30]
4	Develops own understanding of this competency area through Continual Professional Development (CPD). [MCES04]	(A) Identifies potential gaps in own knowledge or development needs in this area, identifying opportunities to address these through continual professional development activities. [MCES04-10A] (A) Performs continual professional development activities to improve their knowledge and understanding in this area. [MCES04-20A] (A) Records continual professional development activities undertaken including learning or insights gained. [MCES04-30A]	Records of Continual Professional Development (CPD) performed and learning outcomes. [MCES04-E10]

Practitioner – Management: Concurrent Engineering			
ID	**Indicators of Competence *(in addition to those at Supervised Practitioner level)***	**Relevant knowledge, experience, and/or behaviors**	**Possible examples of objective evidence of personal contribution to activities performed, or professional behaviors applied**
1	Creates governing concurrency management strategies and uses these to perform concurrent engineering on a project. [MCEP01]	(A) Identifies multi-disciplinary design teams that cover entire life cycle. [MCEP01-10A]	Development and production plan development showing concurrent engineering. [MCEP01-E10]
		(A) Identifies methods for evaluation of product against requirements during development. [MCEP01-20A]	Established production lines executed in parallel with development teams. [MCEP01-E20]
2	Identifies elements which can be developed concurrently on a Systems Engineering project. [MCEP02]	(A) Identifies the schedule for engineering elements so that they can be developed concurrently. [MCEP02-10A]	Project/program schedules showing concurrent engineering elements. [MCEP02-E10]
		(A) Identifies element dependencies and their relationships. [MCEP02-20A]	
		(A) Identifies inputs and outputs of tasks associated with development elements. [MCEP02-30A]	
3	Identifies concurrent interactions within a Systems Engineering life cycle on a project. [MCEP03]	(A) Identifies the concurrent scheduling of engineering tasks. [MCEP03-10A]	Project/program schedules showing concurrent engineering tasks. [MCEP03-E10]
		(A) Identifies concurrent task dependencies and relationships. [MCEP03-20A]	
		(A) Defines inputs and outputs of system development tasks. [MCEP03-30A]	
4	Coordinates concurrent activities and deals with emerging issues on a Systems Engineering project. [MCEP04]	(A) Performs Configuration management on concurrent tasks. [MCEP04-10A]	Configuration control plan. [MCEP04-E10]
		(P) Guides and actively coordinates the concurrent development of interfaces. [MCEP04-20P]	Interface Control Document. [MCEP04-E20]
		(A) Uses change management processes to effect interfaces or system performance. [MCEP04-30A]	Schedule for baselining or interface definition milestones in a project/program schedule. [MCEP04-E30]
		(A) Monitors concurrent tasks to determine progress. [MCEP04-40A]	Task performance or schedule variance interventions. [MCEP04-E40]
		(A) Identifies design review strategy (such as when to hold reviews and maturity of artifacts). [MCEP04-50A]	Resource budget management records. [MCEP04-E50]

(Continued)

ID	Indicators of Competence *(in addition to those at Supervised Practitioner level)*	Relevant knowledge, experience, and/or behaviors	Possible examples of objective evidence of personal contribution to activities performed, or professional behaviors applied
		(A) Performs Interface control on concurrent elements. [MCEP04-60A] (A) Analyzes change requests that effect interfaces or system performance. [MCEP04-70A] (A) Monitors the status of actual resource utilization vs. budget over time, updating as required. [MCEP04-80A] (A) Monitors the status of actual performance vs. budget over time, updating as required. [MCEP04-90A]	Performance budget management records. [MCEP04-E60]
5	Identifies concurrency-related aspects of appropriate management plans for a Systems Engineering project. [MCEP05]	(A) Defines and documents applicable engineering processes. [MCEP05-10A] (A) Identifies and documents applicable life cycle and process tailoring. [MCEP05-20A] (A) Identifies and documents the interfaces to be managed during development. [MCEP05-30A] (A) Identifies and documents system budgets (such as resource and performance) to be managed during development. [MCEP05-40A] (A) Ensures concurrent design approach is identified by management plans. [MCEP05-50A]	Systems Engineering plan concurrency section or comments. [MCEP05-E10]
6	Analyzes concurrency issues and risks on a Systems Engineering project. [MCEP06]	(A) Analyzes project concurrency issues to maintain design integrity. [MCEP06-10A] (A) Uses Change control to address concurrency issues and risks. [MCEP06-20A] (A) Uses Configuration management to address concurrency issues and risks. [MCEP06-30A] (A) Analyzes Interface management approaches for concurrent developments. [MCEP06-40A] (A) Analyzes technical performance measures. [MCEP06-50A] (A) Analyzes system resource budgets. [MCEP06-60A]	Technical notes, reports, email highlighting individual issues or the generic issues and risks. [MCEP06-E10] Periodic project/program report showing awareness of and highlighting these issues and risks. [MCEP06-E20]

7	Guides new or supervised practitioners in concurrent engineering principles in order to develop their knowledge, abilities, skills, or associated behaviors. [MCEP07]	(P) Guides new or supervised practitioners in executing activities that form part of this competency. [MCEP07-10P]	Organizational Breakdown Structure showing their responsibility for technical supervision in this area. [MCEP07-E10]
		(A) Trains individuals to an "Awareness" level in this competency area. [MCEP07-20A]	On-the-job training objectives/guidance etc. [MCEP07-E20]
			Coaching or mentoring assignment records. [MCEP07-E30]
			Records highlighting their impact on another individual in terms of improvement or professional development in this competency. [MCEP07-E40]
8	Maintains and enhances own competence in this area through Continual Professional Development (CPD) activities. [MCEP08]	(A) Identifies potential development needs in this area, identifying opportunities to address these through continual professional development activities. [MCEP08-10A]	Records of Continual Professional Development (CPD) performed and learning outcomes. [MCEP08-E10]
		(A) Performs continual professional development activities to maintain and enhance their competency in this area. [MCEP08-20A]	
		(A) Records continual professional development activities undertaken including learning or insights gained. [MCEP08-30A]	

Lead Practitioner – Management: Concurrent Engineering			
ID	**Indicators of Competence** ***(in addition to those at Practitioner level)***	**Relevant knowledge, experience, and/or behaviors**	**Possible examples of objective evidence of personal contribution to activities performed, or professional behaviors applied**
1	Creates enterprise-level policies, procedures, guidance, and best practice for concurrent engineering, including associated tools. [MCEL01]	(A) Analyzes enterprise need for concurrent engineering policies, processes, tools, or guidance. [MCEL01-10A]	Enterprise level reports and recommendations. [MCEL01-E10]
		(A) Creates enterprise policies, procedures or guidance on concurrent engineering. [MCEL01-20A]	Part of enterprise SE resource pool and proposal or internal R&D reviews. [MCEL01-E20]
		(A) Selects and acquires appropriate tools supporting concurrent engineering. [MCEL01-30A]	
2	Judges the tailoring of enterprise-level concurrency processes and associated work products to meet the needs of a project. [MCEL02]	(A) Evaluates the enterprise concurrent engineering processes against the business and external stakeholder (e.g. customer) needs in order to tailor processes to enable project success. [MCEL02-10A]	Part of enterprise SE resource pool and proposal or internal R&D reviews. [MCEL02-E10]
		(A) Provides constructive feedback on concurrent engineering processes. [MCEL02-20A]	Number of programs supported with successful results. [MCEL02-E20]
		(A) Reviews high value or high-risk program concurrency planning issues across the enterprise. [MCEL02-30A]	Panel reports and white papers generated. [MCEL02-E30]
		(A) Judges the tailoring of enterprise-level concurrency processes and associated work products to meet the needs of a project. [MCEL02-40A]	
3	Coordinates concurrent activities and deals with emerging issues across multiple diverse projects, or across a complex system, with proven results. [MCEL03]	(A) Compares and contrasts the benefits of a range of techniques for performing concurrent engineering across the enterprise. [MCEL03-10A]	Internal or external consultant in the relevant areas. [MCEL03-E10]
		(A) Uses knowledge and experience of the application of different concurrent engineering techniques with successful results. [MCEL03-20A]	Projects that have been improved through activities. [MCEL03-E20]
		(A) Coordinates concurrent activities and deals with emerging issues across multiple diverse projects, or across a complex system, with proven results. [MCEL03-30A]	

4	Guides and actively coordinates interactions within Systems Engineering concurrency issues across multiple diverse projects, or across a complex system. [MCEL04]	(P) Acts as independent advisor to oversee and improve interactions between concurrent projects across the enterprise or across a complex project. [MCEL04-10P] (P) Advises on Systems Engineering concurrency issues across multiple diverse projects, or across a complex system. [MCEL04-20P]	Advice provided regarding concurrency issues across enterprise or on complex projects. [MCEL04-E10]
5	Judges on concurrency issues and risks across multiple diverse projects, or on a complex system. [MCEL05]	(A) Advises on the approach to external stakeholders (e.g. customers) and senior program managers where concurrency issues and risks exist, justifying their proposals. [MCEL05-10A]	Meeting minutes or report showing authored advice. [MCEL05-E10]
6	Judges the suitability of plans for concurrent system developments from across the enterprise. [MCEL06]	(A) Reviews the suitability of plans for concurrent engineering developments from across the enterprise. [MCEL06-10A]	Approval records from multiple management plans. [MCEL06-E10]
7	Persuades key stakeholders to address identified enterprise-level concurrent engineering issues. [MCEL07]	(A) Advises senior program managers on the justification of Concurrent Engineering techniques. [MCEL07-10A] (A) Advises senior managers on the justification for specialized facilities to aid Concurrent Design. [MCEL07-20A] (A) Persuades stakeholders to implement strategies which address enterprise-level concurrency issues. [MCEL07-30A]	Meeting minutes or report showing authored recommendation. [MCEL07-E10] Records indicating changed stakeholder thinking from personal intervention. [MCEL07-E20]
8	Coaches or mentors practitioners across the enterprise in Systems Engineering concurrent engineering in order to develop their knowledge, abilities, skills, or associated behaviors. [MCEL08]	(P) Coaches or mentors practitioners across the enterprise in competency-related techniques, recommending development activities. [MCEL08-10P] (A) Develops or authorizes enterprise training materials in this competency area. [MCEL08-20A] (A) Provides enterprise workshops/seminars or training in this competency area. [MCEL08-30A]	Coaching or mentoring assignment records. [MCEL08-E10] Records of formal training courses, workshops, seminars, and authored training material supported by successful post-training evaluation data. [MCEL08-E20] Listing as an approved organizational trainer for this competency area. [MCEL08-E30]
9	Promotes the introduction and use of novel techniques and ideas in concurrent engineering management across the enterprise, to improve enterprise competence in this area. [MCEL09]	(A) Analyzes different approaches across different domains through research. [MCEL09-10A] (A) Defines novel approaches that could potentially improve the SE discipline within the enterprise. [MCEL09-20A] (P) Fosters awareness of these novel techniques within the enterprise. [MCEL09-30P]	Research records. [MCEL09-E10] Published papers in refereed journals/company literature. [MCEL09-E20] Records showing introduction of enabling systems supporting the new techniques or ideas. [MCEL09-E30]

(Continued)

ID	Indicators of Competence *(in addition to those at Practitioner level)*	Relevant knowledge, experience, and/or behaviors	Possible examples of objective evidence of personal contribution to activities performed, or professional behaviors applied
		(P) Collaborates with enterprise stakeholders to facilitate the introduction of techniques new to the enterprise. [MCEL09-40P] (A) Monitors new techniques after their introduction to determine their effectiveness. [MCEL09-50A] (A) Adapts approach to reflect actual enterprise performance improvements. [MCEL09-60A]	Published papers (or similar) at enterprise level. [MCEL09-E40] Records of improvements made against a recognized process improvement model in this area. [MCEL09-E50]
10	Develops expertise in this competency area through specialist Continual Professional Development (CPD) activities. [MCEL10]	(A) Identifies own needs for further professional development in order to increase competence beyond practitioner level. [MCEL10-10A] (A) Performs professional development activities in order to move own competence toward expert level. [MCEL10-20A] (A) Records continual professional development activities undertaken including learning or insights gained. [MCEL10-30A]	Records of Continual Professional Development (CPD) performed and learning outcomes. [MCEL10-E10]

NOTES	In addition to items above, enterprise-level or independent 3rd-Party-generated evidence may be used to amplify other evidence presented and may include: a. Formally recognized by senior management in current organization as an expert in this competency area b. Evidence of role as Product/System Design Authority or Technical Authority on a complex project with responsibilities in this area or where skills within this competency area were used c. Recognized as an authorizing signatory on behalf of enterprise for formal documentation in this competency area (e.g. policies, processes, and deliverables) d. Formal commendation or award within own enterprise for contribution or item of work successfully performed, which required proficiency in this competency area e. Customer, Supplier, or other external project-specific key Stakeholder accolades for specific work performed in this competency area f. Independently assessed or accredited work in this competency area (e.g. for independent publication or use) g. Formal organizational HR records positively highlighting any specific professional competencies or behaviors identified (if applicable) plus any of the evidence indicators listed at Expert level below

Expert – Management: Concurrent Engineering			
ID	**Indicators of Competence *(in addition to those at Lead Practitioner level)***	**Relevant knowledge, experience, and/or behaviors**	**Possible examples of objective evidence of personal contribution to activities performed, or professional behaviors applied**
1	Communicates own knowledge and experience in Systems Engineering concurrent engineering in order to improve best practice beyond the enterprise boundary. [MCEE01]	(A) Produces papers, seminars, or presentations outside own enterprise for publication in order to share own ideas and improve industry best practices in this competence area. [MCEE01-10A]	Published papers or books etc. on new technique in refereed journals/company literature. [MCEE01-E10]
		(P) Fosters incorporation of own ideas into industry best practices in this area. [MCEE01-20P]	Published papers in refereed journals or internal literature proposing new practices in this competence area (or presentations, tutorials, etc.). [MCEE01-E20]
		(P) Develops guidance materials identifying new (or updating existing) best practice in this competence area. [MCEE01-30P]	Own proposals adopted as industry best practices in this competence area. [MCEE01-E30]
2	Influences key stakeholders beyond the enterprise boundary in order to resolve concurrent engineering issues. [MCEE02]	(P) Acts as independent member of industry panels and professional societies. [MCEE02-10P]	Records showing active role supporting a number of projects successfully. [MCEE02-E10]
		(A) Provides evidence and arguments influencing external stakeholder concurrent engineering activities. [MCEE02-20A]	Publications and presentations used in support of stakeholder arguments. [MCEE02-E20]
		(P) Acts as independent external member of review panels on high-risk concurrent programs or external stakeholder-led reviews. [MCEE02-30P]	Review minutes showing contribution. [MCEE02-E30]
3	Advises organizations beyond the enterprise boundary on the suitability of their approach to concurrent engineering developments. [MCEE03]	(P) Acts as independent reviewer on approval of concurrent engineering plans. [MCEE03-10P]	Records of membership on oversight committee with relevant terms of reference. [MCEE03-E10]
		(A) Advises organizations on their approach to concurrent engineering. [MCEE03-20A]	Review comments or revised document. [MCEE03-E20]
4	Advises organizations beyond the enterprise boundary on complex or sensitive concurrency issues. [MCEE04]	(A) Advises external stakeholders (e.g. customers) on their concurrency issues. [MCEE04-10A]	Meeting minutes or report showing authored advice. [MCEE04-E10]
		(A) Conducts sensitive concurrency-related negotiations. [MCEE04-20A]	

(Continued)

ID	Indicators of Competence *(in addition to those at Lead Practitioner level)*	Relevant knowledge, experience, and/or behaviors	Possible examples of objective evidence of personal contribution to activities performed, or professional behaviors applied
		(A) Conducts complex concurrency-related negotiations taking account of external stakeholders (e.g. customers) background and knowledge. [MCEE04-30A]	
		(P) Arbitrates on marginal decisions which affect concurrency. [MCEE04-40P]	
5	Develops new strategies for concurrent engineering for use beyond the enterprise boundary. [MCEE05]	(A) Develops new methods to improve concurrency efficiency and effectiveness. [MCEE05-10A]	Published papers in refereed journals. [MCEE05-E10]
		(A) Develops new facilities/infrastructure to facilitate concurrency. [MCEE05-20A]	New facility supporting concurrent engineering. [MCEE05-E20]
6	Advises organizations beyond the enterprise boundary on concurrency issues and risks. [MCEE06]	(A) Analyzes engineering management plans for their suitability to facilitate concurrent engineering. [MCEE06-10A]	Meeting minutes or report showing authored advice. [MCEE06-E10]
		(A) Advises external stakeholders (e.g. customers) and senior program managers on best approach where there is concurrency issues and risks. [MCEE06-20A]	Sign-off on multiple SEMPs. [MCEE06-E20]
7	Champions the introduction of novel techniques and ideas in concurrency management, beyond the enterprise boundary, in order to develop the wider Systems Engineering community in this competency. [MCEE07]	(A) Analyzes different approaches across different domains through research. [MCEE07-10A]	Records of activities promoting research and need to adopt novel technique or ideas. [MCEE07-E10]
		(A) Produces reports for the wider SE community on the effectiveness of new techniques after their introduction. [MCEE07-20A]	Records of improvements made to process and appraisal against a recognized process improvement model. [MCEE07-E20]
		(P) Collaborates with those introducing novel techniques within the wider SE community. [MCEE07-30P]	Research records. [MCEE07-E30]
		(A) Defines novel approaches that could potentially improve the wider SE discipline. [MCEE07-40A]	Published papers in refereed journals/company literature. [MCEE07-E40]
		(P) Fosters awareness of these novel techniques within the wider SE community. [MCEE07-50P]	Records showing introduction of enabling systems supporting the new techniques or ideas. [MCEE07-E50]

8	Coaches individuals beyond the enterprise boundary, in the concurrent engineering of Systems Engineering projects in order to further develop their knowledge, abilities, skills, or associated behaviors. [MCEE08]	(P) Coaches or mentors individuals beyond the enterprise boundary, in competency-related techniques, recommending development activities. [MCEE08-10P]	Coaching or mentoring assignment records. [MCEE08-E10]
		(A) Develops or authorizes training materials in this competency area, which are subsequently successfully delivered beyond the enterprise boundary. [MCEE08-20A]	Records of formal training courses, workshops, seminars, and authored training material supported by successful post-training evaluation data. [MCEE08-E20]
		(A) Provides workshops/seminars or training in this competency area for practitioners or lead practitioners beyond the enterprise boundary (e.g. conferences and open training days). [MCEE08-30A]	Records of Training/workshops/seminars created supported by successful post-training evaluation data. [MCEE08-E30]
9	Maintains expertise in this competency area through specialist Continual Professional Development (CPD) activities. [MCEE09]	(A) Reviews research, new ideas, and state of the art to identify relevant new areas requiring personal development in order to maintain expertise in this competency area. [MCEE09-10A]	Records of documents reviewed and insights gained as part of own research into this competency area. [MCEE09-E10]
		(A) Performs identified specialist professional development activities in order to maintain or further develop competence at expert level. [MCEE09-20A]	Records of Continual Professional Development (CPD) performed and learning outcomes. [MCEE09-E20]
		(A) Records continual professional development activities undertaken including learning or insights gained. [MCEE09-30A]	

NOTES	In addition to items above, enterprise-level or independent 3rd-Party-generated evidence may be used to amplify other evidence presented and may include: a. Formally recognized by a reputable external organization as an expert in this competency area b. Evidence of role as independent assessor or reviewer on project outside own organization where skills in this competency area were used c. Evidence of invitation(s) from wider community for contribution of systems engineering expertise in this area (e.g. industry conference panel, government advisory board etc. cross-industry working groups, partnerships, accredited advanced university courses or research, or as part of professional institute) d. Formal commendation beyond the enterprise (e.g. by INCOSE or other recognized authority) for work performed in this competency area e. Independently assessed or accredited work product in this competency area (e.g. for independent publication or use) f. Accolades of expertise in this area from recognized industry leaders

(Continued)

Competency area – Management: Business and Enterprise Integration
Description Businesses and Enterprises are systems in their own right. Systems Engineering is just one of many activities that must occur in order to bring about a successful system development meeting the needs of all its stakeholders. Systems Engineering addresses the needs of all other internal business and enterprise stakeholders, covering areas such as infrastructure, portfolio management, human resources, knowledge management, quality, information technology, production, sales, marketing, commercial, legal, and finance, within and beyond the local enterprise.
Why it matters As businesses and enterprises become larger, more complex and the functions within the enterprise more insular, the interdependencies between individual enterprise functions should be engineered using a systems approach at an enterprise level in order to meet the demands of increased business effectiveness and efficiency.
Possible contributory types of evidence Any combination of the types of evidence may be acceptable (depending on how the Framework is tailored and used). The evidence items identified at each level indicate example work products only. Contributions to work products will generally differ at each proficiency level.
Learning and development The INCOSE Professional Development Portal provides example guidance on how to gain an initial awareness of a competency area and options for developing further competence thereafter.

Awareness – Management: Business and Enterprise Integration		
ID	**Indicators of Competence**	**Relevant knowledge sub-indicators**
1	Explains why a business or enterprise is a system in its own right and describes the business or enterprise "system" using Systems Engineering ideas and terminology. [MBEA01]	(K) Describes analogies between systems and the business infrastructure. [MBEA01-10K] (K) Describes the importance/relevance of interfaces, processes, and methodologies governing operations. [MBEA01-20K] (K) Lists examples of Influences and interactions across the business or enterprise. [MBEA01-30K] (K) Explains how organizational culture may affect business or enterprise integration. [MBEA01-40K]
2	Lists other business or enterprise functions and provides examples. [MBEA02]	(K) Lists business or enterprise business functions affecting Systems Engineering activities (e.g. HR). [MBEA02-10K] (K) Describes key interfaces and links between engineering and stakeholders in the wider enterprise outside engineering (e.g. infrastructure management, human resource management, quality management, knowledge management, portfolio management, and life cycle model management). [MBEA02-20K]

3	Lists work products from business and enterprise functions to the Systems Engineering process and vice versa and can provide examples. [MBEA03]	(K) Lists examples of enterprise processes which link enterprise functions to Systems Engineering. [MBEA03-10K] (K) Lists outputs from Systems Engineering, which have dependencies on other functions. [MBEA03-20K] (K) Lists Inputs to Systems Engineering process, which are dependent on other functions. [MBEA03-30K] (K) Explains how models and frameworks can link business and enterprise concepts. [MBEA03-40K] (K) Explains how allocation of functions and responsibilities can impact Systems Engineering activities. [MBEA03-50K] (K) Explains how a "Gate review" links the wider enterprise to ongoing project activities. [MBEA03-60K] (K) Explains how an enterprise-level peer assessment of project activities can support Systems Engineering. [MBEA03-70K]

Supervised Practitioner – Management: Business and Enterprise Integration			
ID	**Indicators of Competence** *(in addition to those at Awareness level)*	**Relevant knowledge, experience, and/or behaviors**	**Possible examples of objective evidence of personal contribution to activities performed, or professional behaviors applied**
1	Follows a governing process and appropriate tools to plan and control their own business and enterprise integration activities. [MBES01]	(K) Describes the relationship between Systems Engineering processes and enterprise activities. [MBES01-10K] (A) Performs Systems Engineering activities, which support enterprise integration (e.g. risk management, decision management, and data management). [MBES01-20A]	Systems Engineering processes they have helped to integrate within the wider business. [MBES01-E10] Systems Engineering tools they have helped to integrate into the wider business. [MBES01-E20]
2	Describes Systems Engineering work products that support business and enterprise infrastructure and provides examples of why this is the case. [MBES02]	(K) Describes Systems Engineering Work products associated with Enterprise models, architectures and frameworks in the context of processes and functional entities. [MBES02-10K] (K) Describes the business/enterprise infrastructure and how Systems Engineering information is disseminated and coordinated and maintained. [MBES02-20K] (K) Describes the management of interactions between different business and enterprise functions. [MBES02-30K]	Systems Engineering work products they have helped to integrate into the wider business. [MBES02-E10]
3	Describes work products produced elsewhere in the enterprise, which impact Systems Engineering activities and provides examples of why this is the case. [MBES03]	(K) Describes enterprise models, architectures, and frameworks impacting Systems Engineering in the context of processes and functional entities. [MBES03-10K] (K) Describes the business/enterprise infrastructure and how information is disseminated and coordinated with Systems Engineering. [MBES03-20K] (K) Describes interactions between enterprise and business functions. [MBES03-30K]	
4	Prepares inputs to Systems Engineering work products required by other business and enterprise functions. [MBES04]	(A) Prepares information for business and enterprise reports. [MBES04-10A] (A) Ensures information exchange and dissemination across the business or enterprise. [MBES04-20A] (K) Identifies shared working environments and associated business and enterprise tools. [MBES04-30K]	Systems Engineering work products required by the wider business they have helped develop. [MBES04-E10]

5	Analyzes the impact of work products produced by other business and enterprise functions for their impact on Systems Engineering activities, in order to improve integration across the project or enterprise. [MBES05]	(A) Prepares information in support of analysis work performed on work products produced by other enterprise functions for their impact on Systems Engineering activities. [MBES05-10A]	Work products by the wider business reviewed for their impact on Systems Engineering activities. [MBES05-E10]
6	Develops own understanding of this competency area through Continual Professional Development (CPD). [MBES06]	(A) Identifies potential gaps in own knowledge or development needs in this area, identifying opportunities to address these through continual professional development activities. [MBES06-10A] (A) Performs continual professional development activities to improve their knowledge and understanding in this area. [MBES06-20A] (A) Records continual professional development activities undertaken including learning or insights gained. [MBES06-30A]	Records of Continual Professional Development (CPD) performed and learning outcomes. [MBES06-E10]

Practitioner – Management: Business and Enterprise Integration			
ID	**Indicators of Competence** ***(in addition to those at Supervised Practitioner level)***	**Relevant knowledge, experience, and/or behaviors**	**Possible examples of objective evidence of personal contribution to activities performed, or professional behaviors applied**
1	Creates a strategy for Business and Enterprise integration on a project to support SE project and wider enterprise needs. [MBEP01]	(A) Identifies project-specific business and enterprise integration needs that need to be incorporated into SE planning. [MBEP01-10A] (A) Prepares project-specific business and enterprise integration inputs for SE planning purposes, using appropriate processes and procedures. [MBEP01-20A] (A) Prepares project-specific business and enterprise integration task estimates in support of SE planning. [MBEP01-30A] (P) Liaises throughout with stakeholders to gain approval, updating strategy as necessary. [MBEP01-40P]	SE business and enterprise integration work products. [MBEP01-E10]
2	Identifies the needs of the wider business and enterprise in order to ensure integration of ongoing activities (e.g. portfolio sustainment, infrastructure, HR, and knowledge management). [MBEP02]	(A) Identifies wider business or enterprise functions, which require integration into ongoing Systems Engineering projects in order to meet business or enterprise objectives. [MBEP02-10A] (A) Identifies wider processes or procedures, which need to be followed or tailored by Systems Engineering to ensure business and enterprise integration. [MBEP02-20A] (A) Identifies Systems Engineering activities required to ensure business and enterprise integration. [MBEP02-30A]	SE Work products created. [MBEP02-E10]
3	Creates Systems Engineering work products needed by to manage infrastructure across business or enterprise objectives. [MBEP03]	(K) Describes key elements of a successful process, which integrates enterprise infrastructure needs and how this affects different stages of a system life cycle. [MBEP03-10K] (K) Describes key elements of a successful process, which integrates Business and Enterprise Knowledge Management needs and how this affects different stages of a system life cycle. [MBEP03-20K] (A) Uses appropriate standards and governing processes to ensure that enterprise infrastructure needs are incorporated in wider SE planning, tailoring as appropriate. [MBEP03-30A] (A) Uses appropriate standards and governing processes to ensure that Business and Enterprise Knowledge Management needs are incorporated in wider SE planning, tailoring as appropriate. [MBEP03-40A]	SE work products they have developed. [MBEP03-E10] SE work products they have developed. [MBEP03-E20] Other project/program plan or organizational process they have written. [MBEP03-E30] Other project/program plan or organizational process they have written. [MBEP03-E40]

4	Creates Systems Engineering work products needed to initiate or sustain the needs of the wider business or enterprise project portfolio. [MBEP04]	(K) Describes key elements of a successful process which integrates Business portfolio needs and how this affects different stages of a system life cycle. [MBEP04-10K] (A) Uses appropriate standards and governing processes to ensure that business portfolio needs are incorporated in wider SE planning, tailoring as appropriate. [MBEP04-20A] (A) Creates plans and processes for use on a project when ensure business portfolio needs are accommodated on the project. [MBEP04-30A] (P) Liaises throughout with stakeholders to gain approval, updating plans as necessary. [MBEP04-40P]	SE work products they have developed. [MBEP04-E10] Other project/program plan or organizational process they have written. [MBEP04-E20]
5	Creates Systems Engineering work products needed by Human Resources in order to meet business or enterprise objectives. [MBEP05]	(K) Describes key elements of a successful process, which integrates enterprise Human Resources needs and how this affects different stages of a system life cycle. [MBEP05-10K] (A) Uses appropriate standards and governing processes to ensure that enterprise Human Resources needs are incorporated in wider SE planning, tailoring as appropriate. [MBEP05-20A] (A) Creates plans and processes for use on a project when ensure enterprise Human Resources needs are accommodated on the project. [MBEP05-30A] (P) Liaises throughout with stakeholders to gain approval, updating plans as necessary. [MBEP05-40P]	SE work products they have developed. [MBEP05-E10] Other project/program plan or organizational process they have written. [MBEP05-E20]
6	Creates Systems Engineering work products needed by to manage knowledge across business or enterprise objectives in order to support enterprise knowledge management, reuse, or exploitation. [MBEP06]	(K) Describes key elements of a successful process which integrates enterprise Knowledge Management needs and how this affects different stages of a system life cycle. [MBEP06-10K] (A) Uses appropriate standards and governing processes to ensure that enterprise Knowledge Management needs are incorporated in wider SE planning, tailoring as appropriate. [MBEP06-20A] (A) Creates plans and processes for use on a project when ensure enterprise Knowledge Management needs are accommodated on the project. [MBEP06-30A] (P) Liaises throughout with stakeholders to gain approval, updating plans as necessary. [MBEP06-40P]	SE work products they have developed. [MBEP06-E10] Other project/program plan or organizational process they have written. [MBEP06-E20]

ID	Indicators of Competence *(in addition to those at Supervised Practitioner level)*	Relevant knowledge, experience, and/or behaviors	Possible examples of objective evidence of personal contribution to activities performed, or professional behaviors applied
7	Complies with governing plans and processes for business and enterprise integration, interpreting, evolving, or seeking guidance where appropriate. [MBEP07]	(A) Follows defined plans, processes and associated tools to ensure successful business and enterprise integration on a project. [MBEP07-10A]	Enterprise integration plans or work products they have interpreted or challenged to address project issues. [MBEP07-E10]
8	Uses Systems Engineering techniques to contribute to the definition of the business or enterprise. [MBEP08]	(A) Uses systems concepts and system design techniques in support of business or enterprise definition. [MBEP08-10A] (P) Fosters collaboration between disparate functions across the business or enterprise. [MBEP08-20P] (A) Uses Systems Engineering to determine how issues addressed by business and enterprise plans will impact at different levels and phases throughout the life cycle. [MBEP08-30A] (A) Identifies and implements changes to business and organizational practices based upon Systems Engineering techniques. [MBEP08-40A]	Enterprise models/architectures and frameworks. [MBEP08-E10] Technology plan. [MBEP08-E20] Business or Enterprise improvement plan. [MBEP08-E30] Updated practices. [MBEP08-E40]
9	Reviews work products produced by other enterprise functions for their impact on Systems Engineering activities. [MBEP09]	(A) Analyzes and judges work products produced by other enterprise functions for their impact on Systems Engineering activities. [MBEP09-10A]	Comments on work products. [MBEP09-E10]
10	Identifies constraints placed on the Systems Engineering process by the business or enterprise. [MBEP10]	(A) Identifies boundaries within the business or enterprise and the resultant framework of operation, clarifying as necessary. [MBEP10-10A] (A) Communicates information between distinct functions within the business or enterprise. [MBEP10-20A] (A) Identifies business or enterprise constraints affecting the management of information and/or knowledge and acts to address these. [MBEP10-30A]	List of constraints (non-functional requirements on a system). [MBEP10-E10] Records highlighting the organizational conflict and contribution to its resolution. [MBEP10-E20] Allocation of tasks. [MBEP10-E30] Information on time and to the right place. [MBEP10-E40]

11	Guides new or supervised practitioners in Business and Enterprise integration in order to develop their knowledge, abilities, skills, or associated behaviors. [MBEP11]	(P) Guides new or supervised practitioners in executing activities that form part of this competency. [MBEP11-10P]	Organizational Breakdown Structure showing their responsibility for technical supervision in this area. [MBEP11-E10]
		(A) Trains individuals to an "Awareness" level in this competency area. [MBEP11-20A]	On-the-job training objectives/guidance etc. [MBEP11-E20] Coaching or mentoring assignment records. [MBEP11-E30] Records highlighting their impact on another individual in terms of improvement or professional development in this competency. [MBEP11-E40]
12	Maintains and enhances own competence in this area through Continual Professional Development (CPD) activities. [MBEP12]	(A) Identifies potential development needs in this area, identifying opportunities to address these through continual professional development activities. [MBEP12-10A] (A) Performs continual professional development activities to maintain and enhance their competency in this area. [MBEP12-20A] (A) Records continual professional development activities undertaken including learning or insights gained. [MBEP12-30A]	Records of Continual Professional Development (CPD) performed and learning outcomes. [MBEP12-E10]

Lead Practitioner – Management: Business and Enterprise Integration			
ID	**Indicators of Competence** ***(in addition to those at Practitioner level)***	**Relevant knowledge, experience, and/or behaviors**	**Possible examples of objective evidence of personal contribution to activities performed, or professional behaviors applied**
1	Creates enterprise-level policies, procedures, guidance, and best practice for business and enterprise integration, including associated tools. [MBEL01]	(A) Analyzes enterprise need for policies, processes, tools, or guidance to address business or enterprise integration. [MBEL01-10A]	Successful projects/programs with technology advances either in depth and/or broader application. [MBEL01-E10]
		(A) Creates policies, processes, and guidance documents covering business or enterprise integration activities. [MBEL01-20A]	Technology Strategy Document contribution. [MBEL01-E20]
		(A) Selects and acquires appropriate tools supporting business or enterprise integration. [MBEL01-30A]	Records of support for tool introduction. [MBEL01-E30]
2	Judges the tailoring of enterprise-level business and enterprise processes and associated work products to meet the needs of a project. [MBEL02]	(A) Describes enterprise processes and products beyond Systems Engineering. [MBEL02-10A]	Tailoring reports. [MBEL02-E10]
		(A) Selects relevant processes and templates from those available. [MBEL02-20A]	
		(A) Adapts business or enterprise products to meet project needs. [MBEL02-30A]	
3	Coordinates business and enterprise integration across multiple diverse projects or across a complex system, with proven success. [MBEL03]	(P) Acts to facilitate enterprise level resource issues on high-payoff/high-risk projects. [MBEL03-10P]	List of programs supported and contribution. [MBEL03-E10]
			Programs improved through actions in this area. [MBEL03-E20]
4	Judges Systems Engineering work products created for use by other parts of the business and enterprise. [MBEL04]	(P) Acts as independent reviewer in product and milestone reviews across the enterprise. [MBEL04-10P]	Review comments made. [MBEL04-E10]
		(A) Explains the application of business and enterprise integration within multiple context types. [MBEL04-20A]	List of programs supported and contribution. [MBEL04-E20]
		(A) Provides constructive feedback on work products used elsewhere in the business. [MBEL04-30A]	
5	Advises stakeholders across the enterprise regarding activities and work products affecting Systems Engineering. [MBEL05]	(P) Acts as independent reviewer of enterprise Systems Engineering policies, processes, tools, and templates. [MBEL05-10P]	Reports and white papers. [MBEL05-E10]

		(A) Analyzes and judges enterprise-level activities and work products affecting Systems Engineering. [MBEL05-20A] (A) Produces constructive feedback on key infrastructure, portfolio, human resource, quality, and knowledge management stakeholders. [MBEL05-30A]	
6	Advises stakeholders across the enterprise regarding the use of Systems Engineering techniques to contribute to the definition of the business/enterprise. [MBEL06]	(P) Persuades enterprise level stakeholders on the use of Systems Engineering to support wider business or enterprise usage. [MBEL06-10P] (P) Coaches or mentors enterprise level stakeholders in their adoption of Systems Engineering methods, processes, tools and templates. [MBEL06-20P]	Records showing support for business definition activities. [MBEL06-E10]
7	Persuades key stakeholders to address identified enterprise-level business and enterprise integration issues and constraints to reduce project cost, schedule, or technical risk. [MBEL07]	(A) Develops an enterprise technology strategy. [MBEL07-10A] (A) Advises enterprise stakeholders on the impact of business or enterprise impact of decisions made at project level. [MBEL07-20A] (A) Advises stakeholders on multiple projects on the impact of business or enterprise decisions on their projects. [MBEL07-30A] (A) Analyzes enterprise-level business and enterprise integration issues and constraints for their impact on Systems Engineering. [MBEL07-40A] (P) Persuades enterprise level stakeholders to address enterprise-level issues affecting Systems Engineering. [MBEL07-50P] (A) Coordinates improvements in business or enterprise integration. [MBEL07-60A]	Successful projects/programs with technology advances either in depth and/or broader application. [MBEL07-E10] Records indicating changed stakeholder thinking from personal intervention. [MBEL07-E20] Records of strategy changes coordinated across multiple projects. [MBEL07-E30]
8	Coaches or mentors practitioners across the enterprise in business and enterprise integration in order to develop their knowledge, abilities, skills, or associated behaviors. [MBEL08]	(P) Coaches or mentors practitioners across the enterprise in competency-related techniques, recommending development activities. [MBEL08-10P] (A) Develops or authorizes enterprise training materials in this competency area. [MBEL08-20A] (A) Provides enterprise workshops/seminars or training in this competency area. [MBEL08-30A]	Coaching or mentoring assignment records. [MBEL08-E10] Records of formal training courses, workshops, seminars, and authored training material supported by successful post-training evaluation data. [MBEL08-E20] Listing as an approved organizational trainer for this competency area. [MBEL08-E30]

(Continued)

ID	Indicators of Competence *(in addition to those at Practitioner level)*	Relevant knowledge, experience, and/or behaviors	Possible examples of objective evidence of personal contribution to activities performed, or professional behaviors applied
9	Promotes the introduction and use of novel techniques and ideas in business and enterprise integration across the enterprise, to improve the enterprise competence in this area. [MBEL09]	(A) Analyzes different approaches across different domains through research. [MBEL09-10A]	Research records. [MBEL09-E10]
		(A) Defines novel approaches that could potentially improve the SE discipline within the enterprise. [MBEL09-20A]	Published papers in refereed journals/company literature. [MBEL09-E20]
		(P) Fosters awareness of these novel techniques within the enterprise. [MBEL09-30P]	Enabling systems introduced in support of new techniques or ideas. [MBEL09-E30]
		(P) Collaborates and advises enterprise stakeholders to facilitate the introduction of techniques new to the enterprise. [MBEL09-40P]	Published papers (or similar) at enterprise level. [MBEL09-E40]
		(A) Monitors new techniques after their introduction to determine their effectiveness. [MBEL09-50A]	Records of improvements made against a recognized process improvement model in this area. [MBEL09-E50]
		(A) Adapts approach to reflect actual enterprise performance improvements. [MBEL09-60A]	
10	Develops expertise in this competency area through specialist Continual Professional Development (CPD) activities. [MBEL10]	(A) Identifies own needs for further professional development in order to increase competence beyond practitioner level. [MBEL10-10A]	Records of Continual Professional Development (CPD) performed and learning outcomes. [MBEL10-E10]
		(A) Performs professional development activities in order to move own competence toward expert level. [MBEL10-20A]	
		(A) Records continual professional development activities undertaken including learning or insights gained. [MBEL10-30A]	

NOTES	In addition to items above, enterprise-level or independent 3rd-Party-generated evidence may be used to amplify other evidence presented and may include: a. Formally recognized by senior management in current organization as an expert in this competency area b. Evidence of role as Product/System Design Authority or Technical Authority on a complex project with responsibilities in this area or where skills within this competency area were used c. Recognized as an authorizing signatory on behalf of enterprise for formal documentation in this competency area (e.g. policies, processes, and deliverables) d. Formal commendation or award within own enterprise for contribution or item of work successfully performed, which required proficiency in this competency area e. Customer, Supplier, or other external project-specific key Stakeholder accolades for specific work performed in this competency area f. Independently assessed or accredited work in this competency area (e.g. for independent publication or use) g. Formal organizational HR records positively highlighting any specific professional competencies or behaviors identified (if applicable) plus any of the evidence indicators listed at Expert level below

Expert – Management: Business and Enterprise Integration			
ID	**Indicators of Competence *(in addition to those at Lead Practitioner level)***	**Relevant knowledge, experience, and/or behaviors**	**Possible examples of objective evidence of personal contribution to activities performed, or professional behaviors applied**
1	Communicates own knowledge and experience in business and enterprise integration, in order to best practice beyond the enterprise boundary. [MBEE01]	(A) Produces papers, seminars, or presentations outside own enterprise for publication in order to share own ideas and improve industry best practices in this competence area. [MBEE01-10A]	Published papers or books etc. on new technique in refereed journals/company literature. [MBEE01-E10]
		(P) Fosters incorporation of own ideas into industry best practices in this area. [MBEE01-20P]	Published papers in refereed journals or internal literature proposing new practices in this competence area (or presentations, tutorials, etc.). [MBEE01-E20]
		(P) Develops guidance materials identifying new (or updating existing) best practice in this competence area. [MBEE01-30P]	Proposals adopted as industry best practice. [MBEE01-E30]
2	Influences stakeholders beyond the enterprise boundary on business or enterprise issues. [MBEE02]	(P) Acts as independent member of industry panels and professional societies. [MBEE02-10P]	Data from projects supported successfully. [MBEE02-E10]
		(A) Produces evidence and arguments influencing external stakeholder Business and Enterprise integration activities. [MBEE02-20A]	Publications and presentations used in support of stakeholder arguments. [MBEE02-E20]
		(P) Acts as independent reviewer on high-risk Business and Enterprise integration activity. [MBEE02-30P]	
3	Advises organizations beyond the enterprise boundary on the suitability of their approach to business or enterprise integration. [MBEE03]	(A) Analyzes the business or enterprise organization (e.g. assessing a model or otherwise). [MBEE03-10A]	Enterprise business or enterprise activities they have worked on. [MBEE03-E10]
		(A) Assesses the integration of systems engineering activities across a business or enterprise. [MBEE03-20A]	Minutes of meetings associated with business or enterprise integration. [MBEE03-E20]
4	Advises organizations beyond the enterprise boundary on the effectiveness of the business or enterprise as a system. [MBEE04]	(A) Analyzes or reviews potential methods to increase efficiency or enterprise related process improvements. [MBEE04-10A]	Process improvement plans and quantitative results. [MBEE04-E10]
		(A) Selects efficiency programs and enterprise related process improvements. [MBEE04-20A]	

(Continued)

ID	Indicators of Competence *(in addition to those at Lead Practitioner level)*	Relevant knowledge, experience, and/or behaviors	Possible examples of objective evidence of personal contribution to activities performed, or professional behaviors applied
5	Advises organizations beyond the enterprise boundary on developing a Systems Engineering capability within a business or enterprise context. [MBEE05]	(A) Analyzes or reviews business function agility and responsiveness. [MBEE05-10A]	Reports covering integration of business and enterprise. [MBEE05-E10]
		(A) Identifies appropriate Metrics assessing the enterprise Systems Engineering capability. [MBEE05-20A]	Recommendations supplied and taken up. [MBEE05-E20]
			Records of advice provided. [MBEE05-E30]
6	Advises organizations beyond the enterprise boundary on the impact of inputs from other business/enterprise functions on the Systems Engineering process. [MBEE06]	(A) Analyzes the impact of inputs from other business or enterprise functions and determines appropriate response. [MBEE06-10A]	Impact analysis. [MBEE06-E10]
		(A) Identifies enhanced or more efficient interactions and integration of "separate" enterprise functions. [MBEE06-20A]	Report showing uptake of recommendations of analysis. [MBEE06-E20]
			Report indicating impact e.g. time saving and/or quality of the delivery. [MBEE06-E30]
7	Advises organizations beyond the enterprise boundary on the impact of inputs from other business/enterprise functions on the Systems Engineering process. [MBEE06]	(A) Reviews enterprise technology strategies from external organizations. [MBEE07-10A]	Records indicating requests for support fulfilled in this area. [MBEE07-E10]
		(A) Advises external stakeholders on the business or enterprise impact of decisions. [MBEE07-20A]	Records indicating changed stakeholder thinking from personal intervention. [MBEE07-E20]
8	Champions the introduction of novel techniques and ideas in business and enterprise integration, beyond the enterprise boundary, in order to develop the wider Systems Engineering community in this competency. [MBEE08]	(A) Analyzes different approaches across different domains through research. [MBEE08-10A]	Records of activities promoting research and need to adopt novel technique or ideas. [MBEE08-E10]
		(A) Produces reports for the wider SE community on the effectiveness of new techniques after their introduction. [MBEE08-20A]	Records of improvements made to process and appraisal against a recognized process improvement model. [MBEE08-E20]
		(P) Collaborates with those introducing novel techniques within the wider SE community. [MBEE08-30P]	Research records. [MBEE08-E30]

		(A) Defines novel approaches that could potentially improve the wider SE discipline. [MBEE08-40A]	Published papers in refereed journals/company literature. [MBEE08-E40]
		(P) Fosters awareness of these novel techniques within the wider SE community. [MBEE08-50P]	Records showing introduction of enabling systems supporting the new techniques or ideas. [MBEE08-E50]
9	Coaches individuals beyond the enterprise boundary in business and enterprise integration techniques, in order to further develop their knowledge, abilities, skills, or associated behaviors. [MBEE09]	(P) Coaches or mentors individuals beyond the enterprise boundary, in competency-related techniques, recommending development activities. [MBEE09-10P]	Coaching or mentoring assignment records. [MBEE09-E10]
		(A) Develops or authorizes training materials in this competency area, which are subsequently successfully delivered beyond the enterprise boundary. [MBEE09-20A]	Records of formal training courses, workshops, seminars, and authored training material supported by successful post-training evaluation data. [MBEE09-E20]
		(A) Provides workshops/seminars or training in this competency area for practitioners or lead practitioners beyond the enterprise boundary (e.g. conferences and open training days). [MBEE09-30A]	Records of Training/workshops/seminars created supported by successful post-training evaluation data. [MBEE09-E30]
10	Maintains expertise in this competency area through specialist Continual Professional Development (CPD) activities. [MBEE10]	(A) Reviews research, new ideas, and state of the art to identify relevant new areas requiring personal development in order to maintain expertise in this competency area. [MBEE10-10A]	Records of documents reviewed and insights gained as part of own research into this competency area. [MBEE10-E10]
		(A) Performs identified specialist professional development activities in order to maintain or further develop competence at expert level. [MBEE10-20A]	Records of Continual Professional Development (CPD) performed and learning outcomes. [MBEE10-E20]
		(A) Records continual professional development activities undertaken including learning or insights gained. [MBEE10-30A]	

NOTES

In addition to items above, enterprise-level or independent 3rd-Party-generated evidence may be used to amplify other evidence presented and may include:

a. Formally recognized by a reputable external organization as an expert in this competency area
b. Evidence of role as independent assessor or reviewer on project outside own organization where skills in this competency area were used
c. Evidence of invitation(s) from wider community for contribution of systems engineering expertise in this area (e.g. industry conference panel, government advisory board etc. cross-industry working groups, partnerships, accredited advanced university courses or research, or as part of professional institute)
d. Formal commendation beyond the enterprise (e.g. by INCOSE or other recognized authority) for work performed in this competency area
e. Independently assessed or accredited work product in this competency area (e.g. for independent publication or use)
f. Accolades of expertise in this area from recognized industry leaders

Competency area – Management: Acquisition and Supply
Description The purpose of Acquisition is to obtain a product or service in accordance with the Acquirer's requirements. The purpose of Supply is to provide an Acquirer with a product or service that meets agreed needs.
Why it matters All system solutions require agreements between different organizations under which one party acquires or supplies products or services from the other. Systems Engineering helps facilitate the successful acquisition and supply of products or services, in order to ensure that the need is defined accurately, to evaluate the supplier against complex criteria; to monitor the ongoing agreement especially when technical circumstances change; and to support formal acceptance of the product or service.
Possible contributory types of evidence Any combination of the types of evidence may be acceptable (depending on how the Framework is tailored and used). The evidence items identified at each level indicate example work products only. Contributions to work products will generally differ at each proficiency level.
Learning and development The INCOSE Professional Development Portal provides example guidance on how to gain an initial awareness of a competency area and options for developing further competence thereafter.

Awareness – Management: Acquisition and Supply		
ID	**Indicators of Competence**	**Relevant knowledge sub-indicators**
1	Describes the key stages in the acquisition of a system. [MASA01]	(K) Explains the key characteristics of each stage of acquisition, namely: Prepare for, advertise need and specification, establish and maintain agreement, monitor agreement, and accept product or service. [MASA01-10K]
2	Describes the key stages in the supply of a system. [MASA02]	(K) Explains the key characteristics of each stage of supply, namely Prepare for, respond to request, establish and maintain agreement, execute agreement, and deliver and support product or service. [MASA02-10K] (K) Describes types of acquisition or supply contracts (e.g. fixed price, cost plus fixed fee, cost plus incentive fee, time, and materials). [MASA02-20K]
3	Describes legal and ethical obligations associated with acquisition and supply, and provides examples. [MASA03]	(K) Explains key legal and ethical obligations in acquisition and supply, such as acting ethically in requesting/advertising goods and services to ensuring technical and financial obligations can be met. [MASA03-10K]

Supervised Practitioner – Management: Acquisition and Supply			
ID	**Indicators of Competence** ***(in addition to those at Awareness level)***	**Relevant knowledge, experience, and/or behaviors**	**Possible examples of objective evidence of personal contribution to activities performed, or professional behaviors applied**
1	Follows a governing process and appropriate tools to plan and control their own acquisition and supply activities. [MASS01]	(A) Prepares information supporting need requirements preparation and documentation or assessment of acquirer needs and requirements. [MASS01-10A] (A) Maintains status and traceability information for agreement requirements. [MASS01-20A] (A) Prepares information supporting or participates in vendor/acquirer communication. [MASS01-30A]	Acquisition and supply processes used. [MASS01-E10] Requirements traceability information. [MASS01-E20]
2	Prepares inputs to work products associated with acquisition of a system. [MASS02]	(A) Prepares typical work products (e.g. specifications) in support of the acquisition of a system. [MASS02-10A] (A) Prepares information supporting requirements and specification documents. [MASS02-20A] (A) Prepares inputs in support of acquisition, such as potential vendor lists or vendor product data. [MASS02-30A]	Supplier Product/service specifications. [MASS02-E10] Solution analyzes performed. [MASS02-E20] Product or service vendors lists. [MASS02-E30]
3	Identifies potential acquirers of organization systems, products, and services on their program. [MASS03]	(K) Describes the concept of product or service capability. [MASS03-10K] (K) Describes factors influencing acquiring organization need. [MASS03-20K] (A) Prepares information in support of the identification of potential vendors for a program. [MASS03-30A] (A) Prepares information in support of business development efforts in support of product sales. [MASS03-40A]	Vendor product/service data sheets or capability statements. [MASS03-E10] Business solution using vendor products or services. [MASS03-E20]
4	Prepares inputs to work products associated with supply of a system. [MASS04]	(A) Prepares or maintains product description information. [MASS04-10A] (A) Prepares responses to technical specification need statements. [MASS04-20A] (A) Prepares information supporting the development of technical and labor cost proposal inputs. [MASS04-30A]	Product Data or capability statements. [MASS04-E10] Proposal inputs supported. [MASS04-E20] Request for proposal or quotation. [MASS04-E30]
5	Develops own understanding of this competency area through Continual Professional Development (CPD). [MASS05]	(A) Identifies potential gaps in own knowledge or development needs in this area, identifying opportunities to address these through continual professional development activities. [MASS05-10A] (A) Performs continual professional development activities to improve their knowledge and understanding in this area. [MASS05-20A] (A) Records continual professional development activities undertaken including learning or insights gained. [MASS05-30A]	Records of Continual Professional Development (CPD) performed and learning outcomes. [MASS05-E10]

Practitioner – Management: Acquisition and Supply			
ID	**Indicators of Competence** ***(in addition to those at Supervised Practitioner level)***	**Relevant knowledge, experience, and/or behaviors**	**Possible examples of objective evidence of personal contribution to activities performed, or professional behaviors applied**
1	Creates a strategy for Acquisition or Supply on a project to support SE project and wider enterprise needs. [MASP01]	(A) Identifies project-specific Acquisition or Supply needs that need to be incorporated into SE planning. [MASP01-10A]	Acquisition or Supply work products. [MASP01-E10]
		(A) Prepares project-specific Acquisition or Supply inputs for SE planning purposes, using appropriate processes and procedures. [MASP01-20A]	Supplier/subcontractor statements of work. [MASP01-E20]
		(A) Prepares project-specific Acquisition or Supply task estimates in support of SE planning. [MASP01-30A]	Supplier/subcontractor specification work products. [MASP01-E30]
		(P) Liaises throughout with stakeholders to gain approval, updating strategy as necessary. [MASP01-40P]	
2	Creates governing plans, processes, and appropriate tools and uses these to control and monitor Acquisition or Supply on a project. [MASP02]	(K) Identifies the steps necessary to define a process and appropriate techniques to be adopted to establish Acquisition and Supply of system elements. [MASP02-10K]	SE Acquisition or Supply planning work products. [MASP02-E10]
		(K) Describes key elements of a successful Acquisition and Supply process and how decision management activities relate to different stages of a system life cycle. [MASP02-20K]	Planning documents such as the systems engineering management plan, other project/program plan or organizational process. [MASP02-E20]
		(A) Uses Acquisition or Supply tools or uses a methodology. [MASP02-30A]	Acquisition or Supply plans and processes used. [MASP02-E30]
		(A) Uses appropriate standards and governing processes in their Acquisition or Supply planning, tailoring as appropriate. [MASP02-40A]	
		(A) Selects Acquisition or Supply approaches appropriate to each information item. [MASP02-50A]	
		(A) Creates Acquisition or Supply plans and processes for use on a project. [MASP02-60A]	Specifications, statements of work, basis of estimates, proposals. [MASP02-E60]
		(P) Liaises throughout with stakeholders to gain approval, updating plans as necessary. [MASP02-70P]	Request for proposal or quotation. [MASP02-E70]
		(A) Creates need requirements preparation and documentation or assessment of acquirer needs and requirements. [MASP02-80A]	Supply chain reports and analysis. [MASP02-E80]
		(A) Identifies acquisition and supply guidance including milestones, standards, acceptance criteria, and decision gates. [MASP02-90A]	
		(A) Develops approaches to select, negotiate and communicate with vendor/acquirer. [MASP02-100A]	

3	Develops a tender document requesting the supply of a system. [MASP03]	(A) Develops/approves system of interest specification and request for proposal or quotation. [MASP03-10A] (A) Creates solution ensuring parent system needs and acquisition strategy are considered. [MASP03-20A] (A) Defines solutions meeting schedule and cost needs. [MASP03-30A]	Records of a request for proposal or quotation created. [MASP03-E10] System specification. [MASP03-E20]
4	Identifies potential suppliers using criteria to judge their suitability. [MASP04]	(A) Reviews supplier response against request specifications (e.g. tenders) and system requirements. [MASP04-10A] (A) Reviews supplier historical record over cost and ability to provide. [MASP04-20A] (A) Reviews supplier ability to deal with ilities and security. [MASP04-30A]	Supply chain reports and analysis. [MASP04-E10] Proposal evaluation reports. [MASP04-E20]
5	Reviews supplier responses to a tender document and makes formal recommendations. [MASP05]	(A) Analyzes requested information in response to a request for proposal or quotation. [MASP05-10A] (A) Analyzes supplier capability and solution catalog against requested services or product. [MASP05-20A] (A) Reviews approach, schedule, and cost estimate for supplier solution in collaboration with key stakeholders (e.g. architect). [MASP05-30A]	Supplier documentation review records. [MASP05-E10] Technical solution review records. [MASP05-E20] Supplier solution response analyses. [MASP05-E30]
6	Reviews acquirer requests and works with key internal stakeholders to propose a solution that meets acquirer needs. [MASP06]	(A) Prepares requested information in response to a request for proposal or quotation. [MASP06-10A] (A) Analyzes enterprise capability and solution catalog against requested services or product. [MASP06-20A] (A) Develops integrated approach, schedule, and cost estimate for solution in collaboration with key stakeholders (e.g. architect). [MASP06-30A]	Proposal development, cost development inputs, bill of materials. [MASP06-E10] Technical solution proposal. [MASP06-E20] Architecture development and solution trade analyses. [MASP06-E30]
7	Negotiates an agreement with a supplier for a system including acceptance criteria. [MASP07]	(A) Creates subcontract/supplier statement of work including acceptance criteria and requirements. [MASP07-10A] (A) Identifies exit criteria and review milestone agreements with key supplier stakeholders. [MASP07-20A]	Statement of work, schedule and milestones. [MASP07-E10] Program plan. [MASP07-E20] Records demonstrating analysis of supplier enterprise processes and requirements. [MASP07-E30]

(*Continued*)

ID	Indicators of Competence *(in addition to those at Supervised Practitioner level)*	Relevant knowledge, experience, and/or behaviors	Possible examples of objective evidence of personal contribution to activities performed, or professional behaviors applied
8	Negotiates an agreement with an acquirer for a system, including acceptance criteria. [MASP08]	(A) Analyzes statement of work, acceptance milestones, technical requirements in collaboration with key stakeholders, to ensure acceptability. [MASP08-10A] (A) Reviews contract deliverables and schedule highlighting potential modifications as necessary. [MASP08-20A]	Statements of work, schedules, milestones evaluated, counterproposals or similar negotiation documentation. [MASP08-E10] Contract modifications and engineering change proposals. [MASP08-E20] Records demonstrating analysis of enterprise processes and requirements against contractual obligations. [MASP08-E30]
9	Monitors supplier adherence to terms of agreement to ensure compliance. [MASP09]	(A) Analyzes and maintains entrance and exit criteria and establishes communications. [MASP09-10A] (A) Compares status of supplier schedule and cost against original baseline. [MASP09-20A] (A) Monitors technical solution status and evolving risk and opportunity status. [MASP09-30A]	Records of direct communication with vendors. [MASP09-E10] Records showing acting as technical representative for technical requested solution. [MASP09-E20]
10	Maintains an agreement with a supplier to reflect changes on a project. [MASP10]	(A) Analyzes and maintains entrance and exit criteria and establishes communications. [MASP10-10A] (A) Compares status of supplier schedule and cost against original baseline. [MASP10-20A] (P) Negotiates changes as necessary. [MASP10-30P] (A) Maintains a supplier agreement to reflect changes. [MASP10-40A]	Records showing direct communication with vendors. [MASP10-E10] Records showing acting as technical representative for technical requested solution. [MASP10-E20]
11	Maintains an agreement with an acquirer maintaining in accordance with agreement terms and conditions. [MASP11]	(A) Ensures technical compliance against requirements affecting systems engineering activities. [MASP11-10A] (A) Compares status of supplier schedule and cost against original baseline. [MASP11-20A] (A) Identifies necessary changes to acquirer agreement, documenting them for communication to acquirer. [MASP11-30A] (P) Negotiates changes with acquirer as necessary. [MASP11-40P] (A) Monitors technical solution status and evolving risk and opportunity status. [MASP11-50A]	Records detailing active role in direct contact with customer. [MASP11-E10]

12	Guides new or supervised practitioners in Acquisition and Supply in order to develop their knowledge, abilities, skills, or associated behaviors. [MASP12]	(P) Guides new or supervised practitioners in executing activities that form part of this competency. [MASP12-10P] (A) Trains individuals to an "Awareness" level in this competency area. [MASP12-20A]	Organizational Breakdown Structure showing their responsibility for technical supervision in this area. [MASP12-E10] On-the-job training objectives/guidance etc. [MASP12-E20] Coaching or mentoring assignment records. [MASP12-E30] Records highlighting their impact on another individual in terms of improvement or professional development in this competency. [MASP12-E40]
13	Maintains and enhances own competence in this area through Continual Professional Development (CPD) activities. [MASP13]	(A) Identifies potential development needs in this area, identifying opportunities to address these through continual professional development activities. [MASP13-10A] (A) Performs continual professional development activities to maintain and enhance their competency in this area. [MASP13-20A] (A) Records continual professional development activities undertaken including learning or insights gained. [MASP13-30A]	Records of Continual Professional Development (CPD) performed and learning outcomes. [MASP13-E10]

Lead Practitioner – Management: Acquisition and Supply			
ID	**Indicators of Competence** ***(in addition to those at Practitioner level)***	**Relevant knowledge, experience, and/or behaviors**	**Possible examples of objective evidence of personal contribution to activities performed, or professional behaviors applied**
1	Creates enterprise-level policies, procedures, guidance, and for acquisition and supply, including associated tools. [MASL01]	(A) Analyzes enterprise need for acquisition or supply policies, processes, tools or guidance. [MASL01-10A]	Enterprise level reports and recommendations. [MASL01-E10]
		(A) Creates enterprise policies, procedures or guidance on acquisition or supply. [MASL01-20A]	Part of enterprise SE resource pool and proposal or internal R&D reviews. [MASL01-E20]
		(A) Selects and acquires appropriate tools supporting acquisition or supply. [MASL01-30A]	Develops, edits, maintains program supply chain risk management policies. [MASL01-E30]
2	Judges the tailoring of enterprise-level acquisition and supply processes and associated work products to meet the needs of a project. [MASL02]	(A) Evaluates the enterprise Acquisition or Supply processes against the business and external stakeholder (e.g. customer) needs in order to tailor processes to enable project success. [MASL02-10A]	Part of enterprise SE resource pool and proposal or internal R&D reviews. [MASL02-E10]
		(A) Produces constructive feedback on acquisition or supply processes. [MASL02-20A]	Details of programs supported with successful results. [MASL02-E20]
		(A) Reviews high value or high-risk program acquisition or supply planning and review issues across the enterprise. [MASL02-30A]	Reports or white papers generated. [MASL02-E30]
3	Coordinates acquisition and supply across multiple diverse projects or across a complex system, with proven success. [MASL03]	(A) Compares and contrasts the enterprise benefits of a range of techniques for performing acquisition or supply. [MASL03-10A]	Internal or external consultant in the relevant areas. [MASL03-E10]
		(A) Uses knowledge and experience of the application of different acquisition or supply approaches with successful results. [MASL03-20A]	Reports or white papers generated. [MASL03-E20]
4	Identifies opportunities, arising from projects across the enterprise, to supply systems, products, or services in accordance with wider enterprise goals. [MASL04]	(A) Develops strategies supporting business development of supply opportunities in projects across the enterprise. [MASL04-10A]	Business development successes, new customers, or areas through their activities. [MASL04-E10]
		(A) Judges system architecting analyses in projects across the enterprise, to further organizational supply goals. [MASL04-20A]	Papers identifying integration of solutions to meet new needs. [MASL04-E20]
5	Persuades key stakeholders to address identified enterprise-level acquisition and supply issues in order to reduce risk or eliminate issues. [MASL05]	(A) Reviews project/programs to ensure Systems Engineering processes addresses enterprise-level acquisition and supply issues and vice versa. [MASL05-10A]	White papers and reports. [MASL05-E10]
		(A) Advises on high value, high risk, or high opportunity enterprise-level acquisition or supply management review issues. [MASL05-20A]	Records indicating changed stakeholder thinking from personal intervention. [MASL05-E20]
		(A) Persuades stakeholders to adopt a different path on enterprise acquisition or supply issues. [MASL05-30A]	

6	Coaches or mentors practitioners across the enterprise in acquisition and supply in order to develop their knowledge, abilities, skills, or associated behaviors. [MASL06]	(P) Coaches or mentors practitioners across the enterprise in competency-related techniques, recommending development activities. [MASL06-10P]	Coaching or mentoring assignment records. [MASL06-E10]
		(A) Develops or authorizes enterprise training materials in this competency area. [MASL06-20A]	Records of formal training courses, workshops, seminars, and authored training material supported by successful post-training evaluation data. [MASL06-E20]
		(A) Provides enterprise workshops/seminars or training in this competency area. [MASL06-30A]	Listing as an approved organizational trainer for this competency area. [MASL06-E30]
7	Promotes the introduction and use of novel techniques and ideas in Acquisition and Supply, across the enterprise, to improve enterprise competence in this area. [MASL07]	(A) Analyzes different approaches across different domains through research. [MASL07-10A]	Research records. [MASL07-E10]
		(A) Defines novel approaches that could potentially improve the SE discipline within the enterprise. [MASL07-20A]	Published papers in refereed journals/company literature. [MASL07-E20]
		(P) Fosters awareness of these novel techniques within the enterprise. [MASL07-30P]	Records showing introduction of enabling systems supporting the new techniques or ideas. [MASL07-E30]
		(P) Collaborates with enterprise stakeholders to facilitate the introduction of techniques new to the enterprise. [MASL07-40P]	Published papers (or similar) at enterprise level. [MASL07-E40]
		(A) Monitors new techniques after their introduction to determine their effectiveness. [MASL07-50A]	Records of improvements made against a recognized process improvement model in this area. [MASL07-E50]
		(A) Adapts approach to reflect actual enterprise performance improvements. [MASL07-60A]	
8	Develops expertise in this competency area through specialist Continual Professional Development (CPD) activities. [MASL08]	(A) Identifies own needs for further professional development in order to increase competence beyond practitioner level. [MASL08-10A]	Records of Continual Professional Development (CPD) performed and learning outcomes. [MASL08-E10]
		(A) Performs professional development activities in order to move own competence toward expert level. [MASL08-20A]	
		(A) Records continual professional development activities undertaken including learning or insights gained. [MASL08-30A]	

NOTES	In addition to items above, enterprise-level or independent 3rd-Party-generated evidence may be used to amplify other evidence presented and may include: a. Formally recognized by senior management in current organization as an expert in this competency area b. Evidence of role as Product/System Design Authority or Technical Authority on a complex project with responsibilities in this area or where skills within this competency area were used c. Recognized as an authorizing signatory on behalf of enterprise for formal documentation in this competency area (e.g. policies, processes, and deliverables) d. Formal commendation or award within own enterprise for contribution or item of work successfully performed, which required proficiency in this competency area e. Customer, Supplier, or other external project-specific key Stakeholder accolades for specific work performed in this competency area f. Independently assessed or accredited work in this competency area (e.g. for independent publication or use) g. Formal organizational HR records positively highlighting any specific professional competencies or behaviors identified (if applicable) plus any of the evidence indicators listed at Expert level below

Expert – Management: Acquisition and Supply			
ID	Indicators of Competence *(in addition to those at Lead Practitioner level)*	Relevant knowledge, experience, and/or behaviors	Possible examples of objective evidence of personal contribution to activities performed, or professional behaviors applied
1	Communicates own knowledge and experience in acquisition and supply in order to best practice beyond the enterprise boundary. [MASE01]	(A) Produces papers, seminars, or presentations outside own enterprise for publication in order to share own ideas and improve industry best practices in this competence area. [MASE01-10A]	Published papers or books etc. on new technique in refereed journals/company literature. [MASE01-E10]
		(P) Fosters incorporation of own ideas into industry best practices in this area. [MASE01-20P]	Published papers in refereed journals or internal literature proposing new practices in this competence area (or presentations, tutorials, etc.). [MASE01-E20]
		(P) Develops guidance materials identifying new (or updating existing) best practice in this competence area. [MASE01-30P]	Proposals adopted as industry best practice. [MASE01-E30]
2	Influences key acquisition and supply stakeholders beyond the enterprise boundary. [MASE02]	(P) Acts as independent member of industry panels and professional societies. [MASE02-10P]	Records of projects supported successfully. [MASE02-E10]
		(A) Provides evidence and arguments influencing external stakeholder acquisition or supply. [MASE02-20A]	Publications and presentations used in support of stakeholder arguments. [MASE02-E20]
		(P) Acts as independent external member of panels for high-risk acquisition and supply reviews and stakeholder activities. [MASE02-30P]	Management reports indicating contributions made. [MASE02-E30]
3	Advises organizations beyond the enterprise boundary on complex or sensitive acquisition and supply issues. [MASE03]	(A) Advises external stakeholders (e.g. customers) on their acquisition or supply issues. [MASE03-10A]	Meeting minutes or report outlining role as arbitrator. [MASE03-E10]
		(A) Conducts sensitive acquisition or supply negotiations. [MASE03-20A]	
		(A) Conducts complex acquisition or supply negotiations taking account of taking account of external stakeholders' (e.g. customer's) background and knowledge. [MASE03-30A]	
		(P) Arbitrates on acquisition or supply decisions. [MASE03-40P]	
4	Advises organizations beyond the enterprise boundary on the suitability of their approach to acquisition and supply. [MASE04]	(A) Assesses an external organization's approach to acquisition or supply, as independent reviewer. [MASE04-10A]	Records of membership on oversight committee with relevant terms of reference. [MASE04-E10]
		(A) Assesses the acquisition or supply plans of an external organization, as independent reviewer. [MASE04-20A]	Review comments or revised document. [MASE04-E20]
5	Champions the introduction of novel techniques and ideas in acquisition and supply, beyond the enterprise boundary, in order to develop the wider Systems Engineering community in this competency. [MASE05]	(A) Analyzes different approaches across different domains through research. [MASE05-10A]	Records of activities promoting research and need to adopt novel technique or ideas. [MASE05-E10]

		(A) Produces reports for the wider SE community on the effectiveness of new techniques after their introduction. [MASE05-20A]	Records of improvements made to process and appraisal against a recognized process improvement model. [MASE05-E20]
		(P) Collaborates with those introducing novel techniques within the wider SE community. [MASE05-30P]	Research records. [MASE05-E30]
		(A) Defines novel approaches that could potentially improve the wider SE discipline. [MASE05-40A]	Published papers in refereed journals/company literature. [MASE05-E40]
		(P) Fosters awareness of these novel techniques within the wider SE community. [MASE05-50P]	Records showing introduction of enabling systems supporting the new techniques or ideas. [MASE05-E50]
6	Coaches individuals beyond the enterprise boundary in system or element acquisition and supply, in order to further develop their knowledge, abilities, skills, or associated behaviors. [MASE06]	(P) Coaches or mentors individuals beyond the enterprise boundary, in competency-related techniques, recommending development activities. [MASE06-10P]	Coaching or mentoring assignment records. [MASE06-E10]
		(A) Develops or authorizes training materials in this competency area, which are subsequently successfully delivered beyond the enterprise boundary. [MASE06-20A]	Records of formal training courses, workshops, seminars, and authored training material supported by successful post-training evaluation data. [MASE06-E20]
		(A) Provides workshops/seminars or training in this competency area for practitioners or lead practitioners beyond the enterprise boundary (e.g. conferences and open training days). [MASE06-30A]	Records of Training/workshops/seminars created supported by successful post-training evaluation data. [MASE06-E30]
7	Maintains expertise in this competency area through specialist Continual Professional Development (CPD) activities. [MASE07]	(A) Reviews research, new ideas, and state of the art to identify relevant new areas requiring personal development in order to maintain expertise in this competency area. [MASE07-10A]	Records of documents reviewed and insights gained as part of own research into this competency area. [MASE07-E10]
		(A) Performs identified specialist professional development activities in order to maintain or further develop competence at expert level. [MASE07-20A]	Records of Continual Professional Development (CPD) performed and learning outcomes. [MASE07-E20]
		(A) Records continual professional development activities undertaken including learning or insights gained. [MASE07-30A]	

NOTES	In addition to items above, enterprise-level or independent 3rd-Party-generated evidence may be used to amplify other evidence presented and may include: a. Formally recognized by a reputable external organization as an expert in this competency area b. Evidence of role as independent assessor or reviewer on project outside own organization where skills in this competency area were used c. Evidence of invitation(s) from wider community for contribution of systems engineering expertise in this area (e.g. industry conference panel, government advisory board etc. cross-industry working groups, partnerships, accredited advanced university courses or research, or as part of professional institute) d. Formal commendation beyond the enterprise (e.g. by INCOSE or other recognized authority) for work performed in this competency area e. Independently assessed or accredited work product in this competency area (e.g. for independent publication or use) f. Accolades of expertise in this area from recognized industry leaders

Competency area – Management: Configuration Management
Description Configuration Management (CM) manages and controls system elements and configurations over the program life cycle, ensuring the overall coherence of the "evolving" design of a system is maintained in a verifiable manner, throughout the life cycle, and retains the original intent. The Configuration Management activity includes planning; identification; change management and control; reporting; and auditing.
Why it matters Configuration Management ensures that the product functional, performance, and physical characteristics are properly identified, documented, validated, and verified to establish product integrity; that changes to these product characteristics are properly identified, reviewed, approved, documented, and implemented; and that the products produced against a given set of documentation are known. Without Configuration Management, loss of control over the evolving design, development, and operation of a product will occur.
Possible contributory types of evidence Any combination of the types of evidence may be acceptable (depending on how the Framework is tailored and used). The evidence items identified at each level indicate example work products only. Contributions to work products will generally differ at each proficiency level.
Learning and development The INCOSE Professional Development Portal provides example guidance on how to gain an initial awareness of a competency area and options for developing further competence thereafter.

Awareness – Management: Configuration Management		
ID	**Indicators of Competence**	**Relevant Knowledge Sub-indicators**
1	Explains why the integrity of the design needs to be maintained and how configuration management supports this. [MCMA01]	(K) Explains why CM assists robustness. [MCMA01-10K] (K) Explains why CM ensures an understanding of the functional and physical characteristics of a controlled item. [MCMA01-20K] (K) Explains how CM produces a set of coherent and consistent information pertaining to a particular event or item. [MCMA01-30K] (K) Explains why CM enables an organization to return to an earlier build constructed exactly as it was when it was controlled (e.g. in support of testing older build standards). [MCMA01-40K] (K) Explains why CM reduces risk/uncertainty at acceptance. [MCMA01-50K] (K) Describes key interfaces and links between engineering configuration management and stakeholders in the wider enterprise outside engineering (e.g. infrastructure management, human resource management, quality management, knowledge management, portfolio management, and life cycle model management). [MCMA01-60K] (K) Explains why CM provides early indication of future development problems. [MCMA01-70K] (K) Explains why CM supports early identification of variance/inconsistency. [MCMA01-80K] (K) Explains why CM permits design margins or apportioned requirements to be coherently controlled across a suite of related elements or systems. [MCMA01-90K]
2	Describes the key characteristics of a configuration item (CI) including how configuration items are selected and controlled. [MCMA02]	(K) Explains how CM manages and controls of names for critical elements. [MCMA02-10K]

		(K) Explains how CM provides formal control of information relating to a particular product. [MCMA02-20K] (K) Explains how control of identifiers for deliverable items serves to highlight their build standard and/or functional characteristics. [MCMA02-30K]
3	Identifies key baselines and baseline reviews in a typical development life cycle. [MCMA03]	(K) Lists key baselines and baseline reviews in a typical development life cycle. [MCMA03-10K] (K) Explains the purpose of key baselines and baseline reviews in a typical development life cycle. [MCMA03-20K] (K) Explains the concept of a "freeze" and what this means for requirements, start of design, start of production, or start of verification. [MCMA03-30K]
4	Describes the process for changing baselined information and a typical life cycle for an engineering change. [MCMA04]	(K) Describes the change control cycle. [MCMA04-10K] (K) Explains the concepts and differences between change requests, change proposals, and engineering notices (or similar). [MCMA04-20K] (K) Explains the purpose and key participants in a change control board and how change dispositioning works. [MCMA04-30K] (K) Explains the concepts of impact analysis and how this relates to change implementation. [MCMA04-40K]
5	Lists key activities performed as part of configuration management and can outline the key activities involved in each. [MCMA05]	(K) Lists configuration management planning, configuration identification, baseline control and change management, configuration evaluation, release control, and configuration status accounting. [MCMA05-10K] (K) Describes definition of strategy and of life cycle. [MCMA05-20K] (K) Describes identification of system elements and information items, use of unique identifiers and baselines. [MCMA05-30K] (K) Describes change requests, change control board, and impact analysis. [MCMA05-40K] (K) Describes link between CM and reviews and audits. [MCMA05-50K] (K) Describes link between CM and scheduling or planning, verification, documentation. [MCMA05-60K]
6	Explains why change occurs and why changes need to be carefully managed. [MCMA06]	(K) Describes link between change and new requirements, new technology, reduced cost, or improved "-ilities." [MCMA06-10K] (K) Explains topics including "maintaining integrity," "avoiding feature creep," or "ensuring compliance." [MCMA06-20K]
7	Describes the processes and work products used to assist in Change Management. [MCMA07]	(K) Describes where and when change can occur in the life cycle. [MCMA07-10K] (K) Describes the elements of change management. [MCMA07-20K]
8	Describes the meaning of key terminology and acronyms used within Change Management and their relationships. [MCMA08]	(K) Describes a Configuration Management Plan and its link to Change Management plan. [MCMA08-10K] (K) Describes a Change Control Board (CCB) and how this body governs the change. [MCMA08-20K] (K) Describes an Engineering Change Request (ECR) or Engineering Change Proposal (ECP) and outlines the process of origination and processing of each of these. [MCMA08-30K] (K) Describes Configuration Items (CI) and highlights reasons for designation of an item as a CI. [MCMA08-40K]

Supervised Practitioner – Management: Configuration Management			
ID	**Indicators of Competence** ***(in addition to those at Awareness level)***	**Relevant knowledge, experience, and/or behaviors**	**Possible examples of objective evidence of personal contribution to activities performed, or professional behaviors applied**
1	Follows a governing configuration and change management process and appropriate tools to plan and control their own activities relating to maintaining design integrity. [MCMS01]	(A) Performs configuration management activities on own work products as required. [MCMS01-10A] (A) Uses approved configuration management processes and tools to control and manage own configuration management activities. [MCMS01-20A]	Configuration management processes used to control work products created. [MCMS01-E10] Configuration Management Plan used. [MCMS01-E20] Tailored Configuration Management plan created in order to match the principles of a required process. [MCMS01-E30]
2	Prepares information for configuration management work products. [MCMS02]	(A) Prepares information in support of configuration management activities, such as CM planning, configuration item identification, change control, status accounting and auditing, and CCB meeting briefings. [MCMS02-10A] (A) Collates inputs in support of configuration management and change management products to maintain a product baseline. [MCMS02-20A] (A) Performs preparation of engineering change proposals to manage change. [MCMS02-30A]	Configuration management plan, including configuration item identification, change control, status accounting, and auditing. [MCMS02-E10] Configuration management baseline reports they supported. [MCMS02-E20]
3	Describes the need to identify configuration items and why this is done. [MCMS03]	(K) Explains the concept of a "configuration Item" (CI) and why there is a need for configuration items in a design process. [MCMS03-10K] (K) Explains the concept of configuration item identifiers and how the associated numbering system works. [MCMS03-20K] (K) Explains the adverse results of mixing up configuration items. [MCMS03-30K]	
4	Prepares information in support of configuration change control activities. [MCMS04]	(K) Explains why there is a need for baselines, design reviews, etc. [MCMS04-10K] (A) Maintains bi-directional traceability of requirements to design as part of configuration management. [MCMS04-20A] (A) Monitors performance criteria such as parameter budgets, measures of performance, and measures of effectiveness. [MCMS04-30A]	Updates to documentation versions performed in response to changes. [MCMS04-E10] Traceability matrix updates made in response to configuration changes. [MCMS04-E20] Configuration performance parameters tracked as part of configuration management activities. [MCMS04-E30]

5	Prepares material in support of change control decisions and associated review meetings. [MCMS05]	(A) Identifies artifacts affected by a change. [MCMS05-10A] (A) Records materials associated with changes with accuracy. [MCMS05-20A]	Configuration management change reviews supported. [MCMS05-E10] Configuration management records updated as a result of agreed review changes. [MCMS05-E20]
6	Produces management reports in support of configuration item status accounting and audits. [MCMS06]	(A) Identifies measures in use for configuration status accounting. [MCMS06-10A] (K) Explains criteria commonly applied in configuration audits. [MCMS06-20K] (A) Prepares information in support of configuration status accounting and audits. [MCMS06-30A] (A) Prepares information for management reports in support of status accounting and audits. [MCMS06-40A]	Configuration management status accounting reports generated or contributed to. [MCMS06-E10] Configuration management audits they have performed or contributed to. [MCMS06-E20]
7	Identifies applicable standards, regulations, and enterprise level processes on their project. [MCMS07]	(A) Identifies applicable configuration management standards or internal quality system standards other external stakeholder (e.g. customer) standards. [MCMS07-10A] (A) Identifies domain specific standards or regulations. [MCMS07-20A]	Configuration management standards or regulations applied or tailored (e.g. ISO/IEC 15288, 10007 or IEEE 828). [MCMS07-E10]
8	Identifies and reports baseline inconsistencies. [MCMS08]	(K) Explains the concept and benefits of baseline monitoring. [MCMS08-10K] (A) Uses baseline management in their activities. [MCMS08-20A] (A) Uses change management and non-conformance control in their activities. [MCMS08-30A]	Inconsistencies identified and reported as a result of analyzing Configuration Management baselines. [MCMS08-E10]
9	Develops own understanding of this competency area through Continual Professional Development (CPD). [MCMS09]	(A) Identifies potential gaps in own knowledge or development needs in this area, identifying opportunities to address these through continual professional development activities. [MCMS09-10A] (A) Performs continual professional development activities to improve their knowledge and understanding in this area. [MCMS09-20A] (A) Records continual professional development activities undertaken including learning or insights gained. [MCMS09-30A]	Records of Continual Professional Development (CPD) performed and learning outcomes. [MCMS09-E10]

Practitioner – Management: Configuration Management			
ID	**Indicators of Competence** ***(in addition to those at Supervised Practitioner level)***	**Relevant knowledge, experience, and/or behaviors**	**Possible examples of objective evidence of personal contribution to activities performed, or professional behaviors applied**
1	Creates a strategy for Configuration Management on a project to support SE project and wider enterprise needs. [MCMP01]	(A) Identifies project-specific configuration management needs that need to be incorporated into SE planning. [MCMP01-10A] (A) Prepares project-specific configuration management inputs for SE planning purposes, using appropriate processes and procedures. [MCMP01-20A] (A) Prepares project-specific configuration management task estimates in support of SE planning. [MCMP01-30A] (P) Liaises throughout with stakeholders to gain approval, updating strategy as necessary. [MCMP01-40P]	SE configuration management work products. [MCMP01-E10] Configuration Management Plan. [MCMP01-E20]
2	Creates governing configuration and change management plans, processes, and appropriate tools, and uses these to control and monitor design integrity during the full life cycle of a project or system. [MCMP02]	(K) Identifies the steps necessary to define a process and appropriate techniques to be adopted to establish configuration and change management for system elements. [MCMP02-10K] (K) Describes key elements of a successful configuration and change management process and how decision management activities relate to different stages of a system life cycle. [MCMP02-20K] (A) Uses configuration and change management tools or uses a methodology. [MCMP02-30A] (A) Uses appropriate standards and governing processes in their configuration and change management planning, tailoring as appropriate. [MCMP02-40A] (A) Selects configuration and change management method appropriate to each configuration and change item. [MCMP02-50A] (A) Creates configuration and change management plans and processes for use on a project. [MCMP02-60A]	Produces documentation of configuration and change management plans and processes and their application during the full life cycle of a project. [MCMP02-E10] Configuration management process. [MCMP02-E20] Configuration and change management plans and process descriptions used during the full life cycle of a project. [MCMP02-E30]

3	Identifies required remedial actions in the presence of baseline inconsistencies. [MCMP03]	(A) Identifies actual or proposed appropriate corrective or contingent actions to be performed. [MCMP03-10A] (A) Reviews and updates configuration management plan and process as required. [MCMP03-20A] (A) Monitors the status of technical margins over time through configuration management activities. [MCMP03-30A] (A) Maintains technical margins through configuration management actions in response to identified issues. [MCMP03-40A]	Produces updated plans, budgets. [MCMP03-E10]
4	Coordinates changes to configuration items understanding the potential scope within the context of the project. [MCMP04]	(A) Follows controlled steps to effectively manage system change. [MCMP04-10A] (A) Identifies potential changes and their potential impact and acts to minimize this. [MCMP04-20A] (A) Maintains all information associated with all work products affected by changes. [MCMP04-30A] (A) Identifies and controls system baselines. [MCMP04-40A] (A) Maintains bidirectional traceability of requirements throughout life cycle. [MCMP04-50A] (A) Monitors parameter budgets as part of the flow down of requirements through the physical architecture. [MCMP04-60A] (A) Compares planned parameter budgets with actuals, updating budgets as design evolves. [MCMP04-70A] (A) Monitors actual progress and budget against planned progress and allocated budget over time. [MCMP04-80A] (A) Performs and documents design reviews. [MCMP04-90A] (A) Reviews system assumptions over time for their stability, updating as necessary. [MCMP04-100A] (A) Analyzes limiting and out of spec scenarios as well as nominal ones to assure system robustness, dependability, and graceful degradation. [MCMP04-110A] (A) Maintains appropriate margins. [MCMP04-120A]	Produces configuration/change records. [MCMP04-E10] Review minutes. [MCMP04-E20] Budget allocation tables with margin. [MCMP04-E30]

(Continued)

ID	Indicators of Competence *(in addition to those at Supervised Practitioner level)*	Relevant knowledge, experience, and/or behaviors	Possible examples of objective evidence of personal contribution to activities performed, or professional behaviors applied
5	Identifies selection of configuration items and associated documentation by working with design teams justifying the decisions reached. [MCMP05]	(A) Selects rationale for configuration items in collaboration with design stakeholders lead engineers and program management. [MCMP05-10A] (A) Defines configuration item documentation in collaboration with design stakeholders, recording the results. [MCMP05-20A]	Periodic project/program reviews with parameter tracking. [MCMP05-E10] SEMP outlining metrics to be tracked. [MCMP05-E20] Configured design documentation work products (e.g. interface control document, performance models for key user requirements, behavior model for system and system elements, parameter budgets, human computer interface and ergonomic models, whole life cycle cost model, and safety case). [MCMP05-E30]
6	Coordinates change control review activities in conjunction with customer representative and directs resolutions and action items. [MCMP06]	(A) Prepares information (e.g. plans) in support of a change control board (CCB). [MCMP06-10A] (A) Reviews and documents current state of change activities. [MCMP06-20A] (K) Explains impact of proposed changes to stakeholders. [MCMP06-30K] (A) Identifies the direction of proposed changes in collaboration with stakeholders. [MCMP06-40A] (A) Records change decisions and their justification. [MCMP06-50A] (A) Ensures the agreed implementation strategy for changes is folly implemented. [MCMP06-60A]	Produces minutes of CCB meetings. [MCMP06-E10] Produces, reviews, approves CCB meeting briefings, and proposals. [MCMP06-E20]
7	Coordinates configuration status accounting reports and audits. [MCMP07]	(A) Identifies measures for configuration status accounting. [MCMP07-10A] (A) Defines criteria for configuration audit. [MCMP07-20A] (A) Defines reporting schedule. [MCMP07-30A] (A) Defines targets and schedule of audits. [MCMP07-40A]	Documentation covering planning of configuration status accounting activities. [MCMP07-E10] Documentation covering planning of configuration audits. [MCMP07-E20]

		(A) Prepares configuration status accounting reports. [MCMP07-50A] (A) Performs configuration status accounting audits as required. [MCMP07-60A]	
8	Guides new or supervised practitioners in configuration management to develop their knowledge, abilities, skills, or associated behaviors. [MCMP08]	(P) Guides new or supervised practitioners in executing activities that form part of this competency. [MCMP08-10P] (A) Trains individuals to an "Awareness" level in this competency area. [MCMP08-20A]	Organizational Breakdown Structure showing their responsibility for technical supervision in this area. [MCMP08-E10] On-the-job training objectives/guidance etc. [MCMP08-E20] Coaching or mentoring assignment records. [MCMP08-E30] Records highlighting their impact on another individual in terms of improvement or professional development in this competency. [MCMP08-E40]
9	Maintains and enhances own competence in this area through Continual Professional Development (CPD) activities. [MCMP09]	(A) Identifies potential development needs in this area, identifying opportunities to address these through continual professional development activities. [MCMP09-10A] (A) Performs continual professional development activities to maintain and enhance their competency in this area. [MCMP09-20A] (A) Records continual professional development activities undertaken including learning or insights gained. [MCMP09-30A]	Records of Continual Professional Development (CPD) performed and learning outcomes. [MCMP09-E10]

Lead Practitioner – Management: Configuration Management			
ID	**Indicators of Competence** ***(in addition to those at Practitioner level)***	**Relevant knowledge, experience, and/or behaviors**	**Possible examples of objective evidence of personal contribution to activities performed, or professional behaviors applied**
1	Creates enterprise-level policies, procedures, guidance, and best practice for configuration management, including associated tools. [MCML01]	(A) Analyzes enterprise need for Configuration Management policies, processes, tools, or guidance. [MCML01-10A]	Enterprise level reports and recommendations. [MCML01-E10]
		(A) Creates enterprise policies, procedures or guidance on Configuration Management. [MCML01-20A]	Part of enterprise SE resource pool and proposal or internal R&D reviews. [MCML01-E20]
		(A) Selects and acquires appropriate tools supporting Configuration Management. [MCML01-30A]	Enterprise level configuration management plan template. [MCML01-E30]
2	Judges the tailoring of enterprise-level configuration and change management processes and associated work products to meet the needs of a project. [MCML02]	(A) Evaluates the enterprise Configuration Management processes against the business and external stakeholder (e.g. customer) needs in order to tailor processes to enable project success. [MCML02-10A]	Part of enterprise SE resource pool and proposal or internal R&D reviews. [MCML02-E10]
		(A) Provides constructive feedback on Configuration Management processes. [MCML02-20A]	Details of programs supported with successful results. [MCML02-E20]
		(A) Reviews high value or high-risk program Configuration Management planning issues across the enterprise. [MCML02-30A]	Reports and white papers generated. [MCML02-E30]
3	Coordinates configuration management across multiple diverse projects or across a complex system, with proven success. [MCML03]	(A) Compares and contrasts the enterprise benefits of a range of techniques for performing Configuration Management. [MCML03-10A]	Records showing internal or external consultation in the relevant areas. [MCML03-E10]
		(A) Uses knowledge and experience of the application of different CM approaches with successful results. [MCML03-20A]	Reports. [MCML03-E20]
		(A) Uses CM policies and procedures across multiple projects for components of a complex system. [MCML03-30A]	
4	Influences key stakeholders to address identified enterprise-level configuration management issues. [MCML04]	(P) Uses facilitation skills in project/program reviews to ensure Systems Engineering processes addresses enterprise-level Configuration Management issues and vice versa. [MCML04-10P]	White papers and reports. [MCML04-E10]
		(A) Advises on high value, high risk or high opportunity enterprise-level Configuration Management review issues. [MCML04-20A]	Records of enterprise-level meetings. [MCML04-E20]
		(A) Persuades stakeholders to address enterprise-level Configuration management issues. [MCML04-30A]	Records indicating changed stakeholder thinking from personal intervention. [MCML04-E30]

5	Advises stakeholders across the enterprise on remedial actions to address baseline inconsistencies for projects of various size and complexity. [MCML05]	(A) Analyzes causes of baseline inconsistency on projects of various complexity. [MCML05-10A]	Analysis records. [MCML05-E10]
		(P) Fosters agreement between key stakeholders at enterprise level to resolve baseline consistency issues, by promoting a holistic viewpoint. [MCML05-20P]	Minutes of key stakeholder review meetings and actions taken. [MCML05-E20]
			Records demonstrating changed stakeholder viewpoints in this area. [MCML05-E30]
6	Advises stakeholders across the enterprise on major changes and influences them to reduce impact of such changes. [MCML06]	(A) Analyzes impact of major changes. [MCML06-10A]	Analysis records. [MCML06-E10]
		(A) Persuades stakeholders to change strategy in order to reduce the potential impact of change. [MCML06-20A]	Minutes of meetings or other reports that show interactions with key stakeholders in order to reduce the impact of major changes. [MCML06-E20]
			Records showing changed stakeholders viewpoints in this area. [MCML06-E30]
7	Coaches or mentors practitioners across the enterprise in configuration management in order to develop their knowledge, abilities, skills, or associated behaviors. [MCML07]	(P) Coaches or mentors practitioners across the enterprise in competency-related techniques, recommending development activities. [MCML07-10P]	Coaching or mentoring assignment records. [MCML07-E10]
		(A) Develops or authorizes enterprise training materials in this competency area. [MCML07-20A]	Records of formal training courses, workshops, seminars, and authored training material supported by successful post-training evaluation data. [MCML07-E20]
		(A) Provides enterprise workshops/seminars or training in this competency area. [MCML07-30A]	Listing as an approved organizational trainer for this competency area. [MCML07-E30]
8	Promotes the introduction and use of novel techniques and ideas in Configuration Management, across the enterprise, to improve enterprise competence in this area. [MCML08]	(A) Analyzes different approaches across different domains through research. [MCML08-10A]	Research records. [MCML08-E10]
		(A) Defines novel approaches that could potentially improve the SE discipline within the enterprise. [MCML08-20A]	Published papers in refereed journals/company literature. [MCML08-E20]
		(P) Fosters awareness of these novel techniques within the enterprise. [MCML08-30P]	Enabling systems introduced in support of new techniques or ideas. [MCML08-E30]

(Continued)

ID	Indicators of Competence *(in addition to those at Practitioner level)*	Relevant knowledge, experience, and/or behaviors	Possible examples of objective evidence of personal contribution to activities performed, or professional behaviors applied
		(P) Collaborates with enterprise stakeholders to facilitate the introduction of techniques new to the enterprise. [MCML08-40P]	Published papers (or similar) at enterprise level. [MCML08-E40]
		(A) Monitors new techniques after their introduction to determine their effectiveness. [MCML08-50A]	Records of improvements made against a recognized process improvement model in this area. [MCML08-E50]
		(A) Monitors new techniques after their introduction to determine their effectiveness. [MCML08-60A]	
9	Develops expertise in this competency area through specialist Continual Professional Development (CPD) activities. [MCML09]	(A) Identifies own needs for further professional development in order to increase competence beyond practitioner level. [MCML09-10A]	Records of Continual Professional Development (CPD) performed and learning outcomes. [MCML09-E10]
		(A) Performs professional development activities in order to move own competence toward expert level. [MCML09-20A]	
		(A) Records continual professional development activities undertaken including learning or insights gained. [MCML09-30A]	

NOTES In addition to items above, enterprise-level or independent 3rd-Party-generated evidence may be used to amplify other evidence presented and may include:

a. Formally recognized by senior management in current organization as an expert in this competency area
b. Evidence of role as Product/System Design Authority or Technical Authority on a complex project with responsibilities in this area or where skills within this competency area were used
c. Recognized as an authorizing signatory on behalf of enterprise for formal documentation in this competency area (e.g. policies, processes, and deliverables)
d. Formal commendation or award within own enterprise for contribution or item of work successfully performed, which required proficiency in this competency area
e. Customer, Supplier, or other external project-specific key Stakeholder accolades for specific work performed in this competency area
f. Independently assessed or accredited work in this competency area (e.g. for independent publication or use)
g. Formal organizational HR records positively highlighting any specific professional competencies or behaviors identified (if applicable) plus any of the evidence indicators listed at Expert level below

Expert – Management: Configuration Management			
ID	**Indicators of Competence *(in addition to those at Lead Practitioner level)***	**Relevant knowledge, experience, and/or behaviors**	**Possible examples of objective evidence of personal contribution to activities performed, or professional behaviors applied**
1	Communicates own knowledge and experience in configuration management in order to best practice beyond the enterprise boundary. [MCME01]	(A) Produces papers, seminars, or presentations outside own enterprise for publication in order to share own ideas and improve industry best practices in this competence area. [MCME01-10A]	Published papers or books etc. on new technique in refereed journals/company literature. [MCME01-E10]
		(P) Fosters incorporation of own ideas into industry best practices in this area. [MCME01-20P]	Published papers in refereed journals or internal literature proposing new practices in this competence area (or presentations, tutorials, etc.). [MCME01-E20]
		(P) Develops guidance materials identifying new (or updating existing) best practice in this competence area. [MCME01-30P]	Own proposals adopted as industry best practices in this competence area. [MCME01-E30]
2	Persuades individuals beyond the enterprise boundary regarding configuration and change management issues. [MCME02]	(P) Acts as independent member of industry panels and professional societies. [MCME02-10P]	Records of projects supported successfully. [MCME02-E10]
		(A) Provides evidence and arguments influencing external stakeholder configuration management activities. [MCME02-20A]	Publications and presentations used in support of stakeholder arguments. [MCME02-E20]
		(P) Acts as independent external member of review panels for Configuration Management issues. [MCME02-30P]	Records indicating management approval of work performed. [MCME02-E30]
3	Advises organizations beyond the enterprise boundary on the suitability of their approach to configuration management. [MCME03]	(A) Acts as an independent reviewer of external organization configuration management plans. [MCME03-10A]	Records showing membership on review committee. [MCME03-E10]
		(A) Advises external organizations on potential improvements to their configuration management approach. [MCME03-20A]	Review comments or revised document. [MCME03-E20]
4	Advises organizations beyond the enterprise boundary on complex or sensitive configuration and change management issues. [MCME04]	(A) Advises external stakeholders (e.g. customers) on their Configuration management issues. [MCME04-10A]	Meeting minutes or report outlining role as arbitrator. [MCME04-E10]

(Continued)

ID	Indicators of Competence *(in addition to those at Lead Practitioner level)*	Relevant knowledge, experience, and/or behaviors	Possible examples of objective evidence of personal contribution to activities performed, or professional behaviors applied
		(A) Conducts sensitive Configuration management negotiations. [MCME04-20A]	
		(A) Conducts complex Configuration management negotiations taking account of external stakeholders' (e.g. customer's) background and knowledge. [MCME04-30A]	
		(P) Arbitrates on Configuration management decisions. [MCME04-40P]	
5	Champions the introduction of novel techniques and ideas in configuration management, beyond the enterprise boundary, in order to develop the wider Systems Engineering community in this competency. [MCME05]	(A) Analyzes different approaches across different domains through research. [MCME05-10A]	Records of activities promoting research and need to adopt novel technique or ideas. [MCME05-E10]
		(A) Produces reports for the wider SE community on the effectiveness of new techniques after their introduction. [MCME05-20A]	Records of improvements made to process and appraisal against a recognized process improvement model. [MCME05-E20]
		(P) Collaborates with those introducing novel techniques within the wider SE community. [MCME05-30P]	Research records. [MCME05-E30]
		(A) Defines novel approaches that could potentially improve the wider SE discipline. [MCME05-40A]	Published papers in refereed journals/company literature. [MCME05-E40]
		(P) Fosters awareness of these novel techniques within the wider SE community. [MCME05-50P]	Records showing introduction of enabling systems supporting the new techniques or ideas. [MCME05-E50]
6	Coaches individuals beyond the enterprise boundary in Configuration Management, in order to further develop their knowledge, abilities, skills, or associated behaviors. [MCME06]	(P) Coaches or mentors individuals beyond the enterprise boundary, in competency-related techniques, recommending development activities. [MCME06-10P]	Coaching or mentoring assignment records. [MCME06-E10]
		(A) Develops or authorizes training materials in this competency area, which are subsequently successfully delivered beyond the enterprise boundary. [MCME06-20A]	Records of formal training courses, workshops, seminars, and authored training material supported by successful post-training evaluation data. [MCME06-E20]
		(A) Provides workshops/seminars or training in this competency area for practitioners or lead practitioners beyond the enterprise boundary (e.g. conferences and open training days). [MCME06-30A]	Records of Training/workshops/seminars created supported by successful post-training evaluation data. [MCME06-E30]

7	Maintains expertise in this competency area through specialist Continual Professional Development (CPD) activities. [MCME07]	(A) Reviews research, new ideas, and state of the art to identify relevant new areas requiring personal development in order to maintain expertise in this competency area. [MCME07-10A] (A) Performs identified specialist professional development activities in order to maintain or further develop competence at expert level. [MCME07-20A] (A) Records continual professional development activities undertaken including learning or insights gained. [MCME07-30A]	Records of documents reviewed and insights gained as part of own research into this competency area. [MCME07-E10] Records of Continual Professional Development (CPD) performed and learning outcomes. [MCME07-E20]

NOTES

In addition to items above, enterprise-level or independent 3rd-Party-generated evidence may be used to amplify other evidence presented and may include:

a. Formally recognized by a reputable external organization as an expert in this competency area
b. Evidence of role as independent assessor or reviewer on project outside own organization where skills in this competency area were used
c. Evidence of invitation(s) from wider community for contribution of systems engineering expertise in this area (e.g. industry conference panel, government advisory board etc. cross-industry working groups, partnerships, accredited advanced university courses or research, or as part of professional institute)
d. Formal commendation beyond the enterprise (e.g. by INCOSE or other recognized authority) for work performed in this competency area
e. Independently assessed or accredited work product in this competency area (e.g. for independent publication or use)
f. Accolades of expertise in this area from recognized industry leaders

Competency area – Management: Information Management
Description Information Management addresses activities associated with the generation, obtaining, confirming, transforming, retaining, retrieval, dissemination, and disposal of information, to designated stakeholders with appropriate levels of timeliness, accuracy, and security. Information Management plans, executes and controls the provision of information to designated stakeholders that is unambiguous, complete, verifiable, consistent, modifiable, traceable, and presentable. Information includes technical, project, organizational, agreement, and user information.
Why it matters System Engineering requires relevant, timely and complete information during and after the system life cycle to support all aspects of the development; from the analysis of future concepts to the ultimate archiving and potential subsequent retrieval of project data. Information also supports decision-making across every aspect of the development including suppliers and agreements. Information security and assurance are crucial parts of Information Management: ensuring only designated individuals are able to access certain data, while protecting intellectual property and making sure information is available as required in line with the sender's intent.
Possible contributory types of evidence Any combination of the types of evidence may be acceptable (depending on how the Framework is tailored and used). The evidence items identified at each level indicate example work products only. Contributions to work products will generally differ at each proficiency level.
Learning and development The INCOSE Professional Development Portal provides example guidance on how to gain an initial awareness of a competency area and options for developing further competence thereafter.

Awareness – Management: Information Management		
ID	**Indicators of Competence**	**Relevant knowledge sub-indicators**
1	Describes various types of information required to be managed in support of Systems Engineering activities and provides examples. [MIMA01]	(K) Identifies enterprise-wide knowledge captured in training, processes, practices, methods, policies, and procedures. [MIMA01-10K] (K) Describes capture and maintenance of best practices and key personnel expertise and knowledge. [MIMA01-20K]
2	Describes various types of information assets that may need to be managed within a project or system. [MIMA02]	(K) Identifies proposal and financial information, proprietary reports and processes, intellectual property information, sensitive and personal information, external stakeholder (e.g. customer), and classified information. [MIMA02-10K]
3	Identifies different classes of risk to information integrity and can provide examples of each. [MIMA03]	(K) Identifies insufficient information resulting in inefficiencies due to reinventing of the wheel, inconsistent approach across enterprise. [MIMA03-10K] (K) Identifies loss of confidence for example cybersecurity and personnel information violation. [MIMA03-20K] (K) Describes key interfaces and links between engineering information management and stakeholders in the wider enterprise outside engineering (e.g. infrastructure management, human resource management, quality management, knowledge management, portfolio management, and life cycle model management). [MIMA03-30K]

4	Describes the relationship between information management and configuration change management. [MIMA04]	(K) Defines how Information Management and Change management interact to ensure information integrity is maintained across the enterprise when items of information are (for whatever reason) modified. [MIMA04-10K]
5	Describes potential scenarios where information may require modification. [MIMA05]	(K) Describes change management as a means used by information management to keep the information current, consistent, and accurate. [MIMA05-10K]
6	Explains how data rights may affect information management on a project. [MIMA06]	(K) Identifies different classes of information that require respective data rights including enterprise proprietary, intellectual property, financial, personnel, stakeholder, and classified information. [MIMA06-10K] (K) Describes that different types of information require different levels of protection. [MIMA06-20K] (K) Explains Cybersecurity and protection of information, particularly individual records and enterprise sensitive information. [MIMA06-30K]
7	Describes the legal and ethical responsibilities associated with access to and sharing of enterprise and customer information and summarizes regulations regarding information sharing. [MIMA07]	(K) Identifies personnel sensitive information and records are protected by law. [MIMA07-10K] (K) Identifies protection of intellectual property and commercially sensitive information of enterprise, external stakeholder (e.g. customer), vendor, and partner organization. [MIMA07-20K] (K) Identifies international and national law, company regulations, and contractual agreements. [MIMA07-30K]
8	Describes what constitutes personal data and why its protection and management is important. [MIMA08]	(K) Describes different types of personal data that may require protection. [MIMA08-10K] (K) Identifies key principles for personal data management and protection (e.g. lawfulness, fairness and transparency, purpose limitation, data minimization, accuracy, storage limitation, integrity and confidentiality (security), and accountability. [MIMA08-20K] (K) Identifies relevant national or international legislation applicable to the organization in which they work. [MIMA08-30K]

Supervised Practitioner – Management: Information Management			
ID	**Indicators of Competence** ***(in addition to those at Awareness level)***	**Relevant knowledge, experience, and/or behaviors**	**Possible examples of objective evidence of personal contribution to activities performed, or professional behaviors applied**
1	Follows a governing process and appropriate tools to plan and control information management activities. [MIMS01]	(K) Describes principles of planning and controlling information management as defined in the respective processes. [MIMS01-10K] (K) Describes tools used in support of information management. [MIMS01-20K] (A) Follows information management enterprise processes for information management on a particular project. [MIMS01-30A] (A) Uses information management enterprise tools for information management on a particular project. [MIMS01-40A]	Information Management reports to which they have contributed or acted upon. [MIMS01-E10] Information Management plan they have used. [MIMS01-E20] Information Management Plan they have tailored in order to match the principles of a required process. [MIMS01-E30]
2	Prepares inputs to a data dictionary and technical data library. [MIMS02]	(A) Maintains contents of a technical data library. [MIMS02-10A] (A) Uses information to correctly populate or maintain a data dictionary. [MIMS02-20A] (A) Maintains data dictionary as required. [MIMS02-30A]	Data dictionary they have populated or maintained. [MIMS02-E10] Data dictionary they have populated or maintained. [MIMS02-E20]
3	Identifies valid sources of information and associated authorities on a project. [MIMS03]	(K) Describes the method used to identify sources of information. [MIMS03-10K] (A) Prepares information in support of the identification of valid sources of information and associated authorities. [MIMS03-20A]	Information sources and respective authorities they identified as part of the generation of an information management plan. [MIMS03-E10]
4	Maintains information in accordance with integrity, security, privacy requirements, and data rights. [MIMS04]	(A) Uses appropriate information management techniques to obtain, transfer, distribute and maintain, or transform information. [MIMS04-10A] (A) Uses appropriate information security management techniques to maintain the integrity, security, privacy, and access rights of information. [MIMS04-20A] (A) Maintains information over its life cycle, in accordance with integrity, security, privacy requirements, and data rights. [MIMS04-30A]	Information items which they have had to acquire, transfer, distribute, maintain or transformed in accordance with integrity, security, privacy requirements, and data rights on a project. [MIMS04-E10]

5	Identifies information or approaches that requires replanning in order to implement engineering changes on a project. [MIMS05]	(A) Identifies identification and preparation of information required in order to support decision-making. [MIMS05-10A] (A) Prepares information in support of the assessment and re-planning of Systems Engineering activities associated with engineering changes. [MIMS05-20A]	Information packages they have generated to meet the information needs for a specific project change. [MIMS05-E10]
6	Identifies designated information requiring archiving in compliance project requirements on a project. [MIMS06]	(K) Describes rules and methods for identification and archiving of designated information. [MIMS06-10K] (A) Maintains identification and archival of designated information in compliance project requirements. [MIMS06-20A]	Records showing information they archived in accordance with information management requirements. [MIMS06-E10]
7	Identifies information requiring disposal of such as unwanted, invalid, or unverifiable information in accordance with requirements on a project. [MIMS07]	(A) Identifies rules and methods for identification and disposal of information. [MIMS07-10A] (A) Prepares information in support of the identification and disposal of unwanted, invalid or unverifiable information in accordance with requirements. [MIMS07-20A]	Records showing information they disposed of in accordance with information management requirements. [MIMS07-E10]
8	Prepares information management data products to support management reporting at organizational level. [MIMS08]	(A) Prepares data reports relating to entities under configuration management. [MIMS08-10A] (A) Prepares management information in support of organizational information management. [MIMS08-20A]	Data or statistics they created for Information Management reporting. [MIMS08-E10]
9	Prepares inputs to plans and work products addressing information management and its communication. [MIMS09]	(K) Describes purpose and content of typical information management plans. [MIMS09-10K] (K) Describes principles to be applied when communicating matters of information management. [MIMS09-20K] (A) Prepares information in support of information management planning and communication activities. [MIMS09-30A]	Information plans they have either generated or communicated. [MIMS09-E10]

(Continued)

ID	Indicators of Competence *(in addition to those at Awareness level)*	Relevant knowledge, experience, and/or behaviors	Possible examples of objective evidence of personal contribution to activities performed, or professional behaviors applied
10	Records lessons learned and shares beyond the project boundary. [MIMS10]	(A) Records lessons learned and shares beyond the project boundary. [MIMS10-10A] (A) Communicates lessons learned beyond the team or project boundary. [MIMS10-20A]	Records showing Information management lessons learned which were shared with others in the enterprise. [MIMS10-E10]
11	Develops own understanding of this competency area through Continual Professional Development (CPD). [MIMS11]	(A) Identifies potential gaps in own knowledge or development needs in this area, identifying opportunities to address these through continual professional development activities. [MIMS11-10A] (A) Performs continual professional development activities to improve their knowledge and understanding in this area. [MIMS11-20A] (A) Records continual professional development activities undertaken including learning or insights gained. [MIMS11-30A]	Records of Continual Professional Development (CPD) performed and learning outcomes. [MIMS11-E10]

Practitioner – Management: Information Management			
ID	**Indicators of Competence** ***(in addition to those at Supervised Practitioner level)***	**Relevant knowledge, experience, and/or behaviors**	**Possible examples of objective evidence of personal contribution to activities performed, or professional behaviors applied**
1	Creates a strategy for Information Management on a project to support SE project and wider enterprise needs. [MIMP01]	(A) Identifies project-specific Information Management needs that need to be incorporated into SE planning. [MIMP01-10A] (A) Prepares project-specific Information Management inputs for SE planning purposes, using appropriate processes and procedures. [MIMP01-20A] (A) Prepares project-specific Information Management task estimates in support of SE planning. [MIMP01-30A] (P) Liaises throughout with stakeholders to gain approval, updating strategy as necessary. [MIMP01-40P]	SE Information Management work products. [MIMP01-E10]
2	Creates governing plans, processes, and appropriate tools and uses these to control and monitor information management and associated communications activities. [MIMP02]	(K) Identifies the steps necessary to define a process and appropriate techniques to be adopted to establish Information Management for system elements. [MIMP02-10K] (K) Describes key elements of a successful Information management process and how decision management activities relate to different stages of a system life cycle. [MIMP02-20K] (A) Uses Information management tools or uses a methodology. [MIMP02-30A] (A) Uses appropriate standards and governing processes in their Information management planning, tailoring as appropriate. [MIMP02-40A] (A) Selects Information management method appropriate to each information item. [MIMP02-50A] (A) Creates Information management plans and processes for use on a project. [MIMP02-60A]	SE system information management planning work products they have developed. [MIMP02-E10] Planning documents such as the systems engineering management plan or other project/program plan or organizational process they have written. [MIMP02-E20] Information management plans and processes and their application during the full life cycle of a project. [MIMP02-E30] Information management process used. [MIMP02-E40] Information management plans and process descriptions used during the full life cycle of a project. [MIMP02-E50]

(Continued)

ID	Indicators of Competence *(in addition to those at Supervised Practitioner level)*	Relevant knowledge, experience, and/or behaviors	Possible examples of objective evidence of personal contribution to activities performed, or professional behaviors applied
3	Maintains a data dictionary, technical data library appropriate to the project. [MIMP03]	(A) Creates a data dictionary or technical data library appropriate to a specific project. [MIMP03-10A] (A) Uses and maintains a data dictionary or technical data library appropriate to a specific project. [MIMP03-20A] (A) Maintains the Data dictionary to reflect changes. [MIMP03-30A]	Data dictionaries and technical libraries that are appropriate to the respective project. [MIMP03-E10] Communications plan that ensures that the data dictionary or technical data library can be kept current. [MIMP03-E20]
4	Identifies valid sources of information and designated authorities and responsibilities for the information. [MIMP04]	(P) Collaborates with lead engineers and program management to identify critical information. [MIMP04-10P] (A) Performs contract requirements and deliverable analysis to determine key information. [MIMP04-20A] (A) Develops data taxonomies based on project scope, milestones, and typical life cycle. [MIMP04-30A]	Information Management Plan that details the relevant sources of information. [MIMP04-E10] Information Management Plan that differentiates between different kinds of information. [MIMP04-E20]
5	Maintains information artifacts in accordance with integrity, security, privacy requirements, and data rights. [MIMP05]	(A) Identifies activities associated with obtaining, transferring, distributing, and maintaining or transforming information. [MIMP05-10A] (A) Liaises with information security and assurance team to see that the integrity, security, privacy, and access rights for information is obtained across the life cycle. [MIMP05-20A]	Data library employed for data transformation and maintenance of integrity, security, or privacy. [MIMP05-E10]
6	Determines formats and media for capture, retention, transmission, and retrieval of information, and data requirements for the sharing of information. [MIMP06]	(A) Identifies distinguishing properties of data formats and media for the different life cycle stages of respective kinds of information. [MIMP06-10A] (A) Identifies the properties of data representations suitable for sharing information. [MIMP06-20A]	Data library incorporating data formats and media suitable for the information management needs. [MIMP06-E10] Data library that has the properties for being shared successfully. [MIMP06-E20]
7	Selects information archival requirements reflecting legal, audit, knowledge retention, and project closure obligations. [MIMP07]	(K) Describes potential sources for information about information archival requirements like laws and regulations, national and international standards, and company regulations. [MIMP07-10K]	Sources of archival requirements like laws and regulations, standards, or company regulations. [MIMP07-E10]

		(A) Identifies processes in which information with explicit archival requirements are created or modified. [MIMP07-20A]	Processes used for information to which archival requirements apply. [MIMP07-E20]
		(A) Identifies company regulations for information archive which require tailoring to be compliant to archival requirements for information. [MIMP07-30A]	Company regulations that have been tailored to specific information archival requirements. [MIMP07-E30]
		(A) Adapts process descriptions to incorporate information archival requirements. [MIMP07-40A]	Tailored process descriptions incorporating specific information archival requirements. [MIMP07-E40]
8	Prepares managed information in support of organizational configuration management and knowledge management requirements (e.g. sharing lessons learned). [MIMP08]	(K) Explains how lessons learned is collected, documented, and communicated in project teams and the enterprise. [MIMP08-10K]	Lessons learned from information management activities. [MIMP08-E10]
		(A) Communicates lessons learned beyond project boundaries, to other parts of the enterprise. [MIMP08-20A]	
9	Follows security, data management, privacy standards, and regulations applicable to the project. [MIMP09]	(K) Explains the levels or rules for access to various kinds or classes of information that apply in the project. [MIMP09-10K]	Data management plan that details the access rules for different classes of information in the project. [MIMP09-E10]
		(A) Identifies the rules and measures against data loss that apply in the project. [MIMP09-20A]	Data management plan that details the rules and measures against data loss for different kinds of data that are applied in the project. [MIMP09-E20]
		(A) Follows enterprise level data protection requirements. [MIMP09-30A]	
10	Selects and implements information management solutions consistent with project security and privacy requirements, data rights, and information management standards. [MIMP10]	(A) Analyzes project information management need in line with enterprise strategy, acting as an independent reviewer. [MIMP10-10A]	Project information management strategies prepared. [MIMP10-E10]
		(A) Creates project solution for Information management aligned to enterprise information management strategy. [MIMP10-20A]	Information management documentation. [MIMP10-E20]
		(A) Advises on the justification for project tools supporting information management. [MIMP10-30A]	

(*Continued*)

ID	Indicators of Competence *(in addition to those at Supervised Practitioner level)*	Relevant knowledge, experience, and/or behaviors	Possible examples of objective evidence of personal contribution to activities performed, or professional behaviors applied
11	Guides new or supervised practitioners in Information Management to develop their knowledge, abilities, skills, or associated behaviors. [MIMP11]	(P) Guides new or supervised practitioners in executing activities that form part of this competency. [MIMP11-10P] (A) Trains individuals to an "Awareness" level in this competency area. [MIMP11-20A]	Organizational Breakdown Structure showing their responsibility for technical supervision in this area. [MIMP11-E10] On-the-job training records. [MIMP11-E20] Coaching or mentoring assignment records. [MIMP11-E30] Records highlighting their impact on another individual in terms of improvement or professional development in this competency. [MIMP11-E40]
12	Maintains and enhances own competence in this area through Continual Professional Development (CPD) activities. [MIMP12]	(A) Identifies potential development needs in this area, identifying opportunities to address these through continual professional development activities. [MIMP12-10A] (A) Performs continual professional development activities to maintain and enhance their competency in this area. [MIMP12-20A] (A) Records continual professional development activities undertaken including learning or insights gained. [MIMP12-30A]	Records of Continual Professional Development (CPD) performed and learning outcomes. [MIMP12-E10]

Lead Practitioner – Management: Information Management			
ID	**Indicators of Competence *(in addition to those at Practitioner level)***	**Relevant knowledge, experience, and/or behaviors**	**Possible examples of objective evidence of personal contribution to activities performed, or professional behaviors applied**
1	Creates enterprise-level policies, procedures, guidance, and best practice for information management, including associated tools. [MIML01]	(A) Analyzes enterprise need for information management policies, processes, tools, or guidance. [MIML01-10A]	Enterprise level reports and recommendations. [MIML01-E10]
		(A) Creates enterprise policies, procedures or guidance on Information management. [MIML01-20A]	Part of enterprise SE resource pool and proposal or internal R&D reviews. [MIML01-E20]
		(A) Selects and acquires appropriate tools supporting information management. [MIML01-30A]	
2	Judges the tailoring of enterprise-level information management processes and associated work products to meet the needs of a project. [MIML02]	(A) Evaluates the enterprise Information Management processes against the business and external stakeholder (e.g. customer) needs in order to tailor processes to enable project success. [MIML02-10A]	Part of enterprise SE resource pool and proposal or internal R&D reviews. [MIML02-E10]
		(A) Provides constructive feedback on Information Management processes. [MIML02-20A]	Details of programs supported with successful results. [MIML02-E20]
		(A) Reviews high value or high-risk program Information Management planning and review issues across the enterprise. [MIML02-30A]	Reports and white papers generated. [MIML02-E30]
3	Coordinates information management across multiple diverse projects or across a complex system, with proven success. [MIML03]	(A) Compares and contrasts the enterprise benefits of a range of techniques for performing Information Management. [MIML03-10A]	Internal or external consultant in the relevant areas. [MIML03-E10]
		(A) Uses knowledge of the application of different Information Management approaches with successful results. [MIML03-20A]	Reports and white papers generated. [MIML03-E20]
		(A) Uses IM policies and procedures across multiple projects for components of a complex system. [MIML03-30A]	
4	Advises on appropriate information management solutions to be used on projects across the enterprise. [MIML04]	(A) Analyzes information management requirements from projects across the enterprise, maintaining alignment with wider enterprise strategy. [MIML04-10A]	Comments made on strategies proposed. [MIML04-E10]

(Continued)

ID	Indicators of Competence *(in addition to those at Practitioner level)*	Relevant knowledge, experience, and/or behaviors	Possible examples of objective evidence of personal contribution to activities performed, or professional behaviors applied
		(A) Determines project-specific solution option(s) consistent with security and privacy requirements, data rights and information management standards. [MIML04-20A]	Information management documentation. [MIML04-E20]
		(A) Liaises with other enterprise information management stakeholders if decision requires enterprise level actions. [MIML04-30A]	
5	Influences key stakeholders to address identified enterprise-level information management issues. [MIML05]	(P) Uses facilitation skills at project/program reviews to ensure Systems Engineering processes addresses enterprise-level information management issues and vice versa. [MIML05-10P]	White papers and reports. [MIML05-E10]
		(A) Advises on high value, high risk or high opportunity information management milestone review issues. [MIML05-20A]	Records of enterprise-level meetings. [MIML05-E20]
		(A) Persuades stakeholders to address enterprise-level Information management issues. [MIML05-30A]	Records indicating changed stakeholder thinking from personal intervention. [MIML05-E30]
6	Communicates Systems Engineering lessons learned gathered from projects across the enterprise. [MIML06]	(A) Analyzes project/program performance to gather information management lessons learned. [MIML06-10A]	Enterprise level reports and recommendations. [MIML06-E10]
		(A) Creates enterprise level information management lessons learned. [MIML06-20A]	Lessons learned examples. [MIML06-E20]
		(A) Communicates information management lessons learned to key enterprise stakeholders. [MIML06-30A]	
7	Coaches or mentors practitioners across the enterprise in information management in order to develop their knowledge, abilities, skills, or associated behaviors. [MIML07]	(P) Coaches or mentors practitioners across the enterprise in competency-related techniques, recommending development activities. [MIML07-10P]	Coaching or mentoring assignment records. [MIML07-E10]
		(A) Develops or authorizes enterprise training materials in this competency area. [MIML07-20A]	Records of formal training courses, workshops, seminars, and authored training material supported by successful post-training evaluation data. [MIML07-E20]
		(A) Provides enterprise workshops/seminars or training in this competency area. [MIML07-30A]	Listing as an approved organizational trainer for this competency area. [MIML07-E30]

8	Promotes the introduction and use of novel techniques and ideas in Information Management, across the enterprise, to improve enterprise competence in this area. [MIML08]	(A) Analyzes different approaches across different domains through research. [MIML08-10A]	Research records. [MIML08-E10]
		(A) Defines novel approaches that could potentially improve the SE discipline within the enterprise. [MIML08-20A]	Published papers in refereed journals/company literature. [MIML08-E20]
		(P) Fosters awareness of these novel techniques within the enterprise. [MIML08-30P]	Records showing introduction of enabling systems supporting the new techniques or ideas. [MIML08-E30]
		(P) Collaborates with enterprise stakeholders to facilitate the introduction of techniques new to the enterprise. [MIML08-40P]	Published papers (or similar) at enterprise level. [MIML08-E40]
		(A) Monitors new techniques after their introduction to determine their effectiveness. [MIML08-50A]	Records of improvements made against a recognized process improvement model in this area. [MIML08-E50]
		(A) Adapts approach to reflect actual enterprise performance improvements. [MIML08-60A]	
9	Develops expertise in this competency area through specialist Continual Professional Development (CPD) activities. [MIML09]	(A) Identifies own needs for further professional development in order to increase competence beyond practitioner level. [MIML09-10A]	Records of Continual Professional Development (CPD) performed and learning outcomes. [MIML09-E10]
		(A) Performs professional development activities in order to move own competence toward expert level. [MIML09-20A]	
		(A) Records continual professional development activities undertaken including learning or insights gained. [MIML09-30A]	

NOTES In addition to items above, enterprise-level or independent 3rd-Party-generated evidence may be used to amplify other evidence presented and may include:

a. Formally recognized by senior management in current organization as an expert in this competency area
b. Evidence of role as Product/System Design Authority or Technical Authority on a complex project with responsibilities in this area or where skills within this competency area were used
c. Recognized as an authorizing signatory on behalf of enterprise for formal documentation in this competency area (e.g. policies, processes, and deliverables)
d. Formal commendation or award within own enterprise for contribution or item of work successfully performed, which required proficiency in this competency area
e. Customer, Supplier, or other external project-specific key Stakeholder accolades for specific work performed in this competency area
f. Independently assessed or accredited work in this competency area (e.g. for independent publication or use)
g. Formal organizational HR records positively highlighting any specific professional competencies or behaviors identified (if applicable) plus any of the evidence indicators listed at Expert level below

Expert – Management: Information Management			
ID	**Indicators of Competence *(in addition to those at Lead Practitioner level)***	**Relevant knowledge, experience, and/or behaviors**	**Possible examples of objective evidence of personal contribution to activities performed, or professional behaviors applied**
1	Communicates own knowledge and experience in information management, in order to best practice beyond the enterprise boundary. [MIME01]	(A) Produces papers, seminars, or presentations outside own enterprise for publication in order to share own ideas and improve industry best practices in this competence area. [MIME01-10A]	Published papers or books etc. on new technique in refereed journals/company literature. [MIME01-E10]
		(P) Fosters incorporation of own ideas into industry best practices in this area. [MIME01-20P]	Published papers in refereed journals or internal literature proposing new practices in this competence area (or presentations, tutorials, etc.). [MIME01-E20]
		(P) Develops guidance materials identifying new (or updating existing) best practice in this competence area. [MIME01-30P]	Own proposals adopted as industry best practices in this competence area. [MIME01-E30]
2	Influences individuals beyond the enterprise boundary to adopt appropriate information management techniques or approaches. [MIME02]	(P) Acts as independent member of industry panels and professional societies. [MIME02-10P]	Records of projects supported successfully. [MIME02-E10]
		(A) Provides evidence and arguments influencing external stakeholder information management activities. [MIME02-20A]	Publications and presentations used in support of stakeholder arguments. [MIME02-E20]
		(P) Acts as independent member of review panels for external Information Management reviews. [MIME02-30P]	Reports from panels. [MIME02-E30]
3	Advises organizations beyond the enterprise boundary on complex or sensitive information management issues recommending appropriate solutions. [MIME03]	(A) Advises external stakeholders (e.g. customers) on their information management issues. [MIME03-10A]	Meeting minutes or report outlining role as arbitrator. [MIME03-E10]
		(A) Conducts sensitive information management negotiations. [MIME03-20A]	
		(A) Conducts complex information management negotiations taking account of external stakeholders' (e.g. customer's) background and knowledge. [MIME03-30A]	
		(P) Arbitrates on information management decisions. [MIME03-40P]	

4	Advises organizations beyond the enterprise boundary on the suitability of their approach to information management. [MIME04]	(A) Assesses an external organization's approach to Information Management, as independent reviewer. [MIME04-10A]	Records of membership on oversight committee with relevant terms of reference. [MIME04-E10]
		(A) Assesses the Information Management plans of an external organization, as independent reviewer. [MIME04-20A]	Review comments or revised document. [MIME04-E20]
5	Advises organizations beyond the enterprise boundary on security, data management, data rights, privacy standards, and regulations. [MIME05]	(A) Assesses the suitability of an external organizations approach to security, data management, data rights, privacy standards, and regulations, as an independent reviewer. [MIME05-10A]	Records of membership on oversight committee with relevant terms of reference. [MIME05-E10]
		(A) Assesses information management security, data management, data rights, privacy standards, and regulations. [MIME05-20A]	Review comments or revised document. [MIME05-E20]
6	Champions the introduction of novel techniques and ideas in information management, beyond the enterprise boundary, in order to develop the wider Systems Engineering community in this competency. [MIME06]	(A) Analyzes different approaches across different domains through research. [MIME06-10A]	Records of activities promoting research and need to adopt novel technique or ideas. [MIME06-E10]
		(A) Produces reports for the wider SE community on the effectiveness of new techniques after their introduction. [MIME06-20A]	Records of improvements made to process and appraisal against a recognized process improvement model. [MIME06-E20]
		(P) Collaborates with those introducing novel techniques within the wider SE community. [MIME06-30P]	Research records. [MIME06-E30]
		(A) Defines novel approaches that could potentially improve the wider SE discipline. [MIME06-40A]	Published papers in refereed journals/company literature. [MIME06-E40]
		(P) Fosters awareness of these novel techniques within the wider SE community. [MIME06-50P]	Records showing introduction of enabling systems supporting the new techniques or ideas. [MIME06-E50]
7	Coaches individuals beyond the enterprise boundary in information management, in order to further develop their knowledge, abilities, skills, or associated behaviors. [MIME07]	(P) Coaches or mentors individuals beyond the enterprise boundary, in competency-related techniques, recommending development activities. [MIME07-10P]	Coaching or mentoring assignment records. [MIME07-E10]
		(A) Develops or authorizes training materials in this competency area, which are subsequently successfully delivered beyond the enterprise boundary. [MIME07-20A]	Records of formal training courses, workshops, seminars, and authored training material supported by successful post-training evaluation data. [MIME07-E20]
		(A) Provides workshops/seminars or training in this competency area for practitioners or lead practitioners beyond the enterprise boundary (e.g. conferences and open training days). [MIME07-30A]	Records of Training/workshops/seminars created supported by successful post-training evaluation data. [MIME07-E30]

(*Continued*)

ID	Indicators of Competence *(in addition to those at Lead Practitioner level)*	Relevant knowledge, experience, and/or behaviors	Possible examples of objective evidence of personal contribution to activities performed, or professional behaviors applied
8	Maintains expertise in this competency area through specialist Continual Professional Development (CPD) activities. [MIME08]	(A) Reviews research, new ideas, and state of the art to identify relevant new areas requiring personal development in order to maintain expertise in this competency area. [MIME08-10A] (A) Performs identified specialist professional development activities in order to maintain or further develop competence at expert level. [MIME08-20A] (A) Records continual professional development activities undertaken including learning or insights gained. [MIME08-30A]	Records of documents reviewed and insights gained as part of own research into this competency area. [MIME08-E10] Records of Continual Professional Development (CPD) performed and learning outcomes. [MIME08-E20]

NOTES	In addition to items above, enterprise-level or independent 3rd-Party-generated evidence may be used to amplify other evidence presented and may include: a. Formally recognized by a reputable external organization as an expert in this competency area b. Evidence of role as independent assessor or reviewer on project outside own organization where skills in this competency area were used c. Evidence of invitation(s) from wider community for contribution of systems engineering expertise in this area (e.g. industry conference panel, government advisory board etc. cross-industry working groups, partnerships, accredited advanced university courses or research, or as part of professional institute) d. Formal commendation beyond the enterprise (e.g. by INCOSE or other recognized authority) for work performed in this competency area e. Independently assessed or accredited work product in this competency area (e.g. for independent publication or use) f. Accolades of expertise in this area from recognized industry leaders

Competency area – Integrating: Project Management
Description Project management identifies, plans, and coordinates activities required in order to deliver a satisfactory system, product, service of appropriate quality, within the constraints of schedule, budget, resources, infrastructure, available staffing, and technology. project management includes development engineering but covers the complete project (i.e. beyond the engineering boundary), encompassing disciplines such as sales, business development, finance, commercial, legal, human resources, production, procurement and supply chain management, and logistics.
Why it matters Good project management reduces risk; maximizes opportunity; minimizes system, product, or service costs; and improves both the success rate and the return on investment of projects.
Possible contributory types of evidence Any combination of the types of evidence may be acceptable (depending on how the Framework is tailored and used). The evidence items identified at each level indicate example work products only. Contributions to work products will generally differ at each proficiency level.
Learning and development The INCOSE Professional Development Portal provides example guidance on how to gain an initial awareness of a competency area and options for developing further competence thereafter.

Awareness – Integrating: Project Management		
ID	**Indicators of Competence**	**Relevant knowledge sub-indicators**
1	Explains the role the project management function plays in developing a successful system product or service. [IPMA01]	(K) Explains why organizations with a good project management and control are likely to survive better in the longer term than those without. [IPMA01-10K] (K) Explains the reasons why there is a need to perform project management across the full life cycle. [IPMA01-20K] (K) Lists project, external (e.g. customer) and organizational project management stakeholders. [IPMA01-30K] (K) Lists the key interfaces and links between project management stakeholders and the wider enterprise outside engineering development (e.g. finance, HR, Legal, Production, and IT). [IPMA01-40K] (K) Explains the role wider business stakeholders play with regard to project management and the potential impact their decisions may have on Systems Engineering activities on a particular project. [IPMA01-50K]
2	Explains the meaning of commonly used project management terms and applicable standards. [IPMA02]	(K) Lists common project management terms such as sponsor, project charter, and tollgates. [IPMA02-10K] (K) Lists key project management standards, guidance, or regulations applicable to the business. [IPMA02-20K]
3	Explains the relationship between cost, schedule, quality, and performance and why this matters. [IPMA03]	(K) Explains why cost, schedule, performance and quality are important on a project. [IPMA03-10K] (K) Explains the relationship between cost, schedule, performance and quality (e.g. If schedule time is decreased, typically cost will increase, or quality will suffer). [IPMA03-20K]

(Continued)

ID	Indicators of Competence	Relevant knowledge sub-indicators
4	Describes the role and typical responsibilities of a project manager on a project team, within the wider project management function. [IPMA04]	(K) Lists key roles within the project management function. [IPMA04-10K]
		(K) Describes the primary responsibilities and activities of key project management function roles (e.g. Project manager, Project planner) across all stages of a project life cycle (e.g. defining, planning, executing, and closing out a project). [IPMA04-20K]
		(K) Explains the difference between and potential resulting conflicts from the goals of the project management function and those of the Systems Engineering function (e.g. the best technical solution may not be feasible within the cost and schedule constraints). [IPMA04-30K]
5	Describes the differences between performing project management and Systems Engineering management on that project. [IPMA05]	(K) Explains how the project may be a sub-project within the wider enterprise (or program), and that Systems Engineering planning and execution is a sub-component of the project plan. [IPMA05-10K]
		(K) Explains why a project is a temporary effort with a specific objective to meet and how enterprise (or program) operations are ongoing and serve to sustain the business. [IPMA05-20K]
6	Describes the key interfaces between project management stakeholders within the enterprise and the project team. [IPMA06]	(K) Explains the interfaces between key project management function roles and the Systems Engineering team on a project, covering all stages of a systems life cycle. [IPMA06-10K]
		(K) Describes key work products exchanged between project management stakeholders within the enterprise and the project team. [IPMA06-20K]
		(K) Describes the key work products exchanged between the project Systems Engineering team and project management stakeholders within the enterprise across the life cycle. [IPMA06-30K]
7	Describes the wider program environment within which the system is being developed, and the influence each can have on this other. [IPMA07]	(K) Identifies the interfaces between key project management function roles and the Systems Engineering team on a project, covering all stages of a systems life cycle. [IPMA07-10K]
		(K) Describes the influence that the wider project management environment may have on a system is being developed. [IPMA07-20K]
		(K) Describes the influence that a system being developed may have on the wider project management environment within which it is being developed. [IPMA07-30K]

Supervised Practitioner – Integrating: Project Management			
ID	**Indicators of Competence *(in addition to those at Awareness level)***	**Relevant knowledge, experience, and/or behaviors**	**Possible examples of objective evidence of personal contribution to activities performed, or professional behaviors applied**
1	Follows a governing process in order to interface successfully to project management activities. [IPMS01]	(K) Describes the work products required to interface to the project management planning, monitoring and control. [IPMS01-10K]	Work products produced in response to requirements of project management function. [IPMS01-E10]
		(A) Uses approved processes and tools to control and manage own Systems Engineering activities in support of interactions with project management. [IPMS01-20A]	Systems Engineering Management Plan (SEMP) aligning to project management Plan Work products. [IPMS01-E20]
		(A) Uses interfacing work products produced by project management function to guide their own Systems Engineering activities. [IPMS01-30A]	Work products produced by project management function used to control Systems Engineering activities. [IPMS01-E30]
		(A) Uses a governing process and appropriate tools to control their own project management-related tasks. [IPMS01-40A]	
2	Prepares inputs to work products which interface to project management stakeholders to ensure Systems Engineering work aligns with wider project management activities. [IPMS02]	(K) Explains how project management concerns affect the definition of Systems Engineering schedules and resources (money, people, materials/equipment…). [IPMS02-10K]	Records of inputs supplied in support of project plan generation plan generation (e.g. technical work breakdown structure, technical resource plan, and technical development schedule). [IPMS02-E10]
		(K) Describes the input an author of a business project management plan may require from Systems Engineering. [IPMS02-20K]	
		(A) Prepares project management monitoring data to meet the requirements of the project management function. [IPMS02-30A]	
		(A) Collates identified indicators to support the project management function in tracking Systems Engineering activities. [IPMS02-40A]	
		(A) Monitors process performance to support the project management function in achieving the targeted standard of excellence on a project. [IPMS02-50A]	
		(A) Prepares work products for project management which align Systems Engineering work to wider project management activities. [IPMS02-60A]	

(Continued)

ID	Indicators of Competence *(in addition to those at Awareness level)*	Relevant knowledge, experience, and/or behaviors	Possible examples of objective evidence of personal contribution to activities performed, or professional behaviors applied
3	Identifies potential issues with interfacing work products received from project management Stakeholders or produced by Systems Engineering for project management stakeholders taking appropriate action. [IPMS03]	(A) Identifies issues with interfacing work products produced by project management when performing own Systems Engineering activities. [IPMS03-10A]	Records of issues raised and accepted as requiring resolution regarding interfacing Work Products produced by the project management function. [IPMS03-E10]
		(A) Identifies issues with interfacing work products produced by Systems Engineering which support project management activities. [IPMS03-20A]	
		(A) Records concerns regarding issue to ensure affected stakeholders become fully informed. [IPMS03-30A]	
4	Prepares Systems Engineering information for project management in support of wider project initiation activities. [IPMS04]	(K) Describes how projects consist of tasks and the scheduling and resourcing of those tasks. [IPMS04-10K]	Work products produced in support of project task definition. [IPMS04-E10]
		(K) Explains why monitoring and control is important to a project. [IPMS04-20K]	Work products produced in support of project task scheduling. [IPMS04-E20]
		(A) Prepares information supporting project management in the definition and scheduling of tasks and their resourcing. [IPMS04-30A]	Work products produced in support of project task resourcing. [IPMS04-E30]
5	Prepares Systems Engineering Work Breakdown Structure (WBS) information for project management in support of their creation of a wider project WBS. [IPMS05]	(K) Describes the term "Work Breakdown Structure" and how to break down tasks to the level required. [IPMS05-10K]	Work products produced in support of WBS generation. [IPMS05-E10]
		(K) Describes the key characteristics of common WBS models (e.g. US DoD publication MIL-STD 881). [IPMS05-20K]	
		(A) Prepares inputs to the definition of a WBS to support project management activities. [IPMS05-30A]	
6	Prepares Systems Engineering Work Package definitions and estimating information for project management in support of their work creating project-level Work Packages and estimates. [IPMS06]	(K) Explains the purpose of work packages – specific, detailed, and often annually agreed upon tasks, negotiated with the funding sponsor. [IPMS06-10K]	Work products produced in support of Work Package definition. [IPMS06-E10]
		(A) Prepares inputs to the definition of project Work Packages to support project management activities. [IPMS06-20A]	

7	Follows a governing process in order to interface successfully to project management activities. [IPMS07]	(K) Describes the concept, key features and purposes of different types of Integrated Project Teams (IPTs). [IPMS07-10K] (A) Prepares Systems Engineering inputs in support of a multi-disciplinary Project Team to support their activities. [IPMS07-20A]	Systems Engineering work products produced in support of the work of a multi-disciplinary team. [IPMS07-E10]
8	Prepares information used in project management contract reviews for project management on a project. [IPMS08]	(K) Explains why a contract is legally binding and how it is used to help accomplish project execution with the help of external entities. [IPMS08-10K] (K) Describes typical contract deliverable obligations, schedules, review cycles and contract closeout activities. [IPMS08-20K] (K) Describes key aspects of a contract: scope of work, statutory or regulatory requirements, required deliverables and acceptance reviews, etc. [IPMS08-30K] (K) Explains the key aspects of formally closing a project that is complete or being terminated, covering topics such as document acceptance by external stakeholders (e.g. customer), completion of project documentation, document and deliverables transfer, and lessons learned. [IPMS08-40K] (A) Prepares inputs in support of the review of contractual obligations (e.g. deliverables, adherence to contract obligations and interpretation of agreements) to ensure alignment of Systems Engineering activities with contractual requirements. [IPMS08-50A] (A) Prepares inputs to the close out of a project to support project management activities. [IPMS08-60A] (A) Prepares contractual technical inputs in support of the review of contract deliverables. [IPMS08-70A]	Work products produced in support of a contract technical review board or as technical support to the contracting officer. [IPMS08-E10] Work products produced in support of a project close-out review. [IPMS08-E20] Work products produced in support of the review of formal technical deliveries recognized contractually. [IPMS08-E30]
9	Prepares Systems Engineering information for project management in support of wider project termination activities. [IPMS09]	(K) Explains why projects require formal closure and common reasons for closure. [IPMS09-10K] (A) Prepares information supporting project management in the termination or closure of a project. [IPMS09-20A]	Work products produced in support of a project or contract termination. [IPMS09-E10]
10	Develops own understanding of this competency area through Continual Professional Development (CPD). [IPMS10]	(A) Identifies potential gaps in own knowledge or development needs in this area, identifying opportunities to address these through continual professional development activities. [IPMS10-10A] (A) Performs continual professional development activities to improve their knowledge and understanding in this area. [IPMS10-20A] (A) Records continual professional development activities undertaken including learning or insights gained. [IPMS10-30A]	Records of Continual Professional Development (CPD) performed and learning outcomes. [IPMS10-E10]

Practitioner – Integrating: Project Management			
ID	**Indicators of Competence** ***(in addition to those at Supervised Practitioner level)***	**Relevant knowledge, experience, and/or behaviors**	**Possible examples of objective evidence of personal contribution to activities performed, or professional behaviors applied**
1	Follows governing project management plans and processes, and uses appropriate tools to control and monitor project management-related Systems Engineering tasks, interpreting as necessary. [IPMP01]	(K) Describes the purpose of the project management Plan (PMP) and its relationship with SE Plans. [IPMP01-10K]	Systems Engineering work products prepared in accordance with requirements of project management plans. [IPMP01-E10]
		(A) Follows governing project management plans, processes and appropriate tools to control and monitor project-related Systems Engineering tasks. [IPMP01-20A]	
		(A) Uses governing project management plans, processes, and appropriate tools where necessary to ensure outputs meet project management stakeholder needs. [IPMP01-30A]	
		(A) Communicates potential issues with governing project management plans, processes and tools to project management stakeholders. [IPMP01-40A]	
2	Identifies Systems Engineering tasks ensuring that these tasks integrate successfully with project management activities. [IPMP02]	(K) Explains how a large project or task can be broken down into smaller activities, as well as activity sequencing and critical paths. [IPMP02-10K]	Project technical plans and schedules. [IPMP02-E10]
		(K) Explains how to estimate time, resources, and cost associated with any given task. [IPMP02-20K]	Detailed technical work packages. [IPMP02-E20]
		(K) Describes key characteristics of various organizational structures (functional/product/matrix alignment), styles and cultures (centralized or decentralized power/control) and adapts project plans accordingly. [IPMP02-30K]	Project technical management plans contributed to or authored. [IPMP02-E30]
		(A) Adapts project plans to reflect organizational structure, style, and culture. [IPMP02-40A]	Processes and appropriate techniques adopted for the technical planning of a project. [IPMP02-E40]
		(A) Determines tasks, owners and work packages for estimation and execution on a complete project. [IPMP02-50A]	
3	Identifies activities required to ensure integration of project management planning and estimating with Systems Engineering planning and estimating. [IPMP03]	(K) Explains project charters, scope statements, plans, baselines, etc. [IPMP03-10K]	Project technical plan. [IPMP03-E10]

		(K) Describes concepts such as risk management, quality management, configuration management and change control, staffing, communication, etc. and how they are relevant to the project plan. [IPMP03-20K]	Subsidiary plans such as Quality Management Plan, Risk Management Plan, Staffing Management Plan. [IPMP03-E20]
		(A) Prepares Systems Engineering contribution to a Project Plan. [IPMP03-30A]	
4	Develops inputs to a project management plan for a complete project beyond those required for Systems Engineering planning to support wider project or business project management. [IPMP04]	(K) Describes methods commonly used project monitoring and control. (e.g. expert judgment, earned value). [IPMP04-10K]	Chaired a formal decision point meeting. [IPMP04-E10]
		(K) Describes the purpose of performance measurement and recording, including progress on deliverables, expense tracking, the management of staff, equipment, contracts, change requests, etc. [IPMP04-20K]	Data from tracking systems (financial management tools, risk management tools, results of audits…) and resulting analysis of project status. [IPMP04-E20]
		(A) Selects appropriate decision points during program execution to integrate Systems Engineering with project management needs (e.g. design reviews, time-based reviews, and expense-based reviews). [IPMP04-30A]	
		(A) Defines decisions to be made at decision points. [IPMP04-40A]	
5	Develops Systems Engineering inputs for project management status reviews to enable informed decision-making. [IPMP05]	(A) Reviews Systems Engineering information produced in support of project status reviews or decision making. [IPMP05-10A]	Work products which support Systems Engineering decision making on a project. [IPMP05-E10]
		(P) Communicates with project management regarding the status of SE tasks on a project to support decision making. [IPMP05-20P]	Work products defining Systems Engineering status on a project. [IPMP05-E20]
		(A) Prepares Systems Engineering information for use in project management status reviews. [IPMP05-30A]	Records of communications to Project Management regarding Systems Engineering status reviews on a project. [IPMP05-E30]
6	Develops project initiation information required to support Project Start-up by project management on a project. [IPMP06]	(K) Describes contents of a team charter (purpose, scope, objectives, roles, membership, meeting requirements, criteria for termination). [IPMP06-10K]	Working group terms of reference or charter. [IPMP06-E10]
		(K) Explains the purpose of working groups across the project effort. [IPMP06-20K]	Leadership of a working group. [IPMP06-E20]
		(A) Creates working group charters beyond Systems Engineering which work effectively. [IPMP06-30A]	

(Continued)

ID	Indicators of Competence *(in addition to those at Supervised Practitioner level)*	Relevant knowledge, experience, and/or behaviors	Possible examples of objective evidence of personal contribution to activities performed, or professional behaviors applied
7	Develops Systems Engineering information required to support termination of a project by senior management. [IPMP07]	(K) Explains why a project might be terminated (exceeding schedule or budget, significant change to key requirements, sponsor request). [IPMP07-10K] (K) Describes tasks associated with project closure or termination. [IPMP07-20K] (A) Prepares Closeout Systems Engineering information for use by project management. [IPMP07-30A]	Project closure documentation. [IPMP07-E10]
8	Creates working groups extending beyond Systems Engineering. [IPMP08]	(A) Creates working groups (or other collaborate mechanisms) extending beyond systems engineering. [IPMP08-10A] (A) Ensures representatives of disciplines beyond systems engineering (e.g. project management) participate in working group meetings. [IPMP08-20A] (A) Coordinates agreed improvements to working practices in collaboration with project management. [IPMP08-30A]	Working group charter. [IPMP08-E10] List of working group participants. [IPMP08-E20] Improvements made through charter activities. [IPMP08-E30]
9	Guides new or supervised practitioners in finance and its relationship to Systems Engineering, to develop their knowledge, abilities, skills, or associated behaviors. [IPMP09]	(P) Guides new or supervised practitioners in executing activities that form part of this competency. [IPMP09-10P] (A) Trains individuals to an "Awareness" level in this competency area. [IPMP09-20A]	Organizational Breakdown Structure showing their responsibility for technical supervision in this area. [IPMP09-E10] Records of on-the-job training objectives/ guidance. [IPMP09-E20] Assignment as coach or mentor. [IPMP09-E30] Records highlighting their impact on another individual in terms of improvement or professional development in this competency. [IPMP09-E40]
10	Maintains and enhances own competence in this area through Continual Professional Development (CPD) activities. [IPMP10]	(A) Identifies potential development needs in this area, identifying opportunities to address these through continual professional development activities. [IPMP10-10A] (A) Performs continual professional development activities to maintain and enhance their competency in this area. [IPMP10-20A] (A) Records continual professional development activities undertaken including learning or insights gained. [IPMP10-30A]	Records of Continual Professional Development (CPD) performed and learning outcomes. [IPMP10-E10]

Lead Practitioner – Integrating: Project Management			
ID	**Indicators of Competence *(in addition to those at Practitioner level)***	**Relevant knowledge, experience, and/or behaviors**	**Possible examples of objective evidence of personal contribution to activities performed, or professional behaviors applied**
1	Creates enterprise-level policies, procedures, guidance, and best practice in order to ensure Systems Engineering project management activities integrate with enterprise-level Project Management goals. [IPML01]	(A) Creates Enterprise policy and guidance regarding project management integration to ensure SE activities align with project management needs. [IPML01-10A]	Enterprise policy and guidance they have produced. [IPML01-E10]
		(A) Determines requirements for precision level, control thresholds, reporting formats, and periodicity to ensure SE activities align with project management needs. [IPML01-20A]	Work products produced defining Systems Engineering thresholds, reporting formats, and periodicity to ensure SE activities align with project management needs. [IPML01-E20]
		(A) Selects and acquires appropriate tools to align SE activities with project management needs. [IPML01-30A]	Records of tool trade studies and selection. [IPML01-E30]
2	Assesses enterprise-level project management processes and tailoring to ensure they integrate with Systems Engineering needs. [IPML02]	(A) Evaluates enterprise project management processes against Systems Engineering needs in order to enable project success. [IPML02-10A]	Documented tailored process. [IPML02-E10]
		(P) Produces constructive feedback on enterprise level project management processes where these do not align with Systems Engineering approaches or needs. [IPML02-20P]	Project Management Plan feedback and revised plan. [IPML02-E20]
		(A) Assesses project management plans across the enterprise to ensure Systems Engineering activities are aligned. [IPML02-30A]	Records demonstrating project management planning, estimates or other work products evaluated including feedback provided and implemented. [IPML02-E30]
		(A) Assesses project management approaches across the enterprise with regard to their alignment with Systems Engineering activities. [IPML02-40A]	
3	Assesses project management information produced across the enterprise using appropriate techniques for its integration with Systems Engineering data. [IPML03]	(A) Reviews information produced within the project management functions using appropriate techniques to assess its impact on Systems Engineering. [IPML03-10A]	Records demonstrating estimates evaluated including feedback provided and implemented. [IPML03-E10]
		(A) Judges adequacy and correctness of Systems Engineering project management information using appropriate techniques. [IPML03-20A]	

(*Continued*)

ID	Indicators of Competence *(in addition to those at Practitioner level)*	Relevant knowledge, experience, and/or behaviors	Possible examples of objective evidence of personal contribution to activities performed, or professional behaviors applied
4	Judges appropriateness of enterprise-level project management decisions in a rational way to ensure alignment with Systems Engineering needs. [IPML04]	(A) Analyzes project management stakeholders decisions using best practice techniques and based upon Systems Engineering need. [IPML04-10A]	Reports on project management decisions made. [IPML04-E10]
		(A) Determines appropriateness of project management decisions made, with robust justification based upon Systems Engineering need. [IPML04-20A]	Records indicating changed stakeholder thinking from personal intervention. [IPML04-E20]
		(A) Persuades project management stakeholders to change viewpoints and decisions based on the grounds of appropriateness and Systems Engineering need. [IPML04-30A]	
5	Judges conflicts between project management needs and Systems Engineering needs on behalf of the enterprise, arbitrating as required. [IPML05]	(A) Analyzes data associated with the conflict to determine the best strategy at enterprise level. [IPML05-10A]	Original and revised resource or asset plans, showing areas of conflict and how they were resolved. [IPML05-E10]
		(P) Persuades stakeholders to accept deconfliction strategies for personnel or asset scheduling across multiple enterprise projects. [IPML05-20P]	Records of advice provided on Project plans. [IPML05-E20]
6	Guides and actively coordinates complex or challenging relationships with key stakeholders affecting Systems Engineering. [IPML06]	(A) Advises when dealing with complex external stakeholder relationships affecting Systems Engineering. [IPML06-10A]	Records showing communications in complex stakeholder relationships. [IPML06-E10]
		(P) Acts in leading role in discussions with external stakeholders. [IPML06-20P]	Records of advice provided. [IPML06-E20]
7	Persuades key project management stakeholders to address identified enterprise-level project management issues affecting Systems Engineering. [IPML07]	(A) Identifies and engages with key project management stakeholders. [IPML07-10A]	Examples of new project management planning, monitoring, and control techniques introduced. [IPML07-E10]
		(P) Fosters agreement between key stakeholders at enterprise level to resolve project management issues related to systems engineering, by promoting a holistic viewpoint. [IPML07-20P]	Minutes of key stakeholder review meetings and actions taken. [IPML07-E20]
		(P) Persuades key stakeholders to address identified enterprise-level project management issues. [IPML07-30P]	Records indicating changed stakeholder thinking from personal intervention. [IPML07-E30]
8	Coaches or mentors practitioners across the enterprise in the integration of project management with Systems Engineering, in order to develop their knowledge, abilities, skills, or associated behaviors. [IPML08]	(P) Coaches or mentors Systems Engineering or project management practitioners within the enterprise in competency-related techniques, recommending development activities. [IPML08-10P]	Coaching or mentoring assignment records. [IPML08-E10]

		(A) Develops or authorizes enterprise training materials in this competency area. [IPML08-20A]	Records of formal training courses, workshops, seminars, and authored training material supported by successful post-training evaluation data. [IPML08-E20]
		(A) Provides enterprise workshops/seminars or training in this competency area. [IPML08-30A]	Listing as an approved organizational trainer for this competency area. [IPML08-E30]
9	Promotes the introduction and use of novel techniques and ideas across the enterprise, which improve the integration of Systems Engineering and project management functions. [IPML09]	(A) Analyzes different approaches across different domains through research. [IPML09-10A]	Research records. [IPML09-E10]
		(A) Defines novel approaches that could potentially improve the SE discipline within the enterprise. [IPML09-20A]	Published papers in refereed journals/company literature. [IPML09-E20]
		(P) Fosters awareness of these novel techniques within the enterprise. [IPML09-30P]	Records showing introduction of enabling systems supporting the new techniques or ideas. [IPML09-E30]
		(P) Collaborates with enterprise stakeholders to facilitate the introduction of techniques new to the enterprise. [IPML09-40P]	Published papers (or similar) at enterprise level. [IPML09-E40]
		(A) Monitors new techniques after their introduction to determine their effectiveness. [IPML09-50A]	Records of improvements made against a recognized process improvement model in this area. [IPML09-E50]
		(A) Adapts approach to reflect actual enterprise performance improvements. [IPML09-60A]	
10	Develops expertise in this competency area through specialist Continual Professional Development (CPD) activities. [IPML10]	(A) Identifies own needs for further professional development in order to increase competence beyond practitioner level. [IPML10-10A]	Records of Continual Professional Development (CPD) performed and learning outcomes. [IPML10-E10]
		(A) Performs professional development activities in order to move own competence toward expert level. [IPML10-20A]	
		(A) Records continual professional development activities undertaken including learning or insights gained. [IPML10-30A]	

NOTES	In addition to items above, enterprise-level or independent 3rd-Party-generated evidence may be used to amplify other evidence presented and may include: a. Formally recognized by senior management in current organization as an expert in this competency area b. Evidence of role as Product/System Design Authority or Technical Authority on a complex project with responsibilities in this area or where skills within this competency area were used c. Recognized as an authorizing signatory on behalf of enterprise for formal documentation in this competency area (e.g. policies, processes, and deliverables) d. Formal commendation or award within own enterprise for contribution or item of work successfully performed, which required proficiency in this competency area e. Customer, Supplier, or other external project-specific key Stakeholder accolades for specific work performed in this competency area f. Independently assessed or accredited work in this competency area (e.g. for independent publication or use) g. Formal organizational HR records positively highlighting any specific professional competencies or behaviors identified (if applicable) plus any of the evidence indicators listed at Expert level below

Expert – Integrating: Project Management			
ID	**Indicators of Competence** ***(in addition to those at Lead Practitioner level)***	**Relevant knowledge, experience, and/or behaviors**	**Possible examples of objective evidence of personal contribution to activities performed, or professional behaviors applied**
1	Communicates own knowledge and experience in the integration of project management with Systems Engineering, in order to improve Systems Engineering best practice beyond the enterprise boundary. [IPME01]	(A) Produces papers, seminars, or presentations outside own enterprise for publication in order to share own ideas and improve industry best practices in this competence area. [IPME01-10A]	Published papers or books etc. on new technique in refereed journals/company literature. [IPME01-E10]
		(P) Fosters incorporation of own ideas into industry best practices in this area. [IPME01-20P]	Published papers in refereed journals or internal literature proposing new practices in this competence area (or presentations, tutorials, etc.). [IPME01-E20]
		(P) Develops guidance materials identifying new (or updating existing) best practice in this competence area. [IPME01-30P]	Own proposals adopted as industry best practices in this competence area. [IPME01-E30]
2	Advises organizations beyond the enterprise boundary on complex or sensitive project management-related issues affecting Systems Engineering. [IPME02]	(A) Advises external stakeholders (e.g. customers) on their project management plans and issues. [IPME02-10A]	Records of advice provided on project management issues. [IPME02-E10]
		(A) Conducts successful sensitive negotiations regarding the project management of highly complex systems. [IPME02-20A]	Records of advice provided on sensitive project management issues. [IPME02-E20]
		(A) Conducts successful sensitive negotiations making limited use of specialized, project management relevant terminology. [IPME02-30A]	
		(A) Uses a holistic approach to complex issue resolution including balanced, rational arguments on way forward. [IPME02-40A]	
3	Advises organizations beyond the enterprise boundary on the suitability of their approach to project management plans affecting Systems Engineering activities. [IPME03]	(P) Acts as independent external reviewer on external committee (or similar body) which approves such plans. [IPME03-10P]	Records of membership on oversight committee with relevant terms of reference. [IPME03-E10]
		(A) Reviews and advises on project management strategies that has led to changes being implemented. [IPME03-20A]	Review comments or records from revised document. [IPME03-E20]
4	Champions the introduction of novel techniques and ideas to improve the integration of Systems Engineering and project management functions, beyond the enterprise boundary, in order to develop the wider Systems Engineering community in this competency. [IPME04]	(A) Analyzes different approaches across different domains through research. [IPME04-10A]	Records of activities promoting research and need to adopt novel technique or ideas. [IPME04-E10]

		(A) Produces reports for the wider SE community on the effectiveness of new techniques after their introduction. [IPME04-20A]	Records of improvements made to process and appraisal against a recognized process improvement model. [IPME04-E20]
		(P) Collaborates with those introducing novel techniques within the wider SE community. [IPME04-30P]	Research records. [IPME04-E30]
		(A) Defines novel approaches that could potentially improve the wider SE discipline. [IPME04-40A]	Published papers in refereed journals/company literature. [IPME04-E40]
		(P) Fosters awareness of these novel techniques within the wider SE community. [IPME04-50P]	Records showing introduction of enabling systems supporting the new techniques or ideas. [IPME04-E50]
5	Coaches individuals beyond the enterprise boundary, in the relationship between Systems Engineering and project management, to further develop their knowledge, abilities, skills, or associated behaviors. [IPME05]	(P) Coaches or mentors individuals beyond the enterprise boundary, in competency-related techniques, recommending development activities. [IPME05-10P]	Coaching or mentoring assignment records. [IPME05-E10]
		(A) Develops or authorizes training materials in this competency area, which are subsequently successfully delivered beyond the enterprise boundary. [IPME05-20A]	Records of formal training courses, workshops, seminars, and authored training material supported by successful post-training evaluation data. [IPME05-E20]
		(A) Provides workshops/seminars or training in this competency area for practitioners or lead practitioners beyond the enterprise boundary (e.g. conferences and open training days). [IPME05-30A]	Records of Training/workshops/seminars created supported by successful post-training evaluation data. [IPME05-E30]
6	Maintains expertise in this competency area through specialist Continual Professional Development (CPD) activities. [IPME06]	(A) Reviews research, new ideas, and state of the art to identify relevant new areas requiring personal development in order to maintain expertise in this competency area. [IPME06-10A]	Records of documents reviewed and insights gained as part of own research into this competency area. [IPME06-E10]
		(A) Performs identified specialist professional development activities in order to maintain or further develop competence at expert level. [IPME06-20A]	Records of Continual Professional Development (CPD) performed and learning outcomes. [IPME06-E20]
		(A) Records continual professional development activities undertaken including learning or insights gained. [IPME06-30A]	

NOTES	In addition to items above, enterprise-level or independent 3rd-Party-generated evidence may be used to amplify other evidence presented and may include: a. Formally recognized by a reputable external organization as an expert in this competency area b. Evidence of role as independent assessor or reviewer on project outside own organization where skills in this competency area were used c. Evidence of invitation(s) from wider community for contribution of systems engineering expertise in this area (e.g. industry conference panel, government advisory board etc. cross-industry working groups, partnerships, accredited advanced university courses or research, or as part of professional institute) d. Formal commendation beyond the enterprise (e.g. by INCOSE or other recognized authority) for work performed in this competency area e. Independently assessed or accredited work product in this competency area (e.g. for independent publication or use) f. Accolades of expertise in this area from recognized industry leaders

Competency area – Integrating: Finance
Description Finance is the area of estimating and tracking costs associated with the project. It also includes understanding of the financial environment in which the project is being executed.
Why it matters Appropriate funding is the life blood of any system development project. It is important for systems engineers to recognize the importance of cost estimation, budgeting, and controlling project finances and to support the finance discipline in its activities.
Possible contributory types of evidence Any combination of the types of evidence may be acceptable (depending on how the Framework is tailored and used). The evidence items identified at each level indicate example work products only. Contributions to work products will generally differ at each proficiency level.
Learning and development The INCOSE Professional Development Portal provides example guidance on how to gain an initial awareness of a competency area and options for developing further competence thereafter.

Awareness - Integrating: Finance		
ID	**Indicators of Competence**	**Relevant knowledge sub-indicators**
1	Explains the role the finance function plays in developing a successful system product or service. [IFIA01]	(K) Explains why organizations with a good financial management and control are likely to survive better in the longer term than those without. [IFIA01-10K] (K) Explains the reasons why there is a need to estimate, budget, and control costs associated with project execution across the full life cycle. [IFIA01-20K] (K) Lists project, external (e.g. customer) and organizational financial stakeholders. [IFIA01-30K] (K) Explains the role wider enterprise stakeholders play with regard to financial management and the potential impact their decisions may have on Systems Engineering activities on a particular project. [IFIA01-40K]
2	Explains the meaning of commonly used financial terms and applicable standards. [IFIA02]	(K) Describes the key interfaces and links between financial stakeholders and the wider enterprise outside engineering (e.g. infrastructure management, human resource management, quality management, knowledge management, portfolio management, and life cycle model management). [IFIA02-10K] (K) Describes key financial terms such as budget, income statement, balance sheet, cost, profit, return on investment, and cashflow. [IFIA02-20K]
3	Explains how business financial decisions may impact a product or service through its entire life cycle, and vice versa. [IFIA03]	(K) Explains how broader financial activities or decisions can impact the cost of developing, using, maintaining and supporting the product or service and vice versa. [IFIA03-10K] (K) Explains how technical decisions required on a project need to align with wider financial decisions (e.g. make/buy choice of supplier or technology). [IFIA03-20K] (K) Describes the link between risk and financial management. [IFIA03-30K] (K) Describes the relationship between a financial controller within a business and a project. [IFIA03-40K]

4	Explains primary interfaces between the finance function and the Systems Engineering team. [IFIA04]	(K) Describes key roles within the finance function. [IFIA04-10K] (K) Describes the interfaces between key finance function roles and the Systems Engineering team on a project, covering all stages of a systems life cycle. [IFIA04-20K] (K) Describes the primary responsibilities and activities of key finance function roles (e.g. financial manager, financial controller) across all stages of a project life cycle (e.g. defining, planning, executing, and closing out a project). [IFIA04-30K] (K) Explains the difference between and potential resulting conflicts from the goals of the finance function and those of the Systems Engineering function (e.g. price, margin, return on investment, capital investment timing vs process, and technical baseline). [IFIA04-40K]
5	Describes the key work products exchanged between finance stakeholders and the Systems Engineering team. [IFIA05]	(K) Describes key work products exchanged between finance stakeholders within the enterprise and the project team. [IFIA05-10K] (K) Describes the key work products exchanged between the project Systems Engineering team and Financial stakeholders within the enterprise across the life cycle. [IFIA05-20K]
6	Describes the difference between performing financial management on a project or wider enterprise and managing financial resources as part of Systems Engineering activities. [IFIA06]	(K) Describes the scope of project financial management and control and enterprise-level financial management and control. [IFIA06-10K] (K) Describes the difference between financial management (on project or wider enterprise) and managing resources on a project. [IFIA06-20K]
7	Explains how financial management concerns relate to Systems Engineering. [IFIA07]	(K) Explains how financial concerns affect the definition of Systems Engineering schedules and resources (money, people, materials/equipment…). [IFIA07-10K] (K) Describes the input an author of a business finance plan may require from Systems Engineering. [IFIA07-20K]

Supervised Practitioner – Integrating: Finance			
ID	**Indicators of Competence** *(in addition to those at Awareness level)*	**Relevant knowledge, experience, and/or behaviors**	**Possible examples of objective evidence of personal contribution to activities performed, or professional behaviors applied**
1	Follows a governing process in order to interface successfully to financial management activities. [IFIS01]	(K) Describes the work products required to interface to the financial management planning, monitoring and control. [IFIS01-10K] (A) Uses approved processes and tools to control and manage own Systems Engineering activities in support of interactions with finance department. [IFIS01-20A] (A) Uses interfacing work products produced by finance to guide their own Systems Engineering activities. [IFIS01-30A] (A) Uses a governing process and appropriate tools to control their own finance-related tasks. [IFIS01-40A]	Work products created in support of the financial management function, above the level of Systems Engineering planning and execution. [IFIS01-E10] Work products developed in support of financial management showing their alignment with SEMP. [IFIS01-E20] Financial processes and tools used on a project. [IFIS01-E30]
2	Prepares inputs to work products which interface to financial stakeholders to ensure Systems Engineering work aligns with wider financial management activities. [IFIS02]	(K) Explains how financial concerns affect the definition of Systems Engineering schedules and resources (money, people, materials/equipment…). [IFIS02-10K] (K) Describes the input an author of a business finance plan may require from Systems Engineering. [IFIS02-20K] (A) Prepares inputs to interfacing work products required by finance which ensure Systems Engineering work aligns with wider financial management activities. [IFIS02-30A]	Work products for use by finance. [IFIS02-E10] Project cost estimates they have generated or reviewed. [IFIS02-E20]
3	Identifies potential issues with interfacing work products received from Financial Stakeholders or produced by Systems Engineering for financial stakeholders taking appropriate action. [IFIS03]	(A) Identifies issues with interfacing work products produced by finance when performing own Systems Engineering activities. [IFIS03-10A] (A) Identifies issues with interfacing work products produced by Systems Engineering, which support financial management activities. [IFIS03-20A] (A) Records concerns regarding issue to ensure affected stakeholders become fully informed. [IFIS03-30A]	Records of issues they raised in this area. [IFIS03-E10]

4	Prepares inputs to financial cost estimation work products for financial stakeholders ensuring Systems Engineering work aligns with wider financial management activities. [IFIS04]	(K) Explains common financial concerns when estimating costs. [IFIS04-10K] (A) Prepares cost estimation information to assist financial stakeholders in their understanding. [IFIS04-20A]	Plans contributed to relating to the acquisition of project funding. [IFIS04-E10]
5	Uses cost aggregation and analysis techniques to communicate funding information for financial stakeholders during creation or approval of funding requests. [IFIS05]	(A) Uses knowledge of the details of Systems Engineering cost estimating to assist financial stakeholders in their understanding. [IFIS05-10A] (K) Explains cost aggregation and analysis to financial stakeholders to develop funding requests. [IFIS05-20K]	Funding analysis documents they have produced. [IFIS05-E10]
6	Uses system life cycle cost analysis techniques to communicate cost information to financial stakeholders on a project. [IFIS06]	(A) Uses knowledge of system life cycle cost issues to identify system life cycle issues, which affect financial stakeholders. [IFIS06-10A] (A) Uses knowledge of system life cycle cost issues to report system life cycle issues which affect financial stakeholders. [IFIS06-20A] (A) Uses knowledge of system life cycle cost issues to provide inputs to the decisions of financial stakeholders. [IFIS06-30A]	Financial analysis of life cycle costs on a project they have performed or contributed to. [IFIS06-E10] Recommendations they have made in support of financial decision-making on a project. [IFIS06-E20]
7	Uses project performance and expenditure tracking techniques to communicate performance and expenditure tracking information to financial stakeholders on a project. [IFIS07]	(A) Identifies issues with project performance that affect financial stakeholders. [IFIS07-10A] (A) Uses knowledge of project performance to report tracking issues to financial stakeholders. [IFIS07-20A]	Project financial performance data they have tracked. [IFIS07-E10] Project expenditures they have tracked on a project. [IFIS07-E20]
8	Uses financial variance and tolerance data to communicate budget or financial variances to financial stakeholders on a project. [IFIS08]	(A) Uses knowledge of budget and financial reporting to identify budget variance issues which affect financial stakeholders. [IFIS08-10A]	Financial variances they have identified on a project. [IFIS08-E10]

(Continued)

ID	Indicators of Competence *(in addition to those at Awareness level)*	Relevant knowledge, experience, and/or behaviors	Possible examples of objective evidence of personal contribution to activities performed, or professional behaviors applied
		(A) Uses knowledge of budget and financial reporting to identify solutions to budget variances affecting financial stakeholders. [IFIS08-20A] (A) Communicates budget and financial variances to financial stakeholders. [IFIS08-30A]	Recommendations they have made for corrective actions in support of project financial stakeholders. [IFIS08-E20]
9	Develops own understanding of this competency area through Continual Professional Development (CPD). [IFIS09]	(A) Identifies potential gaps in own knowledge or development needs in this area, identifying opportunities to address these through continual professional development activities. [IFIS09-10A] (A) Performs continual professional development activities to improve their knowledge and understanding in this area. [IFIS09-20A] (A) Records continual professional development activities undertaken including learning or insights gained. [IFIS09-30A]	Records of Continual Professional Development (CPD) performed and learning outcomes. [IFIS09-E10]

Practitioner – Integrating: Finance			
ID	**Indicators of Competence** ***(in addition to those at Supervised Practitioner level)***	**Relevant knowledge, experience, and/or behaviors**	**Possible examples of objective evidence of personal contribution to activities performed, or professional behaviors applied**
1	Follows governing finance plans, processes, and uses appropriate tools to control and monitor finance-related Systems Engineering tasks, interpreting as necessary. [IFIP01]	(A) Follows governing financial plans, processes, and appropriate tools to control and monitor finance-related Systems Engineering tasks. [IFIP01-10A] (A) Uses governing financial plans, processes, and appropriate tools where necessary to ensure outputs meet financial stakeholder needs. [IFIP01-20A] (A) Communicates potential issues with governing financial plans, processes, and tools to financial stakeholders. [IFIP01-30A]	Financial work products produced and approved. [IFIP01-E10] Records of review on financial work products produced by others. [IFIP01-E20]
2	Prepares work products required by financial stakeholders to ensure Systems Engineering work aligns with wider financial management activities. [IFIP02]	(A) Analyzes information required by finance for requested work products to ensure understanding. [IFIP02-10A] (A) Prepares work products required by Financial stakeholders, using appropriate processes and procedures. [IFIP02-20A]	Work products for use by finance to which they contributed, or produced. [IFIP02-E10]
3	Creates detailed cost estimating work products required by financial stakeholders to scope the financial aspects of a project. [IFIP03]	(A) Creates detailed cost estimates for financial stakeholders. [IFIP03-10A] (A) Creates supporting evidence to justify cost estimates for financial stakeholders. [IFIP03-20A]	Records showing cost estimates produced or contributed to. [IFIP03-E10] Records showing cost justification documentation or similar supporting evidence produced for cost estimates they have supported. [IFIP03-E20]
4	Analyzes activity costs and scheduling as required by financial stakeholders in order to develop project funding requirements and a cost management plan. [IFIP04]	(A) Analyzes aggregated activity costs and scheduling to develop project funding requirements for financial stakeholders. [IFIP04-10A] (A) Creates a cost management plan for financial stakeholders. [IFIP04-20A]	Records showing a project funding model they produced. [IFIP04-E10] Records showing a cost management plan they produced (or contributed to) for a project. [IFIP04-E20]
5	Analyzes system life cycle cost issues and decisions as required by financial stakeholders in order to make recommendations. [IFIP05]	(A) Analyzes system life cycle cost issues and decisions for financial stakeholders to the program management team. [IFIP05-10A] (A) Develops appropriate recommendations on system life cycle cost issues & decisions for financial stakeholders to the program management team, based on own analysis and review. [IFIP05-20A]	Records showing life cycle cost documentation they created for financial stakeholders. [IFIP05-E10] Records demonstrating analysis, recommendations, or decisions they have contributed to in support of financial management, in order to reduce or control life cycle cost. [IFIP05-E20]

(Continued)

ID	Indicators of Competence *(in addition to those at Supervised Practitioner level)*	Relevant knowledge, experience, and/or behaviors	Possible examples of objective evidence of personal contribution to activities performed, or professional behaviors applied
6	Analyzes project performance and expenditures as required by financial stakeholders in order to determine variance from plans. [IFIP06]	(A) Reviews project performance and expenditures in support of financial stakeholders to the program management team. [IFIP06-10A] (A) Compares project execution against plan, in support of financial stakeholders to the program management team. [IFIP06-20A]	Records demonstrating support of project performance reviews (e.g. details of reviews attended or minute). [IFIP06-E10] Records demonstrating support of financial analysis in support of project progress reviews (e.g. details of reviews attended, or minutes). [IFIP06-E20]
7	Analyzes variances to budget tolerance as required by financial stakeholders in order to identify and implement corrective actions. [IFIP07]	(A) Identifies variances to allocated budgets in support of financial stakeholders to the program management team. [IFIP07-10A] (A) Determines corrective actions in order to stay within tolerance of budgets in support of financial stakeholders to the program management team. [IFIP07-20A] (A) Ensures activities remain within budget tolerances, in support of financial stakeholders to the program management team. [IFIP07-30A]	Records demonstrating a personal contribution to identification of variances to allocated budgets in support of financial stakeholders. [IFIP07-E10] Records demonstrating corrective actions provided in support of project progress reviews. [IFIP07-E20] Records demonstrating personal activities performed in order to implement corrective actions. [IFIP07-E30]
8	Guides new or supervised practitioners in finance and its relationship to Systems Engineering, to develop their knowledge, abilities, skills, or associated behaviors. [IFIP08]	(P) Guides new or supervised practitioners in executing activities that form part of this competency. [IFIP08-10P] (A) Trains individuals to "Awareness" level in this competency area. [IFIP08-20A]	Organizational Breakdown Structure showing their responsibility for technical supervision in this area. [IFIP08-E10] On-the-job training objectives/guidance etc. [IFIP08-E20] Coaching or mentoring assignment records. [IFIP08-E30] Records highlighting their impact on another individual in terms of improvement or professional development in this competency. [IFIP08-E40]
9	Maintains and enhances own competence in this area through Continual Professional Development (CPD) activities. [IFIP09]	(A) Identifies potential development needs in this area, identifying opportunities to address these through continual professional development activities. [IFIP09-10A] (A) Performs continual professional development activities to maintain and enhance their competency in this area. [IFIP09-20A] (A) Records continual professional development activities undertaken including learning or insights gained. [IFIP09-30A]	Records of Continual Professional Development (CPD) performed and learning outcomes. [IFIP09-E10]

Lead Practitioner – Integrating: Finance			
ID	**Indicators of Competence** ***(in addition to those at Practitioner level)***	**Relevant knowledge, experience, and/or behaviors**	**Possible examples of objective evidence of personal contribution to activities performed, or professional behaviors applied**
1	Creates enterprise-level policies, procedures, guidance, and best practice in order to ensure Systems Engineering finance-related activities integrate with enterprise financial goals, including associated tools. [IFIL01]	(A) Creates Enterprise policy and guidance regarding financial management integration which ensure SE activities align with finance needs. [IFIL01-10A]	Enterprise policy and guidance. [IFIL01-E10]
		(A) Determines requirements for financial management activities to ensure SE activities align with finance needs. [IFIL01-20A]	Records of support for tool introduction. [IFIL01-E20]
		(A) Selects and acquires appropriate tools to align SE activities with financial management needs. [IFIL01-30A]	
2	Assesses enterprise-level financial management materials to ensure they integrate with Systems Engineering needs. [IFIL02]	(A) Evaluates enterprise finance processes against Systems Engineering needs in order to enable project success. [IFIL02-10A]	Financial process tailoring to reflect Systems Engineering needs. [IFIL02-E10]
		(A) Develops enterprise-level requirements for financial precision levels, control thresholds, reporting formats and periodicity related to Systems Engineering activities. [IFIL02-20A]	Feedback provided on finance Work Products. [IFIL02-E20]
		(P) Produces constructive feedback on finance plans where these do not align with Systems Engineering approaches or needs. [IFIL02-30P]	Minutes of finance management meetings where they raised Systems Engineering integration issues. [IFIL02-E30]
		(A) Selects and acquires appropriate tools for financial management activities. [IFIL02-40A]	Financial work products evaluated including feedback and revised. [IFIL02-E40]
		(A) Assesses finance approaches across the enterprise with regard to their alignment with Systems Engineering activities. [IFIL02-50A]	
3	Judges tailoring required for enterprise-level Systems Engineering processes in order to ensure that the needs of financial stakeholders are fully integrated. [IFIL03]	(A) Reviews financial planning documents with financial management stakeholders. [IFIL03-10A]	Minutes of financial management meetings on this topic. [IFIL03-E10]
		(A) Assesses financial plans for thorough and robust Systems Engineering activities in the context of the entire project. [IFIL03-20A]	Records of assessment of organizational financial systems or processes from SE standpoint. [IFIL03-E20]

(Continued)

ID	Indicators of Competence *(in addition to those at Practitioner level)*	Relevant knowledge, experience, and/or behaviors	Possible examples of objective evidence of personal contribution to activities performed, or professional behaviors applied
4	Judges appropriateness of enterprise-level financial decisions in a rational way to ensure alignment with Systems Engineering needs. [IFIL04]	(A) Analyzes financial management stakeholder decisions using best practice techniques and based on Systems Engineering needs. [IFIL04-10A]	Reports on financial management decisions made. [IFIL04-E10]
		(A) Determines appropriateness of financial decisions made, with robust justification based on Systems Engineering needs. [IFIL04-20A]	Records indicating changed stakeholder thinking from personal intervention. [IFIL04-E20]
		(P) Persuades financial management stakeholder to change viewpoints and decisions on the grounds of appropriateness. [IFIL04-30P]	
5	Assesses financial information produced across the enterprise using appropriate techniques for its integration with Systems Engineering data. [IFIL05]	(A) Reviews information produced within the financial functions using appropriate techniques to assess its impact on Systems Engineering. [IFIL05-10A]	Records of assessment of financial work products against Systems Engineering data (e.g. estimates and costings). [IFIL05-E10]
		(A) Judges adequacy and correctness of Systems Engineering financial information using appropriate techniques. [IFIL05-20A]	
6	Persuades key financial stakeholders to address identified enterprise-level financial management issues affecting Systems Engineering. [IFIL06]	(A) Identifies and engages with key financial stakeholders. [IFIL06-10A]	Records of issues raised on existing financial management issues and revised or improved techniques adopted to address these issues. [IFIL06-E10]
		(P) Fosters agreement between key stakeholders at enterprise level to resolve financial issues related to Systems Engineering, by promoting a holistic viewpoint. [IFIL06-20P]	Minutes of key stakeholder review meetings and actions taken. [IFIL06-E20]
		(P) Persuades key stakeholders to address identified enterprise-level financial issues. [IFIL06-30P]	Records indicating changed stakeholder thinking from personal intervention. [IFIL06-E30]
7	Coaches or mentors practitioners across the enterprise in the integration of finance with Systems Engineering to develop their knowledge, abilities, skills, or associated behaviors. [IFIL07]	(P) Coaches or mentors Systems Engineering or financial practitioners within the enterprise in competency-related techniques, recommending development activities. [IFIL07-10P]	Coaching or mentoring assignment records. [IFIL07-E10]
		(A) Develops or authorizes enterprise training materials in this competency area. [IFIL07-20A]	Records of formal training courses, workshops, seminars, and authored training material supported by successful post-training evaluation data. [IFIL07-E20]
		(A) Provides enterprise workshops/seminars or training in this competency area. [IFIL07-30A]	Listing as an approved organizational trainer for this competency area. [IFIL07-E30]

8	Promotes the introduction and use of novel techniques and ideas across the enterprise, which improve the integration of Systems Engineering and finance functions. [IFIL08]	(A) Analyzes different approaches across different domains through research. [IFIL08-10A]	Research records. [IFIL08-E10]
		(A) Defines novel approaches that could potentially improve the SE discipline within the enterprise. [IFIL08-20A]	Published papers in refereed journals/company literature. [IFIL08-E20]
		(P) Fosters awareness of these novel techniques within the enterprise. [IFIL08-30P]	Records showing introduction of enabling systems supporting the new techniques or ideas. [IFIL08-E30]
		(P) Collaborates with enterprise stakeholders to facilitate the introduction of techniques new to the enterprise. [IFIL08-40P]	Published papers (or similar) at enterprise level. [IFIL08-E40]
		(A) Monitors new techniques after their introduction to determine their effectiveness. [IFIL08-50A]	Records of improvements made against a recognized process improvement model in this area. [IFIL08-E50]
		(A) Adapts approach to reflect actual enterprise performance improvements. [IFIL08-60A]	
9	Develops expertise in this competency area through specialist Continual Professional Development (CPD) activities. [IFIL09]	(A) Identifies own needs for further professional development in order to increase competence beyond practitioner level. [IFIL09-10A]	Records of Continual Professional Development (CPD) performed and learning outcomes. [IFIL09-E10]
		(A) Performs professional development activities in order to move own competence toward expert level. [IFIL09-20A]	
		(A) Records continual professional development activities undertaken including learning or insights gained. [IFIL09-30A]	

NOTES	In addition to items above, enterprise-level or independent 3rd-Party-generated evidence may be used to amplify other evidence presented and may include: a. Formally recognized by senior management in current organization as an expert in this competency area b. Evidence of role as Product/System Design Authority or Technical Authority on a complex project with responsibilities in this area or where skills within this competency area were used c. Recognized as an authorizing signatory on behalf of enterprise for formal documentation in this competency area (e.g. policies, processes, and deliverables) d. Formal commendation or award within own enterprise for contribution or item of work successfully performed, which required proficiency in this competency area e. Customer, Supplier, or other external project-specific key Stakeholder accolades for specific work performed in this competency area f. Independently assessed or accredited work in this competency area (e.g. for independent publication or use) g. Formal organizational HR records positively highlighting any specific professional competencies or behaviors identified (if applicable) plus any of the evidence indicators listed at Expert level below

Expert – Integrating: Finance			
ID	**Indicators of Competence** ***(in addition to those at Lead Practitioner level)***	**Relevant knowledge, experience, and/or behaviors**	**Possible examples of objective evidence of personal contribution to activities performed, or professional behaviors applied**
1	Communicates own knowledge and experience in the integration of finance needs with Systems Engineering, in order to improve Systems Engineering best practice beyond the enterprise boundary. [IFIE01]	(A) Produces papers, seminars, or presentations outside own enterprise for publication in order to share own ideas and improve industry best practices in this competence area. [IFIE01-10A]	Published papers or books etc. on new technique in refereed journals/company literature. [IFIE01-E10]
		(P) Fosters incorporation of own ideas into industry best practices in this area. [IFIE01-20P]	Published papers in refereed journals or internal literature proposing new practices in this competence area (or presentations, tutorials, etc.). [IFIE01-E20]
		(P) Develops guidance materials identifying new (or updating existing) best practice in this competence area. [IFIE01-30P]	Records of own proposals adopted as industry best practices in this competence area. [IFIE01-E30]
2	Advises organizations beyond the enterprise boundary on the suitability of financial management plans affecting Systems Engineering activities. [IFIE02]	(P) Acts as independent reviewer on external committee (or similar body) which approves such plans. [IFIE02-10P]	Records of advice provided. [IFIE02-E10]
		(A) Reviews and advises on financial management strategies that have led to changes being implemented. [IFIE02-20A]	
3	Advises organizations beyond the enterprise boundary on complex or sensitive Financial matters and their effect on Systems Engineering. [IFIE03]	(A) Advises external stakeholders (e.g. customers) on their financial management plans and issues. [IFIE03-10A]	Records of advice provided on financial management issues. [IFIE03-E10]
		(A) Conducts successful sensitive negotiations regarding the financial management of highly complex systems. [IFIE03-20A]	Records of advice provided on sensitive financial management issues. [IFIE03-E20]
		(A) Conducts successful sensitive negotiations making limited use of specialized, financial management relevant terminology. [IFIE03-30A]	
		(A) Uses a holistic approach to complex financial issue resolution including balanced, rational arguments on way forward. [IFIE03-40A]	
4	Champions the introduction of novel techniques and ideas to improve the integration of Systems Engineering with the finance function, beyond the enterprise boundary, in order to develop the wider Systems Engineering community in this competency. [IFIE04]	(A) Analyzes different approaches across different domains through research. [IFIE04-10A]	Records of activities promoting research and the need to adopt novel technique or ideas. [IFIE04-E10]

		(A) Produces reports for the wider SE community on the effectiveness of new techniques after their introduction. [IFIE04-20A]	Records of improvements made to process and appraisal against a recognized process improvement model. [IFIE04-E20]
		(P) Collaborates with those introducing novel techniques within the wider SE community. [IFIE04-30P]	Research records. [IFIE04-E30]
		(A) Defines novel approaches that could potentially improve the wider SE discipline. [IFIE04-40A]	Published papers in refereed journals/company literature. [IFIE04-E40]
		(P) Fosters awareness of these novel techniques within the wider SE community. [IFIE04-50P]	Records showing introduction of enabling systems supporting the new techniques or ideas. [IFIE04-E50]
5	Coaches individuals beyond the enterprise boundary, in the relationship between Systems Engineering and finance, to further develop their knowledge, abilities, skills, or associated behaviors. [IFIE05]	(P) Coaches or mentors those engaged in Systems Engineering or finance activities beyond the enterprise boundary, in in competency-related techniques, recommending development activities. [IFIE05-10P]	Coaching or mentoring assignment records. [IFIE05-E10]
		(A) Develops or authorizes training materials in this competency area, which are subsequently successfully delivered beyond the enterprise boundary. [IFIE05-20A]	Records of formal training courses, workshops, seminars, and authored training material supported by successful post-training evaluation data. [IFIE05-E20]
		(A) Provides workshops/seminars or training in this competency area for practitioners or lead practitioners beyond the enterprise boundary (e.g. conferences and open training days). [IFIE05-30A]	Records of Training/workshops/seminars created supported by successful post-training evaluation data. [IFIE05-E30]
6	Maintains expertise in this competency area through specialist Continual Professional Development (CPD) activities. [IFIE06]	(A) Reviews research, new ideas, and state of the art to identify relevant new areas requiring personal development in order to maintain expertise in this competency area. [IFIE06-10A]	Records of documents reviewed and insights gained as part of own research into this competency area. [IFIE06-E10]
		(A) Performs identified specialist professional development activities in order to maintain or further develop competence at expert level. [IFIE06-20A]	Records of Continual Professional Development (CPD) performed and learning outcomes. [IFIE06-E20]
		(A) Records continual professional development activities undertaken including learning or insights gained. [IFIE06-30A]	

NOTES	In addition to items above, enterprise-level or independent 3rd-Party-generated evidence may be used to amplify other evidence presented and may include: a. Formally recognized by a reputable external organization as an expert in this competency area b. Evidence of role as independent assessor or reviewer on project outside own organization where skills in this competency area were used c. Evidence of invitation(s) from wider community for contribution of systems engineering expertise in this area (e.g. industry conference panel, government advisory board etc. cross-industry working groups, partnerships, accredited advanced university courses or research, or as part of professional institute) d. Formal commendation beyond the enterprise (e.g. by INCOSE or other recognized authority) for work performed in this competency area e. Independently assessed or accredited work product in this competency area (e.g. for independent publication or use) f. Accolades of expertise in this area from recognized industry leaders

Competency area – Integrating: Logistics
Description Logistics focuses on the support and sustainment of the product once it is transitioned to the end user. It includes areas such as life cycle cost analysis, supportability analysis, sustainment engineering, maintenance planning and execution, training, spares and inventory control, associated facilities and infrastructure, packaging, handling and shipping, and support equipment for the system and its elements.
Why it matters Factoring logistics considerations such as availability, storage and transport, and training needs early in the design effort can significantly reduce total life cycle cost for the system.
Possible contributory types of evidence Any combination of the types of evidence may be acceptable (depending on how the Framework is tailored and used). The evidence items identified at each level indicate example work products only. Contributions to work products will generally differ at each proficiency level.
Learning and development The INCOSE Professional Development Portal provides example guidance on how to gain an initial awareness of a competency area and options for developing further competence thereafter.

Awareness – Integrating: Logistics		
ID	**Indicators of Competence**	**Relevant knowledge sub-indicators**
1	Explains the role the logistics function plays in developing a successful system, product, or service. [ILOA01]	(K) Explains why organizations with a good logistics management and control are likely to survive better in the longer term than those without. [ILOA01-10K] (K) Explains how it is important to have logistics specialists involved in design to identify potential issues early (for example, the use of an unreliable or difficult to maintain part). [ILOA01-20K] (K) Describes reliability and maintainability as factors that influence the operational availability of the system and why they are important requirements. [ILOA01-30K] (K) Explains the reasons why there is a need to maintain a focus on the logistics associated with project across the full life cycle. [ILOA01-40K] (K) Lists project, external stakeholder (e.g. customer) and organizational logistics stakeholders. [ILOA01-50K] (K) Describes the key interfaces and links between logistics stakeholders and the wider enterprise outside engineering (e.g. infrastructure management, human resource management, quality management, knowledge management, portfolio management, and life cycle model management). [ILOA01-60K] (K) Explains the role wider business stakeholders play with regard to logistics management and the potential impact their decisions may have on Systems Engineering activities on a particular project. [ILOA01-70K]
2	Explains the meaning of commonly used logistics terms and applicable standards. [ILOA02]	(K) Explains the meaning of key logistics terms including life cycle cost and how each affect both the system solution and logistics. [ILOA02-10K] (K) Lists key logistics standards or legislation applicable to the business. [ILOA02-20K]

3	Describes key logistics activities and why they are important to the success of a system. [ILOA03]	(K) Lists activities such as sustainment, supply support, maintenance planning, Packaging Handling Storage and Transportation (PHS&T), support equipment, associated maintenance facilities, and infrastructure. [ILOA03-10K] (K) Describes a basic maintenance approach for a system and the associated elements (e.g. trained maintainers, tools and support equipment or facilities, and replacement parts). [ILOA03-20K]
4	Explains primary interfaces between the logistics function and the Systems Engineering team. [ILOA04]	(K) Identifies key roles within the logistics function. [ILOA04-10K] (K) Identifies the interfaces between key logistics function roles and the Systems Engineering team on a project, covering all stages of a systems life cycle. [ILOA04-20K] (K) Describes the primary responsibilities and activities of key logistics function roles (e.g. logistics manager, maintainers, data managers, parts provisioners, reliability engineers…) across all stages of a project life cycle (e.g. defining, planning, executing, and closing out a project). [ILOA04-30K] (K) Explains the difference between and potential resulting conflicts from the goals of the logistics function and those of the Systems Engineering function (e.g. resource, cost capital investment timing vs process and technical baseline). [ILOA04-40K]
5	Describes the key work products exchanged between logistics stakeholders and the Systems Engineering team. [ILOA05]	(K) Describes key work products exchanged between logistics stakeholders within the enterprise and the project team. [ILOA05-10K] (K) Describes the key work products exchanged between the project Systems Engineering team and logistics stakeholders within the enterprise across the life cycle. [ILOA05-20K]
6	Explains the concept and value of life cycle cost and how this affects both the system solution and logistics. [ILOA06]	(K) Explains how the design determines approximately 80% of the system life cycle cost. [ILOA06-10K] (K) Describes the costs associated with operating and maintaining a system such as training, packaging handling, storage and transportation (PHS&T), maintenance, technical data management, and provisioning parts. [ILOA06-20K]
7	Describes the wider logistics environment within which the system is being developed, and the influence each can have on this other. [ILOA07]	(K) Describes key elements of a Life Cycle Support Plan (maintenance and support strategies, metrics, responsible parties, funding required…). [ILOA07-10K] (K) Describes the goals of logistic sustainment (optimize availability, minimize support cost). [ILOA07-20K] (K) Lists examples of process improvements made as a result of logistics related SE tasks. [ILOA07-30K]

(Continued)

Supervised Practitioner – Integrating: Logistics			
ID	**Indicators of Competence** ***(in addition to those at Awareness level)***	**Relevant knowledge, experience, and/or behaviors**	**Possible examples of objective evidence of personal contribution to activities performed, or professional behaviors applied**
1	Follows a governing process in order to interface successfully to logistics management activities. [ILOS01]	(K) Describes the work products required to interface to the Logistics Management Planning, monitoring and control. [ILOS01-10K] (A) Uses approved logistics function processes and tools to control and manage own activities in support of logistics. [ILOS01-20A] (A) Uses interfacing work products produced by logistics to guide their own Systems Engineering activities. [ILOS01-30A] (A) Uses a governing process and appropriate tools to control their own logistics-related tasks. [ILOS01-40A]	Records of inputs to Logistics Management Plan. [ILOS01-E10] Records of inputs to Logistics Management Plan (or similar). [ILOS01-E20] Work products linked to logistics processes and tools applied on a project. [ILOS01-E30]
2	Identifies potential issues with interfacing work products received from logistics Stakeholders or produced by Systems Engineering for logistics stakeholders taking appropriate action. [ILOS02]	(A) Identifies issues with interfacing work products produced by logistics when performing own Systems Engineering activities. [ILOS02-10A] (A) Identifies issues with interfacing work products produced by Systems Engineering which support logistics activities. [ILOS02-20A] (A) Records concerns regarding issue to ensure ensuring affected stakeholders become fully informed. [ILOS02-30A]	Records of issues raised. [ILOS02-E10]
3	Prepares inputs to a supportability analysis on a project to assist logistics stakeholders. [ILOS03]	(K) Describes common factors relevant to supportability assessments. [ILOS03-10K] (A) Uses guidance to perform supportability analysis and calculations. [ILOS03-20A]	Records of inputs to supportability analysis. [ILOS03-E10]
4	Explains how different concepts for maintenance may have different life cycle costs. [ILOS04]	(K) Identifies the key steps to developing a maintenance approach. [ILOS04-10K] (K) Describes in a general sense the difference in costs associated with maintenance approaches, e.g. depot level or Original Equipment Manufacturer (OEM) maintenance is typically more expensive due to transportation costs. [ILOS04-20K] (A) Analyzes different maintenance concepts on a project. [ILOS04-30A]	Records of inputs to activities performed in support of a maintenance or logistics planning team. [ILOS04-E10] Records of inputs to maintenance concept and cost assessment. [ILOS04-E20]

5	Uses recognized analysis techniques to calculate spares, repairs, or supply-related information for logistics stakeholders on a project. [ILOS05]	(A) Identifies provisioning requirements for spare parts and supplies on a system. [ILOS05-10A]	Records of inputs to spares provisioning activity. [ILOS05-E10]
		(A) Monitors spare parts and supply usage for a system. [ILOS05-20A]	Reports from spares and supplies management activities they contributed to in support of system maintenance. [ILOS05-E20]
6	Uses recognized analysis techniques to produce facilities and infrastructure operation and maintenance information for logistics stakeholders on a project. [ILOS06]	(A) Prepares data supporting the analysis of facilities and infrastructure supporting operation and maintenance of a system. [ILOS06-10A]	Records of inputs to a facility assessment report. [ILOS06-E10]
			Records of input to site visit report as part of a logistics assessment. [ILOS06-E20]
7	Uses recognized techniques to produce system engineering information in support of operator or personnel training or simulation activities for logistics stakeholders on a project. [ILOS07]	(K) Explains the different types of training aids and their relative benefits for various system type and operator/maintainer skill sets. [ILOS07-10K]	Records of inputs to work products from user/maintainer/operator training. [ILOS07-E10]
		(A) Develops training aids, simulators and simulations for operators or personnel sustaining the system. [ILOS07-20A]	Records of inputs to training materials created in support of logistics. [ILOS07-E20]
8	Uses recognized techniques to produce system operation and maintenance information for logistics stakeholders on a project. [ILOS08]	(K) Describes types of support equipment required to operate and maintain a system (e.g. replenishable power supplies, maintenance laptops, tools, consumables…). [ILOS08-10K]	Work products from maintenance team activities to which they contributed. [ILOS08-E10]
		(A) Identifies requirements for the support of the acquisition of support equipment. [ILOS08-20A]	Records of inputs to activities identifying or acquiring support equipment. [ILOS08-E20]
		(A) Prepares specifications in support of the acquisition of support equipment. [ILOS08-30A]	
9	Uses recognized techniques to produce system installation, operation, maintenance, and sustainment information for logistics stakeholders on a project. [ILOS09]	(K) Describes different types of technical data (drawings, parts lists, bills of material, software code, installation procedures, operation and maintenance manuals…). [ILOS09-10K]	Records of inputs to logistics technical work products (e.g. drawings, procedures). [ILOS09-E10]
		(A) Prepares analysis and review data for technical data products. [ILOS09-20A]	Records of logistics technical data products reviewed (e.g. redlined manuals). [ILOS09-E20]

(Continued)

ID	Indicators of Competence *(in addition to those at Awareness level)*	Relevant knowledge, experience, and/or behaviors	Possible examples of objective evidence of personal contribution to activities performed, or professional behaviors applied
10	Uses recognized techniques to produce system packaging, handling, storage, and transportation information for logistics stakeholders on a project. [ILOS10]	(K) Describes the type of requirements that are related to packaging, handling, storage and transportation (PHS&T) (e.g. safety, security, environment, and cost). [ILOS10-10K] (A) Prepares inputs to a packaging, handling, storage, and transportation (PHS&T) plan. [ILOS10-20A] (A) Follows a packaging, handling, storage and transportation (PHS&T) plan giving examples of methods for packing, shipping, etc. [ILOS10-30A]	Records of input to packaging, handling, storage and transportation plans. [ILOS10-E10]
11	Develops own understanding of this competency area through Continual Professional Development (CPD). [ILOS11]	(A) Identifies potential gaps in own knowledge or development needs in this area, identifying opportunities to address these through continual professional development activities. [ILOS11-10A] (A) Performs continual professional development activities to improve their knowledge and understanding in this area. [ILOS11-20A] (A) Records continual professional development activities undertaken including learning or insights gained. [ILOS11-30A]	Records of Continual Professional Development (CPD) performed and learning outcomes. [ILOS11-E10]

Practitioner – Integrating: Logistics			
ID	**Indicators of Competence** ***(in addition to those at Supervised Practitioner level)***	**Relevant knowledge, experience, and/or behaviors**	**Possible examples of objective evidence of personal contribution to activities performed, or professional behaviors applied**
1	Follows governing logistics plans, processes, and uses appropriate tools to control and monitor logistics-related Systems Engineering tasks, interpreting as necessary. [ILOP01]	(A) Follows governing logistics plans, processes and uses appropriate tools to control and monitor logistics-related Systems Engineering tasks. [ILOP01-10A]	Work products prepared (or contributed to) as a result of logistics plans, processes, and tools. [ILOP01-E10]
		(A) Uses governing logistics plans, processes and appropriate tools where necessary to ensure outputs meet logistics stakeholder needs. [ILOP01-20A]	
		(A) Communicates potential issues with governing logistics plans, processes and tools to logistics stakeholders. [ILOP01-30A]	
2	Prepares work products required by logistics stakeholders to ensure Systems Engineering work aligns with wider logistics management activities. [ILOP02]	(A) Analyzes information required by logistics for requested work products to ensure understanding. [ILOP02-10A]	Work products prepared (or contributed to) as a result of logistics plans, processes and tools. [ILOP02-E10]
		(A) Prepares work products required by logistics stakeholders, using appropriate processes and procedures. [ILOP02-20A]	Logistics strategy work products they have produced. [ILOP02-E20]
		(A) Identifies project-specific system engineering needs that need to be incorporated into Logistics Management Plans or processes. [ILOP02-30A]	
		(A) Prepares inputs for a governing logistics management process, plan, and associated tools, which reflect project and business strategy and meet the needs of the logistics function. [ILOP02-40A]	
		(A) Prepares process performance data as required by the logistics function in order to achieve the targeted standard of excellence on a project. [ILOP02-50A]	
		(A) Communicates recommendations for corrective actions to be performed and implemented. [ILOP02-60A]	
3	Prepares supportability analysis information required by logistics stakeholders to meet project and enterprise requirements. [ILOP03]	(K) Describes common supportability issues (use of unreliable or difficult to obtain parts, physical designs that complicate preventive maintenance tasks, unique or unusual maintenance tools or fixtures…). [ILOP03-10K]	Supportability Analysis Report or associated information produced. [ILOP03-E10]

(Continued)

ID	Indicators of Competence *(in addition to those at Supervised Practitioner level)*	Relevant knowledge, experience, and/or behaviors	Possible examples of objective evidence of personal contribution to activities performed, or professional behaviors applied
		(A) Analyzes a product baseline design, identifying potential supportability issues or opportunities for improvement/cost reduction and recommends associated design, manufacturing or maintenance process changes. [ILOP03-20A]	
4	Develops maintenance concepts required by logistics stakeholders to ensure alignment with system engineering activities. [ILOP04]	(K) Describes the elements required to implement a maintenance concept and typical approaches along with the benefits or disadvantages of the various approaches. [ILOP04-10K] (A) Determines tasks for generating well documented maintenance procedures. [ILOP04-20A]	Maintenance plans or procedures or information required for this activity. [ILOP04-E10]
5	Develops spares and repair concepts required by logistics stakeholders to ensure alignment with system engineering activities. [ILOP05]	(A) Identifies system requirements for acquisition, catalog, receipt, storage, transferring, issuing and disposal of spares, repair of parts, and supplies sustaining the system. [ILOP05-10A] (A) Defines tasks for acquisition, catalog, receipt, storage, transferring, issuing and disposal of spares, repair of parts and supplies sustaining the system. [ILOP05-20A] (A) Identifies tasking requirements for spares and repairs in order to sustain the system. [ILOP05-30A]	Spares and repair concepts developed or contributed to. [ILOP05-E10] Records of assessment of spares and repair concepts for their impact on Systems Engineering. [ILOP05-E20]
6	Develops facilities infrastructure concepts required by logistics stakeholders to support operation and maintenance of a system across its life cycle. [ILOP06]	(K) Describes considerations for system utilization and support facilities (e.g. location/proximity, power requirements etc., staffing…). [ILOP06-10K] (A) Defines tasks associated with identifying appropriate operations and support facilities and infrastructure. [ILOP06-20A]	Facility Identification or Assessment Report. [ILOP06-E10] Records of review or assessment of operations and support facilities and infrastructure from Systems Engineering perspective. [ILOP06-E20]
7	Develops logistics training products required by logistics stakeholders to maximize the effectiveness of operators and personnel sustaining the system at lowest life cycle cost. [ILOP07]	(K) Describes methods for delivering training for both operators and maintainers (e.g. in person, self-guided via paper or electronic documentation, virtual…). [ILOP07-10K] (A) Determines tasks associated with the development and delivery of system operational and maintenance training to include a training needs analysis. [ILOP07-20A]	Training documentation or review comments. [ILOP07-E10] Training Needs Analysis. [ILOP07-E20]
8	Develops concepts for support equipment in collaboration with logistics stakeholders to sustain the operation and maintenance of a system across its life cycle. [ILOP08]	(K) Describes considerations for identifying appropriate support equipment. [ILOP08-10K]	Records of inputs to development of support equipment strategy or concept. [ILOP08-E10]

		(A) Determines tasks associated with the development and delivery of system operational and maintenance training to include a training needs analysis. [ILOP08-20A]	Support equipment documentation (ground handling or maintenance equipment, tools, calibration equipment, test equipment). [ILOP08-E20]
9	Develops packaging, handling, storage, and transportation required by logistics stakeholders to ensure safe and secure transportation of a system. [ILOP09]	(K) Identifies different methods for packaging, handling storage and transportation of equipment, including transfer of custody when appropriate. [ILOP09-10K]	Packaging or handling procedures, storage or transportation documentation, delivery documentation. [ILOP09-E10]
		(A) Determines tasks associated with the development and delivery of support equipment. [ILOP09-20A]	
10	Develops work products required by logistics stakeholders in order to support the installation, operation, maintenance, and sustainment of the system. [ILOP10]	(K) Explains the purpose of different types of technical data. [ILOP10-10K]	Engineering design data, inspection and calibration procedures, Operator's Manuals, Maintenance Manuals, Installation Procedures. [ILOP10-E10]
		(A) Analyzes issues related to management of data rights (intellectual property, software source code versus executable code, etc.) and addresses these. [ILOP10-20A]	List of contract deliverable technical data with data rights explicitly defined. [ILOP10-E20]
		(A) Reviews technical data, reports and documentation supporting the installation, operation, maintenance and sustainment of the system. [ILOP10-30A]	Records of review of technical data used by logistics activity. [ILOP10-E30]
11	Guides new or supervised practitioners in logistics and its relationship to Systems Engineering, to develop their knowledge, abilities, skills, or associated behaviors. [ILOP11]	(P) Guides new or supervised practitioners in executing activities that form part of this competency. [ILOP11-10P]	Organizational Breakdown Structure showing their responsibility for technical supervision in this area. [ILOP11-E10]
		(A) Trains individuals to an "Awareness" level in this competency area. [ILOP11-20A]	On-the-job training objectives/guidance etc. [ILOP11-E20] Coaching or mentoring assignment records. [ILOP11-E30] Records highlighting their impact on another individual in terms of improvement or professional development in this competency. [ILOP11-E40]
12	Maintains and enhances own competence in this area through Continual Professional Development (CPD) activities. [ILOP12]	(A) Identifies potential development needs in this area, identifying opportunities to address these through continual professional development activities. [ILOP12-10A]	Records of Continual Professional Development (CPD) performed and learning outcomes. [ILOP12-E10]
		(A) Performs continual professional development activities to maintain and enhance their competency in this area. [ILOP12-20A]	
		(A) Records continual professional development activities undertaken including learning or insights gained. [ILOP12-30A]	

Lead Practitioner – Integrating: Logistics			
ID	**Indicators of Competence** ***(in addition to those at Practitioner level)***	**Relevant knowledge, experience, and/or behaviors**	**Possible examples of objective evidence of personal contribution to activities performed, or professional behaviors applied**
1	Creates enterprise-level policies, procedures, guidance, and best practice in order to ensure Systems Engineering logistics-related activities integrate with enterprise logistics goals, including associated tools. [ILOL01]	(A) Creates Enterprise policy and guidance regarding logistics management which ensure SE activities align with logistics needs. [ILOL01-10A]	Enterprise policy and guidance created in this area. [ILOL01-E10]
		(A) Determines requirements for logistics management activities to ensure SE activities align with logistics needs. [ILOL01-20A]	Logistics tools they have used or where they have authorized acquisition. [ILOL01-E20]
		(A) Selects and acquires appropriate tools to align SE activities with logistics management needs. [ILOL01-30A]	
2	Assesses enterprise-level logistics management processes to ensure they integrate with Systems Engineering needs. [ILOL02]	(A) Evaluates enterprise logistics processes against Systems Engineering needs in order to enable project success. [ILOL02-10A]	Records of process developed in this area. [ILOL02-E10]
		(P) Produces constructive feedback on enterprise level logistics processes where these do not align with Systems Engineering approaches or needs. [ILOL02-20P]	Records of comments made and accepted on logistics or Life Cycle Support Plan. [ILOL02-E20]
		(A) Assesses logistics plans across the enterprise to ensure Systems Engineering activities are aligned. [ILOL02-30A]	
		(A) Assesses logistics approaches across the enterprise with regard to their alignment with Systems Engineering activities. [ILOL02-40A]	
3	Judges the appropriateness of enterprise-level logistics decisions in a rational way to ensure alignment with Systems Engineering needs. [ILOL03]	(A) Analyzes logistics stakeholders decisions using best practice techniques based upon Systems Engineering needs. [ILOL03-10A]	Records of inputs to logistics decisions. [ILOL03-E10]
		(A) Determines appropriateness of logistics decisions made, with robust justification, based upon Systems Engineering needs. [ILOL03-20A]	Records of analysis supporting logistics decisions. [ILOL03-E20]
		(P) Persuades logistics stakeholder to change viewpoints and decisions on the grounds of appropriateness based upon Systems Engineering needs. [ILOL03-30P]	

4	Judges the supportability strategies and supportability decisions across the enterprise to ensure they align with Systems Engineering performance, readiness, and life cycle cost needs. [ILOL04]	(P) Acts to remove conflict of personnel or asset scheduling across multiple enterprise projects. [ILOL04-10P]	Records of assessments made on supportability strategies or plans. [ILOL04-E10]
		(A) Describes occasions where they have provided advice on supportability strategies or assessments that has led to changes being implemented. [ILOL04-20A]	Records of advice provided on supportability work products. [ILOL04-E20]
5	Judges logistics plans and decisions across the enterprise to ensure they align with Systems Engineering performance, readiness, and life cycle cost needs. [ILOL05]	(P) Acts to remove conflict of personnel or asset scheduling across multiple enterprise projects. [ILOL05-10P]	Records of assessments made on logistics plans. [ILOL05-E10]
		(A) Describes occasions where they have provided advice on logistics support plans that has led to changes being implemented. [ILOL05-20A]	Records of advice provided on logistics work products. [ILOL05-E20]
6	Assesses enterprise-level logistics work products for their alignment with Systems Engineering. [ILOL06]	(A) Reviews technical logistics work products. [ILOL06-10A]	Records of a work product review. [ILOL06-E10]
		(A) Advises on enterprise logistics work products, resulting in changes to reflect Systems Engineering needs. [ILOL06-20A]	Records of advice provided. [ILOL06-E20]
7	Persuades key logistics stakeholders to address identified enterprise-level logistics management issues affecting Systems Engineering. [ILOL07]	(A) Identifies and engages with key stakeholders. [ILOL07-10A]	Minutes of meetings showing contribution to resolution of identified issues. [ILOL07-E10]
		(P) Fosters agreement between key stakeholders at enterprise level to resolve logistics issues related to Systems Engineering, by promoting a holistic viewpoint. [ILOL07-20P]	Records of contributions made to resolution of logistics issues. [ILOL07-E20]
		(P) Persuades key stakeholders to address identified enterprise-level logistics issues. [ILOL07-30P]	
8	Coaches or mentors practitioners across the enterprise in the integration of logistics with Systems Engineering in order to develop their knowledge, abilities, skills, or associated behaviors. [ILOL08]	(P) Coaches or mentors Systems Engineering or logistics practitioners within the enterprise in competency-related techniques, recommending development activities. [ILOL08-10P]	Coaching or mentoring assignment records. [ILOL08-E10]
		(A) Develops or authorizes enterprise training materials in this competency area. [ILOL08-20A]	Records of formal training courses, workshops, seminars, and authored training material supported by successful post-training evaluation data. [ILOL08-E20]
		(A) Provides enterprise workshops/seminars or training in this competency area. [ILOL08-30A]	Listing as an approved organizational trainer for this competency area. [ILOL08-E30]

(*Continued*)

ID	Indicators of Competence *(in addition to those at Practitioner level)*	Relevant knowledge, experience, and/or behaviors	Possible examples of objective evidence of personal contribution to activities performed, or professional behaviors applied
9	Promotes the introduction and use of novel techniques and ideas across the enterprise, which improve the integration of Systems Engineering and logistics functions. [ILOL09]	(A) Analyzes different approaches across different domains through research. [ILOL09-10A]	Research records. [ILOL09-E10]
		(A) Defines novel approaches that could potentially improve the SE discipline within the enterprise. [ILOL09-20A]	Published papers in refereed journals/company literature. [ILOL09-E20]
		(P) Fosters awareness of these novel techniques within the enterprise. [ILOL09-30P]	Records showing introduction of enabling systems supporting the new techniques or ideas. [ILOL09-E30]
		(P) Collaborates with enterprise stakeholders to facilitate the introduction of techniques new to the enterprise. [ILOL09-40P]	Published papers (or similar) at enterprise level. [ILOL09-E40]
		(A) Monitors new techniques after their introduction to determine their effectiveness. [ILOL09-50A]	Records of improvements made against a recognized process improvement model in this area. [ILOL09-E50]
		(A) Adapts approach to reflect actual enterprise performance improvements. [ILOL09-60A]	
10	Develops expertise in this competency area through specialist Continual Professional Development (CPD) activities. [ILOL10]	(A) Identifies own needs for further professional development in order to increase competence beyond practitioner level. [ILOL10-10A]	Records of Continual Professional Development (CPD) performed and learning outcomes. [ILOL10-E10]
		(A) Performs professional development activities in order to move own competence toward expert level. [ILOL10-20A]	
		(A) Records continual professional development activities undertaken including learning or insights gained. [ILOL10-30A]	

NOTES	In addition to items above, enterprise-level or independent 3rd-Party-generated evidence may be used to amplify other evidence presented and may include: a. Formally recognized by senior management in current organization as an expert in this competency area b. Evidence of role as Product/System Design Authority or Technical Authority on a complex project with responsibilities in this area or where skills within this competency area were used c. Recognized as an authorizing signatory on behalf of enterprise for formal documentation in this competency area (e.g. policies, processes, and deliverables) d. Formal commendation or award within own enterprise for contribution or item of work successfully performed, which required proficiency in this competency area e. Customer, Supplier, or other external project-specific key Stakeholder accolades for specific work performed in this competency area f. Independently assessed or accredited work in this competency area (e.g. for independent publication or use) g. Formal organizational HR records positively highlighting any specific professional competencies or behaviors identified (if applicable) plus any of the evidence indicators listed at Expert level below

Expert – Integrating: Logistics			
ID	**Indicators of Competence** ***(in addition to those at Lead Practitioner level)***	**Relevant knowledge, experience, and/or behaviors**	**Possible examples of objective evidence of personal contribution to activities performed, or professional behaviors applied**
1	Communicates own knowledge and experience in the integration of logistics needs with Systems Engineering, in order to improve Systems Engineering best practice beyond the enterprise boundary. [ILOE01]	(A) Produces papers, seminars, or presentations outside own enterprise for publication in order to share own ideas and improve industry best practices in this competence area. [ILOE01-10A]	Published papers or books etc. on new technique in refereed journals/company literature. [ILOE01-E10]
		(P) Fosters incorporation of own ideas into industry best practices in this area. [ILOE01-20P]	Published papers in refereed journals or internal literature proposing new practices in this competence area (or presentations, tutorials, etc.). [ILOE01-E20]
		(P) Develops guidance materials identifying new (or updating existing) best practice in this competence area. [ILOE01-30P]	Own proposals adopted as industry best practices in this competence area. [ILOE01-E30]
2	Advises organizations beyond the enterprise boundary on the suitability of their approach to logistics management within Systems Engineering. [ILOE02]	(P) Acts as independent reviewer on external committee (or similar body) which approves such plans. [ILOE02-10P]	Records of membership on oversight committee with relevant terms of reference. [ILOE02-E10]
		(A) Reviews and advises on logistics management strategies that has led to changes being implemented. [ILOE02-20A]	Records of review comments made and accepted on logistics management strategy. [ILOE02-E20]
3	Assesses the suitability of Logistics Management Plans affecting Systems Engineering activities. [ILOE03]	(A) Assesses the suitability of Logistics Management Plans for their impact on systems engineering. [ILOE03-10A]	Records of review comments made and accepted on logistics planning documents. [ILOE03-E10]
		(A) Advises on Logistics Management Plan suitability leading to changes being implemented. [ILOE03-20A]	Records of advice provided on logistics planning documents. [ILOE03-E20]
4	Advises organizations beyond the enterprise boundary on complex or sensitive logistics-related issues and its effect on Systems Engineering. [ILOE04]	(A) Advises external stakeholders (e.g. customers) on their Logistics Management Plans and issues. [ILOE04-10A]	Records of advice provided on logistics management issues. [ILOE04-E10]
		(A) Conducts successful sensitive negotiations regarding the logistics management of highly complex systems. [ILOE04-20A]	Records from negotiations showing contribution to problem resolution. [ILOE04-E20]
		(A) Conducts successful sensitive negotiations making limited use of specialized, logistics management relevant terminology. [ILOE04-30A]	Stakeholder approval of Logistics Management Plans. [ILOE04-E30]
		(A) Uses a holistic approach to complex issue resolution including balanced, rational arguments on way forward. [ILOE04-40A]	Records of assessments of problem which include arguments justifying proposed solution. [ILOE04-E40]

(Continued)

ID	Indicators of Competence *(in addition to those at Lead Practitioner level)*	Relevant knowledge, experience, and/or behaviors	Possible examples of objective evidence of personal contribution to activities performed, or professional behaviors applied
5	Champions the introduction of novel techniques and ideas to improve the integration of Systems Engineering and logistics functions, beyond the enterprise boundary, in order to develop the wider Systems Engineering community in this competency. [ILOE05]	(A) Analyzes different approaches across different domains through research. [ILOE05-10A]	Records of activities promoting research and need to adopt novel technique or ideas. [ILOE05-E10]
		(A) Produces reports for the wider SE community on the effectiveness of new techniques after their introduction. [ILOE05-20A]	Records of improvements made to process and appraisal against a recognized process improvement model. [ILOE05-E20]
		(P) Collaborates with those introducing novel techniques within the wider SE community. [ILOE05-30P]	Research records. [ILOE05-E30]
		(A) Defines novel approaches that could potentially improve the wider SE discipline. [ILOE05-40A]	Published papers in refereed journals/company literature. [ILOE05-E40]
		(P) Fosters awareness of these novel techniques within the wider SE community. [ILOE05-50P]	Records showing introduction of enabling systems supporting the new techniques or ideas. [ILOE05-E50]
6	Coaches individuals beyond the enterprise boundary, in the relationship between Systems Engineering and logistics, to further develop their knowledge, abilities, skills, or associated behaviors. [ILOE06]	(P) Coaches or mentors those engaged in Systems Engineering or logistics activities beyond the enterprise boundary, in competency-related techniques, recommending development activities. [ILOE06-10P]	Coaching or mentoring assignment records. [ILOE06-E10]
		(A) Develops or authorizes training materials in this competency area, which are subsequently successfully delivered beyond the enterprise boundary. [ILOE06-20A]	Records of formal training courses, workshops, seminars, and authored training material supported by successful post-training evaluation data. [ILOE06-E20]
		(A) Provides workshops/seminars or training in this competency area for practitioners or lead practitioners beyond the enterprise boundary (e.g. conferences and open training days). [ILOE06-30A]	Records of Training/workshops/seminars created supported by successful post-training evaluation data. [ILOE06-E30]

7	Maintains expertise in this competency area through specialist Continual Professional Development (CPD) activities. [ILOE07]	(A) Reviews research, new ideas, and state of the art to identify relevant new areas requiring personal development in order to maintain expertise in this competency area. [ILOE07-10A] (A) Performs identified specialist professional development activities in order to maintain or further develop competence at expert level. [ILOE07-20A] (A) Records continual professional development activities undertaken including learning or insights gained. [ILOE07-30A]	Records of documents reviewed and insights gained as part of own research into this competency area. [ILOE07-E10] Records of Continual Professional Development (CPD) performed and learning outcomes. [ILOE07-E20]

NOTES	In addition to items above, enterprise-level or independent 3rd-Party-generated evidence may be used to amplify other evidence presented and may include: a. Formally recognized by a reputable external organization as an expert in this competency area b. Evidence of role as independent assessor or reviewer on project outside own organization where skills in this competency area were used c. Evidence of invitation(s) from wider community for contribution of systems engineering expertise in this area (e.g. industry conference panel, government advisory board etc. cross-industry working groups, partnerships, accredited advanced university courses or research, or as part of professional institute) d. Formal commendation beyond the enterprise (e.g. by INCOSE or other recognized authority) for work performed in this competency area e. Independently assessed or accredited work product in this competency area (e.g. for independent publication or use) f. Accolades of expertise in this area from recognized industry leaders

Competency area – Integrating: Quality
Description Quality focuses on customer satisfaction via the control of key product characteristics and corresponding key manufacturing process characteristics.
Why it matters Proactive quality management improves both the quality of the system, product, or service provided, as well as the quality of the project's management processes.
Possible contributory types of evidence Any combination of the types of evidence may be acceptable (depending on how the Framework is tailored and used). The evidence items identified at each level indicate example work products only. Contributions to work products will generally differ at each proficiency level.
Learning and development The INCOSE Professional Development Portal provides example guidance on how to gain an initial awareness of a competency area and options for developing further competence thereafter.

Awareness – Integrating: Quality		
ID	**Indicators of Competence**	**Relevant knowledge sub-indicators**
1	Explains the role the quality function plays in developing a successful system product or service. [IQUA01]	(K) Explains why products developed with a focus on quality have fewer warranty claims, recall actions, etc. [IQUA01-10K] (K) Explains the reasons why there is a need to manage quality across the full life cycle. [IQUA01-20K] (K) Lists project, external stakeholder (e.g. customer) and organizational quality stakeholders. [IQUA01-30K] (K) Describes the key interfaces and links between quality function stakeholders and the wider enterprise outside engineering (e.g. infrastructure management, human resource management, finance, knowledge management, portfolio management, and life cycle model management). [IQUA01-40K] (K) Explains the role wider business stakeholders play with regard to quality management and the potential impact their decisions may have on Systems Engineering activities on a particular project. [IQUA01-50K]
2	Explains the meaning of commonly used quality-related terms and applicable standards. [IQUA02]	(K) Defines the terms "Benchmarking," "cost-benefit analysis," "Cost of quality assessment," "Design of Experiments". [IQUA02-10K] (K) Defines how statistical methods may be used to support a project. [IQUA02-20K] (K) Describes the key aims and characteristics of quality programs such as "Lean Six Sigma," "Total Quality Management (TQM)", "Capability Maturity Model (CMM)". [IQUA02-30K] (K) Lists commonly used quality standards, such as ISO 9000 and associated ISO standards. [IQUA02-40K]
3	Explains primary interfaces between the quality management function and the Systems Engineering team. [IQUA03]	(K) Lists key roles within the quality management function. [IQUA03-10K]

		(K) Describes the interfaces between key quality management function roles and the Systems Engineering team on a project, covering all stages of a systems life cycle. [IQUA03-20K] (K) Describes the primary responsibilities and activities of key quality function roles (e.g. Quality Manager, Quality Controller) across all stages of a project life cycle (e.g. defining, planning, executing, and closing out a project). [IQUA03-30K] (K) Describes the key characteristics of the Quality Assurance (QA) (or Quality Control (QC)) role such as assisting with the development of processes and procedures to ensure quality execution of the project (e.g. fabrication procedures, pre-test checklists, and packaging checklists). [IQUA03-40K] (K) Explains the difference between and potential resulting conflicts from the goals of the quality management function and those of the Systems Engineering function (e.g. quality vs technical performance). [IQUA03-50K]
4	Describes the key work products exchanged between quality management stakeholders and the Systems Engineering team. [IQUA04]	(K) Describes key work products exchanged between quality management stakeholders within the enterprise and the project team. [IQUA04-10K] (K) Describes the key work products exchanged between the project Systems Engineering team and quality management stakeholders within the enterprise across the life cycle. [IQUA04-20K]
5	Explains the difference between quality assurance and quality control. [IQUA05]	(K) Explains the purpose and activities of quality assurance, such as applying planned systematic quality activities to ensure the project employs processes to successfully deliver the required product or service. [IQUA05-10K] (K) Explains the purpose and activities of quality control, such as monitoring and assessing project results to determine conformance with relevant standards and identify and resolve any unsatisfactory results. [IQUA05-20K]
6	Explains how project-level decisions can impact the quality of a system. [IQUA06]	(K) Explains how pressure to meet deadlines may lead to increased errors or rework and how this problem might be addressed. [IQUA06-10K] (K) Explains why using less costly components can negatively affect system performance and reliability resulting in increased costs overall. [IQUA06-20K]
7	Explains the difference between performing quality management on a project or wider enterprise and managing quality as part of Systems Engineering activities. [IQUA07]	(K) Identifies the scope of project quality management and control and enterprise-level quality management and control. [IQUA07-10K] (K) Explains the differences between performing quality activities on a project and performing quality management as part of the wider enterprise and their differing potential impact on Systems Engineering. [IQUA07-20K]
8	Describes the wider quality environment within which the system is being developed, and the influence each can have on this other. [IQUA08]	(K) Describes the influence that the wider quality environment may have on a system being developed. [IQUA08-10K] (K) Describes the influence that a system being developed may have on the wider quality environment within which it is being developed. [IQUA08-20K]

Supervised Practitioner – Integrating: Quality			
ID	**Indicators of Competence** ***(in addition to those at Awareness level)***	**Relevant knowledge, experience, and/or behaviors**	**Possible examples of objective evidence of personal contribution to activities performed, or professional behaviors applied**
1	Follows a governing process in order to interface successfully to quality management activities. [IQUS01]	(K) Describes the work products required to interface to the Quality Management Planning, monitoring and control. [IQUS01-10K] (A) Uses approved quality management processes and tools to control and manage Systems Engineering interface to quality management. [IQUS01-20A] (A) Uses interfacing work products produced by quality to guide their own Systems Engineering activities. [IQUS01-30A] (A) Uses a governing process and appropriate tools to control their own quality-related tasks. [IQUS01-40A]	Quality management work products generated, above the level of Systems Engineering planning and execution. [IQUS01-E10] Quality Management Plan (QMP) they have produced. [IQUS01-E20] Quality processes and tools applied on a project. [IQUS01-E30]
2	Prepares inputs to work products which interface to quality stakeholders to ensure Systems Engineering work aligns with wider quality management activities. [IQUS02]	(K) Explains how quality concerns affect the definition of Systems Engineering processes, schedules and resources (money, people, materials/equipment…). [IQUS02-10K] (K) Describes the input an author of a business quality plan may require from Systems Engineering. [IQUS02-20K] (A) Prepares quality monitoring data (e.g. statistical analysis, sensitivity studies, root-cause analysis) to meet the requirements of the quality function. [IQUS02-30A] (A) Monitors identified quality indicators to support the quality function in achieving the targeted standard of excellence on a project. [IQUS02-40A] (A) Monitors process performance to support the quality function in achieving the targeted standard of excellence on a project. [IQUS02-50A] (A) Prepares inputs to interfacing work products required by quality which ensure Systems Engineering work aligns with wider quality management activities. [IQUS02-60A]	Work products for use by quality which they contributed to or produced. [IQUS02-E10] Reviews of project quality which they have contributed to. [IQUS02-E20]
3	Identifies potential issues with interfacing work products received from quality Stakeholders or produced by Systems Engineering for quality taking appropriate action. [IQUS03]	(A) Identifies issues with interfacing work products produced by quality when performing own Systems Engineering activities. [IQUS03-10A]	Documents highlighting issues raised which they produced. [IQUS03-E10]

		(A) Identifies issues with interfacing work products produced by Systems Engineering which support quality management activities. [IQUS03-20A]	
		(A) Records concerns regarding issue to ensure ensuring affected stakeholders become fully informed. [IQUS03-30A]	
4	Identifies measures of quality which ensure an appropriate standard of excellence is targeted on a project in support of quality function activities. [IQUS04]	(A) Identifies potential measures of quality which ensure an appropriate standard of excellence is targeted on a project in support of quality function requirements. [IQUS04-10A]	Quality Management Plan to which they contributed. [IQUS04-E10]
		(A) Uses quality measures on a project (e.g. volatility, density, rates, time…) to meet the requirements of the quality function. [IQUS04-20A]	Relevant measurable metrics they have identified. [IQUS04-E20]
5	Identifies quality characteristics which ensure an appropriate standard of excellence is targeted on a project in support of quality function activities. [IQUS05]	(A) Identifies quality characteristics which ensure an appropriate "value" is targeted on a project in support of quality function requirements. [IQUS05-10A]	Quality Management Plan to which they contributed. [IQUS05-E10]
		(A) Uses quality characteristics on a project (e.g. performance, reliability, durability, maintainability…) to identify maximum value delivery. [IQUS05-20A]	Relevant quality characteristics they have identified. [IQUS05-E20]
6	Monitors process adherence on a project in support of quality function activities. [IQUS06]	(K) Explains how process tracking can assist in improving quality or effectiveness on a project. [IQUS06-10K]	Quality Management Plan to which they contributed. [IQUS06-E10]
		(A) Monitors process utilization or tailoring usage in support of quality function activities. [IQUS06-20A]	Reports indicating quality metrics they have tracked. [IQUS06-E20]
7	Uses recognized techniques to support verification of product or system conformity for quality stakeholders on a project. [IQUS07]	(A) Uses checklists on a project to provide a repeatable way to provide data which verifies proper execution of processes and meets the requirements of the quality function. [IQUS07-10A]	Quality checklists used to ensure product conformity. [IQUS07-E10]
		(A) Identifies potential acceptance criteria or critical thresholds on a project which meet the requirements of the quality function. [IQUS07-20A]	
8	Uses recognized techniques to perform system root-cause analysis and failure elimination for quality stakeholders on a project. [IQUS08]	(A) Prepares input data for root-cause analysis and elimination of failures to meet the requirements of the quality function. [IQUS08-10A]	Report containing a failure analysis using a well-defined method which they produced (e.g. failure mode effects (and criticality) analysis – FMEA or FMECA – or similar failure or safety analysis document). [IQUS08-E10]
		(A) Prepares inputs to failure effects analysis activities on a project (e.g. Failure Mode Effects and Criticality Analysis (FMECA)) or similar) to meet the requirements of the quality function. [IQUS08-20A]	Report into or root cause of a failure which they produced. [IQUS08-E20]

(Continued)

ID	Indicators of Competence *(in addition to those at Awareness level)*	Relevant knowledge, experience, and/or behaviors	Possible examples of objective evidence of personal contribution to activities performed, or professional behaviors applied
		(A) Prepares potential corrective actions for identified causes of failure to meet the requirements of the quality function. [IQUS08-30A] (A) Performs corrective actions to support the quality function in improving quality. [IQUS08-40A]	
9	Identifies measures of quality which ensure an appropriate standard of excellence is targeted on a project in support of quality function activities. [IQUS09]	(A) Selects measures of quality, which ensure an appropriate standard of excellence is targeted on a project in support of quality function requirements. [IQUS09-10A]	Records showing contribution to a Quality Management (or similar) Plan. [IQUS09-E10]
		(A) Ensures identified quality indicators support the quality function in achieving the targeted standard of excellence on a project. [IQUS09-20A]	Records of metrics identified, proposed, and accepted. [IQUS09-E20]
		(P) Performs associated sampling or statistical analysis. [IQUS09-30P]	List of metrics. [IQUS09-E30]
		(A) Prepares reports on quality measures on a project to meet the requirements of the quality function. [IQUS09-40A]	Records showing project quality data contributions. [IQUS09-E40]
10	Complies with required quality standards to support the quality function in auditing ongoing projects. [IQUS10]	(K) Describes the purpose and benefits of quality audits. [IQUS10-10K]	Records from a quality audit against a recognized standard, which they supported. [IQUS10-E10]
		(K) Explains how correction of deficiencies results in reduced cost of quality and increased acceptance of product or service. [IQUS10-20K] (A) Prepares inputs to a structured independent review as requested by the quality function in order to determine project compliance. [IQUS10-30A]	
11	Develops own understanding of this competency area through Continual Professional Development (CPD). [IQUS11]	(A) Identifies potential gaps in own knowledge or development needs in this area, identifying opportunities to address these through continual professional development activities. [IQUS11-10A] (A) Performs continual professional development activities to improve their knowledge and understanding in this area. [IQUS11-20A] (A) Records continual professional development activities undertaken including learning or insights gained. [IQUS11-30A]	Records of Continual Professional Development (CPD) performed and learning outcomes. [IQUS11-E10]

Practitioner – Integrating: Quality			
ID	**Indicators of Competence** ***(in addition to those at Supervised Practitioner level)***	**Relevant knowledge, experience, and/or behaviors**	**Possible examples of objective evidence of personal contribution to activities performed, or professional behaviors applied**
1	Follows governing quality plans and processes, and uses appropriate tools to control and monitor quality-related Systems Engineering tasks, interpreting as necessary. [IQUP01]	(A) Follows governing quality plans, processes and appropriate tools to control and monitor quality-related Systems Engineering tasks. [IQUP01-10A] (A) Uses governing quality plans, processes and appropriate tools where necessary to ensure outputs meet quality stakeholder needs. [IQUP01-20A] (A) Communicates potential issues with governing quality plans, processes and tools to quality stakeholders. [IQUP01-30A]	Records demonstrating quality plans, processes and tools they defined and used. [IQUP01-E10]
2	Prepares work products required by quality stakeholders to ensure Systems Engineering work aligns with wider quality management activities. [IQUP02]	(A) Analyzes information required by quality for requested work products to ensure understanding. [IQUP02-10A] (A) Identifies project-specific system engineering needs that need to be incorporated into Quality Management Plans or processes. [IQUP02-20A] (A) Prepares inputs for a governing quality management process, plan, or similar document to meet the needs of the quality function reflecting project and business strategy. [IQUP02-30A] (A) Prepares process performance data as required by the quality function in order to achieve the targeted standard of excellence on a project. [IQUP02-40A] (A) Communicates recommendations for corrective actions to be performed and implemented. [IQUP02-50A] (A) Prepares work products required by quality stakeholders, using appropriate processes and procedures. [IQUP02-60A]	Work products produce for use by quality which they contributed to or produced. [IQUP02-E10] Quality strategy work products they have produced. [IQUP02-E20]
3	Identifies alternative mechanisms for measuring quality to support the quality function in achieving the targeted standard of excellence on a project. [IQUP03]	(A) Determines technical performance measures, and uses them to control performance. [IQUP03-10A] (K) Describes key elements of a Quality Management Plan (including both QA and QC elements as well as continuous process improvement). [IQUP03-20K]	Documented process for collecting data. [IQUP03-E10]

(Continued)

ID	Indicators of Competence *(in addition to those at Supervised Practitioner level)*	Relevant knowledge, experience, and/or behaviors	Possible examples of objective evidence of personal contribution to activities performed, or professional behaviors applied
		(K) Describes the concepts of tolerances and control limits. [IQUP03-30K] (K) Describes potential quality metrics for their project (e.g. requirements volatility, defect rate, change proposal review time…). [IQUP03-40K] (K) Describes the Seven Basic Tools of quality (Cause and Effect diagrams, Control charts, flowcharting, histograms, pareto diagrams, run charts, scatter diagrams). [IQUP03-50K] (K) Describes tools or methods used for collecting metric data. [IQUP03-60K] (A) Identifies process improvements as a result of quality management work. [IQUP03-70A]	
4	Identifies mechanisms measuring process performance to support the quality function in achieving the targeted standard of excellence on a project. [IQUP04]	(A) Ensures data tracked in support of process monitoring addresses quality function needs. [IQUP04-10A] (A) Prepares data for quality function recommending actions to improve process performance. [IQUP04-20A]	Results of QA activities. [IQUP04-E10]
5	Guides and actively coordinates Systems Engineering process improvement activities to enable the quality function to achieve its targeted standard of Systems Engineering excellence on a project. [IQUP05]	(A) Monitors process performance on a project. [IQUP05-10A] (A) Determines quality factors that can be used to monitor process performance. [IQUP05-20A] (A) Analyzes process performance on a project looking for trends. [IQUP05-30A] (A) Performs process performance improvement on a project by addressing deficiencies in quality factors. [IQUP05-40A]	Records of process performance tracking and monitoring. [IQUP05-E10]
6	Analyzes design information or test (e.g. verification) results for a product or project to confirm conformance to standards. [IQUP06]	(A) Analyzes a product or project on behalf of quality management, to confirm its conformity to the standards against which it has been designed. [IQUP06-10A]	Records of the result of an analysis against a standard. [IQUP06-E10]
7	Analyzes the root-cause analysis of failures, determining appropriate corrective actions in support of quality function needs. [IQUP07]	(K) Describes the concepts of systemic and single-case errors or causes. [IQUP07-10K]	Root cause analysis report. [IQUP07-E10]

		(A) Performs root cause analysis of a failure using a recognized technique. [IQUP07-20A] (A) Determines root cause of failure and provides recommended corrective actions. [IQUP07-30A]	Failure or fault investigation report and recommendations produced. [IQUP07-E20]
8	Conducts an audit of project practices against recognized quality or project standards to support quality Function needs. [IQUP08]	(A) Conducts a quality audit and documents results. [IQUP08-10A] (A) Performs a work product inspection utilizing a documented process or checklist. [IQUP08-20A]	Quality audit report and findings. [IQUP08-E10]
9	Reviews the results of Quality Management Plans affecting Systems Engineering activities. [IQUP09]	(A) Reviews Systems Engineering aspects of a project Quality Management Plan. [IQUP09-10A] (A) Coordinates implementation of changes to a project Quality Management plan which improves project Systems Engineering. [IQUP09-20A]	Review comments made on plans. [IQUP09-E10]
10	Guides new or supervised practitioners in quality and its relationship to Systems Engineering, to develop their knowledge, abilities, skills, or associated behaviors. [IQUP10]	(P) Guides new or supervised practitioners in executing activities that form part of this competency. [IQUP10-10P] (A) Trains individuals to an "Awareness" level in this competency area. [IQUP10-20A]	Organizational Breakdown Structure showing their responsibility for technical supervision in this area. [IQUP10-E10] On-the-job training objectives/guidance etc. [IQUP10-E20] Coaching or mentoring assignment records. [IQUP10-E30] Records highlighting their impact on another individual in terms of improvement or professional development in this competency. [IQUP10-E40]
11	Maintains and enhances own competence in this area through Continual Professional Development (CPD) activities. [IQUP11]	(A) Identifies potential development needs in this area, identifying opportunities to address these through continual professional development activities. [IQUP11-10A] (A) Performs continual professional development activities to maintain and enhance their competency in this area. [IQUP11-20A] (A) Records continual professional development activities undertaken including learning or insights gained. [IQUP11-30A]	Records of Continual Professional Development (CPD) performed and learning outcomes. [IQUP11-E10]

Lead Practitioner – Integrating: Quality			
ID	**Indicators of Competence** ***(in addition to those at Practitioner level)***	**Relevant knowledge, experience, and/or behaviors**	**Possible examples of objective evidence of personal contribution to activities performed, or professional behaviors applied**
1	Creates enterprise-level policies, procedures, guidance, and best practice in order to ensure Systems Engineering quality-related activities integrate with enterprise-level quality goals including associated tools. [IQUL01]	(A) Creates Enterprise policy and guidance regarding quality management integration to ensure SE activities align with quality needs. [IQUL01-10A]	Records showing their role in embedding Systems Thinking into enterprise policies (e.g. guidance introduced at enterprise level). [IQUL01-E10]
		(A) Determines requirements for precision level, control thresholds, reporting formats and periodicity to ensure SE activities align with quality needs. [IQUL01-20A]	Procedures they have written. [IQUL01-E20]
		(A) Selects and acquires appropriate tools to align SE activities with quality management needs. [IQUL01-30A]	Minutes of enterprise-level reviews. [IQUL01-E30]
2	Assesses enterprise-level quality management processes to ensure they integrate with Systems Engineering needs. [IQUL02]	(A) Evaluates enterprise quality processes against Systems Engineering needs in order to enable project success. [IQUL02-10A]	Records of comments or documents improving the quality interface with Systems Engineering. [IQUL02-E10]
		(P) Provides constructive feedback on enterprise level quality management processes where these do not align with Systems Engineering approaches or needs. [IQUL02-20P]	Feedback provided on quality Plan, accepted and implemented. [IQUL02-E20]
		(A) Assesses quality plans across the enterprise to ensure Systems Engineering activities are aligned. [IQUL02-30A]	Minutes of contributions to quality management meetings covering Systems Engineering integration. [IQUL02-E30]
		(A) Assesses quality approaches across the enterprise with regard to their alignment with Systems Engineering activities. [IQUL02-40A]	Records of feedback provided on quality planning documents. [IQUL02-E40]
3	Judges appropriateness of enterprise-level quality decisions in a rational way to ensure alignment with Systems Engineering needs. [IQUL03]	(A) Analyzes quality stakeholders decisions using best practice techniques and based upon Systems Engineering need. [IQUL03-10A]	Reports on quality decisions made. [IQUL03-E10]
		(A) Determines appropriateness of quality decisions made, with robust justification based upon Systems Engineering need. [IQUL03-20A]	Records indicating changed stakeholder thinking from personal intervention. [IQUL03-E20]
		(P) Persuades quality stakeholders to change viewpoints and decisions based on the grounds of appropriateness and Systems Engineering need. [IQUL03-30P]	

4	Persuades quality stakeholders to address identified enterprise-level quality management issues affecting Systems Engineering. [IQUL04]	(A) Identifies and engages with key stakeholders. [IQUL04-10A]	Records of issues raised on existing quality management issues and revised or improved techniques adopted to address these issues. [IQUL04-E10]
		(P) Persuades stakeholders to address enterprise level issues using a holistic viewpoint. [IQUL04-20P]	Minutes of key stakeholder review meetings and actions taken. [IQUL04-E20]
		(P) Fosters agreement between key stakeholders at enterprise level to resolve quality issues related to Systems Engineering, by promoting a holistic viewpoint. [IQUL04-30P]	Records indicating changed stakeholder thinking from personal intervention. [IQUL04-E30]
5	Assesses quality information produced across the enterprise using appropriate techniques for its integration with Systems Engineering data. [IQUL05]	(A) Reviews information produced within the quality functions using appropriate techniques to assess its impact on Systems Engineering. [IQUL05-10A]	Records of assessments made on Systems Engineering data. [IQUL05-E10]
		(A) Judges adequacy and correctness of Systems Engineering quality information using appropriate techniques. [IQUL05-20A]	Records of assessments made on quality function data. [IQUL05-E20]
6	Reviews quality audit outcomes at enterprise level to establish their impact on system engineering across the enterprise. [IQUL06]	(A) Reviews information produced within the quality functions using appropriate techniques to assess its impact on Systems Engineering. [IQUL06-10A]	Records of comments made on audit data. [IQUL06-E10]
		(A) Judges adequacy and correctness of Systems Engineering quality information using appropriate techniques. [IQUL06-20A]	
7	Promotes continuous improvement in Systems Engineering at the enterprise level to support quality management function initiatives. [IQUL07]	(A) Determines critical capabilities of quality-management systems from an SE perspective. [IQUL07-10A]	Records of quality management improvement inputs. [IQUL07-E10]
		(P) Fosters a culture of Systems Engineering continuous improvement activities at enterprise-level. [IQUL07-20P]	Records of activities supporting quality management Systems Engineering interface improvement at enterprise level. [IQUL07-E20]
8	Assesses quality management plans from projects across the enterprise for their impact on Systems Engineering activities. [IQUL08]	(A) Reviews Systems Engineering aspects of a quality management plans from across the enterprise for their relationship with systems engineering. [IQUL08-10A]	Review comments made on plans. [IQUL08-E10]
		(P) Persuades key stakeholders to implement changes to Quality Management plans which improve Systems Engineering quality or effectiveness. [IQUL08-20P]	

(Continued)

ID	Indicators of Competence *(in addition to those at Practitioner level)*	Relevant knowledge, experience, and/or behaviors	Possible examples of objective evidence of personal contribution to activities performed, or professional behaviors applied
9	Fosters a culture of continuous quality improvement in projects across the enterprise. [IQUL09]	(A) Communicates critical capabilities of quality-management systems from an SE perspective to projects across the enterprise. [IQUL09-10A]	Records indicating input to quality management process improvement across enterprise. [IQUL09-E10]
		(P) Fosters a culture of Systems Engineering quality improvement activities in multiple projects across the enterprise. [IQUL09-20P]	Records of activities supporting quality management Systems Engineering interface improvement across the enterprise. [IQUL09-E20]
10	Coaches or mentors practitioners across the enterprise in the integration of quality with Systems Engineering in order to develop their knowledge, abilities, skills, or associated behaviors. [IQUL10]	(P) Coaches or mentors Systems Engineering or quality management practitioners within the enterprise in competency-related techniques, recommending development activities. [IQUL10-10P]	Coaching or mentoring assignment records. [IQUL10-E10]
		(A) Develops or authorizes enterprise training materials in this competency area. [IQUL10-20A]	Records of formal training courses, workshops, seminars, and authored training material supported by successful post-training evaluation data. [IQUL10-E20]
		(A) Provides enterprise workshops/seminars or training in this competency area. [IQUL10-30A]	Listing as an approved organizational trainer for this competency area. [IQUL10-E30]
11	Promotes the introduction and use of novel techniques and ideas across the enterprise, which improve the integration of Systems Engineering and quality management functions. [IQUL11]	(A) Analyzes different approaches across different domains through research. [IQUL11-10A]	Research records. [IQUL11-E10]
		(A) Defines novel approaches that could potentially improve the SE discipline within the enterprise. [IQUL11-20A]	Published papers in refereed journals/company literature. [IQUL11-E20]
		(P) Fosters awareness of these novel techniques within the enterprise. [IQUL11-30P]	Records showing introduction of enabling systems supporting the new techniques or ideas. [IQUL11-E30]
		(P) Collaborates with enterprise stakeholders to facilitate the introduction of techniques new to the enterprise. [IQUL11-40P]	Published papers (or similar) at enterprise level. [IQUL11-E40]
		(A) Monitors new techniques after their introduction to determine their effectiveness. [IQUL11-50A]	Records of improvements made against a recognized process improvement model in this area. [IQUL11-E50]
		(A) Adapts approach to reflect actual enterprise performance improvements. [IQUL11-60A]	

12	Develops expertise in this competency area through specialist Continual Professional Development (CPD) activities. [IQUL12]	(A) Identifies own needs for further professional development in order to increase competence beyond practitioner level. [IQUL12-10A] (A) Performs professional development activities in order to move own competence toward expert level. [IQUL12-20A] (A) Records continual professional development activities undertaken including learning or insights gained. [IQUL12-30A]	Records of Continual Professional Development (CPD) performed and learning outcomes. [IQUL12-E10]

NOTES	In addition to items above, enterprise-level or independent 3rd-Party-generated evidence may be used to amplify other evidence presented and may include: a. Formally recognized by senior management in current organization as an expert in this competency area b. Evidence of role as Product/System Design Authority or Technical Authority on a complex project with responsibilities in this area or where skills within this competency area were used c. Recognized as an authorizing signatory on behalf of enterprise for formal documentation in this competency area (e.g. policies, processes, and deliverables) d. Formal commendation or award within own enterprise for contribution or item of work successfully performed, which required proficiency in this competency area e. Customer, Supplier, or other external project-specific key Stakeholder accolades for specific work performed in this competency area f. Independently assessed or accredited work in this competency area (e.g. for independent publication or use) g. Formal organizational HR records positively highlighting any specific professional competencies or behaviors identified (if applicable) plus any of the evidence indicators listed at Expert level below

Expert – Integrating: Quality			
ID	**Indicators of Competence** ***(in addition to those at Lead Practitioner level)***	**Relevant knowledge, experience, and/or behaviors**	**Possible examples of objective evidence of personal contribution to activities performed, or professional behaviors applied**
1	Communicates own knowledge and experience in the integration of quality function needs with Systems Engineering, in order to improve Systems Engineering best practice beyond the enterprise boundary. [IQUE01]	(A) Produces papers, seminars, or presentations outside own enterprise for publication in order to share own ideas and improve industry best practices in this competence area. [IQUE01-10A]	Published papers or books etc. on new technique in refereed journals/company literature. [IQUE01-E10]
		(P) Fosters incorporation of own ideas into industry best practices in this area. [IQUE01-20P]	Published papers in refereed journals or internal literature proposing new practices in this competence area (or presentations, tutorials, etc.). [IQUE01-E20]
		(P) Develops guidance materials identifying new (or updating existing) best practice in this competence area. [IQUE01-30P]	Own proposals adopted as industry best practices in this competence area. [IQUE01-E30]
2	Advises organizations beyond the enterprise boundary on the suitability of their approach to Quality Management and the effect of their plans on Systems Engineering activities. [IQUE02]	(P) Acts as independent reviewer on external committee (or similar body) which approves such plans. [IQUE02-10P]	Records of membership on oversight committee with relevant terms of reference. [IQUE02-E10]
		(A) Reviews and advises on quality management strategies, leading to changes being implemented. [IQUE02-20A]	Review comments or revised document to external body on this topic. [IQUE02-E20]
3	Fosters a culture of continuous quality improvement beyond the enterprise boundary. [IQUE03]	(A) Communicates critical capabilities of quality-management systems from an SE perspective for organizations beyond the enterprise boundary. [IQUE03-10A]	Records indicating input to quality management process improvement beyond enterprise. [IQUE03-E10]
		(P) Fosters a culture of Systems Engineering quality improvement activities beyond the enterprise boundary. [IQUE03-20P]	Records of activities supporting quality management Systems Engineering interface improvement beyond enterprise level. [IQUE03-E20]
4	Advises organizations beyond the enterprise boundary on complex or sensitive quality-related issues affecting Systems Engineering. [IQUE04]	(A) Advises external stakeholders (e.g. customers) on their Quality Management Plans and issues. [IQUE04-10A]	Records of advice provided on quality management issues. [IQUE04-E10]
		(A) Conducts successful sensitive negotiations regarding the quality management of highly complex systems. [IQUE04-20A]	Records showing sensitive quality negotiations taking account of customer's background and knowledge, for example in minutes of meetings, position papers, and emails. [IQUE04-E20]

		(A) Conducts successful sensitive negotiations making limited use of specialized quality management relevant terminology. [IQUE04-30A]	
		(A) Uses a holistic approach to complex quality issue resolution including balanced, rational arguments on way forward. [IQUE04-40A]	
5	Champions the introduction of novel techniques and ideas to improve the integration of Systems Engineering and quality functions, beyond the enterprise boundary, in order to develop the wider Systems Engineering community in this competency. [IQUE05]	(A) Analyzes different approaches across different domains through research. [IQUE05-10A]	Records of activities promoting research and need to adopt novel technique or ideas. [IQUE05-E10]
		(A) Produces reports for the wider SE community on the effectiveness of new techniques after their introduction. [IQUE05-20A]	Records of improvements made to process and appraisal against a recognized process improvement model. [IQUE05-E20]
		(P) Collaborates with those introducing novel techniques within the wider SE community. [IQUE05-30P]	Research records. [IQUE05-E30]
		(A) Defines novel approaches that could potentially improve the wider SE discipline. [IQUE05-40A]	Published papers in refereed journals/company literature. [IQUE05-E40]
		(P) Fosters awareness of these novel techniques within the wider SE community. [IQUE05-50P]	Records showing introduction of enabling systems supporting the new techniques or ideas. [IQUE05-E50]
6	Coaches individuals beyond the enterprise boundary, in the relationship between Systems Engineering and quality management, to further develop their knowledge, abilities, skills, or associated behaviors. [IQUE06]	(P) Coaches or mentors those engaged in Systems Engineering or quality management activities beyond the enterprise boundary, in competency-related techniques, recommending development activities. [IQUE06-10P]	Coaching or mentoring assignment records. [IQUE06-E10]
		(A) Develops or authorizes training materials in this competency area, which are subsequently successfully delivered beyond the enterprise boundary. [IQUE06-20A]	Records of formal training courses, workshops, seminars, and authored training material supported by successful post-training evaluation data. [IQUE06-E20]
		(A) Provides workshops/seminars or training in this competency area for practitioners or lead practitioners beyond the enterprise boundary (e.g. conferences and open training days). [IQUE06-30A]	Records of Training/workshops/seminars created supported by successful post-training evaluation data. [IQUE06-E30]

(Continued)

ID	Indicators of Competence *(in addition to those at Lead Practitioner level)*	Relevant knowledge, experience, and/or behaviors	Possible examples of objective evidence of personal contribution to activities performed, or professional behaviors applied
7	Maintains expertise in this competency area through specialist Continual Professional Development (CPD) activities. [IQUE07]	(A) Reviews research, new ideas, and state of the art to identify relevant new areas requiring personal development in order to maintain expertise in this competency area. [IQUE07-10A] (A) Performs identified specialist professional development activities in order to maintain or further develop competence at expert level. [IQUE07-20A] (A) Records continual professional development activities undertaken including learning or insights gained. [IQUE07-30A]	Records of documents reviewed and insights gained as part of own research into this competency area. [IQUE07-E10] Records of Continual Professional Development (CPD) performed and learning outcomes. [IQUE07-E20]

NOTES

In addition to items above, enterprise-level or independent 3rd-Party-generated evidence may be used to amplify other evidence presented and may include:

a. Formally recognized by a reputable external organization as an expert in this competency area
b. Evidence of role as independent assessor or reviewer on project outside own organization where skills in this competency area were used
c. Evidence of invitation(s) from wider community for contribution of systems engineering expertise in this area (e.g. industry conference panel, government advisory board etc. cross-industry working groups, partnerships, accredited advanced university courses or research, or as part of professional institute)
d. Formal commendation beyond the enterprise (e.g. by INCOSE or other recognized authority) for work performed in this competency area
e. Independently assessed or accredited work product in this competency area (e.g. for independent publication or use)
f. Accolades of expertise in this area from recognized industry leaders

SECAG ANNEX B: FRAMEWORK IMPLEMENTATION EXAMPLES

This annex comprises a number of self-contained sections documenting example implementations of the SECF and SECAG.

Note that each example was created within its specific organizational context and thus may or may not be applicable to another content. Thus, these examples are informative rather than guidance.

SECAG ANNEX B1 – USING THE SECF/SECAG FOR CANDIDATE RECRUITMENT AND ASSESSMENT

This is an example of how to create a job announcement for candidate recruitment and a set of candidate assessment questions.

Candidate Recruitment Job Announcement

In this example, the announcement is targeting recent graduates from universities. Most competencies would be found in the Supervised Practitioner proficiency level. The statement from any competency or KSA can become the basis for a statement.

For this example, a system architect is needed. Suppose there are four desired competencies associated with a new system architect job. For this example, the competencies, as listed in SECAG Table B1-1, have been determined by competency managers for the organization.

The individual writing the job description typically begins by filling out a company-provided template using the descriptions of the four desired competencies for the specific proficiency levels desired. The results are presented in SECAG Table B1-2.

Note that the proficiency levels can be selected for any competency at a level desired for that specific competency, and that the levels need not be the same across all competencies. In this example, many competencies are taken from the Supervised Practitioner level, while others are taken from the Practitioner level. The competency and skill statements then are entered in the company template in order for a talent manager to develop the job description. The results are presented in SECAG Table B1-3.

Systems Engineering Competency Assessment Guide: A combined INCOSE Systems Engineering Competency Framework (SECF) and associated Systems Engineering Competency Assessment Guide (SECAG) document, First Edition. INCOSE.

SECAG TABLE B1-1 Competencies desired for example job announcement

Competency number	Name	Description
1	Systems thinking	The application of the fundamental concepts of systems thinking to Systems Engineering. These concepts include understanding what a system is, its context within its environment, its boundaries and interfaces, and that it has a life cycle. System thinking applies to the definition, development, and production of systems within an enterprise and technological environment and is a framework for curiosity about any system of interest.
29	Ethics and professionalism	Professional ethics encompass the personal, organizational, and corporate standards of behavior expected of systems engineers. Professional ethics also encompasses the use of specialist knowledge and skills by systems engineers when providing a service to the public. Overall, competence in ethics and professionalism can be summarized by a personal commitment to professional standards, recognizing obligations to society, the profession, and the environment.
11	Team dynamics	Team dynamics are the unconscious, psychological forces that influence the direction of a team's behavior and performance. Team dynamics are created by the nature of the team's work, the personalities within the team, their working relationships with other people, and the environment in which the team works.
25	System architecting	The definition of the system structure, interfaces, and associated derived requirements to produce a solution that can be implemented to enable a balanced and optimum result that considers all stakeholder requirements (business, technical etc.). This includes the early generation of potential system concepts that meet a set of needs and demonstration that one or more credible, feasible options exist.
7	Communications	The dynamic process of transmitting or exchanging information using various principles such as verbal, speech, body language, signals, behavior, writing, audio, video, graphics, language, etc. Communication includes all interactions between individuals, individuals and groups, or between different groups.

The task statements are listed in multiple groupings to separate the specific technical ones from the more general ones that might apply across many job roles in an organization. An important aspect is that the SECF provides a consistent ability to use competency and task statements for elements for a systems engineering job description, such that the person filling the job should be able to demonstrate the abilities listed.

The job description would then be posted along with other desired information such as education required, location, and salary compensation and other job and employment information.

Candidate Assessment

The duties and responsibilities portion of a position job description is used to assess the qualified candidates to determine the best from among those who applied. Candidates can assess themselves using the respective information in the competency tables with regard to their tasks as identified in the announcement. An assessor in the company would prepare a set of questions to ask the candidates during the job interview. They may also want to consider evidence provided by the candidate, as well. Questions are tailored from the relevant knowledge, experience, and/or behaviors sections of the competency tables shown in SECAG Table B1-4 based on the competencies listed in SECAG Table B1-1.

SECAG TABLE B1-2 Desired skills for example job announcement

Competency	ID number	Proficiency level	Description
Systems thinking	3	Supervised Practitioner	Uses the principles of system partitioning within system hierarchy on a project. [CSTS03]
Ethics and professionalism	10	Supervised Practitioner	Acts ethically when fulfilling own responsibilities. [PEPS10]
Team dynamics	2	Supervised Practitioner	Uses team dynamics to improve their effectiveness in performing team goals. [PTDS02]
Team dynamics	1	Practitioner	Acts collaboratively with other teams to accomplish interdependent project or organizational goals. [PTDP01]
System architecting	2	Supervised Practitioner	Uses analysis techniques or principles used to support an architectural design process. [TSAS02]
System architecting	3	Supervised Practitioner	Develops multiple different architectural solutions (or parts thereof) meeting the same set of requirements to highlight different options available. [TSAS03]
System architecting	4	Supervised Practitioner	Produces traceability information linking differing architectural design solutions to requirements. [TSAS04]
System architecting	5	Supervised Practitioner	Uses different techniques to develop architectural solutions. [TSAS05]
System architecting	8	Supervised Practitioner	Prepares architectural design work products (or parts thereof) traceable to the requirements. [TSAS08]
Communications	2	Practitioner	Uses appropriate communications techniques to ensure a shared understanding of information with all project stakeholders. [PCCP02]
Communications	3	Practitioner	Uses appropriate communications techniques to ensure positive relationships are maintained. [PCCP03]

Employee Performance Assessment and Ratings and Competency Level Assessment

Once a candidate is selected and placed in the position, performance objectives are established and assessed on a regular basis. The performance objectives should initially relate to the skills associated with the tasks for the position. Typically, performance objectives will be more comprehensive than the items used for interviewing a candidate for a job and include as many details regarding specific skills for tasks to be accomplished during a performance rating period as are documented in the actual job description.

Achievement of higher competency levels can be assessed and documented to allow consideration for a promotion to a higher-level position. This is not tied to specific annual performance objectives, but to the competency levels needed to be considered to have achieved a higher-level qualification.

Employee Performance Assessment and Ratings – Example Performance rating is distinct from developing and maintaining competencies required for a position or consideration for promotion to a higher-level position. Performance ratings are based on actual work accomplished in a given year, considering evidence of task completion, progress on

SECAG TABLE B1-3 System architect job announcement in the company template

JOB OVERVIEW	
JOB TITLE	System Architect
DEPARTMENT	Systems Engineering
LOCATION	
SALARY	
GENERAL JOB DESCRIPTION	
The incumbent creates and maintains architectural products throughout the life cycle integrating hardware, software, and human elements; their processes; and related internal and external interfaces that meet user needs and optimize performance. Collaborates with a diverse group of systems engineers, project managers, engineers, and support staff.	
DUTIES AND RESPONSIBILITIES (TASKS)	
Technical • Uses the principles of system partitioning within system hierarchy on a project. [CSTS03] • Uses analysis techniques or principles used to support an architectural design process. [TSAS02] • Develops multiple different architectural solutions (or parts thereof) meeting the same set of requirements to highlight different options available. Develops multiple different architectural solutions (or parts thereof) meeting the same set of requirements to highlight different options available. [TSAS03] • Produces traceability information linking differing architectural design solutions to requirements. [TSAS04] • Uses different techniques to develop architectural solutions. [TSAS05] • Prepares architectural design work products (or parts thereof) traceable to the requirements. [TSAS08] **Ethics and Professionalism** • Acts ethically when fulfilling own responsibilities. [PEPS10] **Teaming and Communications** • Uses team dynamics to improve their effectiveness in performing team goals. [PTDS02] • Acts collaboratively with other teams to accomplish interdependent project or organizational goals. [PTDP01] • Uses appropriate communications techniques to ensure a shared understanding of information with all project stakeholders. [PCCP02] • Uses appropriate communications techniques to ensure positive relationships are maintained. [PCCP03]	

projects, working with colleagues, and other aspects related to how well a person accomplished their work objectives based on their required competencies. Performance ratings are used to review actual performance, and are assessed to determine actual achievement, and determine the level of any bonuses available. Employee and supervisor information is collected and documented to organize the information for the performance review. The employee and supervisor would have a discussion to form the basis to compare the assessment of job performance objectives in the past year by competency using KSAs documented in the job description that apply to the employee's role. Evidence from the employee's job in the past year can also be used to assess performance objectives. In order to determine achievement of performance objectives, the supervisor would conduct an assessment of the employee relative to the performance of the items as identified in SECAG Table B1-5.

The supervisor rates the employee, and both sign the form to document achievement of desired performance levels or to set up expectations for improvement in specific areas.

SECAG TABLE B1-4 Questions to ask a job candidate

Competency	ID number	Proficiency level	Relevant knowledge, experience, and/or behaviors (questions to ask a job candidate)
Systems thinking	3	Supervised Practitioner	Define the following system properties for a system you have worked on: life cycle, context, hierarchy, sum of parts, purpose, boundary, interactions. [CSTS01-10A]
Ethics and professionalism	10	Supervised Practitioner	Describe a situation demonstrating the exercise of responsibilities in an ethical manner. [PEPS10-10K] Describe situations where guidance may be needed to exercise responsibilities in an ethical manner. [PEPS10-20K]
Team dynamics	1	Supervised Practitioner	Explain how team roles were formed for one of your past teams and your role on that team. [PTDS01-10K] Describes how your role interfaced with other roles in your organization. [PTDS01-20A]
Team dynamics	2	Practitioner	Describe a situation where you used team dynamics to improve your effectiveness in achieving team goals. [PTDS02-10A]
System architecting	2	Supervised Practitioner	Describe a systems architectural design, identifying its key architectural features and why they are present. [TSAS02-10K] Describe concepts of abstraction and the benefits of controlling complexity. [TSAS02-20K] Explain the differences between types of architectures. [TSAS02-30K] Describe the advantages of a formal approach for developing a system architecture. [TSAS02-60K]
System architecting	3	Supervised Practitioner	Explain the idea of having multiple "views" of a system model. [TSAS03-10K] Develop a sketch of alternative architectural designs from this set of requirements. [TSAS03-30A]
System architecting	4	Supervised Practitioner	Explain a situation where you identified areas where one of your architectural design solution(s) failed to meet requirements or go beyond identified requirements. [TSAS04-20A]
System architecting	5	Supervised Practitioner	Describe a range of different creativity techniques such as brainstorming, lateral thinking, and TRIZ, highlighting their strengths and weaknesses. [TSAS05-10K] Explain why there is a need for research and data collection to generate concepts. [TSAS05-20K]
System architecting	8	Supervised Practitioner	Describe what architectural design documentation you have prepared. [TSAS08-10A] Prepare a sketch showing multiple "views" of this system model. [TSAS08-20A]
Communications	2	Practitioner	Explain an example of when you used appropriate communications techniques to ensure a shared understanding of information delivered to project stakeholders. [PCCP02-10P] Explain an example of when you used appropriate communications techniques to ensure a common understanding of information received from project stakeholders. [PCCP02-20P]
Communications	3	Practitioner	Explain a situation where you identified differing quality of stakeholder relationships as a result of examining your communications with them. [PCCP03-10A]

SECAG TABLE B1-5 Example competency assessment statements for performance evaluation

Competency	ID number	Proficiency level	KSAs (Can the employee do the following?)
Systems thinking	3	Supervised Practitioner	Explains how system partitioning may be carried out through various techniques such as an analysis of scenarios, functional decomposition, physical decomposition, interface reduction, heritage, etc. [CSTS03-10K] Explains how the process of system partitioning deals with complexity by breaking down the system into realizable system elements. [CSTS03-20K] Explains how partitioning moves from an understanding of high-level purpose, through analysis, to an eventual allocation of identified functions to elements within the system. [CSTS03-30K] Explains the challenges of system partitioning. [CSTS03-40K] Explains the relative merits of different system partitioning approaches. [CSTS03-50K] Explains why hierarchy and partitions are merely constructs but how they impact our solution. [CSTS03-60K] Uses partitioning principles to support system decomposition. [CSTS03-70A]
Ethics and professionalism	10	Supervised Practitioner	Describes situation demonstrating the exercise of responsibilities in an ethical manner. [PEPS10-10K] Describes situations where guidance may be needed to exercise responsibilities in an ethical manner. [PEPS10-20K] Acts ethically when fulfilling own responsibilities, seeking guidance when appropriate. [PEPS10-30P]
Team dynamics	1	Supervised Practitioner	Explains how roles are formed and their role in current team. [PTDS01-10K] Describes how own role interfaces with other roles in the organization. [PTDS01-20A]
Team dynamics	2	Supervised Practitioner	Uses team dynamics to improve their effectiveness in performing team goals. [PTDS02-10A]
System architecting	2	Supervised Practitioner	Describes a systems architectural design, identifying its key architectural features and why they are present. [TSAS02-10K] Describes concepts of abstraction and the benefits of controlling complexity. [TSAS02-20K] Explains the differences between types of architectures. [TSAS02-30K] Describes a set of architectural design principles. [TSAS02-40K] Uses architectural design techniques to support systems architecting work on a project. [TSAS02-50A] Describes the advantages of a formal approach. [TSAS02-60K]

SECAG TABLE B1-5 (Continued)

Competency	ID number	Proficiency level	KSAs (Can the employee do the following?)
System architecting	3	Supervised Practitioner	Explains the idea of having multiple "views" of a system model. [TSAS03-10K] Performs architecture trade-offs in terms of finding an acceptable balance between constraints such as performance, cost, and time parameters. [TSAS03-20A] Develops alternative architectural designs from a set of requirements. [TSAS03-30A] Identifies different architectural design considerations when following different approaches to architectural design. [TSAS03-40A] Uses differing approaches required for different architectural design considerations. [TSAS03-50A]
System architecting	4	Supervised Practitioner	Produces traceability information linking differing architectural design solutions to requirements. [TSAS04-10A] Identifies areas where architectural design solution(s) fail to meet requirements or go beyond identified requirements. [TSAS04-20A]
System architecting	5	Supervised Practitioner	Describes a range of different creativity techniques such as brainstorming, lateral thinking, TRIZ, highlighting their strengths and weaknesses. [TSAS05-10K] Explains why there is a need for research and data collection to generate concepts. [TSAS05-20K] Explains set-based design principles. [TSAS05-30K]
System architecting	8	Supervised Practitioner	Prepares architectural design documentation. [TSAS08-10A] Prepares multiple "views" of a system model. [TSAS08-20A] Prepares decision management or trade-off documentation as part of an architectural solution selection activity. [TSAS08-30A]
Communications	2	Practitioner	Uses appropriate communications techniques to ensure a shared understanding of information delivered to project stakeholders. [PCCP02-10P] Uses appropriate communications techniques to ensure a common understanding of information received from project stakeholders. [PCCP02-20P] Recognizes differing levels of effectiveness of communications with different stakeholders and how these can be improved over time. [PCCP02-30P]
Communications	3	Practitioner	Identifies differing quality (or positivity) of stakeholder relationships as a result of examining communications with them. [PCCP03-10A] Fosters improvements to poor stakeholder relationships over time through improved communications. [PCCP03-20P] Fosters maintenance of good stakeholder relationships through effective communications. [PCCP03-30P]

Employee Competency Assessment Example Suppose that an employee is currently at Supervised Practitioner for a required competency in System Architecting. The employee has already demonstrated behaviors for Supervised Practitioner, so the employee and supervisor concentrate on the assessment of achievement of Practitioner by reviewing the performance of tasks related to the System Architecting Practitioner, as shown in SECAG Table B1-6.

SECAG TABLE B1-6 Practitioner competency assessment

Competency	ID number	Proficiency level	KSAs (Can the employee do the following?)
System architecting	3	Practitioner	Follows defined plans, processes, and associated tools to perform system architecting on a project. [TSAP03-10A] Recognizes situations where deviation from published plans and processes or clarification from others is appropriate in order to overcome complex system architecting challenges. [TSAP03-40P]
System architecting	4	Practitioner	Develops alternative architectural design solutions from a set of requirements. [TSAP04-20A] Uses architectural frameworks in assisting consistency and reusability of architectural design. [TSAP04-30A] Identifies the merits or consideration in different architectural design solutions. [TSAP04-40A] Uses an architectural design tool, methodology, or modeling language. [TSAP04-50A] Uses different architectural approaches to establish preferred solution approaches of different stakeholders. [TSAP04-60A]
System architecting	5	Practitioner	Describes the purpose and potential challenges of reviewing different architectural design solutions. [TSAP05-10K] Performs architecture trade-offs in terms of finding an acceptable balance between constraints such as performance, cost, and time parameters. [TSAP05-20A] Selects preferred options from those available, listing advantages and disadvantages. [TSAP05-30A]
System architecting	6	Practitioner	Lists and describes key characteristics of different analysis techniques (e.g. cost analysis, technical risk analysis effectiveness analysis, or other recognized formal analysis techniques). [TSAP06-10K] Uses techniques for analyzing the effectiveness of a particular architectural solution and selecting the most appropriate solution. [TSAP06-20A] Describes the advantages and limitations of the use of architectural design tools in relation to at least one tool. [TSAP06-30K]
System architecting	7	Practitioner	Identifies areas where discipline implementation or technology constraints dictate partitioning of functionality (e.g. software, hardware, human factors, packaging, and safety). [TSAP07-10A] Develops architectural partitioning between discipline technologies. [TSAP07-20A]
System architecting	8	Practitioner	Analyzes potential options against selection criteria. [TSAP08-10A] Selects credible solutions using criteria. [TSAP08-20A]

The assessment might include evidence for the respective KSAs, or they may require evidence of knowledge by answering questions verbally or by demonstrating the KSA using a model-based systems engineering (MBSE) tool. Once the employee has been verified as meeting the required KSAs, the supervisor can provide their input to a record to verify the employee's level of achievement – the employee is then eligible to be documented as having achieved the higher-level proficiency of Practitioner for System Architecting.

The assessments would be performed for as many competencies and roles for specialties and subspecialties required for the employee.

The overall context for employee development is captured in a comprehensive career development plan, which is a career progression guide that will provide employees with information about the types of skills and enrichment activities needed to further their career goals.

SECAG ANNEX B2 – USING THE SECF/SECAG FOR CAREER AND ORGANIZATIONAL DEVELOPMENT

This example is adapted from a US Office of Personnel Management (OPM) report on development of a systems engineering career competency model for the US Navy.

Career Development

It is the responsibility of employees, managers/supervisors, and the organizations to ensure employee career development. For career development to be effective, everyone must take responsibility and play a role. The following are brief descriptions of the responsibilities involved in career development.

Employee:

- Performs self-assessment.
- Learns about opportunities internal and external to their organization.
- Defines career goals and identifies relevant training and development.
- Discusses career goals with supervisor.
- Prepares an Individual Development Plan (IDP).
- Reviews the Career Path Development Plan.

Manager/Supervisor:

- Defines organizational-level goals for training and development.
- Learns employee goals through discussion and review of IDP.
- Encourages and establishes appropriate timing for employee training and development.
- Follows up with employees after training and development opportunities.
- Coaches/counsels employees.
- Provides feedback on employee progress.
- Ensures employees are provided opportunities to develop required competencies.
- Determines the need for positions based upon the goals and objectives outlined in strategic and business plans.

Organization:

- Provides strategic vision/direction.
- Provides organizational development support.

- Offers individual development tools and programs.
- Analyzes training needs.
- Fosters an environment for career growth.
- Encourages training and development opportunities.

Career development plans contain a comprehensive set of competencies needed to advance through a desired career path, as documented in a Career Development Model (CDM) used to document expectations for proficiency at various career development levels. Career development is based on career path models and career development plans.

Career Path Modeling

Companies establish career paths to guide employees, their supervisors, and the organization as a whole for employee development purposes. Career path models serve as a resource to employees seeking to further develop their professional skills (OPM 2016).

An organization may leverage the content of the SECAG to develop a more specific Career Path Model for their organization. For example, there may be specific organizational SE positions at the Practitioner level and above. The duties associated with those positions can be specified based upon SECF content, either the competencies and/or KSA. Content from the SECAG would indicate which competencies and KSA are important for higher-level positions. This allows employees to better understand the knowledge and skills they would need, in order to be able to compete and qualify for these positions.

An employee can use an organization's Career Path Model to determine their desired career trajectory and determine one or more paths to enable them to reach their career goals. The content of the SECAG would allow them to self-assess their strengths and weaknesses for the various positions they hope to take on. After self-assessing, they can use the enrichment activity suggestions as a starting point for improving on their areas of weakness.

Career Path Modeling Example A career path is a progression of a job comprised of entrance points, grade levels, and exit opportunities and is summarized in SECAG Figure B2-1. This diagram shows progression from "Supervised Practitioner" to "Expert" from the top to the bottom of the diagram.

- *Entrance points* are located on the left side of the career path diagram and are provided for each identified grade level within the systems engineering profession.
- *Proficiency levels* are found in the center of the diagram and range from the lowest possible entry level to the highest achievable systems engineering positions.
- *Exit opportunities* are located on the right side of the diagram and represent opportunities for employees to assume new roles, either within or outside of an organization.

Actual job titles are not included in this systems engineer career path example as these can vary widely from company to company.

Career Development Planning

A Career Path Development Plan is competency-based and includes career paths, proficiency level assessment criteria, and professional development activities.

The purpose of the Career Path Development Plan is to provide all employees with a standard roadmap, from entry through senior levels, for enhancing their professional growth and to assist organizations in developing a highly

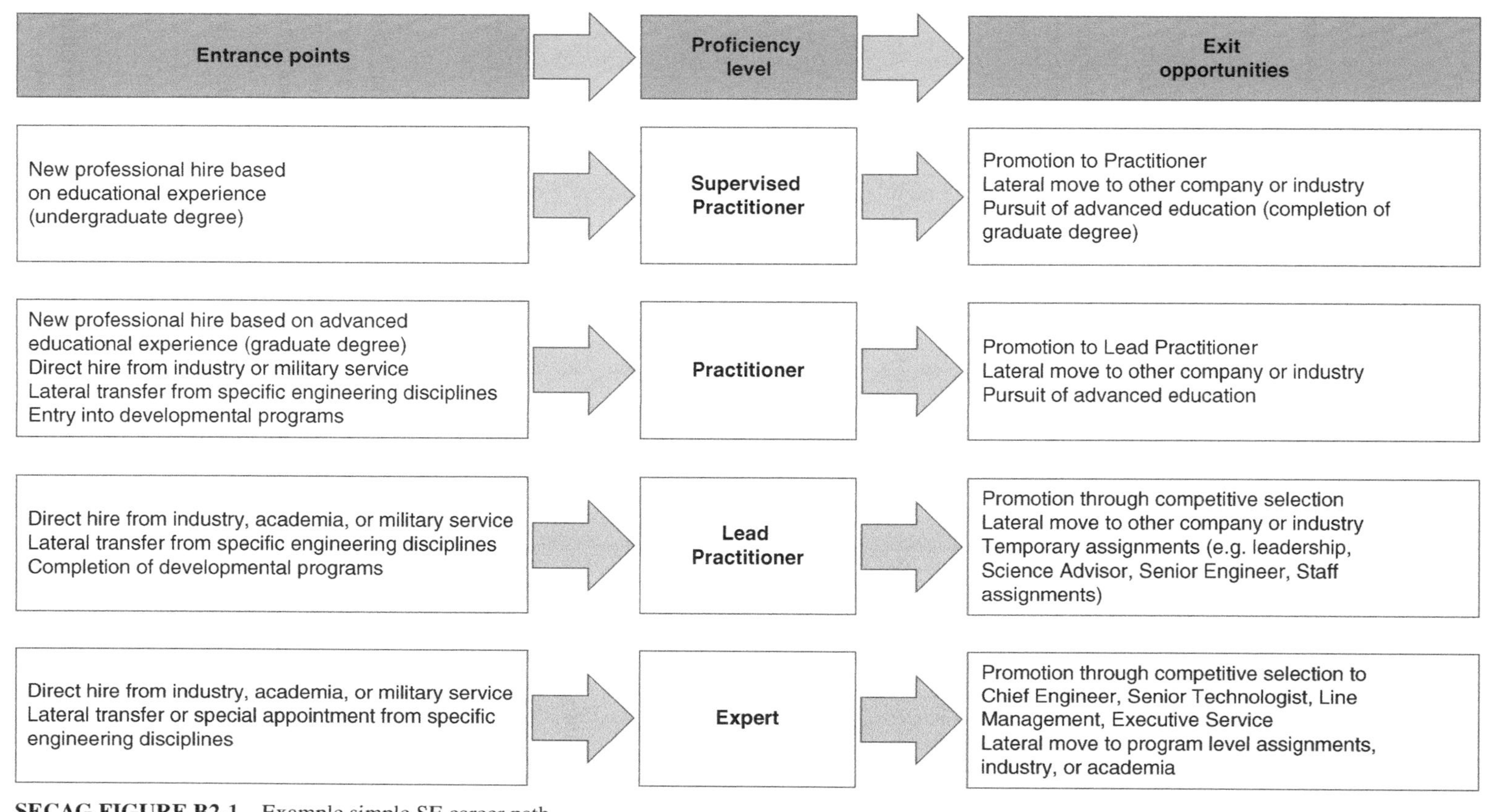

SECAG FIGURE B2-1 Example simple SE career path.

competent and professional systems engineering workforce. More specifically, the Career Path Development Plan objectives are to:

- Display typical *career paths, experience level criteria, and professional development activities* for systems engineers.
- Provide employees and their supervisors with a *single source reference* to assist in determining appropriate education and training.
- Assist supervisors in making *effective use of scarce training resources* by identifying critical competencies and associated training courses.
- Assist employees with *planning and sequencing appropriate career training and development* so they can attend the appropriate courses at the appropriate time in preparation for more senior-level positions.

NOTE: The Career Path Development Plan is NOT a requirement of a promotion process. Instead, it is a career progression guide that provides employees with information about the types of skills and enrichment activities they require in order to progress their career goals.

Career Path Development Plan Example

For this example, a systems engineering role is defined through four competencies deemed to be most beneficial for future development as follows:

- System Architecting
- Requirements Definition
- Systems Modeling and Analysis
- Communications

A competency manager uses the competencies to find sets of KSAs for each competency, as shown in SECAG Table B2-1.

Employees can use the information to:

- Identify competencies required for success as systems engineers.
- Review tasks for their current grade level linked to competencies.
- Compare tasks of their current grade level to those of higher grade levels.

The information can be used to determine training and development needs. For example, an employee can use self-assessments to begin to determine if they need to further develop their skills to successfully perform the tasks linked to a competency. If an employee identifies a need to develop their skills, they can research more formal competency assessments and related professional development opportunities through commercially available training courses, university courses or certificates, professional certification, or on-the-job training for possible developmental opportunities for the competency. Employees should review opportunities available at both their current level and for higher levels to maintain opportunities for advancement.

Organizational Development

Beyond career development, organizations can use the SECAG for workforce assessment, including risk analysis, mission/business case analysis, targeted training investment, and targeted training development.

SECAG TABLE B2-1 System engineer role progression example

Systems Engineer Role Example			
Competency	**Supervised Practitioner**	**Practitioner**	**Lead Practitioner**
System architecting	Uses a governing process using appropriate tools to manage and control their own system architectural design activities. [TSAS01] Uses analysis techniques or principles used to support an architectural design process. [TSAS02] Develops multiple different architectural solutions (or parts thereof) meeting the same set of requirements to highlight different options available. [TSAS03] Uses different techniques to develop architectural solutions. [TSAS05] Compares the characteristics of different concepts to determine their strengths and weaknesses. [TSAS06]	Creates a governing process, plan, and associated tools for systems architecting which reflect project and business strategy. [TSAP02] Uses plans and processes for system architecting, interpreting, evolving, or seeking guidance where appropriate. [TSAP03] Creates alternative architectural designs traceable to the requirements to demonstrate different approaches to the solution. [TSAP04] Analyzes options and concepts in order to demonstrate that credible, feasible options exist. [TSAP05] Identifies the strengths and weaknesses of relevant technologies in the context of the requirement and provides examples. [TSAP09] Monitors key aspects of the evolving design solution in order to adjust architecture, if appropriate. [TSAP10]	Creates enterprise-level policies, procedures, guidance, and best practice for system architectural design including associated tools. [TSAL01] Assesses the tailoring of enterprise-level system architectural design processes to meet the needs of a project. [TSAL02] Advises stakeholders across the enterprise on selection of architectural design and functional analysis techniques to ensure effectiveness and efficiency of approach. [TSAL03] Judges the suitability of architectural solutions across the enterprise in areas of complex or challenging technical requirements or needs. [TSAL04] Assesses system architectures across the enterprise to determine whether they meet the overall needs of individual projects. [TSAL05] Persuades key stakeholders across the enterprise to address identified enterprise-level Systems Engineering architectural design issues to reduce project cost, schedule, or technical risk. [TSAL06] Promotes the introduction and use of novel techniques and ideas in Systems Architecting across the enterprise to improve enterprise competence in this area. [TSAL08]
Requirements definition	Uses a governing process using appropriate tools to manage and control their own requirements definition activities. [TRDS01] Identifies examples of internal and external project stakeholders highlighting their sphere of influence. [TRDS02] Elicits requirements from stakeholders under guidance, in order to understand their need and ensuring requirement validity. [TRDS03] Defines acceptance criteria for requirements, under guidance. [TRDS06] Reviews developed requirements. [TRDS10]	Creates a governing process, plan, and associated tools for Requirements Definition which reflect project and business strategy. [TRDP02] Uses plans and processes for requirements definition, interpreting, evolving, or seeking guidance where appropriate. [TRDP03] Elicits requirements from stakeholders ensuring their validity, to understand their need. [TRDP04] Develops good quality, consistent requirements. [TRDP05] Determines derived requirements. [TRDP06] Determines acceptance criteria for requirements. [TRDP08] Negotiates agreement in requirement conflicts within a requirement set. [TRDP09] Analyzes the impact of changes to requirements on the solution and program. [TRDP10]	Creates enterprise-level policies, procedures, guidance, and best practice for requirements elicitation and management, including associated tools. [TRDL01] Judges the tailoring of enterprise-level requirements elicitation and management processes to meet the needs of a project. [TRDL02] Advises on complex or challenging requirements from across the enterprise to ensure completeness and suitability. [TRDL03] Defines strategies for requirements resolution in situations across the enterprise where stakeholders (or their requirements) demand unusual or sensitive treatment. [TRDL04] Persuades key stakeholders across the enterprise to address identified enterprise-level requirements elicitation and management issues to reduce enterprise-level risk. [TRDL05] Promotes the introduction and use of novel techniques and ideas in Requirements Definition across the enterprise to improve enterprise competence in this area. [TRDL07]

(*Continued*)

SECAG TABLE B2-1 (Continued)

Systems Engineer Role Example			
Competency	**Supervised Practitioner**	**Practitioner**	**Lead Practitioner**
Systems modeling and analysis	Uses modeling and simulation tools and techniques to represent a system or system element. [CSMS01] Analyzes outcomes of modeling and analysis and uses this to improve understanding of a system. [CSMS02] Analyzes risks or limits of a model or simulation. [CSMS03] Uses systems modeling and analysis tools and techniques to verify a model or simulation. [CSMS04] Uses system analysis techniques to derive information about the real system. [CSMS07]	Identifies project-specific modeling or analysis needs which need to be addressed when performing modeling on a project. [CSMP01] Creates a governing process, plan, and associated tools for systems modeling and analysis in order to monitor and control systems modeling and analysis activities on a system or system element. [CSMP02] Determines key parameters or constraints which scope or limit the modeling and analysis activities. [CSMP03] Analyzes a system, determining the representation of the system or system element, collaborating with model stakeholders as required. [CSMP05] Selects appropriate tools and techniques for system modeling and analysis. [CSMP06]	Creates enterprise-level policies, procedures, guidance, and best practice for systems modeling and analysis definition and management, including associated tools. [CSML01] Judges the correctness of tailoring of enterprise-level modeling and analysis processes to meet the needs of a project, on behalf of the enterprise. [CSML02] Advises stakeholders across the enterprise on systems modeling and analysis. [CSML03] Coordinates modeling or analysis activities across the enterprise in order to determine appropriate representations or analysis of complex system or system elements. [CSML04] Adapts approaches used to accommodate complex or challenging aspects of a system of interest being modeled or analyze projects across the enterprise. [CSML05] Assesses the outputs of systems modeling and analysis across the enterprise to ensure that the results can be used for the intended purpose. [CSML06] Coordinates the integration and combination of different models and analyses for a system or system element across the enterprise. [CSML08]
Communications	Follows guidance received (e.g. from mentors) when using communications skills to plan and control their own communications activities. [PCCS01] Uses appropriate communications techniques to ensure a shared understanding of information with peers. [PCCS02] Uses appropriate communications techniques to interact with others, depending on the nature of the relationship. [PCCS04] Uses active listening techniques to clarify understanding of information or views. [PCCS06]	Uses a governing communications plan and appropriate tools to control communications. [PCCP01] Uses appropriate communications techniques to ensure a shared understanding of information with all project stakeholders. [PCCP02] Uses appropriate communications techniques to express alternate points of view in a diplomatic manner using the appropriate means of communication. [PCCP04] Uses full range of active listening techniques to clarify information or views. [PCCP07] Uses appropriate communications techniques to express own thoughts effectively and convincingly in order to reinforce the content of the message. [PCCP06] Uses appropriate feedback techniques to verify success of communications. [PCCP08]	Creates enterprise-level policies, procedures, guidance, and best practice for systems engineering communications, including associated tools. [PCCL01] Uses best practice communications techniques to improve the effectiveness of Systems Engineering activities across the enterprise. [PCCL02] Maintains positive relationships across the enterprise through effective communications in challenging situations, adapting as necessary to achieve communications clarity or to improve the relationship. [PCCL03] Uses effective communications techniques to convince stakeholders across the enterprise to reach consensus in challenging situations. [PCCL04] Uses a proactive style, building consensus among stakeholders across the enterprise using techniques supporting the verbal messages (e.g. nonverbal communication). [PCCL05] Adapts communications techniques or expresses ideas differently to improve effectiveness of communications to stakeholders across the enterprise, by changing language, content, or style. [PCCL06] Reviews ongoing communications across the enterprise, anticipating and mitigating potential problems. [PCCL07]

Workforce Risk Analysis If an organization would like to perform a workforce risk analysis on their ability to perform systems engineering, they could have those in systems engineering roles perform a self-assessment based upon the systems engineering competencies and tasks. A simple answer as to whether the employee feels they could perform the task independently, collected across all systems engineering personnel, would give the organization insight into task types that a limited number, or none, of their employees could accomplish. The result will also indicate the types of tasks that most or all systems engineers can execute.

Workforce Mission/Business Case Analysis The result of a workforce risk analysis can be used to support mission or business case analysis. If a pending work request requires skill in the tasks that only a few systems engineers can perform, leadership may decide not to accept the tasking, lest it not be accomplished properly. Similarly, if there are workforce strengths in some areas, that type of work can be targeted for future business acquisition. There could also be a consideration to rely on contracted assistance for a short time as SE skills are developed for the competencies in need.

Targeted Training Investment If areas of weakness are determined during the course of employee assessment, this can help better target investments in training funds. For example, if a significant number of employees are not skilled at: designing architecture solutions, assessing an architecture's ability to meet requirements, and managing architecture artifacts – a decision could be made to invest corporately in training regarding system architecture development and use.

Targeted Training Development The SECAG also facilitates the development of tailored training and education programs. The competencies and tasks can be used to develop courses or enable efforts with an outside vendor to develop courses or even an entire program of study. The SECAG content can help specify learning outcomes and objectives within systems engineering programs that will ensure the students who finish the course or program have the competencies required to perform successfully in their job.

SECAG ANNEX B3 – ORGANIZATIONAL ROLE DEFINITION

This example is taken from an application within a European aerospace organization.

Introduction

In this example, an engineering organization wishes to have a written and formalized way to address competence assessment and development. A general approach is described, although this has been simplified.

NOTE: All examples of competencies or required proficiency levels for different roles, etc., are for illustrative purposes only. The intent of this section is to describe the process, rather than the content or end result.

Background

The following use cases were defined as the goal of the activity:

1. Describe relevant competences for different roles within the organization.
2. Identify competences and proficiency levels of individual employees including any areas for personal development.

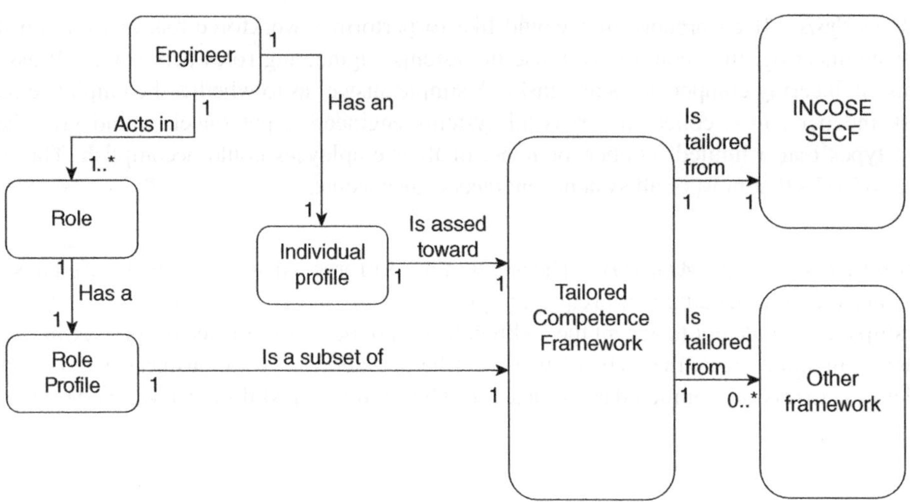

SECAG FIGURE B3-1 Organizational competence framework model.

3. Match an individual's competence with the competence/levels required for a specific role. The aim of this was not to qualify/disqualify an individual, but rather to identify areas where the individual potentially needed mentoring or guidance when assigned to the role.

The organizational competence framework model is illustrated in SECAG Figure B3-1. This figure describes the initial status where only SECF is tailored.

To establish the necessary capability to achieve the goals, the following information needed to be defined:

- Tailored Competence Framework (TCF): This defines the competences the organization wants to address within their competence framework. Furthermore, it ensures that wording and terminology are also aligned to the organization.
- Roles: A list of roles to be described needs to be identified.
- Role Profiles: The required competence areas and associated proficiency levels need also to be defined for each of the identified roles.

Tailored Competence Framework (TCF)

A series of local decisions were made which affected the tailoring:

- For its initial roll-out, the TCF would be limited only to competencies identified in the INCOSE SECF core framework was tailored.
- Professional competencies would be omitted – since there were other means to address those.
- The scope of initial TCF roll-out would be limited to competencies up to Practitioner level. The belief was that this addressed all work necessary to execute projects and processes within the current organization. Guidance and improvement, which are the focus for the Lead Practitioner and Expert levels, were left for future versions of the TCF.
- Initially, around 10 competence areas were identified to form the TCF. As the work with defining the Role Profiles evolved, see below, this list was iterated and extended.

With the above decisions in place, each of the prioritized competence areas were reviewed and tailored.

This tailoring was performed by selecting indicators relevant to the organization.

Each of the selected indicators was evaluated together with the sub-indicators and evidence in the SECAG. The information was rephrased into a set of evidence capable of describing real activities and work products typically produced in the organization. Attention was made to try and make the evidence written in language, and with terminology, known to individuals within the organization.

The result was the set of evidence for each competence area and proficiency levels shown in SECAG Figure B3-2.

Awareness	Supervised practitioner	Practitioner
1. Can describe the differences between iterative and waterfall models in early concept stages 3. Can describe the relationship, differences, and consequences between agile SW and HW development	Has participated in development where different life cycle models has been applicable. Examples: -Linear/waterfall -Incremental development -Reuse of developed components -Agile/scrum	Has been responsible for planning the development for a system/SW/HW and creating a SEMP/SDP/HDP that incorporates e.g.: -Customer life cycle needs and schedule -Corporate processes, methods, and tools -System element maturity (e.g. new technology, reuse, etc.) -Milestones and technical reviews
2. Can describe how the size of a project, complexity of a system, incremental development, and regulatory requirements affect the decision of which life cycle model shall apply	Has applied/used more than one of the processes for stakeholder and system requirements, architecture/design definition, integration, or verification	Has used corporate management system when creating development plans for a system/SW/HW. Has been responsible for tailoring 1 development life cycles to meet the stakeholder needs for a project.
1. Can describe the basis for the corporate management system: Life cycle model, processes, system of interest, system hiearchy and system elements, enabling systems, recursion 2. Can describe the difference between a project life cycle and a system life cycle 3. Can describe how the size of a project, complexity of a system, incremental development, and regulatory requirements affect the decision of which life cycle model shall apply	Has participated in creating, or reviewed, plans for one or several activities regarding how to perform: -requirements, architecture, design, integration, or verification tasks -technical reviews/ milestones for a system.	Has participated in creating the plans for milestones/technical reviews. e.g.: -Which reviews shall be used and how will they apply with respect to different system life cycles, the development model used (e.g. incremental development) -Content and maturity of the information that shall be available at the review
	Can describe how later stages such as Production, Operation, and retirement needs to be adressed in the Development stage. Can describe how activities later in the Development stage needs to be adressed early	Has participated in identifying dependencies or constraints due to different product life cycles (i.e. integration of new and existing/legacy design)
	Works according to the governing plans regarding processes, tools, and reviews in their project (e.g. the SEMP)	

SECAG FIGURE B3-2 Example of the structure of a competence area within the TCF.

Roles and Role Profiles

The work performed in selected roles was the basis for the identified competencies and thus formed the basis for the content of the TCF.

Next, a set of roles were identified based upon the corporate management system but reflecting standard assignments and ways of working in projects. Each role was given a brief description and from that description a set of prioritized competence areas was identified.

Within each role, a selection was then made regarding the expected proficiency level for each competency defined with the role. During this work, the list of competence areas originally prioritized was revised and additional competence areas were added.

Examples of Role Profiles are illustrated in SECAG Figure B3-3.

	Systems requirements engineer	Systems architect	Systems designer	In-service support systems	Technical design authority	Systems engineering leader
Systems thinking	Supervised Practitioner	Practitioner	Supervised Practitioner	Practitioner	Supervised Practitioner	Supervised Practitioner
Life cycles	Supervised Practitioner	Practitioner	Supervised Practitioner	Practitioner	Practitioner	Practitioner
Capability engineering	Practitioner	Practitioner	Supervised Practitioner	Practitioner	Practitioner	Supervised Practitioner
Systems modeling and analysis	Awareness	Practitioner	Practitioner	Awareness	Awareness	Supervised Practitioner
Requirements definition	Practitioner	Supervised Practitioner	Supervised Practitioner	Practitioner	Practitioner	Practitioner
System architecting	Supervised Practitioner	Practitioner	Supervised Practitioner	Supervised Practitioner	Supervised Practitioner	Supervised Practitioner
Design for. ..	Supervised Practitioner	Practitioner	Practitioner	Supervised Practitioner	Practitioner	Supervised Practitioner
Integration	Awareness	Supervised Practitioner	Supervised Practitioner	Practitioner	Awareness	Awareness
Verification	Supervised Practitioner	Awareness	Awareness	Practitioner	Supervised Practitioner	Awareness
Configuration management	Practitioner	Practitioner	Supervised Practitioner	Practitioner	Practitioner	Practitioner
Operation and support				Practitioner		
Planning						Practitioner
Decision management					Practitioner	Practitioner
(Integrate with) Project management						Practitioner
(Integrate with) Logistics				Practitioner		
(Integrate with) Quality					Supervised Practitioner	Supervised Practitioner

SECAG FIGURE B3-3 Example of Roles and Role Profile. A Role Profile is the requirement to be able to act in a role with adequate quality.

When the TCF was established in its first draft version, i.e. the tailored evidence for each proficiency level was defined (SECAG Figure B3-2) and all evidence required for each specific role was collected and validated.

Validation was performed through review and discussions with senior and junior engineers currently working in similar roles. This validation activity resulted in a number of small updates to the originally defined proficiency levels in certain roles and some updates to evidence in the TCF (changes or removals).

Conclusion

At the end of the exercise, the goals originally defined will have been met as follows:

- *Describe relevant competence for different roles*

  For each role, a set of proficiency levels had been defined according to agreed Role Profiles (SECAG Figure B3-3) and for each of those levels, the tailored framework included a description of expected competence. By extracting all evidence items in the TCF identified in a Role Profile, the expected competence for a role could then be defined.

- *Matching an individual's competence with the competence required for a specific role*

  Identification of each individual's competence was achieved through assessment against the finalized TCF. This will have resulted in a record of an individual profile capturing their current proficiency level (SECAG Figure B3-4), enabling this individual profile to be compared with the expected profile for the role.

  NOTE: As expected, in some instances, this identified a competence gap, i.e. the individual did not meet ALL required proficiency levels for a specific role (meets Systems Requirements Engineer but not Technical Design Authority in SECAG Figure B3-4). In such cases, an organization would need to define how to handle these gaps. In the example, it was decided to use the gap as supporting information to guide the individual's manager or team leader to areas where the individual needed support or mentoring.

  For certain critical roles (e.g. Design Authority or roles related to safety), the defined process would normally be more rigorous. For instance, it may require that an individual demonstrates 100% compliance with the required Role Profile before they can take on those roles.

- *Identify individual competence of employees and areas for personal development*

  When performing the assessment of an individual engineer, competence areas with proficiency levels Awareness or Supervised Practitioner could be targeted for professional development, based on the individual's areas of interest. Training and/or future work assignments could be identified as means to increase competence.

  In addition, when comparing the individual profile with different Role Profiles, gaps could be targeted for professional development to support the individual's preferred career path.

	Assessed N N individual profile	Systems requirements engineer role profile	Technical design authority role profile
Systems thinking	Supervised Practitioner	Supervised Practitioner	Supervised Practitioner
Life cycles	Supervised Practitioner	Supervised Practitioner	Practitioner
Capability engineering	Practitioner	Practitioner	Practitioner
Systems modeling and analysis	Awareness	Awareness	Awareness
Requirements definition	Practitioner	Practitioner	Practitioner
System architecting	Supervised Practitioner	Supervised Practitioner	Supervised Practitioner
Design for ...	Supervised Practitioner	Supervised Practitioner	Practitioner
Integration	Supervised Practitioner	Awareness	Awareness
Verification	Supervised Practitioner	Supervised Practitioner	Supervised Practitioner
Configuration management	Practitioner	Practitioner	Practitioner
Operation and support	None		
Planning	None		
Decision management	Awareness		Practitioner
(Integrate with) Project management	None		
(Integrate with) Logistics	Awareness		
(Integrate with) Quality	Supervised Practitioner		Supervised Practitioner

SECAG FIGURE B3-4 Example of an individual profile and comparison with Role Profiles.

SECAG ANNEX B4 – USING THE SECF/SECAG FOR EDUCATIONAL COURSE DEFINITION

This example is based around construction of an introductory course on Systems Integration.

Course Content Definition

Let us imagine a systems engineering faculty member, curriculum developer, or instructor who wishes to develop an introductory systems integration course. Their first step would be to review all competency areas within SECF to select relevant competency areas. There are several candidate areas to consider. In this case, however, the instructor decides to focus their course on the competency entitled "Integration." However, in addition, the instructor decides that managing interfaces is such a significant element of systems integration they will also draw upon the "Interfaces" competency.

As this is an *introductory*-level course, the instructor chooses to select from "Awareness" and "Supervised Practitioner" levels for both of these competencies. Their thinking is that an undergraduate student completing an *introductory*-level course would not be expected to be proficient enough to perform as a full (unsupervised) Practitioner.

The instructor can use the SECAG Competency Area Description and Why it Matters to construct a brief course description. The course title remains Systems Integration, even though aspects of interfaces are included in the course description and learning objectives. The course title and description example includes wording from both the Integration and Interfaces competencies.

Course Title:

Systems Integration

Course Description:

Systems integration is the logical process for assembling a set of system elements and aggregates into the realized system, product, or service that satisfies system requirements, architecture, and design. Systems integration focuses on the testing of interfaces, data flows, and control mechanisms, checking that realized elements and aggregates perform as predicted by their design and architectural solution, since it may not always be practicable or cost-effective to confirm these lower-level aspects at higher levels of system integration. Systems Integration should be planned so that system elements are brought together in a logical sequence to avoid wasted effort. Systematic and incremental integration makes it easier to find, isolate, diagnose, and correct problems. Interfaces occur where system elements interact, for example human, mechanical, electrical, thermal, data, etc. Interface management comprises the identification, definition, and control of interactions across system or system element boundaries. Poor interface definition and management can result in incompatible system elements (either internal to the system or between the system and its environment) which may ultimately result in system failure or project overrun. A system or system element that has not been integrated systematically cannot be relied on to meet its requirements.

The instructor then examines the Indicators of Competence (indicators) and Relevant knowledge, experience, and/or behaviors (sub-indicators) to review these competencies for inclusion in the course learning objectives. All of the items in these two categories are expressed using terms associated with Bloom's taxonomy, making the translation to learning objectives straightforward. The labels "(K)" and "(A)" provide useful information, with "(K)" items related primarily to having students being able to learn and express knowledge of the topic, and "(A)" items related to students having the ability to perform some sort of activity or application of the item.

The instructor reviews the items falling within the Awareness level of the Integration competency area and determines that all indicators identified for this competency area in the SECF are important for a student to understand. Similarly for the Interfaces competency, the instructor determines that all Awareness-level indicators are important.

Moving to the Supervised Practitioner proficiency level, the instructor recognizes that some of the indicators identified in both the Integration and Interfaces competency areas are potentially rather too specific to an organizational implementation, or otherwise seem impractical or inappropriate to cover in their course. As a result, they choose to "tailor out" certain indicators and will not consider them further.

All indicators not tailored out are then used along with the indicators from the Awareness level (e.g. Integration Supervised Practitioner Sub IDs: [TINS02-20A], [TINS02-30A], [TINS03-10A], [TINS03-20A], [TINS04-10A], [TINS04-20A], [TINS04-30A], and [TINS04-50A] and Interfaces Supervised Practitioner Sub IDs [TIFS02-20A], TIFS02-30A], [TIFS03-10K], and [TIFS03-20K]). SECAG Table B4-1 shows the summary of specific items to be included in the course.

Course and Assignment Detailed Construction

The instructor can then use their tailored list of applicable competency indicators when constructing the course in order to make sure all selected indicators are covered within the course.

Once the relevant indicators have been identified, the instructor starts to construct their course.

The familiar statement "Upon course completion the student shall be able to…" can be used, followed by the list of main learning objectives. The indicators provide a useful summary level of learning objectives to communicate expectations to the students, as including a list of learning objectives with all the sub-indicators would be an overwhelming list.

SECAG TABLE B4-1 Systems integration course development example

Competency	ID number	Proficiency level	Indicators of competence	Relevant sub-indicators
Integration	1	Awareness	Explains why integration is important and how it confirms the system design, architecture, and interfaces. [TINA01]	(K) Describes the process of building up the system from smaller elements into larger and larger aggregations, or perhaps in related collections of elements, verifying along the way to ensure system elements work as designed. [TINA01-10K] (K) Explains why, if you put the entire system together and find it does not work as intended, it is difficult to identify clearly what elements may not be performing properly. [TINA01-20K] (K) Explains the potential impact of a system being a "system of systems" on the integration process. [TINA01-30K]
Integration	2	Awareness	Explains why it is important to integrate the system in a logical sequence. [TINA02]	(K) Explains how integration is conducted using a progressive, logical process of assembling system elements, building an element, evaluating it, and then assembling several elements at the next level (system build). [TINA02-10K] (K) Describes alternative integration sequences (top down, bottom up, middle out, etc.) and how they may be assessed in order to define the most appropriate sequence in terms of overall cost and risk. [TINA02-20K] (K) Describes with examples how, if integration is performed in the wrong sequence, rework and extra cost may be incurred (dependency on suppliers, development, new technology, obsolescence, etc.). [TINA02-30K]
Integration	3	Awareness	Explains why planning and management of systems integration is necessary. [TINA03]	(K) Explains why planning for integration should occur at the beginning of the project/program. [TINA03-10K] (K) Explains how a failure to plan could result in a delay to integration; procedures may not be written; the sequences may not have been defined and the environment may not be available. [TINA03-20K] (K) Explains why the integration sequence should be documented. [TINA03-30K] (K) Explains how test requirements may influence the design. [TINA03-40K] (K) Describes key elements of an integration plan (e.g. identification of resources, equipment, and test requirements that influence the design). [TINA03-50K]

SECAG TABLE B4-1 (Continued)

Competency	ID number	Proficiency level	Indicators of competence	Relevant sub-indicators
Integration	4	Awareness	Explains the relationship between integration and verification. [TINA04]	(K) Explains how testing system element performance during integration can help establish the verification of some system requirements. [TINA04-10K] (K) Explains how documenting system element testing can provide evidence in support of verification and potentially even acceptance. [TINA04-20K]
Integration	2	Supervised Practitioner	Prepares inputs to integration plans based upon governing standards and processes including identification of method and timing for each activity to meet project requirements. [TINS02]	(A) Ensures integration plans include measures of successful integration, such as how to verify conformance. [TINS02-20A] (A) Ensures integration plans include all required enterprise or statutory/regulatory process requirements. [TINS02-30A]
Integration	3	Supervised Practitioner	Prepares plans which address integration for system elements (or noncomplex systems) in order to define or scope that activity. [TINS03]	(A) Prepares inputs to integration plan generation documenting the integration sequence, the environment, and approach. [TINS03-10A] (A) Prepares inputs to integration plan generation documenting measures of successful integration and how to verify conformance. [TINS03-20A]
Integration	4	Supervised Practitioner	Records the causes of simple faults typically found during integration activities in order to communicate with stakeholders. [TINS04]	(A) Identifies where results differ from those expected. [TINS04-10A] (A) Records faults appropriately (process, tools used, and method). [TINS04-20A] (A) Analyzes simple faults in a logical manner and initiates the corrective action process. [TINS04-30A] (A) Records corrective actions taken and closes the outstanding fault log. [TINS04-50A]
Interfaces	1	Awareness	Defines key concepts within interface definition and management. [TIFA01]	(K) Describes what an interface is. [TIFA01-10K] (K) Defines different strategies for the successful identification of interfaces. [TIFA01-20K] (K) Describes interface stakeholders. [TIFA01-30K] (K) Describes the importance of ensuring each interface is "owned" by someone and the effects of failing to have this in place. [TIFA01-40K] (K) Explains how interfaces are usually defined. [TIFA01-50K] (K) Describes different strategies for defining the content of interfaces and managing thereafter and when each strategy may be applicable. [TIFA01-60K]

(Continued)

SECAG TABLE B4-1 (Continued)

Competency	ID number	Proficiency level	Indicators of competence	Relevant sub-indicators
Interfaces	2	Awareness	Explains how interface definition and management affects the integrity of the system solution. [TIFA02]	(K) Explains why control of the management of interface development is necessary. [TIFA02-10K] (K) Describes the potential impact on the system of failure to define interfaces properly. [TIFA02-20K] (K) Describes the potential impact on the system of failure to manage interfaces properly. [TIFA02-30K] (K) Explains how and why internal and external interfaces may evolve and subsequently be managed differently. [TIFA02-40K] (K) Explains how different stakeholder types may affect the definition or management of interfaces. [TIFA02-50K] (K) Describes the importance of configuration management when managing interfaces. [TIFA02-60K]
Interfaces	3	Awareness	Identifies possible sources of complexity in interface definition and management. [TIFA03]	(K) Describes different types of interface (functional and physical) covering different domains. [TIFA03-10K] (K) Describes possible sources of interface complexity. [TIFA03-20K]
Interfaces	4	Awareness	Explains how different sources of complexity affect interface definition and management. [TIFA04]	(K) Explains how different interfaces (functional and physical) across different domains are managed differently. [TIFA04-10K] (K) Explains how different sources of interface complexity affect the management of systems interfaces. [TIFA04-20K]
Interfaces	2	Supervised Practitioner	Identifies the properties of simple interfaces in order to define them. [TIFS02]	(A) Identifies simple interfaces for a system. [TIFS02-20A] (A) Defines different parameters of a simple interface (both functional and physical). [TIFS02-30A]
Interfaces	3	Supervised Practitioner	Explains the potential consequences of changes on system interfaces to coordinate and control ongoing development. [TIFS03]	(K) Explains how a change at one end of the interface can impact the other end of the interface. [TIFS03-10K] (K) Explains how system performance may be affected by a change to an interface. [TIFS03-20K]

Upon completion of this course, the student shall be able to:

- Explain why integration is important and how it confirms the system design, architecture, and interfaces.
- Explain why it is important to integrate the system in a logical sequence.
- Explain why planning and management of systems integration is necessary.
- Explain the relationship between integration and verification.
- Prepare plans which address integration for system elements (or noncomplex systems) in order to define or scope that activity.
- Record the causes of simple faults typically found during integration activities in order to communicate with stakeholders.
- Define key concepts within interface definition and management.
- Explain how interface definition and management affects the integrity of the system solution.
- Identify the properties of simple interfaces in order to define them.
- Explain the potential consequences of changes on system interfaces to coordinate and control ongoing development.

The instructor then uses the sub-indicators as a guide for creating course lecture materials as well as knowledge checks and learning assessments.

Knowledge indicators (marked with a "K") can easily become learning objectives assessed with quizzes and/or exams. Note that for sub-indicators marked as "Abilities" (i.e. marked with an "A"), objective evidence provided in the SECAG document can help define course assignments as well as being additional course learning objectives. Learning objectives derived from indicators marked with an "A" can also be used to define learning elements that have a basis in active learning methods, team-based project activities, lab activities, or hands-on learning activities.

For example, "Integration Plan," "Interface Diagram," and "Interface Description Document (IDD)" are all listed in the SECAG as evidence examples demonstrating that an ability exists within the Interfaces competency area. The instructor can use these to construct course assignments, such as creation of an "Integration Plan," "Interface Diagram," and "Interface Design Document (IDD)" for a notional system as elements of their course. SECAG Table B4-2 shows the summary of specific items related to assignment activities to be included in the course.

Course Outcome Verification

With the course design completed, the SECAG document can be used again to check that students attending have gained the correct knowledge or experience from the course.

The tailored set of indicators and sub-indicators which formed the basis of the course design can be used to confirm that the set of indicators defined for the course have been both learned and understood to the required level. For example, Assessment evidence could be verified by a short knowledge test at the end of the course as well as verification that assignments have been completed to a suitable level, in order to check for learning.

SECAG TABLE B4-2 Systems integration course development example

Competency	ID number	Relevant sub-indicators	Possible examples of objective evidence
Integration	2	(A) Ensures integration plans include measures of successful integration, such as how to verify conformance. [TINS02-20A] (A) Ensures integration plans include all required enterprise or statutory/regulatory process requirements. [TINS02-30A]	Integration plan. [TINS02-E10]
Integration	3	(A) Prepares inputs to integration plan generation documenting the integration sequence, the environment, and approach. [TINS03-10A] (A) Prepares inputs to integration plan generation documenting measures of successful integration and how to verify conformance. [TINS03-20A]	Detailed integration plan including specifics on environment, required equipment and tools, procedures, verification measures, etc. [TINS03-E10]
Integration	4	(A) Identifies where results differ from those expected. [TINS04-10A] (A) Records faults appropriately (process, tools used, and method). [TINS04-20A] (A) Analyzes simple faults in a logical manner and initiates the corrective action process. [TINS04-30A] (A) Records corrective actions taken and closes the outstanding fault log. [TINS04-50A]	Documentation of faults, e.g. Trouble Report, Nonconformance Report, and Fault Log. [TINS04-E10] Documentation of fault resolution, e.g. Corrective Action, Engineering Investigation. [TINS04-E20]
Interfaces	2	(A) Identifies simple interfaces for a system. [TIFS02-20A] (A) Defines different parameters of a simple interface (both functional and physical). [TIFS02-30A]	Interface diagram. [TIFS02-E10] Interface definition/description document (IDD). [TIFS02-E20]

ANNEX B5 – USING THE SECF/SECAG FOR "ROUND TRIP" COMPETENCY ASSESSMENT

Introduction

The following case study is based upon an exercise carried out during a five-year program of competency assessment and development for a targeted group of systems engineers within a business unit forming part of an organization established for several years as a "systems integrator" but new to full life cycle systems engineering.

NOTE: The data set should be taken as "representative" rather than actual. The original data set has required modification to protect the anonymity of the organization and the individuals involved and further adjusted to align with the latest version of the SECF/SECAG, since the original assessment was made against an earlier set of similar systems engineering competencies. The activities performed, the conclusions drawn, and the ultimate outcomes discussed are typical of the use of this approach generally.

The case study context was that the organization, based in the Middle East and part of a much larger multinational company, had historically been used by their parent company for local installation and commissioning work within large transportation systems contracts. With the region expected to continue to experience substantial business growth, the parent company had started placing an increased percentage of work locally, hoping to both develop and utilize local systems engineering capability and to respond actively to government expectations of increased local technical content and capability development.

The "new" work placed within the business area had changed from being primarily installation and commissioning to front-end requirements elicitation and systems architecting. In parallel, the parent company had been rolling out a new set of own corporate business processes globally to ensure high-quality development and the new business became exposed to these new ways of working for the first time.

The local team, an expat management team on assignment from the parent company, had been unable to prevent the local business running into difficulties, particularly in their new higher-value prestige full life cycle projects (e.g. overspends, delays, and requirements issues were common) and urgent remedial work was deemed necessary to improve the outlook moving forward.

As part of this, several expat technical leaders were put in place to support the ongoing projects. However, this could only be a short-term solution. One strategic initiative proposed was to engage a small team of systems engineering competency assessment specialists from the parent company to assess a cross-section of the existing systems engineering workforce in order to determine a local strategy to improve project performance.

This case study follows the activities performed and shows the results of the assessment and how performance can be improved through the use of the SECF/SECAG and a strategic "round-trip" competency development approach.

Note that the assessment was focused on systems engineering technical and managerial competencies and was initially carried out on a representative population of 25 engineers. Over the course of the five-year exercise, a small number of these individuals left the business and others were recruited but the case study still reflects a general capability development over that period using this approach.

The parent company HR department was delighted with the outcome of the activity regarding the exercise as vindication of the stable well-funded strategic competency development strategy employed by the local business and a model for use in similar situations elsewhere in the business.

Summary of Work Performed

Right from its inception, the competency development project was conceived as a long-term intervention.

Overall, the following activities were performed in the five-year period:

- SE Role Identification. Work with local stakeholders to identify key roles within the Systems Engineering Department.
- SECF/SECAG Tailoring. Work with local stakeholders to tailor the SECF/SECAG to identify relevant or critical competencies for the organization.
- Work with local key stakeholders to map expected competency levels to identified SE roles.
- Communicate purpose of assessment to local workforce.
- Determine the group of representative individuals to be assessed.
- Perform individual assessments on this group, agreeing outcome both between assessors and with the individuals involved.
- Record assessments in database.
- Analyze data for both organizational and individual patterns.
- Create development pathways for the systems engineering department – both individually for the 25 assessed and organizationally.
- Implement the developments required.
- Reassess the same individual agreeing outcome both between assessors and with the individuals involved.
- Record updated assessments in database.
- Determine value of interventions and any future work required.

A summary of the results of each of these activities is provided in the following sections.

SE Role Identification

The team worked collaboratively with the Director of Engineering to identify a range of roles thought to be both typical of current roles used and reflective of expected role assignments moving forward.

Thus, some of the roles (e.g. IVV Engineer) matched or were close to existing role titles but others (e.g. Systems Architect) were currently not used.

The list of roles identified can be seen as the column headers in SECAG Figure B5-1.

Tailor the Framework to Identify Relevant Competencies for the Organization

This was done through a dialogue with the Engineering Director. A summary of current projects and future prospects was presented by the local team and used to identify the most likely key competencies required by the business now and looking forward.

Not surprisingly, current and historic projects were mainly focused on "right-hand-side of "V" model" whereas moving forward (and new projects) required full life cycle skills. Further detail of this is provided as part of the outcome analysis below.

Although the assessment team suggested that best practice would be to limit the assessment scope to the most critical 6–10 competencies, the local stakeholders were clear that the assessment should evaluate ALL framework competencies in order to characterize the "need" for the organization and were willing to accept the additional cost implications of this requirement.

Map Tailored Competencies to Identified SE Roles

Having identified the competencies required for the business, these were mapped to the SE roles.

Initially, a discussion was held with the technical executive to address:

a. Current role titles used by the business.
b. Future expected role titles (e.g. as the business grew).

These role titles were then assigned idealized required SE proficiency levels.

The selected roles and required proficiency levels are shown in SECAG Figure B5-1.

In Figure B5-1, the letters are as follows:

- N = Not required for this role.
- A = Awareness.
- S = Supervised Practitioner.
- P = Practitioner.
- L = Lead Practitioner.

The color coding is for clarity.

Note that while in theory an "Expert" level ("E") of proficiency exists in the SECF/SECAG, this was deemed to be not required for any of the identified roles. Local stakeholder expectation was that any "expertise" in any particular competency area could be drawn directly from the parent organization for the foreseeable future.

(NOTE: An additional piece of work added "domain-specific" competencies to each identified role, but this activity is not discussed in this case study.)

Local job ID	System design authority	System architect	Senior SE manager	SE manager	WP Mgr	Senior systems engineer	Systems engineer	IVV manager	System IVV engineer	SE process & tools administrator	SE CM manager	Requirements manager	Technical specialist
Role ID	101	102	103	104	105	106	107	108	109	110	111	112	113
Systems and critical thinking	L	P	P	P	A	L	P	P	S	S	P	A	P
Life Cycles	P	P	P	P	A	P	S	P	P	P	P	S	S
Modelling and MBSE	P	L	S	S	A	P	S	S	S	S	A	A	S
Requirements definition	L	P	P	P	A	P	P	P	P	A	A	P	P
Systems architecting	P	L	S	S	N	S	A	A	A	A	A	A	P
Design for...	P	P	P	P	N	P	S	S	P	A	A	A	P
Integration	P	P	P	P	N	P	A	L	P	A	A	A	P
Interfaces	P	P	P	P	N	P	S	P	P	A	A	A	P
Verification and validation	L	S	P	P	A	P	S	L	P	A	A	P	P
Qualification, certification and acceptance	L	S	P	P	A	P	S	L	P	A	P	P	P
Transition	P	S	A	A	N	P	A	P	P	N	A	A	A
Utilization and support	P	P	L	P	N	P	S	S	A	N	A	A	A
Retirement	P	S	P	P	A	P	S	S	A	N	N	A	S
SE planning, monitoring and control	L	S	L	P	S	P	S	P	S	N	P	S	P
Decision management	L	P	P	P	P	P	S	P	P	N	P	A	P
Concurrent engineering	P	S	L	P	P	P	S	P	P	S	S	A	A
Business and enterprise integration	P	S	S	A	A	S	A	A	A	P	S	A	A
Acquisition and supply	L	P	P	S	S	P	S	P	S	S	A	A	S
Information and configuration management	P	P	P	P	S	P	S	P	P	A	S	P	A
Risk and opportunity management	L	P	P	P	P	P	S	P	A	A	A	A	P
Project management	P	A	P	S	P	A	A	P	A	S	S	A	A
Finance	P	A	P	S	P	A	A	P	A	S	S	A	A
Logistics	P	A	P	S	S	A	A	P	S	S	S	A	A
Quality	P	A	P	S	S	A	A	P	S	S	S	A	A

SECAG FIGURE B5-1 Competency "Requirement" vs local role title.

Communicate Purpose of Assessment to Local Workforce

As the presence of the team became visible within the wider local engineering community, rumors had grown (mostly negative) about the purpose of the assessment (e.g. competency assessment = restructuring = redundancies).

The purpose of the exercise had never been one of selecting for restructuring, and to combat this negativity, the assessment team worked with local HR leaders to ensure accurate messaging to the work force. The key messaging included:

- The purpose was to baseline capability in a growing business unit. Restructuring or "redundancies" were not part of this activity.
- Truthful responses to questions were essential – since they minimized the amount of access the assessment team would require to validate evidence claims.
- The outcome would be used to draft individual learning and development plans for the coming years.
- The assessment would follow "Chatham House rules" – in other words, anything said confidentially in the assessment would not be revealed to others outside the assessment process and would not be recorded.
- Although an individual may require development, this cannot be guaranteed but will be prioritized according to business need.
- Individuals were required to prioritize their own attendance at the scheduled time. Project Managers would be instructed to release individuals for the scheduled slot. If this was not possible, the slot would be rescheduled within the two-week period. NOTE: This was important. Individuals sometimes required overnight travel and accommodation as not all individuals were located locally at the time of the assessment. Travel costs were met by HR.

Determine Individuals to Be Assessed

This work was handled locally by the Engineering department. With over 100 people employed at the time, to limit duration and cost, the assessment pool was limited to an initial 25 individuals. They were selected to be representative of the different experience profiles of those currently employed and the roles defined for the assessment. In the end, 24 of the 25 individuals were available for assessment.

It was of course inevitable that some individuals would be selected because of their perceived skills, or failings, but the assessment team requested that they be not made aware of the local view on individual capabilities until all assessments had been completed and formally processed. The local team fully supported this approach and thus the assessments were completed completely independently.

Perform Individual Assessments

The assessments were performed over a period of two weeks by the independent team (generally two but occasionally three assessments each day).

A number of key points should be noted regarding the wider assessment protocol. These are listed below:

- HR managed the scheduling of the assessment cohort with their local projects to ensure impact was limited. This meant that all assessments took place as planned, except one which had to be completed and added at the end of the activity.
- At the start of the assessment, the purpose of the assessment was made clear to the assessee as described above and in particular:
  - This was not to determine their employment status moving forward; more to ensure that individuals received appropriate and necessary training and that their career path aspirations could be assessed and recorded.

- o Truth and openness when questioned was essential in their responses. However, we would respect confidentiality in our conversations (see below).
- o Assessors would go through assessment results with each individuals before taking them further to confirm that the assessee agreed and believed that the proficiency level recorded was fair.
- o The assessment took the form of a one- to two-hour interview with a lead interviewer asking initial questions and the secondary interviewer recording responses and asking follow-up questions as necessary.
- o All competency areas listed for all identified SE roles were assessed. This was done to provide a full baseline for each individual moving forward.
- o At the end of their assessment, the assessee was asked about their own career aspirations and perceived development needs and this was recorded for HR use. This proved very popular with the assesses.
- o Each assessee received a copy of the assessment result for their personal records.

The first two bullets above are important. There had been some degree of hesitation about the purpose of the assessment activity within the workforce. It was known certain projects were not going well and the conclusion was inevitably that the assessment work was a precursor to some form of "culling" of individuals to ensure improvement. We presented a different view: that as a result of new roles and future opportunities, a better understanding of each individual's capabilities and development needs (and wishes) would provide the best way of developing the local business.

NOTE: At the end of the process, the assessment team was briefed by local HR who indicated that general workforce feedback from the exercise was extremely positive: Assessees indicated that the process undertaken had been painless, fair, in depth, and for many, very rewarding personally. This was a great outcome.

Record Assessments in Database

Representative "raw" data from the first cohort assessed is presented in SECAG Figure B5-2.

In Figure B5-2, the letters used have the following meaning:

- A = Assessed as operating at Awareness
- S = Assessed as operating at Supervised Practitioner level
- P = Assessed as operating at Practitioner level
- L = Assessed as operating at Lead Practitioner level
- U = Did not have an awareness of the competency area (Unaware)

The letter "E" was available for assessment proficiency at "Expert" level, but no individual achieved this assessed status.

NOTE: As previously discussed, for the purposes of presenting this case study, assessment data from the original activity has been reanalyzed and assigned to the current SECF/SECAG competency suite in order to maintain anonymity and to aid presentation of this study.

Analyze Data for Both Organizational and Individual Patterns

The "raw" data in SECAG Figure B5-2 was analyzed to produce several reports. Two significant report types are presented here.

- "Deficit Heat Map"(see SECAG Figure B5-3).
- "Individual "Best fit" Role (see SECAG Figure B5-4).

	Assessee name	Person 1	Person 2	Person 3	Person 4	Person 5	Person 6	Person 7	Person 8	Person 9	Person 10	Person 11	Person 12	Person 13	Person 14	Person 15	Person 16	Person 17	Person 18	Person 19	Person 20	Person 21	Person 22	Person 23	Person 24
	Current org role	103	106	113	106	106	106	113	107	106	108	109	109	109	106	109	113	113	108	108	113	103	106	102	106
Subgroup	Site location	S	S	S	S	S	S	S	S	S	S	S	S	S	S	S	S	S	S	S	S	S	S	S	S
Core	Systems and critical thinking	S	S	A	P	S	S	P	A	S	S	P	A	A	A	A	A	S	S	P	P	L	S	P	P
Core	Life Cycles	S	S	A	S	A	S	P	U	S	S	S	A	U	S	A	A	S	S	P	P	L	S	P	P
Core	Modelling and MBSE	A	A	A	S	S	A	A	U	A	A	A	U	U	U	A	A	A	A	S	S	P	S	S	S
Technical	Requirements definition	S	S	A	S	S	S	P	S	P	S	S	A	A	S	S	S	S	P	L	S	L	S	P	P
Technical	Systems architecting	A	U	U	A	U	U	S	U	U	A	A	U	U	A	U	A	A	U	S	U	S	A	S	S
Technical	Design for…	A	A	U	S	A	A	S	A	A	S	A	U	U	A	A	A	A	U	U	S	P	S	S	S
Technical	Integration	S	S	S	S	P	S	P	S	P	S	S	S	P	S	S	S	S	P	L	S	L	S	P	P
Technical	Interfaces	S	S	S	S	P	S	P	S	P	S	S	S	S	S	S	S	S	P	L	S	L	S	P	P
Technical	Verification and validation	P	S	S	S	P	S	P	S	P	S	S	S	S	S	S	S	S	P	L	S	L	S	P	P
Technical	Qualification, certification and acceptance	P	S	S	S	P	S	P	S	P	S	S	S	P	S	S	S	S	P	L	S	L	S	P	P
Technical	Transition	A	A	A	S	P	A	S	A	S	S	S	A	A	S	S	S	S	S	S	S	S	A	P	A
Technical	Utilization and support	P	S	S	S	S	A	P	S	A	P	A	A	P	A	S	S	S	S	S	S	P	A	P	S
Technical	Retirement	A	A	A	S	P	A	A	A	U	S	U	A	U	S	A	S	U	S	A	U	S	A	S	S
SE management	SE planning, monitoring and control	S	A	S	P	P	S	P	U	S	P	P	A	S	A	S	S	S	S	L	S	L	S	P	P
SE management	Decision management	S	S	S	P	A	S	P	U	S	P	P	S	S	A	S	S	S	S	L	P	L	S	P	S
SE management	Concurrent engineering	P	S	S	P	S	A	S	A	S	S	S	S	S	S	P	A	S	P	L	P	P	P	P	P
SE management	Business and enterprise integration	U	U	U	S	U	A	U	U	S	A	A	U	U	U	A	A	S	A	P	S	L	S	P	S
SE management	Acquisition and supply	P	S	S	P	A	A	P	A	S	P	P	S	S	S	P	A	S	P	L	P	S	P	P	P
SE management	Information and configuration management	S	S	A	S	S	S	A	A	S	S	S	S	A	S	A	A	A	S	S	A	S	A	S	S
SE management	Risk and opportunity management	S	S	A	S	S	A	S	A	S	S	A	U	A	U	U	U	U	P	S	A	P	A	S	P
Integrating	Project management	A	S	U	S	A	U	A	A	A	A	A	U	U	A	U	U	U	S	S	U	S	U	A	A
Integrating	Finance	A	A	U	S	U	A	A	A	U	U	A	A	A	U	U	A	A	A	S	U	S	A	A	S
Integrating	Logistics	S	S	S	S	U	A	S	A	S	U	S	U	S	U	S	P	S	S	P	P	P	S	P	P
Integrating	Quality	S	P	A	P	S	U	S	A	A	S	S	U	A	A	A	U	U	S	P	U	P	U	S	A

SECAG FIGURE B5-2 First cohort "Raw" competency assessment data.

Candidate name	Person 1	Person 2	Person 3	Person 4	Person 5	Person 6	Person 7	Person 8	Person 9	Person 10	Person 11	Person 12	Person 13	Person 14	Person 15	Person 16	Person 17	Person 18	Person 19	Person 20	Person 21	Person 22	Person 23	Person 24
Assigned org role ID	103	106	113	106	106	106	113	107	106	108	109	109	109	106	109	113	113	108	108	113	103	106	102	106
Competence area																								
Systems and critical thinking	−1	−2	−2	−1	−2	−2	0	−2	−2	−1	1	−1	−1	−3	−1	−2	−1	−1	0	0	1	−2	0	−1
Life Cycles	−1	−1	−1	−1	−2	−1	1	−2	−1	−1	−1	−2	−3	−1	−2	−1	0	−1	0	1	1	−1	0	0
Modelling and MBSE	−1	−2	−1	−1	−1	−2	−1	−2	−2	−1	−1	−2	−2	−3	−1	−1	−1	−1	0	0	1	−1	−2	−1
Requirements definition	−1	−1	−2	−1	−1	−1	0	−1	0	−1	−1	−2	−2	−1	−1	−1	−1	0	1	−1	1	−1	0	0
Systems architecting	−1	−2	−3	−1	−2	−2	−1	−1	−2	0	0	−1	−1	−1	−1	−2	−2	−1	1	−3	0	−1	−2	0
Design for…	−2	−2	−3	−1	−2	−2	−1	−1	−2	0	−2	−3	−3	−2	−2	−2	−2	−2	−2	−1	0	−1	−1	−1
Integration	−1	−1	−1	−1	0	−1	0	1	0	−2	−1	−1	0	−1	−1	−1	−1	−1	0	−1	1	−1	0	0
Interfaces	−1	−1	−1	−1	0	−1	0	0	0	−1	−1	−1	−1	−1	−1	−1	−1	0	1	−1	1	−1	0	0
Verification and validation	0	−1	−1	−1	0	−1	0	0	0	−2	−1	−1	−1	−1	−1	−1	−1	−1	0	−1	1	−1	1	0
Qualification, certification and acceptance	0	−1	−1	−1	0	−1	0	0	0	−2	−1	−1	0	−1	−1	−1	−1	−1	0	−1	1	−1	1	0
Transition	0	−2	0	−1	0	−2	1	0	−1	−1	−1	−2	−2	−1	−1	1	1	−1	−1	1	1	−2	1	−2
Utilization and support	−1	−1	1	−1	−1	−2	2	0	−2	1	0	0	2	−2	1	1	1	0	0	1	−1	−2	0	−1
Retirement	−2	−2	−1	−1	0	−2	−1	−1	−3	0	−1	0	−1	−1	0	0	−2	0	−1	−2	−1	−2	0	−1
SE planning, monitoring and control	−2	−2	−1	0	0	−1	0	−2	−1	0	1	−1	0	−2	0	−1	−1	−1	1	−1	0	−1	1	0
Decision management	−1	−1	−1	0	−2	−1	0	−2	−1	0	0	−1	−1	−2	−1	−1	−1	−1	1	0	1	−1	0	−1
Concurrent engineering	−1	−1	1	0	−1	−2	1	−1	−1	−1	−1	−1	−1	−1	0	0	1	0	1	2	−1	0	1	0
Business and enterprise integration	−2	−2	−1	0	−2	−1	−1	−1	0	0	0	−1	−1	−2	0	0	1	0	2	1	2	0	1	0
Acquisition and supply	0	−1	0	0	−2	−2	1	−1	−1	0	1	0	0	−1	1	−1	0	0	1	1	−1	0	0	0
Information and configuration management	−1	−1	0	−1	−1	−1	0	−1	−1	−1	−1	−1	−2	−1	−2	0	0	−1	−1	0	−1	−2	−1	−1
Risk and opportunity management	−1	−1	−2	−1	−1	−2	−1	−1	−1	−1	0	−1	0	−3	−1	−3	−3	0	−1	−2	0	−2	−1	0
Project management	−2	1	−1	1	0	−1	0	0	0	−2	0	−1	−1	0	−1	−1	−1	−1	−1	−1	−1	−1	0	0
Finance	−2	0	−1	1	−1	0	0	0	−1	−3	0	0	0	−1	−1	0	0	−2	−1	−1	−1	0	0	1
Logistics	−1	1	1	1	−1	0	1	0	1	−3	0	−2	0	−1	0	2	1	−1	0	2	0	1	2	2
Quality	−1	2	0	2	1	−1	1	0	0	−1	0	−2	−1	0	−1	−1	−1	−1	0	−1	0	−1	1	0

SECAG FIGURE B5-3 First cohort "Deficit" Heat Map.

Individual role deficit map
confidential when completed

Name	Person 2

Type	SubGroup	Competency area	System design authority	System architect	Senior SE manager	SE manager	WP Mgr	Senior systems engineer	Systems engineer	IVV manager	System IVV engineer	SE process & tools administrator	SE CM Manager	Requirements manager	Technical specialist
Competence area	Core	Systems and critical thinking	−2	−1	−1	−1	1	−2	−1	−1	0	0	−1	1	−1
Competence area	Core	Life Cycles	−1	−1	−1	−1	1	−1	0	−1	−1	−1	−1	0	0
Competence area	Core	Modelling and MBSE	−2	−3	−1	−1	0	−2	−1	−1	−1	−1	0	0	−1
Competence area	Technical	Requirements definition	−2	−1	−1	−1	1	−1	−1	−1	−1	1	1	−1	−1
Competence area	Technical	Systems architecting	−3	−4	−2	−2	0	−2	−1	−1	−1	−1	−1	−1	−3
Competence area	Technical	Design for…	−2	−2	−2	−2	1	−2	−1	−1	−2	0	0	0	−2
Competence area	Technical	Integration	−1	−1	−1	−1	2	−1	1	−2	−1	1	1	1	−1
Competence area	Technical	Interfaces	−1	−1	−1	−1	2	−1	0	−1	−1	1	1	1	−1
Competence area	Technical	Verification and validation	−2	0	−1	−1	1	−1	0	−2	−1	1	1	−1	−1
Competence area	Technical	Qualification, certification and acceptance	−2	0	−1	−1	1	−1	0	−2	−1	1	−1	−1	−1
Competence area	Technical	Transition	−2	−1	0	0	1	−2	0	−2	−2	1	0	0	0
Competence area	Technical	Utilization and support	−1	−1	−2	−1	2	−1	0	0	1	2	1	1	1
Competence area	Technical	Retirement	−2	−1	−2	−2	0	−2	−1	−1	0	1	1	0	−1
Competence area	SE management	SE Planning, Monitoring and Control	−3	−1	−3	−2	−1	−2	−1	−2	−1	1	−2	−1	−2
Competence area	SE management	Decision management	−2	−1	−1	−1	−1	−1	0	−1	−1	2	−1	1	−1
Competence area	SE management	Concurrent engineering	−1	0	−2	−1	−1	−1	0	−1	−1	0	0	1	1
Competence area	SE management	Business and enterprise integration	−3	−2	−2	−1	−1	−2	−1	−1	−1	−3	−2	−1	−1
Competence area	SE management	Acquisition and supply	−2	−1	−1	0	0	−1	0	−1	0	0	1	1	0
Competence area	SE management	Information and configuration management	−1	−1	−1	−1	0	−1	0	−1	−1	1	0	−1	1
Competence area	SE management	Risk and opportunity management	−2	−1	−1	−1	−1	−1	0	−1	1	1	1	1	−1
Competence area	Integrating	Project management	−1	1	−1	0	−1	1	1	−1	1	0	0	1	1
Competence area	Integrating	Finance	−2	0	−2	−1	−2	0	0	−2	0	−1	−1	0	0
Competence area	Integrating	Logistics	2	2	2	2	2	2	2	2	2	2	2	2	2
Competence area	Integrating	Quality	3	3	3	3	3	3	3	3	3	3	3	3	3

SECAG FIGURE B5-4 Individual Best fit example.

The "Deficit Heat Map" presents a color-coded analysis of competencies and their assessed proficiency levels against the "expected" proficiency levels for each individual for their current role.

The "Individual Best fit Role" heat map presents a color-coded analysis for a single individual competencies against the "expected" proficiency levels for all defined roles. The individual's current role is highlighted.

On both heat maps, the cell color coding is as follows:

- Green – *meets*, or *exceeds*, the required proficiency level for the role in question. No development required.
- Amber – one level *below* the required proficiency level, so some development required (e.g. amber cell coloring and -1 in cell means assessed as "Unaware" when "Aware" required, "Aware" when "Supervised Practitioner" required, or "Supervised Practitioner" when "Practitioner" required, etc.).
- Red – two or more levels *below* the required proficiency level, so extensive development required (e.g. red cell coloring and -2 in cell means assessed as "Aware" when "Practitioner," and -3 means assessed as "Unaware" when "Practitioner" required, etc.).

Create Organizational and Individual Development Pathways

The analysis above pointed to a number of areas of potential development within the department:

- Although several areas needed to be addressed, looking forward, organizationally the Systems Engineering department would benefit most from interventions in the following areas:
  - Requirements Engineering training and mentoring.
  - Systems Architecting training and mentoring.
  - Basic Project Management and Work Package Management training.
  - Configuration Management training and mentoring.
- Several individuals would need significant individual development. Options included:
  - Development training or mentoring within their current role.
  - Move to a similar role where the competency deficit to overcome was smaller.
  - Discuss possible alternative roles or career paths with individuals.

The above development options for each individual were discussed initially with the local stakeholders who "revealed" their own thoughts on the independent outcomes produced. It was interesting to realize that although the "stars" and "poor performers" within the assessed group generally aligned, one or two individuals had fared better in the assessment than expected while several others had fared substantially worse than expected.

In the end, it was good to see that the local stakeholder team took the advice of the assessment team for development activities rather than reverting to their own positions. This is an important point: the *independence* of the assessment activity from the local team meant that the resultant data set would stand up to independent scrutiny and could be reliably used by the HR team in future discussions with the individuals involved and in their strategic planning.

Implement the Developments Required

The local team decided to fully implement all recommendations. These activities would run alongside additional Project Management training and active recruiting (including long-term overseas assignments) to develop Systems Engineering leadership but also to support Systems Engineering mentoring for individuals on key projects.

It took a further four years to complete the recommended activities.

Reassess the Same Group to Determine Organizational and Individual Improvements Achieved

The assessment team eventually returned five years later and assessed individuals again. Over the course of the intervening period:

- Some individuals had chosen to leave the business.
- Some individuals had changed roles (per recommendations).
- New individuals had been recruited to increase the local talent pool.
- One member of the second assessment team had not participated in the original assessment work but had been trained to the same standard as the original team member.

A similarly representative set of the local engineering department employees (including a few new people to replace leavers but with the majority the same as for the initial assessment) were assessed again.

Clearly for "new" individuals, this was in effect a "baseline" assessment. For those being reassessed, any changes to their own individual competence were of primary interest.

The actual assessment of the second cohort (24 individuals) was performed in the same manner agreeing outcome both between assessors and with the individuals involved. The results for the second cohort are shown in SECAG Figure B5-5.

Record Updated Assessments in Database

The database was updated with the latest data for the second set of individuals following a similar process for the first set.

The updated data from the second Cohort was analyzed to produce heat map reports as before. The "Cohort 2 Deficit Heat Map" from the second group of 25 individuals is presented in SECAG Figure B5-6.

In Figure SECAG B5-6, several points should be noted that while the majority of individuals were still working for the business:

- Two individuals from the first cohort had chosen to leave the business (Person 6 and Person 9) and had been replaced with recruits in "similar" roles (New Person 1 and New Person 3, respectively).
- Two "new" individuals were assessed as part of the "second" cohort. These had been employed to strengthen the architecting team (New Person 2, New Person 4).
- Two individuals from the first cohort had changed roles (per recommendations) to a lower level of responsibility (Person 2 and Person 13).

Determine Value of Interventions and Future Work Required

The data from the second cohort was analyzed and compared explicitly for those individuals who had received a second assessment, but also across the organization for the complete set of individuals.

The local management team was briefed on the second activity as they had been for the first assessments.

For some individuals, further development work was still required but the competence-based approach had become incorporated in the annual assessment regime so requirements and incremental individual experience evidence could be recorded and tracked for all employees over time.

It was recognized that the organizational development training and individual interventions alongside an active recruitment program had improved the systems engineering capability locally and further organizational training was scheduled to reflect the Cohort 2 results.

Assessee name	Person 1	Person 2	Person 3	Person 4	Person 5	New person 1	Person 7	Person 8	New person 3	Person 10	Person 11	New person 4	Person 13	New person 2	Person 15	Person 16	Person 17	Person 18	Person 19	Person 20	Person 21	Person 22	Person 23	Person 24
Current org role	103	107	113	106	106	106	113	107	102	108	109	102	111	106	109	113	113	108	108	113	103	106	102	106
Site Location	S	S	S	S	S	S	S	S	S	S	S	S	S	S	S	S	S	S	S	S	S	S	S	S
Systems and critical thinking	S	P	S	P	P	P	P	S	P	S	P	P	S	P	A	A	S	S	P	P	L	P	P	P
Life Cycles	S	S	A	S	A	S	P	A	S	S	S	P	S	S	A	A	S	S	P	P	L	S	P	P
Modelling and MBSE	A	S	A	S	S	P	A	A	P	A	A	L	A	P	A	A	A	A	S	S	P	S	L	S
Requirements definition	S	S	A	S	S	P	P	S	P	S	S	P	A	P	S	S	S	P	L	S	L	P	P	P
Systems architecting	A	A	A	A	A	S	S	A	P	A	A	L	A	S	A	A	A	A	S	A	S	S	L	P
Design for…	P	S	S	S	S	P	S	A	S	S	A	P	S	P	S	A	S	S	P	S	P	S	P	P
Integration	S	S	S	S	P	S	P	S	P	P	S	P	P	P	S	S	S	P	L	S	L	P	P	P
Interfaces	S	S	S	S	P	P	P	S	P	S	S	P	S	P	S	S	S	P	L	S	L	P	P	P
Verification and validation	P	P	S	S	P	P	P	S	P	S	S	P	S	P	S	S	S	P	L	S	L	P	P	P
Qualification, certification and acceptance	P	P	S	S	P	P	P	S	S	S	S	S	P	P	S	S	S	P	L	S	L	P	P	P
Transition	A	A	A	S	P	A	S	A	S	P	S	S	A	P	S	S	S	S	S	S	S	S	P	A
Utilization and support	P	P	S	S	S	A	P	S	A	P	A	S	P	S	S	S	S	S	S	S	P	S	P	S
Retirement	A	A	A	S	P	A	A	A	A	S	U	S	U	S	A	S	U	S	A	U	S	S	S	S
SE planning, monitoring and control	S	P	S	P	P	P	P	S	P	P	P	S	S	P	S	S	S	S	L	S	L	S	P	P
Decision management	S	S	S	P	A	P	P	U	S	P	P	P	S	P	S	S	S	S	L	P	L	S	P	S
Concurrent engineering	P	S	S	P	S	S	S	A	S	S	S	S	S	P	P	A	S	P	L	P	P	P	P	P
Business and enterprise integration	A	A	A	S	A	S	A	U	A	A	A	L	A	A	A	A	S	A	P	S	L	S	P	S
Acquisition and supply	P	S	S	P	A	P	P	A	S	P	P	P	S	S	P	A	S	P	L	P	S	P	P	P
Information and configuration management	S	S	A	S	S	S	A	A	P	S	S	S	A	S	A	A	A	S	S	A	S	A	S	S
Risk and opportunity management	S	S	A	S	S	P	S	A	P	S	A	P	A	P	S	S	S	P	S	A	P	A	S	P
Project management	A	S	P	S	A	S	A	A	S	A	A	S	A	P	S	S	S	S	S	A	S	A	A	A
Finance	A	A	A	S	A	A	A	A	A	A	A	A	A	S	A	A	A	A	S	A	S	A	A	S
Logistics	S	S	S	S	A	A	S	A	A	A	S	S	S	A	S	P	S	S	P	P	P	S	P	P
Quality	S	P	P	P	S	A	S	A	A	S	S	S	A	P	A	A	A	S	P	A	P	A	S	A

SECAG FIGURE B5-5 Second cohort "Raw" competency assessment data.

Competence area	Person 1	Person 2	Person 3	Person 4	Person 5	New person 1	Person 7	Person 8	New person 3	Person 10	Person 11	New person 4	Person 13	New person 2	Person 15	Person 16	Person 17	Person 18	Person 19	Person 20	Person 21	Person 22	Person 23	Person 24
Assigned org role ID	103	107	113	106	106	106	113	107	102	108	109	102	111	106	109	113	113	108	108	113	103	106	102	106
Systems and critical thinking	–1	0	–1	–1	–1	–1	0	–1	0	–1	1	0	–1	–1	–1	–2	–1	–1	0	0	1	–1	0	–1
Life Cycles	–1	0	–1	–1	–2	–1	1	–1	–1	–1	–1	0	–1	–1	–2	–1	0	–1	0	1	1	–1	0	0
Modelling and MBSE	–1	0	–1	–1	–1	0	–1	–1	–1	–1	–1	0	0	0	–1	–1	–1	–1	0	0	1	–1	0	–1
Requirements definition	–1	–1	–2	–1	–1	0	0	–1	0	–1	–1	0	0	0	–1	–1	–1	0	1	–1	1	0	0	0
Systems architecting	–1	0	–2	–1	–1	0	–1	0	–1	0	0	0	0	0	0	–2	–2	0	1	–2	0	0	0	1
Design for…	0	0	–1	–1	–1	0	–1	–1	–1	0	–2	0	1	0	–1	–2	–1	0	1	–1	0	–1	0	0
Integration	–1	1	–1	–1	0	–1	0	1	0	–1	–1	0	2	0	–1	–1	–1	–1	0	–1	1	0	0	0
Interfaces	–1	0	–1	–1	0	0	0	0	0	–1	–1	0	1	0	–1	–1	–1	0	1	–1	1	0	0	0
Verification and validation	0	1	–1	–1	0	0	0	0	1	–2	–1	1	1	0	–1	–1	–1	–1	0	–1	1	0	1	0
Qualification, certification and acceptance	0	1	–1	–1	0	0	0	0	0	–2	–1	0	0	0	–1	–1	–1	–1	0	–1	1	0	1	0
Transition	0	0	0	–1	0	–2	1	0	0	0	–1	0	0	0	–1	1	1	–1	–1	1	1	–1	1	–2
Utilization and support	–1	1	1	–1	–1	–2	2	0	–2	1	0	–1	2	–1	1	1	1	0	0	1	–1	–1	0	–1
Retirement	–2	–1	–1	–1	0	–2	–1	–1	–1	0	–1	0	0	–1	0	0	–2	0	–1	–2	–1	–1	0	–1
SE planning, monitoring and control	–2	1	–1	0	0	0	0	0	1	0	1	0	–1	0	0	–1	–1	–1	1	–1	0	–1	1	0
Decision management	–1	0	–1	0	–2	0	0	–2	–1	0	0	0	–1	0	–1	–1	–1	–1	1	0	1	–1	0	–1
Concurrent engineering	–1	0	1	0	–1	–1	1	–1	0	–1	–1	0	0	0	0	0	1	0	1	2	–1	0	1	0
Business and enterprise integration	–1	0	0	0	–1	0	0	–1	–1	0	0	2	–1	–1	0	0	1	0	2	1	2	0	1	0
Acquisition and supply	0	0	0	0	–2	0	1	–1	–1	0	1	0	1	–1	1	–1	0	0	1	1	–1	0	0	0
Information and configuration management	–1	0	0	–1	–1	–1	0	–1	0	–1	–1	–1	–1	–1	–2	0	0	–1	–1	0	–1	–2	–1	–1
Risk and opportunity management	–1	0	–2	–1	–1	0	–1	–1	0	–1	0	0	0	0	1	–1	–1	0	–1	–2	0	–2	–1	0
Project management	–2	1	2	1	0	1	0	0	1	–2	0	1	–1	2	1	1	1	–1	–1	0	–1	0	0	0
Finance	–2	0	0	1	0	0	0	0	0	–2	0	0	–1	1	0	0	0	–2	–1	0	–1	0	0	1
Logistics	–1	1	1	1	0	0	1	0	0	–2	0	1	0	0	0	2	1	–1	0	2	0	1	2	2
Quality	–1	2	2	2	1	0	1	0	0	–1	0	1	–1	2	–1	0	0	–1	0	0	0	0	1	0

SECAG FIGURE B5-6 Cohort 2 Deficit Heat Map.

General Conclusions

Some general conclusions can be drawn from this exercise:

- Areas previously with significant deficits in competency had now very much reduced overall organizational deficit. In other words, the areas requiring intervention historically were now much improved.
- Individuals generally had benefitted from their own personal interventions.

Clearly, improvements in the Cohort 2 scores could have been due to a combination of:

- Increased experience just through the passage of "time-on-the-job."
- Ongoing development through the interventions proposed at their initial assessment.
- Increased competency just through osmosis (i.e. "being around" individuals who provided improved local competence or in SECF/SECAG terms, individuals with "Practitioner" or "Lead Practitioner" competency to call upon to provide ad hoc guidance and monitoring).

Whatever the real reason, the general results were very positive. The HR team concluded that the "round trip" assessment activity had been a success, vindicating the significant investment made in its people over the five-year period.

Indeed, the operating position of the local business unit when the second assessment took place was far more positive than when the team first arrived, with significantly improved project performance and effectiveness in all ongoing projects.

It is perhaps of significant note that the same local management team was in place throughout the five-year exercise and the leader of the assessment team was the same for both activities. Both of these inevitably made a significant contribution to the overall stability of the program over its extended timeframe, something which is perhaps unusual and therefore hard to replicate in many businesses.

ANNEX B6 –TAILORING A JOB DESCRIPTION TO ADD SYSTEMS ENGINEERING COMPETENCIES AND SKILLS

This is an example of how to tailor a job description to add systems engineering competencies and skills to an existing job description that currently does not include any systems engineering.

Job Description Tailoring

In this example, a competency manager would like to ensure that some of their engineering jobs require systems engineering competencies as appropriate. In this case, a competency manager updates an existing materials engineer job description to include systems engineering-related competencies and tasks. They start by retrieving the existing job description as shown in SECAG Table B6-1. For this example, the existing job announcement only uses a subset of the tasks required for an individual with a specialty in materials engineering.

Next, the competency manager reviews the job description to highlight any of the tasks that require systems engineering competencies. The following tasks from the existing job description in SECAG Table B6-1 were determined to be related to systems engineering tasks and were targeted for tailoring with language from the SECF/SECAG.

SECAG TABLE B6-1 Example of existing materials engineer job description

Job highlights
The incumbent develops and maintains an internationally recognized research program and leads a diverse group of scientists and support staff in research to develop new bio-based materials and value-added products from traditional as well as new or specialty sources.
Salary:
Location:
Qualifications
• BS degree in applicable Materials Science/Material Engineering, Chemical Engineering, Manufacturing Engineering. Master's Degree a plus. • 5+ years of Sheet Molding/Composite material development, material molding, and leveling technology with proven results. • Strong understanding of batch material processing and process refinement. • Demonstrated problem solving skills. Able to articulate issues across the organization. • Can effectively lead/work with other functions (Design Engineer, Manufacturing Engineer, Maintenance, Quality, and Distribution) for business results.
Responsibilities (Tasks)
• Develops and defines environmentally compliant material substitutes. • Solves complex problems and provides authoritative technical advice and/or instructions in regard to material processes, materials failure analyses, and testing of materials for specification conformance. • Monitors the materials related to the industrial processes in the assigned areas by establishing and monitoring schedules, analysis procedures, and the qualifying criteria for the materials process monitored. • Provides expert advice and instructions concerning the materials aspect of processing or repair of materials, parts, and/or assemblies undergoing periodic rework, overhaul, repair, or modification on all systems. • Represents the company at conferences and meetings. Improves and maintains working relations with industry partners and actively participates in meetings to solve problems and contribute to special development projects. • Develops, assembles, and issues local engineering or process specifications consuming the materials aspects of processing and repair of materials, parts, and/or assemblies undergoing periodic rework, overhaul, repair, modification, as well as processing of new or experimental materials. • Reviews and evaluates correspondence related to material problems for technical accuracy and for development and action. • Presents data in such a manner as to recognize and emphasize both theoretical and practical aspects. • Conducts engineering investigations. • Leads special material research and development projects.

Systems Thinking

- Solve complex problems and provide authoritative technical advice and/or instructions in regard to material processes, materials failure analyses, and testing of materials for specification conformance.

Life Cycle Support

- Develop, assemble, and issue local engineering or process specifications consuming the materials aspects of processing and repair of materials, parts, and/or assemblies undergoing periodic rework, overhaul, repair, modification, as well as processing of new or experimental materials.

SECAG TABLE B6-2 Systems engineering tasks for the job

Task descriptions
Develop holistic research and development of materials over a range of situations throughout the system life cycle.
Ensure the adoption, introduction, and use of novel techniques and ideas related to quality initiatives across the enterprise and integrate materials engineering with enterprise-level quality initiatives, and create enterprise-level policies, guidance, and best practice related to Materials Engineering quality for systems.
Prepare technology plans that address innovation, risk, maturity, readiness levels, and insertion points considering system capability.
Ensure creative or innovative approaches to Materials Engineering are developed and performed across the enterprise.
Create overall processes for modeling and analysis tools for Materials Engineering that address systems aspects.

Quality

- Monitors the materials related to the industrial processes in the assigned areas of the aerospace rework facility serviced by establishing and monitoring schedules, analysis procedures, and the qualifying criteria for the materials process monitored.

Modeling and Analysis

- Leads special material research and development projects.

The competency manager then researches other systems engineering competencies and tasks that the organization would like to add to the job description. A list of additional key systems engineering competencies and tasks required for the job is complied. A list of possible systems engineering tasks to add to this Materials Engineer position is shown in SECAG Table B6-2.

Each of the tasks aligns to one or more competencies. The competency manager develops a final overall list of competencies that cover all of the tasks included for the position to include the systems engineering tasks. The SECF/SECAG is used to identify key competencies required for the job based on the information. A total of six competencies were identified for the job.

- Systems Thinking
- Life Cycles
- Systems Modeling and Analysis
- Quality
- Capability Engineering
- General Engineering

Next, the competency manager uses the SECF/SECAG to find KSA statements and organizes them as the basis for creating tailored tasks to update the job description, as shown in SECAG Table B6-3.

Throughout this process, some of the existing task KSAs were partially updated, rewritten, or new tasks were added to create a comprehensive new job description. The plain text represents items from the original job description. The bold text represents portions from the SECF/SECAG competencies. The updated example job description is shown in SECAG Table B6-4.

SECAG TABLE B6-3 SECF/SECAG information to be used to tailor the original job description

Competency	ID number	Proficiency level	Indicators of competence
Systems thinking	6	Practitioner	Uses appropriate systems thinking approaches to a range of situations, integrating the outcomes to get a full understanding of the whole. [CSTP06]
Life cycles	3	Practitioner	Identifies dependencies aligning life cycles and life cycle stages of different system elements accordingly. [CLCP03]
Systems modeling and analysis	1	Practitioner	Creates a governing process, plan, and associated tools for systems modeling and analysis in order to monitor and control systems modeling and analysis activities on a system or system element. [CSMP02]
Quality	2	Lead Practitioner	Promotes the introduction and use of novel techniques and ideas across the enterprise, which improve the integration of Systems Engineering and quality management functions. [IQUL11] Creates enterprise-level policies, procedures, guidance, and best practice in order to ensure Systems Engineering quality-related activities integrate with enterprise-level quality goals including associated tools. [IQUL01]
Capability engineering	3	Practitioner	Prepares technology plan that includes technology innovation, risk, maturity, readiness levels, and insertion points into existing capability. [CCPP03]
General engineering	6	Lead Practitioner	Fosters creative or innovative approaches to performing general engineering activities across the enterprise. [CGEL06]

SECAG TABLE B6-4 Updated materials engineer position description

Job highlights
The incumbent develops and maintains an internationally recognized research program and leads a diverse group of scientists and support staff in research to develop new bio-based materials and value-added products from traditional as well as new or specialty sources.
Salary:
Location:
Qualifications
• BS degree in applicable Materials Science/Material Engineering, Chemical Engineering, Manufacturing Engineering. Master's Degree a plus. • 5+ years of Sheet Molding/Composite material development, material molding, and leveling technology with proven results. • Strong understanding of batch material processing and process refinement. • Demonstrated problem solving skills. Able to articulate issues across the organization. • Can effectively lead/work with other functions (Design Engineer, Manufacturing Engineer, Maintenance, Quality, and Distribution) for business results.
Responsibilities (Tasks)
• Develops and defines environmentally compliant material substitutes. • **Uses appropriate systems thinking approaches to a range of situations, integrating the outcomes to get a full understanding of the whole** while providing authoritative technical advice and/or instructions in regard to material processes, materials failure analyses, and testing of materials for specification conformance. [CSTP06]

SECAG TABLE B6-4 (Continued)

- Develops and defines environmentally compliant material substitutes.
- **Uses appropriate systems thinking approaches to a range of situations, integrating the outcomes to get a full understanding of the whole** while providing authoritative technical advice and/or instructions in regard to material processes, materials failure analyses, and testing of materials for specification conformance. [CSTP06]
- **Creates and assesses enterprise-level policies, procedures, guidance, and best practice in order to ensure** Materials Engineering **quality-related activities integrate with enterprise-level quality goals including associated tools.** [IQUL01]
- **Promotes the introduction and use of novel techniques and ideas** across the enterprise, **which improve the integration of Systems Engineering and quality management functions.** [IQUL11]
- Provides expert advice and instructions concerning the materials aspect of processing or repair of materials, parts, and/or assemblies undergoing periodic rework, overhaul, repair, or modification on all systems.
- **Prepares** materials **technology plan that includes technology innovation, risk, maturity, readiness levels, and insertion points into existing capability.** [CCPP03]
- Represents the company at conferences and meetings. Improve and maintain working relations with industry partners and actively participates in meetings to solve problems and contribute to special development projects.
- Develops, assembles, and issues local engineering or process specifications consuming the materials aspects of processing and repair of materials, parts, and/or assemblies undergoing periodic rework, overhaul, repair, modification, as well as processing of new or experimental materials **identifying dependencies aligning life cycles and life cycle stages of different system elements accordingly.** [CLCP03]
- Reviews and evaluates correspondence related to material problems for technical accuracy and for development and action.
- Presents data in such a manner as to recognize and emphasize both theoretical and practical aspects.
- Conducts engineering investigations.
- **Fosters creative or innovative approaches to performing** Materials Engineering **activities across the enterprise.** [CGEL06]
- **Creates a governing process, plan, and associated tools for** support of research and development efforts **for systems modeling and analysis in order to monitor and control** materials **modeling and analysis activities on a system or system element.** [CSMP02]

ANNEX B7 – USING THE SECF/SECAG FOR TAILORING A JOB DESCRIPTION TO ADD MODEL-BASED SYSTEMS ENGINEERING COMPETENCIES AND SKILLS

This is an example of how to tailor a job description to add MBSE and digital engineering competencies and skills along with the SECF/SECAG competencies to an existing job description that currently uses a more traditional set of systems engineering competencies and skills.

Model-Based Systems Engineering Job Description Tailoring

In this example, the company would like to update some systems engineer job descriptions to include information related to MBSE and digital engineering competencies, as appropriate. A competency manager creates a new systems architect job description that includes MBSE and digital engineering-related tasks. The current job description is shown in SECAG Table B7-1.

The organization's goal is to include relevant MBSE-related tasks and competencies in the new job description. The competency manager starts with the existing job description task list. Next, they collect a list of key MBSE and digital engineering tasks required for the job. These competencies and tasks are not necessarily explicitly included in the SECF or SECAG. Other sources can be used as a basis to tailor the existing system architect job. The competency manager searches and finds a Digital Engineering Competency Framework (DECF) that includes tables of MBSE architect-related tasks and competencies formatted into the Competency Groups (G), Competency Subgroups (S), and KSA that address desired MBSE information. The DECF competency groups are shown in SECAG Table B7-2.

The DECF Competency Subgroups for the respective Competency Groups are shown in SECAG Table B7-3.

SECAG TABLE B7-1 Current System Architect Job Description

JOB OVERVIEW	
JOB TITLE	System Architect
DEPARTMENT	Systems Engineering
LOCATION	
SALARY	
GENERAL JOB DESCRIPTION	
The incumbent creates and maintains architectural products throughout the life cycle integrating hardware, software, and human elements; their processes; and related internal and external interfaces that meet user needs and optimize performance. Collaborates with a diverse group of systems engineers, project managers, engineers, and support staff.	
DUTIES AND RESPONSIBILITIES (TASKS)	
Technical • Develops an architecture model that includes functional and structural partitioning, interface definitions, design decisions, and requirements traceability. • Develops or modifies architectures to meet organizational goals. • Generates the full architectural design description for a system or program. • Assesses the integrity of the overall architectural model to ensure that it meets the operational/system requirements and the business/mission needs. • Develops various scenarios for system use, functions, and performance in the target environment.` • Collaborates with a disciplinary specialist to synthesize information across multiple users concerning the work tasks, work context, and sociocultural factors that affect the system. • Evaluates disciplinary specialist mitigations, designed by the disciplinary specialist, to gauge the effect on the operational use of the system. • Determines the technical and programmatic areas where agreements need to be reached among stakeholders based upon the systems interfaces and interoperability concerns. • Manages the creation of systems engineering architecture artifacts required as inputs to other functions. • Partitions between discipline technologies to derive discipline-specific requirements. • Recommends partial architectural solutions that meet an important subset of needs quickly, coupled with approaches to seek better or more effective solutions. • Reevaluates the solution design space and recommends adjustments when the solution does not meet customers' needs or fit the situation. • Synthesizes individual solutions into larger solutions to explore new approaches. • Identifies systems interfaces and interoperability concerns. **Management** • Document architecture structure in reports to assist in project reviews and decision-making. **Communication** • Communicate effectively with disciplinary engineers to achieve system-level objectives.	
EDUCATION AND TRAINING	
Degree: Engineering or combination of education and experience. College-level education, training, and/or technical experience that furnished (1) a thorough knowledge of the physical and mathematical sciences underlying engineering, and (2) a good understanding, both theoretical and practical, of systems engineering and techniques and their applications.	
COMPETENCIES	
• System architecting • Modeling and simulation • Project management • Communications	

SECAG TABLE B7-2 Competency groups from the DECF used as the basis for the MBSE competencies

G2	Modeling and simulation	Use of digital models to describe and understand phenomena of interest from initiation of the effort through the entire life cycle maturation. Model literacy – understanding what models are and how they work – is required to move into more advanced skills, from the ability to build a model using appropriate tools, standards, and ontology to creating a modeling environment.	S2
G3	Digital engineering and analysis	Apply traditional engineering methods and processes in a digital environment. Create new engineering processes and methods for a digital environment. Create digital artifacts throughout the project or system life cycle. Use engineering methods, processes, and tools to support the engineering and system life cycle.	S3 S4
G5	Digital enterprise environment	Addresses development of the Digital Engineering environment including hardware and software aspects. Digital Enterprise Environment Management is for management, communications, and planning related to enabling the workforce to manage the adoption of appropriate model-based tools and approaches, techniques, and processes for the operation of digital enterprise environment systems that ensure transformational processes in enterprises occur with pace, high quality, and security. Digital Enterprise Environment Operations and Support within a digital enterprise environment include abilities to operate and support the digital enterprise environment across the enterprise and life cycle. Digital Enterprise Environment Security involves developing policies, standards, processes, and guidelines to ensure the physical and electronic security of digital environments and automated systems.	S7

SECAG TABLE B7-3 Competency subgroups of the respective competency groups from the DECF

S2 Modeling and simulation	C3	Modeling	Modeling is essential to aid in understanding complex systems and system interdependencies and to communicate among team members and stakeholders.
	C4	Simulation	Simulation provides a means to explore concepts, system characteristics, and alternatives; open the trade space; facilitate informed decisions, and assess overall system performance.
S3 Digital systems engineering	C8	Digital architecting	Digital architecture activities use digital models to define a comprehensive digital system model based on principles, concepts, and properties logically related to and consistent with each other. Digital architecture has features, properties, and characteristics that satisfy, as far as possible, the problem or opportunity expressed by a set of system requirements (traceable to mission/business and stakeholder requirements) and life cycle concepts (e.g. operational and support) and which are implementable through digital enterprise-related technologies. Digital architecture competencies relate to the ability to create system digital models and required architectural products and digital artifacts for a system or system-of-systems in accordance with applicable standards and policies.
	C11	Model-based systems engineering processes	Model-based systems engineering is the formalized application of modeling to support system requirements, design, analysis, verification, and validation activities beginning in the conceptual design phase and continuing throughout development and later life cycle phases.

(Continued)

SECAG TABLE B7-3 (Continued)

S4 Engineering management	C12	Digital model-based reviews	Digital model-based reviews define the series and sequence of model-based systems engineering activities that bring stakeholders to the required level of commitment, prior to formal reviews. It utilizes system models, artifacts, and products for analysis of design and technical reviews to execute trade-off and design analyses, prototyping, manufacturing, testing, and sustainment of the system.
S7 Digital enterprise environment management	C20	Management	Management in the digital enterprise environment aims to deliver a framework that ensures transformational processes in enterprises occur with pace, high quality, and security. This is achieved through a set of IT solutions that are designed to make digital businesses fast, seamless, and optimized at every level.
	C21	Communications	Communications include using digital model artifacts from the digital enterprise environment to investigate and manage the adoption of appropriate model-based tools, techniques, and processes for the operation of digital enterprise environment systems and services. Communications also establishes the appropriate guidance to enable transparent decision-making to be accomplished, allowing senior leaders to ensure the needs of principal stakeholders are understood, the value proposition offered by digital enterprise environment is accepted by stakeholders, and the evolving needs of the stakeholders and their need for balancing benefits, opportunities, costs, and risks are embedded into strategic and operational plans.

By collecting tasks as KSA from the DECF, the competency manager created a final list of desired tasks at defined proficiency levels for MBSE.

C3 Modeling – Intermediate

- Create system models for system development efforts in accordance with applicable standards and policies.
- Define the inter-relationships among model elements and diagrams.
- Model current and desired scenarios, as directed.
- Review the system model created by others.
- Select appropriate tools and techniques for system modeling and analysis.
- Use modeling language, concepts, diagrams, and data attributes.

C4 Simulation – Intermediate

- Applies modeling and simulation applications and tools, to cover a full range of modeling situations.
- Integrate modeling capabilities with other product and analytical models including physics-based models.
- Use simulation tools and techniques to represent a system or system element.

C8 Digital Architecting – Basic

- Utilize modeling languages to create or maintain system architectural products based on data provided.
- Comply with style guides to properly develop architectural products.

C8 Digital Architecting – Intermediate

- Provide architecture assessment to make decisions based on the architecture to ensure requirements are met for the system development throughout the life cycle.
- Apply system model and architectural concepts based on different stakeholder views and how they relate.
- Collaborate with disciplinary subject matter experts to create system models and architectural products.
- Create required architectural products for a system or system-of-systems in accordance with applicable standards and policies with minimal or no supervision.

C11 Model-Based Systems Engineering Process – Basic

- Develop digital model artifacts, according to intent.
- Analyze and interpret the results obtained using model-based engineering methods and tools.

C11 Model-Based Systems Engineering Process – Intermediate

- Analyze the system model and architectural products.
- Build models in a digital enterprise environment collaborative modeling environment.
- Ensure digital artifacts are up-to-date, consistent, interoperable, accessible, uncorrupted, and properly and safely stored.
- Generate digital enterprise environment system models.
- Integrate all other model domains and physics-based models with the system model.

C12 Digital Model-Based Reviews – Intermediate

- Conduct model-based reviews and audits, to ensure effective collaboration for system-of-interest evolution.
- Confer with subject matter experts on models produced to gain concurrence on results.
- Review resulting models with stakeholders and gain resolution to resultant issues.

C20 Management – Intermediate

- Produce routine reports to assist in digital environment management activities and decision-making.

C21 Communication – Basic

- Communicate using digital model artifacts from the digital enterprise environment.

The competency `manager tailors and reorganizes the original tasks and competencies using KSA from the DECF integrating them with appropriate indicators of competence from the SECF/SECAG. This information is used to create the updated job description, shown in SECAG Table B7-4. The plain text entries are based on the indicators of competence statements. The bold text portions are derived from the DECF statements. The italicized text are portions retained from the original job description.

SECAG TABLE B7-4 Example of updated system architect description including MBSE SE tasks and competencies

JOB OVERVIEW	
JOB TITLE	System Architect
DEPARTMENT	Systems Engineering
LOCATION	
SALARY	
GENERAL JOB DESCRIPTION	
The incumbent develops and maintains system architectures using model-based methods and collaborates with a diverse group of systems engineers, project managers, disciplinary engineers, and support staff to develop system models using digital modeling tools and environments.	
DUTIES AND RESPONSIBILITIES (TASKS)	
Technical • Creates alternative architectural designs **in a digital enterprise environment modeling environment** traceable to the requirements **in accordance with applicable standards and policies** to demonstrate different approaches to the solution **and reviews resulting models with stakeholders and gains resolution to resultant issues.** [TSAP04] • Uses appropriate analysis techniques to ensure different viewpoints are considered **based on different stakeholder views and how they relate.** [TSAP06] • Elicits derived discipline-specific architectural constraints from specialists to support partitioning and decomposition, **synthesizes individual solutions into larger solutions to explore new approaches, and evaluates disciplinary specialist mitigations, designed by the disciplinary specialist, to gauge the effect on the operational use of the system.** [TSAP07] • Identifies *systems interfaces and interoperability concerns* and the impact on interface definitions as a result of wider changes. [TIFP07] • Defines **and utilizes** appropriate **modeling language, concepts, diagrams, data attributes, and** representations of a system or system element **to create or maintain system architectural products based on data provided.** [CSMP07] • Selects appropriate tools and techniques for system modeling and analysis. [CSMP06] • Analyzes options and concepts **using model-based engineering methods and tools and integrates all other model domains and physics-based models with the system model** in order to demonstrate that credible, feasible options exist. [TSAP05] • Uses the results of system analysis activities to *reevaluate the solution design space* and inform system architectural design *and recommend adjustments when the solution does not meet customers' needs or fit the situation.* [TSAP08] • Monitors key aspects of the evolving design solution *and recommends partial architectural solutions that meet an important subset of needs quickly, coupled with approaches to seek better or more effective solutions* in order to adjust architecture, if appropriate. [TSAP10] **Management** • Follows governing project management plans, processes, and uses appropriate **digital environment management** tools to control and monitor project management-related Systems Engineering tasks, interpreting as necessary. [IPMP01] • Develops **digital model artifacts, according to intent, and ensures digital artifacts are up-to-date, consistent, interoperable, accessible, uncorrupted, and properly and safely stored for use in model-based reviews and audits** as Systems Engineering inputs for project management status reviews to enable informed decision-making. [IPMP05] **Communication** • Uses appropriate communications techniques **using digital model artifacts from the digital enterprise environment** to ensure a shared understanding of information with all project stakeholders. [PCCP02] • Uses appropriate communications techniques to express alternate points of view in a diplomatic manner using the appropriate means of communication **to collaborate with disciplinary subject matter experts to create system models and architectural products and gain concurrence on results.** [PCCP04]	

SECAG TABLE B7-4 (Continued)

EDUCATION AND TRAINING
Degree: Engineering or combination of education and experience. College-level education, training, and/or technical experience that furnished (1) a thorough knowledge of the physical and mathematical sciences underlying engineering, and (2) a good understanding, both theoretical and practical, of the engineering sciences and techniques and their applications to one of the branches of engineering.
COMPETENCIES
• System architecting • Systems modeling and analysis • Interfaces • Project management • Communications

SECAG ANNEX C: SECAG COMMENT FORM

Please submit feedback comment form information to SECAGCompetencyWG@incose.net.

Reviewed document:	INCOSE Systems Engineering Competency Assessment Guide (SECAG)				
Name of submitter:	Given name, family name (e.g. Jo DOE)				
Date of submission:	YYYY-MM-DD (e.g. 2018-04-09)				
Contact info:	Email address (e.g. jo.DOE@anywhere.com)				
Type of submission:	Group, individual				
Group name and number of contributors	Group name if applicable (e.g. INCOSE XYZ Working group)				
Comments	Please provide comment details including precise reference to the document section, paragraph, or line item requiring change. Ideally, comments should be formatted as shown in the table below.				
Comment ID	**Category**	**Section number**	**Specific reference**	**Issue, comment, and rationale**	**Proposed change or new text (mandatory)**
Unique identifier	G, E, TH, TL As follows: • G = general • E = editorial • TH = technical comment, high priority • TL = technical comment, low priority	E.g. section n, table m	E.g. paragraph, line	Please provide rationale so that comment is clear and supportable.	Good quality new or revised text will increase odds of acceptance.

Systems Engineering Competency Assessment Guide: A combined INCOSE Systems Engineering Competency Framework (SECF) and associated Systems Engineering Competency Assessment Guide (SECAG) document, First Edition. INCOSE.

INDEX

Systems Engineering Competency Assessment Guide: A combined INCOSE Systems Engineering Competency Framework (SECF) and associated Systems Engineering Competency Assessment Guide (SECAG) document, First Edition. INCOSE.

Printed and bound by CPI Group (UK) Ltd, Croydon, CR0 4YY

16/04/2025

14658421-0005